MACHINE DESIGN

OTHER BOOKS BY THE SAME AUTHOR

- **Strength of Materials**
  ISBN : 978-93-81068-80-9
  Pages (Approx.) : 600

- **Fluid Mechanics**
  ISBN : 978-93-81068-65-6
  Pages (Approx.) : 680

- **Machine Design**
  ISBN : 978-93-82609-57-5
  Pages (Approx.) : 832

- **Machine Design Data Book**
  ISBN : 978-93-82609-51-3
  Pages (Approx.) : 300

- **Strength of Materials-I**
  ISBN : 978-93-81068-61-8
  Pages (Approx.) : 400

- **Fluid Machinery (Hydraulic Machines)**
  ISBN : 978-93-82609-28-5
  Pages (Approx.) : 500

- **Khanna's Mechanical Objective Type**
  ISBN : 978-93-82609-60-5
  Pages (Approx.) : 1050

- **Khanna's Mechanical Conventional Type**
  ISBN : 978-93-82609-61-2
  Pages (Approx.) : 1050

MACHINE DESIGN

DR. SADHU SINGH

Professor & Head, Mechanical Engineering Department
(Former)
&
Dean, Faculty of Engineering & Technology (Former)
Govind Ballabh Pant University of Agriculture & Technology,
Pantnagar – 263145 (Uttarakhand)
and
(Director Colleges), Punjab Technical University, Jalandhar
(Former)

KHANNA BOOK PUBLISHING CO. (P) LTD.

Publisher of Engineering and Computer Books

4C/4344, Ansari Road, Darya Ganj, New Delhi-110002

Phone : 011-23244447-48 **Mobile:** +91-9910909320

E-mail : contact@khannabooks.com

Website : www.khannabooks.com

Price: ₹ 575.00

ISBN: 978-93-82609-57-5

Machine Design Data Book
by Dr. Sadhu Singh

Revised Edition: 2026

To view complete list of books, Please scan the QR Code:

Published by:
Khanna Book Publishing Co. (P) Ltd.
CIN: U22110DL1998PTC095547

Visit us at: **www.khannabooks.com**
Write us at: **contact@khannabooks.com**

Printed in Bharat.

CONTENTS

1 ENGINEERING DESIGN

Chapter

1.1 DEFINITIONS

Engineering—It may be defined as the profession in which a knowledge of the mathematical and physical sciences gained by study, experience and practice is applied with judgement to develop ways to utilize economically, the materials and forces of nature for the progressive well being of mankind.

Design—To design is to innovate and to create.

Designing—The process of designing is to suggest or outline ways to put together man-made things, or to suggest modifications in man-made things to satisfy optimally (under the given constraints) some specified human needs.

Engineering Design—It has been defined in many ways by various scientists, but the simplest way to define engineering design is :

"An iterative decision making process to conceive and implement optimum systems to solve societies problems and needs."

Engineering design being practical in nature must be concerned with physical realizability, or economic and financial feasibility.

Machine—It may be defined as a combination of kinematic links having unique relationships of motion so as to result in useful work being done in a controlled manner.

Machine Design—It may be defined as a process of achieving a plan for the construction of a machine that works safely, reliably, and well.

Optimal Design—A design which is best for the given objective function under the specified constraints.

Industrial Design—It is a design process in which main considerations are given to aesthetics, economics and production aspects of design.

Empirical Design—It is based upon existing standard design practice and past experience.

1.2 MECHANICAL ENGINEERING DESIGN

Mechanical Engineering Design is defined as the use of scientific principles, technical information and imagination in the description of a mechanical structure, a machine or mechanical system to perform prescribed functions with the maximum economy and efficiency. The description of a mechanical system is in the form of hard copy of drawings.

1.2.1 *Morphology of Mechanical Engineering Design*

The most general design problem can be stated as :

Devise or design : A product, system or process to accomplish the given task or fulfill the given need optimally

Subject to : Certain constraints.

The simplest sequence for problem solving activities is :

1. Analysis
2. Synthesis, and
3. Evaluation.

The collection of time phases of the various steps of mechanical engineering design is called the **morphology of design,** as shown in Fig. 1.1. The various phases are :

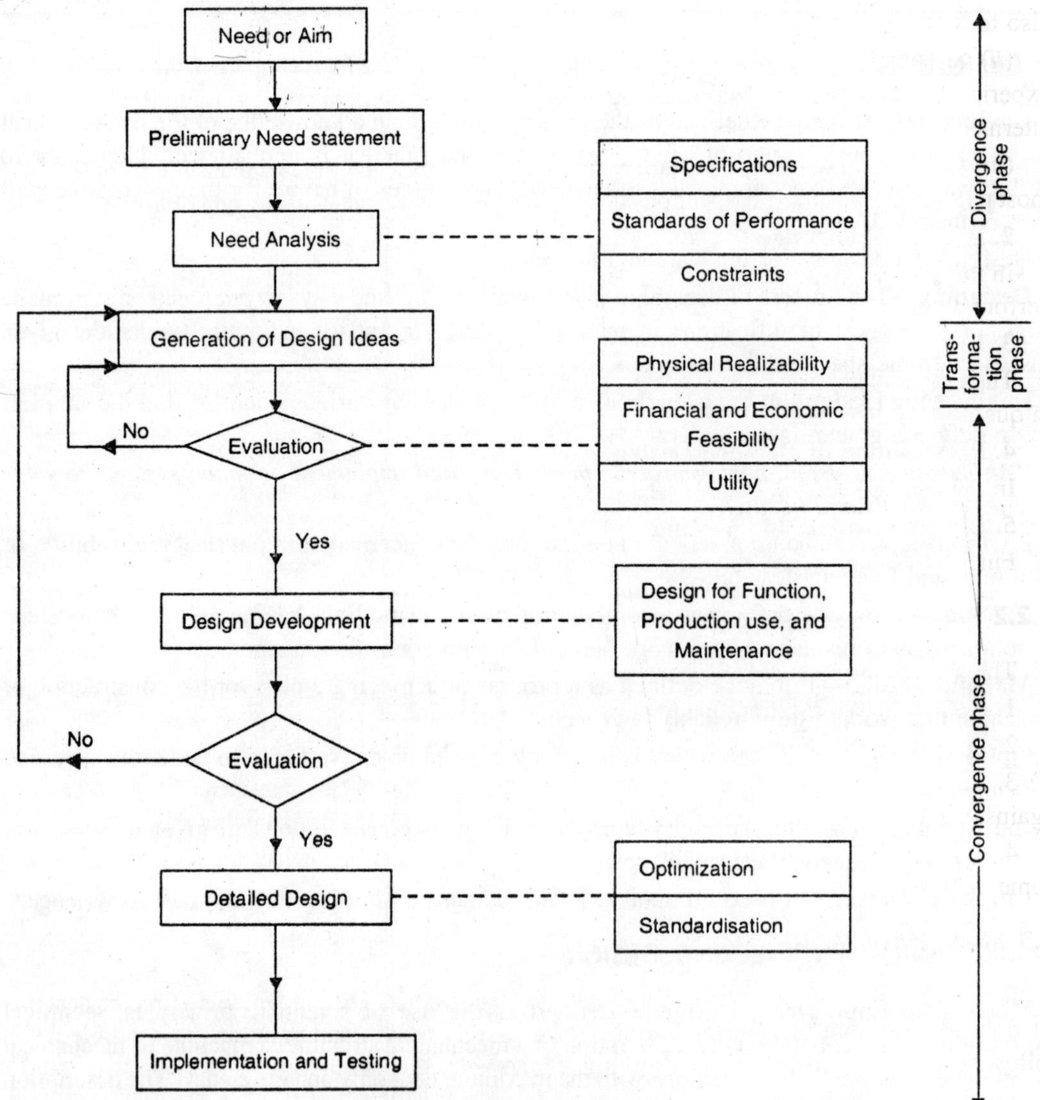

Fig. 1.1 Morphology of Design

1. Feasibility Study

The three phases of design in chronological order are :

(*i*) Divergence phase

(*ii*) Transformation phase, and

(*iii*) Convergence phase.

(*i*) **Divergence phase**— The term divergence refers to the act of extending the boundary of a design situation. This is usually the exploration phase where the designer is to get as much understanding of the problem as possible. The typical divergence methods are : (*a*) Literature survey (*b*) Questionnaire (*c*) Strategy switching, and (*d*) Brain-storming, etc.

The analysis of need consists of exploring the design situation, establishing the specifications of the system desired, deciding on the standards against which the performance of the design can be measured and starting the constraints within which one has to work. The optimality criteria has also to be fixed.

(*ii*) **Transformation phase**—It is creative phase wherein the designer makes use of all his experience , innovative capabilities, insights and genius to conceive of a number of plausible alternate solutions to the problem.

(*iii*) **Convergence phase**—It starts with narrowing down the field of plausible solutions to the most promising one on the basis of physical realizability, anticipated utility and financial feasibility.

2. Preliminary Design

In this phase, the proposed design concepts are synthesized to achieve a pre-specified level of performance.

3. Evaluation of Preliminary Design

The evaluation of preliminary design is based upon reliability, wear and consequent deterioration in quality and performance.

4. Presentation of Detailed Design

In the detailed design, the detailed dimensions, tolerances, finishes, etc. are specified.

5. Implementation and Testing

Finally the design is implemented, a prototype prepared and tested.

1.2.2 *Procedure for Morphological Analysis*

The following procedure may be adopted for morphological analysis :

1. Recognize the essential parameters acceptable for design.
2. Determine the number of possible ways for achieving each of the parameters.
3. Set up a matrix in the left column of which are listed all the parameters and in the rows against each of these parameters are listed the possible methods of achieving them.
4. Study all combinations of sub—solutions for feasibility and select the best on the basis of some valid criterion.

1.3 BRAIN-STORMING

It is the best technique for idea generation. It is a group activity in which a group of designers sit down to generate the greatest number of alternative ideas and solutions to a problem. The following rules should be observed during a brain-storming session :

First Rule : Criticism and judgement, whether favourable or unfavourable are not allowed.

Second Rule : Generate a lot of ideas.

Third Rule : Think wild.

1.3.1 *Scheme for Brain-Storming Session*

The following scheme is quite useful for conducting a brain-storming session.

1. Select a group of designers (around 5) to generate ideas.

2. State the problem as simply as possible.
3. Give about ten minutes to each person to write on a piece of paper as many solution ideas as possible.
4. Allow each person to read one idea in turn and ask the remaining persons to suggest any additional ideas to improve or modify the idea being suggested.
5. Do not allow to criticize or discourage whether an idea is practical or not. Only shortcomings and improvements may be entertained.
6. Encourage humour to release the boredom of the members and to build mutual trust.
7. At the end, find a logical classification of all the ideas generated.
8. Evaluate the design concepts generated and select one which you feel to be the best for further development.

1.4 SKILLS OF A DESIGNER

The basic skills of a designer are :

Inventiveness—It is the ability to think of or discover valuable, useful ideas or concepts for things or processes to accomplish given objectives.

Ability of engineering analysis—It is the capability of the designer to analyse a given component, system, or process using engineering or scientific principles in order to arrive quickly at meaningful answers.

Knowledge and training in an engineering science speciality.

Inter-disciplinary ability.

Mathematical and computational skills.

Thorough knowledge of use of digital computers.

Knowledge of manufacturing processes.

Ability to take appropriate decisions.

Communication skills.

1.5 THE DESIGN PROCESS

The design is the art of devising an original solution to a problem by a combination of principles, resources, and products in design. The design process is the pattern of activities that is followed by the designer in arriving at the solution of a technological problem. The design process is an iterative procedure. It is a step–by– step procedure from some statement of need through identification of the problem, a search for solutions and development of the chosen solution to manufacture, test and use.

1.5.1 Steps in the Design Process

The design process involves six steps as shown in Fig. 1.2.

1. Recognition of Need—It involves the realization by some one that a problem exists for which some feasible solution is to be found. This might be identification of some defect in a current machine designed by an engineer or the perception of a new product marketing opportunity by a salesman.

2. Definition of a Problem—It involves a thorough specification of the item to be designed. This specification will generally include functional and physical characteristics, cost, quality, performance, etc.

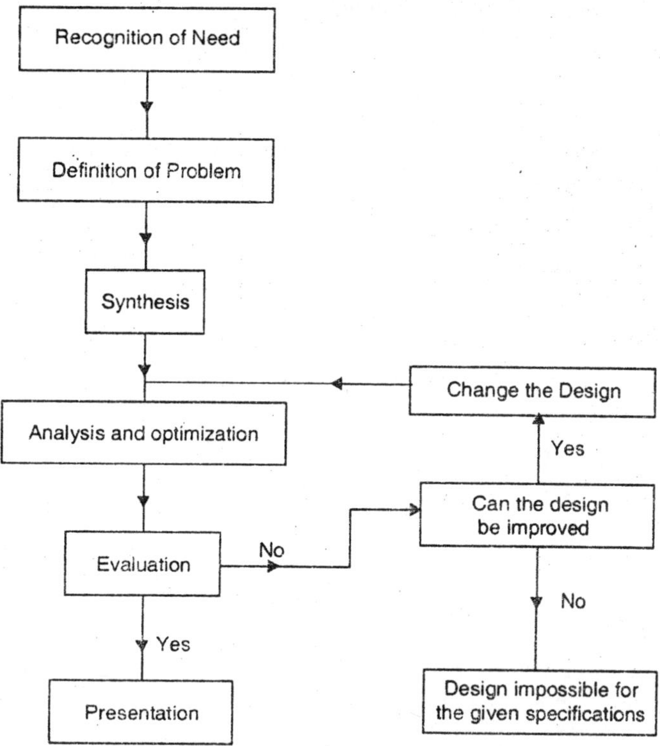

Fig. 1.2 The Design Process

3. Synthesis—During the synthesis phase of design, various preliminary ideas are developed through research of similar products of designs in use.

4. Analysis and Optimization—The resulting preliminary designs are then subjected to appropriate analysis to determine their suitability for the specified design constraints. If these designs fail to satisfy the constraints, they are then redesigned or modified on the basis of the information gained from the analysis. This iterative process continues until the proposed designs meet the specification, or until the designer is convinced that the design is not feasible within the specified constraints. The components, sub-assemblies or sub-systems are then synthesized into the final overall system in a similar iterative manner.

5. Evaluation—The assessment of evaluation of the design against the specifications established during the problem definition phase is then carried out. This often requires the fabrication and testing of prototype mod2el to evaluate operating performance, quality, reliability, etc.

6. Presentation—The final phase in the design process is the presentation of design. This includes documentation of design through drawings, material specifications, assembly lists, and so on.

1.6 CONCEPTUALIZATION OR CREATIVITY

The production of something new is closely related to creativity or inventiveness. An invention is something novel and useful. The secret of inventiveness is to saturate the mind with all the information pertinent to the problem and let it incubate. Then throw up the solution which is further developed and verified. The key for invention is to determine the elements, mechanisms, processes or configuration that in some combination of other result in a design that satisfies the need.

Inventiveness is the ability to get new and useful ideas for accomplishing the objective. It depends upon scientific and technical knowledge and understanding.

1.6.1 Suggestions for Improving Inventiveness

The following suggestions are being made for improving inventiveness.

1. Be curious about how things happen.
2. Don't follow traditional forms and method of solution.
3. Creative solutions can be found by forming new combinations of things, processes or ideas.
4. Simplify the overall system by complicated design of some components.
5. Subdivide the complex design task by subdividing into a number of smaller problems.
6. Keep abreast of the developments in the physical sciences.
7. Inventions not supported by science and analysis should be ignored.
8. Use analogy to get new ideas.
9. Plan the work and have a period of mental rest away from the problem.
10. Brain storming can be a valuable method for getting new ideas.

1.7 AESTHETIC CONSIDERATIONS IN DESIGN

The products are manufactured for a specific purpose and function to meet the requirements of the customers. The contact between the product and the user arises due to the dire necessity of this functional requirement. However, when there is a choice available to the customer of a product having the same qualities of efficiency, durability and cost, the customer is naturally attracted towards the most appealing product. The external appearance is an important feature, which gives grace and luster to the product and dominates the market. This is called aesthetic sense in design. Aesthetic considerations are very important for consumer durables like automobiles, household appliances and audio-visual equipment.

The aesthetic considerations in design have given rise to a separate discipline, known as industrial design. The job of an industrial designer is to create new forms and shapes which are aesthetically pleasing to the mind.

There are five basic forms—step, stream, taper, shear and sculpture. The step form is similar to the shape of a 'skyscraper' or multistorey building. This involves shapes with a vertical rather than a horizontal accent. The stream or streamline form is seen in automobiles and aeroplane structure to reduce the air drag resistance. The taper form consists of tapered blocks interlocked with tapered plinths or cylinders. The shear form has a square outlook, which is ideally suitable for free–standing engineering products. The sculpture form consists of ellipsoids, paraboloids and hyperboloids. The sculpture and stream form are suitable for automobile, while step and shear forms for stationary products.

There is a relationship between functional requirement and appearance of a product. In many cases, functional requirements result in shapes which are aesthetically pleasing. The evolution of streamlined shapes of boeing and racing cars is the result of studies in aerodynamics for reducing air drag. The chromium plating of many household appliances is for the purpose of corrosion resistance rather than appearance.

Selection of proper colour is an important consideration in product aesthetics. The meaning of various colours is as given in Table 1.1.

Table 1.1 Aesthetic meaning of colours

Colour	Aesthetic meaning
Red	Danger – Hazard – Hot
Orange	Possible danger
Yellow	Caution
Green	Safety
Blue	Caution – Cold
Grey	Dull

Other factors affecting aesthetic consideration are :

1. Rigidity and resilience.
2. Tolerance.
3. Surface finish.
4. Motion of individual components.
5. Materials.
6. Manufacturing methods.
7. Vibration and noise.

1.8 ERGONOMIC CONSIDERATIONS IN DESIGN

The word 'ergonomics' is derived from two Greek words : ergon = work and nomos = natural laws. Thus Ergonomics means the natural laws of work. It is the relationship between man and machine and the application of anatomical, physiological principles to solve the problems arising from man–machine relationship. It is also called human engineering.

Anatomy : Study of body dimensions and relations for work design.

Physiology : Study of man and his working environment.

Psychology : Study of adaptive behavior and skills of people.

Ergonomica includes the following factors for study from design considerations :

1. Anatomical factors in design
2. Control of physical work environment.
3. Design of man–machine system.
4. Design of controls and displays.
5. Accidents, fatigue and safety.
6. Work place design.

The fundamental concept in human engineering is the system. A system is composed of humans, machines and other things that interact to accomplish some goal which these same components could not produce independently. The man – machine system is a closed loop system as shown in Fig. 1.3.

Training and experience

M Sensory functions ← Display ← Input

A Data processing MACHINE

N Decision making

Manipulative functions → Control → Output

Environmental : Light, heat, noise, humidity, vibration.

Fig. 1.3 Man–machine system.

The man will receive certain information from the machine either from dials, displays, etc. designed for that purpose or by observation of machine itself. He will process this information and make decisions on what action to take and manipulate controls or attend machine in some other way so as to affect its behaviour in the required manner. The efficiency with which the man functions depends on environmental factors, on his own characteristics such as age, motivation, training and experience as well as efficiency with which the machine provides the information feedback and accepts control measures.

1.8.1 Display Design

Displays are needed by operators, in making decisions and in effecting control responses. The various types of display designs are :

1. Visual displays—Visual displays are classified as :

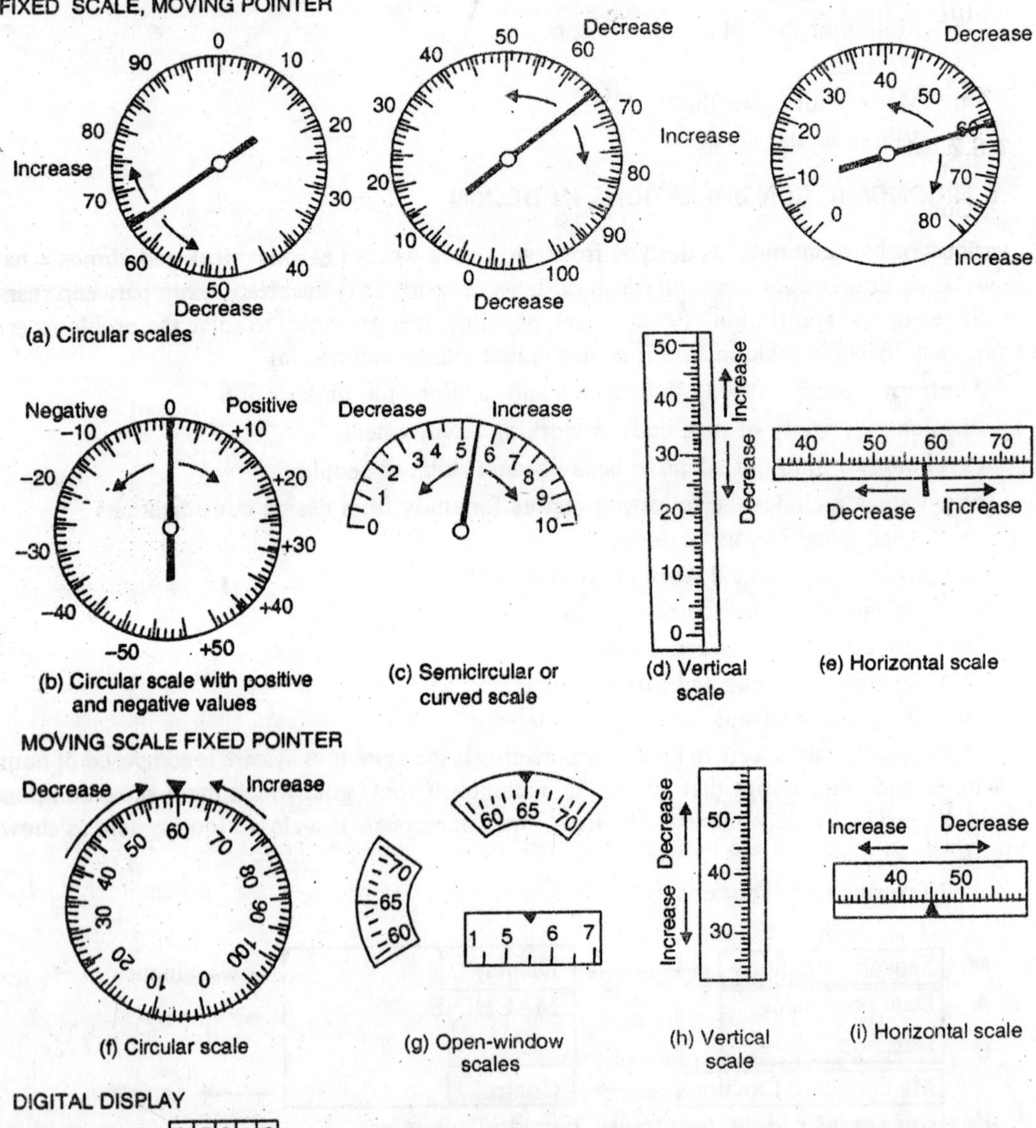

Fig. 1.4 Types of Displays

(*i*) **Quantitative display**—To read a precise numeric value e.g. display for pressure, temperature, speed, angle, etc. Conventional quantitative displays are mechanical devices of the following types.

- Fixed scale with moving points e.g. pressure gauge, automobile speedometer.
- Moving scale with fixed pointer e.g. weighing machine to measure human weight.
- Digital display e.g. digital counters of tape recorder.
- Electronically generated features e.g. digital displays.

The various types of displays are shown in Fig. 1.4.

(*ii*) **Qualitative display**—The display is used to read an approximate value or to indicate rate of change, change in direction, etc. e.g. the increase or decrease in pressure, temperature or speed.

(*iii*) **Check display**—The display gives information about the parameters whether they are normal.

2. Auditory displays—These displays are suitable as warning devices.

1.8.2 Design of Controls

Controls are the means by which information is transmitted to the machine from the man. The various types of controls are :

1. Hand controls : The anthropometric data for the human hand can be used as an aid to design dimensions of hand controls.
2. Hand levers : Levers give a quick control action and can accommodate large forces. They are not suitable for fine adjustments. Levers can provide efficient ON / OFF or step by step control.
3. Hand wheels : Hand wheels provide a controlling torque via both hands and they are used for heavy loads. They can provide good accuracy of adjustments.
4. Cranks : Cranks are intended to provide torque via one hand. Smaller cranks are used for fast control.
5. Knobs : Rotating knobs are recommended for light loading control with either fingers or with whole hand. They are typically used in applications such as instrument control panels.
6. Push buttons and toggle switches : These are essentially used as light load ON / OFF controls and are normally designed for operation by one finger.
7. Joysticks : This is a type of hand control extensively used in computer applications including CNC machines.
8. Foot pedals : These are used for fast action control with medium or heavy loading capacity. They lack in accuracy and range which may be obtained with hand controls. They are used in sitting position.

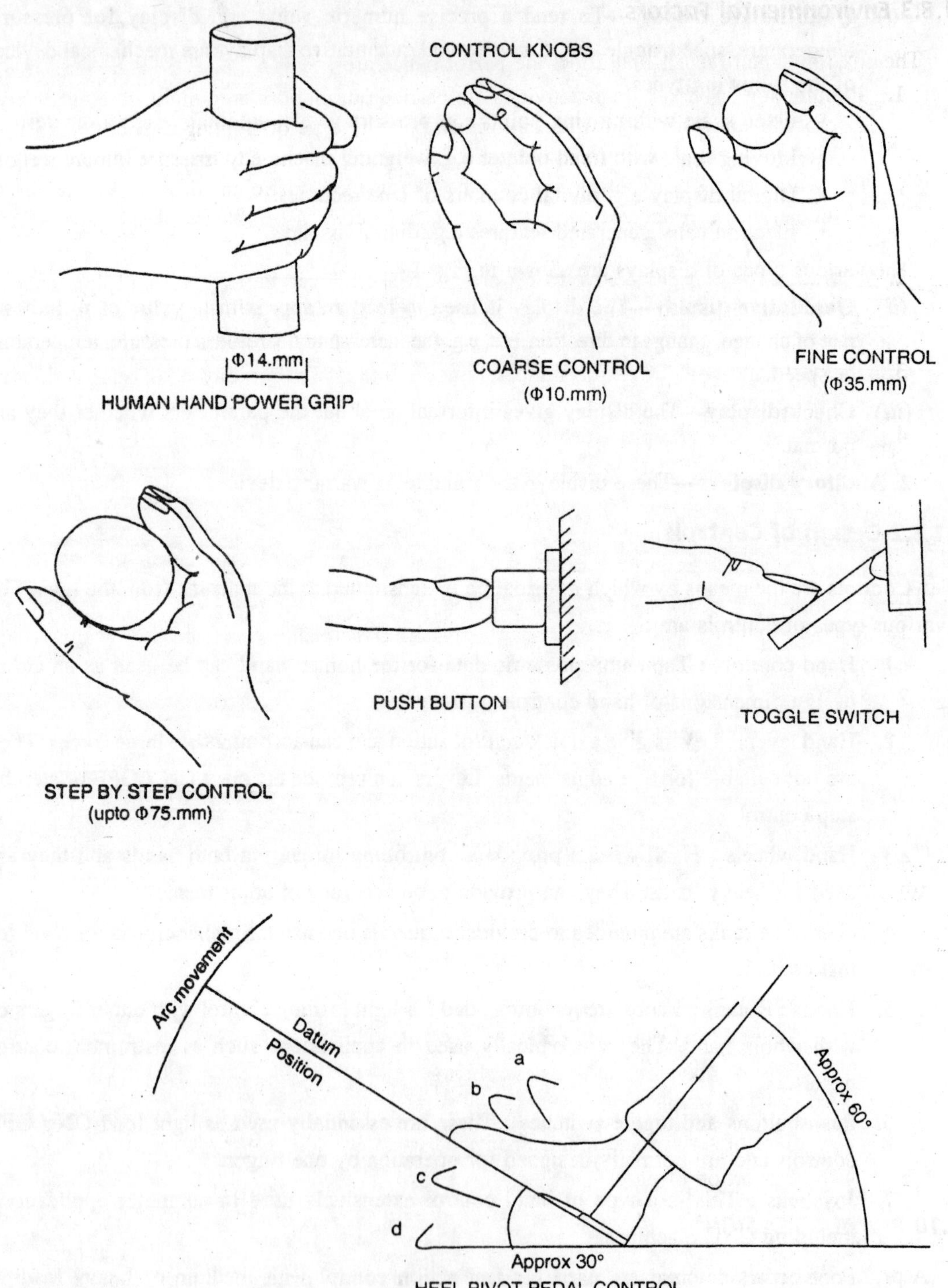

Fig. 1.5 Optimum size for human grip

1.8.3 Environmental Factors

The environmental factors that affect the performance are :

1. Illumination : When human activities are carried out indoors or at night, it is necessary to provide some sort of artificial illumination. The type of lighting depends upon the type of work being performed, the size of the objects, accuracy, speed and duration of the work, etc. The lighting system should provide sufficient brightness, uniform illumination, no reflected glare, and a contrast between brightness of the job and of background.
2. Vibration : Vibrations below 20 Hz are not heard but can be felt. Normally protection from residual vibration is achieved by reducing the force transmitted, by converting vibration energy into thermal energy by using mechanical or hydraulic dampers.
3. Thermal conditions : Comfort temperature = about 25°C, Humidity = 70%, Air velocity = 2 – 2.5 m/s.
4. Ventilation : It is the process of displacement of state air of the building by fresh air to reduce the presence of bad odour, CO_2, concentration, humidity and temperature. Windows, ventilators and exhaust fans can be used for effective ventilation.

1.9 MACHINE DESIGN PROCEDURE

The general procedure for machine design can be outlined as follows :

1. Get thoroughly familiar with what the machine is intended to accomplish and what special requirements or limitations must be considered.
2. Make free hand sketches of various ways the machine might be constructed and make any preliminary calculations which might be required to substantiate your ideas or to establish approximate sizes.
3. Make a layout drawing of the feasible construction of the machine bringing out the details of the construction.
4. Analyze the layout drawings as necessary for the finished design.

While designing, the following important things must be kept in view :

(*i*) Proper functioning of the finished machine.

(*ii*) Cost.

(*iii*) Lubrication.

(*iv*) Strength and rigidity of parts.

(*v*) Ease of manufacture and assembly.

(*vi*) Wear of parts.

(*vii*) Ease of service and replacement of parts.

(*viii*) Proportion of parts.

1.10 PRODUCT DESIGN

A product is an article obtained by the transformation of raw material. It may be a consumer product such as a television, a car or an industrial product like a lathe, a shaping machine, etc. Before starting the manufacturing of a product, it is essential to design it. The designer must consider various factors while designing a product.

1.10.1 Requirements of a Good Design

A good design must fulfil the following essential requirements :

1. Customers satisfaction. The product should satisfy the consumer by fulfilling their need and expectations. In order to achieve this objective, the following points should be kept in view :

(*i*) It should function properly.

(*ii*) It should be of proper quality so as to achieve desired degree of accuracy and reliability.

(*iii*) Easy to use or operate.

(*iv*) Good aesthetic view.

(*v*) It should be of a reasonable price.

2. Adequate profit. The product when manufactured and sold should give adequate profit. For this purpose it must satisfy the following requirements :

(*i*) It should be able to be manufactured at a reasonable price so that it can compete other products in the market.

(*ii*) It should have minimum number of parts.

(*iii*) It should adopt latest technology so that manufacturing requires minimum cost per unit of production.

1.10.2 Factors to be Considered while Designing a Product

While designing a product, the following factors should be considered :

1. Materials—Materials should be cheap and should be able to meet design requirements. During production wastage should be minimized. It should be easily workable and machinable.

2. Manufacturing facility—Product design should commensurate with the facilities available in the factory as regards to equipment, labour and layout.

3. Use of standardisation—The parts used should be of minimum variety and should either be easily available in the market or can be manufactured easily with the machines available. It should have as many standard and interchangeable parts as possible.

4. Aesthetic—The product should be good in appearance and should have attractive colours.

5. Function—It should be able to perform its desired function with desired accuracy, reliability and strength.

6. Ergonomics—It should be easy to use, operate and should cause minimum possible fatigue and provide comfort.

7. Operating conditions—The industrial product should operate with minimum of noise, vibration, heat and other hazards. The operators should be able to operate with available skill.

8. Maintenance—It should be easy to maintain and service.

9. Tolerance—It should not have tight tolerances to minimize rejection.

10. Value engineering—Factors related to the value engineering should be considered thoroughly to reduce the cost of production.

11. Simplification—The simplification techniques should be adopted to avoid manufacturing complications.

12. Economy—All possible ways to reduce the cost should be considered at the design stage.

1.10.3 Factors Affecting Product Design

The following factors need consideration while designing a product.

1. **Technical Factors**
   (*i*) Operating conditions—noise, vibrations, heat, etc.
   (*ii*) Performance—turning speed, accuracy, type of material etc.
   (*iii*) Maintenance aspects.
   (*iv*) Experience of the company.
2. **Industrial Engineering Factors**
   (*i*) Economy in production—materials, workstudy, layout, tooling, etc.
   (*ii*) Appearance—aesthetic look.
   (*iii*) Ergonomics—whether the use of product is comfortable.
3. **Production Engineering Factors**
   (*i*) Standardization.
   (*ii*) Simplification.
   (*iii*) Tolerance.
   (*iv*) Finish.
4. **Customers Satisfaction**

1.10.4 Activities in Product Design And Development

The activities related to designing, developing, producing and marketing are shown in the flow diagram given in Fig. 1.6.

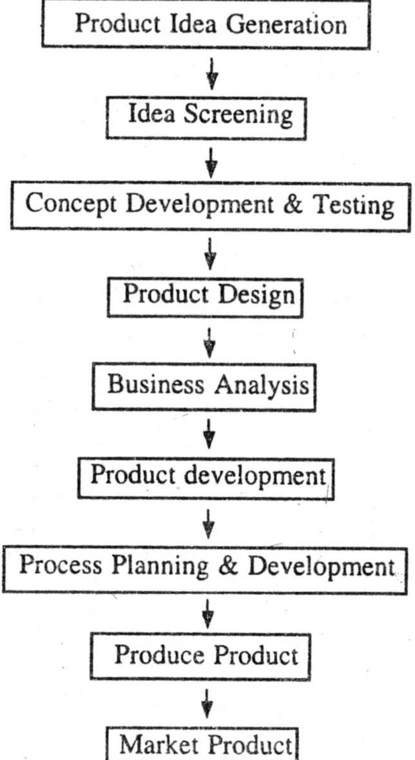

Fig. 1.6 Activities in product design and development

1. Product Idea Generation

The sources of idea generation are :

(*i*) Outside sources—customers, competitive products, foreign products, advertising/distributing agencies, researchers.

(*ii*) Inside sources—sales personnel, marketing personnel, R & D, production department, customer service department.

The idea generation techniques are :

(*i*) Attribute listing of product.

(*ii*) Need / problem identification.

(*iii*) Brainstorming.

2. Idea Screening

Evaluation of idea generated should be done on the following criteria :

(*i*) Product criteria—newness, technical feasibility, servicing requirements, legal considerations, etc.

(*ii*) Market criteria—market size, market share, market growth, distribution features etc.

(*iii*) Financial criteria—profitability, return on investment, cash flow, etc

3. Concept Development and Testing

Attractive, practicable and profitable ideas are then refined into testable product concepts. This is done in the following three stages : (*i*) Product idea—possible product to be offered to the market.

(*ii*) Product concept—a detailed version of the idea expressed.

(*iii*) Product image—customers, imagination about the product.

4. Product Design

The process of designing a product has the following steps :

(*i*) Recognition of Need—It is the realization that a problem exists for which some solution has to be found either by floating a new product or bringing improvements in the existing product.

(*ii*) Definition of Problem—This involves the specifications of the product to be designed and includes physical and functional characteristics, cost, quality and operating performance.

(*iii*) Analysis and Synthesis—The conceptualized product is subjected to analysis, improvement and redesign. The process is repeated until the design is optimized.

(*iv*) Evaluation—It is concerned with measuring the design and comparing it against the specifications established in the problem definition phase. Evaluation generally requires fabrication and testing of a prototype model to assess operative performance, quality, reliability and other characteristics.

(*v*) Presentation of Design—This means documentation of the design through drawings, material specifications etc. of each component / sub–assemblies.

These steps have been shown in Fig. 1.7.

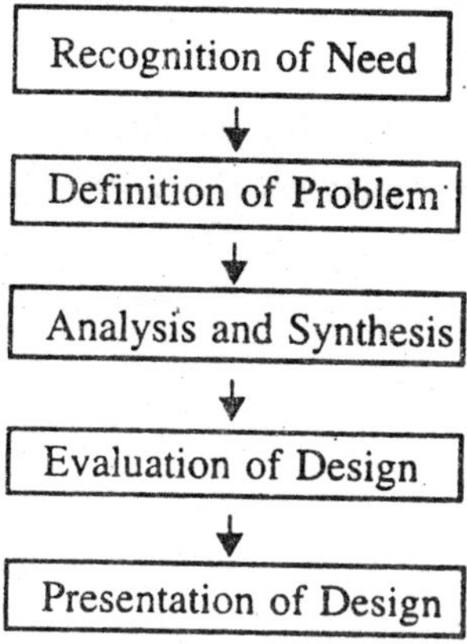

Fig. 1.7 Steps in product design.

5. Business Analysis

At this stage management needs to prepare the sales, cost and profit projections to assess whether it satisfies the company's objectives to yield a satisfactory profit. It is satisfies the objectives, the product can move to the product—development stage.

6. Product Development

After the business test the product is send to R & D and / or engineering division to be developed into a physical product. During this stage the company can come to know whether the product idea can be translated into technically and commercially feasible product. Prototypes are put to rigorous functional and consumer tests.

7. Process Planning and Development

This is related to the production aspects of translating the product design into a real product. This includes the selection of process which is profitable for required quantity of production, utilization of existing equipment, design of jigs and fixtures, etc. Due importance is given for planning and designing for inspection, packaging, transportation, storage and distribution.

8. Produce Product.

In the first stage, it is advisable to produce a batch of products only and observe their performance during its use. If need arises, improvements can be made before taking up the production on full scale.

9. Marketing the Product

The product is then launched in the market as per company's policy.

1.10.5 *Present Scenario of Product Design and Development*

In the present computer age, the product design and development are carried out using the following equipment and software :

1. Graphic workstations for modeling, analysis, optimization, design and NC programming.
2. Input devices like mouse, digitizer, track balls, etc. for data input.
3. Output devices like printer, plotter, colour plotter, inkjet printer etc. for making hard copies of designs.
4. Various software packages for RDBMS modeling analysis etc. and device drivers and communication software.

1.10.6 Steps for Product Development

The following steps are taken before marketing the product.

1. Market research to ascertain the need for a new product.
2. Research to provide base for design.
3. Design the product.
4. Develop the design.
5. Proto–type production.
6. Pre-production.
7. Manufacture.

1.10.7 Need Analysis

Consumption or use of the product usually decides the characteristics for the need. To get the information concerning the market or consumer characteristic, a survey or test-program is planned. Research of the past, present and the future consumption requirements is accomplished to get the realistic estimates of the future demand. Prejudices or pre-conceptions must be disregarded to find all possible characteristics of the need.

A primitive statement of need is prepared from the results of market research and survey. The constraints and limitations are quantified, if possible and criteria based on the need are suggested to judge the results. The final translation of need into statement of project goal or objective takes account or the constraints such as ability to solve, the resources, the time available, and the risk factors.

1.10.8 Traditional Design Methods

The two traditional methods of design are :

1. Design by evolution.
2. Design by drawing.

1. Design by Evolution

Design by evolution means the design and development of a product from earlier forms. For example : bullock cart, rowing boat, plows, musical instruments, cycle, etc. to the prevent developed form are the remarkable features of design by evolution. The salient features of design by evolution are as follows :

(*i*) The craftsmen do not prepare dimensional drawings of their product. They are unable to give satisfactory justification for the design of their product.

(*ii*) These products are developed by trial and error process over many centuries.

(*iii*) The modification in the product is costly because the craftman has to experiment with the product itself.

(*iv*) The essential information about the design of the product and its development are not available. It is passed on from one generation to the next generation of the craftsmen.

(*v*) Technical risks are small.

(*vi*) It can produce discarded features.

(*vii*) The fragmentary information is stored as pattern upon which the craft evolution depends.

Design by evolution has successfully developed some of the complex structures. However, due to its weakness to adapt to sudden changes in needs and to the challenge from mass production system, the method has become obsolete.

2. Design by Drawing

This is the new design approach. The essential features of design by drawing methods are as follows :

(*i*) The dimensions of products are specified in advance of their manufacture.

(*ii*) The complete manufacturing of the product can be subdivided into separate pieces, which can be made by different people. This division of work is not possible with craft evolution.

(*iii*) For a trial and error development of a product, the whole process can be carried out on the drawing board or computer terminal with graphic facility. Actual fabrication of the product is not required.

(*iv*) The design department carries out most of the intellectual activity of product design instead of the shop floor.

1.10.9 Technology Based Developments

Product development is very much dependent upon the technology being used. The technology is bound to change with changing time. The pace of technology has quickened and nowadays most companies are dependent on computer based design and development. Super computers and micro-computers are being used to solve complex problems.

The technology changes very fast and a firm which is unable to cope up with the technological changes may not survive. Changing technology calls for product modification and new developments. The fast changes in technologies affect the product development. Earlier VCP came, then VCR and now CD. Videocassette tapes have no demand nowadays as CDs have become popular due to better audio and video qualities.

From some time industries have witnessed a rapaid growth in the development of harder and difficult to machine materials such as hastalloy, nitralloy, waspalloy, nimonics, carbides, stainless steels, heat resistant steels and many more other high — strength temperature resistant (HSTR) alloys. These materials find wide applications in aerospace, nuclear engineering and other industries, due to their high strength to weight ratio, hardness and heat resisting qualities.

1.11 ROLE OF STANDARDISATION AND STANDARDS

Standardisation aims at establishing mandatory or obligatory norms or standards to which different types, grades, parameters (like length, diameter, etc.) quality characteristics, test methods, rules of marking, packing, storage of finished items, raw materials and semi-finished articles should conform. The main purpose of standardization is to minimize variety so that the number of models may be limited to a definite quantity. Standard items are of much use to a designer to achieve design for manufacture.

Bureau of Indian Standards (BIS) has standardized a number of items. For machine designers the items standardized are of the following categories :

1. Rules for preparing engineering drawings.
2. Symbols and notations for mechanical quantities and their units.
3. Engineering materials, their chemical composition, mechanical properties, method of heat treatment and method of mechanical testing.
4. Dimensions and preferred sizes of various mechine components.
5. Fits and tolerances for various machine elements from assembly considerations.

1.11.1 *Advantages of Standardisation*

Standardisation offers the following advantages :

1. Helps in obtaining design for manufacture.
2. Standardised products have better quality and are more reliable, giving higher service life.
3. Helps in having easy maintenance as standardized products are easily available for replacement.
4. The time and effort needed to manufacture a new machine is reduced remarkably as the standardized products can be easily obtained and assembled.
5. Interchangeability of components is possible due to standards, which reduces labour requirements and manufacturing costs.

1.11.2 *Standards*

There are three types of standards used in machine design, as described below :

1. International Standards, such as the standards prepared by International Standards Organisation (ISO).
2. National Standards, such as :

   I.S. : Bureau of Indian Standards, BIS

   DIN : German

   GOST : Russian

   BS : U.K.

   ASME : American Society of Mechanical Engineers

   AISI : American Iron and Steel Institute

   SAE : Society of Automotive Engineers

   AGMA : American Gear Manufacturer's Association

   ASM : American Society of Metals

   ASTM : American Society of Testing and Materials
3. Company Standards for use in a particular company or a group of sister concerns.

1.12 INTERCHANGEABILITY

Interchangeability of product means that any standardized component shall assemble correctly with the mating component, both being chosen at random. For this purpose, the parts should be as near as possible identical. Interchangeability ensures that the replacement parts can be drawn from the stock with the certainity that they will fit without alteration. Interchangeability of components is possible due to standardization.

1.13 PREFERRED NUMBERS

Preferred numbers form a general basis for standardizing and grading a series of similar dimensions, characteristics or articles. The preferred numbers are arranged in geometric series. The preferred numbers greatly reduce unnecessary variations in the sizes of an article, thus reducing the variety of production and of the material in the store.

The preferred numbers contain the powers of 10, such as 1, 10, 100, ... The intermediate numbers are obtained by dividing each decimal range (1 to 10, 10 to 100, ...) into 5, 10, 20, 40 and 80 geometrically equal steps. The series thus obtained are called the **Basic Series.**

Preferred numbers were first introduced by the French engineer Charles Renard in the 19th century. The geometric series are designated after his name by R5, R10, R20, R40 and R80. The progression ratio of the basic series are :

$$\phi_5 = \sqrt[5]{10} = 1.58$$

$$\phi_{10} = \sqrt[10]{10} = 1.26$$

$$\phi_{20} = \sqrt[20]{10} = 1.12$$

$$\phi_{40} = \sqrt[40]{10} = 1.06$$

$$\phi_{80} = \sqrt[80]{10} = 1.03$$

1.13.1 *Derived Series*

When the steps of a basic series are too close for a particular application, **Derived Series** may be obtained by taking every 2nd, 3rd, 4th or 5th term of a basic series. The derived series are designated as follows :

(*a*) R5/2 (1................10000000) : Derived series comprising every 2nd term of R5 series limited by the term 1 and 1000000, both inclusive.

(*b*) R10/3 (............80........) : Derived series comprising every 3rd term of R10 series, Unlimited in both directions and including the term 80.

(*c*) R20/4 (112..............) : Derived series comprising every 4th term of R20 series and limited at the lower end by the term 112 inclusvie.

(*d*) R40/5 (.................60) : Derived series comprising every 5th term of R40 series and limited at the upper end by the term 60 inclusvie.

1.13.2 *Preferred Value*

According to the accuracy of calculations the following may be defined :

Theoretical Value — The exact value of a term of any of the series. This is of no practical use.

Calculated Value — The value of a term, approximating its theoretical value, expressed to 5 significant digits. This is also of minor importance in practice.

Main or Preferred Value — This is the rounded off value of a term as given in the standard and is used generally. The limits of rounding off are +1.26% and 1.01%.

Theoretical value	Calculated value	Preferred value
$\left(\sqrt[10]{10}\right)^5$	3.1623	3.15
$\left(\sqrt[10]{10}\right)^{11}$	12.589	12.5

Preferred numbers are used in standardizing quantities like lengths, weights, flow rates, force, pressures, allowable loads, revolutions, velocities, gear ratios, etc. Since standardisation aims at minimizing variety, it is advisable to use R5 in preference to R10, R10 in preference to R20, R20 in preference to R40 and R40 in preference to R80. The preferred numbers are given in Table 1.2.

Table 1.2 Preferred numbers

R5	R10	R20	R40
1.00	1.00	1.00	1.00
			1.06
		1.12	1.12
			1.18
	1.25	1.25	1.25
			1.32
		1.40	1.40
			1.50
1.60	1.60	1.60	1.60
			1.70
		1.80	1.80
			1.90
	2.00	2.00	2.00
			2.12
		2.24	2.24
			2.36
2.50	2.50	2.50	2.50
			2.65
		2.80	2.80
			3.00
	3.15	3.15	3.15
			3.35
		3.55	3.55
			3.75
4.00	4.00	4.00	4.00
			4.25
		4.50	4.50
			4.75
	5.00	5.00	5.00
			5.30
		5.60	5.60
			6.00
6.30	6.30	6.30	6.30
			6.70
		7.10	7.10
			7.50
	8.00	8.00	8.00
			8.50
		9.00	9.00
			9.50
10.00	10.00	10.00	10.00

1.13.3 Characteristics of Preferred Series

The preferred series have the following special characteristics :

1. The product or quotient of any two terms of a series is always a term of that series.
2. The integral positive or negative power of any term of a series is always a term of that series.
3. The R series contain 10 and all +ve and –ve integral powers.
4. The number 3.15, which is nearly equal to π would be found among the preferred numbers of R10, R20 and R40. It follows that the length of circumference and the area of a circle, the diameter of which is a preferred number may also be expressed by preferred numbers.
5. The R40 series includes the numbers 3000, 1500, 1000, 750 and 375, which are the synchronous speeds of electrical machines for the standard frequency of 50 Hz.

1.13.4 Number of Steps

Let Range ratio, R_n = Maximum value of a quantity / Minimum value of the quantity

Common ratio $= \phi$

Number of steps $= z$

Then $$R_n = \phi^{(z-1)}$$

Taking ln of both sides, we get

$$\ln R_n = (z-1)\ln\phi$$

$$z = 1 + \ln R_n/\ln\phi$$

Example 1.1. A gear box is to have nine speeds in the range of 30 rpm to 1000 rpm. Calculate the preferred speeds.

Solution.

Given : $N_{max} = 1000$ rpm, $N_{min} = 30$ rpm, Number of steps, $z = 9$

$$N_{max}/N_{min} = \phi^{z-1}$$

$$1000/30 = \phi^{-1} = \phi^0$$

$$\phi = (1000/30)^{1/8} = 1.55$$

The speeds of gear box are :

$$N_1 = 30 \text{ rpm}$$

$$N_2 = \phi N_1 = 1.55 \times 30 = 46.5 \text{ rpm}$$

$$N_3 = \phi N_2 = 1.55 \times 46.5 = 72 \text{ rpm}$$

$$N_4 = \phi N_3 = 1.55 \times 72 = 112 \text{ rpm}$$

$$N_5 = \phi N_4 = 1.55 \times 112 = 173 \text{ rpm}$$

$$N_6 = \phi N_5 = 1.55 \times 173 = 268 \text{ rpm}$$

$$N_7 = \phi N_6 = 1.55 \times 268 = 416 \text{ rpm}$$

$$N_8 = \phi N_7 = 1.55 \times 416 = 645 \text{ rpm}$$

$$N_9 = \phi N_8 = 1.55 \times 645 = 1000 \text{ rpm}$$

Example 1.2 A manufacturer is interested in manufacturing 5 different models of generating sets ranging from 10 kW to 100 kW capacities. Specify the capacities of models.

If he further wishes to expand to 10 models within the same range, what will be the capacities of the additional models ?

Solution :

Given : $P_{max} = 100$ kW, $P_{min} = 10$ kW, $z = 5$

$$P_{max} / P_{min} = \phi^{z-1}$$

$$100/10 = \phi^{5-1} = \phi^4$$

$$\phi = (100/10)^{1/4} = 1.778$$

The speed of gear box are :

$$P_2 = 10 \text{ kW}$$

$$P_1 = \phi P_1 = 1.778 \times 10 = 17.78 \cong 18 \text{ kW}$$

$$P_3 = \phi P_2 = 1.778 \times 17.78 = 31.6 \cong 31.5 \text{ kW}$$

$$P_4 = \phi P_3 = 1.778 \times 31.6 = 56.2 \cong 56 \text{ kW}$$

$$P_5 = \phi P_4 = 1.778 \times 56.2 = 100 \text{ kW}$$

For $z = 10$, we have

$$P_{max}/P_{min} = \phi^{z-1}$$

$$100/10 = \phi^{10-1} = \phi^9$$

$$\phi = (100/10)^{1/9} = 1.2915$$

The speeds of gear box are :

$$P_1 = 100 \text{ kW}$$

$$P_2 = \phi P_1 = 1.2915 \times 10 = 12.915 \approx 13 \text{ kW}$$

$$P_3 = \phi P_2 = 1.2915 \times 12.915 = 16.68 \approx 16.5 \text{ kW}$$

$$P_4 = \phi P_3 = 1.2915 \times 16.68 = 21.54 \approx 21.5 \text{ kW}$$

$$P_5 = \phi P_4 = 1.2915 \times 21.54 = 27.82 \approx 28 \text{ kW}$$

Capacities of additional models :

$$P_6 = \phi P_5 = 1.2915 \times 27.82 = 35.93 \approx 36 \text{ kW}$$

$$P_7 = \phi P_6 = 1.2915 \times 35.93 = 46.4 \approx 46.5 \text{ kW}$$

$$P_8 = \phi P_7 = 1.2915 \times 46.4 = 59.93 \approx 60 \text{ kW}$$

$$P_9 = \phi P_8 = 1.2915 \times 59.93 = 77.4 \approx 77.5 \text{ kW}$$

$$P_{10} = \phi P_9 = 1.2915 \times 77.4 = 99.96 \approx 100 \text{ kW}$$

EXERCISES

1.1 A manufacturer is interested to produce 6 different models of trucks ranging from 50 kW to 400 kW power. Specify the paper capacities of the models.

1.2 A lathe machine is to be manufactured having minimum and maximum spindle speeds of 30 rpm to 900 rpm. If the speeds are to be in R10 series, determine the speeds.

1.3 The headstock gear box of a lathe have an arrangement for 9 spindle speeds from 20 to 800 rpm. Design the speeds.

1.4 The geared headstock of a machine is driven by an electric motor running at 3000 rpm. The headstock is to have 9 spindle speeds rangine from 60 to 750 rpm. Design the speeds.

1.5 (*i*) What do you understand by "Mechanical Engineering Design" ?

(*ii*) Explain the importance of aesthetic considerations in Machine Design.

(*iii*) What are preferred numbers ? How will you find the number belonging to R_{10} series ?

1.6 (*a*) What are Preferred Numbers ? Explain their application in standardization. Give examples.

(*b*) A manufacturer is interested in manufacturing 4 different models of generating sets ranging from 5 kW to 50 kW capacities. Specify the capacities of models. If he further wishes to expand to 8 models within the same range, what will be the capacities of the additional models ?

1.7 Explain the importance of preferred size in design process.

1.8 (*a*) What are preferred numbers ? Explain their advantages.

A manufacturer is interested to start the business with five different models of a machine ranging from 7.5 kW to 75 kW. Specify the power capacity of models.

(*b*) Explain what do you understand from "Mechanical Engineering Design". Define standardization. Explain various standards used in Machine Design.

1.9 (*i*) What do you understand by "Mechanical Engineering Design" ?

(*ii*) Discuss the basic requirments for machine elements and machines.

1.10 (*a*) What are the different stages in the design of a machine element, explain.

(*b*) With suitable example, explain the significance of preferred numbers in design.

(*c*) What is the importance of standardization in design, discuss with suitable examples.

2 SELECTION OF ENGINEERING MATERIALS

Chapter

2.1 INTRODUCTION

It is the responsibility of the design engineer to specify the material to be used for each element of the machine which he has designed. The material from which a part is to be made depends upon what the part is to do, how many parts are needed, and what material and processes are available to make it economically viable. The selection of a material depends primarily upon its properties and many more factors which a designer has to consider for a given situation.

2.2 SELECTION OF ENGINEERING MATERIALS

There are many factors which have to be considered while selecting a material. The most important to consider are those that are essential to the function of the part. Equal importance must be given to the properties, availability, manufacturing considerations, and economics aspects.

1. Properties

Selection of a material for a given application must be based on the following logical steps :

(*i*) Establishing the performance requirements that the material must meet.

(*ii*) Selecting the appropriate material for evaluation.

(*iii*) Selecting the material that best meets the performance requirements and economic constraints.

The properties of materials are :

(*a*) Mechanical :

(*i*) Tensile/compressive stength

(*ii*) Shear strength

(*iii*) Endurance limit

(*iv*) Impact strength

(*v*) Creep strength

(*vi*) Stiffness

Fatigue. The reduced tendency of a material to offer resistance to the applied stresses due to fluctuating cyclic loads is called fatigue. This phenomenon leads to failure of the material much below the tensile strength.

Endurance Limit. The maximum stress below which a material can presumably withstand an infinite number of cycles of stress is called endurance limit or fatigue strength.

Creep. It may be defined as the slow and progressive deformation of a material with time under a constant stress.

Creep strength. It may be defined as the highest stress that a material can withstand for a specified length of time without excessive deformation.

Stiffness. It is defined as the load per unit deflection or deformation.

(*b*) Tribological
- (*i*) Friction
- (*ii*) Wear, and
- (*iii*) Lubrication

(*c*) Physical
- (*i*) Density
- (*ii*) Coefficient of expansion
- (*iii*) Melting point
- (*iv*) Electrical properties
- (*v*) Magnetic properties
- (*vi*) Flammability

(*d*) Chemical
- (*i*) Composition
- (*ii*) Bonding structure
- (*iii*) Environmental resistance

(*e*) Dimensional
- (*i*) Flatness
- (*ii*) Surface finish
- (*iii*) Stability
- (*iv*) Critical size

(*f*) Technological
- (*i*) Machinability
- (*ii*) Weldability
- (*iii*) Castability
- (*iv*) Forgability
- (*v*) Bendability
- (*vi*) Malleability
- (*vii*) Ductility
- (*viii*) Hardenability

2. Availability

When initially considering the use of a material, the following points should be considered :

(*i*) Is the material on hand ?

(*ii*) If answer is No; can we get it in one to two weeks, and so on ?

(*iii*) If Yes, how much quantity is required ?

(*iv*) What is the nature of the supplier ? It is not advisable to use a proprietary material that is available only from one supplier.

(*v*) Is there any special method required for fabrication of raw material ? In that case it is essential for the designer to establish a time line for procurement of materials.

3. Economics

Cost is used as a prime selection factor for materials. The cost of a material in a machine component consists of the initial cost, processing cost and the maintenance cost.

4. Manufacturing Considerations

The manufacturing processes, such as casting, forging, rolling, extrusion, welding and machining govern the selection of a material. Machinability of materials is an important consideration in their selection. The shape of the components also plays an important role in the selection of the manufacturing process. Casting method may be preferable over machining methods for complicated shapes.

2.3 CLASSIFICATION OF ENGINEERING MATERIALS

The engineering materials may be classified as shown in Fig. 2.1.

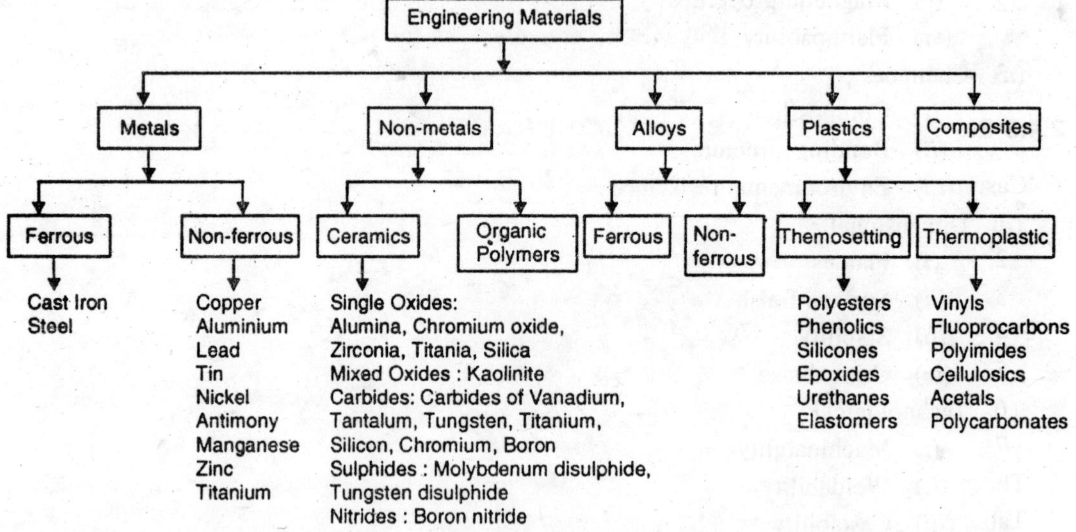

Fig. 2.1 Classification of engineering materials.

Metals are those in which there are extra valence electrons in the outermost shell of their structure. Metals can be ferrous or non-ferrous. In ferrous metals, carbon is one of the constituents, like steels and cast irons. Non-ferrous metals are those in which there are elements other than carbon like aluminium, copper, tin, manganese, etc. Non-metals are those in which there are no extra valence electrons in the outermost shell, like carbon, boron etc. Alloys are obtained by combining two or more than two ferrous metals, or ferrous and non-ferrous metals. The plastics and composites shall be discussed later on.

2.4 CAST IRONS

Cast iron is an alloy of iron, carbon and silicon. The percentage of carbon lies between 1.8 to 3 percent approximately. The carbon is present either in the form of graphite or iron carbide.

2.4.1 Advantages of Cast Irons

Cast irons offer the following advantages from design point of view :

1. It is very hard and brittle. Therefore, it has good wear properties.
2. It is very strong in compression and weak in tension.

3. It has good vibration absorption properties.
4. It is easily machinable.
5. It has good castability.
6. It is available at a reasonably low cost.
7. It has low notch sensitivity.

2.4.2 Applications of Cast Irons

Cast irons have following applications :

1. It is generally used for machine housings and frames, tables and guides, pulleys, highly loaded gears, sprockets, hydraulic cylinders, anvils and dies.
2. I.C. engine cylinder blocks and cylinder head, piston and piston rings, crank case, exhaust manifolds, flywheel, brake drum, pressure plate for clutch and brake shoes.
3. Hydraulic system pipes and pipe fittings, valve bodies and dome covers.

2.4.3 BIS System of Designation of Cast Irons

Cast irons may be classified as follows :

1. Grey Cast Iron
2. Malleable Cast Iron
3. Spheroidal or Nodular Graphite Iron
4. Austenitic Flake Graphite Iron
5. Austenitic Spheroidal or Nodular Graphite Iron
6. Abrasion Resistance Iron
7. Corrosion Resistant Iron

The BIS system of designation of cast irons and their applications are given in Table 2.1.

Table 2.1 BIS designation of cast irons and their applications.

Cast Irons	Designation	Applications
1. Grey	FG followed by the tensile strength in MPa, e.g. FG 350 indicates Grey cast iron of tensile strength 350 N / mm^2	Machine tool bodies, automobile parts, pipe fittings, agricultural implements.
2. Malleable	Black heart, white heart, and Perlitic malleable cast irons are designated respectively by BM, WM and PM followed by the tensile strength in MPa.	Gear housing, brake pedals, automotive and agricultural parts.
3. Spheroidal or Nodular Graphite	SG followed by the tensile strength in MPa and oblique percentage elongation, e.g. SG 600 / 2 indicates a spheroidal and nodular graphite iron having tensile strength of 600 MPa and elongation of 2%.	Crankshafts, pistons, guides, control levers of heavy machines, wrenches and dies.

Cast Irons	Designation	Applications
4. Austenitic Flake Graphite	AFG followed by the name and percentage of important elements of chemical composition.	Pressure covers for turbine-generator sets, housings for switch gear, insulator flanges, bushings, piston rings, pump components, valve castings, machine tools.
5. Austenitic Spheroidal or Nodular Graphite	ASG followed by the name and percentage of important elements of chemical composition.	-do-
6. Abrasion Resistant	ABR followed by the minimum tensile strength in MPa and the name and percentage of important elements of chemical composition.	Stone crushers, cement plants mixtures, rolling mills
7. Corrosion Resistant	CSC followed by the minimum tensile strength in MPa and the name and percentage of important elements of chemical composition.	Industrial furnaces, acid plants, sea going ships.

2.4.4 Effect of Alloying Elements

The effect of alloying elements on cast iron is given in Table 2.2.

Table 2.2 Effect of alloying elements on cast iron.

Alloying Element	Percentage	Effects
Nickel	0.25-5%	Refines grain structure, increases strength and toughness, increases resistance to corrosion. Acts as a graphitizer.
Chromium	10-30%	Refines grain structure, increases strength, hardness and resistance to corrosion, increases wear resistance and heat resistance. Tends to prevent graphitization.
Copper	Upto 1%	Increases fluidity, imparts corrosion resistance and improves toughness and hardness. Improves machinability slightly.
Molybdenum	0.25-1.25%	Increases the strength and toughness. Improves high-temperature strength.
Vanadium	0.10-0.50%	Increases strength and resistance to fatigue stresses. Reduces graphitization.
Boron	Upto 0.05%	Increases surface hardness.

2.4.5 Effect of Impurities on Iron

The impurities generally present in cast iron are : Sulphur, Manganese, Phosphorus and Silicon. Their effects on cast iron are given in Table 2.3.

Table 2.3 Effect of impurities on cast iron.

Impurity	**Percentage**	**Effects**
Sulphur	Upto 0.1%	Lowers fluidity during pouring, decreases strength and makes the metal more brittle.
Manganese	Upto 0.75%	Tends to whiten and harden cast iron.
Phosphorus	Upto 1.0%	Increases flowability of grey cast iron, increases brittleness.
Silicon	Upto 2.5%	Makes cast iron soft and easily machinable, decreases blow holes in castings and helps in producing sound castings.

Table 2.4 Typical Properties of Grey Iron Castings.

Properties	*Grade*						
	FG 150	*FG 200*	*FG 220*	*FG 260*	*FG 300*	*FG 350*	*FG 400*
Tensile strength, N/mm^2	150	200	220	260	300	350	400
Compressive strength, N/mm^2	600	720	768	864	960	1080	1200
Shear strength, N/mm^2	173	230	253	299	345	403	460
Torsional strength, N/mm^2	173	260	253	299	345	403	460
Modulus of Elasticity :							
Compression Tension, GN/m^2	100	114	120	128	135	140	145
Modulus of Rigidity, GN/m^2	40	46	48	51	54	56	58
Poisson's Ratio	0.26	0.26	0.26	0.26	0.26	0.26	0.26
Fatigue limit, N/mm^2 (Wohler, unnotched, 8.4 mm dia)	68	90	99	117	135	149	152
Density, kg/m^3	7050	7100	7150	7250	7250	7300	7300

IS : 210 – 1978

Table 2.5 Properties of Spheroidal Graphite or Nodular Graphite Iron

Grade	*Tensile strength N/mm^2*	*Proof strees, 0.2% N/mm^2*	*Elongation %*	*Impact strength Charpy Nm*	*Brinell Hardness*	*Predominant Structure*
SG 800/2	800	480	2	–	245.350	Pearlite or tempered
SG 700/2	700	440	2	–	225–300	Pearlite
SG 600/3	600	370	3	–	190–270	Pearlite and Ferrite
SG 500/7	500	320	7	–	170–240	Ferrite and Pearlite
SG 400/12	400	250	12	6	max-200	Ferrite
SG 370/17	370	230	17	15	max-180	Ferrite

IS : 1865–1974

Table 2.6 Tensile Properties of White Heart Malleable Iron Castings

Grade* Designation	Sectional Thickness Represented mm	Diameter of Test Bar mm	Tensile strength N/mm^2	0.5 per cent proof stress. Min N/mm^2	Elongation (L_0 = 3D) Min %
WM 410	8 and under	9	350	200	10
	over 8 upto and including 13	12	390	230	6
	over 13	15	410	250	4
WM 340	8 and under	9	270	–	7
	over 8 upto and including 13	12	310	–	4
	over 13	15	340	–	3

IS : 2107–1977

As per IS : 4843–1968

Table 2.7 Tensile Properties of Blackheart Malleable Iron Castings

Grade* Designation	Brinell Hardness HB Max	Dia of test bar mm	Tensile strength Min N/mm^2	0.5 per cent proof stress, Min N/mm^2	Elongation (L_0 = 3D) %
BM 340	150	15	340	200	12
BM 310	150	15	310	190	10
BM 290	160	15	290	–	6

* As per IS : 4843 – 1968

IS : 210–1977

Table 2.8 Tensile Properties of Pearlitic Malleable Iron Castings

Grade* Designation	Brinell Hardness HB Max	Diameter of Test Bar mm	Tensile strength Min N/mm^2	0.5 per cent Proof stress, Min N/mm^2	Elongation (L_0 = 3D) %
PM 690	241–285	15	690	540	2
PM 590	200–248	15	590	420	3
PM 540	180–230	15	540	350	4
PM 490	170–229	15	490	310	5
PM 440	149–201	15	440	270	7

* As per IS : 4843–1968

IS : 2640 – 1977

2.5 STEELS

Carbon steels are alloys of iron and carbon only. It contains carbon upto a maximum of 1.8%. It is obtained by oxidizing the impurities in molten pig iron and then adding the required amount of carbon. The carbon exists in steel in the combined form as iron carbide.

2.5.1 Classification of Steels

Steels for machine applications may be classified as follows :

1. Carbon steel—used for wire for ropes, cold rolled sheets and strip, plates for general engineering use.
2. Alloy steel—used for bearings.
3. Stainless steel—used for spring washers, helical springs, valves for I.C. engines, cold forged rivets, bars and flats, plate, sheet and strips.
4. Special steel—used for pressure vessels, spring strips, volute, laminated and helical springs for railways and automobiles, naval vessels.

Low carbon or mild steel : C = 0.05 – 0.30%

Medium carbon steel : C = 0.30 – 0.60%

High carbon steel : C = 0.60 – 1.5%

2.5.2 Designation of Steel

Steels contain varying amount of carbon and other alloying elements. To distinguish one type of steel from another it is desirable that some system of naming them may be evolved. The Bureau of Indian Standards (BIS) have recommended a system of naming various types of steels. This is called the system of designation for steels. All types of steels are designated as follows :

1. On the Basis of Mechanical Properties

Carbon and low alloy steels where the main criterion in the selection and inspection is the tensile or yield stress are designated in the following order :

(*i*) Symbol "Fe" or "FeE" depending on whether the steel has been specified on the basis of minimum tensile strength or yield stress.

(*ii*) Figure indicating the minimum tensile strength or yield strength in N / mm^2. If no minimum tensile or yield strength is guaranteed the figure shall be 00.

(*iii*) Chemical symbols for elements the presence of which characterize the steel.

(*iv*) Symbol indicating special characteristics covering method of deoxidation, degreepf purity, etc.

(*v*) Symbol indicating applications, if necessary. For example, Fe 410 Cu K indicates, killed steel containing copper as alloying elements with a minimum tensile strength of 410 N / mm^2.

2. On the Basis of Chemical Composition

(*a*) Unalloyed steels. They are designated in the following order :

(*i*) Figure indicating 100 times the average percentage of carbon content.

(*ii*) Letter "C".

(*iii*) Figure indicating 10 times the average percentage of manganese content and rounded off to the nearest integer following the rule for rounding off.

(*iv*) Symbol indicating special characteristics including guaranteed hardenability for which symbol "G" is used at the end of the designation.

Examples. 1. 55C4 indicates a plain carbon steel with 0.55% carbon and 0.4% manganese. 2. A steel with 0.35 – 0.45% carbon and 0.7 – 0.9% manganese is designated as 40C8.

(*b*) **Unalloyed tool steels.** They are designated in the following order :

(*i*) Figure indicating 100 times the average percentage of carbon.

(*ii*) Symbol "T" for tool steel.

(*iii*) Figure indicating 10 times the average percentage of manganese content.

Examples : 1. 50T4 indicates an unalloyed tool steel with 0.50% carbon and 0.4% manganese.

2. A tool steel with 0.35 – 0.45% carbon and 0.7 – 0.9% manganese is designated as 40T8.

(*c*) **Unalloyed free cutting steels.** They are designated in the following order :

(*i*) Figure indicating 100 times the average percentage of carbon.

(*ii*) Letter "C".

(*iii*) Figure indicating 10 times the average percentage of manganese.

(*iv*) Symbol "S", "Se", "Te" or "Pb" depending on the element present which makes the steel free cutting, following by the figure indicating 100 times the percentage content of the element. For phosphorized steels the symbol P is included.

(*v*) Symbol indicating special characteristics.

(*d*) **Low and medium alloy steels.** They are designated in the following order :

(*i*) Figure indicating 100 times the average percentage of carbon.

(*ii*) Chemical symbol for alloying elements each followed by the figure for its average percentage content multiplied by 100.

(*iii*) When the alloying element is less than one percent, it is written upto two decimal places underlined by a bar.

Examples : 1. Consider an alloy steel with the following composition :

Carbon = 0.12 – 0.18%

Silicon = 0.10 – 0.35%

Manganese = 0.40 – 0.60%

Chromium = 0.50 – 0.80%

Average percentage of carbon = 0.15, which is denoted by 15

Silicon and manganese are not important alloying elements and are deleted.

The significant element chromium has average percentage 0.65%, which is less than 1%. Hence, it is denoted as Cr$\underline{65}$. Therefore, complete designation of steel is 15Cr$\underline{65}$

2. A steel designated by 40Cr$\underline{14}$ has average carbon 0.4% and chromium 0.14%.

3. A steel having chemical composition : C = 0. 15 – 0.25%, Si = 0.10 – 0.50%, Mn = 0.30 – 0.50 %, Ni = 1.5 – 2.5%, Cr = 16 – 20% has major average alloying elements Cr = 18% and Ni = 2%. C = 0.20%. Hence its designation is : 20Cr18Ni2

(*e*) **High Alloy steels.** They are designated in the following order :

(*i*) Letter "X".

(*ii*) Figure indicating 100 times the average percentage of carbon content.

(*iii*) Chemical symbol for alloying elements each followed by the figure for its average percentage content rounded off to the nearest integer.

(*iv*) Chemical symbol to indicate specially added elements to attain the desired properties.

(*v*) Symbol indicating specific characteristics.

(*f*) **Alloy tool steels.** They are designated as follows :

(*i*) Letter "T" for low alloy and medium alloy.

(*ii*) Letter "XT' for high alloy and as given for low and medium and high alloy steels.

(*g*) **Free cutting alloy steels.** They are designated as follows :

As stated above plus a symbol of the additional element present followed by the figure indicating 100 times its content.

2.5.3 Applications of Steels

The applications of steels are given in Table 2.9.

Table 2.9 Applications of steels.

Designation of steel	Applications	Ultimate tensile strength S_{ut}, MPa	Yield strength S_{yt} MPa
1. Plain carbon steels			
7C4	Automobile bodies and hoods.	320	176
10C4	Cam and camshaft, worm, gudgeon pin, chain wheel and spindle.	340	187
30C8	Socket, tie rod, yoke, lever and rocker arm.	500	400
40C8	Transmission shaft, crankshaft, spindle, connecting rod, stud and bolt.	580	380
45C8	Transmission shaft, machine tool spindle, bolt and gear of large dimensions.	630	380
50C4	Transmission shaft, worm, gear and cylinder.	760	460
55C8	Gear, cam, sprocket, cylinder and key.	720	460
60C4	Machine tool spindle, hardened bolt and pinion.	750	480
65C6	Flat and coil springs.	750	480
2. Free cutting steels			
10C8S10	Bushes, tappets, gudgeon pin, pawls and worms.	460	253
14C14S14	Pinions, motorcycle and bicycle components.	520	286
25C12S14	Bolts and studs.	560	308
40C10S18	Bolts, crankshaft and connecting rod.	600	330
40C15S12	High strength bolts, axles and shafts.	640	352

Designation of steel	Applications	Ultimate tensile strength S_{ut}, MPa	Yield strength S_{yt} MPa
3. Alloy steels			
40Cr1	Gear, connecting rod and axle.	680	520
40Ni3	Turbine blades and high strength bolts.	780	580
35Ni1Cr60	Crankshaft, connecting rod, gear shaft, cam shaft and gears.	680	520
30Ni14Cr1	High strength gears.	1510	1270
40Ni2Cr1Mo15	Transmission shaft, axle, gear and bolts.	780	580
27Mn2	Welded parts, spindle and shaft.	580	430
37Mn2	Transmission shaft and axle.	580	430
50Cr1V23	Leaf and coil springs.	1860	1760
4. Stainless steels			
04Cr13	Hydraulic pumps, valves and steam applications.	440	242
05Cr17		440	242
15Cr25V20		490	270
02Cr18Ni11	Cold press work	440	242
04Cr18Ni10		490	270
10Cr17Ni7		590	324
04Cr17Ni12Mo2		490	270
02Cr17Ni12Mo2		440	242
12Cr13	Springs and high temperature applications	590	324
20Cr13		690	380
30Cr13		780	429
15Cr16Ni2		830	456

2.5.4 Effect of Alloying Elements on Steel

Table 2.10 Eeffect of alloying elements on steel.

Alloying Element	Percentage	Effects
Chromium	Upto 30%	Increases toughness and hardness, increases resistance to abrasion, wear, corrosion and oxidation.
Nickel	Upto 50%	Increases strength and toughness without decreasing ductility.

Alloying Element	Percentage	Effects
Silicon	Upto 0.8%	Acts as deoxidizing agent, produces high magnetic permeability.
Molybdenum	0.20 – 0.70%	Increases hardness and improves machining properties.
Vanadium	0.2 – 0.5%	Increases toughness and strength, increases life of tools, springs and members subjected to high temperature.
Tungsten	0.4 – 22%	Widely used in tool steels, increases toughness and hardness
Cobalt	0.15 – 0.30%	Increases hot hardness.
Copper	0.15 – 0.30%	Improves resistance to atmospheric corrosion.
Aluminium	1 – 5%	Used as a deoxidizer, restricts grain growth.
Sulphur	0.06 – 0.30%	Increases machinability.

2.5.5 *Reasons for Use of Alloy Steels in Machine Design*

The alloy steels consist of iron plus carbon and varying percentages of the following alloying metals :

Tungsten (W), Molybdenum (Mo), Chromium (Cr), Vanadium (V), Cobalt (Co), Nickel (Ni), Silicon (Si), Copper (Cu), Aluminium (Al) and Sulphur (S).

These alloying metals are added to plain carbon steels to improve one or the other properties so that the part may function to the desired expectancy. The properties to be improved upon are the following :

1. Toughness and hardness.
2. Abrasion, wear, corrosion and oxidation resistance.
3. Tensile strength and toughness without decreasing ductility.
4. Machinability.
5. High temperature strength.
6. Electrical and magnetic properties.
7. Hot hardness
8. Grain structure.

2.5.6 *Special Alloy Steels*

1. (*a*) Stainless Steels. These steels have high resistance to corrosion against atmospheres of air and water vapour, sea water, solutions of acids or salts, etc. There are three types of stainless steels. (*a*) Austenitic stainless steels—These are chrome-nickel steels having the following chemical composition :

$$C = 0.12 - 0.14\%$$

$$Cr = 17 - 20\%$$

$$Ni = 8 - 11\%$$

The most common steel in this group is the 18 – 8 stainless steel containing Cr = 18% and Ni = 8%. These steels are used in food processing and handling machinery, kitchen equipment, dairy plants, textile machinery, surgical and dental instruments, and nuclear engineering.

(*b*) **Ferritic stainless steels.** These steels mainly contain chromium from 14 – 30%. A thin film of chromium oxide formed on the surface of steel protects it against corrosion.

(*c*) **Martensitic stainless steels.** It contains chromium from 12-14%. These steels are used for valves, screens, pump shafts, cutlery, bolts, nuts and various other parts in chemical and petroleum industries, steam turbine blades, jet engine compressor blades, carburetor parts, instrument parts, surgical instruments, ball bearings, bushings, valve seats, etc.

2. Killed Steel. Steel that has been deoxidized with a strong deoxidizing agent, such as Silicon or Aluminium, in order to eliminate a reaction between carbon and oxygen during solidification. This steel has fewer blow holes and are suitable for sound forgings and heavy rolled sections.

3. High Speed Steel (HSS). This is a high carbon steel tool material to which the various alloying elements added are : Tungsten (W), Molybdenum (Mo), Chromium (Cr), Vanadium (V) and Cobalt (Co). These alloying elements improve the hardness, toughness and wear resistance properties of steel. They are capable of retaining their hardness upto 600°C and can be operated at much higher cutting speeds as compared to alloy steels.

The 18 – 4 – 1 H.S.S. contains : W = 18%, Cr = 4% and V = 1%. Cobalt = 5 – 8% is added to increase red hardness. Carbon is about 0.75%. This material is designated as T – 4. The BIS designation of this steel is :

T 75 W 18 Co 5 Cr 4 V 1 Mo 70.

The most commonly used grades of HSS are : M – 1, M – 2, M – 7, M – 10, T – 1 and T – 2. The composition of these high speed steel tool materials is given in Table 2.6.

Table 2.11 High Speed steels

Designation	**Percentage of constituents**					
	C	**W**	**Mo**	**Cr**	**V**	**Co**
M – 1	0.80	1.75	8.50	3.75	1.15	-
M – 2	0.85	6.00	5.00	4.00	2.00	-
M – 7	1.00	1.75	8.75	4.00	2.00	-
M – 10	0.90	-	8.00	4.00	2.00	-
T – 1	0.75	18.00	-	4.00	1.00	-
T – 2	0.85	18.00	-	4.00	2.00	-
T – 4	0.75	18.00	-	4.00	1.00	5

4. Spring Steel. The cold coiled springs are made from circular section steel wire patented and cold drawn, unalloyed. The hot coiled springs are made from Grades 1, 2 and 3 steel wire of following chemical composition :

50Cr4V2, 60Cr4V2, 55Si7, 60Si7 and 65Si7

The railway rolling stock springs are made from hot rolled steel of following grades :

98C6, 113C6, 55Si7, 60Si7, 50Cr4V2 and 60Cr4V2

5. Free Cutting Steels

These are carbon steels and carbon-manganese steels with a small percentage of sulphur. Sulphur improves the machinability of these steels.

Table 2.12 Mechanical Properties of Steels Specified by Tensile and/or Yield Properties

Designation	Tensile Strength *N/mm²*	Yield Stress *Min* *N/mm²*	Elongation percent, *Min* *Gauge Length* = $5.65\sqrt{A_0}$
Fe 290	290	170	27
Fe E 220	290	220	27
Fe 310	310	180	26
Fe 230	310	230	26
Fe 330	330	200	26
Fe E 250	330	250	26
Fe 360	360	220	25
Fe E 270	360	270	25
Fe 410	410	250	23
Fe E 310	410	310	23
Fe 490	490	290	21
Fe E 370	490	370	21
Fe 540	540	320	20
Fe E 400	540	400	20
Fe 620	620	380	15
Fe E 460	620	460	15
Fe 690	690	410	12
Fe E 520	690	520	12
Fe 770	770	460	10
Fe E 580	770	580	10
Fe 870	870	520	8
Fe E 650	870	650	8

IS : 1570 (Part–I)–1978

Table 2.13 Typical Uses of Steels Specified by Tensile Properties

Steel Designation	*Typical Uses*
Fe 290	Structural steel sheets for plain drawn parts, tubes for oil well casing, steam, water and air passages, cycle, motor cycle and automobile tubes, rivet bars and wire.
Fe 310 Fe 330	Steels for locomotive carriage and car structures, screw stock
Fe 360	Structural steel for chemical pressure vessels.

Steel Designation	*Typical Uses*
Fe 380	Structural steel for ships, chemical vessels, air receivers and fasteners.
Fe 410	Structural steel for bridges and building construction, railway rolling stock, screw spikes, and oil well casing.
Fe 430	Structural steel for railway stock, pressure vessels, fasteners and valve fittings for compressed gas cylinders.
Fe 460	Structural steel for railway rolling stock, pressure parts of marine and land boilers and rivets for air receivers.
Fe 490	Structural steel for mines, forgings for marine engines, and machine parts.
Fe 510	Steel for railway wheels, disc wheel centres for railway and electric tramway cars, and seamless tubes.
Fe 540	High tensile steel for locomotive, carriage wagon and tramway axles, bolts, and seamless and welded tubes.
Fe 570	High tensile steel for bridges and general construction and bars and wire for concrete reinforcement.
Fe 620	High tensile steel for tramway axles and seamless tubes.
Fe 650 Fe 770 Fe 870	High tensile for locomotive, carriage and wagon wheels and tyres, and machine parts for heavy loading.

Table 2.14 Typical Uses of Carbon Steels

Steel Designation	*Typical Uses*
4 C 2	Dead soft steel generally used in electrical industry.
5 C 3, 7 C 3, 10 C 4	Used for forming and deep drawing. Rimming quality used for automobile body and fender stock, hoods, lamps, oil pans. They are also used for heading wire and rivets and low carbon wire products. Killed quality is used for forgings and heat treating applications.
10 C4, 14 C 6	Case hardening steels are used for making cam shafts, cams, light duty gears, worms, gudgeon pins, selector forks, spindles, pawls, ratchets, chain wheels, tappets etc.
15 C 4	Used for lightly stressed parts. It is suitable where cold web, such as bending are riveting may be necessary.
15 C 8, 20 C 8, 25 C 4 and 25 C 8	General purpose steels for low stressed components
30 C 8	Used for cold formed shift and brake levers. Hardened and tempered steel is used to make parts such as socket, tie rods, yolk adjustable control lever cable, shaft fork and rear hub, sprocket, lever, cams, rocker arms and bushes, tubes. Also used for tubes for aircraft, automobile and bicycle.
35 C 4	Steel for low stressed parts, automobile tubes and fasteners.

Steel Designation	*Typical Uses*
35 C 8	Steel for making low stressed parts in machine structure, cycle and motor cycle frame tubes fish plates for rails and fasteners.
40 C 8	Steel for crank, shafts, spindle, automobile axle beams, push rods, connecting rods, studs, bolts, lightly stressed gears etc.
45 C 8	Steel for spindle of machine tools, bigger gear, bolts and shafts.
50 C 8	Steel for making keys, shafts, cylinders, machine components required for moderate power resistance. Hardened steel is used for a large pitch worms and gears.
50 C 12	Rail steel. Also used for making spike bolts, gear shafts, rocking levers and cylinder liners.
55 C 6 and 55 C	Steels used for making gears, cylinders, cams, keys, crankshafts, sprockets and machine parts requiring moderate wear resistance for which toughness is not of primary importance.
60 C 6	Steel for making spindles for machine tools, hardened screws and nuts. Couplings, crankshaft, axles and pinions.
65 C 6	High tensile structure steel for making locomotive carriage and wagon tyres. Used also to make engine valve springs, small washers and thin stamped parts.
70 C 6	Steel for making baffle springs, shock absorbers, rail tyres, unhardened gears and worms.
75 C 6	Used for making light flat spirings formed from annealed stock. When properly heat treated it is used for making plough shears, rake teeth scrappers and cultivator's shovels.
80 C 6 and 85 to 6	Used for making flat and coil springs for automobiles and railway vehicles. After suitable heat treatment these steels are also used for making scraper blades, discs and spring tooth harrows.
98 C 6 and 113 C 6	In oil hardened and tempered condition it is used for coil or spiral springs.

Table 2.15 Typical Uses of Alloy Steels

Steel Designation	*Typical Uses*
10 C 8 S 11	Small parts to be cyanided or carborised.
14 C 14 S 14	Parts where good machinability and finish are important.
25 C 12 S 14	Bolts, studs and other heat treated parts of small section.
40 C 10 S 18	Heat treated bolts, engine shafts, connecting rods, miscellaneous gun carriage and small parts not under high stresses and wear.
13 C 10 S 25	Components under light stresses and not under shock. Suitable for components for production on automatic lathes.
40 C S 12	Heat treated axles, shafts, small crankshafts and other vehicle parts.
55 Si 7	Oil hardening spring steel for road and railway vehicles.
20 Mn 6, 27 Mn 6	Welded structures, crankshaft steering levers, shafting spindles, etc.

Steel Designation	*Typical Uses*
37 Mn 6	Axles, shafts, crankshafts, connecting rods, etc.
35 Mn 6 Mo 3	General engineering components such as crankshafts, bolts, wheel studs, axleshafts, levers and connecting rods.
35 Mn 6 Mo 4	General engineering components such as crankshafts, connecting rods, axle shafts, levers, bolts and wheel studs.
17 Mn 5 Cr 4	Small gear wheels and shafts, cardan joints and steering regulators.
40 Cr 4	Gears, connecting rods, stud axles, steering arms, wear resistant plates for earthmoving and concrete handling equipment.
50 Cr 4	Spring steel.
105 Cr 4	Roller bearing steels.
105 Cr 5, 50 Cr 4 V 2	Laminated, coil and volute springs.
40 Cr 4 Mo 3	Axle shafts, crankshafts, connecting rods, gears, high tensile bolts and studs, propeller shaft joints, etc.
15 Cr 4 Mo 6	Boiler superheater tubes, plates, pipefittings, forging and bolting material for petroleum and chemical industries.
15 Cr 13 Mo 6, 25 Cr 20 Mo 6	Crankshafts, cylinder liner for aero and automobile engines, gears and machine parts with high surface hardness and wear resistance.
40 Cr 13 Mo 10 V 2	Components requiring high tensile properties, heavy stress and severe wear.
40 Cr 7 Al 10 Mo 2	Components requiring maximum surface hardness of a nitraded case and high core strength.
40 Ni 14	Parts requiring high toughness, parts, working at low temperatures, heavy forgings, turbine blades, severely stressed screws bolts and nuts.
16 Ni 4-Cr 3	Gear box and transmission components, cams, camshafts, etc.
35 Ni 5 cr 2	Crankshafts, connecting rods, gear, shafts, chain parts, clutches, flexible shafts for planetary gears, camshaft of aircrafts and heavy vehicles.
35 Ni 6 Cr 5	Highly stressed gears, components requiring high tensile strength.
15 Ni 5 Cr 4 Mo 1,	Heavy duty gears, heavy vehicles and automobile transmission components.
15 Ni 7 Cr 4 Mo 2	
40 Ni 6 Cr 4 Mo 2	General machine parts such as bolts, nuts, gear, axles, connecting rods, shafts etc.
40 Ni 6 Cr 4 Mo 3	High strength machine parts, collects, spindles, screws, high tensile bolts and stubs, gears, pinions, axle shafts, tappets crankshafts, connecting rods, boring bars, arbours etc.
31 Ni 10 Cr 3 Mo 6	Highly stressed bolts and nuts, shafts, crankshafts, connecting rods, gears, axles mandrel bars, components for aircraft power units, airframes and under carriages, for low temperature service.
40 Ni 10 Cr 3 Mo 6	Highly stressed gears, connecting rods, crankshaft, components for all types of aircraft power units, axles, airframes and under carriages, die blocks, mandrel bars, etc., for low temperature service.

Table 2.16 Mechanical Properties (Core) of Case Hardening Steels in the Refined and Quenched Condition

Steel Designation	*Tensile Strength* N/mm^2	*Minimum Elongnation Percentage (gauge length* $= 5.65\sqrt{S_0}$)	*Izod Impact Value, Min (if specified)* *N.m.*	*Limiting ruling section, mm*	*Brinell Hardness Max.* *HB.*
10 C 4	490.3	17	54	15	130
14 C 6	490.3	17	54	over 15 upto 30	143
10 C 8 S 11	490.3	17	54	30	143
14 C 14 S 14	588.4	17	40.2	30	154
11 Mn 6	588.4	17	54	30	154
15 Cr 3	588.4	13	47	30	170
17 Mn 5 Cr 4	784.5	10	34.3	30	207
20 Mn Cr 5	980.7	8	37.3	30	217
16 Ni 3 Cr 2	686.5	15	40.2	90	184
16 Ni 4 Cr 3	833.6	12	40.2	30	217
	784.5	–	–	60	217
13 Ni 13 Cr 3	833.6	12	47	60	229
	784.5	–	–	100	229
15 Ni 16 Cr 5	1223.9	9	34.3	30	241
	1176.8	–	–	60	
	1127.8	–	–		
20 Ni 8 Mo 3	833.6	12	60.8	30	
	686.5	–	–	80	207
10 Ni Cr Mo 2	882.6	11	40.2	30	
	784.5	–	–	60	213
	735.5	–	–	90	
15 Ni 5 Cr 4 Mo 2	980.7	9	40.2	30	
	931.6	–	–	90	217
15 Ni 7 Cr 4 Mo 2	1078.7	9	34.3	30	
	980.7	–	–	60	217
	931.6	–	–	90	
20 Ni 8 Cr 6 Mo 2	1323.9	9	34.3	30	
	1176.8	–	–	60	224
	1127.8	–	–	90	

IS : 4432 – 1967

Table 2.17 Mechanical Properties of Alloy Steel Forgings for General Industrial Use

Designation	*N/mm²*	*Tensile Strength Condition Percentage* $= 5.65\sqrt{S_0}$	*Elognation Percent, min (gauge length*	*Izod Impact Value, Min N.m*	*Brinell Hardness HB*	*Limiting Ruling section, mm*
20 Mn 6	H + T	588.4 – 735.5	18	47	217 Max	63
15 Cr 3	R + Q	588.4 Min	13	47	–	–
17 Mn 5 Cr 4	R + Q	784.5 Min	10	34.3	–	–
20 Mn Cr 5	R + Q	980.7 Min	8	37.3	217 Max	60
21 Cr 4 Mo 3	A	588.4 – 735.5	–	–	217 Max	100
07 Cr 4 Mo 6	N + T	441.3	–	–	–	–
10 Cr 9 Mo 10	N + T	490.3 – 627.6	20	53.9	143 – 187	–
13 Ni 13 Cr 3	R + Q	833.6 Min	12	47	–	–
	(core properties)					
15 Ni 16 Cr 5	R+ Q	1323.9	9	43.3	–	–
15 Ni 5 Cr 4 Mo 1	R + Q	980.7 Min	9	40.2	–	–
15 Ni 7 Cr 4 Mo 2	R + Q	1078.7 Min	9	34.3	–	–
16 Ni 8 Cr 6 Mo 2	R + Q	1323.9 Min	9	34.3	–	–
36 Si 7	H + T	784.5 – 931.6	–	–	217 Max	100
37 Mn 6	H + T	686.5 – 833.6	18	47	248 Max	100
35 Mn 6 Mo 3	H + T	784.5 – 931.6	16	54	217 Max	100
40 Cr 4	H + T	686.5 – 833.6	18	54	210 – 248	100
40 Cr 4 Mo 3	H + T	784.5-931.6	16	54	229-177	100
35 Ni 5 Cr 2	H + T	784.5-931.6	16	54	229-177	100
40 Ni 6 Cr 4 Mo 3	H + T	980.7 – 1127.8	13	47	285 – 341	100
40 Ni 10 Cr 3 Mo 6	H + T	1520 Min	8	40.2	444 Min	100
25 Cr 20Mo 6	H + T	1078.7-1225.8	12	40.2	311-365	100
55 Si 7	H + T	1274.9 – 1471	–	–	235 Max	–
50 Cr 4 V 2	A	784.5 Max	–	–	235 Max	563

IS : 4367–1957

Code : H = Hardened, T = Tempered
R = Refined, Q = Quenched
A = Annealed, N = Normalized.

Table 2.18 Mechanical Properties of Steels in the Hardened and Tempered Condition

Steel Designation	*Tensile strength N/mm²*	*0.2 Per cent Proof Stress, Min N/mm²*	*Elogation Min* $L_0 = 5.65\sqrt{A}$ *J*	*Izod Impact Value, J*	*Limiting Ruling Section mm*
30 C 8	600 – 750	400	18	55	30
35 C 8	600 – 750	400	18	55	63
40 C 8	600 – 750	380	18	41	63
40 C 8	700 – 850	480	17	35	30

Steel Designation	Tensile strength N/mm²	0.2 Per cent Proof Stress, Min N/mm²	Elogation Min $L_0 = 5.65\sqrt{A}$	Izod Impact Value, J	Limiting Ruling Section mm
45 C 8	700 – 850	480	17	35	30
	600 – 750	380	17	41	100
50 C 8	700 – 850	480	15	35	30
	700 – 850	460	15	–	63
55 C 8	800 – 950	540	13	–	30
	700 – 850	460	15	–	63
41 C 10 S 18	800 – 950	540	13	–	30
	600 – 750	380	18	41	60
40 C 15 S 12	700 – 850	480	17	49	30
	700 – 850	500	18	35	60
	600 – 750	440	18	48	30
	700 – 850.	500	16	48	15
27 C 15	600 – 750	440	18	48	63
	700-850	500	16	48	30
37 C 15	600 – 750	440	18	48	100
	700 – 850	540	18	48	63
	800 – 950	600	16	48	30
	900 – 1050	700	15	41	15
35 Mn 6 Mo 3	700 – 850	540	18	55	150
	800 – 950	600	16	55	100
	900 – 1050	700	15	55	63
	1000 – 1150	800	13	48	30
35 Mn 6 Mo 4	800 – 950	600	16	55	150
	900 – 1050	700	16	55	100
	1000 – 1150	800	15	48	63
55 Cr 3	900 – 1050	660	12	35	63
	1000 – 1150	740	10	17	63
40 Cr 4	700 – 850	540	18	55	100
	800 – 950	600	16	55	63
	900 – 1050	700	15	55	30
40 Cr 4 Mo 3	700 – 850	540	18	55	150
	800 – 950	600	16	55	100
	900 – 1050	700	15	55	63
	1000–1150	800	13	48	30
15 Cr 13 Mo 6	700 – 850	540	18	55	150
25 Cr 15 Mo 6					
	800 – 950	600	16	55	150
	900 – 1050	700	15	55	150
	1000-11500	800	13	48	150
	1100 – 1250	880	12	41	100

Steel Designation	Tensile strength N/mm²	0.2 Per cent Proof Stress, Min N/mm²	Elogation Min $L_0 = 5.65\sqrt{A}$ J	Izod Impact Value, J	Limiting Ruling Section mm
1550 Min	1300	8	14	63	
40 Cr 13 Mo	1350 Min	11230	8	21	63
v 2	1550 Min	1330	8	14	30
40 Cr 7 Al 10 Mo 2	700 – 850	540	18	55	150
	980 – 950	600	16	55	100
	800 – 1050	700	15	48	63
40 Ni 14	800 – 950	600	16	55	150
	900 – 1050	700	15	55	100
35 Ni 5 Cr 2	700 – 850	540	18	55	150
	800 – 950	600	16	55	100
	900 – 1050	700	15	55	63
30 Ni 13 Cr 5	1550 Min	1300	8	14	63 (air hardened) 150 (oil hardened)
40 Ni 6 Cr 4 Mo 2	800 – 950	600	16	55	150
	900 – 1050	700	15	55	100
	1000 – 1150	800	13	48	63
	1100 – 1250	880	11	41	30
40 Ni Cr 4	800-950	600	16	55	150
Mo 3	900-1050	700	15	55	150
	1000-1150	800	18	48	100
	1100-1250	880	11	41	63
	1200 – 1350	1000	10	30	30
	1550 Min	1300	6	11	30
30 Ni 10 Cr 3 Mo 6	900 – 1050	700	15	55	150
	1000 – 1150	800	12	48	150
	1100 – 1250	880	11	41	100
	1200 – 1350	1000	10	35	63
	1550 Min	1300	8	14	63
40 Ni 10 Cr 3 Mo 6	1000 – 1150	800	12	48	150
	1100 – 1250	880	11	41	150
	1200 – 1350	1000	10	35	150
	1550 Min	1300	8	14	100

IS : 5517 – 1978

2.6 HEAT TREATMENT METHODS

Tempering. It is the process of heating hardened steel to a temperature below 723°C, holding it there for the proper time, and then cooling slowly.

Annealing. It is the process to obtain softness, improve machinability, increase ductility and toughness, relieve internal stresses, reduce or eliminate structural inhomogeneity, refine grain size, and to prepare steel for subsequent heat treatment.

Normalising. It consists in heating the steel to a temperature from 40 – 50°C above 785°C, holding at this temperature for a short time, and subsequent cooling in air. It is used to eliminate coarse-grained structure obtained in rolling and forging to increase the strength of medium carbon steels and reduce internal stresses.

Hardening. It consists in heating steel to a temperature above the critical point, holding at this temperature, and then quenched in water, oil, or molten salt bath. Austenite converts directly into martensite. Hardening must be followed by tempering to reduce brittleness, relieve internal stresses, and increase wear resistance.

Surface Hardening. It is a selective heat treatment method in which the surface layers of a metal are hardened to a certain depth whilst a relatively soft core is maintained.

Induction Hardening. It is a surface hardening process in which high frequency current is accomplished by the thermal effect of current induced in the part being heated. The part is placed in an alternating magnetic field set up by the high frequency current. It does not produce any scaling and deformation is considerably reduced. It is extensively used for crankshafts, camshafts, gears and many other automobile and tractor parts.

Flame Hardening. It is carried out by heating by an oxyacetylene flame having a temperature of 3000 – 3200°C and subsequent quenching hardens the surface layer. It is used for large cast gears, worms, shafts, steel guideways, mill rolls, etc.

Case Carburising. It is the process of saturating the surface layer of steel with carbon to a concentration of 0.75 – 1.2% and subsequent quenching. It increases fatigue strength of materials.

Nitriding. It is the process of saturating the surface of steel with nitrogen by holding it for a prolonged period at a temperature from 480 – 650°C in an atmosphere of ammonia. It increases the surface hardness, wear resistance, fatigue strength and corrosion resistance.

Cyaniding. It is a process in which both carbon and nitrogen are added to the surface layer of steel by heating the steel in liquid or solid media to increase its hardness and wear resistance.

2.7 NON-FERROUS METALS AND ALLOYS

Non-ferrous metals and alloys offer the following advantages :

1. Low density.
2. Attractive colours.
3. Softness and facility for cold working.
4. Good formability.
5. Resistance to corrosion.
6. Ease of fabrication.
7. Excellent machinability.
8. High thermal conductivity.
9. Excellent bearing properties.

2.7.1 *Copper and Copper Alloys*

Copper is a valuable metal having red colour. Commercial copper is a tough, ductile and malleable metal. It has high electrical conductivity and is used extensively for wire, cable and other electrical and electronic apparatus which must conduct the current. It is also a good conductor of heat and highly resistant to corrosion by liquids. It is used in water heating apparatus, water pipes, vessels in brewery, chemical and petroleum plants, condenser tubes and solar water heaters.

The alloys of copper form an important group of materials with a wide variety of properties. Three groups of copper alloys are used in industry. These are : Brasses, bronzes and monel metal.

1. Brass. Brass is an alloy of copper and zinc. The useful range is 10-30% Zn and 70-90% Cu. It is a good bearing material. By adding small quantities of other elements like aluminium, iron, manganese and tin, the strength of brass may be greatly increased. To improve the machinability of brass, 1-2% of lead is added. Small amounts of tin are added to brass to increase its hardness and resistance against corrosive action of sea water.

2. Bronze. It is an alloy of copper and tin. The useful range is 5-25% Sn and 75-95% Cu. Tin bronzes are divided into two groups : Wrought bronzes containing upto 8% Sn, and Casting bronzes containing over 8% Sn. Tin bronze is comparatively hard, resists surface wear and can be cast into shape or rolled into wire, rods and sheets very easily. It is used in hydraulic fittings, pump linings, in making utensils, bearings and bushes etc.

The bronze alloys are : Phosphor bronze, silicon bronze, aluminium bronze, manganese bronze and beryllium bronze.

***(i)* Phosphor Bronze.** When bronze contains phosphorus, it is called phosphor bronze. Its composition may vary as given below :

Cu = 80 – 90%; Sn = 10 – 20%; P = 15% (max.)

Phosphor bronze is used for all average bearings in which wearing qualities are desired. It is used for pump parts, linings, propellers, heavily loaded bearings, worm gears, gears, nuts for machine lead screws and springs.

***(ii)* Gun metal.** It contains Cu = 88%, Sn = 10% and Zn = 2%. It is used extensively for casting boiler fittings, bushes, bearings, glands, etc.

***(iii)* Babbit metal.** An alloy of tin, lead, copper and antimony is called babbit meal. It is the most common bearing material. The composition of common babbit metals and their uses are given in Table 2.19.

Table 2.19 Common Babbit Metals

Percentage Composition				Uses
Tin	Antimony	Copper	Lead	
93	3.5	3.5	-	Big end bearings, motor and aero engines
86	10.5	3.5	-	Main bearings, motor and aero engines
80	11	3	6	Bearings for heavy loads and high speeds
60	10	1.5	28.5	Bearings for engines and electric machines, railways and tramways.
40	10	1.5	48.5	Heavy pressure and bearing speed
20	15	1.5	63.5	Medium pressure and speed
5	15	-	80	Long bearing with medium load

***(iv)* Monel metal.** Monel metal is the most important nickel-copper alloy, having high strength, toughness and excellent corrosion resistance. It contains Ni = 67% and Cu=30% with small amounts of iron, manganese, silicon and carbon. It is a white, tough and ductile metal that can be easily machined, welded and heat treated. It is used in the form of rod, sheet, wire and welded tubing. It is widely used for structural and machine parts which must have a high resistance to corrosion and have high strength at elevated temperatures, as in the case of steam turbine blades, high temperature valves, impeller of centrifugal pump, and for springs subjected to temperatures above 200°C.

2.7.2 Aluminium and Aluminium Alloys

Aluminium is silvery white in colour. It is used where a light non-corrosive metal is necessary as in aircraft, missile and automobile components. Due to its good electrical conductivity, it is used for electrical wires and overhead cables. Its high resistance to corrosion and its non-toxicity makes it a useful metal for cooking utensils. It can be beaten into foil which is widely used for reflectors, mirrors and telescopes.

Aluminium alloying elements are : cooper, silicon, manganese, magnesium, iron, zinc and nickel. Aluminium alloys can be classified as : cast and wrought. Casting alloys are used in foundry practice and wrought alloys are used to make great variety of articles by rolling, forming, drawing, forging and press working.

Duralumin. It is a wrought aluminium alloy having the following composition :

Cu = 2.2 – 5.2%; Mg = upto 1.75%; Mn = upto 1.%; Si = upto 1%; Fe = upto 1%. It has good strength and corrosion resistance. It is used in aircraft, manufacture of pistons, vanes, screws, housings, motors, pump impellers, rings, disks and other parts.

2.8 NON-METALLIC MATERIALS

The commonly used non-metallic materials are : Wood, glass, rubber, leather, carbon and plastics.

1. Wood. It is used where light weight parts subjected to moderate shock loading are required, e.g. in circuit breaker operating rods where non-metallic bearing material is desirable.

2. Glass. It is considered to be a weak and brittle material but glass with newer compositions may be used for design components. Glass parts can be moulded by heating and blowing. Glass is used for all parts of centrifugal for acids, piping for mechanical process, pipe fittings, heat exchangers and linings for tanks and fittings.

3. Rubber. By definition, a rubber is a substance that has at least 200% elongation in a tensile test and is capable of returning rapidly and forcibly to its original dimensions when load is removed. There are five types of rubber—natural rubber, neoprene, silicone, polyurethane and fluoroelastomers. It is an excellent material for seats and diaphragms, for water lubricated bearings, for parts subjected to vibrations as vibration damper, flexible couplings and bearings, and for tubes and hose pipes.

4. Leather. It is flexible and can withstand considerable wear. It is used for belts and as a packing or a washer.

5. Carbon. It gives low friction losses and low wear rates against metals. It is used in chemical handling equipment, pump, rotors, vanes and gears. Clutch plates and rings of metal and cork have been replaced by carbon parts. Self lubricated carbon bearings are being used.

6. Plastics. Plastics are synthetic materials processed by heat and pressure.

(*a*) Advantages

(*i*) Low specific gravity resulting in light weight construction.

(*ii*) Good resistance to shocks and vibrations.

(*iii*) Low coefficient of friction, resistance to wear and self-lubricating property.

(*iv*) Resistance to corrosive atmosphere.

(*v*) Low cost.

(*b*) Limitations.

(*i*) Low strength and rigidity, low heat resistance and low dimensional stability.

(*ii*) Low temperature creep behaviour.

(*iii*) High thermal expansion coefficient.

(*iv*) Embrittlement with age.

(*c*) Types of Polymers

(*i*) Chain of Addition Polymers. They consists of a *Monomer* and an *Initiator*. The monomer contains double bond that can be opened. The initator is an organic or inorganic substance or a catalyst. Hydrocarbons from straight chains (aliphatic hydrocarbons) or Benzene rings (aromatic hydrocarbons). Other polymer molecules may contain N, O, S, P or Si, in the side chains while the monovalent Cl, F, or Br may replace H.

(*ii*) Step – reaction of Condensation Polymers – Two dissimilar monomers are joined into short groups followed by the polymerization and cross – linking of these groups; a byproduct of low molecular weight is often released.

(*iii*) Linear Polymers – Straight chain Covalent bonds exists only between molecules which are held to each other by secondary bonds that could be broken and reformed during deformation.

(*iv*) Cross – linked Polymers – Covalent bonds exist between molecules and no permanent deformation is possible.

(*d*) Types of Plastics

There are two types of plastics : Thermoplastics and Thermosetting plastics.

(*i*) Thermoplastics. They are linear straight chain polymers which can be softened repeatedly without undergoing change in chemical composition, e.g. polyimides, cellulosics, acetals, polycarbonates, PVC, nylon, plexiglass, Teflon, etc.

(*ii*) Thermosetting plastics. They are cross – linked polymers which undergo a chemical change with application of heat and pressure and cannot be resoftened, *e.g.* polysters, phenolics, silicones, epoxies, urethanes, elastomers, araldite, catalin, duraplex, beckosol, etc.

(*e*) Applications of Plastics

Table 2.20 Applications of plastics

Name of Plastic	Applications
(*a*) Thermoplastics	
Polyethlene	Blow – moulded bottles, automotive gas, tanks, drums, trash and garbage cans, household, and automotive parts, extruded pipes.
Polypropylene	Blow – moulded bottles, closures, automotive parts, appliances, housewares and toys.
Polystyrene	Toys, displays and housewares, protective packing materials, insulated drinking cups and floatation devices.
ABS (Arcylonitrile Butadiene Styrene)	Automotive parts, telephones, luggage, pipe fitting and conduit.
Polyvinyl chloride (PVC)	Window frames, gutters, pipes, wrie insulation and bottles, garden hose, boots, tubing, packing and raincoats.
Polyesters	Gears, rollers, bearings, housings, for pumps and appliances, impellers, pulleys, switch parts, automotive components, electrical/electronic components.
Polyimides (Nylons)	Fibres, automotive components, gears, cams, pulleys, rollers, boat propellers.
Polyacetals	Gears, bearings, bushings, automobile door handles, plumbing valves.
Polycarbonates	Saftey glazing, safety shields, non breakable windos, automobile tail lights, poer tool housings.
Polyimides	Gear, bushings, seals, insulators, structural components.
PTFE (Polytetrafluoroethylene)	Nonstick coating on cookwares, non-lubricated bearings, chemical resistant pipe fittings, valves, pump parts, gaskets, seals and packings.
PCTFE	Pipe and pump parts, valve diaphragms, coatings for corrosive industries.
(*b*) Thermosetting Plastics	
Epoxy	Tools and dies, adhesives and coatings, laminates for aircraft, photoelastic models.
Melamine – formaldehyde	Table wares, electric insulation, automotive ignition parts, cutlery handles, jars and bowls.
Phenol – formaldehyde	Industrial electric parts, automotive electrical components.
Alkyd	Boats, tanks, bins, housing truck, trailer and tractor components, ducts, shrouds vaulting poles, athletic goods.
Urea – formaldehyde	Lighting appliances, switch plates, reflectors, housings closures, containers, handles, knobs, tableware, and piano keys.

2.9 COMPOSITE MATERIALS

Composites are cohesive structures made by physically combining two or more compatible materials, differing in composition and characteristics and some times in form. The fibre reinforced composites are only those heterogeneous materials which are prepared by association and bonding in a single structure of material possessing quite different properties and when these are complementary, give a composite material possessing additional and / or superior properties to individual components either alone or mixed together. These are ideal materials for structural applications where high strength – to – weight and stiffness – to – weight ratios are required. The most important class of composite materials is fibre – reinforced plastics (FRP). The matrix material is invariably a plastic and the reinforcing materials are : glass, carbon, boron, silica, ceramic, metals, and organic material like bamboo or jute fibres.

The composites as a class of engineering materials provide almost unlimited potential for higher strength, stiffness, and corrosion resistance over the pure material system of metals, ceramics and polymers.

2.10 WEIGHTED FACTOR METHOD OF MATERIAL SELECTION

The weighted point method is a systematic method for the selection of materials. It consists of the following five steps :

1. Prepare a list of desirable properties.
2. Assign values to the desired properties. The desirable properties could be yield strength, endurance limit, hardness, corrosion resistance etc.
3. The desirable properties are divided into two groups : "Go – no – go" parameters and "discriminating" parameters. The "Go – no – go" parameters are the constraints. If a material is not available or cannot be fabricated into given shape, it is totally rejected.
4. The discriminating parameters are the properties of material which can be given quantitative values. The weightage depends upon the importance of that particular property in the given application. The weightage factor may vary from 1 to 5, with 1 for the poorest and 5 for the best.
5. Assign a rating to each property to the candidate material ranging from 1 to 5, depending upon how closely it meets the requirements. These ratings are multiplied by the weighing factors for each property. These numbers are finally added and materials are arranged in descending order of their total points.

The main drawback of this method is the skill and judgement required to assign the weightage. Consider eight different types of steel with the desirable properties listed in Table 2.21.

Table 2.21 Weightage Factors for Comparative properties of Cast steels.

Sl. No	Material	Abrasive wear resistance	Damping capability	Corrosion resistance	Weldability	Machinability	Castability	Forgability
Rating factor⟶		3	2	3	2	4	1	5
1.	Loc C	P	P	P	E	F	F	F
2.	Medium C	F	P	P	G	F	F	F
3.	High C	G	VP	VP	P	F	F	G
4.	Low alloy	F	P	P	G	F	F	G
5.	High alloy	P	VP	F	VP	VP	P	P
6.	Heat resistant	VP	P	G	F	F	VP	VP
7.	Corrosion resistant	G	G	F	F	F	F	G
8.	Wear resistant	E	P	F	F	VP	F	P

E = Excellent = 5 points, G = good = 4, F = fair = 3, P = poor = 2, VP = very poor = 1

Table 2.22 Multiplication of weightage factor and rating factor.

Sl. No.	Material wear resistance	Abrasive capability	Damping resistance	Corrosion	Weldability	Machinability	Castability	Forgability	Total
1.	Low C	6	4	6	10	12	3	15	56
2.	Medium C	9	4	6	8	12	3	15	57
3.	High C	12	2	3	4	12	3	20	56
4.	Low alloy	9	4	6	8	12	3	20	62
5.	High alloy	6	2	9	2	4	2	10	35
6.	Heat resistant	3	4	12	6	12	1	5	43
7.	Corrosion resistant	12	8	9	6	12	3	20	68
8.	Wear resistant	15	4	9	6	4	3	10	51

Descending order of materials : 7 – 4 – 2 – 3 – 1 – 8 – 6 – 5

We find that material number 7 is found to be best based on weighted point method.

EXERCISES

2.1 Designate the following materials

(*a*) Grey cast iron with an ultimate tensile strength of 350 MPa.
(*b*) Plain carbon steel with a minimum tensile strength of 340 MPa.
(*c*) Plain carbon steel with a minimum yield strength of 250 MPa.
(*d*) Plain carbon steel with 0.45% carbon and 0.8% manganese.
(*e*) Alloy steel with carbon = 0.40 – 0.50% and chromium = 0.90 – 1.20%.

2.2 What do you understand by the following designation of materials ?

(*i*) FG 400 (*ii*) 15C8
(*iii*) 45C15S15 (*iv*) 45Cr1V$\underline{23}$
(*v*) 23Mn2.

2.3 Recommend suitable materials for the following components :

(*a*) Valve spring (*b*) Gasket
(*c*) Screw jack spindle (*d*) Helical springs
(*e*) Screw jack nut.

2.4 (*i*) What do you mean by the term "Mechanical properties of materials" ?
(*ii*) What is the main reason for the use of alloy steels in machine design ?
(*iii*) Discuss the important non – metallic materials used in engineering practices.

2.5 How will you designate the following materials ?

(*i*) Grey cast – iron with an ultimate tensile strength of 300 N / mm^2.
(*ii*) Plain carbon steel with a minimum tensile strength of 320 N / mm^2.
(*iii*) Plain carbon steel with a minimum yield strength of 200 N / mm^2.
(*iv*) Alloy steel with composition :
Carbon = 0.35% to 0.45%
Chromium = 0.9% to 1.20%
(*v*) Plain carbon steel with 0.4% carbon and 0.8% manganese.
(*vi*) Cast steel with a minimum tensile strength of 450 N / mm^2 and a yield strength of 230 N / mm^2.

2.6 What do you understand by the following designation of materials ?
40C8; 40Cr1; 35Ni1Cr$\underline{60}$; 50Cr1V$\underline{23}$; FG350.

2.7 (*i*) Define creep and fatigue.
(*ii*) What are the reasons for the use of alloy steels in machine parts ?
(*iii*) How do you classify materials for engineering use ?

2.8 (*i*) Write short notes on high speed tool steel and spring steel.
(*ii*) Discuss the effect of Si, Mn, S and P on cast iron.
(*iii*) Explain the uses of wood and glass in place of steel in machine design.

2.9 (*i*) Give two typical applications of each of the following alloys :
(*a*) Monel metals (*b*) Babbit metal.
(*ii*) Explain hardening and tempering.

2.10 Discuss the system of designation of steels. What do you understand from the following designations of materials :

(*i*) FG 350 (*ii*) 10C4 (*iii*) 50Cr1V$\underline{23}$.

2.11 (*i*) Which alloy steel is suitable for springs and why ?
(*ii*) Give the materials which are commonly used for the following parts and give the reason also. Shaft, Screw, Piston, Connecting rod.

2.12 (*a*) Explain the BIS method of specifying cast iron and its major types.
(*b*) Briefly discuss the weighted factor method of material selection.
(*c*) Enumerate advantages and disadvantages of plastic materials over metals.

DESIGN FOR STATIC LOAD

Chapter

3.1 INTRODUCTION

A static load may be defined as that load which does not vary in magnitude and direction with respect to time. It is a gradually applied load. To determine the stresses in machine members under static loading, simple formulae of strength of materials may be used. The allowable or safe design stress is based on eiher the ultimate tensile strength for brittle materials or yield strength for ductile materials to give a reasonable factor of safety.

3.2 REVIEW OF STRESSES IN MACHINE MEMBERS

A machine component may be subjected to the following stresses :

(a) Simple Stresses

1. Tension / Compression

Let P = load applied in tension / compression, N

A = area of cross – section normal to P, mm^2

Then tensile / compressive stress, σ_t or $\sigma_c = \dfrac{P}{A}$ N/mm^2 or MPa ....(3.1)

2. Simple Shear

Let F = shear force acting on the member, N

A = area under shear, *i.e.* parallel to F, mm^2

Then shear stress, $\tau = \dfrac{F}{A}$ MPa ....(3.2)

3. Bending

Let M = bending moment applied to the machine member, N.m

z = section modulus of the cross – section, m^3

Using bending formula, $\dfrac{M}{I} = \dfrac{\sigma}{y} = \dfrac{E}{R}$

where I = area moment of inertia about the bending axis, m^4

$= \dfrac{\pi d^4}{64}$ for a circular shaft of diameter d

y = distance of the farthermost fibres from the neutral axis, m

E = Young's modulus of elasticity of the material of machine member, N/m^2

R = radius of bent member, m

$$z = \text{section modulus in bending} = \frac{I}{y} = \frac{\pi d^3}{32} \text{ for a circular shaft of diameter } d\text{, m}^3$$

Then bending stress, $\sigma_b = \frac{M}{z} = \frac{32M}{\pi d^3}$ N/m² ....(3.3)

The bending stress is zero at the neutral axis and maximum at the outermost fibres of the component.

The bending stress varies linearly across the cross – section of the component.

4. Torsion

Let T = torque applied to the member, N.m

z_p = polar section modulus, m³

Using the torsion formula, $\frac{T}{J} = \frac{\tau}{r} = \frac{G\theta}{L}$

Where J = polar moment of inertia, m⁴

$= \frac{\pi d^4}{32}$ for a circular shaft of diameter d

τ = shear stress, N/m²

r = distance of the fibres from the centre of shaft, m

G = shear modulus of rigidity, N/m²

θ = angle of twist, rad

L = length of the shaft, m

$$z_p = \text{section modulus in torsion} = \frac{J}{r} = \frac{\pi d^3}{16} \text{ for a circular shaft of diameter } d\text{, m}^3$$

Then torsional shear stress, $\tau = \frac{T}{z_p} = \frac{16T}{\pi d^3}$ N/m² ....(3.4)

The torsional shear stress is zero at the centre of the shaft and maximum at the outermost fibres. It varies linearly through the shaft cross – section.

***(b)* Eccentric Axial Loading**

Consider a member subjected to an eccentric axial load P at an eccentricity e, as shown in Fig. 3.1 (*a*). According to the principle of statics, the eccentric load P can be replaced by a parallel force P passing through the centroid along with a couple P × e as shown in Fig. 3.1 (*b*) and (*c*).

In Fig. 3.1 (*b*), the force P causes a simple tensile stress of magnitude $\frac{P}{A}$. In Fig. 3.1 (*c*), the couple causes bending stress of magnitude $\frac{Pe}{z}$. The resultant stresses at the cross – section are obtained by the principle of superposition of stresses. They are given by :

$$\sigma = \frac{P}{A} \pm \frac{Pe}{z} \quad ...(3.5)$$

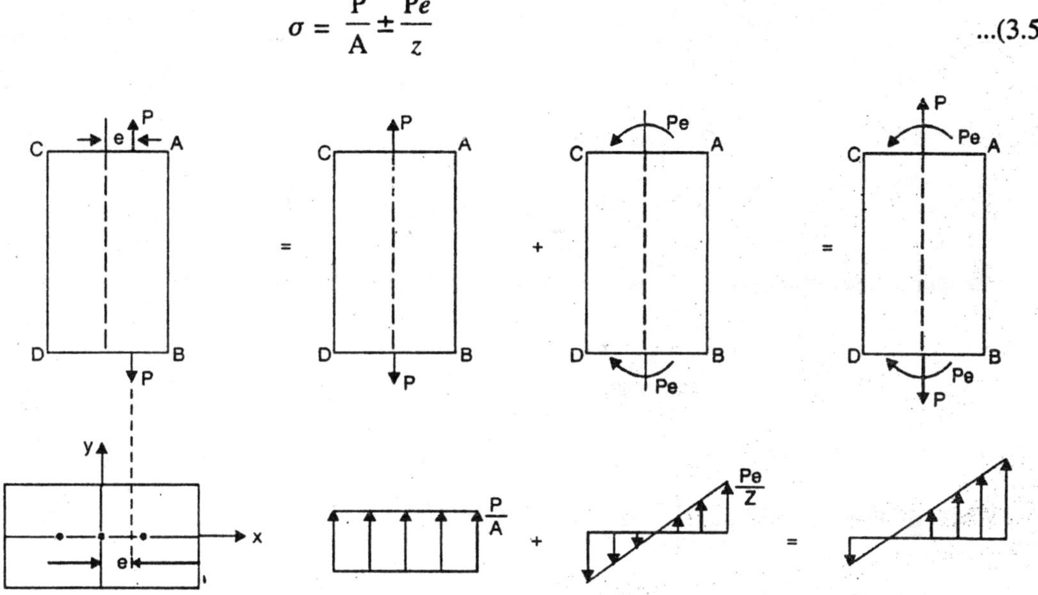

Fig. 3.1 Eccentric axial loading

On the face AB, the resultant stress is,

$$(\sigma_r)_{AB} = \frac{P}{A} + \frac{Pe}{z} \quad ...(3.6a)$$

Which is tensile in nature.

On the face CD, the resultant stress is,

$$(\sigma_r)_{cd} = \frac{P}{A} - \frac{Pe}{z} \quad ...(3.6b)$$

Which may be tensile or compressive depending upon the magnitude of bending stress.

(c) Principal Stresses

However complex the state of stress may be at a point in three-dimensional system, there are always three mutually perpendicular planes in a body on which the normal stresses are either maximum or minimum. These planes are called the principal planes and the normal stresses acting on them are called the principal stresses.

In a two – dimensional system, there are two principal stresses. If σ_x, σ_y and τ_{xy} are the stresses, then the principal stresses are given by :

$$\sigma_{1,2} = \left(\frac{1}{2}\right)(\sigma_x + \sigma_y) \pm \left(\frac{1}{2}\right)[(\sigma_x - \sigma_y)^2 + 4\tau^2_{xy}]^{1/2} \quad ....(3.7)$$

Maximum shear stress, $\tau_{max} = \pm \left(\frac{1}{2}\right)[(\sigma_x - \sigma_y)^2 + 4\tau^2_{xy}]^{1/2}$ ...(3.8)

For a shaft subjected to bending moment M and torque T, the stresses are :

Bending stress, $\sigma_x = \dfrac{32M}{\pi d^3}$

$\sigma_y = 0$

Shear stress, $\tau_{xy} = \dfrac{16T}{\pi d^3}$

Principal stresses are, $\sigma_{1,2} = \left(\dfrac{1}{2}\right)(\sigma_x + \sigma_y) \pm \left(\dfrac{1}{2}\right) [(\sigma_x - \sigma_y)^2 + 4\tau^2_{xy}]^{1/2}$

$$= \left[\frac{16}{\pi d^3}\right] [M \pm (M^2 + T^2)^{1/2}] \quad .....(3.9)$$

Maximum shear stress, $\tau_{max} = \left[\dfrac{16}{\pi d^3}\right] [\pm (M^2 + T^2)^{1/2}]$ ....(3.10)

3.3 SIMPLE AND PRINCIPAL STRAINS

If a machine member of area of cross – section A, length L and Young's modulus E is subjected to an axial load P, then the change in length of the member is,

$$\delta L = \frac{PL}{AE}$$

Longitudinal strain, $\varepsilon = \dfrac{\text{Change in length}}{\text{Original length}}$

$$= \frac{\delta L}{L}$$

Poisson's ratio, $\nu = \dfrac{\text{Lateral strain}}{\text{Longitudinal strain}}$ ...(3.11)

Where lateral strain $= -\dfrac{\delta d}{d}$

and longitudinal strain $= \dfrac{\delta L}{L}$

Since the lateral strain is opposite in sense to the longitudinal strain, therefore, Poisson's ratio is always negative.

If σ_1, σ_2 and σ_3 are the principal stresses at a point in a body, then the principal strains are :

$$\varepsilon_1 = \left(\frac{1}{E}\right) [\sigma_1 - \nu(\sigma_2 + \sigma_3)] \quad ...(3.12)$$

$$\varepsilon_2 = \left(\frac{1}{E}\right)[\sigma_2 - \nu(\sigma_3 + \sigma_1)]$$

$$\varepsilon_3 = \left(\frac{1}{E}\right)[\sigma_3 - \nu(\sigma_1 + \sigma_2)]$$

3.4 RELATIONSHIP BETWEEN ELASTIC CONSTANTS

Let G = shear modulus
K = bulk modulus
E = Young's modulus of elasticity
ν = Poisson's ratio

Then
$$G = \frac{E}{2(1+\nu)} \quad \text{....(3.13)}$$

$$K = \frac{E}{3(1-2\nu)} \quad \text{....(3.14)}$$

3.5 FACTOR OF SAFETY

The terms factor of safety or design factor are commonly used synonymously to indicate the ratio of estimated strength to computed stress. It is defined as the ratio of maximum strength to the working stress. In case of ductile materials, it is based upon the yield strength of the material. In the case of brittle materials, it is based upon the ultimate strength of the material. Thus

$$\text{Factor of safety, } n = \frac{\text{Ultimate strength}}{\text{Working stress}}, \text{ for brittle materials.} \quad \text{....(3.15a)}$$

$$= \frac{\text{Yield strength}}{\text{Working stress}}, \text{ for ductile materials} \quad \text{....(3.15b)}$$

The factor of safety depends upon the various factors as listed below :

1. Variation in material properties.
2. Effect of size in stating material strength properties.
3. Type of loading.
4. Effect of machining and forming processes.
5. Effect of heat treatment upon the physical properties of the material.
6. Effect of wear upon the functions and life of a machine member.
7. Effect of time and environment in which the device is expected to operate.
8. Specific requirements for life and reliability.
9. Overall concern for human safety.

For steady loads, the factor of safety "n" may be calculated from the following formula :

$$n = a \times b \times c \quad \text{....(3.16)}$$

Where Factor, 'a' = elastic ratio

$$= \frac{\text{Ultimate strength}}{\text{Yield strength}}, \text{ for static loading}$$

$$= \frac{\text{Ultimate strength}}{\text{Endurance limit}}, \text{ for dynamic loading}$$

Factor, 'b' = shock or service factor

$$= \frac{\text{Stress produced by a shock or suddenly applied load}}{\text{Stress produced by a gradual load}}$$

$$= \left[1 + \left(1 + \frac{2h}{\delta_{st}}\right)^{1/2}\right] \text{ for linear impact}$$

$$= \left[1 + \left\{1 + \frac{2h}{r\theta_{st}}\right\}^{1/2}\right] \text{ for angular impact}$$

Where h = height fallen by the load

δ_{st}, θ_{st} = static linear or angular deflection

For sudden loading, $h = 0$ and $b = 2$

b = 1.25 – 1.5 for light shocks

= 2 – 3 for heavy shocks

= 5 for extreme cases

Factor, 'c' = real margin of safety to account for unreliable material and to account for unforeseen causes of failure

Average values of 'c' are :

c = 1.5 – 2 for ductile materials of uniform structure

= 2.5 – 3.0 for brittle materials

= 3 – 4 for timber

Suggested values for factor of safety are given in Table 3.1.

Table 3.1 Suggested values for factor of safety

Working conditions	Factor of safety
1. Exceptionally reliable materials subjected to well defined loads under controlled conditions	1.25 – 1.5
2. Well known materials subjected to known loads under reasonably constant environmental conditions	1.5 – 2
3. Average materials subjected to known loads under ordinary environmental conditions.	2 – 2.5
4. Less tried or brittle materials under average conditions of environment and load.	2.5 – 3
5. Untried materials under average conditions of environment and load.	3 - 4
6. Better known materials under uncertain environment and load.	3 – 4
7. Repeated loads.	1 – 6
8. Impact loads.	3 – 6
9. Brittle materials.	Twice the value in 1 to 6

3.6 WORKING STRESS

While designing machine components, it is desirable to keep the induced stress below the ultimate tensile strength (S_{ut}) or tensile yield strength (S_{yt}) of the material. This stress is known as the working (or design, or safe, or allowable or permissible) stress.

$$\text{Design stress,} \quad \sigma_d = \frac{S_{ut}}{n}, \text{ for brittle materials} \quad ...(3.17a)$$

$$= \frac{S_{yt}}{n}, \text{ for ductile materials.} \quad ...(3.17b)$$

$$\text{Margin of safety} = n - 1 \quad ....(3.18)$$

$$\text{Factor of utilization} = \frac{1}{n} \quad ....(3.19)$$

3.7 MODES OF FAILURE

A static load is defined as a force which is gradually applied to a machine component and which remains unaltered over a span of time. The engineering materials can be broadly classified into two groups : ductile and brittle materials. A material is considered ductile if the tensile strain before fracture is relatively high as compared to a brittle material. As a thumb rule if the tensile strain is more than 5% then the material is considered ductile otherwise brittle. For example structural steel, copper and aluminium etc. are ductile materials whereas cast iron is brittle in nature.

The ductile materials are distinguished by their clearly defined yield point in the simple tensile test stress–strain diagram where as brittle materials have no clearly defined yield point. For materials having no clearly defined yield point the offset method is used to determine the yield point stress. In this method, a straight line parallel to the linear portion of the stress – strain diagram is drawn at 0.2% of maximum strain to intersect the stress – strain curve. Then a line parallel to the strain – axis is drawn at this point to locate the yield strength of the material on the stress axis.

By failure of a material we do not mean total fracture but whenever the stress exceeds beyond a pre-specified stress level, like the yield strength or ultimate strength, then the material is said to have failed. The only failures of concern to the designer are those in which the design can be assumed to be faulty. The general procedure in this case is to check the balance between the stress and the strength. In the broadest sense, failure of a machine part means any action resulting from service conditions, causing it to cease to function satisfactorily.

There are five types of structural damage which mahine parts may experience as listed below :

1. Excessive elastic deformation.
2. Elastic buckling.
3. Plastic deformation.
4. Creep, and
5. Fracture.

There are three common type of fractures observed in materials : (*i*) Grain boundary fractures (*ii*) Cleavage fractures across the grains, and (*iii*) Shear fracture across the grains. In iron, the grain boundary fractures are rare at ordinary temperatures, shear fractures are associated with ductile behaviour, and brittle behaviour is characterized by the cleavage mode of fracture. Cleavage fractures follow the cube faces of ferrite crystals whereas in the shear fractures, translation always occurs in the direction having the minimum inter-atomic distance and usually along the planes having the greatest atomic density. Consequently, either of the two modes of fracture can be identified by a microscopic examination. The shear mode of fracture is associated with a grey and silky appearance of the fracture surface whereas the cleavage mode causes the surface to be bright and granular.

The factors influencing the type of failure for metals are given in Table 3.2.

Table 3.2 Factors favourable for fracture.

Factors favourable to shear fracture	Factors favourable to cleavage fracture
1. High shear stresses.	1. High tensile stresses.
2. Low tensile stresses.	2. Low shear stresses.
3. High temperature.	3. Low temperatures.
4. Slow rate of loading.	4. High rate of loading.
	5. High stress concentration.
	6. Repeated loads.

3.8 THEORIES OF FAILURE

When a material is subjected to a single type of stress, *i.e.* axial, bending or torsional stress, then it is a simple matter to anticipate the likely occurrence of failure. However, when the material is subjected to a complex system of stresses then it is difficult to predict the failure of the material straight away. In order to predict the failure of the material under combined stresses, the following theories of failure have been postulated :

(*a*) For brittle materials :

(*i*) Maximum normal stress theory.

(*ii*) Mohr's theory.

(*b*) For ductile materials :

(*i*) Maximum shear stress theory.

(*ii*) Maximum distortion energy theory.

1. Maximum Normal Stress Theory

According to this theory failure occurs in a multi-axial system when either a principal tensile stress reaches the uniaxial tensile ultimate strength σ_{ut}, or a principal compressive stress reaches the uniaxial ultimate compressive strength σ_{uc}. According to this theory, the strength of the material depends upon only one of the principal stresses (i.e. the maximum tensile or compressive stress) and is entirely independent of the other two principal stresses.

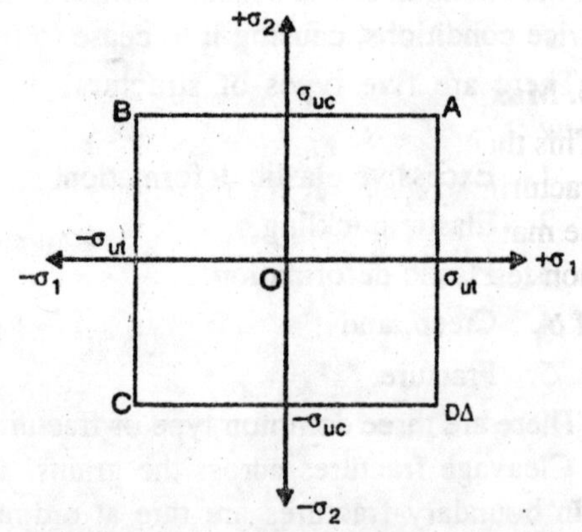

Fig. 3.2 Maximum normal stress theory

If $\sigma_1 > \sigma_2 > \sigma_3$ are the three principal stresses at a point, then according to this theory, we have

$$\sigma_1 = \sigma_{ut} \text{ or } \sigma_1 = \sigma_{uc} \quad ...(3.20)$$

This theory is representented graphically in Fig. 3.5

This theory has been represented for the two-dimensional stress system ($\sigma_3 = 0$) in Fig. 3.2. The maximum normal stress heory is represented by the square ABCD if it is assumed that $\sigma_{ut} = \sigma_{uc}$.

2. Mohr's Theory

This theory suggests that principal Mohr's circles be drawn representing every available test condition and that the envelope of these circles be taken as the envelope of any and all principal Mohr's circles representing stress stateson the verge of failure.

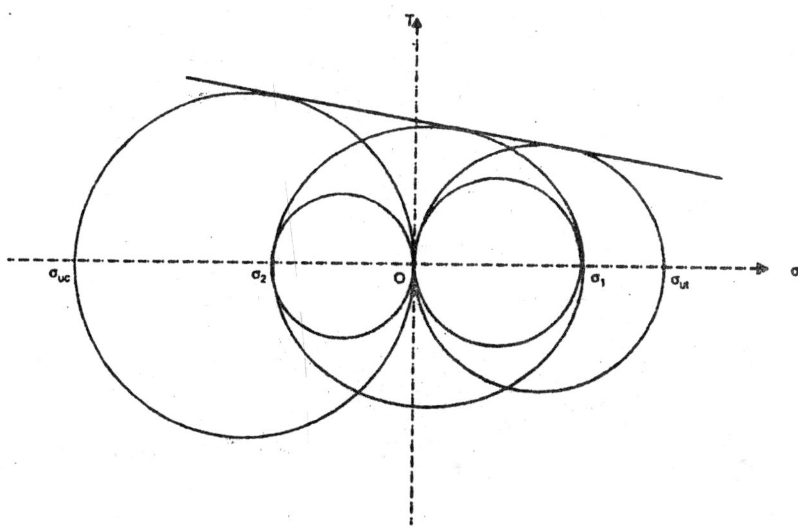

Fig. 3.3 Mohr's theory

The Mohr's theory is illustrated in Fig. 3.3 Circles of σ_{ut} are drawn as shown. Any other stress state for which the Mohr's circle (which envelops circles of diameters σ_1 and σ_2) is tangent to the line of tangency of the σ_{ut} and σ_{uc} circles, represents a condition of failure. The following condition holds good for the first quadrant.

$$\frac{\sigma_1}{\sigma_{ut}} + \frac{\sigma_2}{\sigma_{uc}} = 1 \quad ....(3.21)$$

3. Maximum Shear Stress Theory

This theory states that a material subjected to any combination of stresses shall fail (by yielding or fracturing) whenever the maximum shear stress exceeds the shear strength (yield or ultimate) of the material. The shear strength is usually assumed to be determined from the standard uniaxial tension test.

If $\sigma_1 > \sigma_2 > \sigma_3$ and $\sigma_3 = 0$, then according to this theory, we have

$$\frac{(\sigma_1 - \sigma_2)}{2} = \frac{\sigma_y}{2}$$

or

$$(\sigma_1 - \sigma_2) = \sigma_y \quad ....(3.22)$$

This theory is represented graphically in Fig. 3.4. The major drawback of this theory is that the intermediate stress is totally ignored.

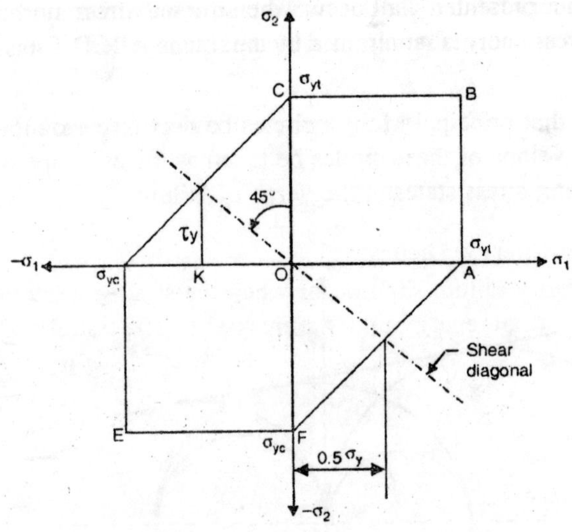

Fig. 3.4 Maximum shear stress theory

For the torsion case, $\sigma_2 = -\sigma_1 = \tau_y$ Fig. 3.4

$$\tau_y = 0.5\,\sigma_y \quad \text{....(3.23)}$$

4. Maximum Distortion Energy Theory

According to this theory failure by yielding occurs when the energy of distortion in a multi-axial stress system exceeds the energy of distortion of a simple tensile test. Thus

$$(\sigma_1 - \sigma_2)^2 + (\sigma_2 - \sigma_3)^2 + (\sigma_3 - \sigma_1)^2 = 2\sigma^2{}_y \quad \text{...(3.24)}$$

If $\sigma_3 = 0$, then

$$\sigma_1^2 + \sigma_2^2 - \sigma_1\sigma_2 = \sigma^2{}_y \quad \text{...(3.25)}$$

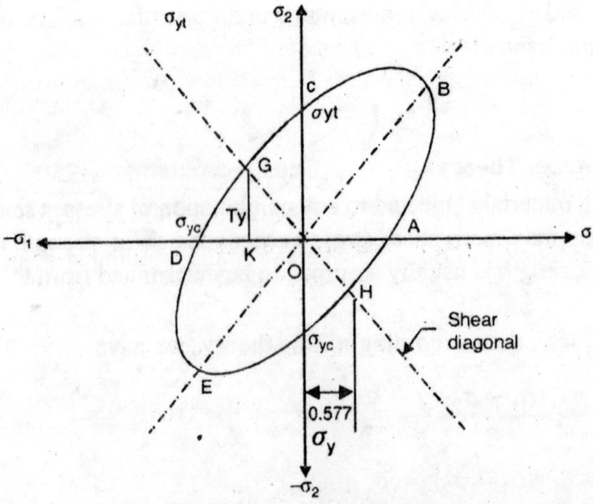

Fig. 3.5 Maximum distortion energy theory.

For the torsion case,

$$\sigma_2 = -\sigma_1 = \tau$$

$$\tau_y = 0.577\,\sigma_y \quad ...(3.26)$$

5. Maximum Strain Theory

According to this theory failure shall occur when the maximum normal strain under the three principal strains excees the normal strain in a uniaxial tensile test. Thus

$$\left(\frac{1}{E}\right) = [\sigma_1 - \nu(\sigma_2 + \sigma_3)] = \frac{\sigma_y}{E}$$

Or

$$\sigma_1 - \nu(\sigma_2 + \sigma_3)] = \sigma_y \quad ....(3.27)$$

6. Maximum Strain Energy Theory

According to this theory failure shall occur when the strain energy under the three principal stresses shall exceed the strain energy under a uniaxial tensile test. Thus

$$\left[\frac{1}{2E}\right][\sigma_1^2 + \sigma_2^2 + \sigma_3^2 - 2\nu(\sigma_1\sigma_2 + \sigma_2\sigma_3 + \sigma_3\sigma_1)] = \frac{\sigma_y^2}{2E}$$

or

$$\sigma_1^2 + \sigma_2^2 + \sigma_3^2 - 2\nu(\sigma_1\sigma_2 + \sigma_2\sigma_3 + \sigma_3\sigma_1) = \sigma_y^2 \quad ...(3.28)$$

3.8.1 Selection of a Failure Theory

In engineering practice many implications must be considered in connection with the selection and use of a failure theory. Situations involving excessive elastic deflection, buckling, creep, impact and fatigue all require considerations in addition to those covered above in art. 3.8.

For the prediction of failure by yielding of ductile materials, distortion energy theory is generally considered to give the most accurate results. For the common case of biaxial stresses, the Mohr's theory is preferred. The maximum shear stress theory is simple to apply for both biaxial and triaxial stresses, but it is more conservative.

For predicting the fracture of brittle materials, Mohr's theory can be used for biaxial stress states. This theory gives quite accurate but conservative results. The maximum normal stress theory gives results which are somewhat on the non-conservative side for biaxial tension/ compression.

Example 3.1 *An offset link subjected to a force of 30 kN is shown in Fig. 3.6. It is made of grey cast iron FG 350 and the factor of safety is 2.5. Determine the dimensions of the cross-section of the link.*

Solution. Working stress for the link, $\sigma_w = \dfrac{\text{Tensile strength}}{\text{Factor of safety}} = \dfrac{350}{2.5} = 140\text{ MPa}$

The cross-section is subjected to direct tensile stress and bending stresses. The top fibres are highly stressed.

Area of cross-section, $A = t \times 2.5t = 2.5t^2\text{ mm}^2$

Moment of inertia about N – A, $I = \dfrac{t \times (2.5t)^3}{12} = 1.3t^4\text{ mm}^4$

Direct stress, $\sigma_d = \frac{P}{A} = \frac{30,000}{2.5t^2} = \frac{12,000}{t^2}$ N/mm^2

Eccentricity of load, $e = (10 + t)$ mm

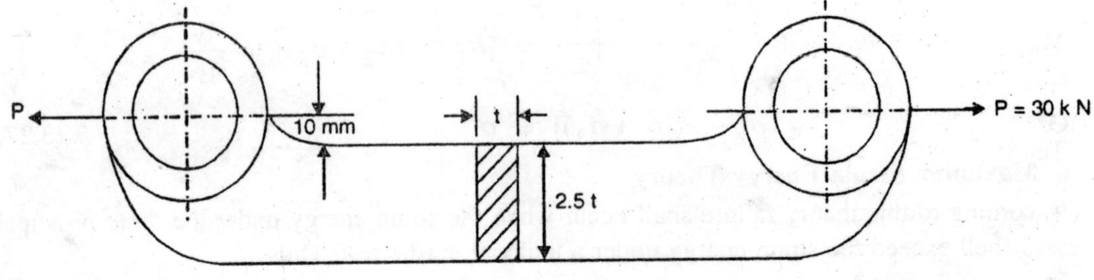

Fig. 3.6

Bending moment, $M = Pe = 30,000 \times (10 + t) = 300,000 + 30,000t$ N.mm

Bending tensile stress, $\sigma_t = \frac{Mt}{I} = \frac{30,000\,(10+t)\times t}{1.3t^4} = \frac{23076.92\,(10+t)}{t^3}$ N/mm^2

Maximum tensile stress at the top fibres $= \sigma_d + \sigma_t = \frac{12,000}{t^2} + \frac{23077\,(10+t)}{t^3} = 140$

$$140t^3 = 12,000t + 230770 + 23077t$$

or $\quad t^3 - 79.12t - 1648.36 = 0$

We solve the cubic equation by hit and trial method

t	15	14
$f(t)$	539.82	– 12.05

Adopt $t = 14$ mm

Example 3.2 *A wall bracket with a rectangular cross-section is shown in Fig. 3.7. The depth of cross-section is twice the width. The force P = 6 kN acting on the bracket is inclined at 30° with the horizontal. The material of the bracket is grey cast iron FG 250 and the factor of safety is 2.5. Determine the dimensions of the cross-section of the bracket using the maximum normal stress theory of failure.*

Solution. Working stress for the link, $\sigma_w = \frac{\text{Tensile strength}}{\text{Factor of safety}} = \frac{250}{2.5} = 100$ MPa

The stress is maximum at point A which is subjected to the combined effect of direct tensile stress and tensile bending stresses. Resolving the force P into its horizontal and vertical components P_H and P_V respectively, we have

$$P_H = P\cos 30° = 6000\cos 30° = 5196.15 \text{ N}$$
$$P_V = P\sin 30° = 6000\sin 30° = 3000 \text{ N}$$

Direct stress $\sigma_d = \dfrac{P_H}{A} = \dfrac{5196.15}{(2t^2)} = \dfrac{2598.1}{t^2}$ N/mm^2

Bending moment at A, $M = P_H \times 100 + P_V \times 250$
$= 5196.15 \times 100 + 3000 \times 250 = 1269615$ N.mm

Moment of inertia, $I = \dfrac{t \times (2t)^3}{12} = 0.667t^4$ mm^4

Tensile bending stress, $\sigma_t = \dfrac{My}{I} = \dfrac{1269615 \times t}{(0.667t^4)} = \dfrac{1903470}{t^3}$ N/mm^3

Total tensile stress at $A = \sigma_d + \sigma_t$

$$= \frac{2598.1}{t^2} + \frac{1903470}{t^3}$$

$= 100$

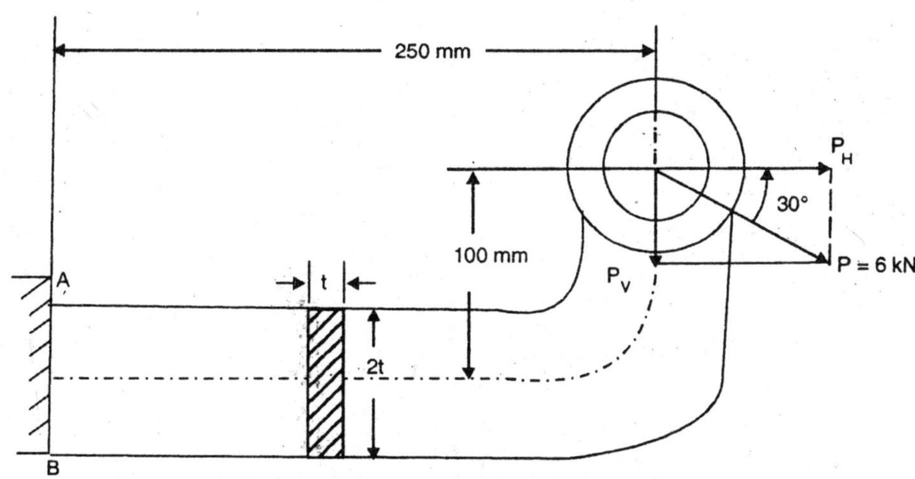

Fig. 3.7

or $t^3 - 25.981t - 19034.7 = 0$

We solve the cubic equation by hit and trial method.

t	30	29	27
$f(t)$	7185.87	4600.85	– 53.2

Adopt $t = 27$ mm

Example 3.3 *AC-frame subjected to a force of 20 kN is shown in Fig. 3.8. It is made of grey cast iron FG 300 and the factor of safety is 3. Determine the dimensions of the cross-section of the frame.*

Solution.

Working stress for the frame, $\sigma_w = \dfrac{\text{Tensile strength}}{\text{Factor of safety}} = \dfrac{300}{3} = 100$ MPa

Area of cross section, $A = 4t^2$ mm²

Direct stress, $\sigma_d = \frac{P}{A} = \frac{20,000}{4t^2} = \frac{5000}{t^2}$ N/mm²

Bending moment, $M = Pe = 20,000 \times 6t = 120,000t$ N.mm

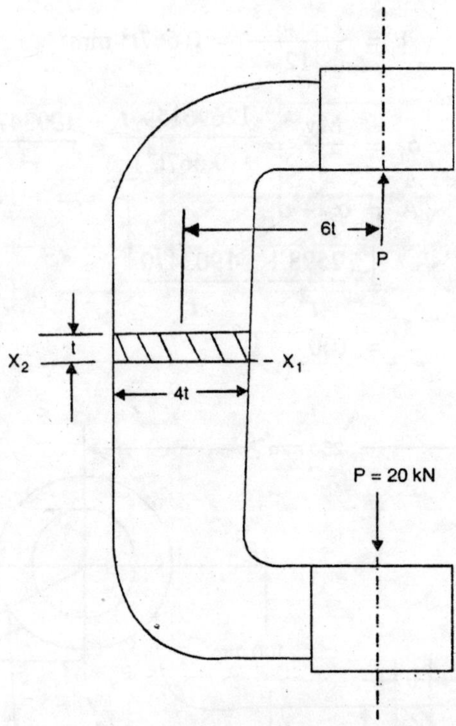

Fig. 3.8

Section modulus at $X_1 - X_2$, $z = \frac{I}{y} = \frac{t \times (4t)^2}{6} = \frac{8t^3}{3}$ mm³

Bending stress, $\sigma_b = \frac{M}{z} = \frac{120,000t \times 3}{(8t^3)} = \frac{45,000}{t^2}$ N/mm²

Maximum tensile stress at $X_1 = \sigma_d + \sigma_b$

$$= \frac{5000}{t^2} + \frac{45,000}{t^2} = \frac{50,000}{t^2}$$

$$= 100$$

$$t = 22.36 \text{ mm} \approx 23 \text{ mm}$$

Example 3.4 *A cast-iron link as shown in Fig. 3.9 is required to carry a tensile load of 15 kN. If the tensile and compressive stresses in the link are limited to 20 MPa and 60 MPa respectively, determine the dimensions of the cross-section of the link at the middle section.*

Solution.

Area of cross-section :

$$A_1 = 2a \times a = 2a^2 \text{ mm}^2$$

$$A_2 = 2 \times a \times \frac{a}{2} = a^2$$

$$A = A_1 + A_2 = 3a^2 \text{ mm}^2$$

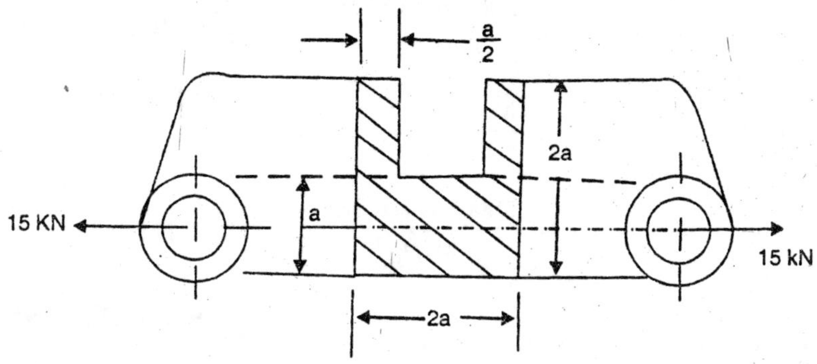

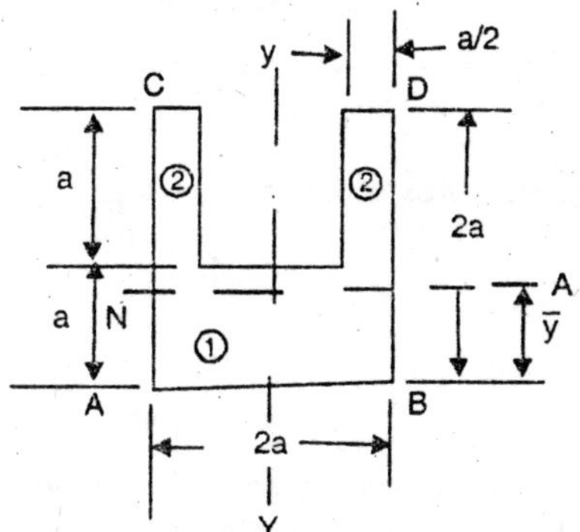

Fig. 3.9

Location of Neutral Axis :

$$3a^2y = \frac{2a^2 \times a}{2} + a^2 \times \left(a + \frac{a}{2}\right) = 2.5a^3$$

$$y = 0.833a \text{ mm}$$

Calculate moment of inertia about neutral axis N – A :

$$I = \left[\frac{2a \times a^3}{12 + 2a^2\,(0.833 - 0.5)^2 a^2}\right] + 2\left[\left(\frac{a}{2}\right)\left(\frac{a^3}{12}\right) + \left(\frac{a^2}{2}\right)(1.5 - 0.833)^2 a^2\right]$$

$$= 0.9167\,a^4 \text{ mm}^4$$

$$y_t = 0.833\,a \text{ mm},\; y_c = 2a - 0.833a = 1.167a \text{ mm}$$

Eccentricity, $e = (0.833 - 0.5)\,a = 0.333a$ mm

Bending moment, $M = Pe = 15 \times 10^3 \times 0.333a = 5000a$ N.mm

AB fibres are under tension and CD under compression.

Direct stress, $\sigma_d = \frac{P}{A} = \frac{15,000}{3a^2} = \frac{5000}{a^2}$ N/mm²

Tensile bending stress, $\sigma_t = \frac{My_t}{I}$

$$= \frac{5000a \times 0.833a}{0.9167a^4} = \frac{4545.1}{a^2} \text{ N/mm}^2$$

Compressive bending stress, $\sigma_c = \frac{My_c}{I}$

$$= \frac{5000a \times 1.167a}{0.9167a^4} = \frac{6365.2}{a^2} \text{ N/mm}^2$$

Resultant tensile stress in fibres at AB $= \sigma_d + \sigma_t = \frac{(5000 + 4545.1)}{a^2} = \frac{9545.1}{a^2}$ N/mm²

$$= 20$$

$$a = 21.84 \approx 22 \text{ mm}$$

Resultant compressive stress in fibres at CD $= \sigma_c - \sigma_d = \frac{(6365.2 - 5000)}{a^2} = \frac{1365.2}{a^2}$ N/mm²

$$= 60$$

$$a = 4.77 \approx 5 \text{ mm}$$

Adopt $a = 22$ mm

Example 3.5 *A shaft 8 m long and weighing 2KN/m is fitted with a pulley 1.4 m diameter and a gear as shown in Fig. 3.10. The driving gear, situated vertically below, delivers 75 kW to the shaft while driving the shaft at 600 rpm. The pulley ratio of belt tensions is 2 and it delivers the power from the shaft to another pulley situated vertically below it. Determine the required diameter of the shaft if the allowable shear stress is 56 MPa.*

Solution.

Given : l = 8m, P = 75 kW, w_s = 2kN/m = 3.5 kN, d_p 1.4 m, W_g = 0.3 kN, d_g = 0.5 m, N = 600 rpm, $F_1/F_2 = 2$, τ = 56 MPa

Forces due to gear :

Power transmitted, $P = \frac{2\pi n T}{60 \times 10^3}$

Torque transmitted, $T = \frac{75 \times 60 \times 10^3}{2\pi \times 600} = 1193.7$ N.m

Tangential force on gear on the shaft, $F_t = \frac{2T}{d_g} = \frac{2 \times 1193.7}{0.5} = 4774.8$ N

Let pressure angle of gear, $\alpha = 20°$

Radial force on gear, $F_r = F_t \tan \alpha = 4774.8 \times \tan 20° = 1738$ N ↑

Total force at gear location $= F_r - W_g = 1738 - 300 = 1438$ N ↑

Force at pulley location :

Let F_1 and F_2 be the tensions in the belt on the tight and slack sides respectively. Then

$$\frac{F_1}{F_2} = 2$$

Also $$\frac{(F_1 - F_2)\, d_p}{2} = T$$

$$F_2 = \frac{1193.7 \times 2}{1.4} = 1705.25 \text{ N}$$

$$F_1 = 3410.5 \text{ N}$$

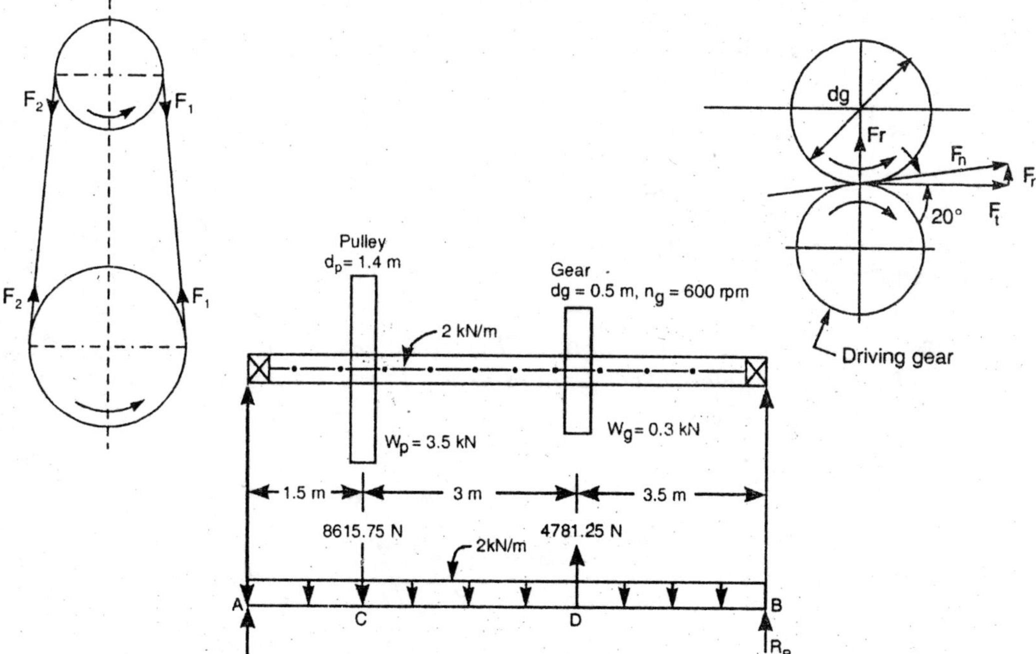

Fig. 3.10

Load on pulley due to belt load, $F = F_1 + F_2 = 3410.5 + 1705.25 = 5115.75$ N ↓

Total load at pulley location $= F + W_p = 5115.75 + 3500 = 8615.75$ N

Bearing Reactions :

The loading diagram on the shaft is shown in Fig. 3.10 (*b*).

Taking moments about B, we have

$$R_A \times 8 = 8615.75 \times 6.5 - 1438 \times 3.5 + 2000 \times 8 \times 4 = 114969$$

$$R_A = 14371 \text{ N}$$

Maximum Bending Moment :

Consider a section at a distance *x* nearer to B The bending moment at this location is,

$$M_x = 14371 \times x - 8615.75\,(x - 1.5) + 1438\,(x - 4.5) - 1000x^2$$

For M_x to be maximum, $\frac{dM_x}{dx} = 0$

$$\frac{dM_x}{dx} = 14371 - 8615.75 + 1438 - 2000x = 0$$

$$x = 3.6\text{m}$$

$$M_{max} = 14371 \times 3.6 - 8615.75\,(3.6 - 1.5) - 1000 \times (3.6)^2$$

$$= 20682 \text{ N.m}$$

Equivalent Torque :

$$T_e = [M^2 + T^2]^{1/2}$$

$$= [(20682)^2 + (1193.7)^2]^{1/2} = 20716 \text{ N.m}$$

$$= \left(\frac{\pi}{16}\right) \times d^3 \times \tau$$

$$\text{Shaft diameter, } d = \left[\frac{20716 \times 10^3 \times 16}{(\pi \times 56)}\right]^{1/3} = 123.5 \approx 125 \text{ mm}$$

Example 3.6 *A bracket as shown in Fig. 3.11 carries a load P. The cross-section at 50 mm from the fixed end is rectangular with dimensions 50 m × 15 mm. If the maximum stress is limited to 60 MPa, find the value of P.*

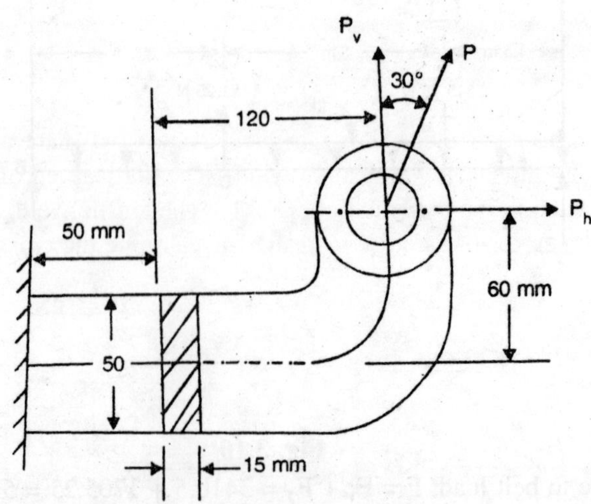

Fig. 3.11

Solution.

Given : t = 15 mm, b = 50 mm, σ_{max} = 60 MPa

Area of cross-section, A = tb = 15 × 50 = 750 mm^2

$$\text{Section modulus, } z = \frac{tb^2}{6} = 15 \times \frac{50^2}{6} = 6250 \text{ mm}^3$$

Resolving the force P into its horizontal and vertical components P_H and P_V respectively, we have

$$P_H = P \sin 30° = P \sin 30° = 0.5P \text{ N}$$
$$P_V = P \cos 30° = P \cos 30° = 0.866P \text{ N}$$

$$\text{Direct stress} \quad \sigma_d = \frac{P_H}{A} = \frac{0.5P}{750} = 6.667 \times 10^{-4} \times P \text{ N mm}^2 \text{ (tensile)}$$

Horizontal bending moment at the cross – section, $M_H = P_H \times 60 = 0.5P \times 60 = 30P$ N.mm

$$\text{Horizontal bending stress, } (\sigma_b)_H = \frac{M_H}{z} = \frac{30P}{6250} = 4.8 \times 10^{-4} \text{ P N/mm}^2$$

Which is tensile on top fibres and compressive on bottom fibres.

Vertical bending moment at the cross – section, MV = PV × 120 = 0.866P × 120 = 103.92P N.mm

$$\text{Vertical bending stress, } (\sigma_b)_v = \frac{M_v}{z} = \frac{103.92P}{6250} = 166.3 \times 10^{-4} \text{ P N/mm}^3$$

Which is compressive on the top fibres and tensile on the bottom fibres.

Maximum stress at the section occurs at the bottom fibres, $\sigma_{max} = \sigma_d - (\sigma_b)_H + (\sigma_b)_V$

$$= (6.667 - 4.8 + 166.3) \times 10^{-4} \times P$$
$$= 168.167 \times 10^{-4} \times P$$
$$= 60$$
$$P = 3567.9 \text{ N}$$

EXERCISES

3.1 The frame of a hacksaw blade shown in Fig. 3.12 is of U-shape and rectangular in cross – section having thickness 3 times the width. The vertical leg of the frame is 150 mm. The initial tension in the blade is to be 300 N. The tensile yield strength of the frame is 380 MPa and a factor of safety 3 is desired. Determine the cross-section.

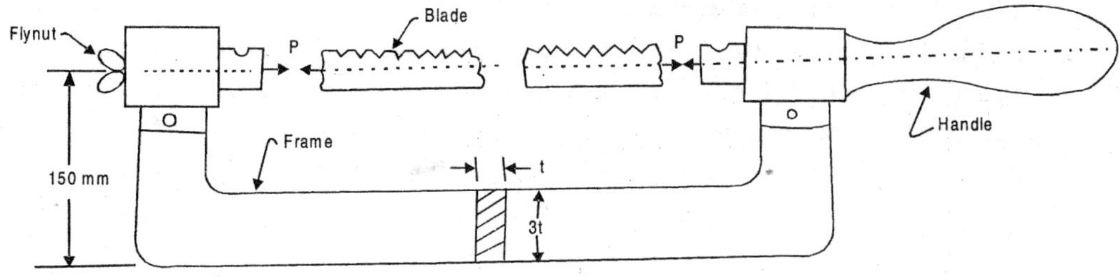

Fig. 3.12 Hand Hacksaw

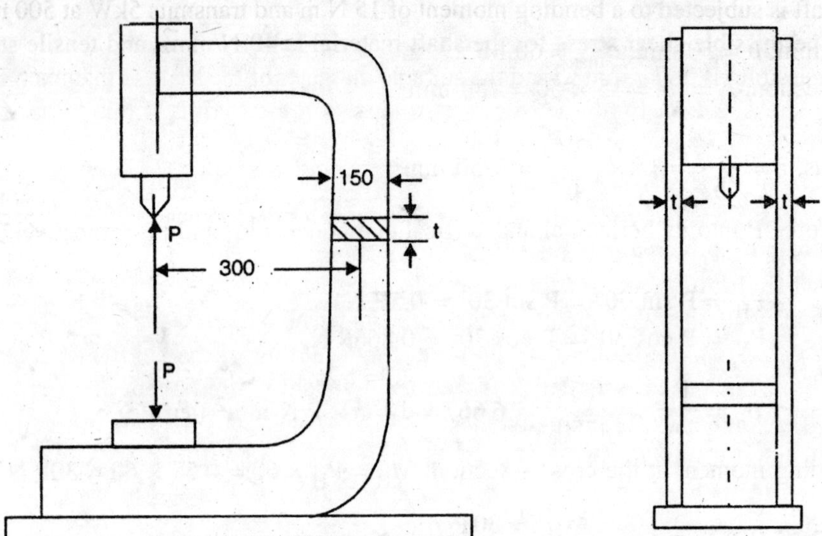

Fig. 3.13 Frame of hydraulic press

3.2 The frame of a hydraulic press shown in Fig. 3.13 consists of two identical steel plates 180 mm wide. The maximum load on the plates is 25 kN at an eccentricity of 300 mm. The tensile yield strength of frame is 400 MPa. Calculate the thickness of plates if a factor of safety 2.5 is desired.

3.3 A shaft is subjected to the combined action bending and torsion. The maximum bending moment is 400 N.m and the maximum torque is 300 N.m. Calculate the diameter of the shaft based on maximum shear stress theory if the permissible shear stress in the shaft is 80 MPa.

3.4 A 10 m long shaft weighing 2.5 kN/m is fitted with a pulley A of 1.25 m diameter at a distance of 1.5 m and a B gear of 500 mm diameter at a distance of 5 m from the left hand bearing. The driving gear situated vertically below delivers 75 kW to the gear B. The shaft runs at 600 rpm. The ratio of belt tensions on the pulley is 2.5. Pulley A drives another pulley situated vertically below it. Calculate the diameter of the shaft if the allowable shear stress is 60 MPa. The shock and fatigue factors in bending and torsion are 1.5 and 1.25 respectively.

3.5 A cylindrical shaft, made of steel of yield strength 760 MPa, is subjected to a static load consisting of bending moment of 15 kN.m and a torsional moment of 25 kN.m Determine suitable diameter of shaft using maximum shear stress theory and maximum strain energy theory, assuming a factor of safety 2.5.

(Take E = 200 GPa and Poisson's ratio = 0.25)

3.6 A line shaft rotating at 200 rpm is to transmit 200 kW. The allowable shear stress for the material of the shaft is 42 MPa. If the shaft carries a central load of 900 N and is simply supported between bearings 3 m apart, determine the diameter of the shaft. The maximum tensile or compressive stress is not to exceed 56 MPa.

3.7 (*i*) What is the criterion of failure for ductile material subjected to static load ?

(*ii*) Write about the various theories of failure.

3.8 (*a*) A shaft is subjected to a bending moment of 15 N.m and transmits 5kW at 500 rpm. The **permissible** shear stress for the shaft material is 40 N/mm^2, and tensile stress (permissible) is 58 N/ mm^2. Find the suitable diameter of the shaft so that maximum shear stress and maximum tensile stress induced in the shaft is within permissible limit.

(*b*) What is the necessity of having more than one theories of failure ? State maximum shear stress theory and show that it is more suitable for ductile material in comparison to maximum principal stress theory.

3.9 A mild steel shaft of 4 cm diameter is subjected to a bending moment of 15,000 N.cm and a torque T. If the yield point of the steel in tension is 20,000 N/cm^2, find the maximum value of this torque T without causing yielding of the shaft according to (*i*) maximum principal stress theory (*ii*) the maximum shear stress theory.

3.10 A line shaft rotating at 200 rpm is to transmit 25 kW. The allowable shear stress for the material of the shaft is 4200 N/cm^2. If the shaft carries a central load of 900 N and is simply supported between bearings 3 m apart, determine the diameter of the shaft. The maximum tensile or compressive stress is not to exceed 5600 N/cm^2.

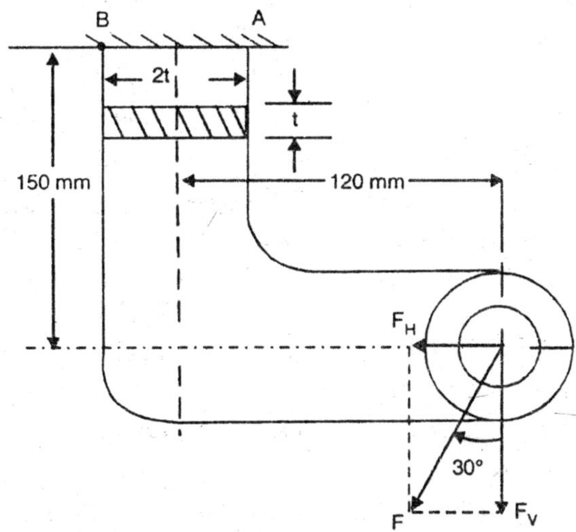

Fig. 3.14

3.11 Figure 3.14 shows a hanger with rectangular cross-section with the proportions indicated. A force F of magnitude 10 kN acts at 30° to the vertical as shown. The permissible stress in the hanger mateial is 60 MPa. Find the cross section of hanger.

3.12 A shaft is required to transmit 15 kW at 200 rpm. It is supported on two bearings 600 mm apart and two gears are keyed to it. A 18 teeth, 14.5° involute, 5 mm module pinion is located at 120 mm to the right of right bearing and delivers power to a gear directly below the shaft. A 72 tooth, 14.5° involute, 5 mm module gear is located at 150 mm to the right of the left bearing and receive power from the gear, directly aboue it. Determine the shaft diameter if permissible normal stress is 80 MPa.

DESIGN FOR VARIABLE LOAD

Chapter

4.1 INTRODUCTION

A variable load may be defined as the force which does not remain constant with respect to time but fluctuates between two extremes. Transmission shafts carrying gears, pulleys, etc are subjected to loads which vary throughout the cycle. Therefore, the design of machine components under variable loads cannot be carried out by the simple formulae of strength of materials. Also the machine members have geometric and other types of discontinuities which further complicate the analysis for design. We shall study all these aspects for the safe design of machine components.

4.2 STRESS-CONCENTRATION

The elementary formulae used in design are based on the assumption that the cross-section of the machine member is constant without any discontinuities. Such conditions are however hardly ever attained throughout the stressed region of actual machine members. There are always present some discontinuities in the cross-section in the form of holes, notches, grooves, fillet radius, keyway, etc. These are called stress raisers. The stress level at the stress raisers is higher than that of the stress calculated by the normal stress formulae.

Stress – concentration may be defined as the localization of maximum stress at the location of these discontinuities. The cause of this highly localized stress is termed as stress raiser. Stress – concentration is measured by the stress – concentration factor, defined by :

$$K_t = \text{Maximum stress / Nominal stress} = \frac{\sigma_{max}}{\sigma_{nom}} \quad \text{... (4.1)}$$

Where subscript 't' stands for 'theoretical'.

K_t is called the geometric or theoretical stress – concentration factor which is based on the usual assumptions in theory of elasticity, i.e. the material is homogeneous, isotropic and linearly elastic. The nominal stress is based on the knowledge from strength of materials. The nominal stress may be either based upon the gross or net area of cross – section. For axial loading the theoretical stress – concentration factor is denoted by K_{ta}, for bending by K_{tb} and for torsion by K_{ts}.

Consider a long thin flat plate of finite width w and thickness t with a central circular hole of diameter d subjected to uniaxial tensile force P, as shown in Fig. 4.1.

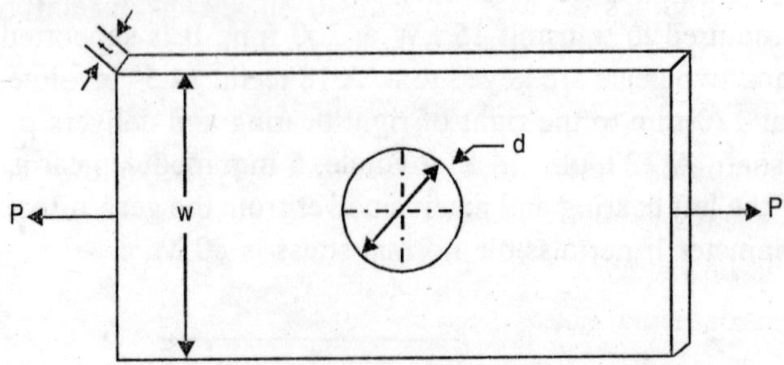

Fig. 4.1 Thin plate with a circular hole.

Gross area of cross – section of plate, $A_g = wt$

Net area of plate, $A_n = (w - d)\,t$

Gross nominal tensile stress in plate, $\sigma_g = \dfrac{P}{A_g}$

Net nominal tensile stress in plate, $\sigma_n = \dfrac{P}{A_n}$

Stress – concentration factor based on gross area, $K_{tg} = \dfrac{\sigma_{max}}{\sigma_g}$

Stress – concentration factor based on net area, $K_{tn} = \dfrac{\sigma_{max}}{\sigma_n}$

$$\frac{K_{tn}}{K_{tg}} = \frac{\sigma_g}{\sigma_n} = \frac{A_n}{A_g}$$

$$= \frac{w-d}{w}$$

$$= 1 - \frac{d}{w}$$

$$= 1 - \lambda \qquad \text{... (4.2)}$$

Where $\lambda = d/w$ is a fraction by which area is decreased.

The stress – concentration factors in the form of charts for various types of stress raisers are given at the end of this chapter.

4.2.1 Causes of Stress-Concentration

The principal cause of stress – concentration is the presence of some sort of a discontinuity. The principal sources of such discontinuities to produce disturbances in a uniform stress field and most commonly encountered are :

1. Geometric discontinuities as represented by holes, notches, keyways, grooves, threads and fillet radii, etc.
2. Load discontinuities, such as those due to concentrated and discontinuously distributed surface loads.
3. Material discontinuities, such as those due to inclusions or reinforcements by materials whose elastic properties differ from the surrounding matrix.

4.2.2 Determination of Stress - Concentration Factors

The stress – concentration factors may be determined by the following methods :

1. Mathematical calculations.
2. Measurement on models.
3. Measurement on actual parts.

Mathematical calculations are generally limited to relatively simple two – dimensional problems. For stress raisers not solvable by mathematical analysis, stress distribution pattern may be evaluated from experimental measurements on models. Measurements of local stresses on actual parts may be made for which it is difficult to simulate by models. The experimental methods generally used are : photoelasticity, strain gauges, brittle coatings and analogous methods.

4.2.3 Mitigation of Stress - Concentration

The following methods may be used to mitigate or reduce the stress – concentration :

1. Removal of undesirable material, [Fig. 4.2 (a)].

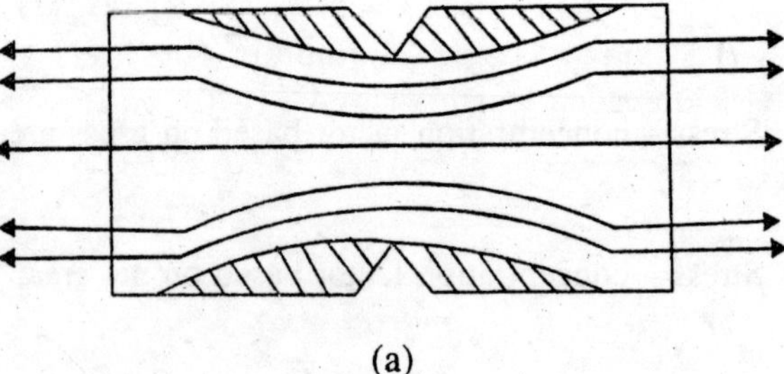

(a)

2. Added grooves, [Fig. 4.2 (b)].

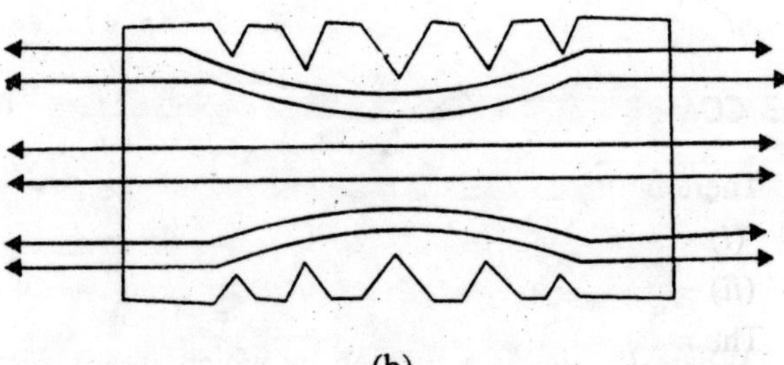

(b)

3. Drilled holes, [Fig. 4.2 (c)]

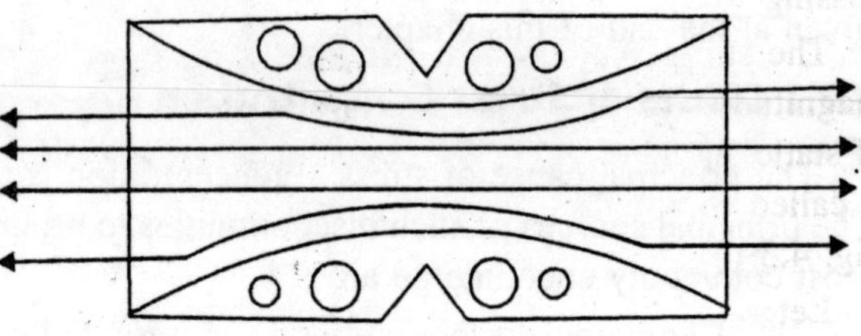

(c)

4. Using large fillet radius, [Fig. 4.2 (d)].

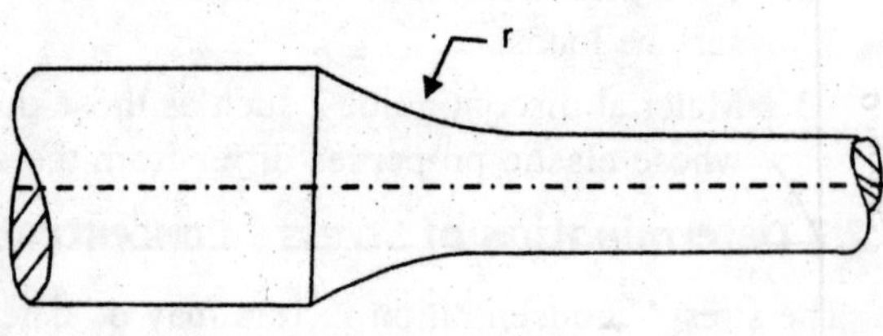

(d)

5. Combining grooves and fillet radii, [Fig. 4.2 (e)].

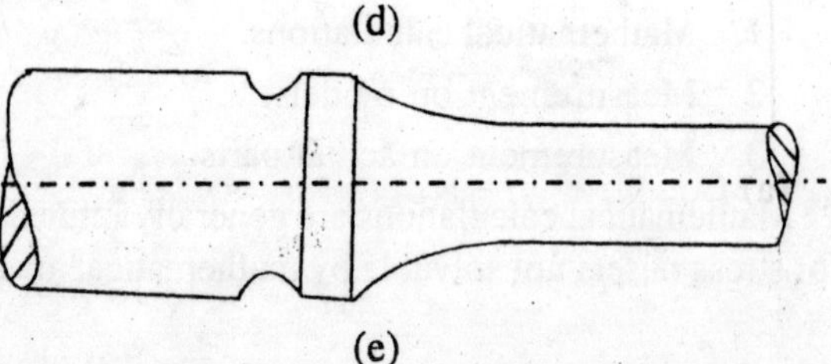

(e)

6. Using undercut shoulders, [Fig. 4.2 (f)].

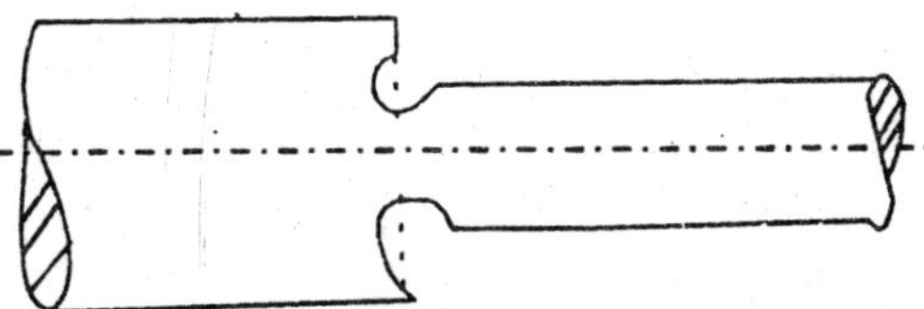

(f)

7. Using larger diameter for threaded length, [Fig. 4.2 (g)].

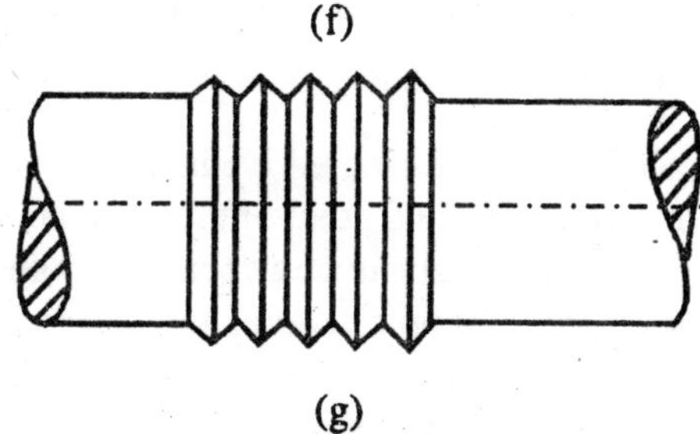

(g)

Fig.4.2 Methods of reducing stress-concentration.

4.3 COMPLETELY REVERSED AND FLUCTUATING STRESSES

There are two types of problems in fatigue design :

(*i*) Components subjected to completely reversed stresses, and

(*ii*) Components subjected to fluctuating stresses.

The mean stress is zero in case of completely reversed stresses. The stress distribution consists of tensile stresses for the first half cycle and compressive stresses for the second half cycle, passing through zero.

The stresses induced in machine components due to forces which are not static but vary in magnitude with respect to time are called fluctuating stresses. These stresses are the combination of static and alternating stresses. The fluctuating stress in which the mean or static stress is zero is called completely reversed stress, as shown in Fig. 4.3 (a). The fluctuating stress is shown in Fig. 4.3 (b).

Let σ_{max} = maximum stress

σ_{min} = minimum stress

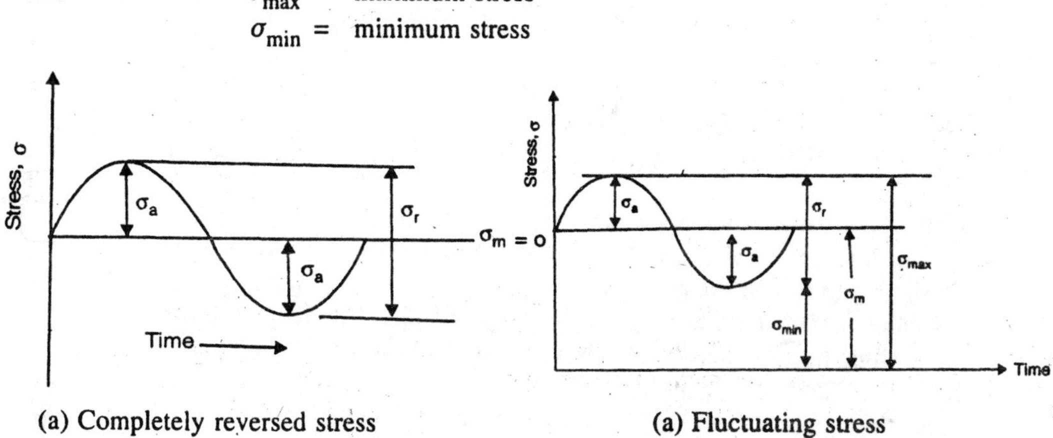

(a) Completely reversed stress

(a) Fluctuating stress

Fig. 4.3 Types of fatigue stresses

Then

Mean stress, $\sigma_m = \dfrac{\sigma_{max} + \sigma_{min}}{2}$ ... (4.3)

Alternating stress, $\sigma_a = \dfrac{\sigma_{max} - \sigma_{min}}{2}$ ... (4.4)

Stress range, $\sigma_r = \sigma_{max} - \sigma_{min}$ ... (4.5)

4.4 FATIGUE FAILURES

Fatigue failures are progressive in nature which are caused due to repeatedly imposing stresses generally below the yield point of the material. Fatigue failures occur without perceptible yielding or necking. They originate at a localized point and progress gradually through a portion of the member. Ultimate fatigue failures take place because the reduced cross – section can no longer withstand the load. It has been found that the magnitude of the stress causing fatigue failure decreases as the number of stress cycles increases. This phenomenon of decreased resistance of materials to fluctuating stresses is called fatigue. The differences between static and fatigue failures are listed in Table 4.1.

Table 4.1 Static v's fatigue failures.

Static failures	Fatigue failures
1. The load is gradually applied and there is sufficient time for elongation of fibres.	1. The fatigue failure begins with a crack at some point in the material.
2. In ductile materials, there is considerable plastic flow prior to fracture.	2. The crack spreads due to fluctuating stresses.
3. The fracture is silky fibrous structure due to the stretching of crystals at the fractured surface.	3. Fatigue failure may be either due to region indicating slow growth of crack with a fine fibrous appearance or region of sudden fracture with a coarse granular appearance.
4. Static failures give warning well in advance.	4. The fatigue failure is sudden and total.

4.5 FATIGUE CONSIDERATIONS IN DESIGN

Fatigue. The reduced tendency of a material to offer resistance to the applied stresses is called fatigue.

The resistance to fatigue failures can be improved by the following methods :

1. Eliminating stress raisers.
2. Designing to take stress raisers into account when cannot be eliminated.
3. Avoiding surface tears such as those caused by punching, shearing, etc.
4. Controlling heat treatment to prevent decarburization, oxidation, etc.
5. Improving fabricating and fastening details in the design.
6. Protecting the part from corrosion, erosion, etc.
7. Guarding against the formation of surface defects in service.
8. Minimize tensile mean stresses during assembly.

4.6 ENDURANCE LIMIT AND WOHLER'S CURVE

Endurance limit (S_e). It may be defined as the maximum value of completely reversed bending stress that the standard specimen of a material can sustain for an infinite number of cycles without fatigue failure.

Fatigue strength (S_f). It may be defined as the limiting stress below which a material will withstand a specified number of cycles of stress without fracture.

Fatigue life (N). It is the number of cycles of stress that can be sustained by a material prior to failure of a specified nature for a stated condition.

The endurance limit of a material is determined by applying to several specimens repeated bending loads which cause completely reversed stresses of known values, and recording the number of stress reversals each specimen did endure before it failed. Each specimen is subjected to a lower stress than was the preceding one, and it breaks after a greater number of cycles. Finally, a stress is reached that does not cause failure, regardless of how many reversed cycles are applied. This test is performed on R.R. Moore rotating beam fatigue testing machine. The number of stress cycles to determine the endurance limit are very large as compared to the stress. Therefore, the experimental results are plotted on a log – log graph sheet with the stress as ordinates and the number of cycles as abscissas, as shown in Fig. 4.4. This plot is caled the S – N or Wohler's curve.

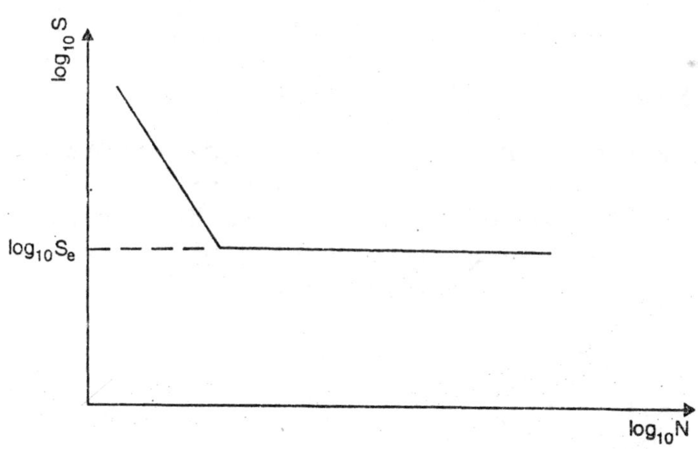

Fig. 4.4 S—N or Wohler curve

The uncorrected endurance limit in bending or axial loading is represented by S'_e and by S'_{es} in torsional shear. The endurance limit of some of the materials as a ratio of the ultimate tensile strength is given in Table 4.2

Table 4.2 Ratios of $\frac{S'_e}{S_{ut}}$

Materials	$\frac{S'_e}{S_{ut}}$
1. Wrought iron at 10^6 cycles	0.4
2. Cast steel at 10^6 cycles	0.5
3. Cast iron at 10^6 cycles	0.4 – 0.5
4. Wrought aluminium alloys at 5×10^8 cycles	0.4
5. Copper alloys at 10^8 cycles	0.25 – 0.50

4.7 FATIGUE STRESS - CONCENTRATION FACTOR

Experimental results confirm that reductions in fatigue strength caused by stress raisers are less than the predicted reduction based on theoretical stress – concentration factor, K_t. To deal with the situation, it is customary to use the so called fatigue stress – concentration factor, K_f, defined as :

$$K_f = \frac{\text{Fatigue limit of un- notched specimen (axial or bending)}}{\text{Fatigue limit of notched specimen (axial or bending)}} \quad \text{... (4.6a)}$$

$$\text{Fatigue strength reduction (or fatigue notch) factor, } k_f = \frac{1}{K_f} \quad \text{... (4.6b)}$$

4.8 NOTCH SENSITIVITY

The effect of a notch on the fatigue strength of a part varies considerably with materials and notch geometry. It has been observed that it is usually less than the effect that would be predicted by using the value of theoretical stress – concentration factor. This general phenomenon is called 'notch sensitivity'. Notch sensitivity may be considered as a measured of the degree to which the theoretical effect of stress – concentration is obtained. It is defined as :

$$q = \frac{K_f - 1}{K_t - 1} \quad \text{... (4.7)}$$

or $$K_f = 1 + q\,(K_t - 1) \quad \text{... (4.8)}$$

The values of q ranges from 0 to 1. For a completely notch sensitive material, $q = 1$ and experiences the full effect of K_t. For a completely non – notch sensitive material, $q = 0$ and experiences no effect of the stress raiser.

Similarly for torsional loading, we can define that,

$$q = \frac{K_{fs} - 1}{K_{ts} - 1} \quad \text{... (4.8)}$$

Where $$K_{fs} = \frac{\text{Fatigue limit of un- notched specimen (torsion)}}{\text{Fatigue limit of notched specimen (torsion)}} \quad \text{... (4.9)}$$

K_{ts} = Stress – concentration factor (shear stress)

Thus $$K_{fs} = 1 + q\,(K_{ts} - 1) \quad \text{... (4.10)}$$

The values of notch sensitivity factor q can be obtained from Fig. 4.5 for axial and reversed bending, and for reversed torsion from Fig. 4.6.

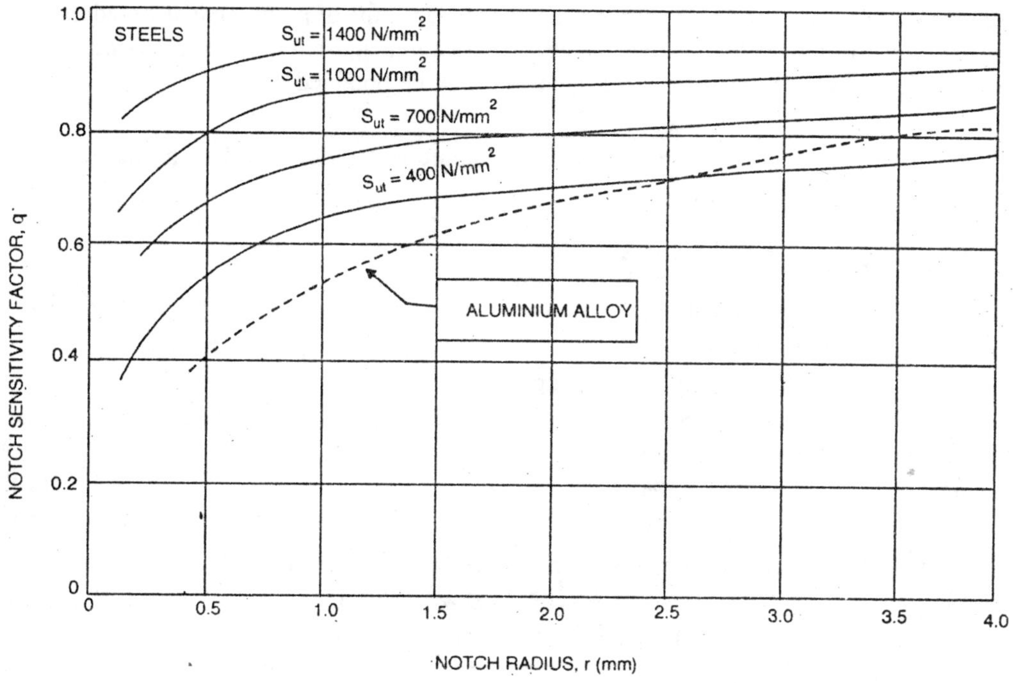

Fig. 4.5 Notch sensitivity charts for reversed bending and reversed axial stresses

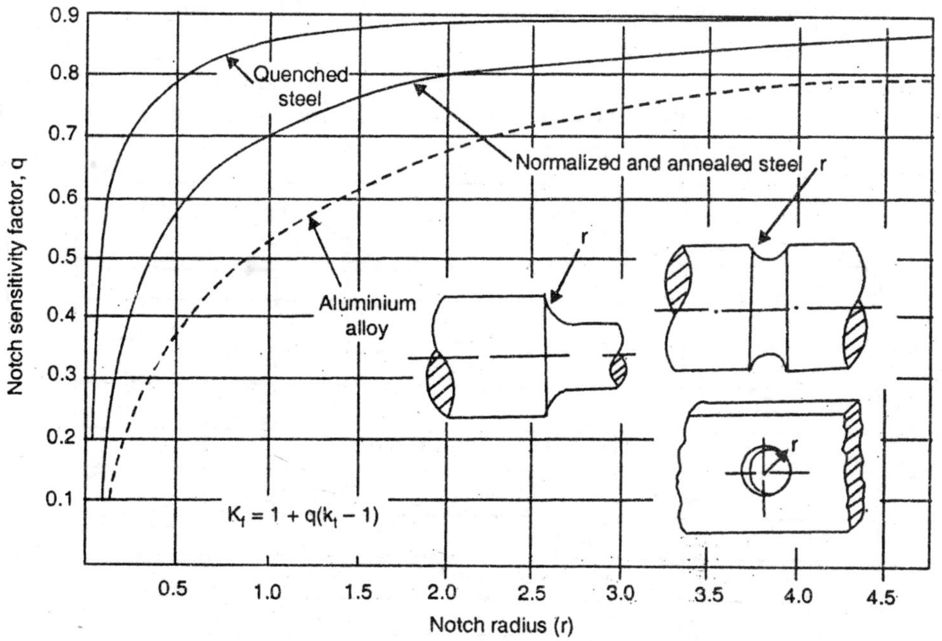

Fig. 4.6 Notch sensitivity factor (q)

4.9 FACTORS AFFECTING ENDURANCE LIMIT

The endurance limit of a component is different from the endurance limit of rotating beam specimen as determined experimentally in the R.R. Moore machine due to a number of factors. The derating factors are used to determine the endurance limit of a component by adjusting its endurance limit to account for loading effects, part features and environmental conditions. The corrected endurance limit of a machine component may be obtained as :

$$S_e = k_a k_b k_c k_d k_e k_f k_g S'_e \quad \text{... (4.11)}$$

Where S'_e = rotating beam (uncorrected) theoretical endurance limit of a specimen

k_a = surface finish factor

k_b = size factor

k_c = reliability factor

k_d = temperature factor

k_e = load factor

k_f = fatigue strength reduction factor

k_g = miscellaneous correction factor for press – fitted collar, residual stress, corrosion, impact, etc.

For steel, S_{ut} = 3.5 BHN MPa ... (4.12)

$S'_e = 0.5\ S_{ut} = 1.75$ BHN ... (4.13)

For fluctuating torsional shear stresses, endurance limit is given by :

$S'_{es} = 0.5\ S'_e$, based on maximum shear stress theory ... (4.14*a*)

$= 0.577\ S'_e$, based on distortion energy theory ... (4.14*b*)

1. Surface Finish Factor, k_a

The surface finish of a part may affect its endurance limit in three ways :

(*i*) by introducing stress – concentration resulting from surface roughness,

(*ii*) by altering the physical properties of the surface layer of the material, and

(*iii*) by introducing residual stresses.

The surface finish factor is defined as the ratio between the endurance limit obtained with arbitrary surface finish and that obtained with the standard mirror polished finish test specimen. The surface finish factor may be obtained from Figs. 4.7 or 4.8.

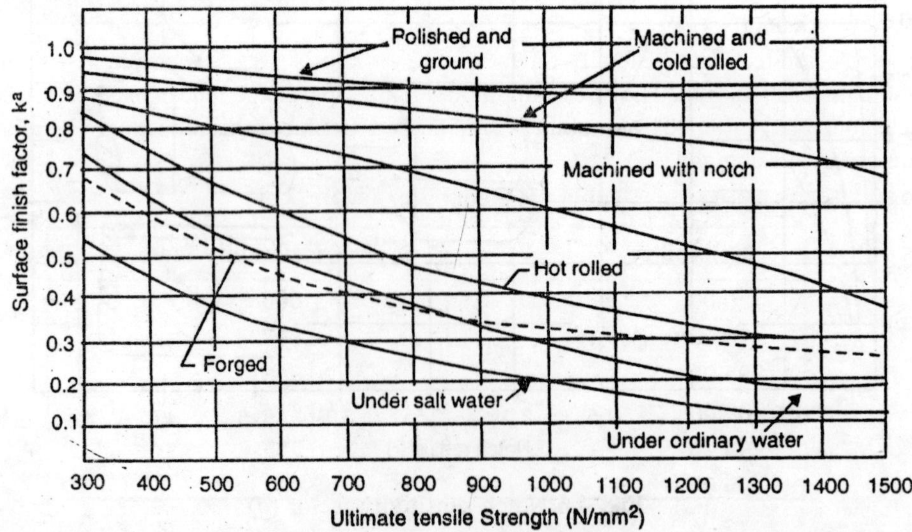

Fig. 4.7 Surface finish factor (k_a)

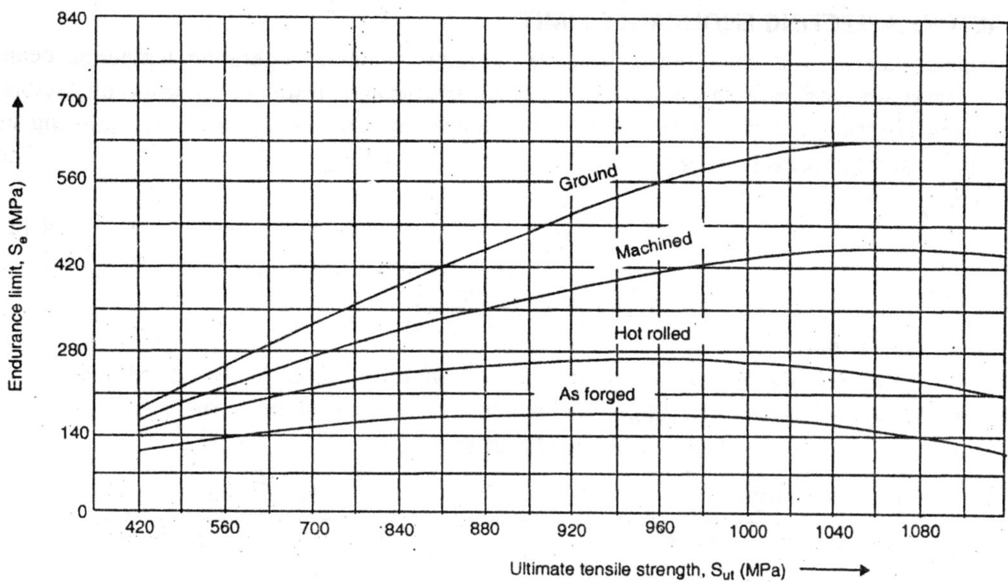

Fig. 4.8 Endurance limit of steel corresponding to ultimate tensile strength.

2. Size factor, k_b'

The endurance limit under bending and torsional loads tends to decrease as size increases. The size factor may be assumed as given in Table 4.3.

Table 4.3 Size factor, k_b

Type of loading	Size factor, k_b
1. Reversed bending and torsional loads :	
(*i*) $d \leq 0.75$ mm	1.00
(*ii*) $7.5 < d \leq 50$ mm	0.85
(*iii*) $d > 50$ mm	0.75
2. Reversed axial loads for $d < 50$ mm	1.0

3. Reliability Factor, k_c

The rotary beam test data generally reflect a survival rate of 50%. To ensure that more than 50% of the parts will survive. The stress amplitude that the part experiences must be lower than the tabulated endurance limit. The reliability factor is used to achieve this reduction. The reliability factors are given in Table 4.4.

Table 4.4 Reliability factor, k_c

Reliability, %	Reliability factor, k_c
50.00	1.000
90.00	0.897
95.00	0.868
99.00	0.814
99.90	0.753
99.99	0.702

4. Temperature Factor, k_d

The endurance limit is specified at room temperature, whereas operating temperatures of machine parts are generally higher. A decline in endurance limit occurs with increasing temperature. To account for reduction in the endurance limit of steels for temperature above 71°C, the temperature factor is defined as :

$$k_d = \frac{344}{273 + T} \quad \text{... (4.15)}$$

where T = temperature, °C

5. Load Factor, k_e

The load constant is the ratio of the endurance limit under the given loading conditions to the endurance limit under reversed bending. The load factor may be obtained from Table 4.5.

Table 4.5 Load factor, k_e

Type of loading	Load factor, k_e
1. Reversed or rotating bending	1.0
2. Reversed axial loading with no bending	0.9
3. Reversed axial loading with indeterminate bending	0.6 – 0.85
4. Reversed torsion :	
(*i*) Ductile materials	0.577
(*ii*) Brittle materials	0.80

6. Fatigue strength reduction factor; k_f

$$k_f = \frac{1}{K_f} \quad \text{... (4.16)}$$

where $K_f = 1 + q(K_t - 1)$

7. Miscellaneous correction factors.

In addition to the above stated factors, there are more factors like impact factor press – fitted collar factor, residual stresses, corrosion factor, etc., which have to be taken into account depending upon the conditions prevailing in a given problem.

4.10 DESIGN FOR FINITE (OR LIMITED) AND INFINITE (OR UNLIMITED) LIFE

1. Completely Reversed Stresses

For completely reversed stresses, $\sigma_m = 0$, $\tau_m = 0$ and only σ_a and τ_a are present.

The design problems for completely reversed stresses are divided into two groups :

(*i*) Design for infinite life, and

(*ii*) Design for finite life.

(*i*) Design for Infinite life. When a component is to be designed for infinite life (number of cycles ≥ 10^6), the endurance limit becomes the criterion for failure. The stresses induced in such components should be lower than the endurance limit in order to withstand the infinite number of cycles. Such components are designed with the help of the following equations :

Induced stress in bending, $$\sigma_a = \frac{S_e}{n} \quad \text{... (4.17)}$$

Induced stress in torsion, $\tau_a = \dfrac{S_{es}}{n}$ ... (4.18)

S_e, S_{es} = corrected endurance strengths in bending and torsion respectively.

σ_a, τ_a = amplitude (or alternating) normal and shear stresses respectively.

n = factor of safety

***(ii)* Design for finite life.** When a component is to be designed for finite life (number of cycle $< 10^6$), the $\log_{10} S_f$ - $\log_{10}$ N curve for steels, as shown in Fig. 4.9, can be used. It consists of a straight line AB drawn from $\log_{10}(0.9S_{ut})$ at $\log_{10}(10^3) = 3$ cycles to $\log_{10}(S_e)$ at $\log_{10}(10^6) = 6$ cycles. The design procedure for such problems is as follows :

1. Locate point A with coordinates $[3, \log_{10}(0.9\ S_{ut})]$.

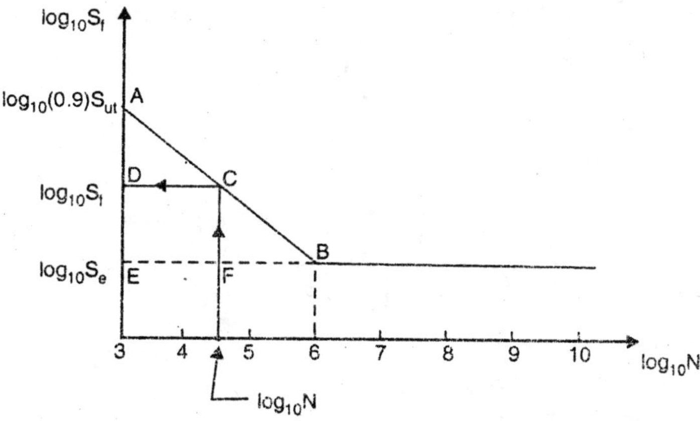

Fig. 4.9 S—N curve for finite life

2. Locate point B with coordinated $[6, \log_{10}(Se)]$.
3. Join AB, which is used as a criterion of failure for finite life problems.
4. For the desired life N cycles of the machine component, draw a vertical line passing through $\log_{10}(N)$ on the abscissa, intersecting line AB at C.
5. Draw a line CD parallel to the abscissa. The ordinate at point D gives the endurance strength corresponding to N cycles. This value of fatigue strength S_f is used for the design calculations.

For finite life,

$$S_f = \frac{10^b}{N^m} \quad \text{... (4.19)}$$

For $10^3 \leq N \leq 10^6$, the values of b and m are given by :

$$b = \log\left[\frac{(0.9S_{ut})^2}{S_e}\right] \quad \text{... (4.20)}$$

$$m = (1/3)\log\left[\frac{0.9S_{ut}}{S_e}\right] \quad \text{... (4.21)}$$

For $N \geq 10^6$; $S_f = S_e$ ... (4.22)

2. Fluctuating Stresses

When a component is subjected to fluctuating stresses, the stresses are resolved into two parts : mean stress, σ_m and alternating stress amplitude, σ_a. A σ_m - σ_a diagram can be plotted to predict the failure under fluctuating stresses. In this diagram, the mean stress is plotted on the abscissa, with tensile stresses to the right of origin and compressive stresses to the left of origin. The alternating stress amplitude is plotted on the ordinate.

(*i*) Ductile Materials under Axial or Bending Loads

The σ_m - σ_a diagram for axial and bending loading for ductile materials having tensile stresses is shown in Fig. 4.10. If the stress is purely static, $\sigma_a = 0$ and a mean stress of S_{yt} will initiate yielding and a mean stress of S_{ut} will cause fracture. If the stress is completely reversed, $\sigma_m = 0$ and the criterion of failure is endurance limit S_e, which is plotted on the ordinate. When the component is subjected to both types of stresses σ_m and σ_a, the failure criterion can be obtained as follows :

(*a*) Soderberg line Fig. 4.10

This line is obtained by joining S_e and S_{yt} by a straight line. The equation of Soderberg line is :

$$\frac{\sigma_m}{S_{yt}} + \frac{\sigma_a}{S_e} = 1 \quad ....(4.23)$$

(*b*) Goodman line

This line is obtained by joining S_e and S_{ut} by a straight line. The equation of Goodman line is :

$$\frac{\sigma_m}{S_u} + \frac{\sigma_a}{S_e} = 1 \quad ....(4.24)$$

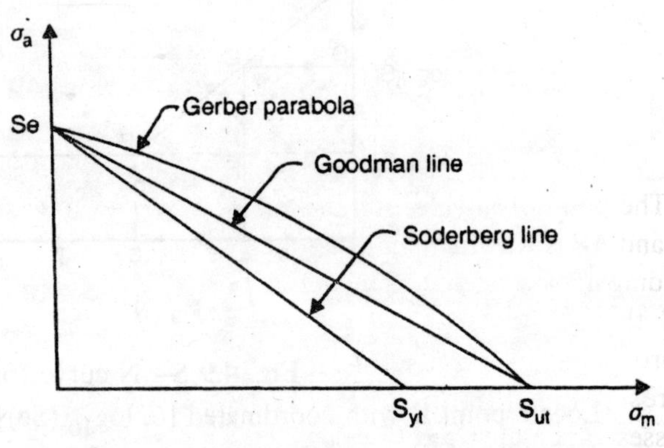

Fig. 4.10 σ_m— σ_a **diagrams**

(*c*) Gerber parabola

This is obtained by joining S_e and S_{ut} by a parabolic curve. The equation of Gerber parabola is :

$$\left(\frac{\sigma_m}{S_{ut}}\right)^2 + \frac{\sigma_a}{S_e} = 1 \quad .....(4.25)$$

Soderberg line is used as a very conservative approximation and has the feature that its use automatically predicts the possibility of failure by general yielding. Its use is recommended sparingly. Goodman line is used where a generalized curve must be used because of lack of data. Gerber parabola is used when some small eccentricities of axial loads exist.

(*d*) Modified Goodman Diagram for Axial or Bending Loads

In Fig. 4.11, the Goodman line AB is the criterion for fatigue fracture, and the line CD the criterion of static yielding. The equation of line CD is :

$$\sigma_m + \sigma_a = S_{yt}$$

All points below AED correspond to fluctuating stresses which should cause neither fatigue fracture nor plastic deformation. Area OAED represents the region of safety for components subjected to fluctuating stresses. Line CD is inclined at 45° to the abscissa. While solving a problem, a line OF with slope $\tan\theta = \dfrac{\sigma_a}{\sigma_m}$ is constructed.

Since $$\frac{\sigma_a}{\sigma_m} = \frac{\left(\dfrac{P_a}{A}\right)}{\left(\dfrac{P_m}{A}\right)} = \frac{P_a}{P_m}$$

Therefore $$\tan\theta = \frac{P_a}{P_m}$$

Similarly $$\tan\theta = \left(\frac{M_a}{M_m}\right)$$

Fig. 4.11 Modified Goodman diagram

The point of intersection of line OF and AE is X, which indicates the dividing line between the safe region and the region of failure. The coordinates of point X(S_m, S_a) represent the limiting values of stresses which are used to calculate the dimensions of the component. The design stresses are :

$$\sigma_{dm} = \frac{S_m}{n} \text{ and } \sigma_{da} = \frac{S_a}{n}$$

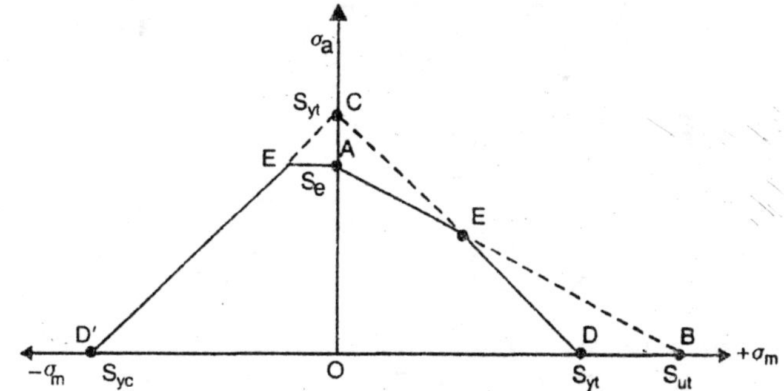

Fig. 4.12 Complete Goodman diagram

The complete diagram for both tensile and compressive stresses is shown in Fig. 4.12.

(*ii*) Ductile Materials under Torsional Loads

The various failure criterion are shown in Fig. 4.13.

(*a*) Soderberg line

This line is obtained by joining S_{es} and S_{ys} by a straight line. The equation of Soderberg line is :

$$\frac{\tau_m}{S_{ys}} + \frac{\tau_a}{S_{es}} = 1 \qquad ....(4.26)$$

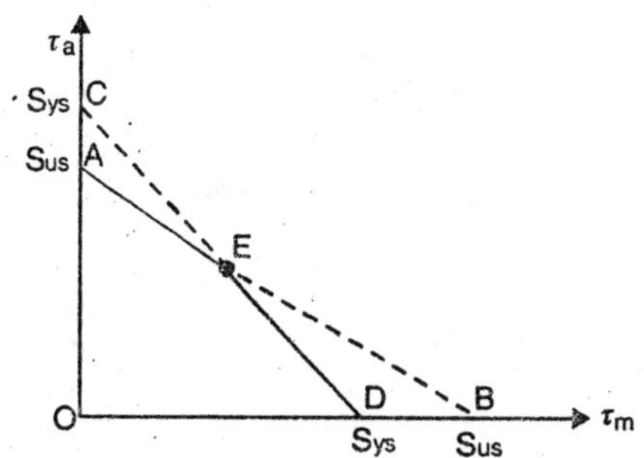

Fig. 4.13 τ_m–τ_a diagram for torsion

(b) Goodman line

This line is obtained by joining S_{es} and S_{us} by a straight line. The equation of Goodman line is :

$$\frac{\tau_m}{S_{us}} + \frac{\tau_a}{S_{es}} = 1 \quad \text{....(4.27)}$$

(c) Gerber parabola

This is obtained by joining S_{es} and S_{us} by a parabolic curve. The equation of Gerber parabola is :

$$\left(\frac{\tau_m}{S_{us}}\right)^2 + \frac{\tau_a}{S_{es}} = 1 \quad \text{....(4.28)}$$

(d) Modified Goodman Diagram for Torsional Loads

The modified Goodman diagram for fluctuating torsional shear stresses is shown in Fig. 4.14. In this diagram, the torsional yield strength S_{ys} is plotted on both the axes and a line AB is constructed to join these two points. This line is inclined at 45° to the abscissa. A line CD is drawn through S_{es} on the ordinate and parallel to the abscissa to intersect line AB at D. The area OCDB represents the region of safe design.

To solve problem, a line OE with slope

$$\tan\theta = \frac{\tau_a}{\tau_m} = \frac{T_a}{T_m} \text{ is constructed.}$$

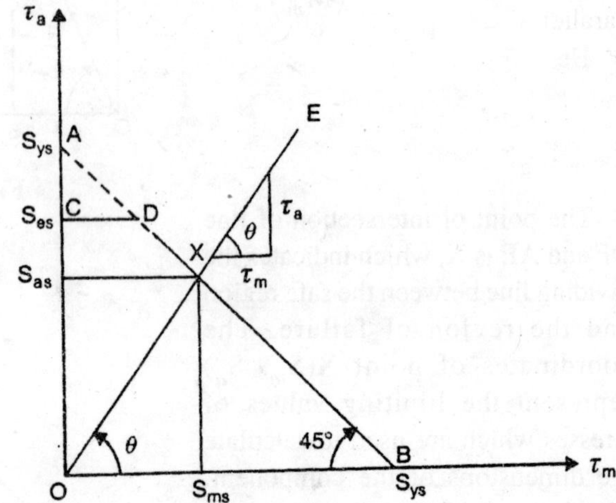

Fig. 4.14 Modified Goodman diagram for torsion.

The point of intersection of line OE and DB is X, which indicates the dividing line between the safe region and the region of failure. The coordinates of pont X (S_{ms}, S_{as}) represent the limiting values of stresses which are used to calculate the dimensions of the component. The design stresses are :

$$\tau_{da} = \frac{S_{as}}{n} \text{ and } \tau_{dm} = \frac{S_{ms}}{n}$$

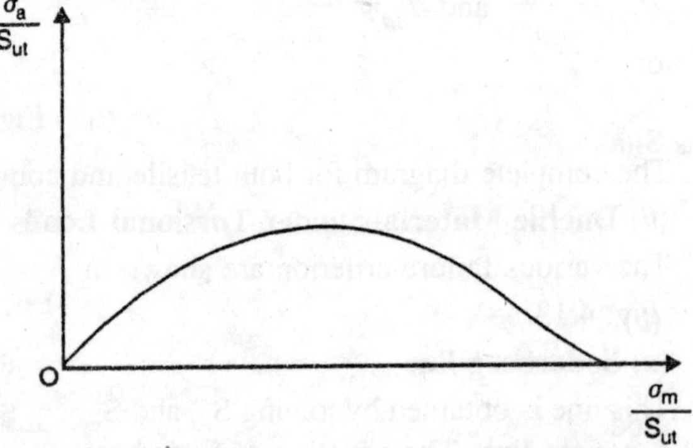

Fig. 4.15 σ_m— σ_a diagram for brittle materials

(iii) $\sigma_m - \sigma_a$ Diagram for Brittle Materials

The $\sigma_m - \sigma_a$ diagram for brittle materials for axial loading is shown in Fig. 4.15. In this case on yielding criteria is involved and a compressive mean stress permits great increase in alternating stress. The diagram for bending loads would be similar. For torsional loads $S_{us} = 0.8\ S_{ut}$ and $S_{es} = 0.8\ S_e$. The Goodman line for torsional loads would be :

$$\frac{\tau_m}{S_{us}} + \frac{\tau_a}{S_{es}} = 1 \qquad ...(4.29)$$

4.11 WORKING STRESSES

When a machine component is subjected to fluctuating stresses then the failure of the material shall take place with stress values lying along a line of failure, i.e. Soderberg line or Goodman line. The fatigue stress — concentration factor K_f is applied only to the alternating component of stress in case of ductile materials. However, in case of brittle materials, the fatigue stress — concentration factor is applied to both mean as well as alternating components of stress.

(*i*) Ductile Materials

(*a*) Axial or Bending Stresses

The Soderberg and Goodman diagrams are shown in Fig. 4.16 by AB and FB respectively. Consider a point E well below the line AB to have a factor of safety *n*. Through E draw a line CL parallel to the line AB. For the Soderberg line, Δs CDE and AOB are similar.

Hence

$$\frac{DE}{CD} = \frac{OB}{OA}$$

$$\frac{JO}{(OC - OD)} = \frac{OB}{OA}$$

$$K_f \sigma_a = S_e$$

$$\frac{K_f \sigma_a}{\left(\frac{S_{yt}}{n}\right) - \sigma_m} = \frac{S_e}{S_{yt}}$$

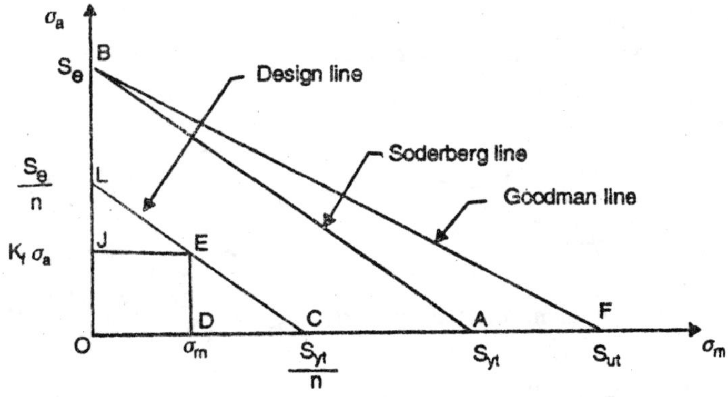

Fig. 4.16 Design line determination for ductile materials.

or
$$\frac{\sigma_m}{S_{yt}} + K_f\left(\frac{\sigma_a}{S_e}\right) = \frac{1}{n} \qquad ....(4.30)$$

Similarly for the Goodman line

$$\frac{\sigma_m}{S_{ut}} + K_f\left(\frac{\sigma_a}{S_e}\right) = \frac{1}{n} \qquad ....(4.31)$$

(*b*) Torsional Stresses

Soderberg line :
$$\frac{\tau_m}{S_{ys}} + K_{fs}\left(\frac{\tau_a}{S_{es}}\right) = \frac{1}{n} \qquad ....(4.32)$$

Goodman line :
$$\frac{\tau_m}{S_{us}} + K_{fs}\left(\frac{\tau_a}{S_{es}}\right) = \frac{1}{n} \qquad ....(4.33)$$

***(ii)* Brittle Materials**

***(a)* Axial or Bending Stresses**

Soderberg line : $$K_t\left[\frac{\sigma_m}{S_{yt}}+\frac{\sigma_a}{S_e}\right] = \frac{1}{n}$$ ....(4.34)

Goodman line : $$K_t\left[\frac{\sigma_m}{S_{ut}}+\frac{\sigma_a}{S_e}\right] = \frac{1}{n}$$ ....(4.35)

***(b)* Torsional Stresses**

Soderberg line : $$K_{ts}\left[\frac{\tau_m}{S_{ys}}+\frac{\tau_a}{S_{es}}\right] = \frac{1}{n}$$ ...(4.36)

Goodman line : $$K_{ts}\left[\frac{\tau_m}{S_{us}}+\frac{\tau_a}{S_{es}}\right] = \frac{1}{n}$$ ...(4.37)

Where $S_{us} = 0.8\ S_{ut}$ and $S_{es} = 0.8\ S_{us}$

***(iii)* Equivalent Stresses**

***(a)* Axial or Bending Stresses**

(i) Soderberg line

$$\text{Equivalent mean stress, } \sigma_{me} = \frac{S_{yt}}{n} = \sigma_m + K_f\left(\frac{S_{yt}}{S_e}\right)\sigma_a$$ ...(4.38)

$$\text{Equivalent alternating stress, } \sigma_{ae} = \left(\frac{S_e}{S_{yt}}\right)\sigma_m + K_f\,\sigma_a$$ ...(4.39)

(ii) Goodman line

$$\text{Equivalent mean stress, } \sigma_{me} = \frac{S_{ut}}{n} = \sigma_m + K_f\left(\frac{S_{ut}}{S_e}\right)\sigma_a$$ ...(4.40)

$$\text{Equivalent alternating stress, } \sigma_{ae} = \left(\frac{S_e}{S_{ut}}\right)\sigma_m + K_f\,\sigma_a$$ ....(4.41)

***(b)* Torsional Stresses**

Equivalent mean shear stress,

$$\tau_{me} = \frac{S_{ys}}{n} = \tau_m + K_{fs}\left(\frac{S_{ys}}{S_{es}}\right)\tau_a$$ ...(4.42)

Equivalent alternating shear stress,

$$\tau_{ae} = \left(\frac{S_{es}}{S_{ys}}\right)\tau_m + K_{fs}\,\tau_a$$ ...(4.43)

Caution : While calculating the corrected endurance limit, the fatigue reduction factor should be ignored as it is being accounted for in calculating working stresses.

4.12 COMBINED FLUCTUATING STRESSES

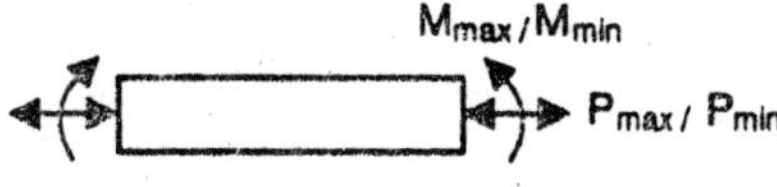

Fig. 4.17

(a) Soderberg Equation :

1. Aixal and bending loading (Fig. 4.17)

$$\frac{P_{eq}}{A}+\frac{M_{eq}}{z} = \frac{S_{yt}}{n} \quad ....(4.44)$$

Where
$$P_{eq} = P_m+\left(\frac{S_{yt}}{S_{ae}}\right)P_a$$

$$M_{eq} = M_m+\left(\frac{S_{yt}}{S_e}\right)M_a \quad ...(4.45)$$

$$S_{ae} = 0.8\ S_e \quad ...(4.46)$$

2. Two — dimensional normal stresses (Fig. 4.18)

$$\sigma_{xm} = \frac{(\sigma_x)_{\max}+(\sigma_x)_{\min}}{2}$$

$$\sigma_{xa} = \frac{(\sigma_x)_{\max}-(\sigma_x)_{\min}}{2}$$

$$\sigma_{ym} = \frac{(\sigma_y)_{\max}+(\sigma_y)_{\min}}{2}$$

$$\sigma_{ya} = \frac{(\sigma_y)_{\max}-(\sigma_y)_{\min}}{2}$$

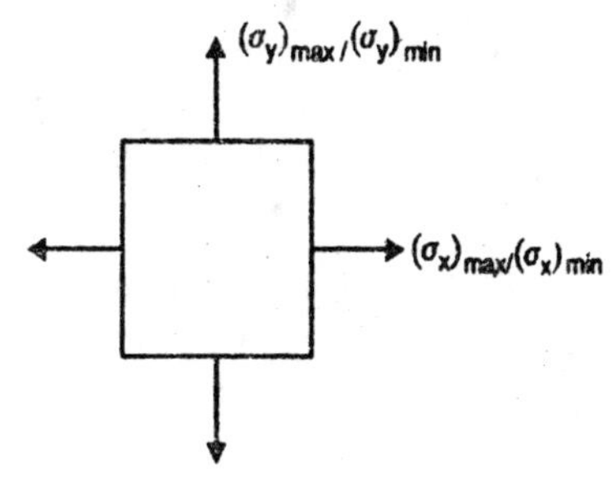

Fig. 4.18

$$\sigma_m = \left[\sigma_{xm}^2+\sigma_{ym}^2-\sigma_{xm}\,\sigma_{ym}\right]^{1/2}$$

$$\sigma_a = \left[\sigma_{xa}^2+\sigma_{ya}^2-\sigma_{xa}\,\sigma_{ya}\right]^{1/2}$$

Soderberg equation using maximum distortion energy theory gives,

$$\sigma_m+\left(\frac{S_{yt}}{S_e}\right)\sigma_a = \frac{S_{yt}}{n} \quad ....(4.47)$$

3. Two-dimensional normal stresses and shear stress (Fig. 4.19)

$$\tau_{xym} = \frac{(\tau_{xy})_{\max}-(\tau_{xy})_{\min}}{2}$$

$$\tau_{xya} = \frac{(\tau_{xy})_{max} - (\tau_{xy})_{min}}{2}$$

$$\sigma_m = \left[\sigma_{xm}^2 + \sigma_{ym}^2 - \sigma_{xm}\,\sigma_{ym} + 3\tau_{xym}^2\right]^{1/2}$$

$$\sigma_a = \left[\sigma_{xa}^2 + \sigma_{ya}^2 - \sigma_{xa}\,\sigma_{ya} + 3\tau_{xya}^2\right]^{1/2}$$

Fig. 4.19

Soderberg equation using maximum distortion energy theory gives,

$$\frac{\sigma_m}{S_{yt}} + \frac{\sigma_a}{S_e} = \frac{1}{n} \qquad \text{....(4.48)}$$

If stress — concentration factors are given in bending and torsion, then for ductile materials, the stress – concentration factors are applied only to the alternating components of stress.

$$\sigma_m = \left[\sigma_{xm}^2 + \sigma_{ym}^2 - \sigma_{xm}\,\sigma_{ym} + 3\tau_{xym}^2\right]^{1/2}$$

$$\sigma_a = \left[(K_{tb}\,\sigma_{xa})^2 + (K_{tb}\,\sigma_{ya})^2 - K_{tb}^2\,\sigma_{xa}\,\sigma_{ya} + 3(K_{ts}\,\tau_{xya})^2\right]^{1/2} \qquad \text{...(4.49)}$$

Example 4.1 A stepped thin flat plate with a circular hole is subjected to a uniaxial tensile force of 6 kN as shown in Fig. 4.20. The material of plate is FG 200 and a factor of safety of 2.5 is desired. Determine the thickness of plate.

Solution.

Given : P = 6 kN, r = 5 mm, d = 40 mm, D = 50 mm, d = 10 mm, w = 40 mm, S_{ut} = 200 MPa, n = 2.5

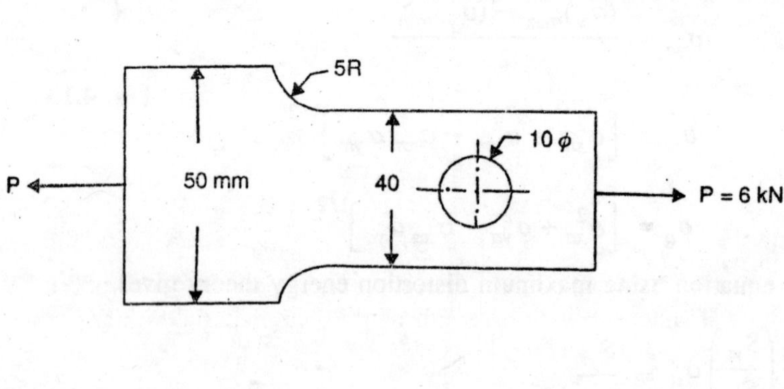

Fig. 4.20

Fillet :

$$\frac{r}{d} = \frac{5}{40} = 0.125 \text{ and } \frac{D}{d} = \frac{50}{40} = 1.25$$

For a stepped plate under tension, $K_t = 1.75$ (Fig. 4.35)

Maximum stress at the fillet,

$$\sigma_{max} = \frac{K_t P}{dt}$$

$$= \frac{1.75 \times 6000}{40 \times t} = \frac{262.5}{t} \text{N/mm}^2$$

Hole :

$$\frac{d}{w} = \frac{10}{40} = 0.25$$

For a finite width plate under tension with a hole, $K_t = 2.42$ (Fig. 4.33)

Maximum stress at the hole,

$$\sigma_{max} = \frac{K_t P}{(w - d)\,t}$$

$$= \frac{2.42 \times 6000}{(40 - 10)t}$$

$$= \frac{484}{t}$$

We observe that the stress at the hole is maximum, which shall be the basis for design to calculate the plate thickness.

Design stress,

$$\sigma_d = \frac{S_{ut}}{n} = \frac{200}{2.5} = 80 \text{ MPa}$$

$$\frac{484}{t} = 80$$

$$t = 6.05 \approx 6 \text{ mm}$$

Example 4.2 A non – rotating stepped axle supports a transverse load of 3 kN at the centre as shown in Fig. 4.21. The tensile yield strength of the axle material is 300 MPa and a factor of safety of 3 is desired. Determine the diameter of the axle.

Solution :

Given :

$$\frac{r}{d} = 0.1,\ \frac{D}{d} = 1.2,\ P = 3 \text{ kN},\ S_{yt} = 300 \text{ MPa},\ n = 3$$

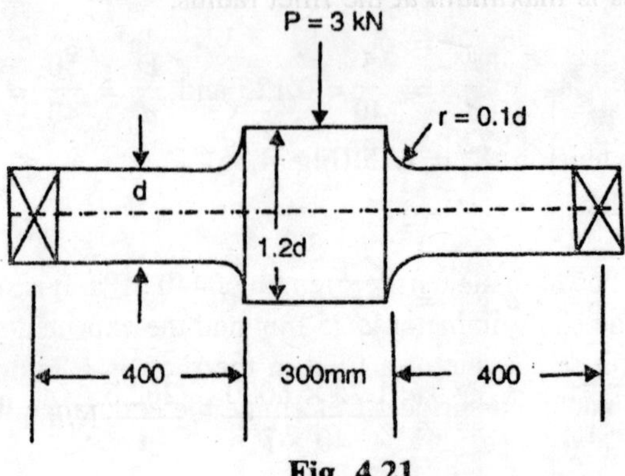

Fig. 4.21

Design stress,

$$\sigma_d = \frac{S_{yt}}{n} = \frac{300}{3} = 100 \text{ MPa}$$

For a circular member with a fillet radius under bending, K_t = 1.62 (Fig. 4.40)
Bearing reactions due to symmetry of loading = 1.5 kN
Bending moment at the fillet, M_f = 1.5 × 400 = 600 kN.mm
Bending moment at the centre,

$$M_c = \frac{PL}{4} = \frac{3 \times 1100}{4} = 825 \text{ kN.mm}$$

Maximum bending stress at the fillet,

$$\sigma_{max} = \frac{K_t \times 32 M_f}{\pi d^3}$$

$$= \frac{1.62 \times 32 \times 600 \times 10^3}{\pi \times d^3}$$

$$= \left[\frac{32 \times 10^3}{\pi d^3}\right] \times 972 \text{ MPa}$$

Maximum bending stress at the centre,

$$\sigma_{max} = \frac{32 M_c}{\pi d^3}$$

$$= \frac{32 \times 825 \times 10^3}{\pi \times d^3}$$

$$= \left[\frac{32 \times 10^3}{\pi d^3}\right] \times 825 \text{ MPa}$$

We observe that the stress is maximum at the fillet radius.

$$\sigma_{max} = \sigma_d$$

$$\left[\frac{32 \times 10^3}{\pi d^3}\right] \times 972 = 100$$

$$d = 46.3 \text{ mm}$$

Example 4.3 A steel shaft has ultimate tensile strength of 440 MPa. It is subjected to completely reversed torsional stresses. The shaft diameter is 25 mm and the expected reliability is 95%. The theoretical stress concentration factor due to a fillet in the shaft is 1.8 and the notch sensitivity factor is 0.86. The shaft has machined surface. Determine the endurance limit.

Solution.

Given :S_{ut} = 440 MPa, d = 25 mm, R = 95%, K_t = 1.8, q = 0.86

$S'_e = 0.5\ S_{ut} = 0.5 \times 440 = 220$ MPa

For machined surface, surface finish factor for S_{ut} = 440 MPa, k_a = 0.82 (Fig. 4.7)

For d = 25 mm, size factor, k_b = 0.85 (Table 4.3)

For R = 95%, reliability factor, k_c = 0.868 (Table 4.4)

Temperature factor, $k_d = 1.0$

For torsion, load factor,

$$k_e = 0.577 \quad \text{(Table 4.5)}$$

$$K_f = 1 + q\,(K_t - 1)$$

$$= 1 + 0.86\,(1.8 - 1) = 1.688$$

Fatigue strength reduction factor,

$$k_f = \frac{1}{K_f} = \frac{1}{1.8} = 0.592$$

Edurance limit,

$$S_e = k_a\, k_b\, k_c\, k_d . k_e\, k_f\, S'_e$$

$$= 0.82 \times 0.85 \times 0.868 \times 1.0 \times 0.592 \times 220$$

$$= 45.73 \text{ MPa}$$

Example 4.4 A steel plate having ultimate tensile strength of 440 MPa is hot – rolled and normalized. It is 60 mm wide and has a central circular hole of 12 mm diameter is shown in Fig. 4.22. It is subjected to a completely reversed axial load of 20 kN. The notch sensitivity factor is 0.8 and the expected reliability is 90%. The factor of safety is 2. Determine the thickness of plate.

Solution.

Given :S_{ut} = 440 MPa, w = 60 mm, a = 12 mm, P = 20 kN, q = 0.8, R = 90%, n = 2

$$S'_e = 0.5\ S_{ut} = 0.5 \times 440 = 220 \text{ MPa}$$

For hot rolled and normalized condition, surface finish factor for S_{ut} = 440 MPa, k_a = 0.67 (Fig. 4.7)

Assuming t < 50 mm, size factor, k_b = 0.85 (Table 4.3)

For R = 90%, reliability factor, k_c = 0.897 (Table 4.4)

Temperature factor,

$$k_d = 1.0$$

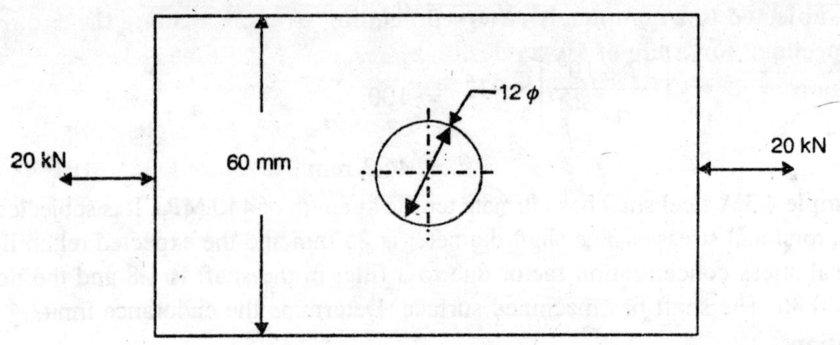

Fig. 4.22

For axial loading, load factor,

$$k_e = 0.9 \quad \text{(Table 4.5)}$$

For the circular hole,

$$\frac{a}{w} = \frac{12}{60} = 0.2$$

$$K_t = 2.50 \quad \text{(Fig. 4.33)}$$

$$K_f = 1 + q\,(K_t - 1)$$

$$= 1 + 0.8\,(2.50 - 1) = 2.216$$

Fatigue strength reduction factor,

$$k_f = \frac{1}{K_f} = \frac{1}{2.2} = 0.454$$

Endurance limit,

$$S_e = k_a\, k_b\, k_c\, k_d\, k_e\, k_f\, S'_e$$

$$= 0.67 \times 0.85 \times 0.897 \times 1.0 \times 0.9 \times 0.454 \times 220$$

$$= 45.94 \text{ MPa}$$

Design stress,

$$\sigma_d = \frac{S_e}{n} = \frac{45.94}{2} = 22.97 \text{ MPa}$$

Stress in plate at hole location,

$$\sigma_{max} = \frac{P}{(w - a)\,t}$$

$$= \frac{20{,}000}{(60 - 12)\,t}$$

$$= 22.97$$

$$t = 18.14 \approx 20 \text{ mm}$$

Example 4.5 A rotating beam specimen made of steel having ultimate tensile strength 650 MPa is subjected to a completely reversed bending stress. Calculate the endurance strength of the specimen for a life of 10^5 cycles.

Solution :

Given :

$$S_{ut} = 650 \text{ MPa}, N = 10^5 \text{ cycles}$$
$$S_e' = 0.5\, S_{ut} = 0.5 \times 650 = 325 \text{ MPa}$$
$$0.9\, S_{ut} = 0.9 \times 650 = 585 \text{ MPa}$$
$$\log S_e' = \log 325 = 2.512$$
$$\log (0.9 S_{ut}) = \log 585 = 2.767$$
$$\log 10^5 = 5$$

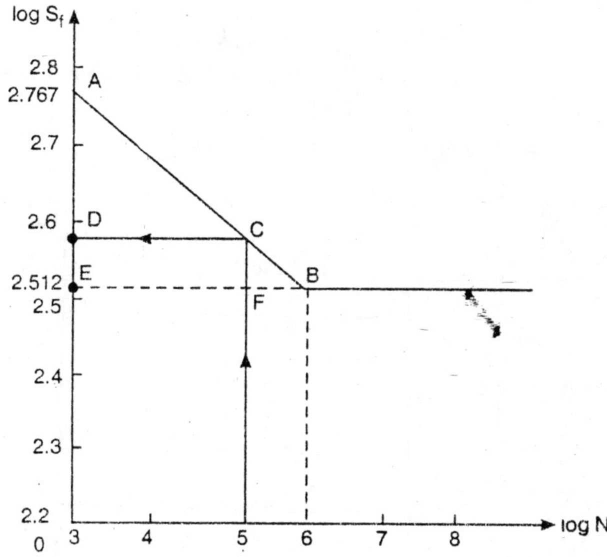

Fig. 4.23

The log S_f -- log N curve is shown in Fig. 4.23.

$$\log S_f = OD = OE + ED = OE + CF$$
$$= OE + (BF / BE)\, AE$$
$$= 2.512 + \left[\frac{6-5}{6-3}\right] \times (2.767 - 2.512)$$
$$= 2.597$$
$$S_f = \text{antilog } 2.597 = 395.37 \text{ MPa.}$$

Example 4.6 A cantilever beam made of cold drawn steel having ultimate tensile strength 540 MPa is subjected to a completely reversing load of 1500 N as shown in Fig. 4.24 (a). The notch sensitivity factor at the fillet may be taken as 0.85 and the expected reliability is 90%. Determine the diameter of the beam for a life of 10^4 cycles.

Solution :

Given : $S_{ut} = 540$ MPa, N = 10^4 cycles, $q = 0.85$, R = 90%, P = 1500 N

$$S_e' = 0.5\, S_{ut} = 0.5 \times 540 = 270 \text{ MPa}$$
$$0.9\, S_{ut} = 0.9 \times 540 = 486 \text{ MPa}$$

For cold drawn steel and S_{ut} = 540 MPa, surface finish factor, $k_a = 0.79$ (Fig. 4.7)

Assuming $d < 50$ mm, size factor, $k_b = 0.85$ (Table 4.3))

For reliability, R = 90%, reliability factor, $k_c = 0.897$ (Table 4.4))

Temperature factor, $k_e = 1.0$

For $\dfrac{r}{d} = 0.25$ and $\dfrac{D}{d} = 1.5$, $K_t = 1.33$ under bending. (Fig. 4.40)

$$K_t = 1 + q\,(K_t - 1) = 1 + 0.85\,(1.33 - 1) = 1.2805$$

Fatigue strength reduction factor, $k_f = \dfrac{1}{K_f} = 1/1.2805 = 0.781$

$$\begin{aligned} S_e &= k_a\, k_b\, k_c\, k_d\, k_e\, k_f\, S_e' \\ &= 0.79 \times 0.85 \times 0.897 \times 1.0 \times 1.0 \times 0.781 \times 220 \\ &= 127 \text{ MPa} \\ \log S_e &= \log 127 = 2.1038 \end{aligned}$$

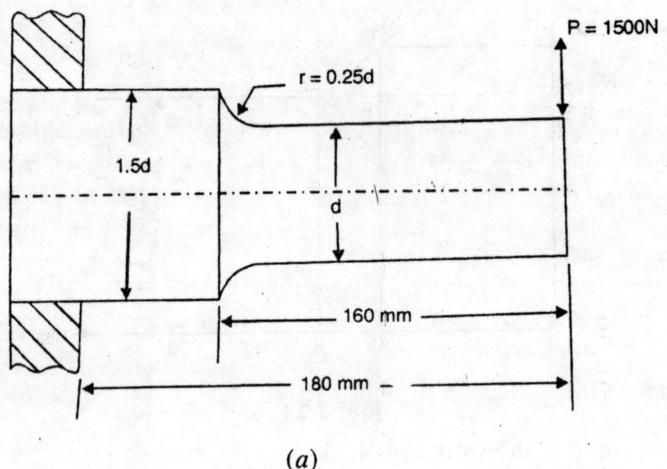

(a)

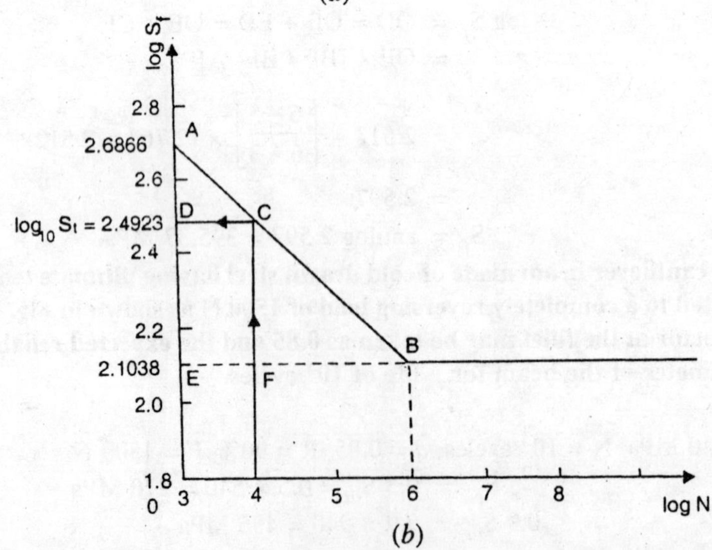

(b)

Fig. 4.24

$$\log (0.9\ S_{ut}) = \log 486 = 2.6866$$

$$\log 10^4 = 4$$

The log S_f – log N curve is shown in Fig. 4.24 (*b*).

$$OD = OE + ED = OE + CF$$

$$= OE + (BF / BE)\ AE$$

$$= 2.1038 + \left[\frac{(6-4)}{(6-3)}\right] \times (2.6866 - 2.1038)$$

$$= 2.4923$$

$$S_f = \text{antilog } 2.4923 = 310.69 \text{ MPa}$$

The failure may occur at B due the fillet instead of at A even thought the bending moment is more at A.

$$M_B = 1500 \times 160 = 240{,}000 \text{ N.mm}$$

$$\frac{32 M_B}{\pi d^3} = S_f$$

$$\frac{32 \times 240{,}000}{\pi d^3} = 310.69$$

$$d = 20 \text{ mm}$$

Example 4.7 A cantilever beam made of cold drawn steel having ultimate tensile strength 600 MPa and tensile yield strength 380 MPa, as shown in Fig. 4.25 (*a*), is subjected to a force P acting at the free end. The force varies from – 100 N to + 200 N. The expected reliability is 90% and the factor of safety is 2. The notch sensitivity factor at the fillet is 0.9. Determine the diameter d at the file cross-section.

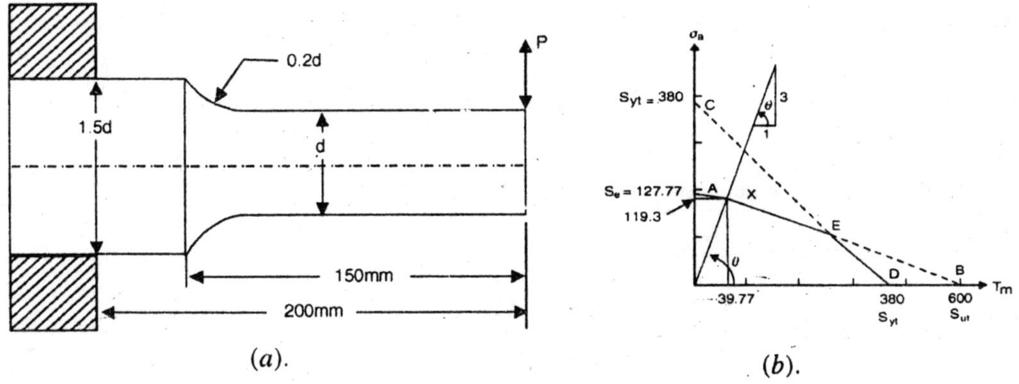

Fig. 4.25

Solution.

Given : $S_{ut} = 600$ MPa, $S_{yt} = 380$ MPa, $q = 0.9$, R = 90%, $n = 2$, $P_{min} = -100$ N, $P_{max} = 200$ N

$S_e' = 0.5\ S_{ut} = 0.5 \times 600 = 300$ MPa

For cold drawn steel and $S_{ut} = 600$ MPa, surface finish factor, $k_a = 0.76$ (Fig. 4.7)

Assuming $d < 50$ mm, size factor, $k_b = 0.85$ (Table 4.3)

For reliability, R = 90%, reliability factor, $k_c = 0.897$

Temperature factor, $k_d = 1.0$ (Table 4.4)

For bending, load factor, $k_e = 1.0$ (Table 4.5)

For $\frac{r}{d} = 0.2$ and $\frac{D}{d} = 1.5$, $K_t = 1.42$ under bending. (Fig. 4.40)

$$K_f = 1 + q\,(K_t - 1) = 1 + 0.9\,(1.42 - 1) = 1.36$$

Fatigue strength reduction factor,

$$k_f = \frac{1}{K_f} = \frac{1}{1.36} = 0.7353$$

$$S_e = k_a\,k_b\,k_c\,k_d\,k_e\,k_f\,S_e$$
$$= 0.76 \times 0.86 \times 0.897 \times 1.0 \times 1.0 \times 0.7353 \times 220$$
$$= 127.77 \text{ MPa}$$

$$M_{max} = 200 \times 150 = 30{,}000 \text{ N.mm}$$
$$M_{min} = -100 \times 150 = 15{,}000 \text{ N.mm}$$

$$M_m = \frac{M_{max} + M_{min}}{2} = \frac{(30{,}000 - 15{,}000)}{2} = 7500 \text{ N.mm}$$

$$M_a = \frac{M_{max} - M_{min}}{2} = \frac{(30{,}000 + 15{,}000)}{2} = 22500 \text{ N.mm}$$

$$\tan\theta = \frac{M_a}{M_m} = \frac{22500}{7500} = 3$$

$$\theta = 71.565°$$

The equation of Goodman line drawn in Fig. 4.25 (*b*) is,

$$\frac{\sigma_m}{S_{ut}} + \frac{\sigma_a}{S_e} = 1$$

Also $$\frac{\sigma_a}{\sigma_m} = 3$$

Solving the above two equations, we have

$$\frac{\sigma_a}{600} + \frac{3\sigma_m}{127.77} = 1$$

$$\sigma_m = 39.767 \text{ MPa}$$
$$\sigma_a = 119.3 \text{ MPa}$$

Design alternating stress, $$\sigma_{da} = \frac{\sigma_a}{n} = \frac{119.3}{2} = 59.65 \text{ MPa}$$

$$= \frac{32M_a}{\pi d^3}$$

$$= \frac{32 \times 22500}{\pi d^3}$$

$$d = 15.66 \approx 16 \text{ mm}$$

Example 4.8 A cold drawn steel transmission shaft having ultimate strength in tension 500 MPa and yield strength 300 MPa is subjected to a fluctuating torque which varies from – 200 N.m to + 500 N.m. The factor of safety is 2 and the expected reliability is 90%. Neglecting the effect of stress - concenfration, determine the diameter of the shaft.

Solution.

Given : $S_{ut} = 500$ MPa, $S_{yt} = 300$ MPa, R = 90%, $n = 2$, $T_{min} = -200$ N.m., $T_{max} = +500$ N.m.

$S'_e = 0.5\ S_{ut} = 0.5 \times 500 = 250$ MPa

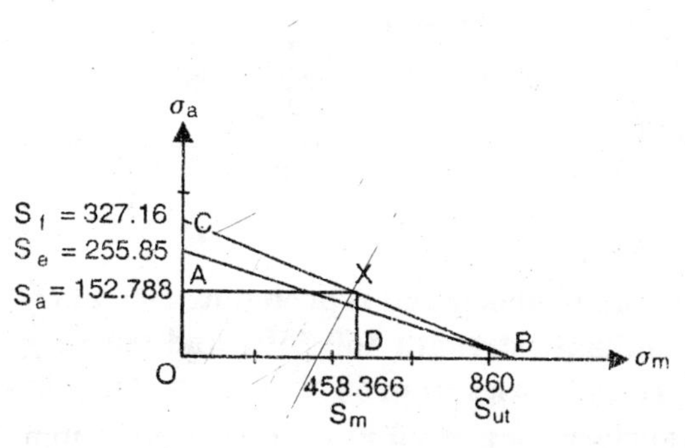

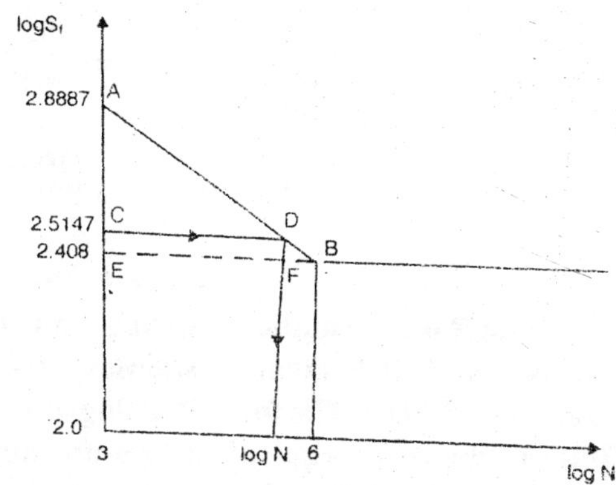

Fig. 4.27

For cold drawn steel and $\sigma_{ut} = 500$ MPa, surface finish factor, $k_a = 0.80$ (Fig. 4.7)

Assuming $d < 50$ mm, size factor, $k_b = 0.85$ (Table 4.3)

For reliability, R = 90%, reliability factor, $k_c = 0897$ (Table 4.4)

Temperature factor, $k_d = 1.0$

For reversed torsional, load, load factor, $k_e = 0.577$ (Table 4.5)

$$S_{es} = k_a\, k_b\, k_c\, k_d\, k_e\, k_f\, S'_e$$

$$= 0.80 \times 0.85 \times 0.897 \times 1.0 \times 0.577 \times 250 \text{ MPa}$$

$$= 87.99 \text{ MPa}$$

Based on distortion energy theory, yield stress in shear,

$$S_{ys} = 0.577,\ S_{yt} = 0.577 \times 300 = 173.1 \text{ MPa}$$

$$T_m = \frac{T_{max} + T_{min}}{2} = \frac{500 - 200}{2} = 150 \text{ N.m}$$

$$T_a = \frac{T_{max} - T_{min}}{2} = \frac{500 + 200}{2} = 350 \text{ N.m}$$

$$\tan\theta = \frac{T_a}{T_m} = \frac{350}{150} = 2.3333$$

$$\theta = 66.8^\circ$$

The modified Goodman diagram for torsion is drawn in Fig. 4.26.

$$S_{as} = S_{es} = 87.99 \text{ MPa}$$

$$S_{ms} = \frac{S_{as}}{2.3333} = \frac{87.99}{2.3333} = 37.71 \text{ MPa}$$

Design alternating stress, $\tau_{da} = \frac{S_{as}}{n} = \frac{87.99}{2} = 43.995$ MPa

$$= \frac{16T_a}{\pi d^3}$$

$$= \frac{16 \times 350{,}000}{\pi d^3}$$

$$d = 34.34 \approx 35 \text{ mm}$$

Example 4.9 A cantilever spring made of 10 mm diameter rod is protuding 500 mm from the fixed end. It is made of stainless steel of ultimate strength 860 MPa and tensile yield strength 690 MPa. The force P acting at the freee end varies from 60 N to 120 N. The surface finish of the rod is equivalent to a machined surface. Neglected any stress – concentration effects, determine the life of the spring for 50 % reliability.

Solution.

Given : S_{ut} = 860 MPa, S_y = 690 MPa, P_{min} = 60 N, P_{max} = 120 N, R = 50%

$S'_e = 0.5\ S_{ut} = 0.5 \times 870 = 430$ MPa

For machined surface and S_{ut} = 860 MPa, surface finish factor, $k_a = 0.7$ (Fig. 4.7)

Assuming $d < 50$ mm, size factor, $k_b = 0.85$ (Table 4.3)

For reliability, R = 50%, reliability factor, $k_c = 1.0$ (Table 4.4)

Temperature factor, $k_d = 1.0$

For bending, load factor, $k_e = 1.0$ (Table 4.5)

$$S_e = k_a\, k_b\, k_c\, k_d\, k_e\, k_f\, S'_e$$

$$= 0.70 \times 0.85 \times 1.0 \times 1.0 \times 1.0 \times 250$$

$$= 255.85 \text{ MPa}$$

$$M_{max} = 120 \times 500 = 60{,}000 \text{ N.mm}$$

$$M_{min} = 60 \times 500 = 30{,}000 \text{ N.mm}$$

$$M_m = \frac{M_{max} + M_{min}}{2} = \frac{60{,}000 + 30{,}000}{2} = 45{,}000 \text{ N.mm}$$

$$M_a = \frac{M_{max} - M_{min}}{2} = \frac{60{,}000 - 30{,}000}{2} = 15{,}000 \text{ N.mm}$$

$$\sigma_m = \frac{32M_m}{\pi d^3} = \frac{32 \times 45{,}000}{\pi \times 10^3} = 458.366 \text{ MPa}$$

$$\sigma_a = \frac{32M_a}{\pi d^3} = \frac{32 \times 15{,}000}{\pi \times 10^3} = 152.788 \text{ MPa}$$

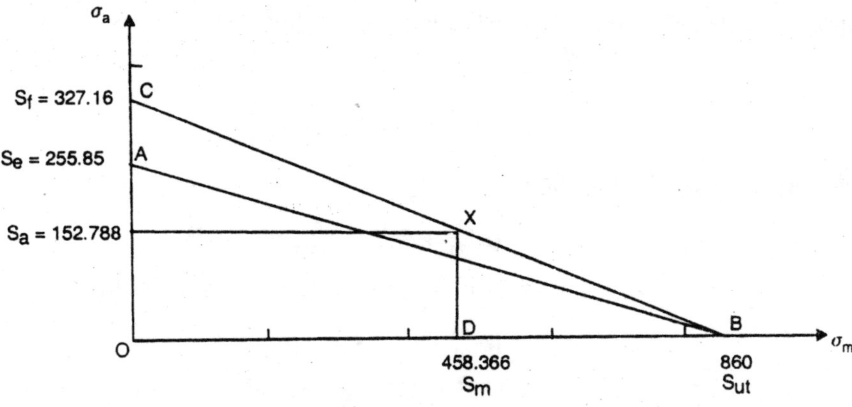

Fig. 4.27

Draw the Goodman line as shown in Fig. 4.27 (*a*). Locate point x (4578.366, 152.788). Join BX and produce to intersect ordinate at C. Then OC = S_f.

Triangles XDB and BOC are similar

$$OC = S_f = \frac{XD \times OB}{BD} = \frac{152.788 \times 860}{(860 - 458.366)} = 327.16 \text{ MPa}$$

$$\log (0.9\, S_{ut}) = \log (0.9 \times 860) = 2.8887$$

$$\log (S_e) = \log (255.85) = 2.408$$

$$\log (S_f) = \log (327.16) = 2.5147$$

Locate these points in the log S_f – log N diagram shown in Fig. 4.27 (*b*). Draw a line CD parallel to the abscissa to interect line AB at D. Drop perpendicular from D on the abscissa. Now triangles ACD and AEB are similar.

$$CD = EF = BE \times \frac{AC}{AE}$$

$$= \frac{(6-3) \times (2.8887 - 2.5147)}{2.8887 - 2.408} = 2.334$$

Then
$$\log N = 3 + 2.334 = 5.334$$

$$N = 215822 \text{ cycles}$$

Example 4.10 A machine component having ultimate tensile strength in tension 600 MPa is subjected to two – dimensional stresses. The tensile stress in the x – direction varies from + 50 MPa to + 100 MPa and the tensile stress in the y – direction varies from + 20 MPa to + 70 MPa. The corrected endurance limit of the material is 280 MPa. Determine the factor of safety used by the designer.

Solution.

Given : S_{ut} = 600 MPa, S_e = 280 MPa, $(\sigma_x)_{min}$ = + 50 MPa, $(\sigma_x)_{max}$ = + 100 MPa, $(\sigma_y)_{min}$ = + 20 MPa, $(\sigma_y)_{max}$ = + 70 MPa

$$\sigma_{xm} = \frac{100 + 50}{2} = 75 \text{ MPa}$$

$$\sigma_{xa} = \frac{100 - 50}{2} = 25 \text{ MPa}$$

$$\sigma_{ym} = \frac{70 + 20}{2} = 45 \text{ MPa}$$

$$\sigma_{ya} = \frac{70 - 20}{2} = 25 \text{ MPa}$$

According to distortion energy theory,

$$\sigma_m = [\sigma_{xm}^2 + \sigma_{xm}^2 - \sigma_{xm}\,\sigma_{ym}]^{1/2}$$

$$= [75^2 + 45^2 - 75 \times 45]^{1/2} = 65.38 \text{ MPa}$$

$$\sigma_a = [\sigma_{xa}^2 + \sigma_{ya}^2 - \sigma_{xa}\,\sigma_{ya}]^{1/2}$$

$$= [25^2 + 25^2 - 25 \times 25]^{1/2} = 25 \text{ MPa}$$

$$\tan\theta = \frac{\sigma_a}{\sigma_m} = \frac{25}{65.38} = 0.38238$$

$$\theta = 20.92°.$$

The modified Goodman diagram is shown in Fig. 4.28.

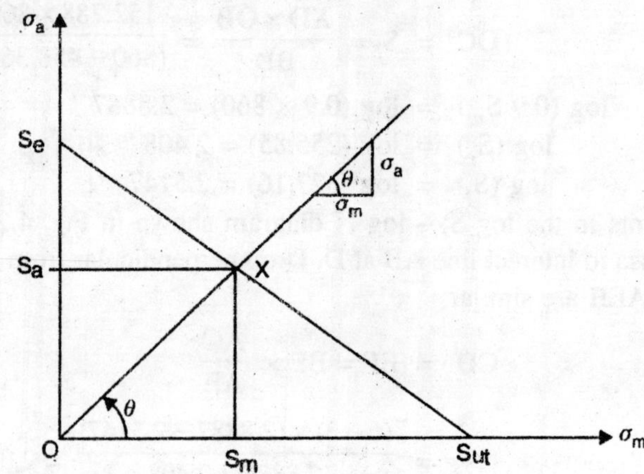

Fig. 4.28

$$\frac{S_m}{S_{ut}} + \frac{S_a}{S_e} = 1$$

$$\frac{S_m}{600} + 0.38238 \times \frac{S_m}{280} = 1$$

$$S_m = 329.78 \text{ MPa}$$

$$S_a = 126.1 \text{ MPa}$$

Factor of safety, $$n = \frac{S_a}{\sigma_a} = \frac{126.1}{25} = 5.04.$$

Example 4.11 A shaft made of steel having ultimate tensile strength 600 MPa and tensile yield strength 400 MPa is subjected to fluctuating bending moment varying from 200 N.m to 500 N.m and torque varying from 60 N.m to 180 N.m. The corrected endurance limit of the shaft material is 200 MPa. Using a factor of safety of 2, determine the diameter of the shaft.

Solution.

Given : $S_{ut} = 600$ MPa, $S_{yt} = 400$ MPa, $S_e = 200$ MPa, $M_{min} = +\,200$ N.m., $M_{max} = +\,500$ N.m., $T_{min} = +\,60$ N.m., $T_{max} = +\,180$ N.m., $n = 2$

$$M_m = \frac{M_{max} + M_{min}}{2} = \frac{500 + 200}{2} = 350 \text{ N.m}$$

$$M_a = \frac{M_{max} - M_{min}}{2} = \frac{500 - 200}{2} = 150 \text{ N.m}$$

$$T_m = \frac{T_{max} + T_{min}}{2} = \frac{180 + 60}{2} = 120 \text{ N.m}$$

$$T_a = \frac{T_{max} - T_{min}}{2} = \frac{180 - 60}{2} = 60 \text{ N. m}$$

$$\sigma_{xm} = \frac{32 M_m}{\pi d^3} = \left[\frac{32 \times 10^3}{\pi d^3}\right] = 350 \text{ MPa}$$

$$\sigma_{xa} = \frac{32 M_a}{\pi d^3} = \left[\frac{32 \times 10^3}{\pi d^3}\right] \times 150 \text{ MPa}$$

$$\tau_{xym} = \frac{16 T_m}{\pi d^3} = \left[\frac{32 \times 10^3}{\pi d^3}\right] \times 60 \text{ MP}$$

$$\tau_{xya} = \frac{16\, T_a}{\pi d^3} = \left[\frac{32 \times 10^3}{\pi d^3}\right] \times 30 \text{ MPa}$$

$$\sigma_a = [\sigma_{xa}^2 + 3\tau_{xya}^2]^{1/2}$$

$$= \left[\frac{32 \times 10^3}{\pi d^3}\right] [350^2 + 3 \times 60^2]^{1/2}$$

$$= \left[\frac{32 \times 10^3}{\pi d^3}\right] \times 365.1 \text{MPa}$$

$$\sigma_a = [\sigma_{xa}^2 + 3\tau_{xya}^2]^{1/2}$$

$$= \left[\frac{32 \times 10^3}{\pi d^3}\right] [150^2 + 3 \times 30^2]^{1/2}$$

$$= \left[\frac{32 \times 10^3}{\pi d^3}\right] \times 158.745 \text{ MPa}$$

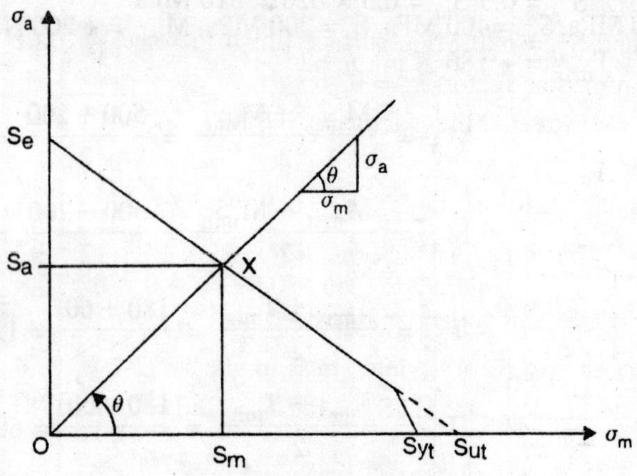

Fig. 4.29

$$\tan\theta = \frac{\sigma_a}{\sigma_m} = \frac{158.745}{365.1} = 0.4348$$

$$\theta = 23.5°$$

The modified Goodman diagram is shown in Fig. 4.29.

$$\frac{S_m}{S_{ut}} + \frac{S_a}{S_e} = 1$$

$$\frac{S_m}{600} + \frac{S_a}{200} = 1$$

Also $$\frac{S_a}{S_m} = 0.4348$$

$$\frac{S_m}{600} = \frac{0.4348 \times S_m}{200} = 1$$

$$S_m = 260.37 \text{ MPa}$$

$$S_a = 113.21 \text{ MPa}$$

Hence $$\sigma_a = \frac{S_a}{n}$$

$$= \left[\frac{32 \times 10^3}{\pi d^3}\right] \times 158.745 = \frac{113.21}{2}$$

$$d = 30.57 \approx 31 \text{ mm.}$$

Example 4.12 A solid shaft of steel Fe 620 (S_{ut} = 620 N/mm^2 and S_{yt} = 380 N/mm^2) is subjected to an alternating torque, which varies from – 200 N.m to + 400 N.m. The shaft is ground and the expected reliability is 90%. Neglecting stress – concentration, calculate the shaft diameter for infinite life. The factor of safety is 2.5. Use the distortion energy theory of failure.

Solution.

Given : S_{ut} = 620 MPa, S_{yt} = 380 MPa, R = 90%, n = 2.5, T_{min} = – 200 N.m. T_{max} = + 400 N.m

$$S'_e = 0.5\ S_{ut} = 0.5 \times 620 = 310 \text{ MPa}$$

For ground shaft and S_{ut} = 620 MPa, surface finish factor, k_a = 0.92 (Fig. 4.7)

Assuming d < 50 mm, size factor, k_b = 0.85 (Table 4.3)

For reliability, R = 90%, reliability factor, k_c = 0.897 (Table 4.4)

Temperature factor, k_d = 1.0 (Table 4.5)

For reversed torsional load, load factor, k_e = 0.577

$$S_{es} = k_a\, k_b\, k_c\, k_d\, k_e\, k_f\, S'_e$$
$$= 0.92 \times 0.85 \times 0.897 \times 1.0 \times 0.577 \times 1.0 \times 310$$
$$= 125.47 \text{ MPa}$$

Based on distortion energy theory, yield stress in shear,

$$S_{ys} = 0.577 S_{yt} = 0.577 \times 380 = 219.26 \text{ MPa}$$

$$T_m = \frac{T_{max} + T_{min}}{2} = \frac{400 - 200}{2} = 100 \text{ N.m}$$

$$T_a = \frac{T_{max} - T_{min}}{2} = \frac{400 + 200}{2} = 300 \text{ N.m}$$

$$\tan\theta = \frac{\tau_a}{\tau_m} = \frac{T_a}{T_m} = \frac{300}{100} = 3$$

$$\theta = 71.56°$$

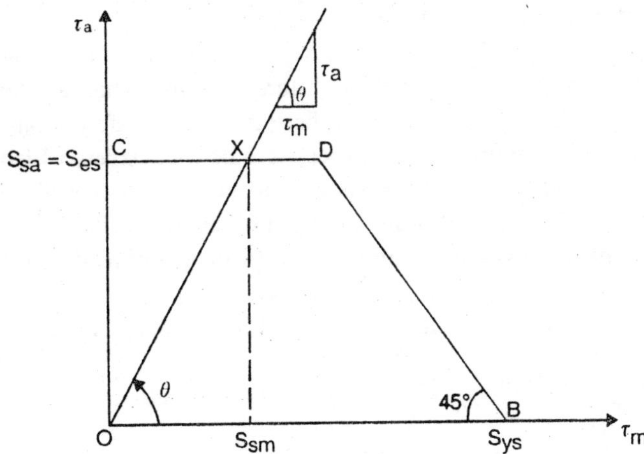

Fig. 4.30

The modified Goodman diagram for torsion is drawn in Fig. 4.30,

$$S_{as} = S_{es} = 125.47 \text{ MPa}$$

$$S_{ms} = \frac{S_{as}}{2.3333} = \frac{125.47}{3} = 41.823 \text{ MPa}$$

$$\text{Design alternating stress, } \tau_{da} = \frac{S_{as}}{n} = \frac{125.47}{2.5} = 50.188 \text{ MPa}$$

$$= \frac{16T_a}{\pi d^3}$$

$$= \frac{16 \times 300{,}000}{\pi d^3}$$

$$d = 31.22 \approx 32 \text{ mm}$$

Example 4.13 (*i*) Define and differentiate between failure stress, design stress and working stress.

(*ii*) What is meant by "Stress—Concentration" ? How does you take it into consideration in case of a component subjected to dynamic loading ?

(*iii*) What information do you obtain from "Soderberg diagram" ?

Solution.

(*i*) **Failutre.** It is the stress at which the plastic deformation (yielding) becomes sufficiently large to destroy the engineering usefulness of the part even though no fracture (separation) has occurred.

Design stress. For ductile materials, it is the ratio of tensile yield strength of the material to the factor of safety. For brittle materials, it the ratio of ultimate tensile strength to the factor of safety.

Working stress. It is defined under combined effect of stready (or mean) and alternating stresses when the failure of the material will take place with stress values lying along a line of failure like the Soderberg or Goodman line.

(*ii*) Refer to art 4.1.

Ductile materials. For dynamic loading, we have the fatigue stress – concentration factor, K_f. It is applied only to the alternating component of stress.

Brittle materials. It is applied to both the mean and alternating components.

(*iii*) Soderberg diagram helps us to fix the failure criterion when a material is subjected to fluctuating stresses. The region below the Soderberg line is the safe region.

Example 4.14 A machine component is subjected to bending stress which fluctuates between 300 N / mm² tensile and 150 N/mm² compressive in a cyclic manner. Using Goodman formula, calculate the minimum required ultimate strength of the material. Take factor of safety as 1.5 and endurance limit in reversed bending as 50% of ultimate strength.

Solution.

Given : $\sigma_{max} = +300 \text{ N/mm}^2$, $\sigma_{min} = -150 \text{ N/mm}^2$, $n = 1.5$, $S_e = 0.5\, S_{ut}$

$$\sigma_m = \frac{\sigma_{max} + \sigma_{min}}{2} = \frac{300 - 150}{2} = 75 \text{ N/mm}^2$$

$$\sigma_a = \frac{\sigma_{max} - \sigma_{min}}{2} = \frac{300 + 150}{2} = 225 \text{ N/mm}^2$$

The Goodman failure line is :

$$\frac{\sigma_m}{S_{ut}} + \frac{\sigma_a}{S_e} = \frac{1}{n}$$

$$\frac{75}{S_{ut}} + \frac{225}{0.5S_{ut}} = \frac{1}{1.5}$$

$$S_{ut} = 787.5 \text{ N/mm}^2$$

Example 4.15 Determine the size of the piston rod subjected to a total load having cyclic fluctuation from 15 kN in compression to 25 kN in tension. The endurance limit is 360 MPa and yield strength is 400 MPa. Take impact factor 1.25, factor of safety = 1.5, surface finish factor = 0.88 and stress concentration factor = 2.25.

Solution.

Given : P_{min} = – 15*k*N, P_{max} = 25 *k*N, S'_e = 360 MPa, S_y = 400 MPa, K_i = 1.25, *n* = 1.25, k_a = 0.88, K_t = 2.25

$$P_m = \frac{P_{max} + P_{min}}{2} = \frac{25-15}{2} = 5\ kN$$

$$P_a = \frac{P_{max} - P_{min}}{2} = \frac{25+15}{2} = 20\ kN$$

Let d = diameter of piston rod, mm

$$\sigma_m = \frac{4P_a}{\pi d^2} = \frac{4 \times 5000}{\pi d^2} = \frac{20,000}{\pi d^2}\ \text{N/mm}^2$$

$$\sigma_a = \frac{4P_9}{\pi d^2} = \frac{4 \times 20,000}{\pi d^2} = \frac{80,000}{\pi d^2}\ \text{N/mm}^2$$

Let $q = 0.8$

$$K_f = 1 + q\,(K_t - 1)$$
$$= 1 + 0.8\,(2.25 - 1) = 2.0$$

Fatigue strength reduction factor, $k_f = \dfrac{1}{K_f} = \dfrac{1}{2} = 0.5$

Let $d < 50$ mm, size factor, $k_b = 0.85$

Load factor, $k_e = 0.9$ for axial loading with no bending

Impact reduction factor, $k_i = \dfrac{1}{K_i} = \dfrac{1}{1.25} = 0.8$

$$S_e = k_a\, k_b\, k_e\, k_i\, k_f\, S'_e$$
$$= 0.88 \times 0.85 \times 0.9 \times 0.8 \times 0.5 \times 360$$
$$= 96.94\ \text{MPa}$$

Using Soderberg failure equation, we have

$$\frac{\sigma_m}{S_y} + \frac{\sigma_a}{S_e} = \frac{1}{n}$$

$$\left(\frac{1}{\pi d^2}\right)\left[\frac{20,000}{400} + \frac{80,000}{96.94}\right] = \frac{1}{2.25}$$

$$d = 25\ \text{mm}$$

Example 4.16 A circular rod of ductile material has endurance strength of 280 N/mm^2 ultimate strength 350 N/mm^2. The member is subjected to a variable axial load varying from + 700 kN (tensile) to – 300 kN (compressive). Take fatigue stress concentration factor as 1.8 and factor of safety as 2.0. Find the suitable diameter of the rod.

Solution.

Given : $P_{min} = -300$ kN, $P_{max} = +700$ kN, $S'_e = 280$ MPa, $S_y = 350$ MPa, $K_f = 1.8$, $n = 2$

$$P_m = \frac{P_{max} + P_{min}}{2} = \frac{700 - 300}{2} = 200 \text{ kN}$$

$$P_a = \frac{P_{max} - P_{min}}{2} = \frac{700 + 300}{2} = 500 \text{ kN}$$

Let d = diameter of piston rod, mm

$$\sigma_m = \frac{4P_m}{\pi d^2} = \frac{4 \times 600}{\pi d^2} = \frac{800}{\pi d^2} \text{ kN/mm}^2$$

$$\sigma_a = \frac{4P_a}{\pi d^2} = \frac{4 \times 500}{\pi d^2} = \frac{2000}{\pi d^2} \text{ kN/mm}^2$$

Fatigue strength reduction factor, $k_f = \frac{1}{K_f} = \frac{1}{1.8} = 0.555$

Let $d < 50$ mm, size factor, $k_b = 0.85$ (Table 4.3)

Load factor, $k_e = 0.9$ for axial loading with no bending (Table 4.4)

Let the surface be cold rolled and machined, so that surface finish factor for $S_u = 350$ MPa,

$k_a = 0.93$ (Fig. 4.7)

$$S_e = k_a\, k_b\, k_e\, k_f\, S_e'$$
$$= 0.93 \times 0.85 \times 0.9 \times 0.555 \times 280$$
$$= 110.56 \text{ MPa}$$

Using Goodman failure equation, we have

$$\frac{\sigma_m}{S_u} + \frac{\sigma_a}{S_e} = \frac{1}{n}$$

$$\left(\frac{1}{\pi d^2}\right)\left[\frac{800}{350} + \frac{2000}{110.56}\right] \times 10^3 = \frac{1}{2}$$

$$d = 113.89 \approx 114 \text{ mm}$$

Example 4.17 Use Soderberg formula to determine the thickness of a 110 mm wide plate for safe continuous operation, if the plate is to be subjected to a tensile load that has a maximum value of 250,000 N and minimum value of 10,000 kgf. The properties of plate material are as follows :

Endurance limit = 22,500 N/cm^2

Yield stress = 30,000 N/cm^2

The factor of safety = 1.5

Solution.

Given : $w = 110$ mm, $P_{max} = 250$ kN, $P_{min} = 10{,}000$ kgf or 98.1 kN, $S_e = 225$ N/mm², $S_y = 300$ N/mm², $n = 1.5$

Let t = thickness of plate, mm

$$P_m = \frac{P_{max} + P_{min}}{2} = \frac{250 + 98.1}{2} = 174.05 \, kN$$

$$P_a = \frac{P_{max} - P_{min}}{2} = \frac{250 - 98.1}{2} = 75.95 \, kN$$

$$\sigma_m = \frac{P_m}{wt} = \frac{174.05}{110t} = 1.5823 / t \, kN/mm^2$$

$$\sigma_a = \frac{P_a}{wt} = \frac{75.95}{110t} = 0.6905 / t \, kN/mm^2$$

Using Soderberg failure criterion,

$$\frac{\sigma_m}{S_y} + \frac{\sigma_a}{S_e} = \frac{1}{n}$$

$$\left(\frac{10^3}{t}\right)\left[\frac{1.5823}{300} + \frac{0.6905}{225}\right] = \frac{1}{1.5}$$

$$t = 12.5 \text{ mm}$$

Example 4.18 A pulley is mounted on a shaft midway between two bearings. The bending moment at the pulley varies from 200 kN. mm to 600 kN. mm and a torsional moment varies from 60 kN. mm to 180 kN. mm. The shaft material has an ultimate strength of 550 MPa and yield strength of 400 Mpa. If factor of safety is 1.5 and the stress concentration factors for the keyway in the pulley in bending and torsion is 1.5 and 1.2 respectively, find the shaft size for infinite life.

Solution.

Given : $M_{min} = 200$ kN.mm, $M_{max} = 600$ kN. mm, $T_{min} = 60$ kN.mm, $T_{max} = 180$ kN.mm, $S_u = 550$ MPa,

$S_y = 400$, MPa, n = 1.5, $K_{tb} = 1.5$, $K_{ts} = 1.2$

$S_e = 0.5 \, S_u = 0.5 \times 550 = 275$ MPa

Let d = diameter of shaft, mm

$$M_m = \frac{M_{max} + M_{min}}{2} = \frac{600 + 200}{2} = 400 \, kN. \, mm$$

$$M_a = \frac{M_{max} - M_{min}}{2} = \frac{600 - 200}{2} = 200 \, kN/mm^2$$

$$\sigma_{xm} = \frac{32M_m}{\pi d^3} = \left[\frac{32}{\pi d^3}\right] \times 400 \, kN / mm^2$$

$$\sigma_{xa} = \frac{32M_a}{\pi d^3} = \left[\frac{32}{\pi d^3}\right] \times 200 \ kN / mm^2$$

$$T_m = \frac{T_{max} + T_{min}}{2} = \frac{180 + 60}{2} = 120 \ kN. \ mm$$

$$T_a = \frac{T_{max} - T_{min}}{2} = \frac{180 - 60}{2} = 60 \ kN.mm$$

$$\tau_{xym} = \frac{16T_m}{\pi d^3} = \left[\frac{32}{\pi d^3}\right] \times 60 \ kN/min^2$$

$$\tau_{xya} = \frac{16T_a}{\pi d^3} = \left[\frac{32}{\pi d^3}\right] \times 30 \ \ kN/mm^2$$

Using distortion energy theory.

Stress – concentration factors are applied only to the alternating stress components for ductile materials.

$$\sigma_m = [(\sigma_{xm})^2 + 3\ (\tau_{xym})^2]^{1/2}$$

$$= \left[\frac{32}{\pi d^3}\right] [400)^2 + 3(60)^2]^{1/2}$$

$$= \left[\frac{32}{\pi d^3}\right] \times 413.28 \ kN/\ mm^2$$

$$\sigma_a = [(K_{tb}\sigma_{xa})^2 + 3\ (K_{ts}\ \tau_{xya})^2]^{1/2}$$

$$= \left[\frac{32}{\pi d^3}\right] [(1.5 \times 200)^2 + 3\ (1.2 \times 30)^2]^{1/2}$$

$$= \left[\frac{32}{\pi d^3}\right] \times 306.411 \ kN/\ mm^2$$

Soderberg equation gives,

$$\sigma_m + \left(\frac{S_y}{S_e}\right)\sigma_a = \frac{S_y}{n} \left[\frac{32}{\pi d^3}\right]\left[413.28 + \left(\frac{550}{275}\right)306.411\right] \times 10^3 = \frac{400}{1.5}$$

$$d = 34 \text{ mm}$$

Example 4.19 A flat bar as shown in Fig. 4.31 is subjected to an axial load of 50 kN. If the stress in the bar is limited to 100 N / mm², determine the thickness of the bar.

Solution. Hole. a = 25 mm, w = 150 mm

$$\frac{a}{w} = \frac{25}{150} = 0.167, \ K_t = 2.56 \qquad \text{(Fig. 4.33)}$$

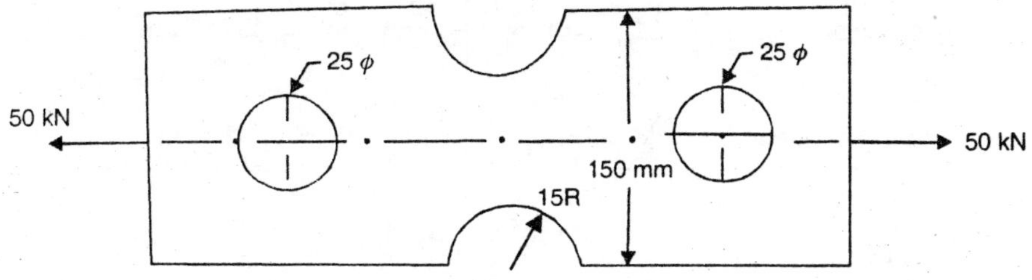

Fig. 4.31

$$\text{Net maximum stress} = \frac{K_t P}{(w-a)t}$$

$$= \frac{2.56 \times 50 \times 10^3}{(150-25)\,t}$$

$$= 100$$

$$t = 10.24 \text{ mm}$$

Notch : $d = 150 - 30 = 120$ mm, D = 150 mm

$$\frac{r}{d} = \frac{15}{120} = 0.125, \quad \frac{D}{d} = \frac{150}{120} = 1.25$$

$$K_t = 2.3$$ (Fig. 4.46)

$$\text{Maximum stress} = \frac{2.3 \times 50 \times 10^3}{[(150-30)\,t]}$$

$$= 100$$

$$t = 9.58 \text{ mm}$$

Adopt, $t = 11$ mm

Example 4.20 A circular bar of 500 mm length is supported freely at its two ends. It is acted upon by a central concentrated cyclic load having a minimum value of 20 kN and a maximum value of 50 kN. Determine the diameter of bar by taking a factor of safety of 1.5, size effect of 0.85, surface finish factor of 0.9. The material properties of bar are given by : ultimate strength of 650 MPa, yield strength of 500 MPa and endurance strength of 350 MPa.

Solution.

Given : L = 500 mm, $P_{min} = 20$ *k*N, $P_{max} = 50$ *k*N, $n = 1.5$, $k_a = 0.9$, $k_b = 0.85$, $S_{ut} = 650$ MPa, $S_{yt} = 500$ MPa, $S'_e = 350$ MPa

Corrected endurance strength, $S_e = k_a k_b S'_e = 0.9 \times 0.85 \times 350 = 267.75$ MPa

$$P_m = \frac{P_{max} + P_{min}}{2} = \frac{50+20}{2} = 35 \, kN$$

$$P_a = \frac{P_{max} - M_{min}}{2} = \frac{50-20}{2} = 15 \, kN$$

Mean bending moment, $M_m = \frac{P_m L}{4} = \frac{35 \times 500}{4} = 4375$ N.m

Alternating bending moment, $M_a = \frac{P_a L}{4} = \frac{15 \times 500}{4} = 1875$ N.m

Mean bending stress, $\sigma_m = \frac{M_m}{z} = \frac{32 M_m}{\pi d^3} = \frac{32 \times 4375}{\pi d^3}$

Alternating bending stress, $\sigma_a = \frac{M_a}{z} = \frac{32 M_a}{\pi d^3} = \frac{32 \times 1875}{\pi d^3}$

Let $K_f = 1.0$

Soderberg criteria of failure gives,

$$\frac{\sigma_m}{S_{yt}} + \frac{\sigma_a}{S_e} = \frac{1}{n}$$

$$\left[\frac{32 \times 10^3}{\pi d^3}\right]\left[\frac{4375}{500} + \frac{1875}{267.75}\right] = \frac{1}{1.5}$$

$$d = 62.2 \approx 63 \text{ mm}$$

EXERCISES

4.1 A component machined from a plate made of steel is having ultimate tensile strength 650 MPa. The component has a step from 120 mm to 60 mm with a fillet radius of 5 mm. It is subjected to a completely reversed axial force of 60 kN. The surface of the component may be assumed to be machined and the expected reliability is 90%. The notch sensitivity is 0.85. If the factor of safety desired is 2.5, determine the thickness of the plate for unlimited life.

4.2 A forged steel bar of 63 mm diameter is subjected to a reversed bending stress of 300 MPa. Calculate the life of the bar for 90% reliability if it is made of steel having ultimate tensile strength equal to 650 MPa.

4.3 A 30 mm diameter forged steel shaft having 600 ultimate tensile strength has a step to 20 mm diameter with a fillet radius of 5 mm. The notch sensitivity factor is 0.85. Calculate the endurance strength of the shaft if it is subjected to reversed bending moment.

4.4 A solid circular shaft is made of steel having S_{ut} = 650 MPa and S_{yt} = 380 MPa. It is subjected to an alternating torsional moment varying from – 150 N.m. to + 350 N.m. The shaft is ground and polished and the expected reliability is 90%. The stress – concentration factor due to keyway is 1.3 and the notch sensitivity is 0.90. Calculate the shaft diameter for unlimited life. A factor of safety of 2.5 is desired. Use maximum distortion energy theory of failure.

4.5 A solid circular shaft 20 mm in diameter is subjected to bending stress varying from – 20 MPa to + 40 MPa and torsional shear stress varying from - 15 MPa to + 30 MPa. The material of the shaft has S_{ut} = 540 MPa and S_{yt} = 360 MPa. The corrected endurance strength of shaft material is 200 MPa. Calculate the factor of safety.

4.6 Determine the size of a piston rod made of hot rolled carbon steel (S_{ut} = 560 MPa, S_{yt} = 350 MPa) subjected axial load fluctuations from 20 kN in compression to 30 kN in tension. Minimum factor of safety desired is 1.5. Impact factor has been estimated to be 1.25. Assume suitable data not given.

4.7 A cantilever shown in Fig. 4.32 is of diameter 1.5d for the first 100 mm and diameter d for the remaining 150 mm with a fillet radius of $\frac{r}{d}$ = 0.20. It is subjected to an axial load P which varies from – 200 N to + 500 N and also a transverse load F at its free end which varies from 100 N up and 150 N down. Determine the diameter for a minimum safety margin of 2. The shaft is made of 40C8 steel.

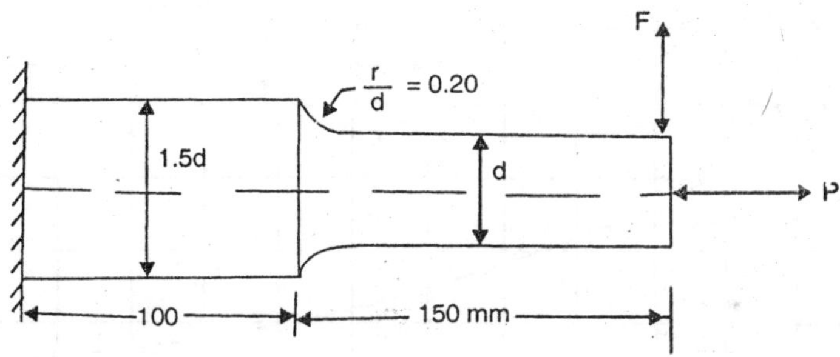

Fig. 4.32

4.8 A pulley is keyed to a shaft midway between two bearings. The shaft is made of cold drawn steel for which ultimate tensile strength is 550 MPa and the yield strength in tension is 400 MPa. The bending moment at the pulley varies from – 150 N.m to + 400 N.m and the torque on the shaft varies from – 50 N.m to + 150 N.m. Determine the diameter of the shaft for infinite life. The stress concentration factor for the keyway at the pulley in bending and torsion are 1.6 and 1.3 respectively. Assume the following values :

Factor of safety = 1.5; Load correction factors = 1.0 in bending and 0.6 in torsion; Size factor = 0.85, Surface finish factor = 0.88.

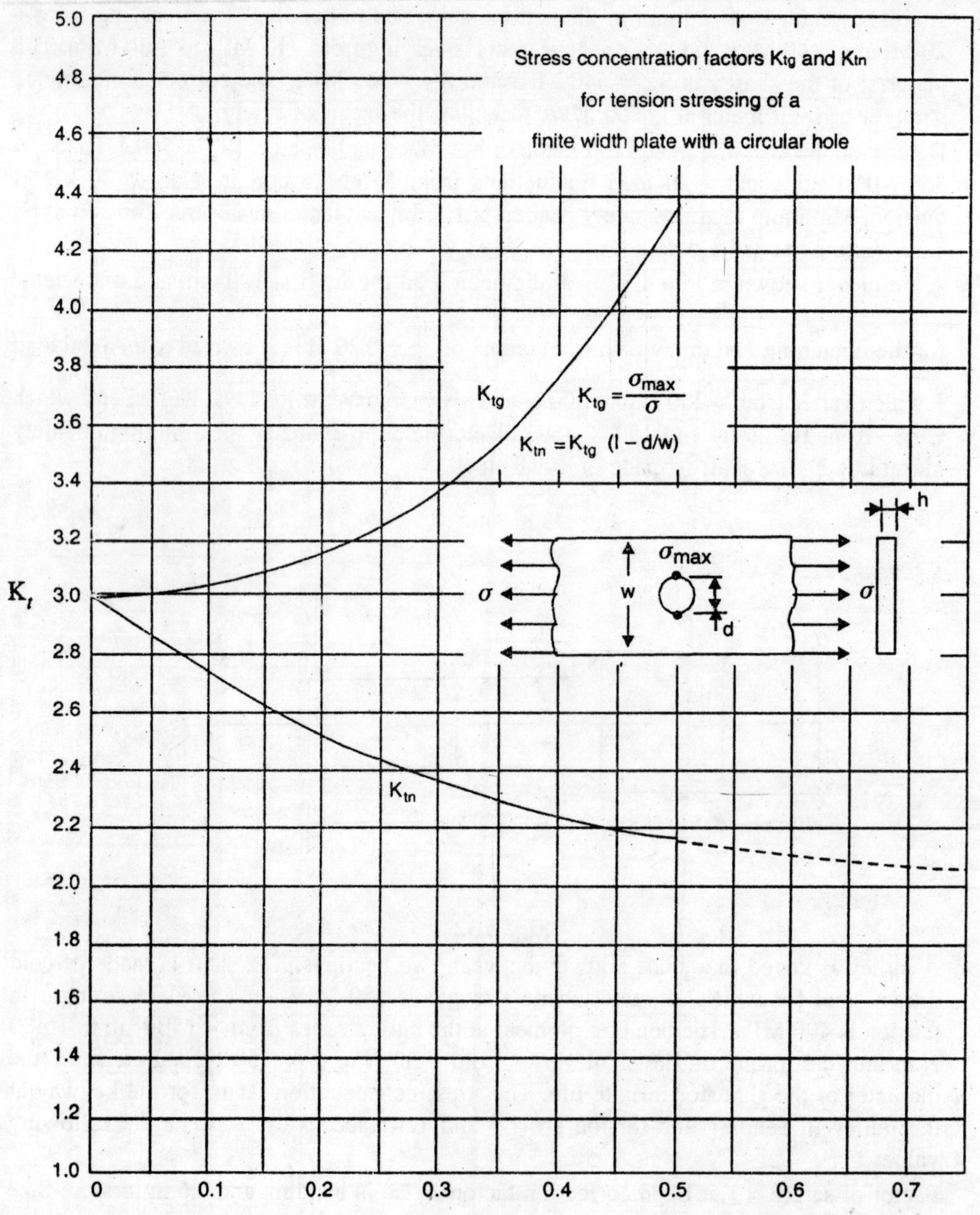

$\frac{d}{w}$	0.05	0.1	0.15	0.20	0.25	0.30	0.35	0.40	0.45	0.50	0.55
K_{tn}	2.83	2.69	2.59	2.50	2.43	2.37	2.32	2.26	2.22	2.17	2.13

Fig. 4.33

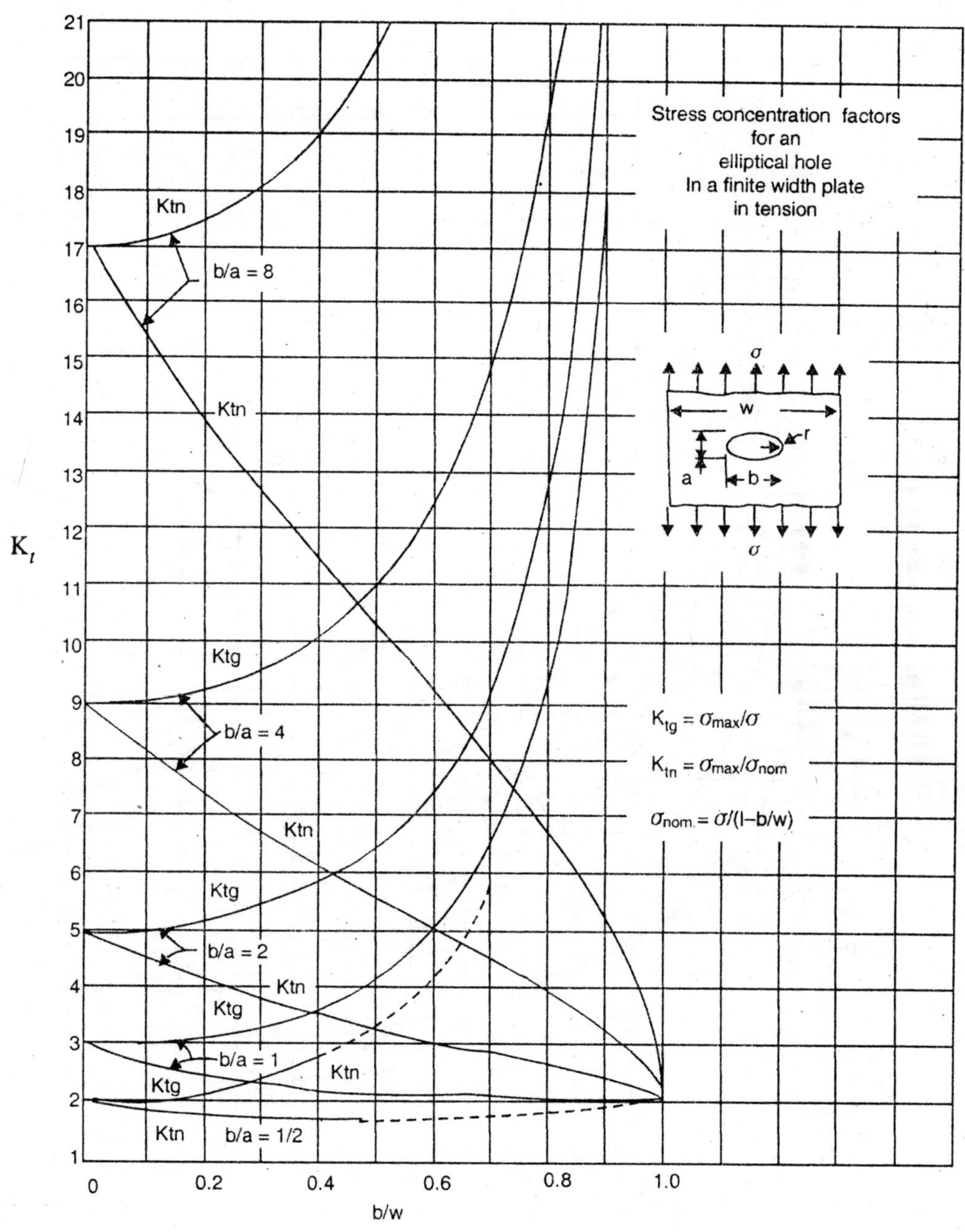
Stress concentration factors
for an
elliptical hole
In a finite width plate
in tension
$K_{tg} = \sigma_{max}/\sigma$
$K_{tn} = \sigma_{max}/\sigma_{nom}$
$\sigma_{nom} = \sigma/(1-b/w)$
σ
w
r
a
b
Ktn
b/a = 8
Ktg
b/a = 4
b/a = 2
b/a = 1
b/a = 1/2
K_t
b/w
0
0.2
0.4
0.6
0.8
1.0
1
2
3
4
5
6
7
8
9
10
11
12
13
14
15
16
17
18
19
20
21

Fig. 4.34

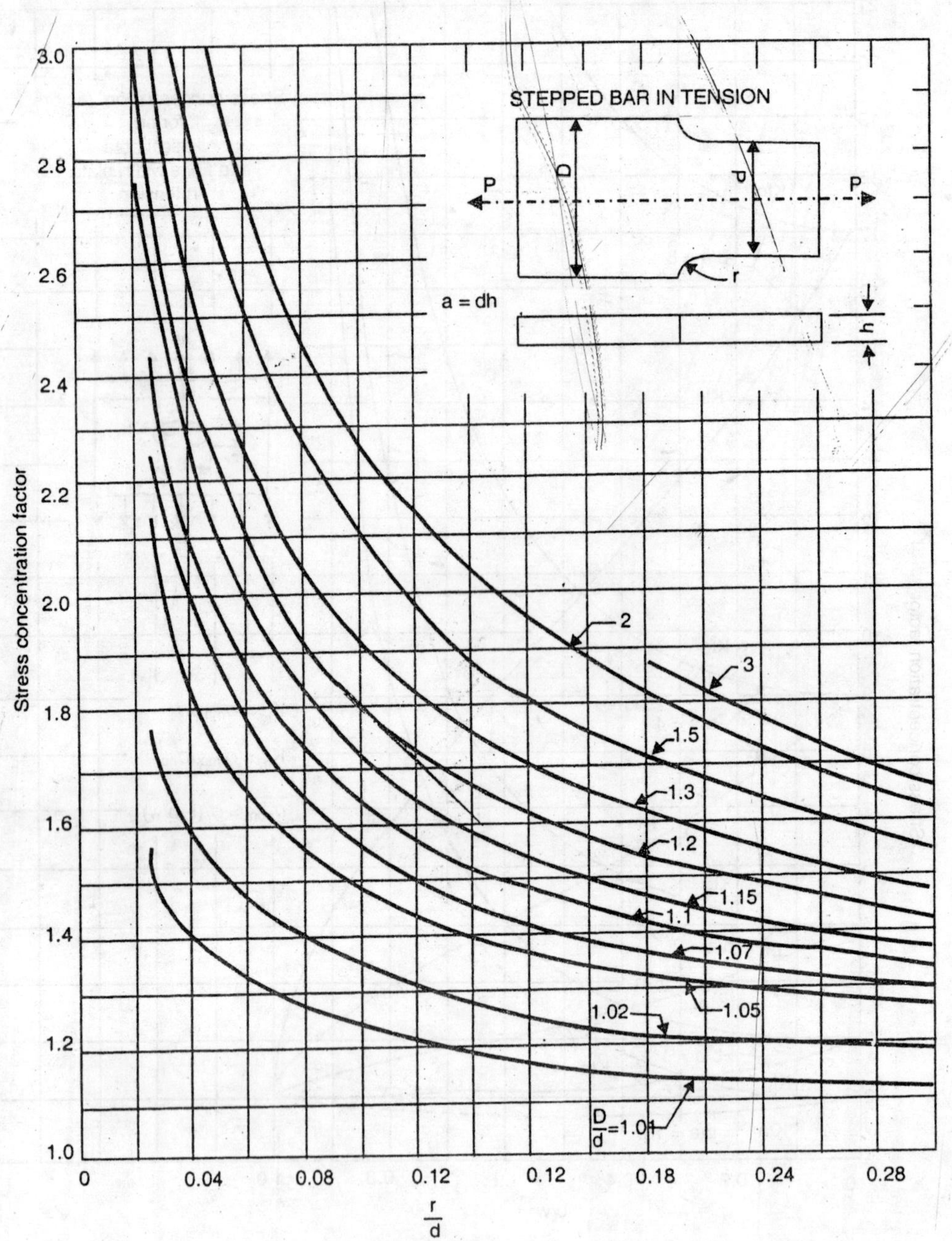
STEPPED BAR IN TENSION
D
d
P
P
r
a = dh
h
Stress concentration factor
3.0
2.8
2.6
2.4
2.2
2.0
1.8
1.6
1.4
1.2
1.0
0
0.04
0.08
0.12
0.12
0.18
0.24
0.28
r/d
2
3
1.5
1.3
1.2
1.15
1.1
1.07
1.05
1.02
D/d = 1.01

Fig. 4.35

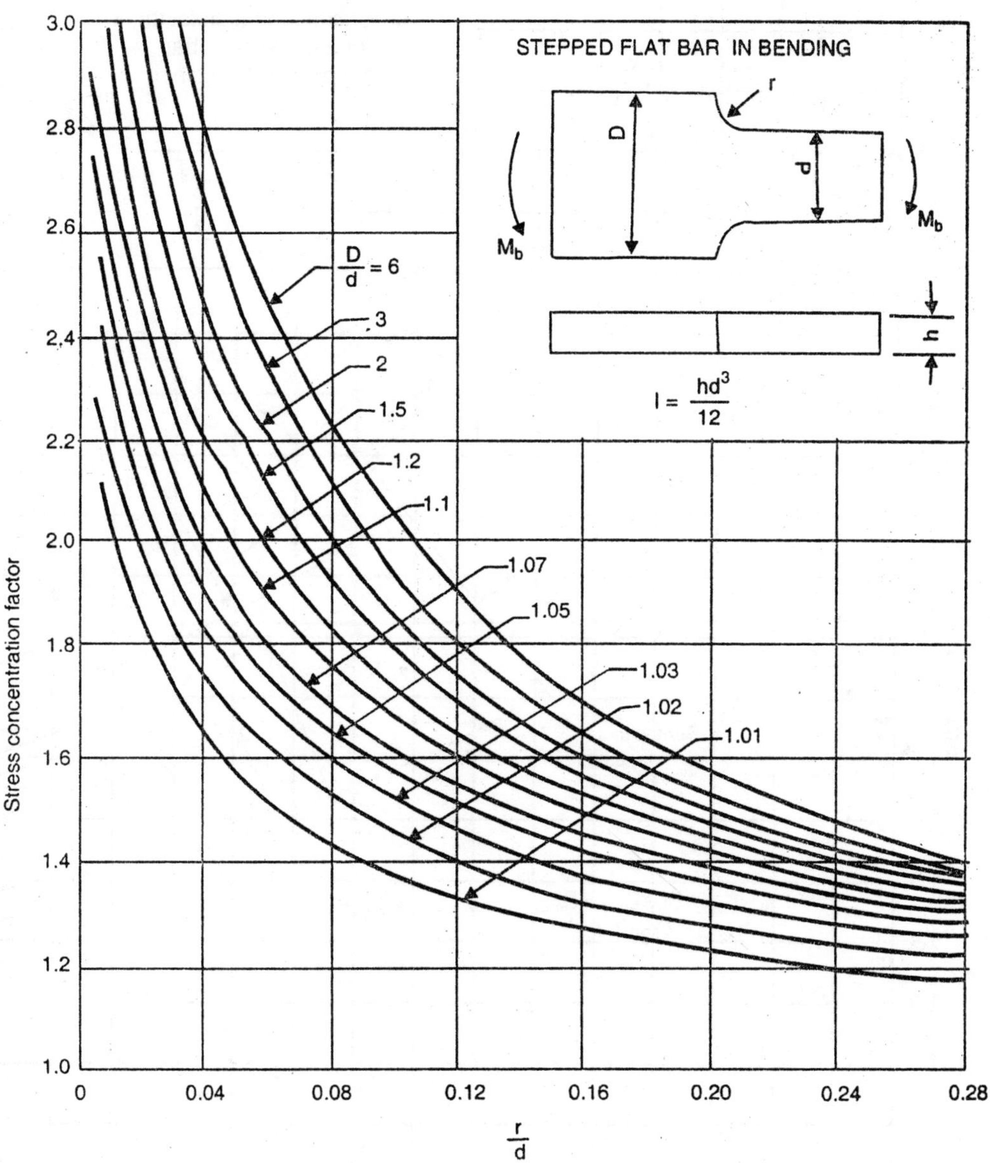
STEPPED FLAT BAR IN BENDING
r
D
d
M_b
M_b
h
$I = \frac{hd^3}{12}$
$\frac{D}{d} = 6$
3
2
1.5
1.2
1.1
1.07
1.05
1.03
1.02
1.01
Stress concentration factor
3.0
2.8
2.6
2.4
2.2
2.0
1.8
1.6
1.4
1.2
1.0
0
0.04
0.08
0.12
0.16
0.20
0.24
0.28
$\frac{r}{d}$

Fig. 4.36

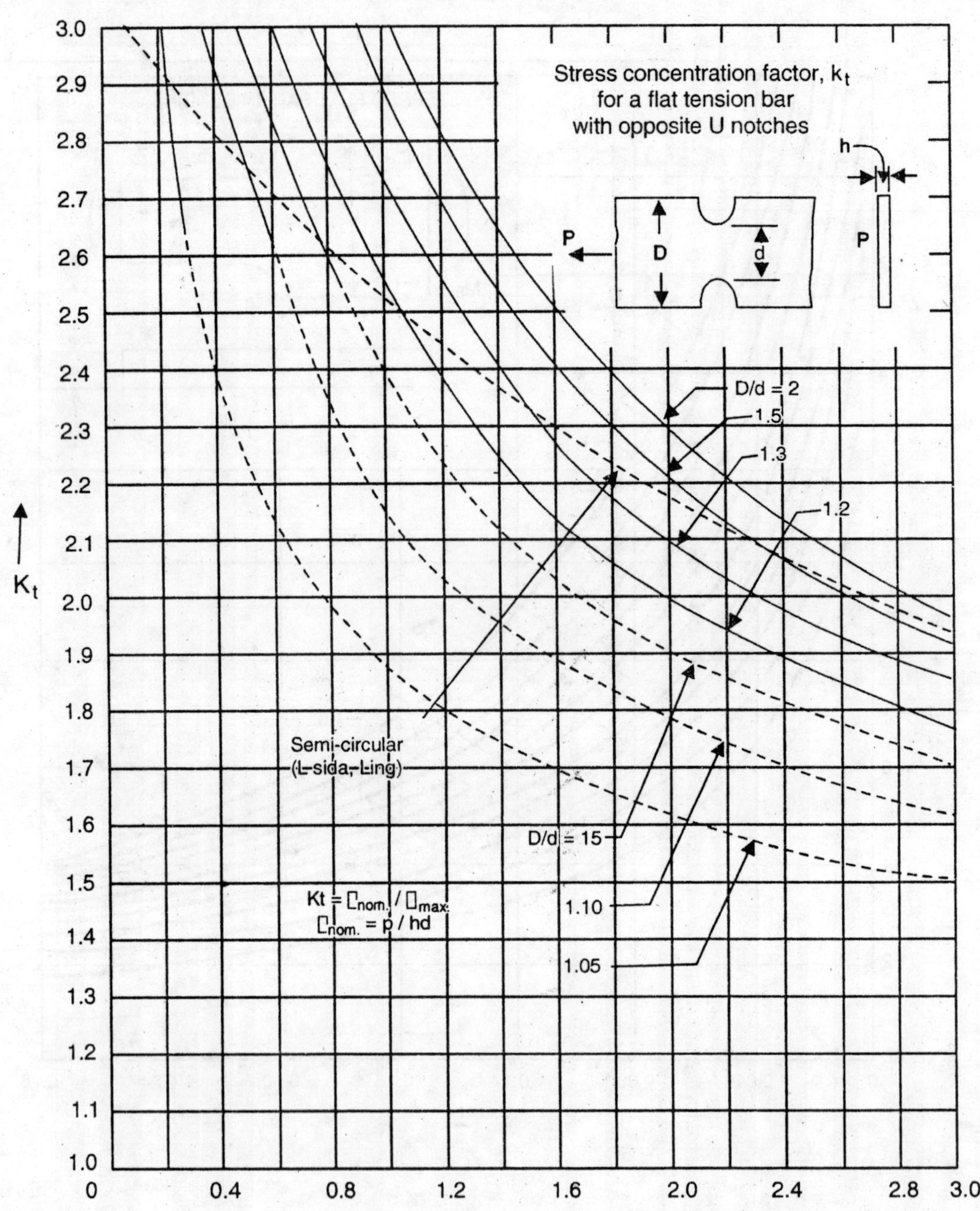
Stress concentration factor, k_t
for a flat tension bar
with opposite U notches
h
P
D
d
P
D/d = 2
1.5
1.3
1.2
Semi-circular
(L-sida, Ling)
D/d = 15
1.10
1.05
Kt = □nom. / □max.
□nom. = p / hd
K_t
3.0
2.9
2.8
2.7
2.6
2.5
2.4
2.3
2.2
2.1
2.0
1.9
1.8
1.7
1.6
1.5
1.4
1.3
1.2
1.1
1.0
0
0.4
0.8
1.2
1.6
2.0
2.4
2.8
3.0

Fig. 4.37

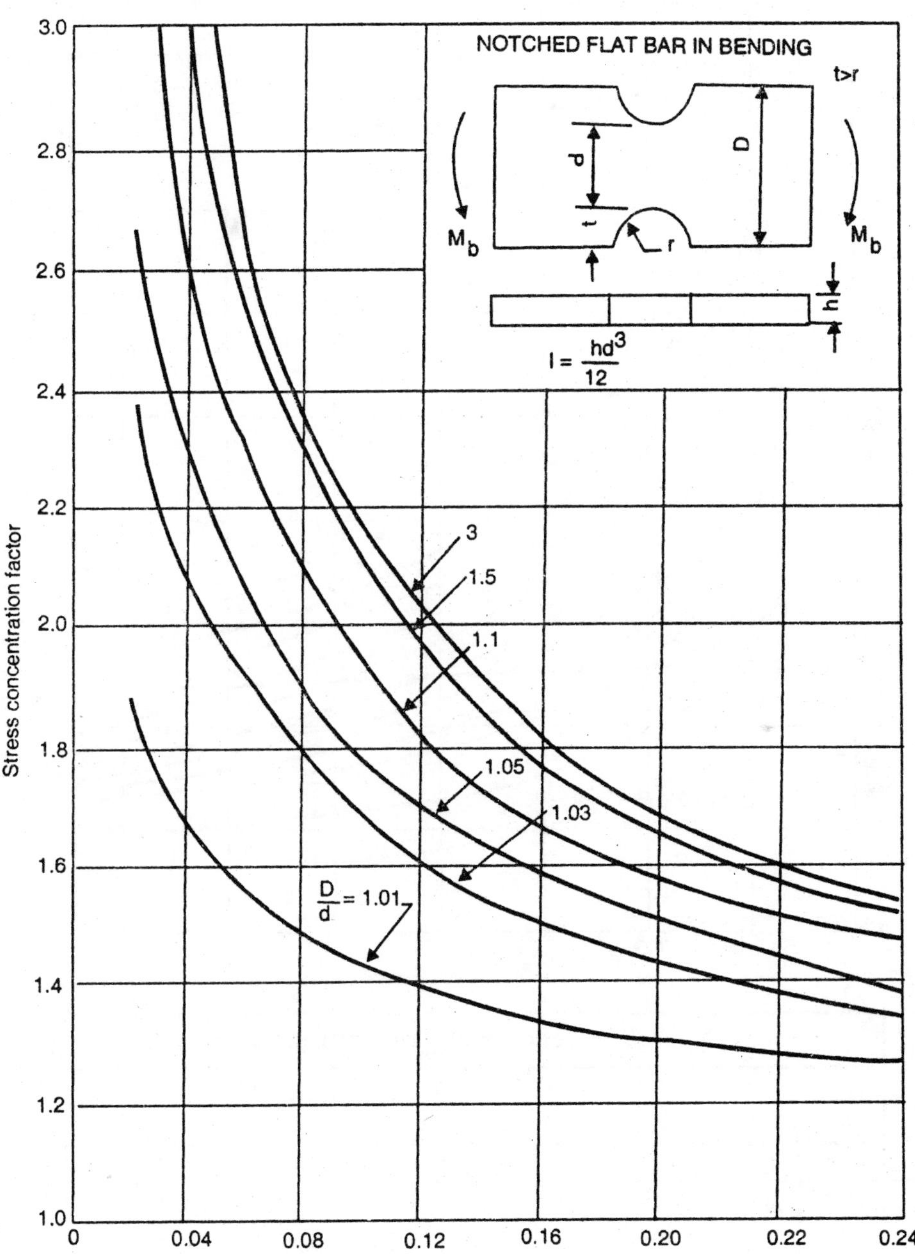
NOTCHED FLAT BAR IN BENDING
t>r
D
d
t
r
M_b
M_b
h
$I = \frac{hd^3}{12}$
3
1.5
1.1
1.05
1.03
$\frac{D}{d}$ = 1.01
Stress concentration factor
3.0
2.8
2.6
2.4
2.2
2.0
1.8
1.6
1.4
1.2
1.0
0
0.04
0.08
0.12
0.16
0.20
0.22
0.24

Fig. 4.38

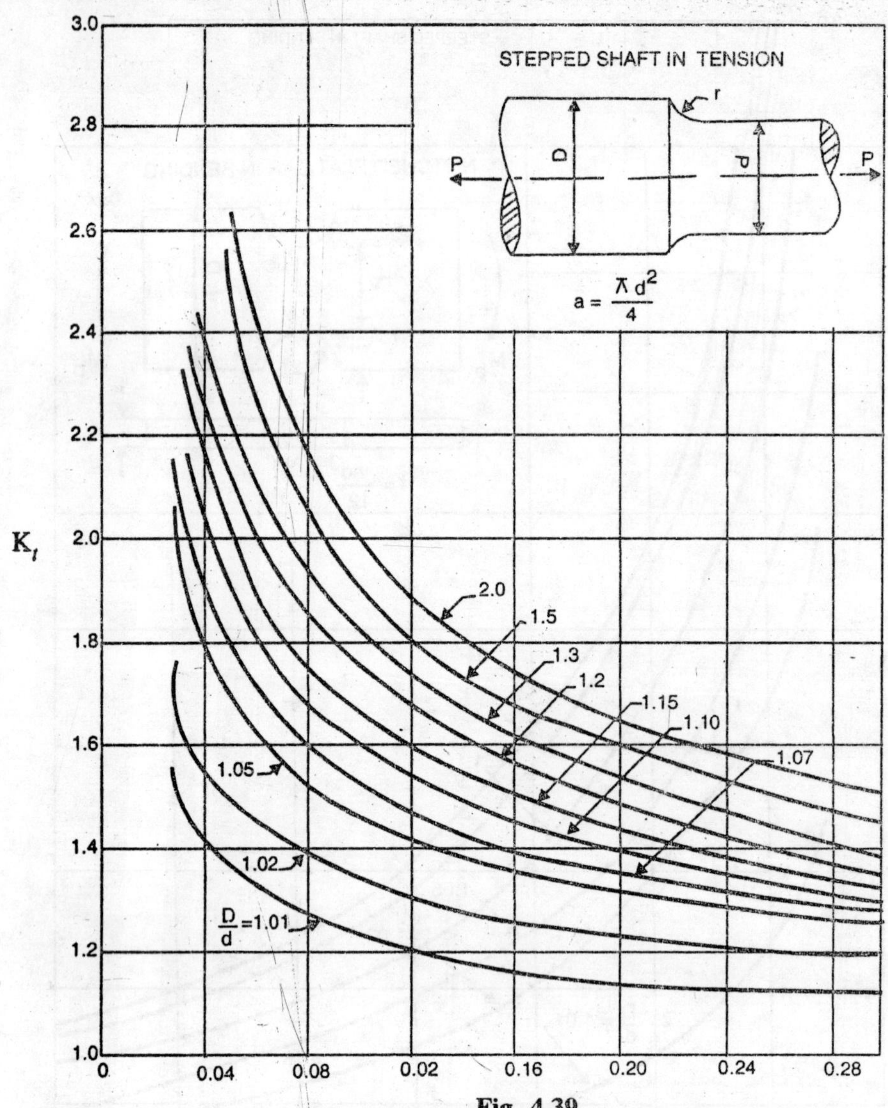

Fig. 4.39

$\frac{D}{d}$	Theoretical stress concentratio factor (K_t)									
	$\frac{r}{d}$									
	0.08	0.10	0.12	0.16	0.18	0.20	0.22	0.24	0.28	0.30
1.01	1.27	1.24	1.21	1.17	1.16	1.15	1.15	1.14	1.13	1.13
1.02	1.38	1.34	1.30	1.26	1.24	1.23	1.22	1.21	1.19	1.19
1.05	1.53	1.46	1.42	1.36	1.34	1.32	1.30	1.28	1.26	1.25
1.10	1.65	1.56	1.50	1.43	1.39	1.37	1.34	1.33	1.130	1.28
1.15	1.73	1.63	1.56	1.46	1.43	1.40	1.37	1.35	1.32	1.31
1.20	1.82	1.68	1.62	1.51	1.47	1.44	1.41	1.38	1.35	1.34
1.50	2.03	1.84	1.80	1.66	1.60	1.56	1.53	1.50	1.46	1.44
2.00	2.14	1.94	1.89	1.74	1.68	1.64	1.59	1.56	1.50	1.47

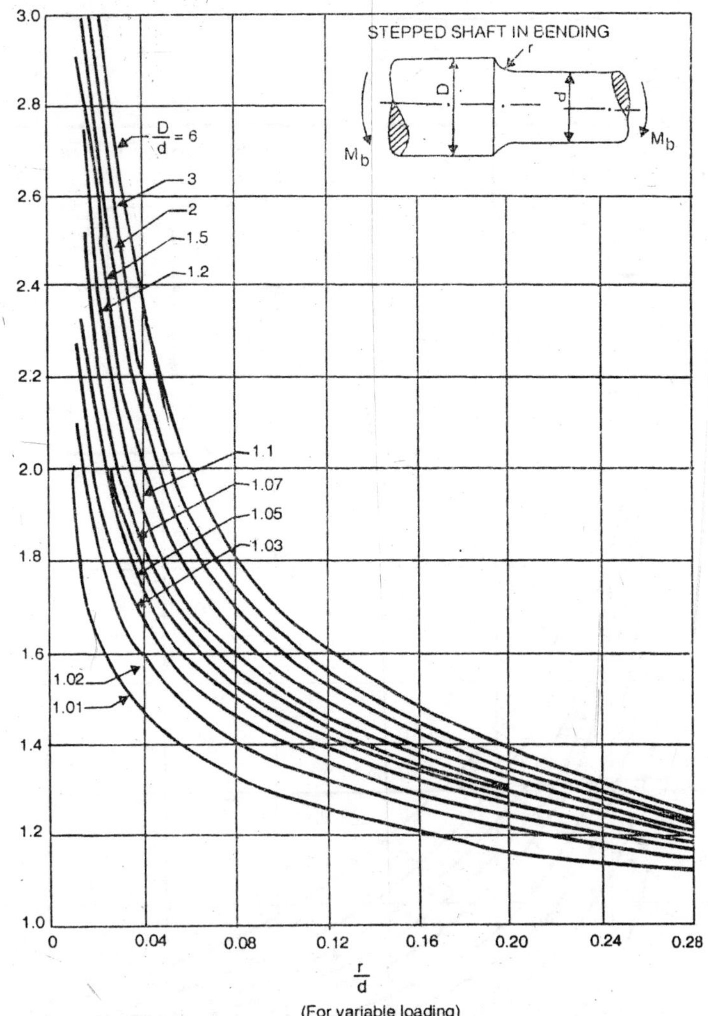

Fig. 4.40

$\frac{D}{d}$	Theoretical stress concentration factor (K_t)									
	$\frac{r}{d}$									
	0.02	0.04	0.08	0.10	0.12	0.16	0.20	0.24	0.28	0.30
1.01	1.85	1.61	1.42	1.36	1.32	1.24	1.20	1.17	1.15	1.14
1.02	1.97	1.72	1.50	1.44	1.40	1.32	1.27	1.23	1.21	1.20
1.05	2.20	1.88	1.60	1.53	1.48	1.40	1.34	1.30	1.27	1.25
1.10	2.36	1.99	1.66	1.58	1.53	1.44	1.38	1.33	1.28	1.27
1.20	2.52	2.10	1.72	1.62	1.56	1.46	1.39	1.34	1.29	1.28
1.50	2.75	2.20	1.78	1.68	1.60	1.50	1.42	1.36	1.31	1.29
2.00	2.86	2.32	1.87	1.74	1.64	1.53	1.43	1.37	1.32	1.30
3.00	3.00	2.45	1.95	1.80	1.69	1.56	1.46	1.38	1.34	1.32
6.00	3.04	2.58	2.04	1.87	1.76	1.60	1.49	1.41	1.35	1.33

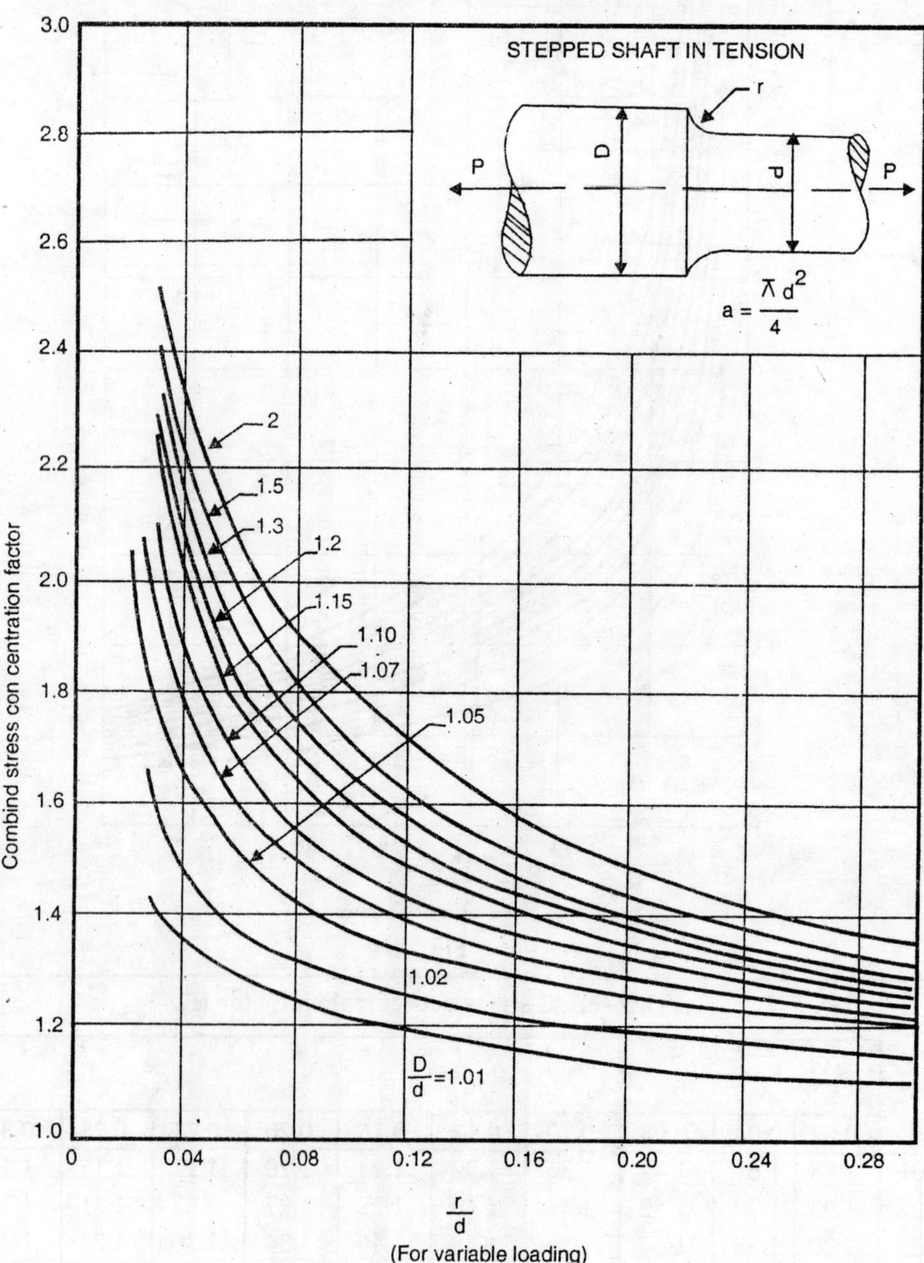
STEPPED SHAFT IN TENSION
r
D
P
d
P
$a = \frac{\pi d^2}{4}$
Combind stress con centration factor
3.0
2.8
2.6
2.4
2.2
2.0
1.8
1.6
1.4
1.2
1.0
2
1.5
1.3
1.2
1.15
1.10
1.07
1.05
1.02
$\frac{D}{d}$=1.01
0
0.04
0.08
0.12
0.16
0.20
0.24
0.28
$\frac{r}{d}$
(For variable loading)

Fig. 4.41

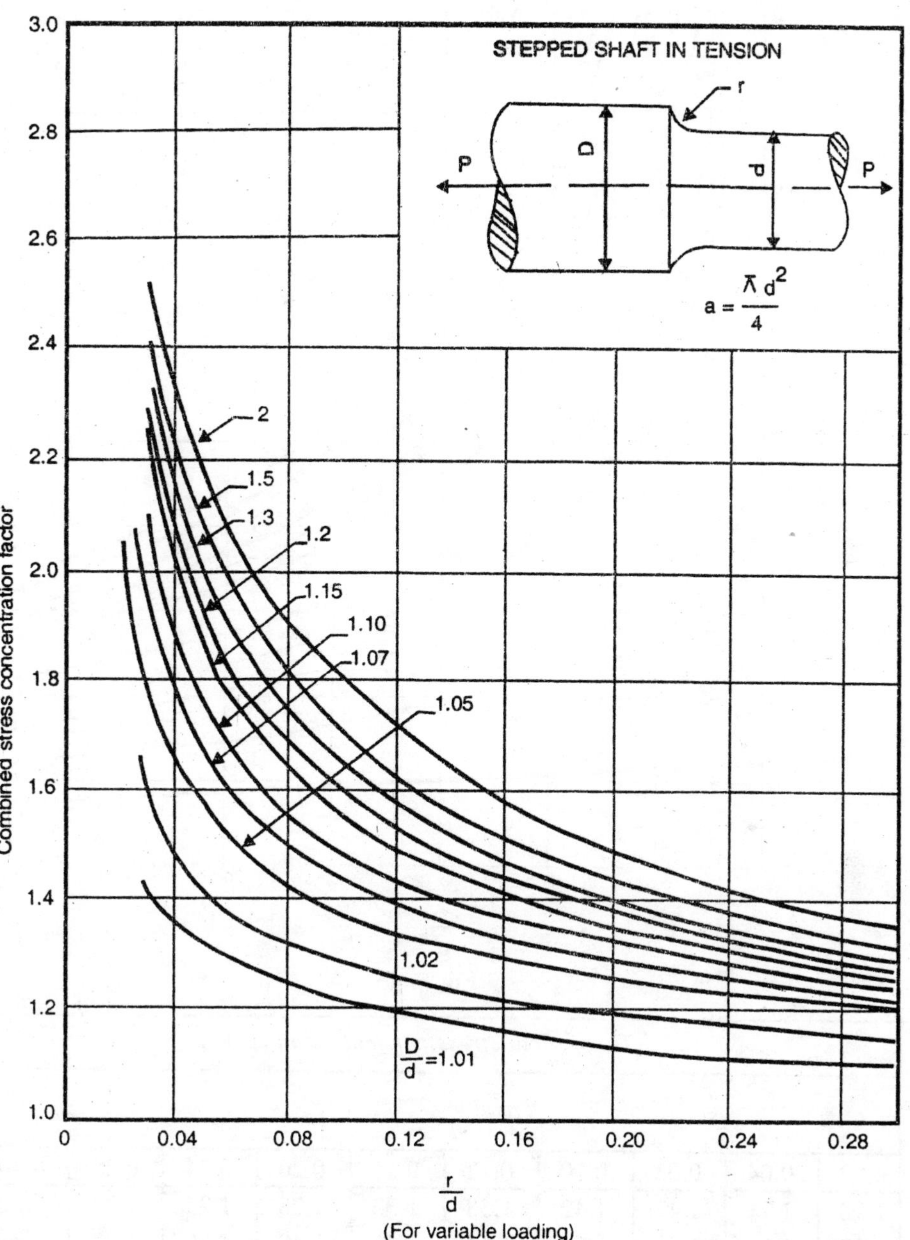

$\frac{D}{d}$	$\frac{r}{d}$									
	0.02	0.04	0.08	0.10	0.12	0.16	0.20	0.24	0.28	0.30
1.09	1.54	1.32	1.19	1.16	1.15	1.12	1.11	1.10	1.09	1.09
1.20	1.98	1.67	1.40	1.33	1.28	1.22	1.18	1.15	1.13	1.13
1.33	2.14	1.79	1.48	1.41	1.35	1.28	1.22	1.19	1.17	1.16
2.00	2.27	1.84	1.53	1.46	1.40	1.32	1.26	1.22	1.19	1.18

Fig. 4.42

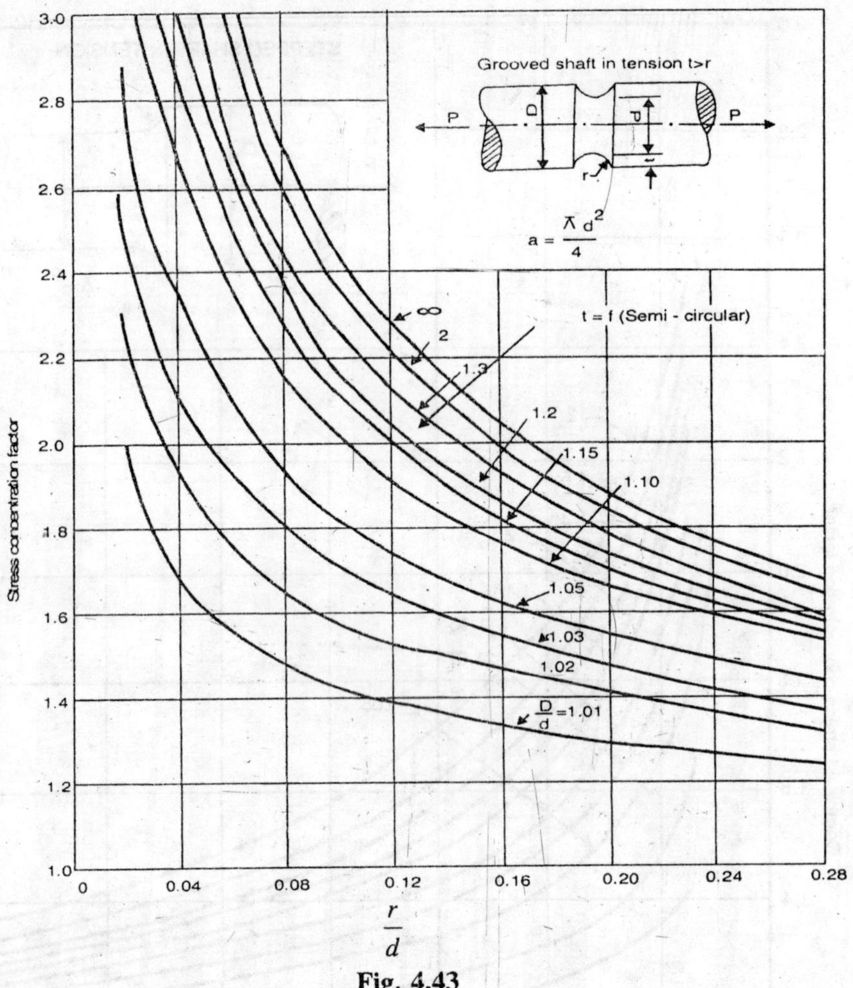

Fig. 4.43

$\frac{D}{d}$	*Theoretical stress concentration* (K_t)									
	$\frac{r}{d}$									
	0.02	0.04	0.08	0.10	0.12	0.16	0.20	0.24	0.28	0.30
1.01	1.98	1.71	1.47	1.42	1.38	1.33	1.28	1.25	1.23	1.22
1.02	2.30	1.94	1.66	1.59	1.54	1.45	1.40	1.36	1.33	1.31
1.03	2.60	2.14	1.77	1.69	1.63	1.53	1.46	1.41	1.37	1.36
1.05	2.85	2.36	1.94	1.81	1.73	1.61	1.54	1.47	1.43	1.41
1.10	...	2.70	2.16	2.01	1.90	1.75	1.70	1.57	1.50	1.47
1.20	...	2.90	2.36	2.17	2.04	1.86	1.74	1.64	1.56	1.54
1.30	....	...	2.46	2.26	2.11	1.91	1.77	1.67	1.59	1.56
1.50	...	...	2.54	2.33	2.16	1.94	1.79	1.69	1.61	1.57
2.00	...	...	2.61	2.38	2.22	1.98	1.83	1.72	1.63	1.59
∞	...	...	2.69	2.44	2.26	2.03	1.86	1.74	1.65	1.61

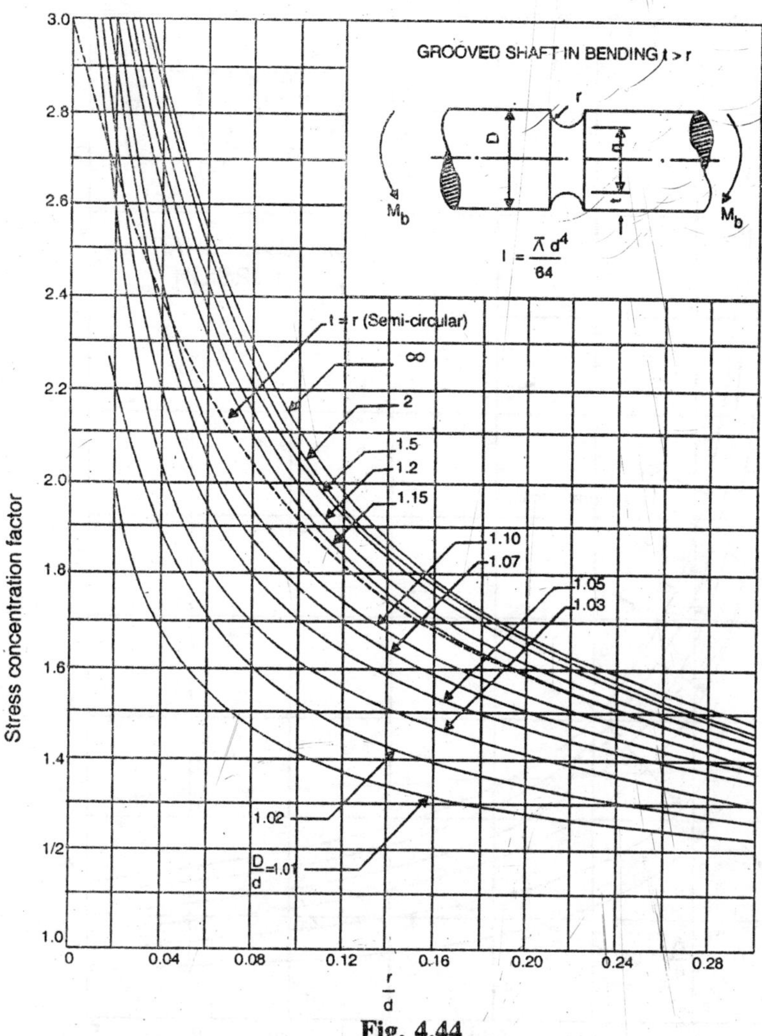

Fig. 4.44

$\frac{D}{d}$	*Theoreticai stress concentration (K_t)*									
	$\frac{r}{d}$									
	0.02	0.04	0.08	0.10	0.12	0.16	0.20	0.24	0.28	0.30
1.01	1.74	1.68	1.47	1.41	1.38	1.32	1.27	1.23	1.22	1.20
1.02	2.28	1.89	1.64	1.53	1.48	1.40	1.34	1.30	1.26	1.25
1.03	2.46	2.04	1.68	1.61	1.55	1.47	1.40	1.35	1.31	1.28
1.05	2.75	2.22	1.80	1.70	1.63	1.53	1.46	1.40	1.35	1.33
1.12	3.20	2.50	1.97	1.83	1.75	1.62	1.52	1.45	1.38	1.34
1.30	3.40	2.70	2.04	1.91	1.82	1.67	1.57	1.48	1.42	1.38
1.50	3.48	2.74	2.11	1.95	1.84	1.69	1.58	1.49	1.43	1.40
2.00	3.55	2.78	2.14	1.97	12.86	1.71	1.59	1.55	1.44	1.41
∞	3.60	2.85	2.17	1.98	1.88	1.71	1.60	1.51	1.45	1.42

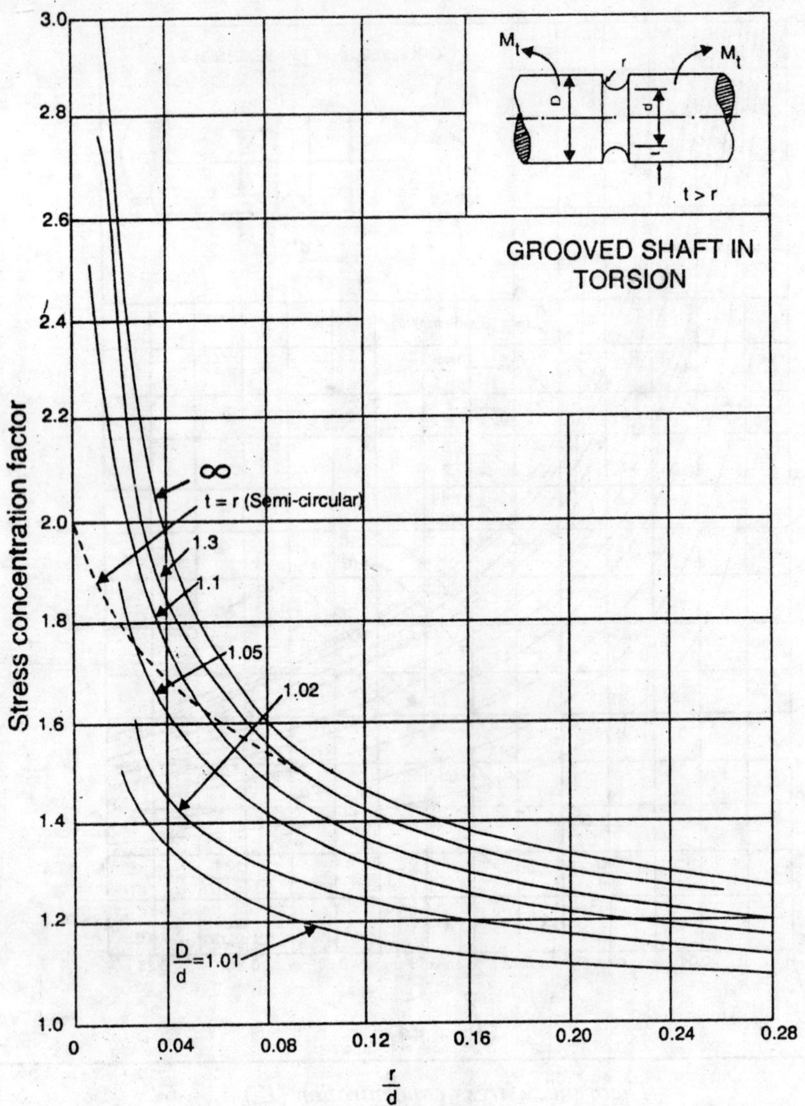

Fig. 4.45

$\frac{D}{d}$	$\frac{r}{d}$									
	0.02	0.04	0.08	0.10	0.12	0.16	0.20	0.24	0.28	0.30
1.01	1.50	1.30	1.22	1.20	1.18	1.16	1.13	1.12	1.12	1.12
1.02	1.62	1.45	1.31	1.27	1.23	1.20	1.18	1.16	1.15	1.16
1.05	1.88	1.61	1.40	1.35	1.32	1.26	1.22	1.20	1.18	1.17
1.10	2.05	1.73	1.47	1.41	1.37	1.31	1.26	1.24	1.21	1.20
1.20	2.26	1.83	1.53	1.46	1.41	1.34	1.27	1.25	1.22	1.21
1.30	2.32	1.89	1.55	1.48	1.43	1.35	1.30	1.26	1.22	1.22
2.00	2.40	1.93	1.58	1.50	1.45	1.36	1.31	1.26	1.23	1.22
∞	2.50	1.96	1.60	1.51	1.46	1.38	1.32	1.27	1.24	1.24

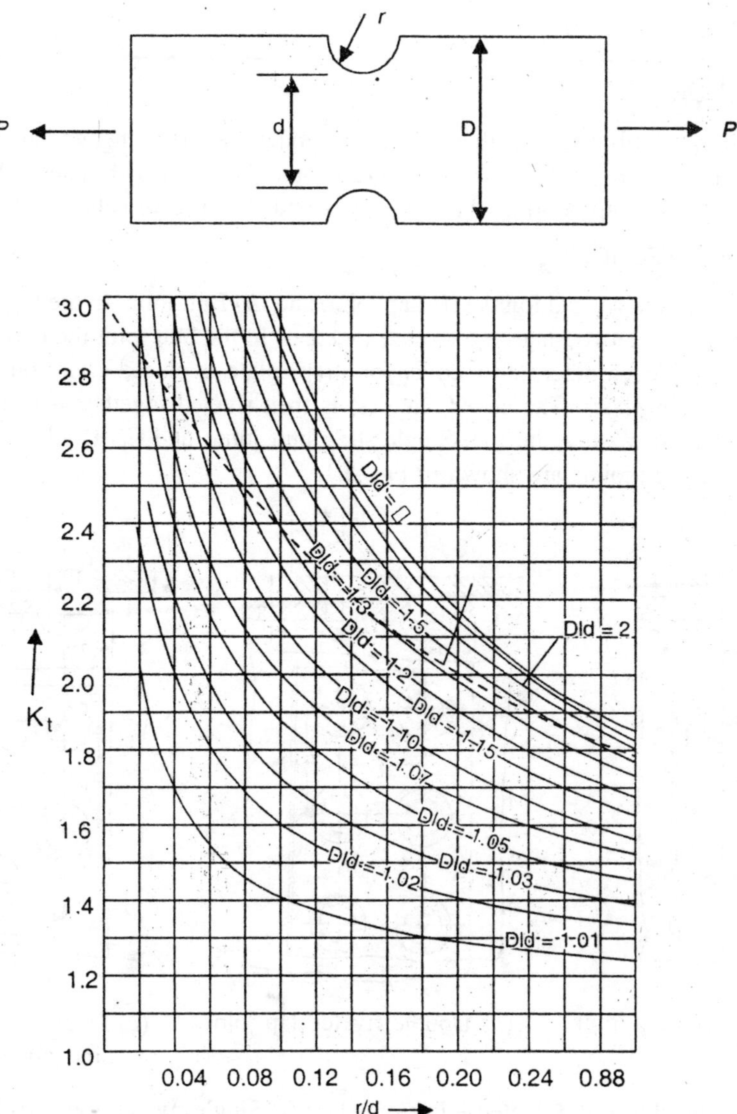

Fig. 4.46 Theoretical stress concentration factor (k_t) *for a notched flat bar in tension*

5 RIVETED JOINTS

Chapter

5.1 INTRODUCTION

A rivet is a short cylindrical bar with a head to one end, tail on the other end and shank or body in between. The rivets are used to make permanent joints between two plates. They are widely used in structural work, ship building, bridges, tanks and pressure vessels.

5.2 TYPES OF RIVETED JOINTS

There are two types of riveted joints : (*i*) Lap joint, and (*ii*) Butt joint. In the lap joint, one plate overlaps the other and the two plates are riveted together. In the butt joint the two plates are kept in alignment touching each other and a cover (or strap) plate is placed either on one side or on both sides of the main plates. The cover plate (s) is then riveted together with the main plate. Accordingly the butt joints may be : (*i*) Single strap butt joint, and (*ii*) Double strap butt joint. Various types of riveted joints are shown in Fig. 5.1.

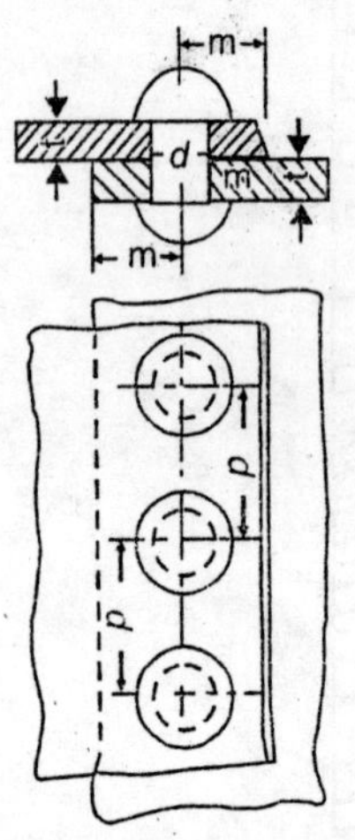

(*a*) Single riveted lap joint

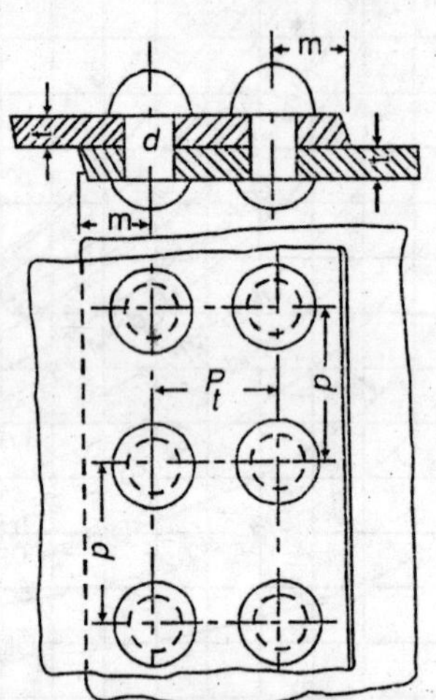

(*b*) Double riveted lap joint

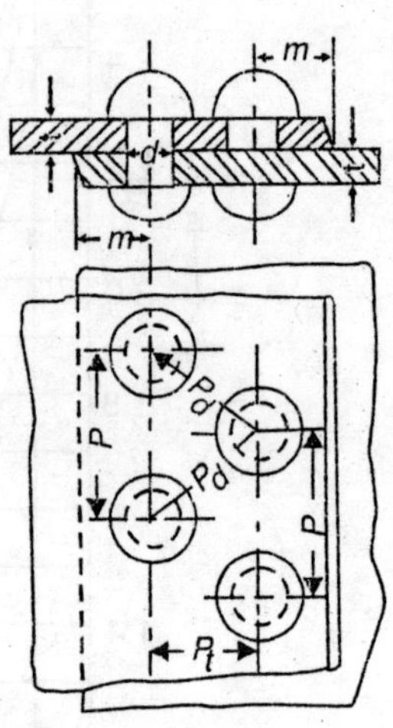

(*c*) Double riveted lap joint (zig-zag)

In addition to the above, the riveted joints may be : (*i*) Single riveted joint (*ii*) Double riveted joint (*iii*) Triple riveted joint, and so on. In the double and higher riveted joints, the joint may be : (*i*) Chain riveted, and (*ii*) Zig – zag riveted. In the chain riveted joint, the rivets in the various rows are opposite to each other. On the other hand, in the zig – zag riveted joint, the rivets in the adjacent rows are stagerred in such a way that every rivet is in the middle of the two rivets of the opposite row. Lap joints are used to transmit small loads, whereas butt joints are used to transmit heavy loads.

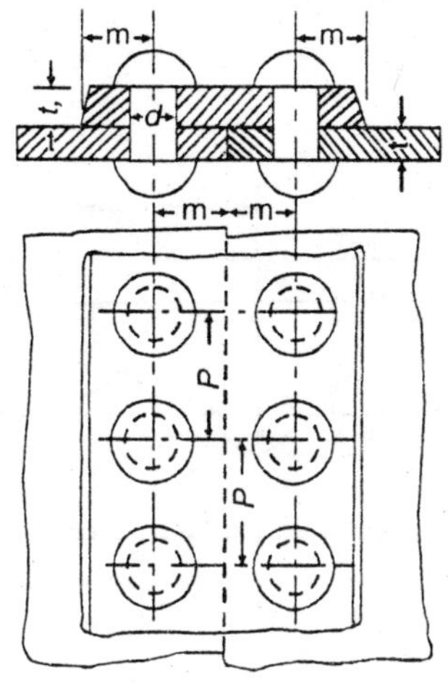

(*d*) Single riveted butt joint with single cover plate

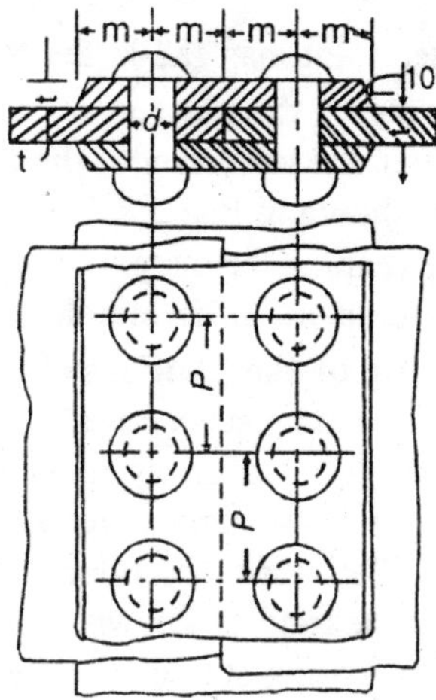

(*e*) Single riveted but joint with double cover plates

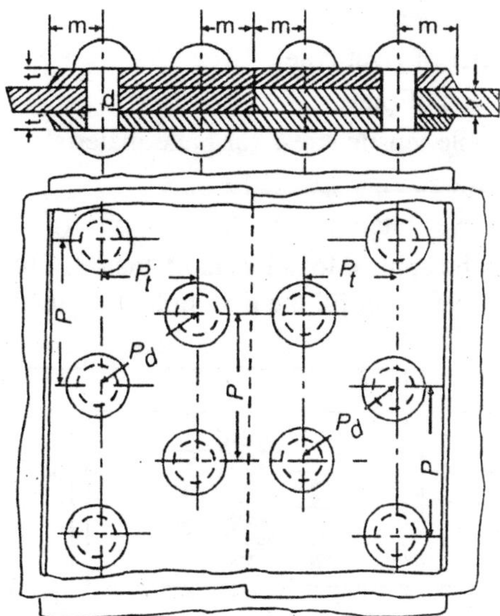

(*f*) Double riveted butt joint with double cover plates

Fig. 5.1 Types of riveted joints.

5.3 TERMINOLOGY

Pitch, p. It is the distance from the centre of one rivet to the centre of the next rivet measured parallel to the seam.

Back or transverse pitch, p_t – It is the perpendicular distance between the center lines of the adjacent rows.

Diagonal pitch, p_d – It is the distance between the centres of rivets in the adjacent rows of zig – zag riveted joint.

Margin, m – It is the distance between the centre of rivet hole to the nearest edge of the plate.

5.4 MODES OF FAILURE OF A RIVETED JOINT

A riveted joint may fail in the following ways :

1. Tearing of the plate at an edge. A joint may fail due to tearing of the plate at an edge, as shown in Fig. 5.2. This can be avoided by keeping the margin, $m = 1.5d_0$, where d_0 = diameter of rivet hole.

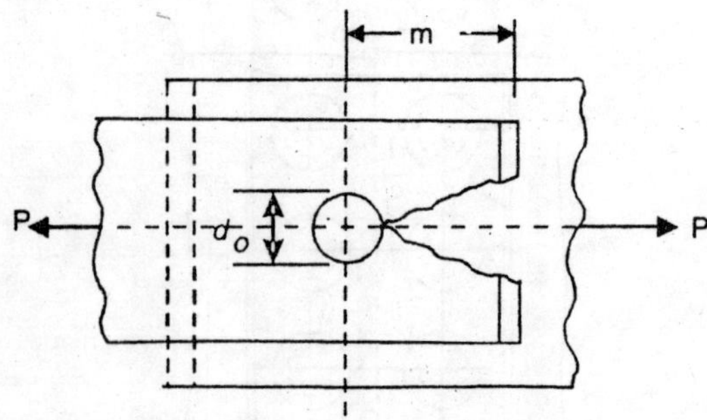

Fig. 5.2 Tearing of plate at an edge

2. Tearing of the plate across a row of rivets. The main plate or cover plate may tear off across a row of rivets due to tensile stresses set up in the main plate, as shown in Fig. 5.3. The resistance offered by the plate against tearing is known as tearing resistance or tearing strength of the plate. It is determined across one pitch length only as every rivet is responsible for that much length.

Let p = pitch of rivets

d_0 = diameter of rivet hole

t = thickness of plate

σ_t = allowable tensile stress for plate material.

Tearing area per pitch length, $A_t = (p - d_0)\,t$

Tearing strength of plate per pitch length, $P_t = A_t \times \sigma_t = (p - d_0)\,t\sigma_t$ ...(5.1)

The joint shall fail when the applied load P is more than P_t. The rivet hole diameter is taken for all calculations because the rivet shall fill the rivet hole completely after the joint has been made.

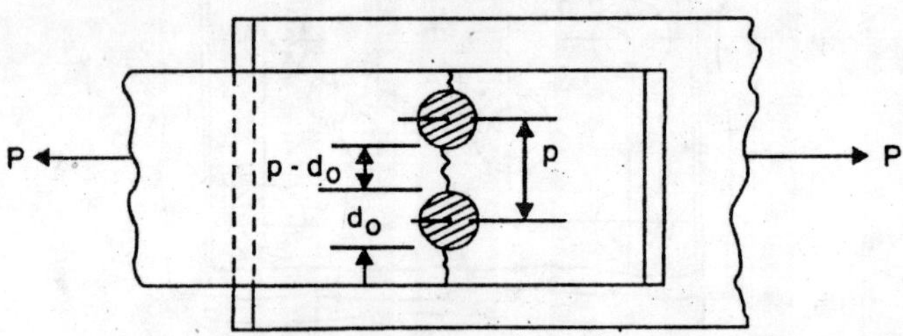

Fig. 5.3 Tearing of plate across the rows of rivets

3. Shearing of the rivets. The axis of the rivets is perpendicular to the load line. As a result of this the rivets can be either in single shear or double shear, as shown in Fig. 5.4. The rivets shall be in single shear in case of lap joint or butt joint with single cover strap. In case of butt joint with double cover straps, the rivets shall be in double shear.

Let d_0 = diameter of rivet hole

n = number of rivets per pitch length

τ = allowable shear stress for rivets material

z = number of shears

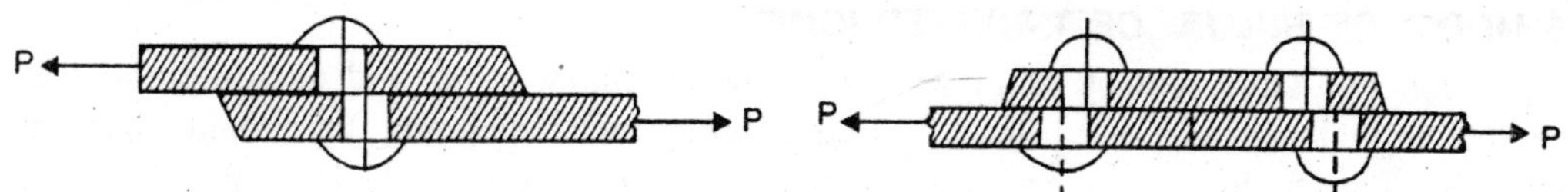

(*a*) Shearing off rivet in lap joint (*b*) Shearing off rivet in a single strap butt joint

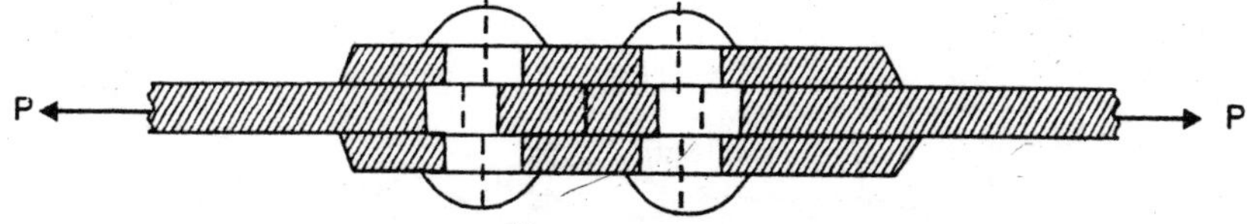

(*c*) Shearing off rivet in a double strap butt joint.

Fig. 5.4 Shearing of rivets.

Then Shearing area, $A_s = n \times z \times \frac{\pi}{4 \times d^2_0}$ ...(5.2)

Where z = 1 for single shear
= 2 for double shear
= 1.875 for double shear as per Indian Boiler Regulations (I.B.R.)

Shearing strength of rivets, $P_s = A_s \times \tau$ ...(5.3)

The joint shall fail if the applied load P is more than P_s.

4. Crushing of the plate or rivets. The crushing of the rivets is shown in Fig. 5.5 as a result of which the hole becomes of oval shape and the joint becomes loose. This type of failure of rivets is called crushing or bearing. The resistance offered by the rivets is called crushing or bearing strength. If σ_c is the allowable crushing stress for the rivets, then

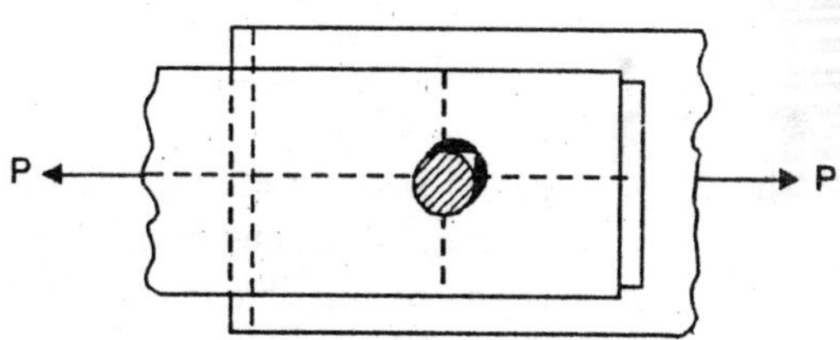

Fig. 5.5 Crushing of a rivet.

Crushing or projected area of n rivets, $A_c = ndt$

Crushing strength of rivets, $P_c = A_c \times \sigma_c$ ...(5.4)

The joint shall fail when the applied load P is more than P_c.

5. Strength of the joint. The strength of the joint is defined as the maximum value of the load it can transmit. It is the smallest value of P_t, P_s and P_c.

6. Strength of the solid plate. The strength of the solid plate per pitch length is calculated after ignoring the rivet hole.

Strength of solid plate, $P = p \times t \times \sigma_t$ ...(5.5)

7. Efficiency of the joint. It is defined as the ratio of the strength of riveted joint to the strength of the solid plate.

$$\text{Efficiency, } \eta = \frac{\text{Least value of Pt, Ps and } P_c}{P} \quad ...(5.6)$$

5.5 DESIGN OF BOILER JOINTS

The boiler is fabricated by making a riveted longitudinal joints and circumferential joints. The longitudinal joint is a butt joint with two cover straps to maintain the circular shape of the boiler. The longitudinal joints are staggered. The circumferential joint is a lap joint which is required to obtain the desired length of the boiler. For this purpose one ring overlapping the other alternately is used. The longitudinal and circumferential joints of a boiler are shown in Fig. 5.6.

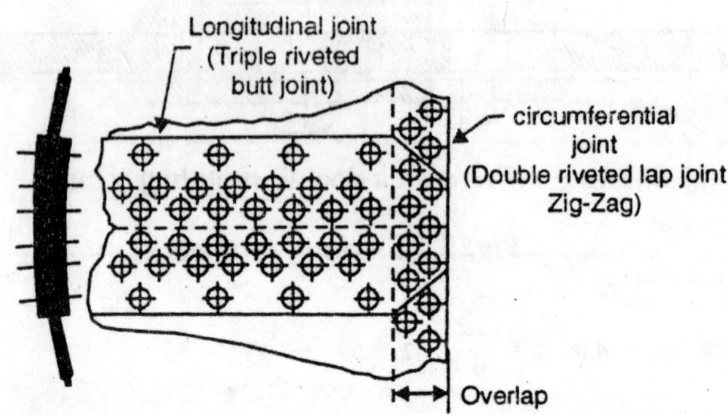

Fig. 5.6 Circumferential and longitudinal joint of a boiler

5.5.1 Longitudinal Butt Joint

(i) Plate thickness

The thickness of boiler shell plate is given by the thin cylindrical pressure vessel theory under internal pressure as :

$$t = \frac{p_s D}{2\sigma_t \eta_l} + c \qquad \text{...(5.7)}$$

where
p_s = steam pressure, MPa gauge
D = shell diameter, mm
σ_t = permissible tensile stress for shell plates material at operating temperature, MPa (see Table 5.1 for tensile strength of boiler plates and rivets.)
η_l = efficiency of longitudinal joint (see Table 5.3)
c = corrosion allowance ≈ 1 mm

The boiler plate thickness should not be less than 7 mm. For standard sizes of plates, refer to Table 5.5.

Table 5.1 Mechanical properties of boiler plates and rivets.

Grade	Tensile strength, MPa	Yield strength as percentage of minimum tensile strength
1	410 – 490	55
2	510 – 610	50

Table 5.2 Recommended type of joint for pressure vessels.

Diameter of shell, mm	Thickness of shell, mm	Type of joint
600 – 1800	6 – 13	Double riveted
900 – 2100	13 – 25	Triple riveted
1500 – 2700	16 – 40	Quadruple riveted

Table 5.3 Range of expected efficiencies for riveted joints.

Type of joint Lap joints :	Efficiency, %	Type of joint Butt joints :	Efficiency, %
Single	45 – 65	Single	60 – 70
Double	63 – 77	Double	75 – 85
Triple	75 – 85	Triple	80 – 90
		Quadruple	85 – 95

Table 5.4 Recommended factor of safety for boiler joints.

Type of Joint	Factor of safety based on ultimate strength	
	Hand riveting	Machine riveting
Lap joint	4.75	4.50
Single strap butt joint	4.75	4.50
Single riveted butt joint with two equal cover straps	4.75	4.50
Double riveted butt joint with two equal cover straps	4.25	4.00

Table 5.5 Standard nominal size of steel plates.

Thickness, mm	5, 6, 8, 12, 14, 16, 18, 20, 22, 25, 28, 32, 36, 40, 45, 50, 56 and 63
Width, mm	900, 1000, 1100, 1200, 1250, 1400, 1500, 1600, 1800, 2000, 2200 and 2500
Length, mm	220, 2500, 2800, 3200, 3600, 4000, 4500, 5000, 5600, 6300, 7100, 8000, 9000, 10000, 11000 and 12000.

(*ii*) Rivet hole diameter.

Rivet hole diameter is given by Unwin's formula,

$$d_0 = 6\sqrt{t} \quad \text{for } t \geq 8 \text{ mm} \qquad \text{...(5.8 } a)$$

for $t \leq 8$ mm, calculate rivet hole diameter as follows :

$$d_0 = \sqrt{50\,t} - 4 \text{ mm for lap joints} \qquad \text{...(5.8 } b)$$

$$d_0 = \sqrt{50\,t} - 5 \text{ mm for single riveted butt joints} \qquad \text{...(5.8 } c)$$

$$d_0 = \sqrt{50\,t} - 6 \text{ mm for double riveted butt joints} \qquad \text{...(5.8 } d)$$

(*iii*) Diameter of rivets.

$$\begin{aligned}\text{Diameter of rivet, } d &= d_0 - 1 \text{ mm for } 12 \text{ mm} \leq d \leq 24 \text{ mm} \\ &= d_0 - 1.5 \text{ mm for } 27 \text{ mm} \leq d \leq 36 \text{ mm} \\ &= d_0 - 2 \text{ mm for } 39 \text{ mm} \leq d \leq 48 \text{ mm}\end{aligned}$$

Standard diameter of rivets for boilers are : 12, 14, 16, 18, 20, 22, 24, 27, 30, 33, 36, 39, 42 and 48 mm.

(iv) Pitch of rivets.

The pitch of rivets should not be less than $2d_0$ to enable the rivet heads to be formed. The maximum pitch of rivets allowed in the longitudinal joint prescribes by IBR is given by :

$$p = C \times t + 40 \text{ mm}$$

where C = coefficient as given in Table 5.6. ...(5.9)

Table 5.6 Value of coefficient C.

Number of rivets per pitch	Lap joint	Single strap butt joint	Double strap butt joint
1	1.31	1.53	1.75
2	2.62	3.06	3.50
3	3.47	4.05	4.63
4	4.17	–	5.52
5	–	–	6.00

(v) Transverse pitch.

Transverse pitch, $p_t = 2d_0$ for chain riveting ...(5.10 *a*)

$= 0.33p + 0.67\, d_0$ for zig – zag riveting ...(5.10 *b*)

In case of zig – zag riveting, if the outermost row has half the number of rivets of inner rows, the distance of outermost row from the inner row should not be less than $(0.2\,p + 1.15\,d_0)$. The transverse pitch of inner rows having same number of rivets should not be less than $(0.165\,p + 0.67\,d_0)$. In all the above distances, p is the pitch of rivets in the outer rows.

(vi) Thickness of butt strap, t_1.

(*a*) Minimum thickness of butt strap = 10 mm

(*b*) $t_1 = 1.125t$, for ordinary chain riveting with single butt strap. ...(5.11 *a*)

$= 1.125t\left[\dfrac{p-d_0}{p-2d_0}\right]$, for single butt straps, every alternate rivet in outer rows being omitted. ...(5.11 *b*)

$= 0.625\,t$, for double butt straps of equal widths having chain riveting. ...(5.11 *c*)

$= 0.625\,t\left[\dfrac{p-d_0}{p-2d_0}\right]$, for double butt straps of equal width having every alternate rivet in outer rows being omitted.

(*c*) For unequal width of butt straps, the thickness of butt strap will be;

$t_1 = 0.75t$, for wide trap on the inside, and ...(5.12 *a*)

$= 0.625t$, for narrow strap on the outside. ...(5.12 *b*)

(vii) Margin.

$m = 1.5d_0$ ...(5.13)

5.5.2 Circumferential Lap Joint

(i) Thickness of boiler shell and diameter of rivets.

It shall be same as for the longitudinal joint.

(*ii*) **Number of rivets.**

Shearing resistance of the rivets in single shear,

$$P_s = n \times \frac{\pi}{4} \times d^2 \times \tau$$

Total shearing load due to steam pressure, $P_s = \frac{\pi}{4} \times D^2 \times p_s$

$$n \times \frac{\pi}{4} \times d^2{}_0 \times \tau = \frac{\pi}{4} \times D^2 \times p_s$$

$$n = \left(\frac{D}{d_0}\right)\left(\frac{p_s}{\tau}\right) \quad ...(5.14)$$

(*iii*) **Pitch of rivets.**

Let η_c = efficiency of circumferential joint

$= 0.5\ \eta_l$

Then $\eta_c = \frac{p_c - d_0}{p_c}$

(*iv*) **Number of rows.**

$$\text{Number of rows of rivets, N} = \pi \frac{D + t}{p_c} \quad ...(5.15)$$

(*v*) **Overlap.**

$$\text{Overlap} = (N - 1)\ p_b + m$$

Example 5.1 A steam boiler of 1.5 m inside diameter is to be designed for a steam pressure of 2.5 MPa. Design the longitudinal and circumferential joints assuming the following allowable stresses in steel plates and rivets :

Tension= 80 MPa, shear = 60 MPa, crushing = 120 MPa

Solution.

Given : $p_s = 2.5$ MPa, D = 1.5 m, $\sigma_t = 80$ MPa, $\tau = 60$ MPa, $\sigma_c = 120$ MPa

1. Longitudinal Joint

(*i*) **Plate thickness.**

From Table 5.2 we observe that the joint shall be triple riveted diamond butt joint with double strap unequal cover plates and efficiency, $\eta_1 = 80\%$. The thickness of boiler shell plate :

$$t = \frac{p_s D}{2\sigma_t \eta_1} + c$$

$$= \frac{2.5 \times 1.5 \times 10^3}{2 \times 80 \times 0.8} + 1$$

$$= 30.29 \approx 32 \text{ mm}$$

(*ii*) Diameter of rivet hole for $t > 8$ mm, $d_0 = 6\sqrt{t} = 6\sqrt{32} = 33.94 \approx 34.5$ mm

Diameter of rivet, $d = d_0 - 1.5 = 34.5 - 1.5 = 33$ mm

(*iii*) Pitch of rivets.

Tearing resistance of plate per pitch length,

$$P_t = (p - d_0)\, t\, \sigma_t = (p - 34.5) \times 32 \times 80 = 2560\,(p - 34.5)\ \text{N}$$

Number of rivets per pitch length are five out of which four are in double shear and one in single shear.

Shearing resistance of rivets, $P_s = (4 \times 1.875 + 1) \times \frac{\pi}{4} \times 34.5^2 \times 60$

$= 476758$ N

For $P_t = P_s$, we have

$$2560\,(p - 34.5) = 476758$$

$$p = 220.7 \text{ mm}$$

As per IBR, $p_{max} = Ct + 40$

where IBR = Indian Boiler Regulations.

For double strap butt joint with 5 rivets per pitch length, C = 6 (see Table 5.6)

$$p_{max} = 6 \times 32 + 40 = 232 \text{ mm}$$

Adopt $p = 220$ mm

Pitch of rivets in the inner row, $p_i = \frac{p}{2} = 110$ mm.

(*iv*) Distance between rows of rivets.

According to IBR, the distance between the outer row and adjacent row,

$$= 0.2p + 1.15\, d_0$$

$$= 0.2 \times 220 + 1.15 \times 34.5 = 83.5 \approx 85 \text{ mm}$$

Distance between the inner rows for zig – zag riveting,

$$= 0.165p + 0.67d_0$$

$$= 0.165 \times 220 + 0.67 \times 34.5 = 59.4 \approx 60 \text{ mm}$$

(*v*) Thickness of butt straps.

Thickness of wide butt strap, $t_1 = 0.75t = 0.75 \times 32 = 24$ mm

Thickness of narrow butt strap, $t_2 = 0.625\, t = 0.625 \times 32 = 20$ mm

Wide butt strap is put on the inside of boiler shell.

(*vi*) Margin, $m = 1.5d_0 = 1.5 \times 34.5 = 51.75 \approx 52$ mm

(*vii*) Efficiency of joint.

Tearing resistance of plate in the outer row,

$P_t = (p - d_0)\, t\sigma_1 = 220 - 34.5) \times 32 \times 80 = 474880$ N

Shearing resistance of rivets, $P_s = 476758$ N

Crushing resistance of rivets, $P_c = nd_0 t\sigma_c = 5 \times 34.5 \times 32 \times 120 = 662400$ N

Considering tearing of the plate between the rivets in the second row when the rivets in the outermost shears, we have

Combined tearing and shearing resistance

$$= (p - 2d_0)\, t\sigma_t + \frac{\pi}{4} \times d^2{}_0 \times \tau$$

$$= (220 - 2 \times 34.5) \times 32 \times 80 + \frac{\pi}{4} \times 34.5^2 \times 60 = \mathbf{442649\ N}$$

Strength of the joint $= 442649$ N

Strength of solid plate, P $= pt\sigma_t = 220 \times 32 \times 80 = 563200$ N

Efficiency of joint, $\eta_l = \dfrac{442649}{563200} = 0.786$ or 78.6% ≈ 80% as assumed.

2. Circumferential Joint

(*i*) $t = 32$ mm, $d_0 = 34.5$ mm, $d = 33$ mm

(*ii*) Number of rivets.

$$n = \left(\frac{D}{d_0}\right)^2 \left(\frac{p_s}{\tau}\right) = \left(\frac{1500}{34.5}\right)^2 \left(\frac{2.5}{60}\right) = 78.7 \approx 80$$

(*iii*) Pitch of rivets.

Assuming double riveted lap joint with zig – zag riveting, number of rivets per row

$$= \frac{n}{2} = \frac{80}{2} = 40$$

Pitch of rivets, $p_c = \dfrac{\pi(D + t)}{\text{Number of rivets per row}}$

$$= \frac{\pi(1500 + 32)}{40} = 120.32 \approx 120 \text{ mm}$$

(*iv*) Efficiency of joint, $\eta_c = \dfrac{p_c - d_0}{p_c} = \dfrac{120 - 34.5}{120} = 0.7125$ or 71.25%

(*v*) Distance between rows of rivets.

For zig – zag riveting, distance between the rows of rivets

$= 0.33p_c + 0.67\, d_0$

$= 0.33 \times 120 + 0.67 \times 34.5 = 62.7 \approx 65$ mm.

(*vi*) Margin, $m = 1.5\, d_0 = 1.5 \times 34.5 = 51.75 \approx 52$ mm

5.6 STRUCTURAL RIVETED JOINT - LOZENGE JOINT

The triple riveted double strap butt Lozenge joint shown in Fig. 5.7. It is a joint of uniform strength.

Let w = width of plate

t = thickness of plate

d_0 = diameter of rivet hole.

The sizes of rivets for general purposes are given in Table 5.7. Permissible stresses for aluminium and steel rivets are given in Table 5.8.

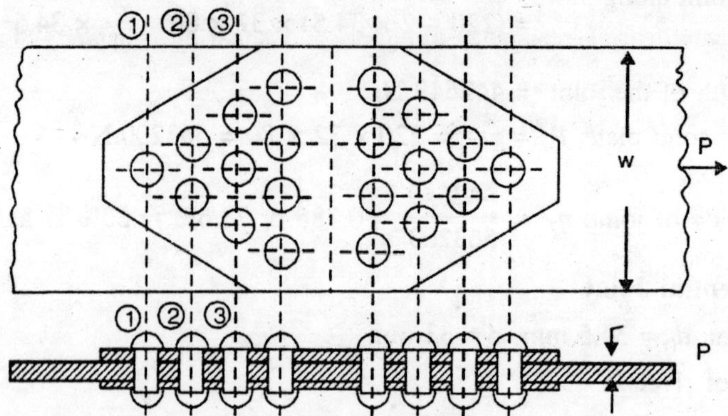

Fig. 5.7 "Lozenge" or diamond joint.

Table 5.7 Sizes of rivets for general purposes

Diameter of rivet, *d*, mm	12, 14, 16, 18, 20, 22, 24, 27, 30, 33, 36, 39, 42 and 45

Diameter of rivet hole, mm :

$$d_0 = d + 1.5 \text{ mm for } 12 \text{ mm} \le d \le 24 \text{ mm}$$
$$= d + 2 \text{ mm for } 27 \text{ mm} \le d \le 50$$

(*i*) Diameter of rivets.

Diameter of river hole, $d_0 = 6\sqrt{t}$

Diameter of rivets, $d = d_0 - \text{clearance}$

(*ii*) Number of rivets.

Tearing resistance of the plate in the outermost row, $P_t = (w - d_0)\, t\sigma_1$

Shearing resistance of rivets, $P_s = 1.75 \times \frac{\pi}{4} \times d^2{}_0 \times \tau$

Crushing resistance of one rivet, $P_c = dt\sigma_c$

Number or rivets, $n = \dfrac{P_t}{\text{Least of } P_s \text{ and } P_c}$

(*iii*) Thickness of butt straps.

$$t_1 = 1.25t \text{ for single cover strap}$$
$$= 0.75 \text{ for double cover strap}$$

(*iv*) Efficiency of joint.

At section 1 – 1, there is only one rivet hole.

Resistance of the joint against tearing along 1 – 1,

$$P_1 = (w - d_0)\, t\sigma_t$$

Resistance of joint along section 2 – 2,

$$P_2 = \text{tearing resistance along } 2 - 2 + \text{shearing resistance of one rivet at } 1 - 1$$

$$= (w - 2d_0)\, t\sigma_t + 1.75 \times \frac{\pi}{4} \times \tau$$

Resistance of joint along 3 – 3,

p_3 = tearing resistance along 3 – 3 + shearing resistance of one rivet at 1 – 1

$$= (w - 3d_0)\, t\sigma_t + 3 \times 1.75 \times \frac{\pi}{4} \times \tau$$

Strength of solid plate, $P = wt\sigma_t$

Shear resistance of all rivets, $P_s = n \times 1.75 \times \frac{\pi}{4} \times d^2_0 \times \tau$

Crushing resistance of all rivets, $P_c = ndt\sigma_c$

Efficiency of joint, η = Least of P_1, P_2, P_3, P_s and $\frac{P_c}{P}$

(*v*) Pitch of rivets.

$p = 3d_0 + 5$ mm

(*vi*) Margin, $m = 1.5d_0$

(*vii*) Distance between rows of rivets = $2.5d_0$

Table 5.8 Permissible stresses for aluminium and steel rivets in N/mm².

Material	Process	Shear	Tension
Aluminium			
64430-W	Cold driven	62	Non recommended
53000-O	Cold driven	54	-do-
	Hot driven	49	-do-
Steel	Power driven (shop)	100	77
	Power driven (field)	93	62

IS : 8147-1976

Example 5.2 A tie bar consisting of flat 400 mm wide and 20 mm thick is connected to a gusset plate of the same thickness by a double cover butt joint. Design an economical joint assuming the following permissible stresses :

Tension = 90 MPa, Shear = 60 MPa and crushing = 140 MPa.

Solution.

Given : w = 400 mm, t = 20 mm, σ_t = 90 MPa, τ = 60 MPa, σ_c = 140 MPa

For the design to be economical, the Lozenge joint is chosen.

1. Diameter of rivet.

$$d_0 = 6\sqrt{t} = 6\sqrt{20} = 26.83 \text{ mm}$$

The standard diameter of rivet hole, d_0 = 29 mm and the corresponding diameter of rivet, d = 27 mm.

2. Number of rivets.

Maximum tearing strength of plate, $P_t = (w - d_0)\, t\sigma_t$

$= (400 - 29) \times 20 \times 90 = 667800$ N

Shearing resistance of one rivet in double shear, $P_s = 1.75 \times \frac{\pi}{4} \times d^2_0 \times \tau$

$$= 1.75 \times \frac{\pi}{4} \times 29^2 \times 60 = 69355 \text{ N}$$

$$\text{Number of rivets, } n = \frac{P_t}{P_s} = \frac{667800}{69355} = 9.628 \approx 10$$

3. The arrangement of rivets is shown in Fig. 5.8.
4. Thickness of butt straps.

   Inner strap, $t_1 = 0.75t = 0.75 \times 20 = 15$ mm

   Outer strap, $t_2 = t_1 = 15$ mm
5. Efficiency of joint.

   Section 1 – 1 :

$$P_1 = (w - d_0)\, t\sigma_t$$
$$= (400 - 29) \times 20 \times 90$$
$$= 667800 \text{ N}$$

Section 2 – 2 :

$$P_2 = (w - 2d_0)\, t\sigma_t + \frac{\pi}{4} \times d^2_0 \times \tau$$
$$= (400 - 2 \times 29) \times 20 \times 90 + 1.75 \times \frac{\pi}{4} \times 29^2 \times 60$$
$$= 684955 \text{ N}$$

Section 3 – 3 :

$$P_3 = (400 - 3 \times 29) \times 20 \times 90 + 3 \times 69355$$
$$= 771465 \text{ N}$$

Section 4 – 4 :

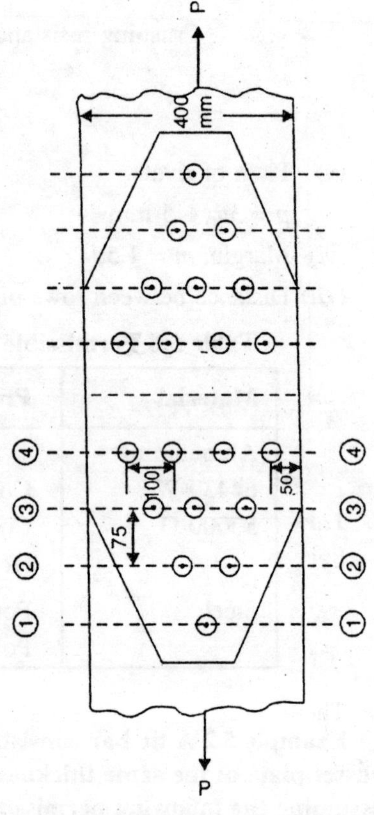

Fig. 5.8

$$P_4 = (400 - 4 \times 29) \times 20 \times 90 + 6 \times 69355 = 927330 \text{ N}$$

Shearing resistance of all the rivets, $P_s = 10 \times 69355 = 693550$ N

Crushing resistance of all the rivets, $P_c = 10 \times 29 \times 20 \times 140 = 812000$ N

Strength of the joint = $P_1 = 667800$ N

Strength of solid plate, $P = wt\sigma_t = 400 \times 20 \times 90 = 720000$ N

$$\text{Efficiency of joint, } \eta = \frac{667800}{720000} = 0.9275 \text{ or } 92.75\%.$$

6. Pitch of rivets, $p = 3d_0 = 3 \times 29 + 5 = 92 \approx 100$ mm
7. Margin, $m = 1.5d_0 = 1.5 \times 29 = 43.5 \approx 50$ mm

   $w = 3p + 2m = 3 \times 100 + 2 \times 50 = 400$ mm
8. Distance between the rows = $2.5d_0 = 2.5 \times 29 = 72.5 \approx 75$ mm

5.7 ECCENTRICALLY LOADED RIVETED JOINT

Consider the eccentrically loaded riveted joint, as shown in Fig. 5.9 (*a*) Let

P = eccentric load

e = eccentricity of load

n = number of rivets in the group

The following procedure may be adopted for the design of eccentrically loaded riveted joint :

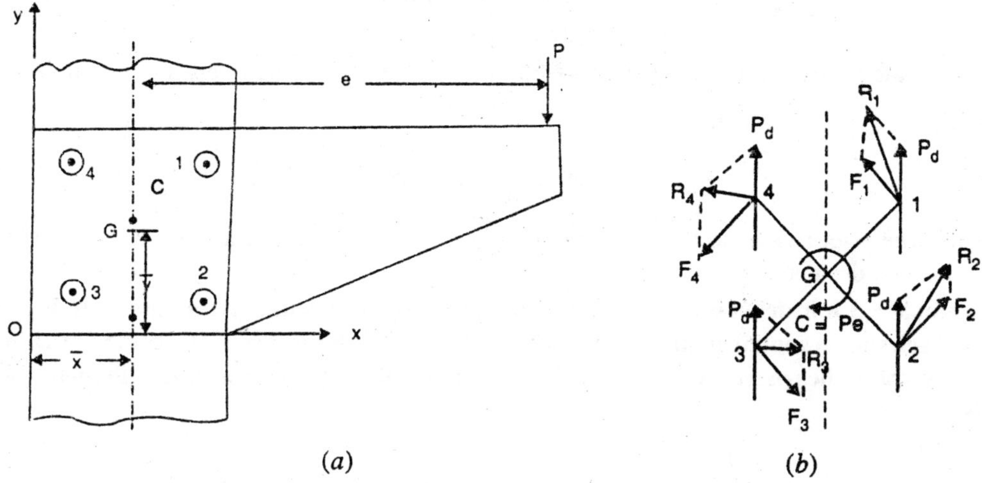

Fig. 5.9 Eccentrically loaded riveted joint

1. Locate the centre of gravity of the rivet group.

Let A = area of cross – section of each rivet

x_i = distance of ith rivet from y-axis

y_i = distance of ith rivet from x-axis

Then

$$x = \frac{A \Sigma x_i}{(n A)}$$

$$y = \frac{A \Sigma y_i}{(n A)}$$

The forces acting on the rivets are shown in Fig. 5.9 (*b*).

2. Direct primary shear force in each rivet, $P_d = \dfrac{P}{n}$

3. Couple about C.G., $C = P \times e$

4. Let F_1 = secondary shear force in each rivet

l_i = perpendicular distance of F_i from C.G.

The secondary shear force in each rivet is proportional to the perpendicular distance from the C.G. Therefore, we have

$$\frac{F_1}{l_1} = \frac{F_2}{l_2} = \text{.........} = \frac{F_i}{l_i}$$

5. Equate the couple to the resisting moment of all the rivets.

$$C = \Sigma F_i l_i$$

$$= \left(\frac{F_1}{l_1}\right) \Sigma l_i^2$$

6. Calculate the resultant shear force in each rivet due to the primary shear and secondary shear.

$$R_i = [P_d^2 + F_i^2 + 2P_d F_i \cos \theta_i]^{1/2}$$

Where θ_i = angle between P_d and F_i

7. Find the maximum resultant shear force in a rivet. Then, as the rivets are in double shear, we have

$$R_{max} = 1.75 \times \frac{\pi}{4} \times d_0^2 \times \tau$$

Note. The rivet nearest to the load line shall be subjected to maximum resultant shear force, which may be designed.

Example 5.3 A bracket is riveted to a column by 6 rivets of same size as shown in Fig. 5.10 (*a*). It carries an eccentric load of 54 kN at a distance of 200 mm from the centre of column. If the allowable shear stress in the rivets is limited to 120 MPa, determine the diameter of the rivets.

Solution.

Given : $n = 6$, $P = 54$ kN, $e = 200$ mm, $\tau = 120$ MPa

Primary or direct shear load in the rivets, $P_d = \frac{54}{6} = 9$ kN

The C.G. of the rivet group lies in the centre. The forces acting on the rivets are shown in Fig. 5.10 (*b*).

Couple, $C = P \times e = 54 \times 200 = 10800$ kN. mm

$$l_1 = l_3 = l_4 = l_6 = [50^2 + 100^2]^{1/2} = 111.8 \text{ mm}$$

$$l_2 = l_5 = 50 \text{ mm}$$

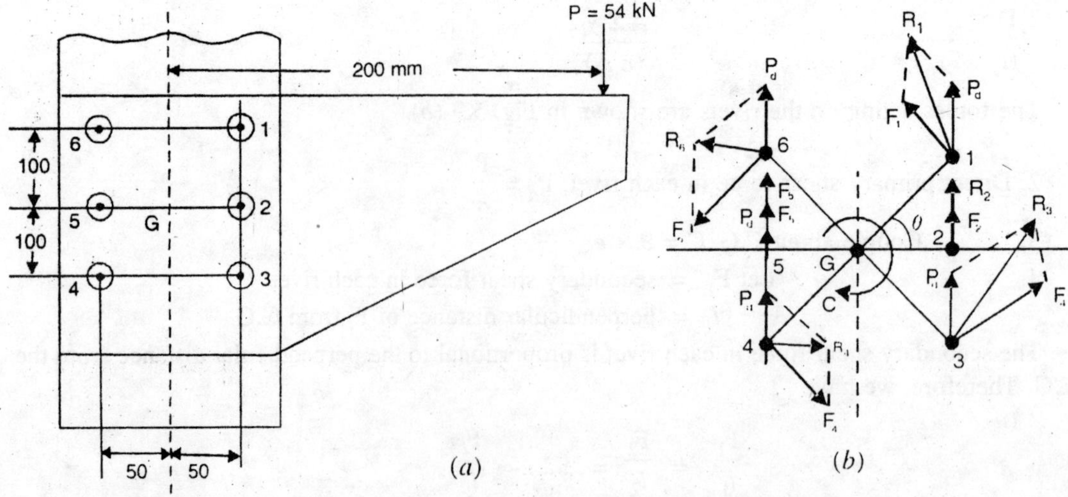

Fig. 5.10

$$\left(\frac{F_1}{l_1}\right)[4l^2_1 + 2l^2_2] = C$$

$$\left(\frac{F_1}{111.8}\right)[4 \times (111.8)^2 + 2 \times 50^2] = 10800 \times 10^3$$

$$F_1 = 21955 \text{ N}$$

Rivet 1 is nearest to the load line and shall be subjected to maximum shear load.

$$\cos\theta_1 = \frac{50}{111.8} = 0.44723$$

$$\theta_1 = 63.434°$$

Resultant force in rivet 1,

$$R_1 = [P_d^2 + F_1^2 + 2P_dF_1\cos\theta_1]^{1/2}$$

$$= [9000^2 + 21955^2 + 2 \times 9000 \times 21955 \times 0.44723]^{1/2}$$

$$= 27198 \text{ N}$$

$$= \frac{\pi}{4} \times d^2_0 \times 120$$

$$d_0 = 16.98 \approx 17.5 \text{ mm}$$

$$d = 16 \text{ mm}$$

Example 5.4 Design a double riveted butt joint with two cover plates for longitudinal seam of a boiler shell, 0.75 m diameter, to carry a maximum steam pressure of 1.55 N/mm^2. Assume joint efficiency as 75%, allowable tensile stress in the plate 90 MPa, compressive stress in plate and rivet 140 MPa, and shear stress in plate and rivet 56 MPa.

Solution.

Given : $D = 0.75$ m, $p_s = 1.55 \dfrac{\text{N}}{\text{mm}^2}$, $\eta_l = 0.75$, $\sigma_t = 90$ MPa, $\sigma_c = 140$ MPa, $\tau = 56$ MPa

Design of Longitudinal joint

(*i*) Plate thickness.

The thickness of boiler shell plate :

$$t = \frac{p_s D}{2\sigma_t \eta_1} + c$$

$$= \frac{1.55 \times 0.75 \times 10^3}{2 \times 90 \times 0.75} + 1$$

$$= 9.61 \approx 10 \text{ mm}$$

(*ii*) Diameter of rivet hole for $t > 8$ mm, $d_0 = 6\sqrt{t} = 6\sqrt{10} = 18.97 \approx 19.5$ mm

Diameter of rivet, $d = d_0 - 1.5 = 19.5 - 1.5 = 18$ mm

(*iii*) Pitch of rivets.

Tearing resistance of plate per pitch length,

$$P_t = (p - d_0)\, t\, \sigma_t = (p - 19.5) \times 10 \times 90 = 900\,(p - 19.5) \text{ N}$$

Number of rivets per pitch length are 2, which are in double shear.

Shearing resistance of rivets, $P_s = 2 \times 1.875 \times \frac{\pi}{4} \times 19.5^2 \times 56$

$$= 61716 \text{ N}$$

For $P_t = P_s$

$$900\,(p - 19.5) = 61716$$

$$p = 89.18 \text{ mm}$$

As per IBR, $p_{max} = Ct + 40$

For double strap butt joint wth 2 rivets per pitch length, C = 3.5

$$p_{max} = 3.5 \times 10 + 40 = 75 \text{ mm}$$

Adopt $p = 75$ mm.

(*iv*) Distance between rows of rivets.

According to IBR, distance between the rows for zig – zag riveting,

$$= 0.33\,p + 0.67\,d_0$$

$$= 0.33 \times 75 + 0.67 \times 19.5 = 37.8 \approx 40 \text{ mm}$$

(*v*) Thickness of butt straps.

Thickness of butt strap, $t_1 = 0.625\,t = 0.625 \times 10 = 6.25 \approx 7$ mm

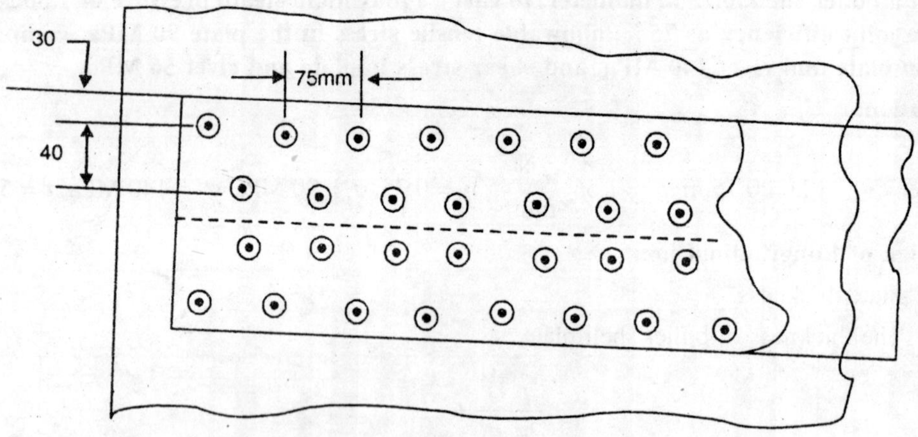

Fig. 5.11 Double riveted zig-zag double strap longitudinal butt joint.

(*vi*) Margin, $m = 1.5d_0 = 1.5 \times 19.5 = 29.25 \approx 30$ mm

(*vii*) Efficiency of joint.

Tearing resistance of plate in the outer row,

$$P_t = (p - d_0)\, t\sigma_t = (75 - 19.5) \times 10 \times 90 = 49950 \text{ N}$$

Shearing resistance of rivets, $P_s = 61716$ N

Crushing resistance of rivets, $P_c = nd_0\, t\sigma_c = 2 \times 19.5 \times 10 \times 140 = 54600$ N

Strength of the joint – 49950 N

Strength of solid plate, P= $pt\sigma_t = 75 \times 10 \times 90 = 67500$ N

$$\text{Efficiency of joint, } \eta_1 = \frac{49950}{67500} = 0.74 \text{ or } 74\% \approx 75\% \text{ as given}$$

The joint is shown in Fig. 5.11.

Example 5.5 A bracket is attached to a column by means of five rivets as shown in Fig. 5.12. Find the suitable diameter of rivets for the riveted connection. Assume suitable values of permissible stresses.

Solution.

Given : $n = 5$, P = 45 kN, $\alpha = 30°$

$$\text{Primary or direct shear load in the rivets, } P_d = \frac{45}{5} = 9 \text{ kN}$$

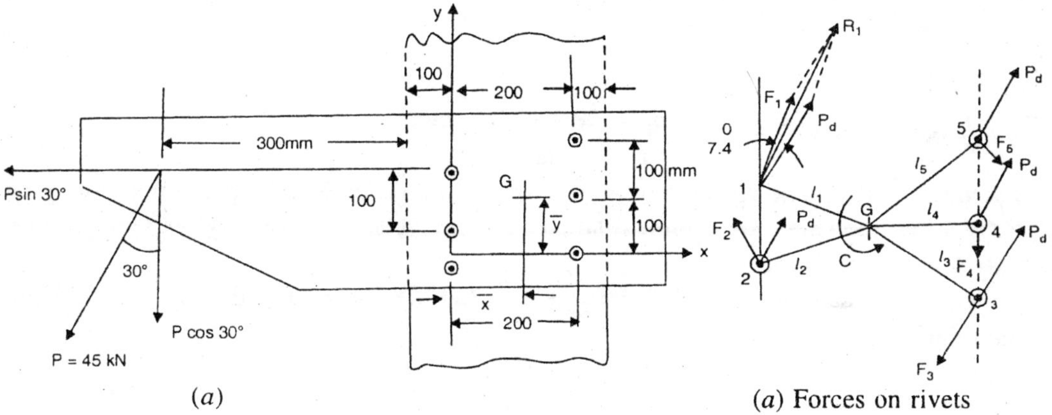

(*a*) (*a*) Forces on rivets

Fig. 5.12

The C.G. of the rivet group :

$$5x = 3 \times 200$$
$$x = 120 \text{ mm}$$
$$5y = 1 \times 50 + 1 \times 150 + 1 \times 100 + 1 \times 200 = 500$$
$$y = 100 \text{ m}$$
$$e_x = 300 + 100 + 120 = 520 \text{ mm}$$
$$e_y = 100 \text{ mm}$$
$$l_1 = (120^2 + 50^2)^{1/2} = 130 \text{ mm}$$
$$l_2 = l_1$$
$$l_3 = (100^2 + 80^2)^{1/2} = 128.06 \text{ mm}$$
$$l_4 = 80 \text{ mm}$$
$$l_5 = l_3$$

$$2F_1\, l_1 + 2F_3 l_3 + F_4\, l_4 = P \cos \alpha \times e_x + P \sin \alpha \times e_y$$

$$\left(\frac{F_1}{130}\right) \quad [2 \times 130^2 + 2 \times (128.06)^2 + 80^2]$$

$$= [45 \cos 30° \times 520 + 45 \sin 30° \times 100] \times 10^3$$

$$F_1 = 40096 \text{ N}$$

Rivet 1 is nearest to the load line and shall be subjected to maximum shear load.

$$\tan \beta = \frac{120}{50} = 2.4$$

$$\beta = 67.4°$$

$$\theta_1 = \beta - (90° - \alpha)$$

$$= 67.4 - (90° - 30°) = 7.4°$$

$$\cos \theta_1 = \cos 7.4° = 0.99167$$

Resultant force in rivet 1,

$$R_1 = [P^2{}_d + F^2{}_1 + 2\, P_d F_1 \cos \theta_1]^{1/2}$$

$$= [9000^2 + 40096^2 + 2 \times 9000 \times 40096 \times 0.99167]^{1/2}$$

$$= 49035 \text{ N}$$

Let allowable shear stress in rivets, τ = 120 MPa

$$49035 = \frac{\pi}{4} \times d^2{}_0 \times 120$$

$$d_0 = 20.4 \approx 21.5 \text{ mm}$$

$$d = 20 \text{ mm}$$

Example 5.6 A double riveted lap joint with chain riveting is to be made for joining two plates of 6 mm thickness. Determine diameter of the rivets, distance between rows of rivets and pitch of rivets.

Indicate how the joint will fail ? Assume σ_t = 220 MPa, τ = 100 MPa, σ_c = 150 MPa.

Solution.

Given : t = 6 mm, σ_t = 220 MPa, τ = 100 MPa, σ_c = 150 MPa.

There are 2 rivets per pitch length and they are in single shear.

Shearing strength of rivets, $P_s = 2 \times \frac{\pi}{4} \times d^2{}_0 \times \tau$

Crushing strength of rivets, $P_c = 2 d_0 t \sigma_c$

Equating shearing and crushing strength, we have

$$d_0 = \left(\frac{4t}{\pi}\right)\left(\frac{\sigma_c}{\tau}\right)$$

$$= \left(\frac{4 \times 6}{\pi}\right)\left(\frac{150}{100}\right) = 11.46 \approx 13.5 \text{ mm}$$

Diameter of rivet, d = 12 mm

Tearing strength of plate per pitch length, $P_t = (p - d_0)\, t\sigma_t = (p - 13.5) \times 6 \times 220$

Shearing strength of rivets per pitch length, $P_s = 2 \times \frac{\pi}{4} \times (13.5)^2 \times 100$

For $P_t = P_s$, we have

$$(p - 13.5) \times 6 \times 220 = 2 \times \frac{\pi}{4} \times (13.5)^2 \times 100$$

$$p = 35.2 \approx 35 \text{ mm}$$

Back pitch, $p_b = 0.6p = 0.6 \times 35 = 21$ mm

Tearing strength of plate per pitch length, $P_t = (35 - 13.5) \times 6 \times 220 = 28380$ N

Shearing strength of rivets per pitch length, $P_s = 28628$ N

Crushing strength of rivets per pitch length, $P_c = 2 \times 13.5 \times 6 \times 150 = 24300$ N

Strength of joint = 24300 N

Strength of solid plate per pitch length,

$$P = pt\sigma_t = 35 \times 6 \times 220 = 46200 \text{ N}$$

$$\text{Efficiency of joint} = \frac{24300}{46200} = 0.526 \text{ or } 52.6\%$$

The joint shall fail in crushing of rivet in the rivet hole.

Example 5.7 A bracket is supported by means of 4 rivets of same size as shown in Fig. 5.13. Determine the diameter of rivets if the maximum permissible shear stress is 150 N/mm².

Solution.

Given : P = 20 kN, e = 80 mm, τ = 150 MPa

The C.G. lies at 45 mm from top rivet.

Direct shear force, $P_d = \frac{P}{4} = \frac{20}{4} = 5$ kN

Couple, $C = P \times e = 20 \times 80 = 1600$ kN. mm

$$l_1 = l_4 = 45 \text{ mm}, l_2 = l_3 = 15 \text{ mm}$$

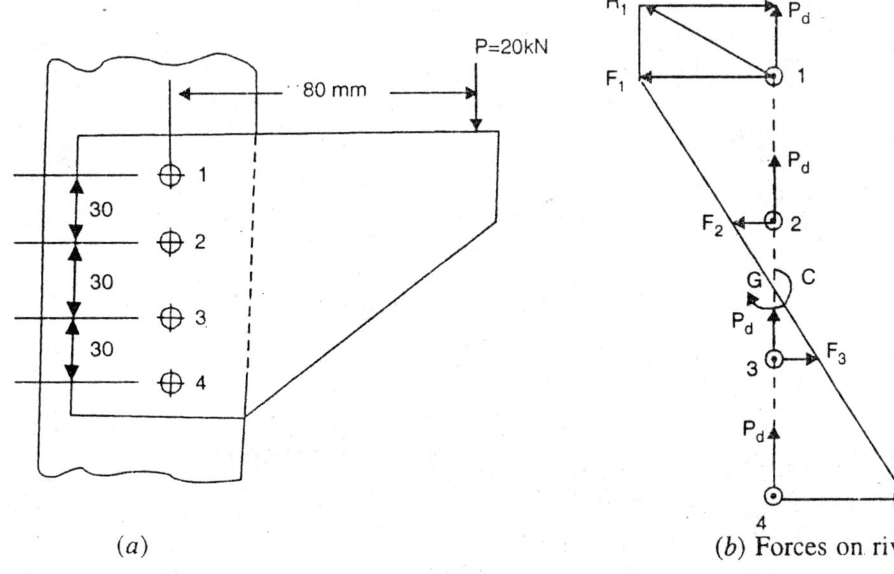

(a) (b) Forces on rivets

Fig. 5.13

$$\left(\frac{F_1}{45}\right)[2 \times 45^2 + 2 \times 15^2] = 1600 \times 10^3$$

$$F_1 = 16,000 \text{ N}$$

Rivet 1 is nearest to the load line and shall be subjected to maximum shear stress.

$$R_1 = [P_d^2 + F^2{}_1]^{1/2} = [5000^2 + 16,000^2]^{1/2} = 16763 \text{ N}$$

$$= \frac{\pi}{4} \times d^2{}_0 \times 150$$

$$d_0 = 11.928 \approx 13.5 \text{ mm}$$

$$d = 12 \text{ mm.}$$

EXERCISES

5.1 Design a double riveted butt joint with two cover plates for longitudinal seam of a boiler shell, 1.2 m diameter, to carry a maximum steam pressure of $\frac{1.5 \text{ N}}{\text{mm}^2}$. Assume joint efficiency as 78%, allowable tensile stress in the plate 90 MPa, compressive stress in plate and rivet 140 MPa, and shear stress in plate and rivet 56 MPa.

5.2 What are the various modes in which a riveted joint may fail ?

5.3 A bracket is attached to a column by means of seven rivets as shown in Fig. 5.14. Find the suitable diameter of rivets for the riveted connection. The bracket plate is 20 mm thck. Permissible shear stress is 60 MPa and crushing stress is 120 MPa.

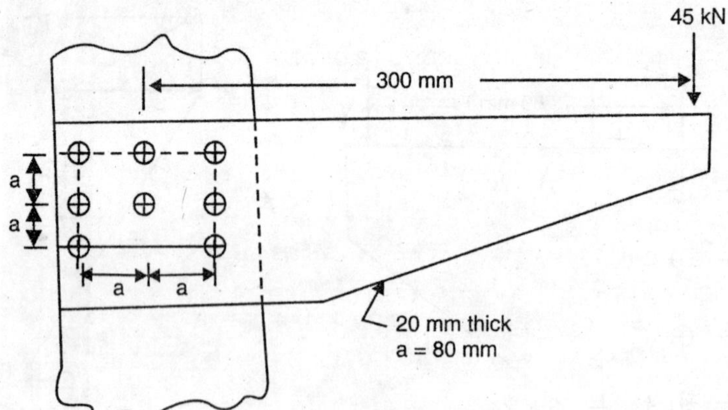

Fig. 5.14

5.4 A double riveted lap joint with chain riveting is to be made for joining two plates of 6 mm thickness. Determine diameter of the rivets, distance between rows of rivets and pitch of rivets. Indicate how the joint will fail ? Assume $\sigma_t = 200$ MPa, $\tau = 100$ MPa, $\sigma_c = 140$ MPa.

5.5 A bracket is supported by means of 9 rivets of same size as shown in Fig. 5.15. Determine the diameter of rivets if the maximum permissible shear stress is $\frac{45\text{ N}}{\text{mm}^2}$.

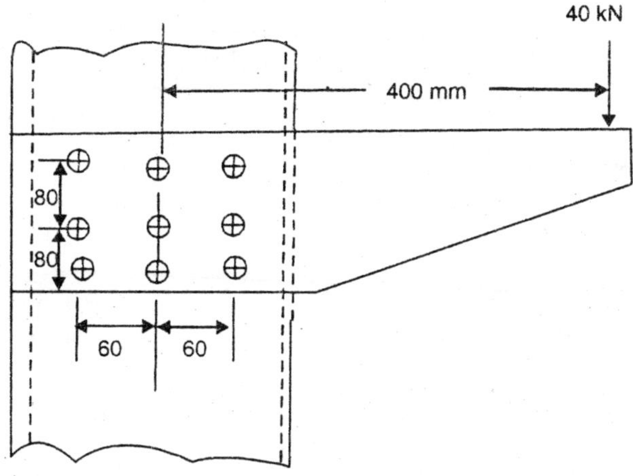

Fig. 5.15

5.6 Calculate the main dimensions of the longitudinal and circumferential joints of a steam boiler whose inner diameter is 2.5 m and the pressure of steam is 20 bar. Assume suitable data from the Data Handbook.

5.7 Two steel plates 250 mm × 10 mm are connected by a diamond butt riveted joint with single strap. The plates are required to carry 250 kN load. The allowable stresses are 120 MPa in tension, 80 MPa in shear and 210 MPa in crushing. Design the joint.

5.8 Design an economical butt joint with two cover straps to connect two steel plates 300 mm × 12 mm. Assume suitable data from the Data Book.

5.9 A plate is joined to a structure member as shown in Fig. 5.16. The cross – section area of each rivet is 315 mm^2. Estimate the effect on the maximum shearing stress induced in the rivets if rivet D is not used.

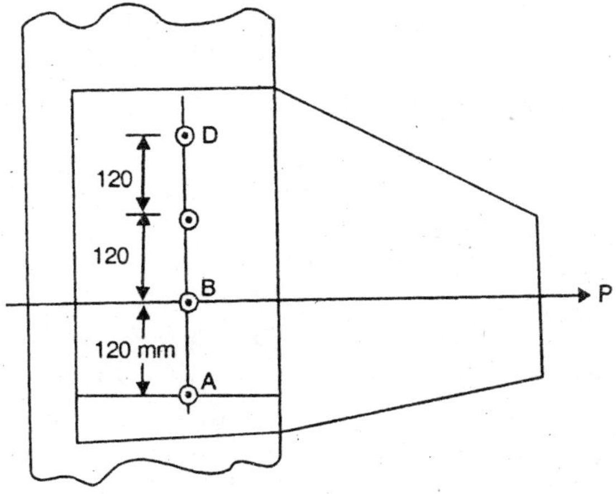

Fig. 5.16

5.10 A steel plate 20 mm thick is attached to a girder by means of four rivets of 20 mm diameter as shown in Fig. 5.17. Determine the distance 'e' which permits the application of a 50 kN load to limit the shear stress to 45 MPa in rivets and crushing stress to 70 MPa in rivets and plate.

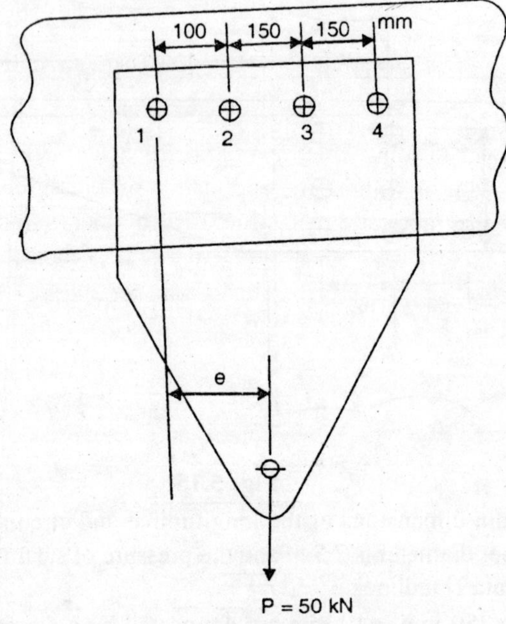

Fig. 5.17

6 WELDED JOINTS

Chapter

6.1 INTRODUCTION

A welded joint is a permanent joint which is obtained by the fusion of the edges of the two parts to be joined together, with or without the application on pressure and filler metal. Welded joints are widely used for the fabrication of structures, boilders, ship building and machine components.

6.2 ADVANTAGES AND DISADVANTAGES OF WELDED JOINTS

The advantages and disadvantages of welded joints over riveted joints are :

Advantages

1. Lighter in weight.
2. Easier to make alterations and additions in existing structures.
3. Higher efficiency.
4. Gives pleasing look, being smoother in appearance.
5. No weakening effect due to drilling of holes.
6. Higher strength and rigidity.
7. Faster process.

Disadvantages

1. Introduces distortion and residual stresses due to heating.
2. Requires highly skilled labour and supervision.
3. Possibility of cracks due to excessive expansion.
4. More difficult process.

6.3 TYPES OF WELDED JOINTS

There are two types of welded joints as explained below :

Lap (or fillet) joint. It is obtained by overlapping the plates and then welding the edges of the plates.

The cross-section of the fillet weld is approximately a triangle. If h is the side of the weld triangle then throat thickness. $t = 0.707h$ in a simple weld and $0.85t$ in a reinforced weld. h is called the size of the weld.

Butt joint. It is obtained by placing the plates edge to edge and then welding. The plate edges may be square, single-V, double-V, single-U and double-U

The various types of welded joints are shown in Fig. 6.1.

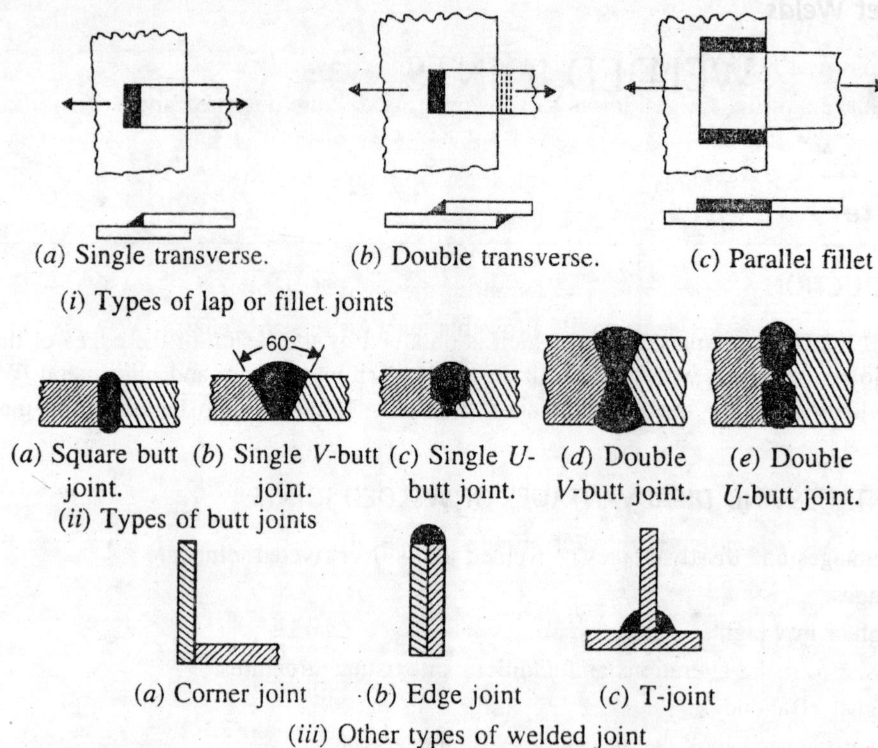

(a) Single transverse. (b) Double transverse. (c) Parallel fillet

(i) Types of lap or fillet joints

(a) Square butt joint. (b) Single V-butt joint. (c) Single U-butt joint. (d) Double V-butt joint. (e) Double U-butt joint.

(ii) Types of butt joints

(a) Corner joint (b) Edge joint (c) T-joint

(iii) Other types of welded joint

Fig. 6.1 Types of welded joints

6.4 STRESSES IN WELDED JOINTS

The stresses for welded joints are given in Table 6.1.

Table 6.1 Stresses for welded joints.

Type of weld	**Bare electrode**		**Coated electrode**	
	Steady load MPa	Fatigue load MPa	Steady load MPa	Fatigue load MPa
Fillet weld (all types)	80	21	98	35
Butt welds :				
Tension	90	35	110	55
Compression	100	35	125	55
Shear	55	21	70	35

6.5 DESIGN OF WELDED JOINTS

Assumptions. The following assumptions are made in the design of welded joints :

1. The load is distributed uniformly over the entire length of the weld.
2. The cross-section of the fillet weld is a right angled triangle with equal sides of length h.
3. The minimum section is the hypotenuse with throat thickness, $t = 0.707h$.
4. The area which resists the applied load is always the throat area.

6.5.1 Fillet Welds

1. Transverse Fillet Welds

The transverse fillet welded joints as shown in Fig. 6.2 are designed under tension. Let

h = leg or size of the weld
= thickness of plate
t = throat thickness
= $h \sin 45° = 0.707h$
l = length of the weld
σ_t = allowable tensile stress in weld
P = applied tensile force

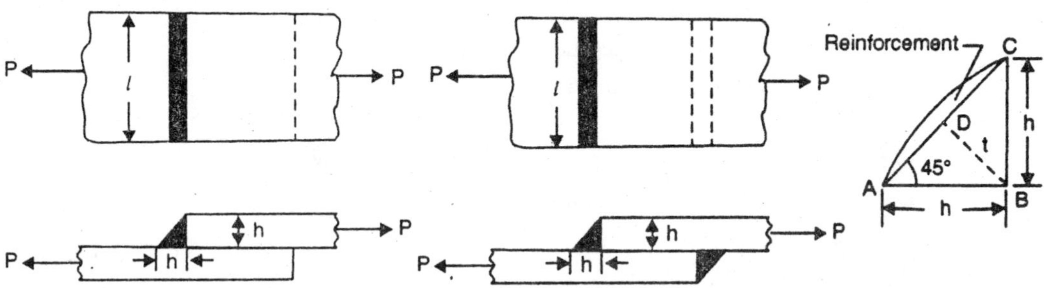

(*a*) Single transverse fillet weld (*b*) Double transverse fillet weld (*b*) Enlarged view of fillet weld

Fig. 6.2 Transverse fillet welds

Minimum area of weld, $A = t \times l = 0.797\ hl$

Tensile strength of single fillet weld (Fig. 5.13a),

$$P = A\sigma_t = 0.707h \times l \times \sigma_t \quad \text{... (6.1a)}$$

For double fillet weld (Fig. 5.13b),

$$P = 2A\sigma_t = 1.41h \times l \times \sigma_t \quad \text{... (6.1b)}$$

Fig. 5.13(c) shows the reinforced fillet weld.

2. Parallel Fillet Welds

The parallel fillet welded joints as shown in Fig. 6.3(a) are designed under shear. For the double parallel fillet weld, we have

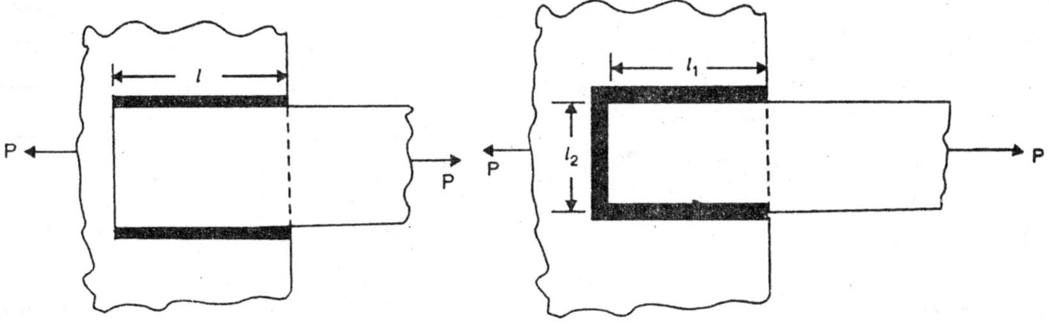

(*a*) Double parallel fillet weld (*b*) Combination of transverse and parallel filled welds.

Fig. 6.3

$$P = 2A_{\tau} = 1.414h \times l \times t \qquad ...(6.2)$$

For the combined single transverse and double parallel fillet welds as shown in Fig. 6.3(b), we have

$$P = 0.707h \times l_t \times \sigma_t + 1.414h \times l_2 \times t \qquad ...(6.3)$$

3. Circular Fillet Weld under Torsion

A circular fillet weld under torsion is shown in Fig. 6.4. Let

d = diameter of the rod

T = torque applied to the rod

J = polar moment of inertia of weld section

$= \pi\,[(d + 2t)^4 - d^4] / 32$

$= \pi\,[\{(d + 2t)^2 + d^2\}\,\{(d + 2t)^2 - d^2\}] / 32$

$= \pi\,[(2d^2 + 4t^2 + 4dt)\,(4t^2 + 4dt)] / 32$

$= \pi\,[(2d^2 + 4dt)\,(4dt)] / 32$, neglecting t^2 terms

$= \pi\,[8d^3t] / 32$, neglecting t^2 term

$= \pi t d^3/4$

$z_p = 2J/d = \pi t d^2/2$

Shear stress induced in the rod material,

$\tau = T/z_p = 2T/(\pi t d^2)$

$= 2T/(\pi \times 0.707h \times d^2)$

$= 2.83T/(\pi \times h \times d^2) \qquad ...(6.4)$

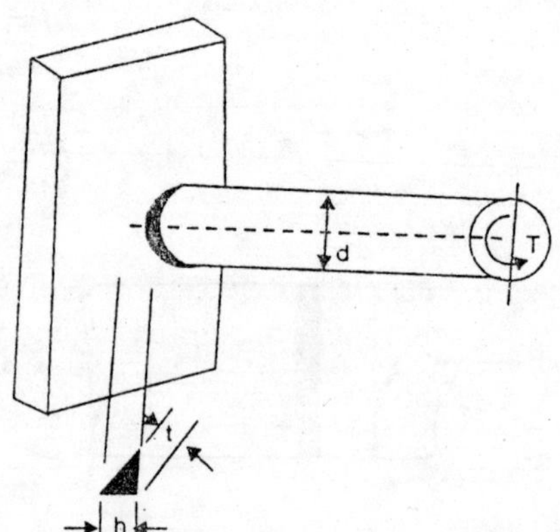

Fig. 6.4 Circular fillet weld under torsion

4. Circular Fillet Weld under Bending Moment

A circular fillet weld under bending is shown in Fig. 6.5 Let

d = diameter of the rod

M = bending moment applied to the rod

$= Pl$

z = section modulus of weld section

$= \pi t d^2/4$

Bending stress induced in the rod material,

$\sigma_b = M/z = 4M/(\pi t d^2)$

$= 4M/(\pi \times t \times d^2)$

$= 4M/(\pi \times 0.707h \times d^2)$

$= 5.66M/(\pi \times h \times d^2) \qquad ...(6.5)$

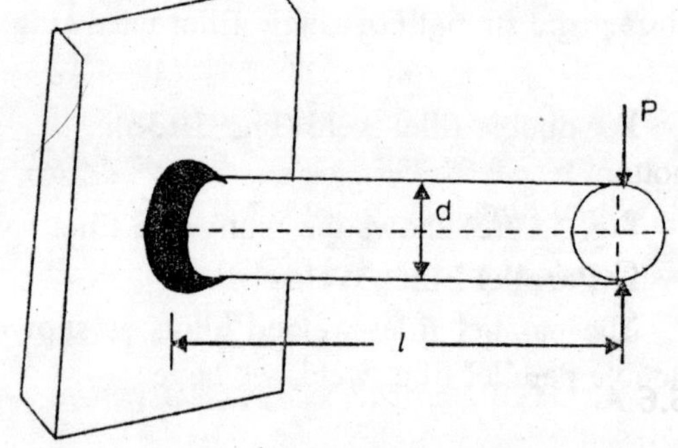

Fig. 6.5 Circular fillet weld under bending

5. Long Fillet Weld under Torsion

Consider a vertical plate welded to a horizontal plate by two indentical fillet welds as shown in Fig. 6.6 and subjected to torque T.

Polar moment of inertia of weld section,

$J = 2 \times t \times l^3/12 = t \times l^3/6$

Shear stress in weld,

$t = T \times (1/2)/(t \times l^3/6)$

$= 3T/(t \times l^2)$

$= 3T/(0.707h \times l^2)$

$= 4.242T/(h \times l^2) \qquad ...(6.6)$

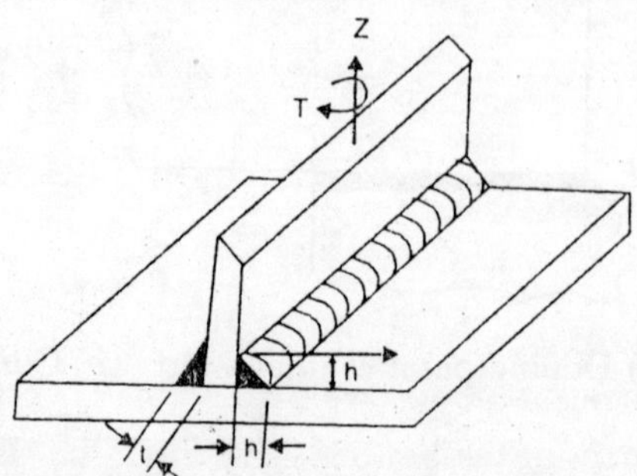

Fig. 6.6 Long fillet weld under torsion

6.5.2 Butt Welds

1. Butt Welds under Tension

(*i*) **Single V-butt joint.** Consider the single V-butt welded joint as shown in Fig. 6.7(a).

Tensile strength of the weld, $P = t \times l \times \sigma_t$ ... (6.7)

Where l = length of weld

= width of plate

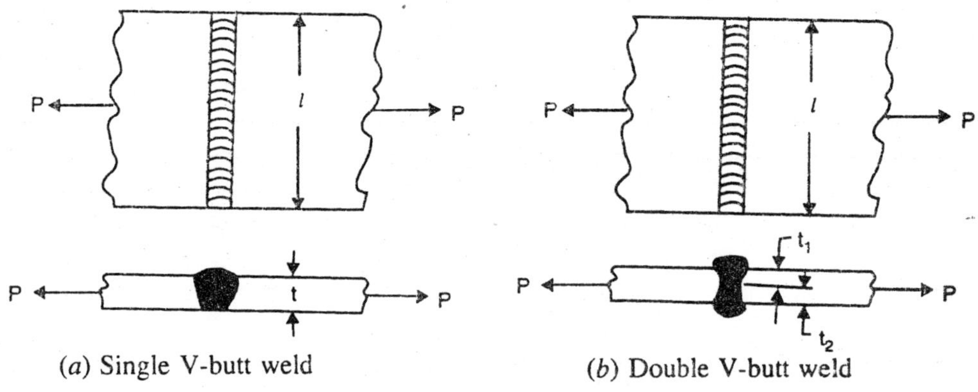

(*a*) Single V-butt weld (*b*) Double V-butt weld

Fig. 6.7 Butt joints under tension

(*ii*) **Double V-butt joint**

Tensile strength of double V-butt joint as shown in Fig. 6.7(b),

$$P = (t_1 + t_2)\, l \times \sigma_t \quad ... (6.8)$$

Where t_1, t_2 = throat thicknesses at top and bottom respectively.

2. Butt Weld under Shear

Consider a butt weld under shear as shown in Fig. 6.8.

Shear stress in weld, $t = P/(hl)$ ... (6.9)

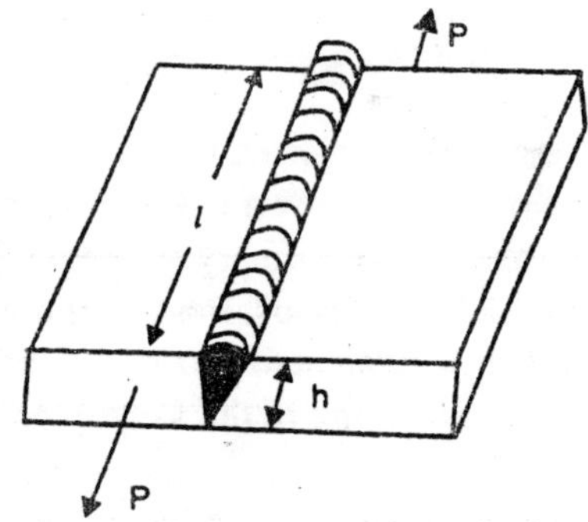

Fig. 6.8 Butt weld in shear

6.6 AXIALLY LOADED UNSYMMETRICAL WELDED SECTIONS

Consider unsymmetrical sections like angle, channel, T – sections welded on the flange edges and loaded axially as shown in Fig. 6.9. Let

l_1 = length of weld at top

l_2 = length of weld at the bottom

$l = l_1 + l_2$ = total length of weld

a_1 = distance of top weld from C.G. axis

a_2 = distance of bottom weld from C.G. axis

P = axial load

f = force of resistance offered by the weld per unit length

Moment of the top weld about the C.G. axis $= f \times l_1 \times a_1$

Moment of the bottom weld about the C.G. axis $= f \times l_2 \times a_2$

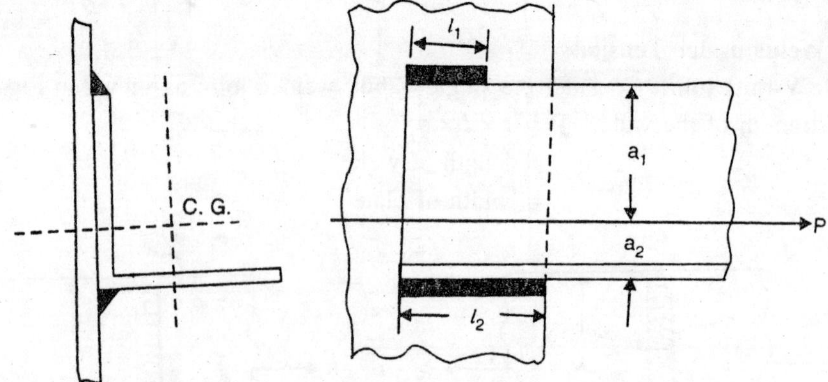

Fig. 6.9 Axially loaded unsymmetrical welded section

For equilibrium of the weld,

$$f \times l_1 \times a_1 = f \times l_2 \times a_2$$

or
$$l_1 \times a_1 = l_2 \times a_2$$

Also
$$l = l_1 + l_2$$

Thus
$$l_1 = la_2/(a_1 + a_2) \quad \text{... (6.10a)}$$

$$l_2 = la_1/(a_1 + a_2) \quad \text{... (6.10b)}$$

6.7 ECCENTRICALLY LOADED WELDED JOINTS

6.7.1 *Welded Joint under Shear*

A bracket subjected to an eccentric load P at eccentricity *e* and welded to the support by means of two welds is shown in Fig. 6.10(a). The design of welded joint under eccentric load consists of determination of primary and secondary shear stresses. In such problems, the first step is to locate the C.G. of the welds, treating the weld as a line. Let G be the C.G. of two welds. The eccentric load P is equivalent to a load P of the same magnitude and sense as the eccentric load and a couple C = Pe (cw) lying in the same plane. The load P at G produces primary shear stress,

$$\tau_1 = P/A$$

Where A = throat area of weld

The couple C produces secondary shear stress,

$$\tau_2 = Cr/J = Per/J$$

Where r = distance of a point in weld from G

J = polar moment of inertia of all welds about G.

The primary and secondary shear stresses along with the resultant shear stress are shown in Fig. 6.10 (*b*)

The secondary shear stress at any point in the weld is proportional to the distance from the C.G. Thus

$$\frac{\tau_2}{r_2} = \frac{\tau}{r} = \text{constant}$$

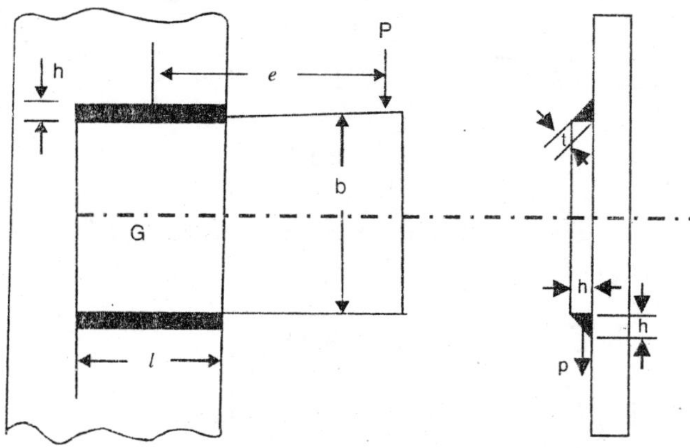

(*a*) Eccentrically loaded weld in shear

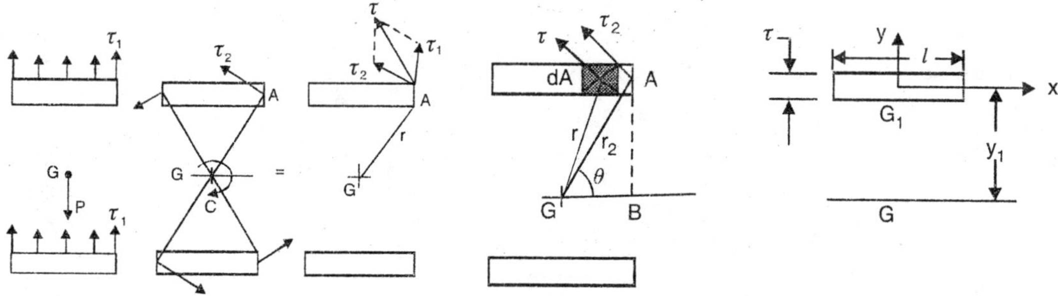

(*b*) Primary and secondary shear stresses (*c*) Eccentrically load welded joint under shear

Fig. 6.10 Eccentrically loaded welded joint under shear.

Where τ_2 = shear stress at the farther most point A.

$$\tau = \frac{\tau_2 \times r}{r_2}$$

Consider an elementary section of the weld having area dA at a distance r from G as shown in Fig. 6.10 (*c*).

Shear force on this elementary section = $\tau \times d$A

Turning moment of this shear force about G,

$$dT = \tau \times dA \times r = \left(\tau_2 \times \frac{r}{r_2}\right) \times dA \times r$$

$$= \left(\frac{\tau_2}{r_2}\right) \times dA \times r^2$$

Total turning moment over the whole weld area,

$$T = \int \left(\frac{\tau_2}{r_2}\right) \times dA \times r^2$$

$$= \left(\frac{\tau_2}{r_2}\right) \int dA \times r^2$$

$$= \left(\frac{\tau_2}{r_2}\right) \times J$$

Where J = polar moment of inertia of the throat area about G.

Now $T = C = P \times e$

$$\tau_2 = P \times e \times \frac{r_2}{J}$$

The polar moment of inertia of the throat area about the C.G. can be obtained by the use of parallel axis theorem as shown in Fig. 6.10 (*d*).

$$I_{xx} = \frac{l \times t^3}{12},\ I_{yy} = \frac{t \times l^3}{12} \text{ and } J = I_{zz} = I_{xx} + I_{yy}$$

Now $I_{xx} << I_{yy}$ as t is small.

Thus $J = I_{yy}$

$$J = \frac{tl^3}{12} = \frac{tl\left(l^2\right)}{12} = \frac{A\,l^2}{12}$$

$$J_G = J + Ay_1^2 = A\left[\frac{l^2}{12} + y_1^2\right]$$

If there are number of welds with polar moment of inertias J_1, J_2, J_3, ... about the C.G., then total polar moment of inertia about C.G. is given by,

$$J = J_1 + J_2 + J_3 + \ldots\ldots$$

For the welded joint shown in Fig. 6.10. (*a*),

$$J = 2A\left[\frac{l^2}{12} + \frac{b^2}{4}\right]$$

Resultant shear stress,

$$\tau = \left[\tau_1^2 + \tau_2^2 + 2\tau_1\tau_2\cos\theta\right]^{\frac{1}{2}} \qquad \text{....(6.11)}$$

where θ = angle between τ_1 and τ_2.

$$\cos\theta = \frac{GB}{AG} = \frac{\left(\frac{l}{2}\right)}{\left[\left(\frac{l}{2}\right)^2 + \left(\frac{b}{2}\right)^2\right]^{\frac{1}{2}}}$$

The polar moment of inertia and section modulus of welds is given in Table 6.2.

Table 6.2 Polar moment of inertia and section modulus of welds.

Sl. No.	Type of weld	Polar moment of inertia, J	Section modulus, z_p
1.	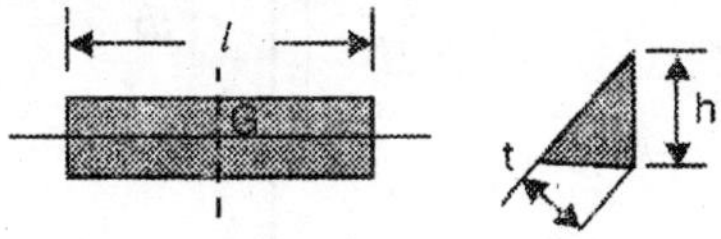	$\frac{tl^3}{12}$	
2.	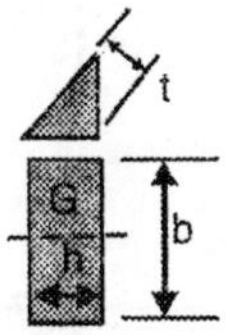	$\frac{tb^3}{12}$	$\frac{tb^2}{6}$
3.	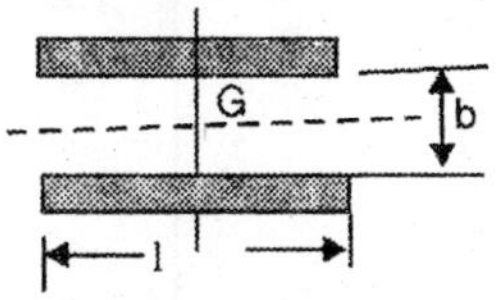	$\frac{tl\left(3b^2+l^2\right)}{6}$	tbl
4.	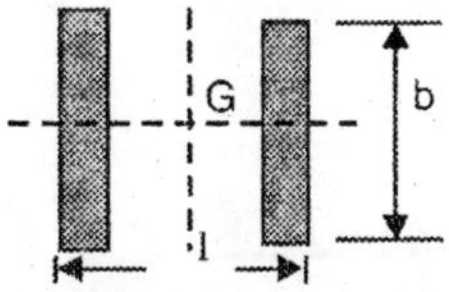	$\frac{tb\left(b^2+3l^2\right)}{6}$	$\frac{tb^2}{3}$
5.	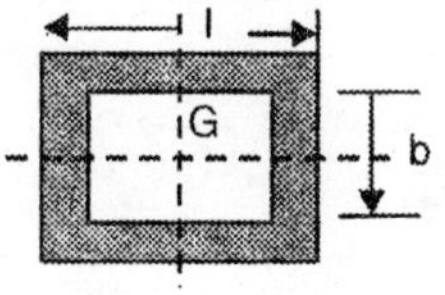	$\frac{t\,(b+l)^3}{6}$	$\frac{t\left(bl+b^2\right)}{3}$
6.	$\bar{x} = \frac{l^2}{2(l+b)}$ 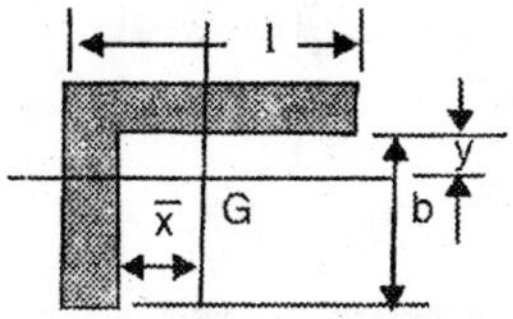 $\bar{y} = \frac{b^2}{2(l+b)}$	$t\left[\frac{(b+1)^4-6b^2t^2}{12\,(t+b)}\right]$ $= \frac{b^2}{2\,(l+b)}$	$t\left[\frac{4lb+b^2}{6}\right]$ top $\left[\frac{b^2\,(4lb+b)}{6(2l+b)}\right]$ (bottom)

Sl. No	Type of weld	Polar moment of inertia, J	Section modulus, z_p
7.	$\bar{x} = \frac{l^2}{2l+b}$	$t\left[\frac{(b+2l)^3}{12} - \frac{l^2\left(b+l^2\right)}{b+2l}\right]$	$t\left(tb + \frac{b^2}{6}\right)$
8.		$\frac{\pi t d^3}{4}$	$\frac{\pi t d^2}{4}$

6.7.2 Welded Joint under Bending

Consider a cantilever beam of rectangular section welded to a support by means of two fillet welds as shown in Fig. 6.11 and subjected to bending load P at an eccentricity *e*.

$$\text{Primary shear stress, } \tau_1 = \frac{P}{A}$$

Where A = throat area of welds

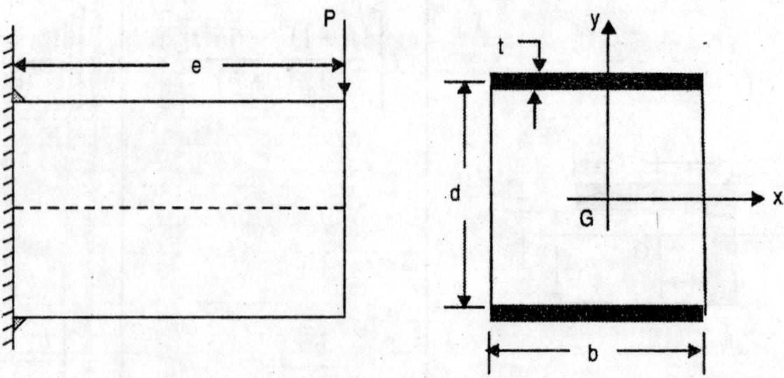

Fig. 6.11 Welded joint subjected to bending moment

Bending moment on the welds, M = P*e*

$$\text{Bending stress in the welds } \sigma_b = \frac{My}{I}$$

Where y = distance of the point in the weld from the neutral axis
I = moment of inertia of all welds based on throat area

$$I_{xx} = \frac{bt^3}{12} + bt \times \left(\frac{d}{2}\right)^2$$

The first term is negligible. Thus

$$I_{xx} = \frac{btd^2}{4}$$

$$I = 2I_{xx} = \frac{btd^2}{2}$$

$$\sigma_b = \frac{\left(\frac{\text{Ped}}{2}\right)}{\left(\frac{btd^2}{2}\right)}$$

$$= \frac{\text{Pe}}{btd}$$

Resultant shear stress in the welds, $\tau = \left[\left(\frac{\sigma_b}{2}\right)^2 + \tau^2{}_1\right]^{\frac{1}{2}}$ ... (6.12)

6.8 WELDED JOINT SUBJECTED TO FLUCTUATING FORCES

Stress – concentration is an important consideration while designing welded joints under fatigue loading. The fatigue stress – concentration factors K_f for welded joints are given in Table 6.3.

Table 6.3 Fatigue stess — concentration factors for welded joints.

Type of joint	Fatigue stress – concentration factor, K_f
Reinforced butt welds	1.2
Toe of transverse fillet weld	1.5
End of parallel fillet weld	2.7
T – butt joint with sharp corner	2.0

The design procedure as explained in chapter 4 may be adopted to solve a problem.

Example 6.1 A plate 100 mm wide and 100 mm thick is joined with another plate by a single transverse and a double parallel fillet weld as shown in Fig. 6.12. The maximum tensile and shear stresses are 70 MPa and 56 MPa respectively. Find the length of each parallel fillet weld if the joint is subjected to static loading.

Solution.

Given : $b = 100$ m, $h = 10$ mm, $\sigma_t = 70$ MPa, $\tau = 65$ MPa

Static loading :

Effective length of transverse weld,

$$l_1 = b - h = 100 - 10 = 90 \text{ mm}$$

Let l_2 = length of each parallel fillet weld

Maximum load the plate can carry,

$$P = \sigma_1\, bh = 70 \times 100 \times 10 = 70{,}000 \text{ N}$$

Load carried by single transverse weld,

$$P_1 = 0.707\, hl_1\sigma_t = 0.707 \times 10 \times 90 \times 70 = 44541 \text{ N}$$

Load carried by double parallel fillet weld,

$$P_2 = 1.414\, hl_2\tau = 1.414 \times 10 \times l_2 \times 56 = 791.84\, l_2 \text{ N}$$

Now $P = P_1 + P_2$

$$70{,}000 = 44541 + 791.84\, l_2$$

$$l_2 = 32.2 \text{ mm}$$

Adding 10 mm for starting and stopping of weld run, we have

$$l_2 = 32.2 + 10 = 42.2 \text{ mm.}$$

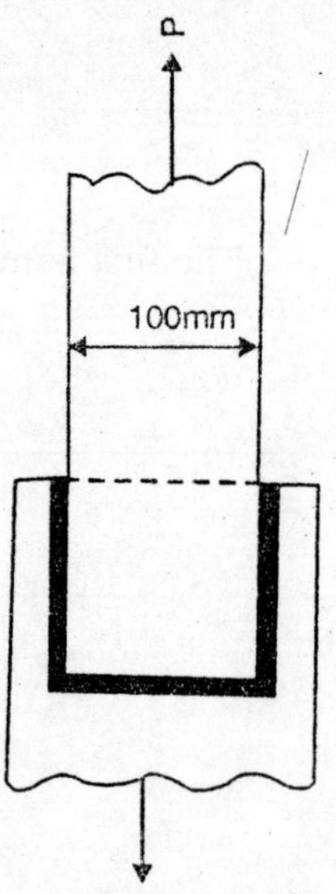

Fig. 6.12

Example 6.2 A 200 × 100 × 10 mm angle is to be welded to a steel plate by fillet welds as shown in Fig. 6.13. If the angle is subjected to a static load of 200 kN, find the length of welds at the top and bottom. The allowable shear stress may be taken as 60 MPa.

Solution.

Given : $a_1 + a_2 = 200$ mm, P = 200 kN, $\tau = 60$ MPa

Let l_1 = length of weld at the top

l_2 = length of weld at the bottom

$l = l_1 + l_2$ = total length of weld

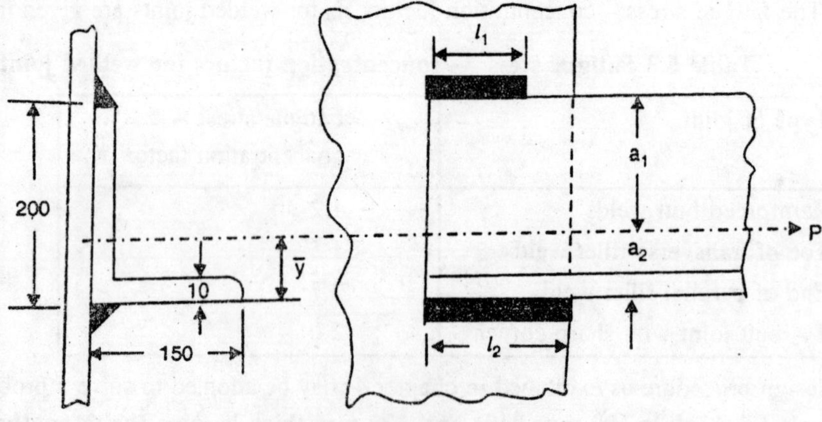

Fig. 6.13

$$(190 \times 10 + 150 \times 10)\, \bar{y} = 190 \times 10 \times 105 + 150 \times 10 \times 5$$

$$\bar{y} = 60.88 \text{ mm}$$

$$a_1 = 200 - 60.88 = 139.12$$

$$a_2 = 60.88 \text{ mm}$$

$$h = 10 \text{ mm}$$

$$P = 0.707\ hl\tau$$

$$200.000 = 0.707 \times 10 \times l \times 60$$

$$l = 471.5 \text{ mm}$$

$$l_1 = \frac{la_2}{a_1 + a_2} = 471.5 \times \frac{60.88}{200} = 143.5 \text{ mm}$$

$$l_2 = 471.5 - 143.5 = 328 \text{ mm}$$

Example 6.3 A circular shaft 50 mm in diameter is welded to a support by means of a fillet weld as shown in Fig. 6.14. Determine the size of the weld if the permissible shear stress in the weld is limited to 80 MPa.

Solution.

Given : $d = 50$ mm, $\tau = 80$ MPa

Primary shear stress, $\tau_1 = \frac{P}{A} = \frac{P}{\pi dt}$

$$= \frac{10,000}{\pi \times 50 \times t} = \frac{63.662}{t} \text{ N/mm}^2$$

$$I_{xx} = \pi tr^3 = \pi \times t \times 25^3 = 49087.4\ t \text{ mm}^3$$

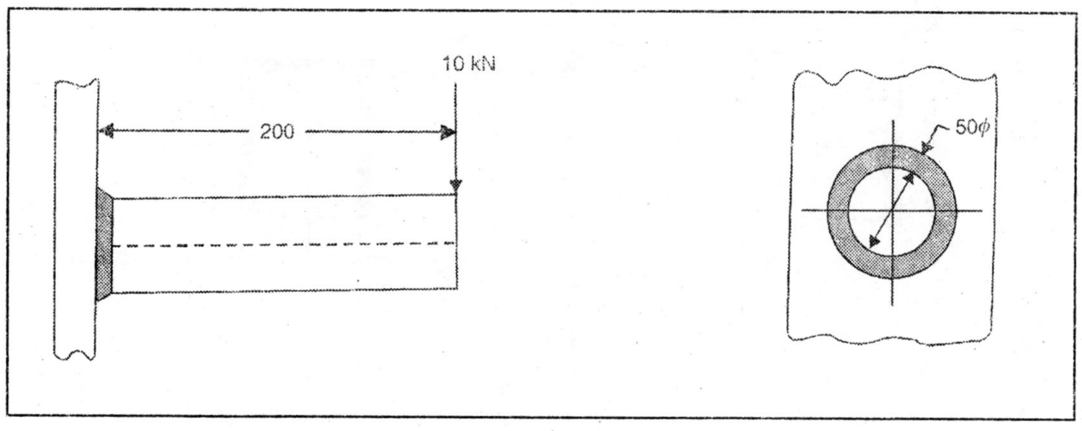

Fig. 6.14

Bending stress, $\sigma_b = \frac{My}{I_{xx}} = \left[\frac{10,000 \times 200 \times 25}{49087.4t}\right] = \frac{1018.6}{t} \text{ N/mm}^2$

Maximum shear stress, $\tau = \left(\frac{1}{t}\right)\left[\left(\frac{1018.6}{2}\right)^2 + (63.662)^2\right]^{\frac{1}{2}}$

$$= \frac{513.26}{t}$$

$$= 80$$

$$t = 6.4 \text{ mm}$$

$$h = \frac{6.4}{0.707} = 9.07 \approx 9 \text{ mm}$$

Example 6.4 A rectangular cross – section bar is welded to a support by means of fillet welds as shown in Fig. 6.14 Determine the size of the welds if the permissible shear stress in the weld is limited to 70 MPa.

Solution.

Given : P =20 kN, τ = 70 MPa, e = 400 mm, l = 100 mm, b = 150 mm

Let h = size of welds

t = throad thickness

$$A = t\,(2b + 2l)$$

$$= 0.707h\,(2b + 2l)$$

$$= 1.414h\,(b + l)$$

$$= 1.414\,(150 + 100)\,h = 353.5\,h \text{ mm}^2$$

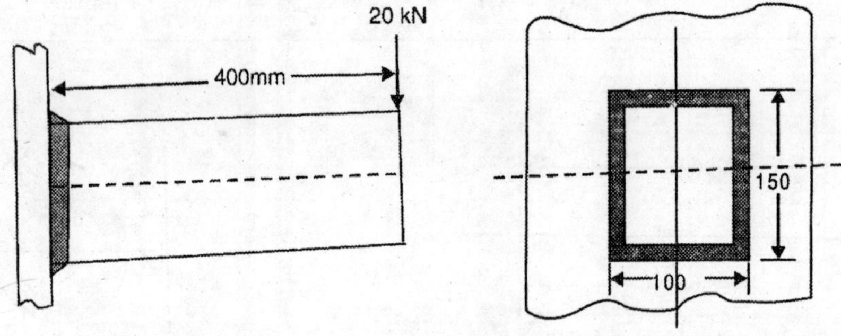

Fig. 6.15

Direct shear stress, $\tau = \frac{P}{A} = \frac{20{,}000}{353.5h} = \frac{56.577}{h}$ N/mm^2

Bending moment, $M = Pe = 20{,}000 \times 400 = 8 \times 10^6$ N.mm

Section modulus, $z = t\left(bl + \frac{b^2}{3}\right) = 0.707h\left[150 \times 100 + \frac{150^2}{3}\right]$

$$= 15907.5\,h \text{ mm}^3$$

Bending stress, $\sigma_b = \frac{M}{z} = \frac{8 \times 10^6}{15907.5h} = \frac{502.9}{h}$ N/mm^2

Maximum shear stress, $$\tau = \left[\left(\frac{\sigma_b}{2}\right)^2 + \tau^2\right]^{\frac{1}{2}}$$

$$= \frac{\left[\left(\frac{502.9}{2}\right)^2 + (56.577)^2\right]^{1/2}}{h}$$

$$= \frac{257.74}{h}$$

$$= 70$$

$$h = 3.68 \approx 4 \text{ mm}$$

Example 6.5 A bracket carrying a load of 20 kN is to be welded as shown in Fig. 6.16. Find the size of weld required if the allowable shear stress is not to exceed 75 MPa.

Solution.

Given : P = 20 kN, e = 150 mm, τ = 75 MPa, b = 100 mm, l = 60 mm

Throat area, A = $2tl = 2 \times 0.707 \times h \times 60 = 84.84$ h mm

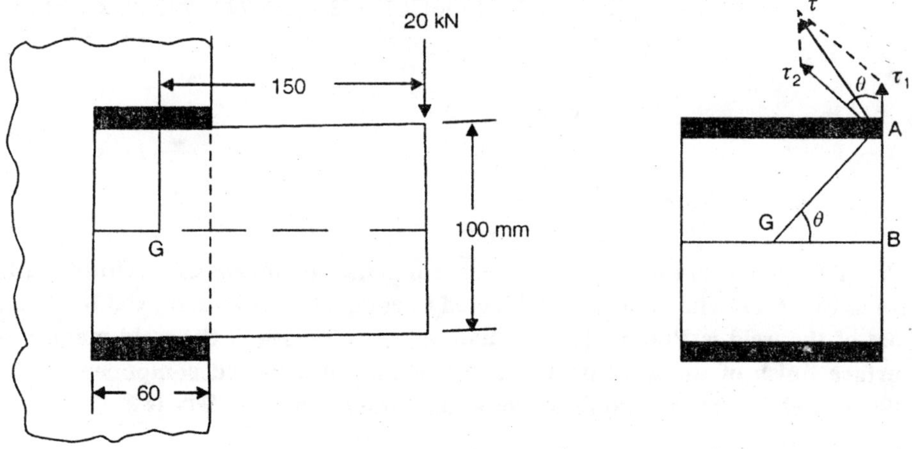

Fig. 6.16

Primary shear stress, $$\tau_1 = \frac{P}{A} = \frac{20{,}000}{84.84\,h} = \frac{235.74}{h} \text{ N/mm}^2$$

Polar moment of inertia of throad about weld C.G.,

$$J = \frac{tl\left(3b^2 + l^2\right)}{6} = 0.707\,h \times 60\,\frac{\left[3 \times 100^2 \times 60^2\right]}{6}$$

$$= 237552h \text{ mm}^4$$

Maximum radius of the weld,

$$r_2 = \left[\left(\frac{l}{2}\right)^2 + \left(\frac{b}{2}\right)^2\right]^{\frac{1}{2}}$$

$$= \left[30^2 + 50^2\right]^{\frac{1}{2}} = 58.3 \text{ mm}$$

Secondary shear stress, $\tau_2 = \dfrac{\text{Per}_2}{\text{J}}$

$$= \frac{20{,}000 \times 150 \times 58.3}{237552\, h} = \frac{736.26}{h} \text{ N/mm}^2$$

$$\cos\theta = \frac{l}{2r_2} = \frac{30}{58.3} = 0.5146$$

Resultant shear stress, $\tau = \left[\tau_1^2 + \tau_2^2 + 2\tau_1\tau_2\cos\theta\right]^{\frac{1}{2}}$

$$= \frac{\left[(235.74)^2 + (736.26)^2 + 2 \times 235.74 \times 736.26 \times 0.5146\right]^{\frac{1}{2}}}{h}$$

$$= \frac{881.07}{h}$$

$$= 75$$

$$h = 11.74 \approx 12 \text{ mm}$$

Example 6.6 Two plates 20 mm thick are welded together by means of a reinforced butt weld as shown in Fig. 6.17. The plates are subjected to completely reversed axial force ± 100 kN. The throat of the weld is 20 mm. The ultimate tensile strength of the weld metal is 440 MPa. The surface finish of the weld is equivalent to that of a forged component and the reliability is 90%. Determine the length of the weld if the factor of safety is 2.

Solution.

Given : $t = 20$ mm, $P_{max} = 100$ kN, $P_{min} = -100$ kN, $S_{ut} = 440$ MPa, R = 90%, $n = 2$

$S'_e = 0.5 S_{ut} = 0.5 \times 440 = 220$ MPa

Surface finish factor for forged surface and

$S_{ut} = 440$ MPa, $k_a = 0.5$

Fig. 6.17

For a 20 mm depth cross – section, size factor,

$k_b = 0.85$

For R = 90%, reliability factor, $k_c = 0.897$

Temperature factor, $k_b = 1.0$

For reversed axial loading without bending, load factor, $k_e = 0.9$

For a reinforced weld, $K_f = 1.2$ so that fatigue strength reduction factor,

$$k_f = \frac{1}{K_f} = \frac{1}{1.2}$$

Corrected endurance limit, $S_e = k_a k_b k_c k_d k_e k_f S'_e$

$$= 0.5 \times 0.85 \times 0.897 \times 1.0 \times 0.9 \times \frac{1}{1.2} \times 220 = 62.9 \text{ MPa}$$

$$\sigma_t = \frac{S_e}{n} = \frac{62.9}{2} = 31.45 \text{ MPa}$$

$$\frac{P}{hl} = \sigma_t$$

$$\frac{100,000}{20 \times l} = 31.45$$

$$l = 159 \text{ mm}$$

EXERCISES

6.1 What are the assumptions made in the design of welded joints ?

6.2 A plate 100 mm wide and 13 mm thick is to be welded to another plate by means of parallel fillet welds. The plates are subjected to a load of 50 kN. Find the length of the weld so that the maximum stress does not exceed 56 N/mm^2.

6.3 Two plates are welded by means of fillet weld as shown in Fig. 6.18. The leg dimension of the welds is 15 mm and the permissible shear stress at the throat cross – section is 90 MPa. Determine the length of the weld. P = 40 kN.

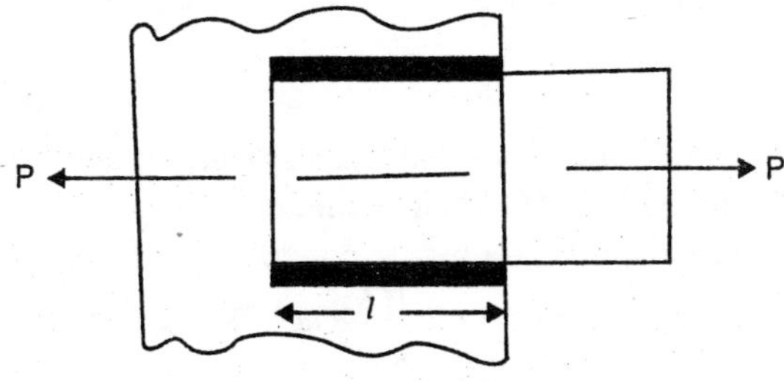

Fig. 6.18

6.4 A bracket as shown in Fig. 6.19 is welded to a plate with the same size welds. The permissible force per mm of weld length is 1.25 kN. Calculate the lengths l_1 and l_2.

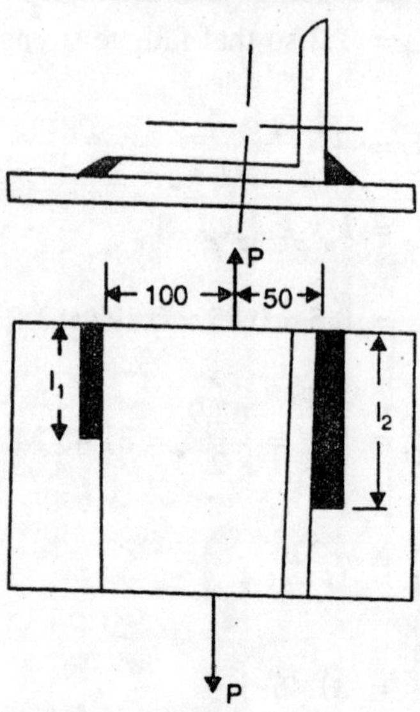

Fig. 6.19

6.5 A welded connection of steel plates as shown in Fig. 6.20 is subjected to an eccentric load of 15 kN. Determine the throat thickness of the weld if the permissible stress is limited to 90 MPa. Fig. 6.20

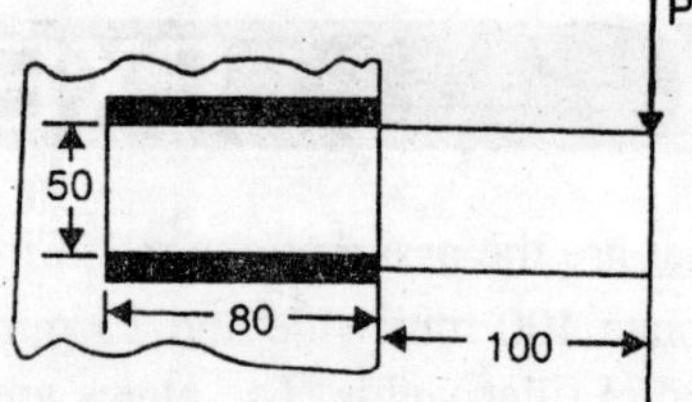

Fig. 6.20

6.6 A solid circular shaft 30 mm diameter is welded to a support by means of a fillet weld as shown in Fig. 6.21. Determine the leg dimension of the weld if the permissible shear stress is 90 MPa.

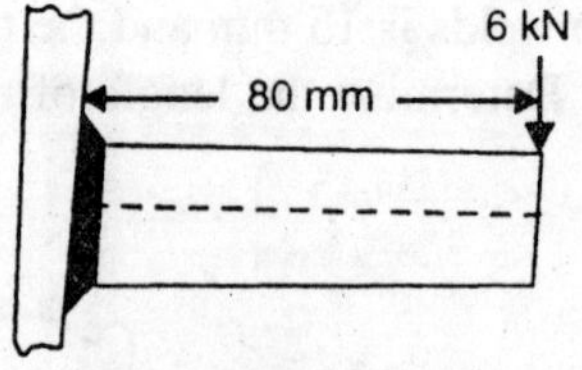

Fig. 6.21

6.7 A butt welded joint with ground and flush surface is subjected to an axial tensile load which varies from 30 kN to 80 kN. The thickness of plates welded is 10 mm. Determine the length of weld required for a life of 2×10^6 cycles and factor of safety of 2. Assume S'_{ut} = 410 MPa, $S_{yt} = 0.7\ S_{ut}$, $S'_e = 0.5\ S_{ut}$.

SCREWED JOINTS

Chapter

7.1 INTRODUCTION

A screwed thread is formed by cutting continuous helical groove on a cylindrical surface. The screw thread may be single start or multi – start and either right hand or left hand. A screwed joint consists of a bolt and a nut assembly. It is a temporary type of fastening. The screwed joints are widely used where the machine parts are required to be readily connected to disconnected without damage to the machine or the fastening.

7.2 ADVANTAGES AND DISADVANTAGES OF SCREWED JOINTS

The advantages and disadvantages of screwed joints are :

Advantages :

1. Highly reliable in operation.
2. Easy to assemble and disassemble.
3. Adaptable to various operating conditions.
4. Relatively cheaper to produce.

Disadvantages :

1. Stress – concentration in the threaded portion.
2. Lower strength as compared to riveted and welded joints.

7.3 TERMINOLOGY

The nomenclature of a screw thread is shown in Fig. 7.1

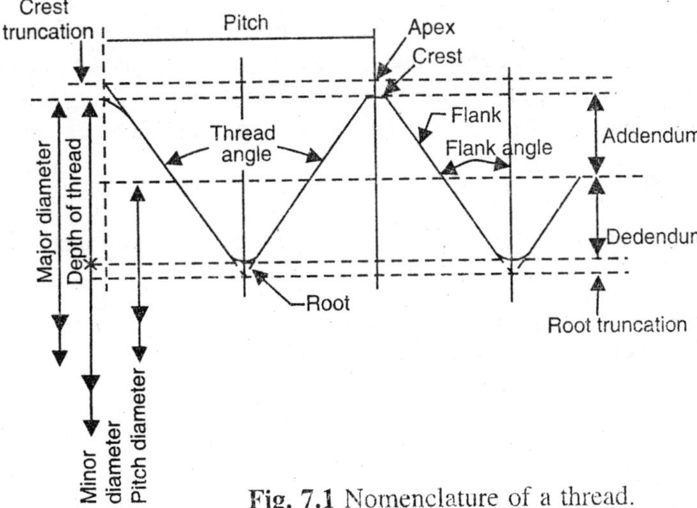

Fig. 7.1 Nomenclature of a thread.

External thread. This is a thread formed on the external surface of a cylinder.

Internal thread. This is a thread formed on the internal surface of a hollow cylinder.

Flank. It is the straight portion of the surface on either side of the thread form.

Crest. It is that part of the surface of a thread which connects the adjacent flanks at the top.

Root. It is that part of the surface of a thread which connects the adjacent flanks at the bottom.

Thickness of thread. This is the distance measured parallel to the axis between the flanks of a thread at the pitch line.

Major diameter. It is the largest diameter of an external or internal screw thread. It is also called outside or nominal diameter.

Minor diameter. It is the smallest diameter of an external or internal screw. It is also called core or root diameter.

Pitch diameter. It is the diameter of an imaginary cylinder on a cylindrical screw thread the surface of which would pass through the thread at such points as to make equal the width of the thread and the width of the spaces between the threads.

Pitch. It is the distance between two corresponding points on the same helix.

Lead = Pitch × number of starts

Angle of the thread. It is the angle included by the flanks of the thread.

7.4 FORMS OF SCREW THREADS

There are two main types of screw thread forms :

(*a*) V – threads

(*b*) Square threads

V – threads are stronger than the square threads for the same depth. V – thread screw has twice the amount of material resisting the shearing action at the root of the thread as compared to the screw with square threads. V – threads offer greater frictional resistance to motion than the square threads. In V – threads, a tensile force, in the direction of the axis, has a bursting action on the nut, which causes friction. The frictional force increases with the increase in thread angle. In square threads, there is no component of tensile force perpendicular to the axis and is, therefore, the ideal thread for power transmission. However, square threads are costly to produce as they cannot be cut by dies.

High frictional resistance of V – threads is an advantage in preventing a nut from slackening back. Therefore, V – threads are used to tighten pieces together as in the cases of bolts and nuts, studs, top bolt, screws, etc. Square threads are suitable for power transmission or feed mechanism of machine tools like lead screw of lathe, feed mechanism of a milling machine, screw jack, etc.

7.4.1 Forms of V - Threads

The various forms of V – threads are shown in Fig. 7.2 (*a*) to (*f*). These threads are :

(*a*) Trapezodial threads

(*b*) Saw tooth threads

(*c*) British Association threads

(*d*) American National or Seller's threads

(*e*) Pipe threads

(*f*) I.S.O. metric screw threads.

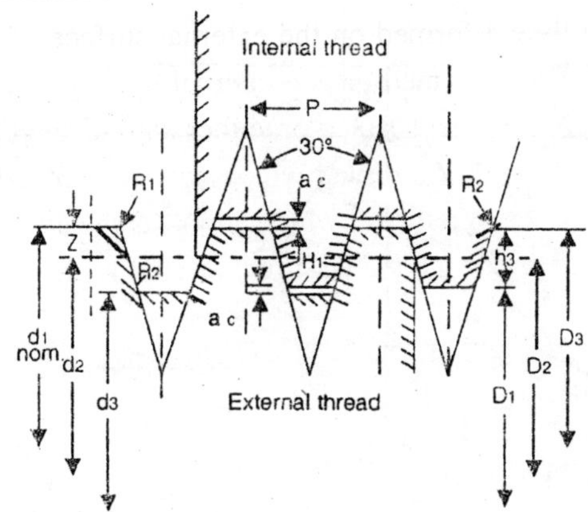

$H_1 = 0.5\ P$ $\qquad$ R_1 max. $= 0.5\ a_c$

$H_3 = h_3 = 0.5\ P + a_c = H_1 + a_c$ $\qquad$ R_2 max. $= a_c$

$z = 0.25\ P = \dfrac{H_1}{2}$ $\qquad$ $D_1 = d - 2H_1 = d - P$

$d_3 = d - 2H_3$ $\qquad$ $D_4 = d + 2a_c$

$d_2 = D_2 = d - 2z = d - 0.5\ P$

(*a*) Trapezoidal thread

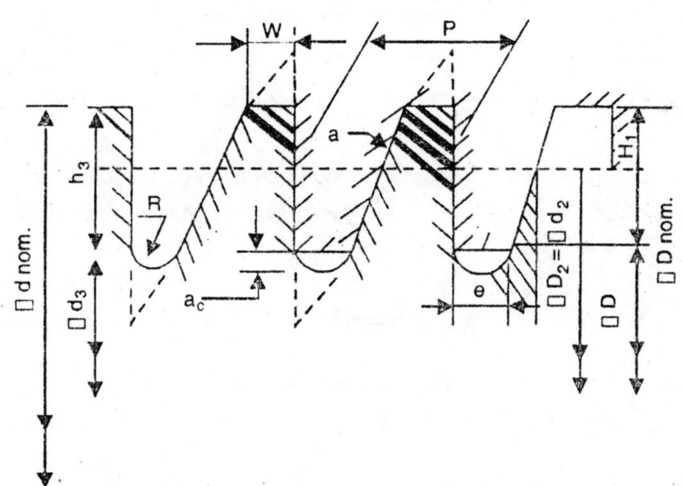

$H_1 = 0.75\ P$ $\qquad$ $R = 0.12427\ P$

$h_3 = H_1 + a_c = 0.86777\ P$ $\qquad$ $D_1 = d - 2H_1 = d - 1.5\ P$

$a = 0.1\sqrt{P}$ $\qquad$ $d_3 = d - 2h_3$

$a_c = 0.11777\ P$ $\qquad$ $d_2 = D_2 = d - 0.75\ P$

$W = 0.26384\ P$ $\qquad$ $s = 0.3149\ A_0$ where A_0 = Basic deviation (Upper deviation) for external thread in the pitch diameter)

$e = 0.26384\ P - 0.1\sqrt{P} = W - a$

(*b*) Saw-tooth thread

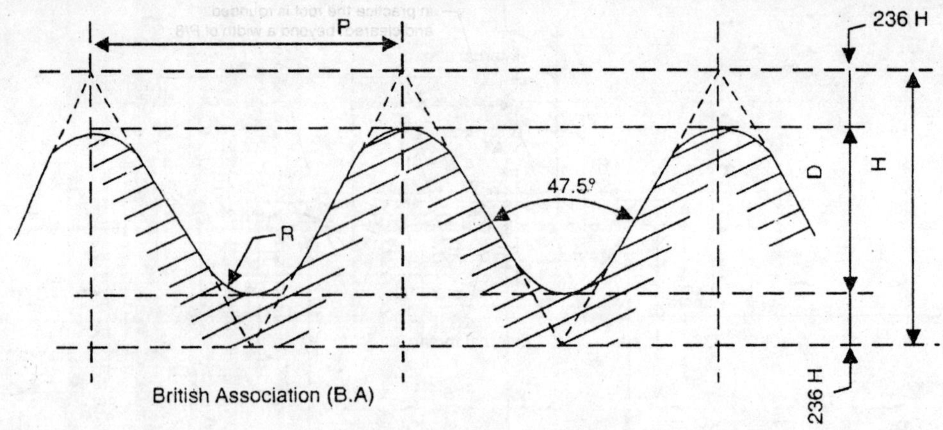

H =1.136 P, D = 0.6, R = $\frac{2 \times P}{11}$ = 0.18 P approx.

$$P = \frac{1}{TPl}$$

(*c*) B.A. thread

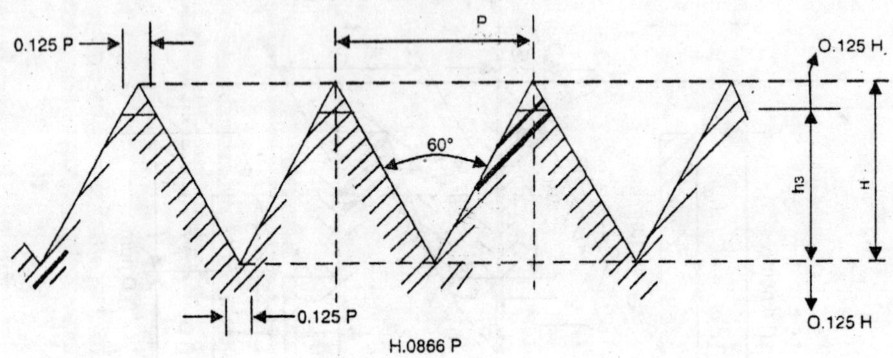

(*d*) Seller's thread.

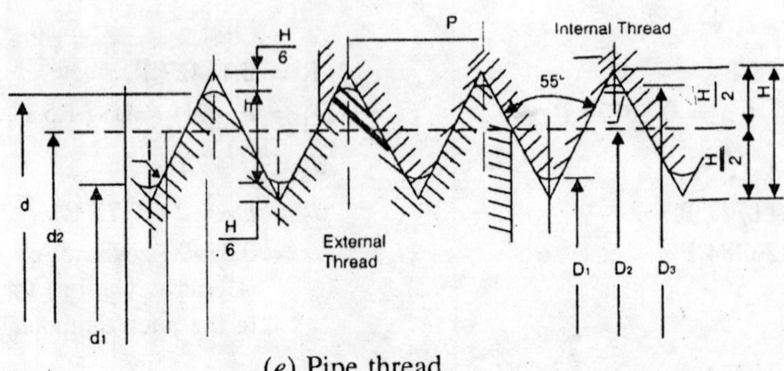

(*e*) Pipe thread.

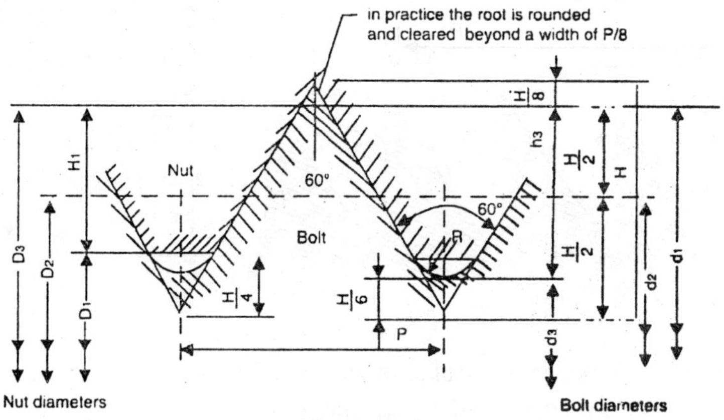

P = Pitch, H = 0.86603 P, D = d = Major diameter

$$H = 0.866025\ P \qquad H_1 = \frac{D - D_1}{2} = 0.625\ H = 0.54127\ P$$

$$D_1 = d_2 - \frac{H}{2} = d - 2\ H_1 = d - 1.08253\ P \qquad h_3 = \frac{d - d_1}{2} = \frac{17}{24}\ H = 0.61343\ P$$

$$D_2 = d_2 = d - 0.75\ H = d - 0.64952\ P \qquad R = \frac{H}{6} = 0.14434\ P,$$

$$d_3 = d_2 - \frac{H}{3} = d - 1.22687\ P$$

Fig. 7.2 Forms of V-threads

P
90°
H
0.5 P
0.5 P
H=0.5 P

(a) Square thread

f = 0.3707 P
P
29°
H
0.3707 P
H= 0.5P

(b) ACME thread

P/2
P
R
R= 1/4 P

(c) Geometry of knuckle thread

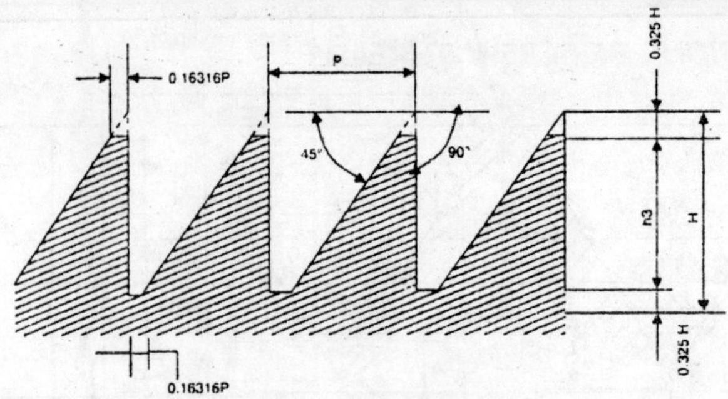

(*d*) Geometry of Buttress thread.

Fig. 7.3 Forms of square thread

7.4.2 FORMS OF SQUARE THREADS

The various forms of square threads are shown in Fig. 7.3 (*a*) to (*d*). These screw threads are :

(*a*) Square threads
(*b*) Acme threads
(*c*) Knuckle threads
(*d*) Buttress threads

7.5 DESIGNATION OF THREADED FASTENERS

Sizes of screw threads are designated by the letter M followed by diameter and pitch. The diameter and pitch being separated by the sign ×. When there is no mention of the pitch, it shall mean that a coarse pitch will be used. Thus a fine series screw of major diameter 8 mm and pitch 1 mm shall be designated as : M8 × 1.

A coarse series screw of major diameter 56 mm and pitch 1 mm shall be designated as : M6

To specify the tolerances on the screw threads, the following tolerance grades are used :

'7' for fine grade

'8' for normal (medium) grade

'9' for coarse grade.

The tolerance position is specified as follows :

'H' for unit thread

'*d*' for bolt thread with allowance, and

'*h*' for bolt thread without allowance.

7.6 STANDARD DIMENSIONS OF SCREW THREADS

The dimensions of screw threads used for fastening purpose are given in Tables 7.1 to 7.5

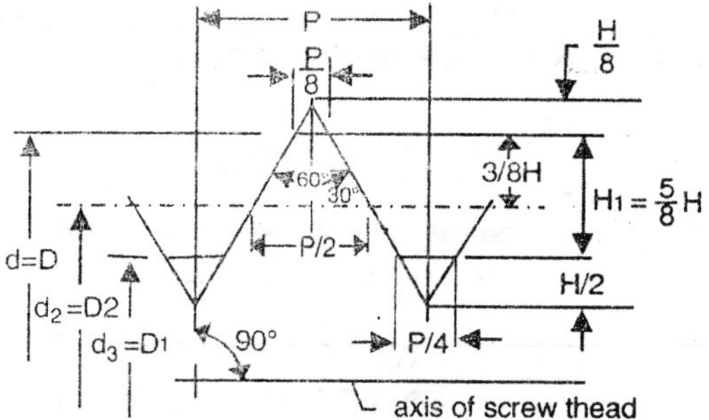

Basic profile for ISO Metric Screw Threads.

Table 7.1 Dimensions for Screw Threads for Bolts and Nuts (Diameter range M2.5 to M36)

All dimensions in mm

Series	Size	Pitch P	Major Dia d = D	Pitch dia $d_1 = D_2$	Minor dia Bolt d_3	Minor dia Nut d_1	Depth of Thread h_3	Maximum depth of engagement H_1	Stress area* A_s mm^2
Coarse series	M2.5	0.45	2.5	2.208	1.948	2.013	0.276	0.244	3.39
	M3	0.5	3	2.675	2.387	2.459	0.307	0.271	5.03
	M4	0.7	4	3.545	3.141	3.242	0.429	0.379	8.78
	M5	0.8	5	4.480	4.019	4.134	0.491	0.433	14.2
	M6	1	6	5.350	4.773	4.918	0.613	0.541	20.1
	M8	1.25	8	7.188	6.466	6.647	0.767	0.677	36.6
	M10	1.5	10	9.026	8.160	8.376	0.920	0.812	58.0
	M12	1.75	12	10.863	9.853	10.106	1.074	0.947	84.3
	M16	2	16	14.701	13.546	13.835	1.227	1.083	157
	M20	2.5	20	18.376	16.933	17.294	1.534	1.353	245
	M24	3	24	22.051	20.320	20.752	1.840	1.624	353
	M30	3.5	30	27.727	25.706	26.211	2.147	1.894	561
	M33	3.5	33	30.727	28.706	29.211	2.147	1.894	694
Fine series	M36	4	36	33.402	31.093	31.67	2.454	2.165	817
	M8X1	1	8	7.350	6.773	6.918	0.613	0.541	39.2
	M10 × 1.25	1.25	10	9.188	8.466	8.647	0.767	0.677	61.2
	M12 × 1.25	1.25	12	11.188	10.466	10.647	0.767	0.677	92.1
	M16 × 1.5	1.5	16	15.026	14.16	14.376	0.920	0.812	167
	M20 × 1.5	1.5	20	19.026	18.16	18.376	0.920	0.812	272
	M24 × 2	2	24	22.701	21.546	21.835	1.227	1.083	384
	M30 × 2	2	30	28.701	27.546	27.835	1.227	1.083	621
	M36 × 3	3	36	34.051	32.32	35.752	1.840	1.629	865

IS : 4218–1976

$$* \text{ Stress area} = \frac{\pi}{4}\left(\frac{d_2 + d_3}{2}\right)^2$$

Designation : A fine series screw of major diameter 8 mm and pitch 1 mm shall be designated as : M × 81.

A coarse series screw of major diameter 6 mm and pitch 1 mm shall be designated as : M6.

Table 7.2 Design Dimensions for Screw Threads for Bolts and Nuts (Diameter range M 42 to M 150)

All dimensions in mm

Size	Pitch P	Allowance μmm	Major dia d = D	Pitch data $d_2 = D_2$	Minor dia		Depth of thread h_3	Maximum depth of engagement H_1	Radius r	Stress area A_s mm^2
					Bolt d_3	Nut D_1				
M 42	4.5	63	42	39.077	36.416	37.129	2.760	2.436	0.650	1120
(M 45)	4.5	63	45	42.077	39.416	40.129	2.760	2.436	0.650	1300
M 48	5	71	48	44.752	41.795	42.587	3.067	2.706	0.722	1470
(M 52)	5	71	52	48.752	45.795	46.587	3.067	2.706	0.722	1760
M 56	5.5	75	56	52.428	49.177	50.046	3.374	2.977	0.794	2030
(M 60)	5.5	75	60	56.428	53.177	54.046	3.374	2.977	0.794	2360
M 64	6	80	64	60.103	56.559	57.505	3.681	3.248	0.866	2680
(M 68)	6	80	68	64.103	60.559	61.505	3.681	3.248	0.866	3060
M 92	6	80	72	68.103	64.559	65.505				3460
(M 76)	6	80	76	72.103	68.559	69.505				3890
M 80	6	80	80	76.103	72.559	73.505				4340
(M 85)	6	80	85	81.103	77.559	78.505				4950
M 90	6	80	90	86.103	62.559	83.505				5590
(M 95)	6	80	95	91.103	87.559	88.505				6270
M 100	6	80	100	96.103	92.559	93.505	3.681	3.248	0.866	7000
(M 105)	6	80	105	101.103	97.559	98.505				7760
M 110	6	80	110	106.103	102.559	103.505				8560
(M 115)	6	80	115	111.103	107.559	108.505				9400
M 120	6	80	120	116.103	112.559	113.505				10300
M 125	6	80	125	121.103	117.559	118.505				11200
(M 130)	6	80	130	126.103	122.559	123.505				12100
M 140	6	80	140	136.103	132.559	133.505				14200
(M 150)	6	80	150	146.103	142.559	143.505				16400

Note : Second preference sizes are shown in brackets.

$$A_s = \frac{\pi}{4}\left(\frac{d_2 + d_3}{2}\right)^2$$

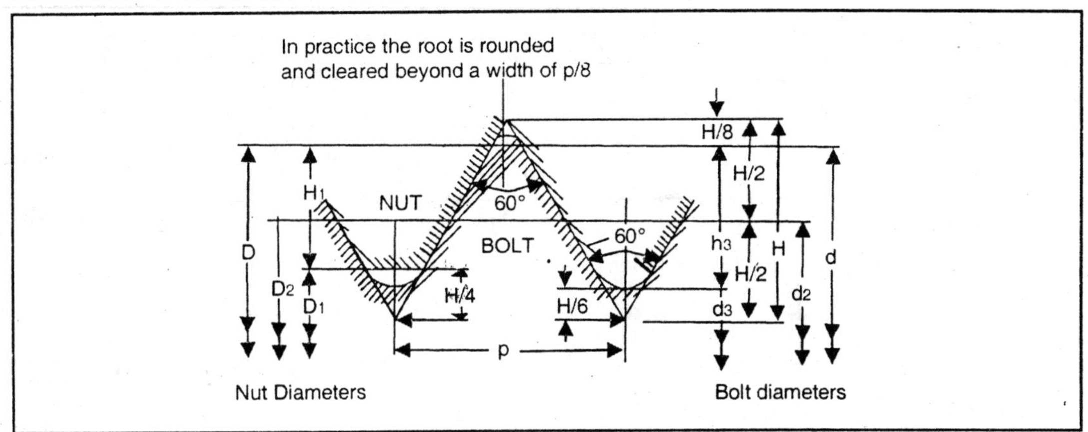

Table 7.3 Basic Dimensions for Design Profiles of ISO Metric Screw Threads

Basic Dia, mm	Pitch P, mm	Major Diameter d_1 mm	Pitch Diameter d_2, mm	Minor Diameter, mm		Lead Angle at Basic Pitch Diameter		Tensile Stress Area, A_c mm^2
				External Threads, d_1	Internal Threads, D_1	deg	min	
1	0.25	1.0	0.837620	0.693283	0.729367	5	27	0.46
	0.20	1.0	0.870096	0.754626	0.783494	4	11	0.53
2	0.40	2.0	1.740192	1.509252	1.566987	4	11	2.07
	0.25	2.0	1.837620	1.693283	1.729367	2	29	2.45
2.5	0.45	2.5	2.207716	1.947909	2.012861	3	43	3.39
	0.35	2.5	2.272668	2.070596	2.121114	2	20	3.70
3.0	0.50	3.0	2.675240	2.386565	2.458734	3	24	5.03
	0.35	3.0	2.772668	2.570596	2.621114	2	18	5.61
4.0	0.70	4.0	3.545337	3.141191	3.242228	3	36	8.78
	0.50	4.0	3.675240	3.386565	3.458734	2	29	9.79
5.0	0.80	5.0	4.480385	4.018505	4.133975	3	15	14.2
	0.50	5.0	4.675240	4.386565	4.458734	2	57	16.1
6.0	1.00	6.0	5.350481	4.773131	4.917468	3	24	20.1
	0.75	6.0	5.512861	5.079848	5.188101	2	29	22.0
7.0	1.00	7.0	6.350481	5.773131	5.917408	2	52	28.9
	0.75	7.0	6.512861	6.079848	6.188101	2	6	31.3
8.0	1.25	8.0	7.188101	6.466413	6.646835	3	10	36.6
	1.00	8.0	7.350481	6.773131	6.917468	2	29	39.2
10	1.50	10.0	9.025721	8.15969696	8.376202	3	2	58.0
	1.25	10.0	9.188101	8.466413	8.646835	2	28	61.2
	1.00	10.0	9.350481	8.773131	8.917468	1	57	64.5

Table 7.3 continued

Basic Dia, mm	Pitch P, mm	Major Diameter d_1 mm	Pitch Diameter d_2, mm	Minor Diameter, mm		Lead Angle at Basic Pitch Diameter		Tensile Stress Area, A_c mm^2
				External Threads, d_1	Internal Threads, D_1	deg	min	
12	1.75	12	10.863342	9.852979	10.105569	2	56	84.3
	1.50	12	11.025721	10.159686	10.376202	2	29	88.1
	1.25	12	11.188101	10.466413	10.646835	2	2	92.1
	1.00	12	11.350481	10.773131	10.917468	1	36	96.1
14	2.00	14	12.700962	11.546261	11.834936	2	52	115
	1.50	14	13.025721	12.159696	12.376202	2	6	125
	1.25	14	13.188101	12.466413	12.646835	1	44	129
16	2.00	16	14.700962	13.546261	13.834936	2	29	157
	1.50	16	15.025721	14.159696	14.276202	1	49	167
18	2.50	18	16.376202	14.932827	15.293671	2	47	192
	2.00	18	16.700962	15.546261	15.834936	2	11	204
	1.50	18	17.025721	15.159696	16.376202	1	36	216
20	2.50	20	18.376202	16.932827	17.293671	2	29	245
	2.00	20	18.700962	17.516261	17.834936	1	57	258
	1.50	20	19.025721	18.159696	18.376202	1	26	272
22	2.50	22	20.376202	18.932827	19.293671	2	14	303
	2.00	22	20.700962	19.546261	19.834936	1	46	318
	1.50	22	21.025721	20.159696	20.376202	1	18	333
24	3.00	24	22.051443	20.319392	20.752405	2	49	353
	2.00	24	22.700962	21.556261	21.834936	1	39	384
	1.50	24	23.025721	22.15969	22.376202	1	11	401
25	3.00	25	23.051443	21.319392	21.752405	1	36	385
30	3.50	30	27.726683	25.705957	26.211139	2	18	361
	3.00	30	28.051443	26.319392	26.752405	1	57	581
	2.00	30	28.700962	27.546261	27.834936	1	16	621
	1.50	30	29.025721	28.159696	28.376202	0	57	642
35	1.50	35	34.055721	33.159696	33.376202	0	48	860
42	4.5	42	39.072114	36.479088	37.128607	2	6	1120
	4.0	42	39.401924	37.092523	37.669873	1	51	1150
	3.0	42	40.051443	38.319392	38.752405	1	22	1210
	2.0	42	40.700962	39.546261	39.834836	0	52	1260
	1.5	42	41.025771	40.159696	40.376202	0	40	1290
45	4.5	45	42.077164	39.479088	40.128607	1	57	1300
	4.0	45	42.401924	40.092523	40.669873	1	43	1340
	3.0	45	43.051443	41.319392	41.752405	1	16	1340
	2.0	45	43.700962	42.546261	42.834936	0	50	1460
	1.5	45	44.025771	43.159696	43.376202	0	37	1490

Table 7.3 continued

Basic Dia, mm	Pitch P, mm	Major Diameter d_1 mm	Pitch Diameter d_2, mm	Minor Diameter, mm		Lead Angle at Basic Pitch Diameter		Tensile Stress Area, A_s mm^2
				External Threads, d_1	Internal Threads, D_1	deg	min	
52	5.0	52	48.752405	45.865653	46.587341		29	1830
	4.0	52	49.401924	47.092523	47.669873	1	29	1830
	3.0	52	50.051443	48.319392	48.752405	1	6	1900
	2.0	52	50.700962	49.546261	49.834936	0	43	1970
	1.5	52	51.025721	50.159696	50.376202	0	32	2010
60	5.5	60	56.427645	53.252219	54.046075	1	47	2360
	4.0	60	57.401924	55.092523	55.669873	1	16	2890
	3.0	60	58.051443	56.319292	56.752405	0	57	2570
	2.0	60	58.700962	57.546261	57.834936	0	37	2650
	1.5	60	59.025721	58.159696	58.376202	0	28	2700
72	6	72	68.102886	64.638784	66.504809	1	36	3460
	4	72	69.401924	67.092523	67.669873	1	3	3660
	3	72	70.051443	68.319392	68.752405	0	47	3760
	2	72	70.700962	69.546261	69.834936	0	31	3860
80	6	80	76.102886	72.638724	73.504809	1	26	4340
	4	80	77.401924	75.092523	75.669873	0	57	4570
	3	80	78.051443	76.319392	76.752405	0	42	4680
	2	80	78.700962	77.546261	77.834936	0	28	47890
90	6	90	86.102886	82.638784	83.504809	1	16	5590
	4	90	87.401924	85.092523	85.669873	0	59	5840
	3	90	88.051449	86.319292	86.752405	0	37	5970
	2	90	88.700962	87.546261	87.834936	0	25	5970
100	6	100	96.102886	92.638784	93.504809	1	8	7000
	4	100	97.401924	95.092523	95.669873	0	45	7280
	3	100	98.05114	96.319392	96.752405	0	33	7420
	2	100	98.00962	97.546261	97.834936	0	22	7560
110	6	110	106.102886	102.638784	103.504809	1	2	8560
	4	110	107.401924	105.092523	105.669873	0	41	8870
	3	110	108.051443	106.319392	106.752405	0	30	9020
120	6	120	116.102886	112.638784	113.504819	0	57	10300
	4	120	117.401924	115.092523	115.669873	0	37	10600
	3	120	118051443	116.319392	116.752405	0	28	10800
150	6	150	146.102886	142.538784	143.504809	0	45	16400
	4	150	147.401924	145.092523	145.669873	0	30	16800
	3	150	148.051443	146.319392	146.7525405	0	22	17000

Table 7.3 continued

Basic Dia, mm	Pitch P, mm	Major Diameter d_1 mm	Pitch Diameter d_2, mm	Minor Diameter, mm		Lead Angle at Basic Pitch Diameter		Tensile Stress Area, A_c mm^2
				External Threads, d_1	Internal Threads, D_1	deg	min	
160	6	160	156.102886	152.638784	153.504809	0	42	18700
	4	160	157.401924	155.092523	155.669873	0	28	19200
	3	160	158.051443	156.319392	156.775405	0	21	19400
180	6	180	176.102886	172.638784	173.504809	0	37	23900
	4	180	177.401924	175.092523	175.669873	0	25	24400
	3	180	178.051443	176.319392	176.752405	0	18	24700
200	6	200	196.102886	192.638784	193.504809	0	33	29700
	4	200	197.401924	195.092523	195.669873	0	22	30200
	3	200	198.051453	196.319392	196.752504	0	17	30500
250	6	250	246.102886	242.638784	294.504809	0	27	46900
	4	250	247.401924	245.092523	245.669873	0	18	47600
	3	250	248.051443	246.319392	246.752405	0	13	48000
300	6	300	296102886	295.638784	293.504809	0	22	68100
	4	300	297.401924	292.092523	295.669873	0	15	68900

IS : 4218—1967 (Part III)]

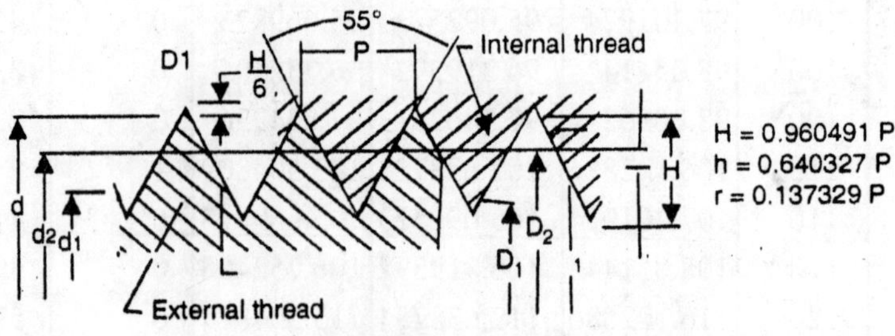

Figure For Table 7.4

Table 7.4 Dimensions for Pipe Threads for Fastening Purposes
(All dimensions in mm)

Size	Pitch	Major dia $d = D$	Pitch dia $d_2 = D_2$	Minor dia $d_1 = D_1$	Depth of thread h	Radius r ≃
1/16	0.907	7.723	7.142	6.561	0.581	0.125
1/18	0.907	9.728	9.147	8.566	0.581	0.125
1/4	1.337	13.157	12.301	11.445	0.856	0.184
3/8	1.337	16.662	15.806	14.490	0.856	0.184
1/2	1.814	20.953	19.793	18.631	1.162	0.249
5/8	1.814	22911	21.749	20.587	1.62	0.249
3/4	1.814	26.441	25.279	24.117	1.162	0.249
7/0	1.814	30.201	29.039	27.877	1.162	0.249
1	2.309	33.249	31.291	30.291	1.479	0.317
$1\frac{1}{8}$		37.897	36.418	34.939		
1¼	2.309	41.910	40.431	38.952	1.479	0.371
1½		47.803	46.324	44.845		0.317
1¾		53.746	52.267	50.788		
2	2.309	59.614	58.135	56.656	1.479	0.317
2¼		65.710	64.231	62.752		
2½		75.184	73.705	72.226		
2¾	2.309	81.534	80.055	78.576	1.479	0.317
3		87.884	86.405	84.926		
3½		100.330	98.851	97.372		
		113.030	111.551	100.072		
4	2.309				1.479	0.317
4½		125.730	124.251	122.772		
5		138.430	136.951	135.472		
5½	2.309	151.130	149.651	148.172	1.479	0.317
6		163.830	162.351	160.872		

IS : 2643 (Part-I)—1975

Table 7.5 Preferred pitches of screw threads in mm

Screw thread form	Preferred Pitches
For Bolts and Nuts	Coarse series—0.45, 0.5, 0.7, 0.8, 1, 1.25, 1.5, 1.75, 2, 2.5, 3, 3.5 and 4. Fine series—1, 1.25, 1.5, 2 and 3.
Square	Fine series—2, 3, 4 and 6. Normal series—5, 6, 7, 8, 9, 10, 12, 14 and 16 Coarse series—8, 10, 12, 14, 16, 18, 20, 22 and 24.
Knuckle	2.54, 3.175, 4.233 and 6.35.
Saw tooth	2, 3, 4, 5, 6, 7, 8, 9, 10, 11, 12, 14, 16, 18, 20, 22, 24, 28, 32, 36, 40 and 44.
ISO Metric trapezoidal	1.5, 2, 3, 4, 5, 6, 7, 8, 9, 10, 12, 14, 16, 18, 20, 22, 24, 28, 32, 36, 40 and 44.
Pipe	0.907, 1.337, 1.814 and 2.309.
ISOMetric	0.20, 0.25, 0.30, 0.35, 0.40, 0.45, 0.50, 0.6, 0.7, 0.75, 0.80, 1.0, 1.25, 1.5, 1.75, 2.0, 2.5, 3, 3.5, 4, 4.5, 5, 5.5 and 6.

7.7 MECHANICAL PROPERTIES OF BOLTS, SCREW AND STUDS

The mechanical properties of bolts, screws and studs are given in Table 7.6.

Table 7.6 Mechanical properties of Bolts, Screw and Studs.

Mechanical property		4.6	4.8	5.6	5.9	6.8	8.8		9.8	10.9	12.9
							<M16	M16			
Tensile strength	Nom	400		500		600	800	800	900	1000	1200
S_{ut} MPa	Min.	400	420	500	520	600	800	830	900	1040	1220
Yield strength,	Nom.	240	320	300	400	480					
S_{yt} MPa	Min	240	340	300	420	480					
Stress at Premanent	Nom.	–					640	640	720	900	1080
Set limit, $S_{0,2}$, MPa	Min						640	660	720	940	1100

The values of factor of safety of bolts, screws and studs are given in Table 7.7

Table 7.7 Values of Factor of Safety for Bolts, Screws and Studs

Material	Steady load			Variable load	
	d_0, mm				
	6 – 16	16 – 30	30 – 60	6 – 16	16 – 30
Carbon steels	4 – 5	2.5 – 4	1.5 – 2.5	8 – 12	8
Alloy steels	5 – 7	3 – 5	3	6 – 10	7

7.7 STRESSES IN SCREWED JOINTS

7.7.1 Static Loading

The following stresses are induced in screwed joints :

1. Internal stresses due to screwing up forces.
2. Stresses due to external forces.
3. Combined stresses due to (1) and (2).

(*a*) Initial stresses

1. Torisonal shear stress caused by the frictional resistance of thread during tightening.

Due to friction, the turning moment acts on the screw in a plane normal to the screw axis. The useful force P_i is exerted by the thread in the direction of the axis. If there were no friction, this force would exert a pressure N normal to the surface of the thread a shown in Fig. 7.4 (*a*). The angle between P_i and N is the helix angle of the threads.

Now, $$\tan \alpha = \frac{p}{\pi d_m} \quad \text{... (7.1)}$$

where p = pitch of threads

d_m = main pich diameter of threads.

Becuase of friction the resultant force R will be inclined to the normal by the amount of the angle of friction ϕ, and will have a component H_1 in the plane normal to the screw axis. The component H_1 is overcome and balanced by an opposite force H_0, where

$$H_1 = H_0 = P_i \tan(\alpha + \phi)$$

$$= \frac{P_i(\tan\alpha + \tan\phi)}{1 - \tan\alpha \tan\phi}$$

$$= \frac{P_i(\tan\alpha + \mu)}{1 - \mu\tan\alpha} \quad \text{... (7.2)}$$

where $\tan\phi = \mu$ = coefficient of friction

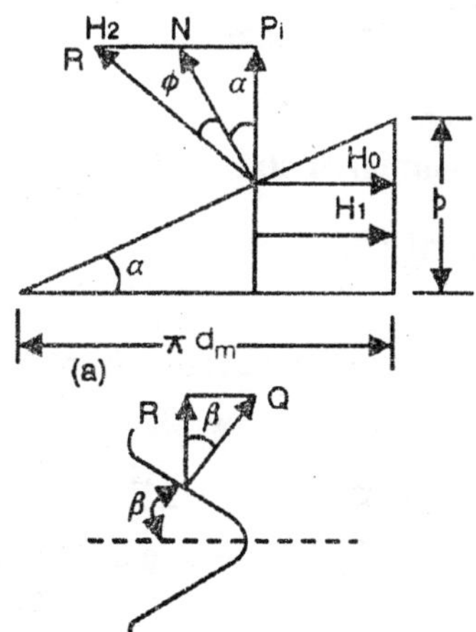

(*b*) Forces on V-thread.

Fig. 7.4 Initial stress duel to friction in screw threads.

For triangular threads as shown in Fig. 7.4 (*b*),

$$Q = \frac{R}{\cos\beta}$$

where 2β = angle of threads, and the corrected tangential resistance becomes,

$$H_1 = P_i\left[\frac{\tan\alpha + \mu\sec\beta}{1 - \mu\tan\alpha\sec\beta}\right] \quad \text{... (7.3)}$$

When the nut or a screw with head is turned, the additional frictional force with which the nut or head bears is given by :

$$H_2 = P_i \frac{\mu_c d_c}{d_m} \quad \text{.... (7.4)}$$

Where μ_c = coefficient of friction between the collar (nut or bolt head) and the surface on which the collar bears.

d_c = mean collar diameter (nut or bolt head)

Total frictional force,
$$H = H_1 + H_2 = P_i\left[\frac{\tan\alpha + \mu\sec\beta}{1-\mu\tan\alpha\sec\beta} + \frac{\mu_c d_c}{d_m}\right] \quad \text{... (7.5)}$$

Frictional torque,
$$T_f = \frac{H d_m}{2} \quad \text{... (7.6)}$$

2. Initial tensile stresses due to screwing up forces – The initial tension in a bolt is given empirically by :

$$P_i = 2805d \text{ N, for a leak proof joint.} \quad \text{... (7.7a)}$$
$$= 1402d \text{ N, for an ordinary joint} \quad \text{... (7.7b)}$$

Where d = nominal diameter of bolt, mm

Initial tensile stress,
$$\sigma_i = \frac{P_i}{A_s} \quad \text{... (7.8)}$$

Where A_s = stress area of bolt.

3. Torsional stress—Initial torsional shear stress due to tightening,

$$\tau_i = \frac{16T_f}{\pi d^3} \quad \text{...(7.9)}$$

4. Shear stress across the threads—It is given by :

$$\tau_s = \frac{P_i}{\pi d_c b n} \quad \text{...(7.10)}$$

Where b = width of thread section at the root

n = number of threads in engagement

5. Crushing stress on threads—The crushing stress is given by :

$$\sigma_c = \frac{4P_i}{\pi(d^2 - d^2{}_c)\,n} \quad \text{....(7.11)}$$

Where d = major diameter

6. Bending stress—The bending stres is induced when the surface of the part connected by the bolt under the head or nut are not perfectly perpendicular to the bolt axis. The bolt will be subjected to a bending moment M that is constant over the whole free length l of the bolt as shown in Fig. 7.5 (*a*). The unit elongation of the outer fibres of the bolt as shown in Fig. 7.5 (*b*) is :

$$\text{Strain, } \varepsilon = \frac{\Delta dx}{dx} = \frac{r.d\theta}{R.d\theta} = \frac{r}{R}$$

$$\text{Stress, } \sigma = \varepsilon \,.\, E = \frac{rE}{R}$$

From Fig. 7.5 (*a*), we have

$$\frac{l}{R} = \frac{2a}{D}$$

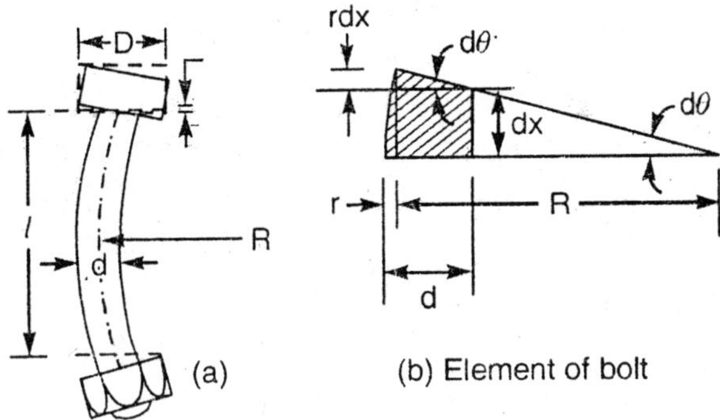

Fig. 7.5 Bolt under bendng

and $r = \frac{d}{2}$, $D = 2d$

Bending stress in bolt, $\sigma_b = \frac{aE}{(2l)}$ ...(7.12)

Where A = difference in height between the extreme corners of the nut or head
l = length of shank of the bolt
E = Young's modulus of elasticity of bolt material

***(b)* Stress due to external forces**

1.Tensile stress—This stress is induced when the load acts in the direction of the bolt axis. It is given by :

$$\sigma_t = \frac{P}{A_s}$$ ....(7.13)

2. Shear stress—This stress is induced in the bolt when the load acts perpendicular to the bolt axis.

$$\tau = \frac{4P}{\pi d^2 n}$$ ...(7.14)

3. Combined tension and shear stress—In such a case, the maximum shear stress I given by :

$$\tau_{max} = \left[\left(\frac{\sigma_t}{2}\right)^2 + \tau^2\right]^{1/2}$$ ...(7.15)

and the maximum principal stress is given by :

$$\sigma_{max} = \frac{\sigma_t}{2} + \left[\left(\frac{\sigma_t}{2}\right)^2 + \tau^2\right]^{1/2}$$ ...(7.16)

7.7.2 Fluctuating Loading

In many applications, the external forces acting on the bolted assembly fluctuates between two limits. The endurance strength is the criterion of failure in these cases. The endurance limit is determined as explained in art. 4.8. The surface finish of the bolt is usually considered as equivalent to a machined surface. The values of fatigue stress – concentration factors, K_f are given in Table 7.8. The Soderberg or modified Goodman diagram may be used to determine the bolt size.

Table 7.8 Fatigue Stress – concentration Factors K_f for Screw Threads

Material	Turned	Rolled
Cast steel	3 – 4.5	1.5 – 4
Alloy steel	5.5 – 6	3 – 5.5

7.8 EXTERNAL LOAD WITH PRELOAD

7.8.1 Axial Loads

Assume that a flexible gasket is used in a bolted joint as shown in Fig. 7.6. Let

P_i = initial load due to tightening of the bolt.

P = external load

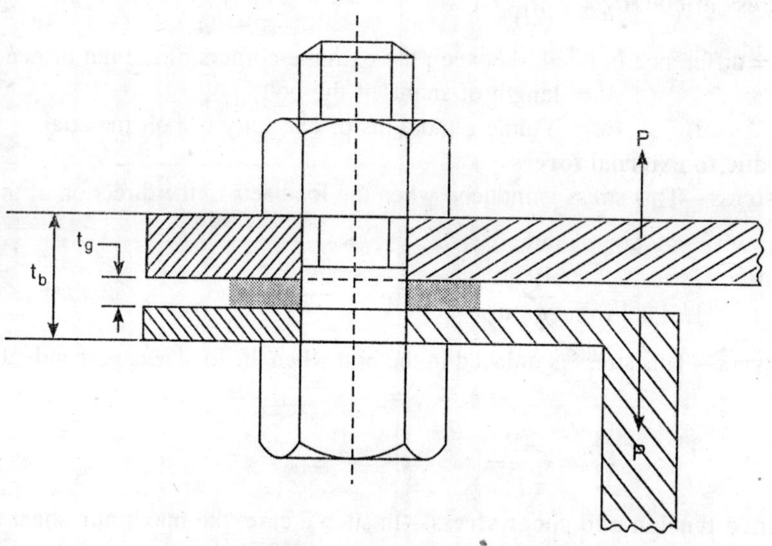

Fig. 7.6. Bolt joint with flexible gasket.

P_g = compression load on the gasket

P_b = combined load on the bolt

k_b, k_g = stiffness of bolt and gasket respectively, N/mm

δ_b = elongation of bolt, mm/N = $\dfrac{1}{k_b}$

δ_g = compression of gasket, mm/N = $\dfrac{1}{k_g}$

n = total number of bolts

Initial compression of gasket = $P_i \delta_g$

Initial elongation of bolt = $P_i \delta_b$

When the load P is applied, increase in length of bolt or gasket = $(P_b - P_i)\,\delta_b$

Final deformation of gasket, $P_g \delta_g = P_i \delta_g - (P_b - P_i)\,\delta_b$

Final compressive load on gasket, $P_g = P_i - \dfrac{(P_b - P_i)\,\delta_b}{\delta_g}$

Combined load on bolt, $$P_b = P_g + P = \frac{P_i - (P_b - P_i)\,\delta_b}{\delta_g} + P$$

$$= P_i + \left[\frac{\delta_g}{\delta_g + \delta_b}\right] P$$

$$= P_i + CP \qquad \text{...(7.17)}$$

Where $$C = \frac{\delta_g}{\delta_g + \delta_b} = \frac{k_b}{k_b + k_g}$$

Hence $$P_g = P_i - \left[\frac{\delta_g}{\delta_g + \delta_b}\right] P \qquad \text{...(7.18)}$$

Now $$\delta_b = \frac{l_b}{A_b E_b} = \frac{1}{k_b}$$

Where l_b, A_b = length and area of bolt respectively.

For the gasket and flange,

$$\delta_g = \frac{t_g}{A_g E_g} + \frac{t_f}{A_f E_f} = \frac{1}{k_g}$$

For a thick gasket, the second term is very small. Hence

$$C = \frac{t_g A_b E_b}{t_g A_b E_b + l_b A_g E_g} \qquad \text{...(7.19)}$$

The value of P should be such that P_g remains compressive. If the value of P becomes such that P_g becomes tensile then Eqs (7.17) and (7.18) do not apply and the bolt will carry the total load P.

7.8.2 Metal - to - metal Joint

For this case, when $E_b = E_f$, $t_f = l_b$ and $A_f >> A_b$, then $C \rightarrow 0$, and the external load does not affect the stress in the bolt as long as $P < P_i$. For a tight joint, $P < P_i$ by about 20%. If $P > P_i$, the joint will separate and the stress will be due to the external load alone.

7.8.3 Gasket Joint

The values of C for a gasket joint are given in Table 7.9.

Table 7.9 Values of C

Type of joint	Factor C
1. Soft, elastic gasket with studs	1.00
2. Soft gasket with through bolts	0.90
3. Copper – asbestos gasket	0.60
4. Soft copper gasket with long through bolts	0.50
5. Soft copper corrugated gasket	0.40
6. Lead gasket with studs	0.10
7. Narrow copper ring	0.01
8. Metal – to – metal joint	0.00

7.9 ECCENTRIC LOADING ON BOLTS

7.9.1 Eccentric Load Parallel to the Axis of Bolts

Consider a bracket having a rectangular base bolted to a wall by means of four bolts as shown in Fig. 7.7. In this case the bolts are under tension. The direct load in each bolt, $P_d = \frac{P}{n}$, where n = number of bolts. The load P tends to rotate the bracket about the edge AB. Let w be the load in a bolt per unit distance due to the turning effect of the bracket. Let P_1, P_2 be the loads on each of the bolts at distances l_1 and l_2 respectively from the tilting edge. Then

Load on each bolt at distance l_1, $\quad P_1 = wl_1$

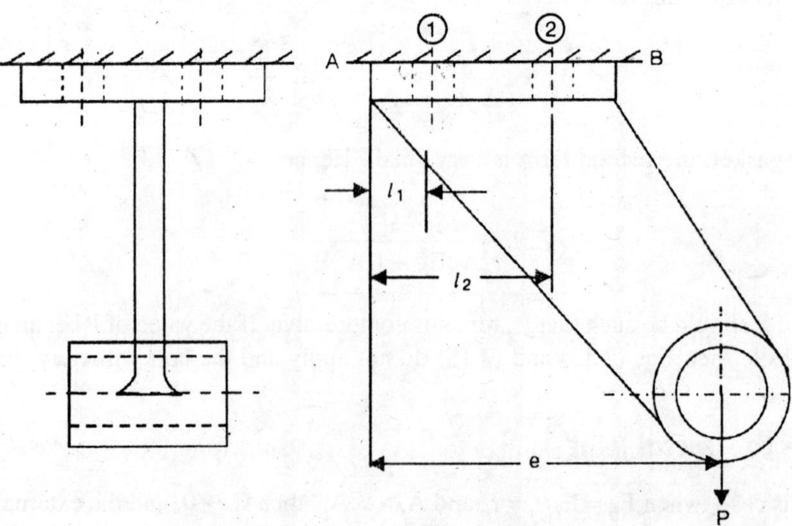

Fig. 7.7 Eccentric Load Parallel to the axis of bolts.

$$\text{Moment of this load about the tilting edge} = wl_1 \times l_1 = wl^2_1$$

$$\text{Similarly moment of each bolt at distance } l_2 = wl^2_2$$

$$\text{Total moment of the load on bolts about the tilting edge} = 2w\,(l^2_1 + l^2_2)$$

$$\text{Moment of due to load P about the tilting edge} = Pe$$

$$2w\,(l^2_1 + l^2_2) = Pe$$

$$w = \frac{Pe}{2\,(l^2_1 + l^2_2)}$$

It may be observed that the most heavily loaded bolts are those which are located at the greatest distance from the tilting edge. Thus

$$P_2 = wl_2 = \frac{Pel_2}{2\,(l^2_1 + l^2_2}$$

Total tensile load on the heavily loaded bolt,

$$P_t = P_d + P_2$$

$$= \frac{P}{4} + \frac{Pel_2}{2\,(l^2_1 + l^2_2)} \qquad ..(7.20)$$

If d_c is the core diameter of the bolt, then

$$\left(\frac{\pi}{4}\right) d^2{}_c\,\sigma_t = P_t$$

Or

$$d_c = \left[\frac{4P_t}{\pi\sigma_t}\right]^{1/2} \qquad ...(7.21)$$

7.9.2 Eccentric Load Acting Perpendicular to the Axis of Bolts.

Consider a wall bracket carrying an eccentric load perpendicular to the axis of the bolts on rectangular base as shown in Fig. 7.8. In this case the bolts are under direct shear.

Direct shearing load in each bolt, $P_s = \dfrac{P}{n}$

The load P tends to cause the bracket to rotate clockwise about the edge AB. This produces tensile load in the bolts. The moment Pe of load P must be balanced by the sum of the moments of all the bolts about the edge AB. Let P_1, P_2, P_3 be the loads on the bolts, then

$$Pe = 2\,(P_1l_1 + P_2l_2 + P_3l_3)$$

Also the forces in the bolts are proportional to the distance from the edge AB. Hence

$$P_2 = \frac{P_1l_2}{l_1},\; P_3 = \frac{P_1l_3}{l_1}$$

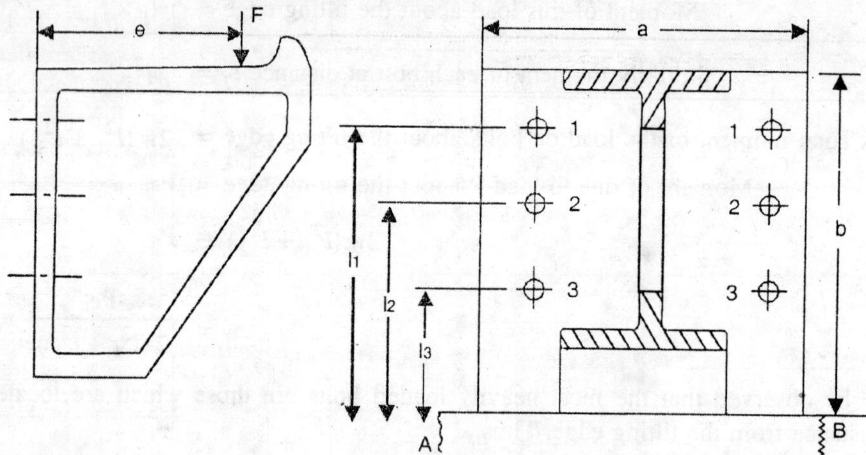

Fig. 7.8 Eccentric load perpendicular to axis of bolts

Hence tensile load in the most heavily loaded bolt, $P_1 = \dfrac{Pel_1}{2(l^2_1 + l^2_2 + l^2_3)}$

The bolts are subjected to the combined action of shear and tension. The maximum tensile load in bolt 1 becomes :

$$(P_l)_{max} = \left(\frac{1}{2}\right)[P_1 + (P^2_1 + 4P^2_s)^{1/2}] \quad ....(7.22)$$

and the maximum shear load,

$$(P_s)_{max} = \left(\frac{1}{2}\right)(P^2_1 + 4P^2_s)^{1/2} \quad ...(7.23)$$

$$\left(\frac{\pi}{4}\right)d^2_c\,\sigma_t = (P_t)_{max}$$

or

$$\left(\frac{\pi}{4}\right)d^2_c\,\tau = (P_s)_{max}$$

7.9.3 Eccentric Load on a Bracket with Circular Base

Consider round flange bearing having four bolts as shown in Fig. 7.9. Let

a = radius of the column flange

b = radius of the bolt pitch circle.

e = distance of the load from the tilting edge

l_1, l_2, l_3, l_4 = distance of bolt centres from the tilting edge.

Then

$$P_1 = \frac{Pel_1}{l^2_1 + l^2_2 + l^2_3 + l^2_4}$$

$$l_1 = a - b\cos\alpha$$

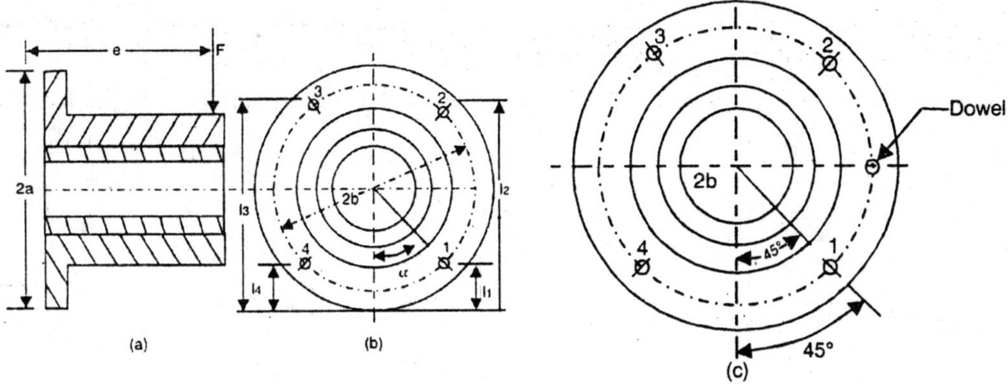

Fig. 7.9 Eccentric load on a Bracket with Circular Base.

$$l_2 = a + b \sin \alpha$$

$$l_3 = a + b \cos \alpha$$

$$l_4 = a - b \sin \alpha$$

$$P_1 = P\left[\frac{e(a - b\cos\alpha)}{4a^2 + 2b^2}\right]$$

For n bolts, equally spaced, the general expression is,

$$P_1 = P\left[\frac{2e(a - b\cos\alpha)}{(2a^2 + b^2)\,n}\right] \quad ...(7.24)$$

Where $\alpha = \frac{180°}{n}$

For P_1 to be maximum, $\cos \alpha = -1$ or $\alpha = 180°$

$$P_{max} = P\left[\frac{2e(a + b)}{(2a^2 + b^2)\,n}\right] \quad ....(7.25)$$

Eq. (7.25) gives the absolute maximum value and should be used if the direction of load P can change with relation to the bolts, as in the case of a pillar crane. If the direction of P is fixed the maximum load on the bolts can be reduced by locating the bolts so that two of the bolts are equally stressed, as shown in Fig. 7.9 (*c*). In this case,

$$\alpha = 180° - \frac{180°}{(2n)}$$

and

$$P_{max} = P\left[\frac{2e\left(a + b\cos\left(\frac{180°}{n}\right)\right)}{(2a^2 + b^2)\,n}\right] \quad ....(7.26)$$

7.9.4. Eccentric Load Acting in the Plane containing the Bolts

Consider the bracket bolted to the frame and loaded as shown in Fig. 7.10. This problem can be solved in the same way as explained for rivets in art. 5.7.

For the bracket shown in Fig. 7.10 (*a*) the forces on the bolts is shown in Fig. 7.10 (*b*).

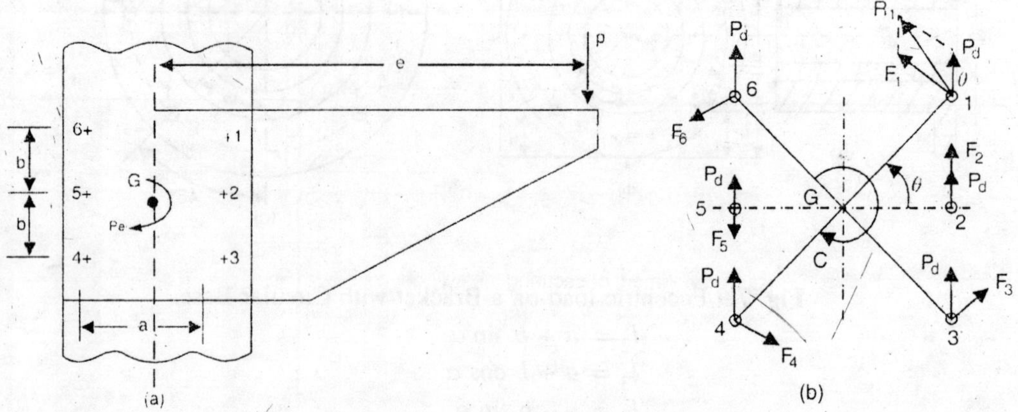

Fig. 7.10 Coplanar eccentric load.

Direct load in each bolt, $P_d = \frac{P}{6}$

Couple acting about the C.G. of the bolt group, $C = Pe$

Resisting couple of all the bolts $= \left(\frac{F_1}{l_1}\right)[4l^2_1 + 2l^2_2]$

$$F_1 = \frac{Pel_1}{4l^2_1 + 2l^2_2}$$

$$\tan \theta = \frac{2b}{a}$$

Maximum resultant force on bolt 1, $R_1 = [P^2_d + F^2_1 + 2P_dF_1]^{1/2}$

The bolts are under shear. Let τ = permissible shear stress in bolts.

Then $\tau \times A_s = R_1$

Where A_s = required stress area of bolt

7.10 DESIGN OF CYLINDER COVER

The cylinder cover may be secured by means of bolts or studs as shown in Fig. 7.11. The various parts of the cylinder may be designed as follows :

1. Cylinder

The cylinder may be designed as a thick pressure vessel. The thickness of cylinder wall is given by,

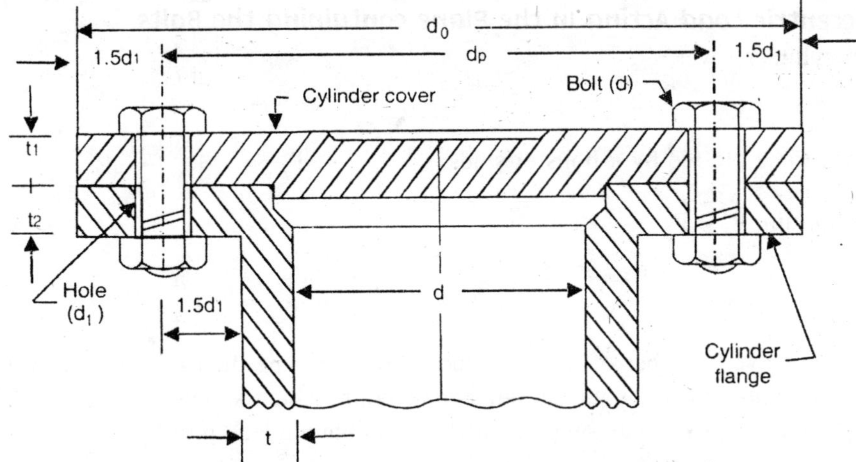

(a) Arrangement of securing the cylinder cover with bolts

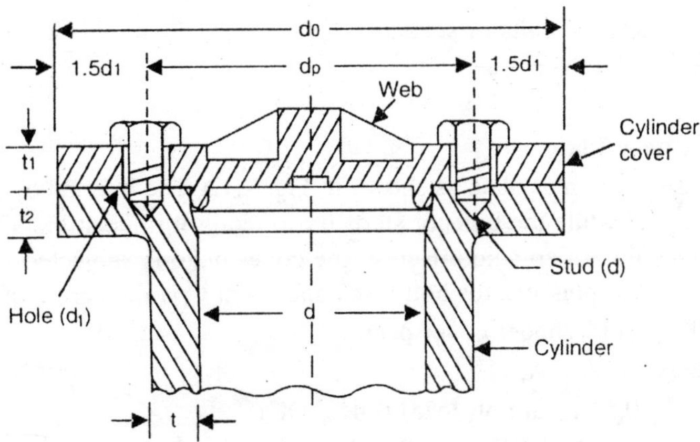

(b) Arrangement of securing the cylinder cover with studs.

Fig. 7.11 Cylinder head designs.

$$t = r\left[\left\{\frac{\sigma_\theta + p}{\sigma_\theta - p}\right\}^{1/2} - 1\right] \qquad \text{...(7.27)}$$

where r = inner radius of cylinder

σ_θ = permissible hoop stress in cylinder wall

p = internal pressure

2. Bolts or Studs

Let d = inner diameter of cylinder

d_c = core diameter of bolts or studs

n = number of bolts or studs

σ_t = permissible tensile stress in bolts or studs

Force in cylinder cover, $$P_c = \left(\frac{\pi}{4}\right) d^2 p$$

Resistance force offered by n bolts or studs, $$P_b = n\left(\frac{\pi}{4}\right) d_c^2\, \sigma_t$$

For equilibrium, $P_c = P_b$

$$\left(\frac{\pi}{4}\right) d^2 p = n\left(\frac{\pi}{4}\right) d_c^2\, \sigma_t \quad \text{....(7.28)}$$

Assume some even number of bolts or studs and determine their core diameter. Otherwise assume $d_c > 16$ mm and calculate the number of bolts, which should be even. The effect of screwing up force on gasket may be accounted for to obtain a leak proof joint.

The circumferential pitch of bolts or studs, $p = 20\sqrt{d_1}$ to $30\sqrt{d_1}$ ....(7.29)

where d_1 = diameter of hole in mm for bolt or stud.

Pitch circle diameter of bolts or studs, $d_p = d + 2t + 3d_1$ ...(7.30)

Outside diameter of cover, $d_0 = d_p + 3d_1 = d + 2t + 6d_1$ ...(7.31)

3. Cylinder Cover Plate

Consider the semi-circular plate as shown in Fig. 7.12. The internal cylinder pressure tries to lift the cylinder cover while the bolts or studs try .o retain it in position. The centres of pressure of these two forces do not coincide. Hence, the cover plate is subjected to bending stress. The point B is the centre of pressure for bolt force and point C is the centre of internal pressure.

Let t_1 = thickness of cylinder cover plate

Bending moment at A – A,

$$M = 0.5\ (\text{Total bolt load})\ (OB - OC)$$
$$= 0.5\ P_b\ (0.318\ d_p - 0.212\ d_p)$$
$$= 0.053\ P_b d_p$$

Section modulus of cover plate,

$$z_c = \frac{w t_1^2}{6}$$

Where w = width of cover plate

$= d_0 - 2d_1$

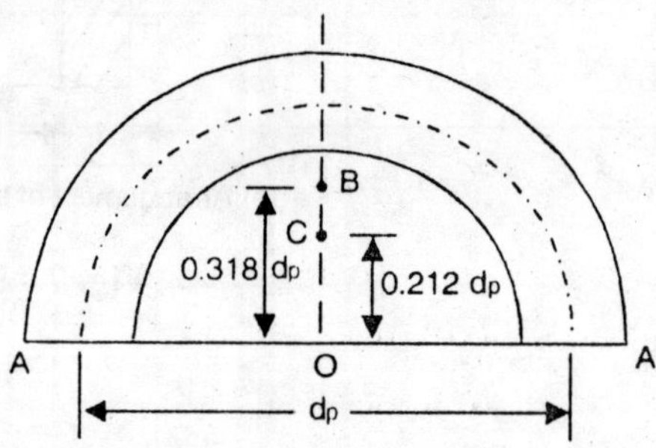

Fig. 7.12 Semi-cover plate of a cylinder

Bending stress in cover plate, $\sigma_{bc} = \dfrac{M}{z_c}$ ....(7.32)

Knowing the allowable bending stress in cover plate material, its thickness t_1 can be determined.

4. Cylinder Flange

The thickness t_2 of cylinder flange may be determined from bending consideration. The portion of the cylinder flange under the influence of one bolt is shown in Fig. 7.13. The force in the bolt produces bending stress at the section $x - x$. The eccentricity of force from section $x - x$ is,

$$e = 0.5\ d_p - (0.5 d_1 + t)$$

Bending moment,

$$M_x = P_b \times \frac{e}{n}$$

Radius of the section

$x - x$, $r_x = 0.5d + t$

Width of the section $x - x$,

$$W_x = \frac{2\pi r_x}{n}$$

Section modulus,

$$z_x = \frac{w_x t_2^2}{6}$$

Bending stress in cylinder flange,

$$\sigma_{bf} = \frac{M_x}{z_x} \quad \text{....(7.33)}$$

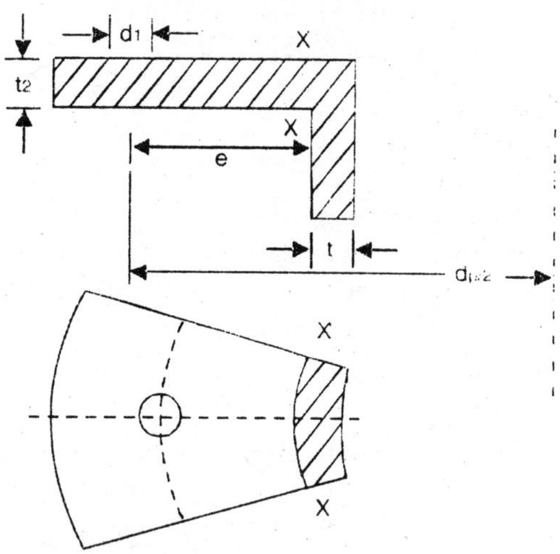

Fig. 7.13 A portion of cylinder flange.

Knowing the permissible tensile stress in flange, its thickness t_2 can be determined.

Example 7.1 The cylinder head of a steam engine is subjected to a steam pressure of 0.8 N/mm². It is held in position by 12 bolts. A soft copper gasket is used to make the joint leak proof. The effective diameter of cylinder is 300 mm. Find the size of the bolts so that the stress in the bolts is not to exceed 100 N/mm².

Solution.

Given $p = 0.8$ N/mm², $n = 12$, $d = 300$ mm, $\sigma_{tb} = 100$ N/mm²

Bolt load, $P_b = P_i + CP$

Let d_b = nominal diameter of bolts

Where P_i = initial tension due to tightening

$= 2805\, d_b$ N

C = 0.5 for soft copper gasket with long through bolts (Table 7.9)

Maximum load on 12 bolts,

$$P_{max} = \frac{\pi d^2 p}{4} = \frac{\pi \times 300^2 \times 0.8}{4} = 56548 \text{ N}$$

$$\text{Load on each bolt, } P = \frac{P_{max}}{n} = \frac{56548}{12} = 4712.4 \text{ N}$$

$$\text{Total bolt load, } P_b = P_i + CP$$

$$= 2805\, d_b + 0.5 \times 4712.4$$

$$= 2805\, d_b + 2356.2 \text{ N}$$

$$\text{Allowable load in the bolt} = \sigma_{tb} \times A_s$$

$$= 100 \times \left(\frac{\pi}{4}\right) \times (0.84 d_b)^2$$

$$= 55.42\, d_b^2$$

$$55.42\, d_b^2 = 2805d + 2356.2$$

or $$d_b^2 - 50.6\, d_b - 42.51 = 0$$

$$d_b = [50.6 + \{(50.6)^2 + 4 \times 42.51\}^{1/2}]$$

$$= 51.4 \text{ mm}$$

We select M52 bolt size with 5 mm pitch.

Example 7.2 The outline of a drill press is shown in Fig. 7.14. The column is secured to the bottom block by 8 bolts as shown (two bolts on each side of column on each location). If allowable tensile stress for the bolt material iş 84 MPa and the external load P = 20 kN acting as shown in figure, find the suitable size of the bolt.

Solution.

Given : $P = 20$ kN, $e = 1700$ mm, $n = 8$, $\sigma_t = 84$ MPa

Distances of the bolts from the tilting edge are :

$$l_1 = l_2 = 80 \text{ mm}$$

$$l_3 = l_8 = 230 \text{ mm}$$

$$l_4 = l_7 = 380 \text{ mm}$$

$$l_5 = l_6 = 530 \text{ mm}$$

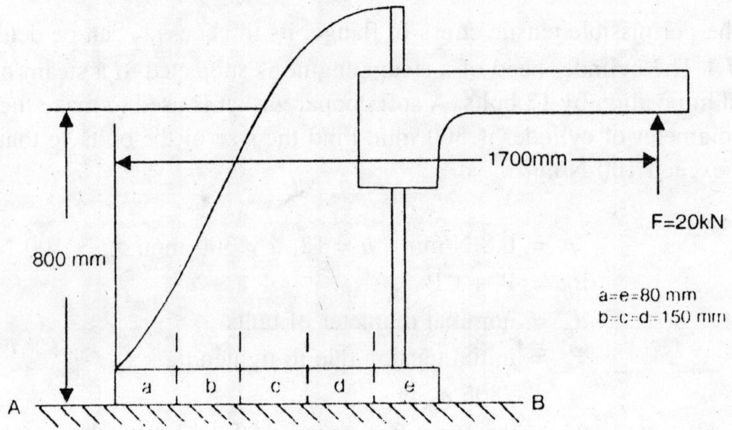

Fig. 7.14

$$P_5 = \frac{Pel_5}{2(l_1^2 + l_3^2 + l_4^2 + l_5^2)}$$

$$= \frac{20,000 \times 1700 \times 530}{2(80^2 + 230^2 + 380^2 + 530^2)}$$

$$= 18592 \text{ N}$$

Direct load, $$P_d = \frac{P}{n} = \frac{20,000}{8} = 2500 \text{ N}$$

Tensile load in the most heavily loaded bolt $= P_d + P_5$

$$= 2500 + 18592 = 21092 \text{ N}$$

Required stress area of bolt, $A_s = \dfrac{21092}{84} = 251.1 \text{ mm}^2$

M 20 × 2 ISO metric threads (fine series) provide, $A_s = 258 \text{ mm}^2$

Hence, we select M 20 × 2 bolt size.

Example 7.3 The inspection hole in the cover plate of a pressure vessel is of 110 mm diameter. The maximum inside pressure is 5 MPa. Determine the plate thickness and size of bolts. Assume allowable tensile stress for mild steel plate as 55 MPa and for bolt material as 40 MPa.

Solution.

Given : $d = 110$ mm or $r = 55$ mm, $p = 5$ MPa, $\sigma_t = 55$ MPa, $\sigma_{tb} = 40$ MPa

Thickness of pressure vessel considering it as a thick cylinder,

$$t = r\left[\left\{\frac{\sigma_t + p}{\sigma_t - p}\right\}^{1/2} - 1\right]$$

$$= 55\left[\left\{\frac{55+5}{55-5}\right\}^{1/2} - 1\right] = 5.247 \approx 6 \text{ mm}$$

Number of bolts, $n = \left(\dfrac{d}{d_c}\right)^2 \times \left(\dfrac{p}{\sigma_{tb}}\right)$

Let nominal diameter of bolts, $d_b = 20$ mm

Then core diameter of bolt, $d_{cb} = 16.933$ mm (Table 7.1)

$$n = \left(\frac{110}{16.933}\right)^2 \times \left(\frac{5}{40}\right) = 5.27 \approx 6$$

Bolt hole diameter, $d_1 = d_b + 1 = 20 + 1 = 21$ mm

Pitch circle diameter of bolts, $d_p = d + 2t + 3d_1 = 110 + 2 \times 6 + 3 \times 21 = 185$ mm

Circular pitch of bolts, $p_c = \dfrac{\pi d_p}{n} = \dfrac{\pi \times 185}{6} = 96.86$ mm

For a leak proof joint,

$$p_c = 20\sqrt{d_1} \text{ to } 30\sqrt{d_1}$$

$$= 20\sqrt{21} = 91.65 \text{ mm to}$$

$$30\sqrt{21} = 137.5 \text{ mm}$$

The calculated circular pitch lies within the specified limits. Hence, M20 bolt size is satisfactory.

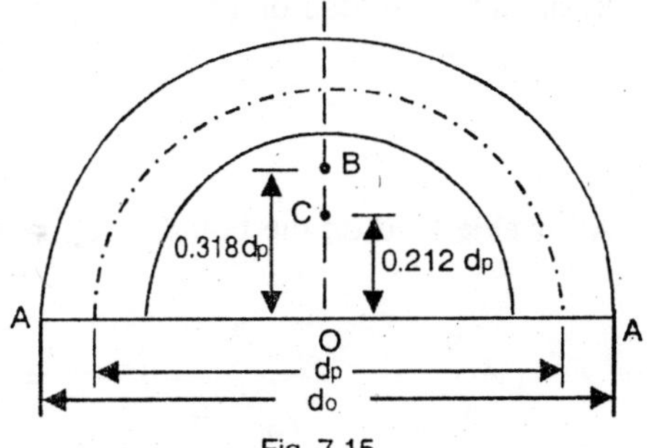

Fig. 7.15

Fig. 7.15

Cover plate :

The geometry of cover plate is shown in Fig. 7.15.

Total load on the bolts, $P = n\left(\frac{\pi}{4}\right)d^2{}_c\sigma_{tb}$

$$= 6 \times \left(\frac{\pi}{4}\right) \times (16.933)^2 \times 40 = 54047 \text{ N}$$

Bending moment at A – A, $M = 0.053\ P \times d_p = 0.053 \times 54047 \times 185 = 529927$ N.mm

Outside diameter of cover plate, $d_0 = d_p + 3d_1 = 185 + 3 \times 21 = 248$ mm

Width of plate, $w = d_0 - 2d_1 = 248 - 2 \times 21 = 206$ mm

Section modulus of plate, $z = \frac{wt_1^2}{6} = 206 \times \frac{t_1^2}{6} = 34.3\ t^2{}_1\ \text{mm}^3$

Where t_1 = thickness of plate

$$M = \sigma_t \times z$$

$$529927 = 55 \times 34.3\ t^2{}_1$$

$$t_1 = 16.75 \approx 18 \text{ mm}$$

Example 7.4 The cylinder head of an air compressor is subjected to 0.6 MPa. It is held in position by **means of** 10 bolts. A soft copper gasket is used to make the joint leak proof. The effective diameter of cylinder is 240 mm. Find the size of the bolts so that the stress in the bolts does not exceed 100 MPa.

Solution.

Given : $p = 0.6$ MPa, $n = 10$, $d = 240$ mm, $\sigma_{tb} = 100$ MPa

Maximum load on all the bolts,

$$P_{max} = \left(\frac{\pi}{4}\right)d^2 p = \left(\frac{\pi}{4}\right) \times 24^2 \times 0.6 = 27143 \text{ N}$$

External load on each bolt, $P = \frac{P_{max}}{n} = \frac{27143}{10} = 2714.3$ N

Let d_b = nominal diameter of bolts

Initial load due to tightening of nuts, $P_i = 2805\ d_b$ N

For soft copper gasket with long through bolts, C = 0.5 (Table 7.9)

Resultant axial load on bolts, $P_b = P_i + CP$

$$= 2805\ d_b + 0.5 \times 2714.3$$

$$= 2805\ d_b + 1357.15 \text{ N}$$

Allowable load on the bolt, $P_a = \left(\frac{\pi}{4}\right)d^2{}_c\sigma_{tb}$

$$= \left(\frac{\pi}{4}\right)(0.84\ d_b)^2\sigma_{tb}$$

$$= \left(\frac{\pi}{4}\right)(0.84\ d_b)^2 \times 100$$

$$= 55.42\ d^2{}_b$$

For $P_a = P_b$, we have

$$55.42\, d_b^2 = 2805\, d_b + 1357.15$$

or

$$d_b^2 - 50.6\, d_b - 24.4 = 0$$

$$d_b = 0.5\,[50.6 + \{(50.6)^2 + 4 \times 24.4\}^{1/2}]$$

$$= 51.08 \approx 52 \text{ mm}$$

We select M52 bolt size.

Example 7.5 A steam engine of effective diameter 260 mm is subjected to a steam pressure of 1.25 MPa. The cylinder head is connected by 8 bolts having tensile yield strength 330 MPa and endurance limit 240 MPa. The bolts are tightened with an initial preload of 1.5 times the steam load. A soft copper gasket is used to make the joint leak proof. Assuming a factor of safety 2, determine the size of bolts required.

Solution.

Given : $d = 260$ mm, $p = 1.25$ MPa, $i = 8$, $S_{yt} = 330$ MPa, $S_e = 240$ MPa, $P_i = 1.5P$, $n = 2$

Load due to steam pressure acting on the bolts, $P = \left(\frac{\pi}{4}\right) d^2 p = \left(\frac{\pi}{4}\right) \times 260^2 \times 1.25 = 66366$ N.

Initial tightening load on bolts, $P_i = 1.5 \times P = 1.5 \times 66366 = 99549$ N

Bolt load, $P_b = P_i + CP$

$C = 0.5$ for soft copper gasket with long through bolts (Table 7.9)

$$P_b = 99549 + 0.5 \times 66366 = 132732 \text{ N}$$

Load per bolt, $P_{max} = \dfrac{P_b}{i} = \dfrac{132732}{8} = 16591$ N

Minimum load on bolts, $P_{min} = \dfrac{P_i}{i} = \dfrac{99549}{8} = 12443$ N

Mean load on bolts, $P_m = 0.5\,(P_{max} + P_{min})$

$= 0.5\,(16591 + 12443) = 14517$ N

Alternating load on bolts, $P_a = 0.5\,(P_{max} - P_{min})$

$= 0.5\,(16591 - 12443) = 2074$ N

Mean stress, $\sigma_m = \dfrac{P_m}{A_s} = \dfrac{14517}{A_s} \dfrac{\text{N}}{\text{mm}^2}$

Alternating stress, $\sigma_a = \dfrac{P_a}{A_s} = \dfrac{2074}{A_s} \dfrac{\text{N}}{\text{mm}^2}$

Soderberg failure criterion gives,

$$\frac{\sigma_m}{S_{yt}} + \frac{\sigma_a}{S_e} = \frac{1}{n}$$

$$\left(\frac{1}{A_s}\right)\left[\frac{14517}{330} + \frac{2074}{240}\right] = \frac{1}{2}$$

$$A_s = 105.625 \text{ mm}^2$$

M16 size bolt has stress area = 157 mm^2, which is selected.

Example 7.6 A bracket as shown in Fig. 7.7 supports a load of 20 kN. Determine the size of bolts if the allowable tensile stress in the bolts is 60 MPa. $l_1 = 100$ mm, $l_2 = 300$ mm, $e = 600$ mm.

Solution.

Given : P = 20 kN, $\sigma_{tb} = 60$ MPa, $l_1 = 100$ mm, $l_2 = 300$ mm, $e = 600$ mm

Direct tensile load in bolts, $P_d = \frac{P}{n} = \frac{20}{4} = 5$ kN

Load in heavily loaded bolt, $P_1 = \frac{Pel_2}{2(l_1^2 + l_2^2)}$

$$= 20 \times 600 \times \frac{300}{2(100^2 + 300^2)}$$

$$= 18 \text{ kN}$$

Maximum tensile load on heavily loaded bolt,

$$P_{max} = P_d + P_1 = 5 + 18 = 23 \text{ kN}$$

Allowable load on bolts, $P_a = \sigma_{tb} \times A_s = 60\ A_s$

For $P_a = P_{max}$, we have

$$A_s = \frac{23{,}000}{60} = 383.3 \text{ mm}^2$$

For M30 size bolt, $A_s = 561$ mm^2, which is selected.

Example 7.7 A bracket is loaded as shown in Fig. 7.16. Determine the size of the bolts if permissible tensile stress for the bolt and arm material is 82 MPa. Also find the cross – section of the arm of the bracket.

Solution.

Given : P = 15 kN, $e = 500$ mm, $l_1 = 50$ mm, $l_2 = 50 + 400 = 450$ mm, $n = 4$, $\sigma_{tb} = 82$ MPa

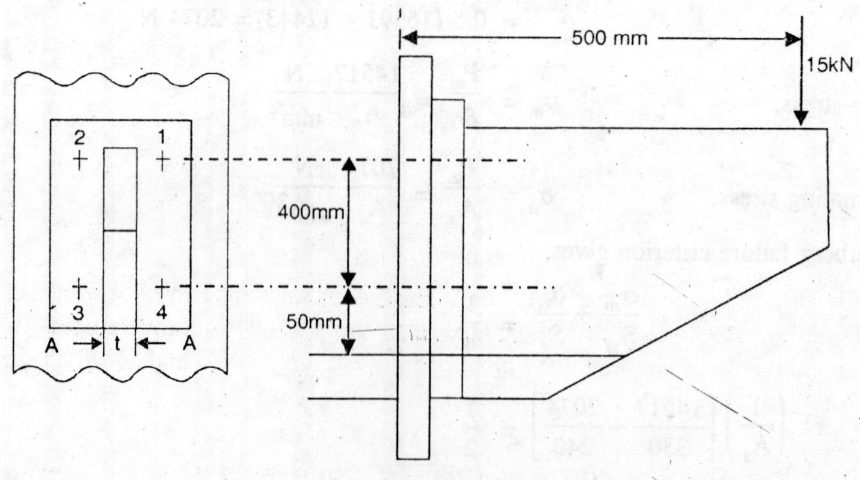

Fig. 7.16

Bolt size :

Direct shear load on each bolt, $P_s = \frac{p}{n} = \frac{15}{4} = 3.75$ kN

Bolts 1 and 2 are heavily loaded. Tensile load in bolts at 1 and 2 positions,

$$P_1 = \frac{Pel_2}{2(l^2_1 + l^2_2)}$$

$$= 5 \times 500 \times \frac{450}{2(50^2 + 450^2)} = 8.232 \text{ kN}$$

Equivalent tensile load, $P_{te} = 0.5\,[P_1 + (P^2_1 + 4P^2_s)^{1/2}]$

$$= 0.5\,[8.232 + \{(8.232)^2 + 4\,(3.75)^2\}^{1/2}]$$

$$= 9.683 \text{ kN}$$

Let A_s = stress area of bolt required, mm^2

$$\sigma_{tb} \times A_s = P_{te}$$

$$A_s = \frac{9683}{82} = 118 \text{ mm}^2$$

We select M16 bolt giving $A_s = 157$ mm^2.

Arms of Bracket :

Let t = thickness of arm

b = width of arm

= 300 mm (assumed)

Section modulus of arm, $z = \frac{tb^2}{6}$

Bending moment in arm, $M = Pe = 15 \times 500 = 7500$ N.m

Bending stress in arm, $\sigma = \frac{M}{z}$

$$82 = 7500 \times 10^3 \times \frac{6}{(t \times 300^2)}$$

$$t = 6 \text{ mm}$$

Example 7.8 Determine the size of the bolts and the thickness of arm for the bracket as shown in Fig. 7.17. It carries a load of 30 kN inclined at an angle of 60° with the vertical. The material of the bracket and bolts is the same for which safe stresses are 75 MPa and 45 MPa in tension and shear respectively.

Solution.

Given : P = 30 kN, σ_t = 75 MPa, τ = 45 MPa, l_1 = 50 mm, l_2 = 50 + 200 = 250 mm

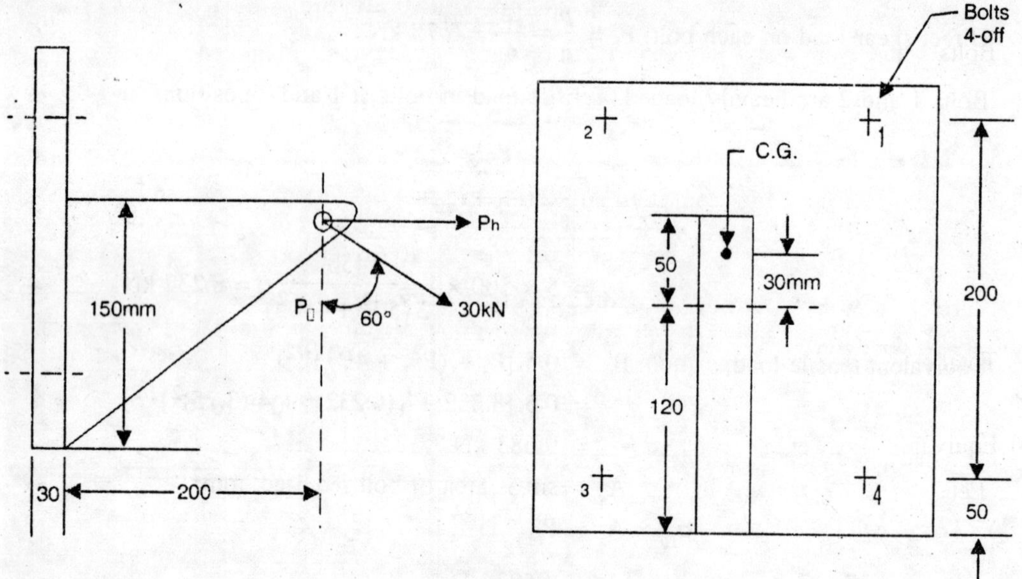

Fig. 7.17

Bolt size :

Horizontal component of load, $P_h = 30 \sin 60° = 25.98$ kN

Vertical component of load, $P_v = 30 \cos 60° = 15$ kN

Horizontal component :

The horizontal component is parallel to the axes of the bolts.

Direct tensile load in each bolt, $$P_{th} = \frac{P_h}{n} = \frac{25.98}{4} = 6.495 \text{ kN}$$

C.G. of bolts from the bottom $$= 50 + \frac{200}{2} = 150 \text{ mm}$$

Distance of C.G. of bolts from line of action of P_h, $y = 150 - 120 = 30$ mm

Twisting moment of the bolts,

$$T_h = P_h \times y = 25.98 \times 30 = 779.4 \text{ N.m (ccw)}$$

Vertical component :

The vertical load is perpendicular to the axes of the bolts.

Shear load in each bolt, $$P_{sv} = \frac{P_v}{n} = \frac{15}{4} = 3.75 \text{ kN}$$

Distance of line of action of P_v from edge of bracket, $x = 200$ mm

Twisting moment on the bolts,

$$T_v = P_v \times x = 15 \times 200 = 3000 \text{ N.m (cw)}$$

Net twisting moment on the bolts,

$$T = T_v - T_h = 3000 - 779.4 = 2220.6 \text{ N.m (cw)}$$

Bolts 1 and 2 are heavily loaded. Load on each heavily loaded bolt,

$$P_{t1} = \frac{Tl_2}{2(l^2_1 + l^2_2)}$$

$$= 2220.6 \times 10^3 \times \frac{250}{2(50^2 + 250^2)}$$

$$= 4270 \text{ N}$$

Total tensile load on upper bolts,

$$P_t = P_{th} + P_{t1} = 6495 + 4270 = 10765 \text{ N}$$

Equivalent tensile load,

$$P_{te} = 0.5\,[P_t + (P^2_t + 4P^2_{sv})^{1/2}]$$

$$= 0.5\,[10765 + (10765^2 + 4 \times 3750^2)^{1/2}] = 11942 \text{ N}$$

Bolt stress area required, $A_s = \frac{P_{te}}{\sigma_t} = \frac{11942}{75} = 159.22 \text{ mm}^2$

M16 × 1.5 gives $A_s \doteq 167 \text{ mm}^2$.

We select M16 × 1.5 bolt size

Arm size :

Width of arm, $b = 120 + 50 = 170$ mm

Let t = thickness of arm, mm

Area of cross – section $A_a = 170t \text{ mm}^2$

Section modulus, $z = \frac{tb^2}{6} = t \times \frac{170^2}{6} = 4817t \text{ mm}^3$

Horizontal load :

Tensile stress, $\sigma_{t1} = \frac{P_h}{A_a} = \frac{25980}{(170\,t)} = \frac{152.82}{t} \text{N}/mm^2$

Horizontal bending moment about C.G. of arm,

$$M_h = P_h\left(120 - \frac{170}{2}\right)$$

$$= 25.98 \times 35 = 909.3 \text{ N.m}$$

Bending tensile stress, $\sigma_{t2} = \frac{M_h}{z} = \frac{909300}{4817t} = \frac{188.8}{t} \text{N}/mm^2$

Vertical load :

Direct shear stress, $\tau = \frac{P_v}{A_a} = \frac{15000}{170t} = \frac{88.23}{t} \text{N}/mm^2$

Vertical bending moment, $M_v = P_v(200 + 30) = 15 \times 230 = 3450$ N.m

Bending tensile stress, $\sigma_{t3} = \frac{M_v}{z} = \frac{3450 \times 10^3}{4817t} = \frac{716.2}{t} \, N/mm^2$

Total tensile stress in arm, $\sigma_t = \sigma_{t1} + \sigma_{t2} + \sigma_{t3} = \frac{(152.82 + 188.8 + 716.2)}{t}$

$$= \frac{1057.82}{t} \; N/mm^2$$

Maximum tensile stress, $(\sigma_t)_{max} = 0.5\,[\sigma_t + (\sigma^2{}_1 + 4\tau^2)^{1/2}]$

$$= \left(\frac{0.5}{t}\right)[1057.82 + \{(1057.82)^2 + 4 \times (88.23)^2\}^{1/2}]$$

$$= 75$$

$$t = 14.2 \approx 15 \text{ mm}$$

Maximum shear stress, $\tau_{max} = 0.5\,(\sigma^2{}_1\, t + 4\tau^2)^{1/2}$

$$= \left(\frac{0.5}{t}\right)[(1057.82)^2 + 4\,(88.23)^2]^{1/2}$$

$$= \frac{536.2}{t}$$

$$= \frac{536.2}{15} = 35.74 \text{ MPa} < 45 \text{ MPa}$$

Hence safe.

Example 7.9 The base of a pillar crane is fastened to the foundation by 8 bolts spaced equally on a bolt circle of diameter 1.5 m. The diameter of the pillar base is 2 m. Determine the size of the bolts when the crane carries a load of 80 kN at a distance of 5 m from the centre of base. The allowable stress in the bolt material is 100 MPa in tension.

Solution.

Given : $n = 8$, $2b = 1.5$ m, $2a = 2$m, P = 80 kN, $e = 5 - b = 4.25$ m, $\sigma_{tb} = 100$ MPa

The pillar crane is shown in Fig. 7.18

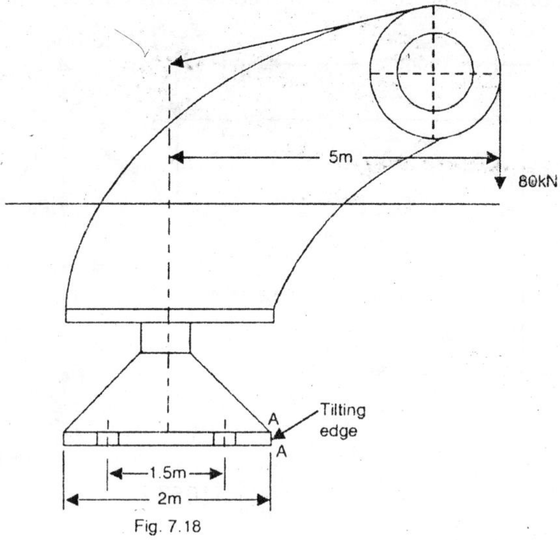

Fig. 7.18

Maximum load on the heavily stressed bolt,

$$P_{max} = P \times 2e \times \frac{(a+b)}{n(2a^2+b^2)}$$

$$= 80 \times 2 \times 4.25 \frac{(1+0.75)}{8(2\times 1+0.75^2)}$$

$$= 58.05 \text{ kN}$$

Stress area of bolt required, $A_s = \frac{P_{max}}{\sigma_{tb}} = \frac{58050}{100} = 580.5 \text{ mm}^2$

M30 × 3 gives $A_s = 581$ mm^2. We select M30 × 3 bolt size.

Example 7.10 A bracket is bolted to a column by 6 bolts of equal size as shown in Fig. 7.19 (*a*). Determine the size of the bolts if maximum stress is limited to 150 MPa.

Solution.

Given : $P = 50$ kN, $e = 150$ mm, $n = 6$

Direct load in each bolt, $P_d = \frac{P}{n} = \frac{50}{6} = 8.33$ kN

The forces in the bolts is shown in Fig. 7.19 (*b*).

$$l_1 = l_3 = l_4 = l_6 = [37.5^2 + 75^2]^{1/2} = 83.85 \text{ mm}$$

$$l_2 = l_5 = 37.5 \text{ mm}$$

Couple acting on the bolts, $C = Pe = 50 \times 150 = 7500$ N.m

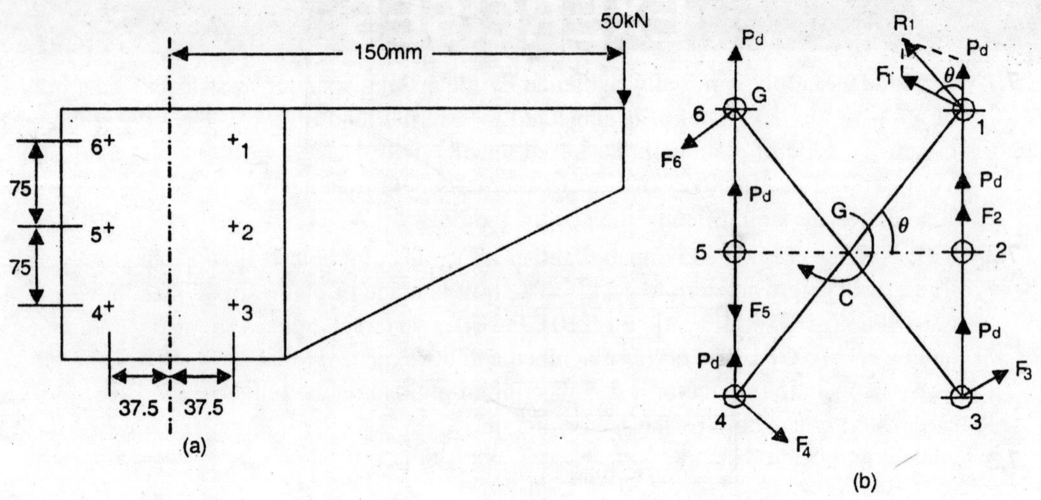

Fig. 7.19

$$\text{Couple due to the forces in the bolts} = \left(\frac{F_1}{l_1}\right)[4l_1^2 + 2l_2^2]$$

$$= \left(\frac{F_1}{83.85}\right)[4 \times (83.85)^2 + 2 \times (37.5)^2]$$

$$= 368.942\ F_1$$

$$= 7500 \times 10^3$$

$$F_1 = 20328 \text{ N}$$

$$\tan\theta = \frac{75}{37.5} = 2$$

$$\theta = 63.4°$$

Bolt 1 is most heavily loaded. Resultant load in bolt 1,

$$R_1 = [P_d^2 + F_1^2 + 2P_dF_1 \cos\theta]^{1/2}$$

$$= [(8330)^2 + (20328)^2 + 2 \times 8330 \times 20328 \times \cos 63.4°]^{1/2}$$

$$= 28661 \text{ N}$$

Stress area of bolt required, $A_s = \dfrac{R_1}{\sigma_{tb}} = \dfrac{28661}{150} = 191 \text{ mm}^2$

M18 × 2.5 size bolt gives $A_s = 192 \text{ mm}^2$. Hence, we select M18 × 2.5 mm size bolt.

EXERCISES

7.1 A bolted assembly is initially tightened by means of a spanner so as to induce a preload of 3 kN in the bolt. Then it is subjected to external load of 6 kN. The bolt with coarse threads is made of plain carbon steel of tensile yield strength 450 MPa and the factor of safety is 3. The effective stiffness of the parts held together by the bolt is 3 times the stiffness of the bolt. Specify the size of the bolt.

7.2 Two circular plates are clamped together by means of a bolt. It is subjected to a vertical force P varying from zero to 15 kN. The bolt is made of plain carbon steel having S_{ut} = 630 MPa, S_{yt} = 380 MPa, and E = 210 GPa. The two circular plates are made of aluminium having E = 70 GPa. The bolt is machined and the expected reliability is 90%. The fatigue stress - concentration factor is 2.3. The initial preload on the bolt is 5 kN. Determine the size of the bolt with 3 as the factor of safety.

7.3 A steel plate is attached to a column by four bolts as shown in Fig. 7.20. The tensile strength of bolt material is 400 MPa and the factor of safety is 2. Calculate the diameter of the bolts.

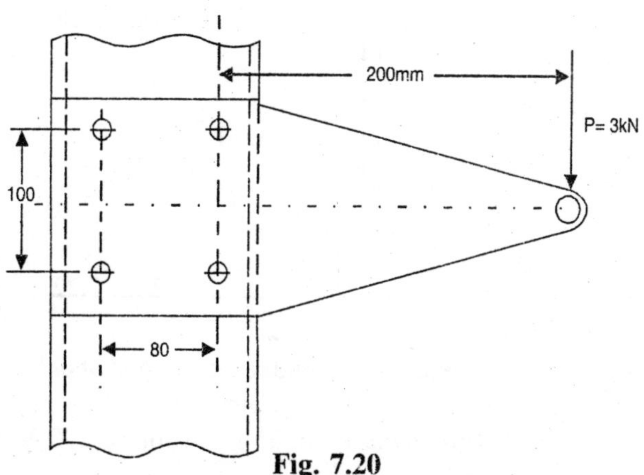

Fig. 7.20

7.4 Determine the size of the bolts and the thickness of the arms for the bracket as shown in Fig. 7.21, if it carries a load of 50 kN. The material of the bracket and the bolts is the same for which safe stresses can be assumed as 70, 50 and 105 MPa in tension, shear and compression respectively.

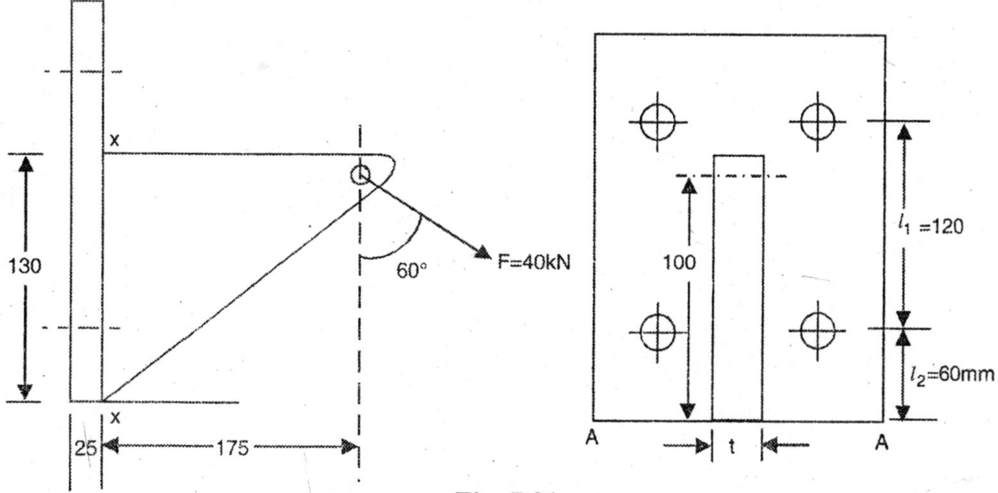

Fig. 7.21

7.5 An eccentric bearing plate connected by bolts is loaded as shown in Fig. 7.22. Locate the bolt with the greatest force and find its diameter if the permissible shear stress in bolts is 40 MPa.

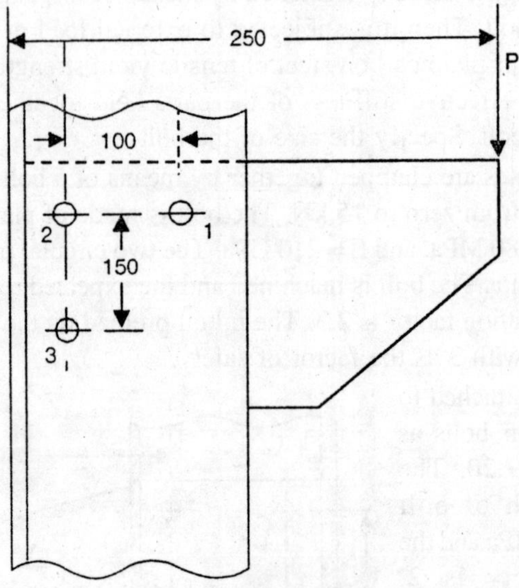

Fig. 7.22

7.6 A load of 50 kN is suspended from a plate held by three bolts as shown in Fig. 7.23. The thickness of the plate is 12 mm and l = 250 mm, $l_1 = l_2$ = 100 mm. The bolts are made of steel with an elastic limit in tension of 250 MPa. Taking a factor of safety of 2, find the size of the bolts.

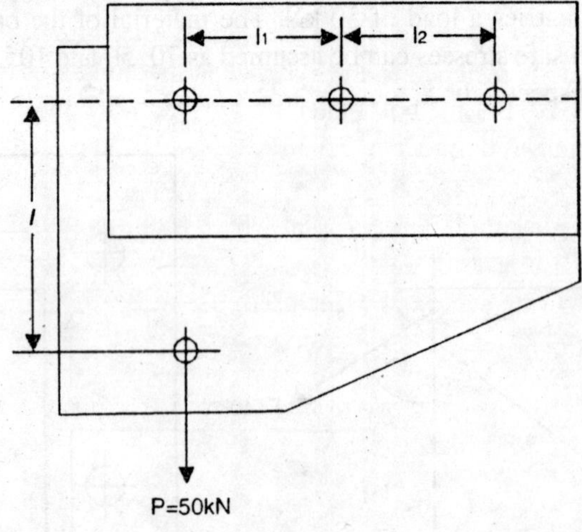

Fig. 7.23

7.7 The base of a pillar crane is fastened to the foundation by eight bolts spaced equally on a bolt circle of 1.6 m diameter as shown in Fig. 7.24. The diameter of the pillar base is 2 m. Determine the size of bolts when the crane carries a load of 100 kN at a distance of 5 m from the centre of the base. The allowable stress for the bolt material is 100 MPa.

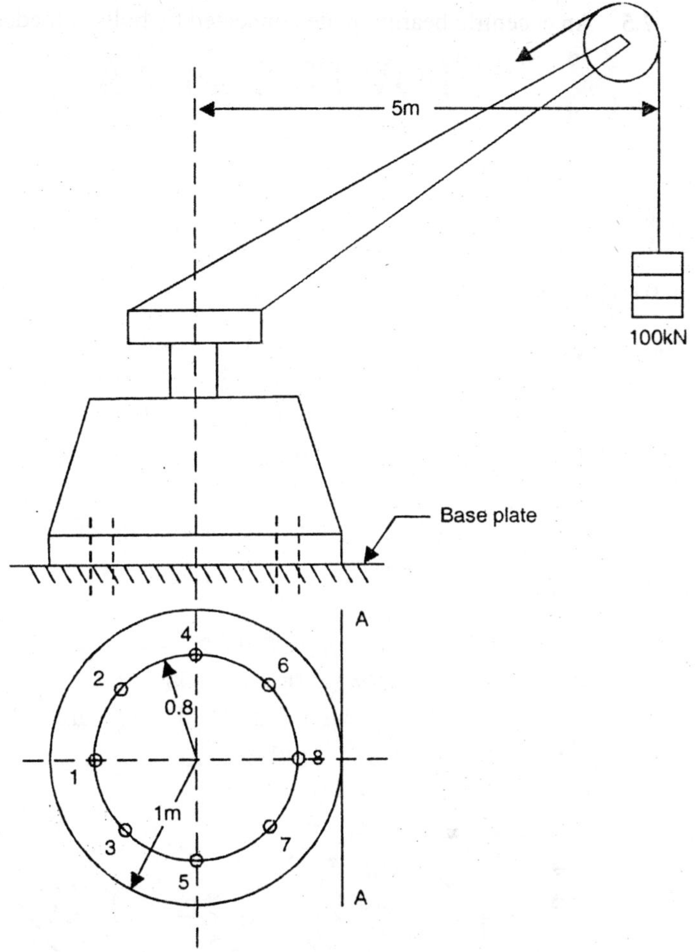

Fig. 7.24

7.8 A cylinder head is fastened to the cylinder of a compressor of 100 mm diameter using 8 bolts of size M16. For the bolt material S_{yt} = 450 MPa and S_{es} = 210 MPa. The maximum pressure of fluid in cylinder is 3.5 MPa. Calculate the factor of safety in the following cases :

(*a*) A soft gasket (k_g = 1) is used which requires initial compressive load of about 70 kN for sealing.

(*b*) An O – ring seal (k_g = 0) is used.

8 POWER SCREWS

Chapter

8.1 INTRODUCTION

A power screw is a mechanical device to convert rotary motion into translational motion and for transmitting power. The main applications of power screws are :

1. To raise or lower the load, e.g. screw jack.
2. To obtain accurate motion in machining operation, e.g. lead screw of a lathe.
3. To clamp a workpiece, e.g. bench vice.
4. To load a specimen, e.g. universal testing machine.

8.2 TYPES OF SCREW THREADS

The following thread forms are used for power screws :

1. Square 2. A4 cme 3. ISO metric trapezoidal and 4. Buttress

These threads have been shown in Fig. 8.1. The guidelines for the selection of a proper thread profile for the power screws are as follows :

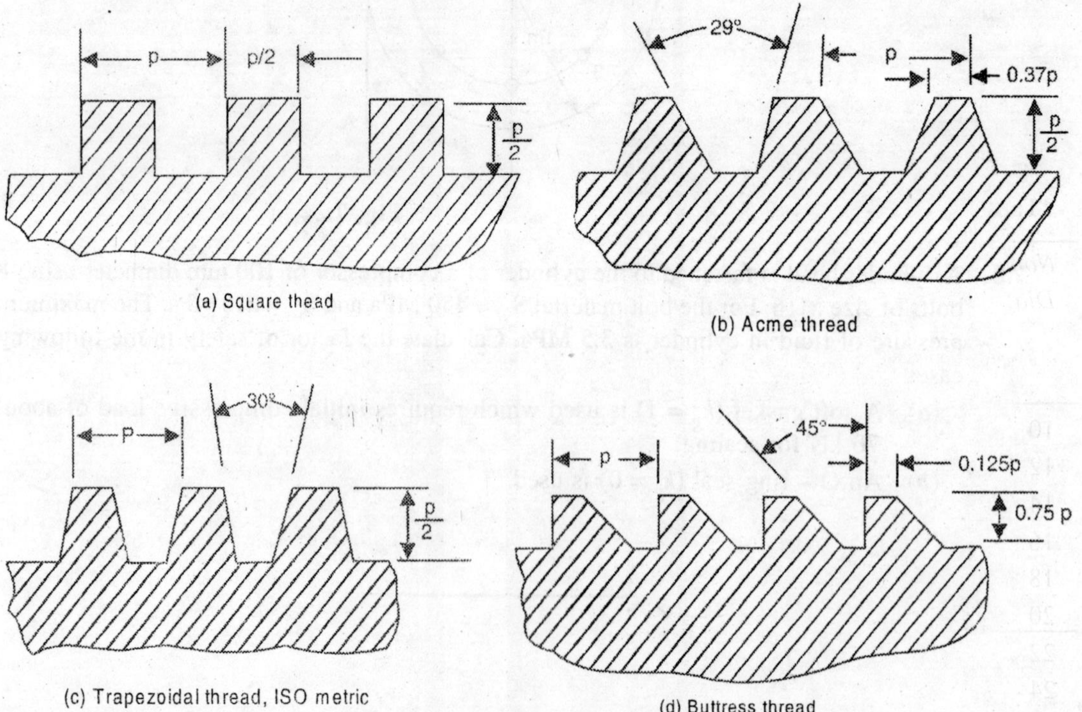

(a) Square thead

(b) Acme thread

(c) Trapezoidal thread, ISO metric

(d) Buttress thread

Fig. 8.1 Threads for power screw

(*i*) The efficiency of square threads is more than that of other types of threads.

(*ii*) Square threads are difficult to manufacture. They are mainly machined on lathe. It is not possible to manufacture the threads on milling machine. On the other hand, acme and trapezoidal threads can be efficiently manufactured on thread milling machine.

(*iii*) The strength of acme and trapezoidal threads is more as compared to square thread due to more thread thickness.

(*iv*) It is not possible to compensate wear in square threads by using a split-type nut. However, it is possible to compensate axial wear in case of acme and trapezoidal threads.

(*v*) Buttress threads can transmit power only in one direction whereas square, acme and trapezoidal threads can transmit power in both directions.

(*vi*) Square threads are used for screw jack and other clamping devices. Acme and trapezoidal threads for the lead screw and other power transmission devices in machine tools. Buttress threads are used in vices where force is applied only in one direction.

The dimensions of screw threads for power screws are gien in Tables 8.1. to 8.4.

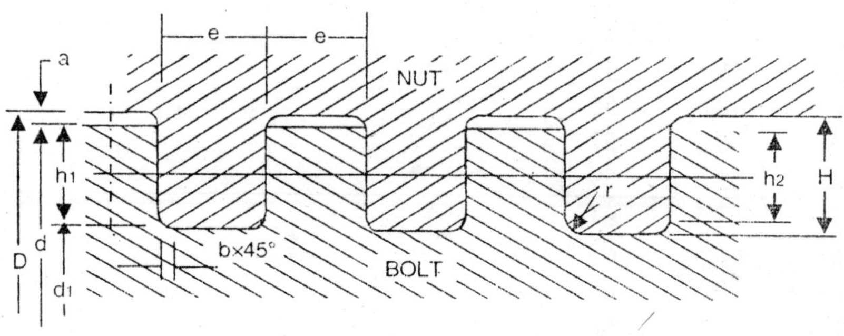

Table 8.1 Basic Dimesion for Square Threads-Fine Series (Dimensions in mm)

Nom., Dia.	*Major Dia*		*Mino Dia*	*Pitch*								*Area of Core.*
	Bolt	*Nut*										
	d	*D*	d_1	*P*	*e*	*r*	h_2	*b*	h_1	*a*	*H*	mm^2
10	10	10.5	8									50.3
12	12	12.5	10									78.5
14	14	14.5	12	2	1	0.12	.75	0.25	1	0.25	1.25	113
16	16	16.5	14									154
18	18	18.5	16									201
20	20	20.5	18									254
22	22	22.5	19									284
24	24	24.5	21									346
26	26	26.5	23									415
28	28	28.5	25									491
30	30	30.5	27									573

Table 8.1 continued

Nom., Dia.	*Major Dia*		*Mino Dia*	*Pitch*								*Area of Core.*
	Bolt	*Nut*										
	d	D	d_1	P	e	r	h_2	b	h_1	a	H	mm^2
32	32	32.5	29									661
(34)	34	34.5	31									755
36	36	36.5	33									855
(38)	38	38.5	35									962
40	40	40.5	37									1075
42	42	42.5	39	3	1.5	0.12	1.25	0.25	1.5	0.25	1.75	1195
44	44	44.5	41									1320
(46)	46	46.5	43									1452
48	48	48.5	45									1590
50	50	50.5	47									1735
52	52	52.5	49									1886
55	55	55.5	52									2124
(58)	58	58.5	55									2376
60	60	60.5	57									2552
(62)	62	62.5	59									2734
65	65	65.5	61									2922
(68)	68	68.5	64									3217
70	70	70.5	66									3421
(72)	72	72.5	68									3632
75	75	75.5	71									3959
(78)	78	78.5	74									4301
80	80	80.5	76									4536
(82)	82	82.5	78									4778
85	85	85.5	81	4	2	0.12	1.75	.25	2	0.25	2.25	5153
(88)	88	88.5	84									5542
90	90	90.5	86									5809
92	92	92.5	88									5809
95	95	95.5	91									6082
(98)	98	98.5	94									6940
100	100	100.5	96									7238
(105)	105	105.5	101									8012
110	110	110.5	106									8825
(115)	115	115.5	109									9331
120	120	120.5	114									10207
(125)	125	125.5	119									11122
130	130	130.5	124	6	3	0.25	2.5	3		0.25	3.25	12076
(135)	135	135.5	129									13070
140	140	140.5	134									14103
(145)	145	145.5	139									15175
150	150	150.5	144									16286

Table 8.2 Basic Dimension for Square Threads-Normal Series (All Dimensions in mm)

Nom., Dia.	*Major Dia*		*Mino Dia*	*Pitch*								*Area of Core.*
	Bolt	*Nut*										
	d	*D*	d_1	*P*	*e*	*r*	h_2	*b*	h_1	*a*	*H*	mm^2
22	22	22.5	17									227
24	24	24.5	19									284
26	26	26.5	21	5	2:5	0.25	2	0.5	2.5	0.25	2.75	346
28	28	28.5	23									415
30	30	30.5	24									452
32	32	32.5	26									531
(34)	34	34.5	28	6	3	0.25	2.5	0.5	3	3.25	616	
36	36	36.5	30									707
(38)	38	38.5	31									755
40	40	40.5	33									855
(42)	42	42.5	35	7	3.5	0.25	3	0.5	3.5	0.25	3.75	962
44	44	44.5	37									1075
(46)	46	46.5	38									1134
48	48	48.5	40									1257
50	50	50.5	42	8	4	0.25	3.5	0.5	4	0.25	4.25	1385
52	52	52.5	44									1521
55	55	55.5	46									1662
(58)	58	58.5	49									1886
60	60	60.5	51	9	4.5	0.25	4	0.5	4.5	0.25	4.75	2043
(62)	62	62.5	53									2206
65	65	65.5	55									2376
(68)	68	68.5	58									2642
70	70	70.5	60									2827
(72)	72	72.5	62									3019
75	75	75.5	65	10	5	0.25	4.5	0.5	5	0.25	5.25	3318
(78)	78	78.5	68									3632
80	80	80.5	70									3848
(82)	82	82.5	72									4072
85	85	85.5	73									4185
(88)	88	88.5	76									4536
90	90	90.5	78									4778
(92)	92	92.5	80									5027

Table 8.2 continued

Nom., Dia.	*Major Dia*		*Mino Dia*	*Pitch*								*Area of Core.*
	Bolt	*Nut*										
	d	*D*	d_1	*P*	*e*	*r*	h_2	*b*	h_1	*a*	*H*	mm^2
95	95	95.5	83	12	6	0.25	5.5	0.5	6	0.25	6.25	5411
(98)	98	98.5	86									5809
100	100	100.5	88									6082
(105)	105	105.5	93									6793
110	110	110.5	98									7543
(115)	115	116	101									8012
120	120	121	106									8825
(125)	125	126	111									9677
130	130	131	116	14	7	0.5	6	1	7	0.5	7.5	10568
(135)	135	136	121									11499
140	140	141	126									12469
(145)	145	146	131									13478
150	150	151	134	16	8	0.5	7	1	8	0.5	8.5	14013

Table 8.3 Basic Dimension for Square Threads-Normal Series (All Dimensions in mm)

Nom., Dia.	*Major Dia*		*Mino Dia*	*Pitch*								*Area of Core.*
	Bolt	*Nut*										
	d	*D*	d_1	*P*	*e*	*r*	h_2	*b*	h_1	*a*	*H*	mm^2
22	22	22.5	14									164
24	24	24.6	16									201
26	26	25.5	18	8	4	0.25	3.5	0.5	4	0.25	4.25	254
28	28	28.5	20									314
30	30	30.5	20									314
32	32	32.5	22									380
(34)	34	34.5	24	10	5	0.25	4.5	0.5	5	0.25	5.25	452
36	36	36.5	26									531
(38)	38	38.5	28									616
(40)	40	40.5	28									616
(42)	42	42.5	30									707
44	44	44.5	32									804
(46)	46	46.5	34	12	6	0.25	5.5	0.5	6	0.25	6.25	908
48	48	48.5	36									1018
50	50	50.5	38									1134

Table 8.3 continued

Nom., Dia.	*Major Dia*		*Mino Dia*	*Pitch*								*Area of Core.*
	Bolt	*Nut*										
	d	*D*	d_1	*P*	*e*	*r*	h_2	*b*	h_1	*a*	*H*	mm^2
52	52	52.5	40									1257
55	55	56	41									1320
(58)	58	59	44									1521
60	60	61	46	14	7	0.5	6	1	7	0.5	7.5	1662
(62)	62	63	48									1810
65	65	66	49	16	8	0.5	7	1	8	0.5	8.5	1886
(68)	68	69	52									2124
70	70	71	54									2290
(72)	72	73	56									2463
75	75	76	59	16	8	0.5	7	1	8	0.5	8.5	2734
(78)	78	79	62									3019
80	80	81	64									3217
(82)	82	83	66									3421
85	85	86	67									3526
(88)	88	89	70									3848
90	90	91	72									4072
(92)	92	93	74	18	9	0.5	8	1	9	0.5	9.5	4301
95	95	96	77									4657
(98)	98	99	80									5027
100	100	101	80									5027
(105)	105	106	85	20	10	0.5	9	1	10	0.5	10.5	5675
110	110	111	90									6362
(115)	115	116	93									6793
120	120	121	98	22	11	0.5	10	1	11	0.5	11.5	7543
(125)	125	126	103									8332
130	130	131	108									9161
(135)	135	136	111									9677
140	140	141	116	24	12	0.5	11	1	12	0.5	12.5	10568
(145)	145	146	121									11499
150	150	151	126									12469

Note : Diameters within brackets are of second preference.

Table 8.4 ISO Metric Trapezoidal Screw Threads (Dimensions as per IS : 7008—1973)

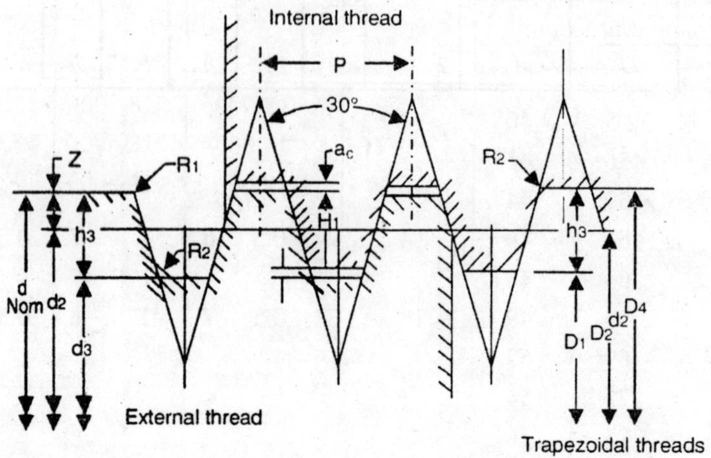

$H_1 = 0.5\ P$

$H_3 = h_3 = 0.5\ P + a_c = H_1 + a_c$

$z = 0.25\ P = H_1/2$

$d_3 = d - 2H_3$

$d_2 = D_2 = d - 2z = d - 0.5\ P$

$s = 0.26795\ es$

$R_1\ max. = 0.5\ a_c$

$R_2\ max. = a_c$

$D_1 = d - 2H_1 = d - P$

$D_4 = d + 2a_c$

Trapezoidal threads (All dimensions in mm)

Nominal Diameter d			Pitch P	Pitch Diameter	Major Diameter	Minor Diameter	
Column 1	Column 2	Column 3		$d_2 = D_2$	D_4	d_3	D_1
8			1.5	7.250	8.300	6.200	6.500
			1.5	8.250	9.300	7.200	7.500
	9		2	8.000	9.500	6.500	7.000
			1.5	9.250	10.300	8.200	8.500
10			2	9.000	10.500	7.500	8.000
			2	10.000	11.500	8.500	9.000
	11		3	9.500	11.500	7.500	8.000
12			3	10.500	12.500	8.500	9.000
			2	13.000	14.500	11.500	12.000
	14		3	12.500	14.500	10.500	11.000
			2	15.000	16.500	13.500	14.000
16			4	14.000	16.500	11.500	12.000

Table 8.4 continued

Nominal Diameter d			Pitch P	Pitch Diameter	Major Diameter	Minor Diameter	
Column 1	Column 2	Column 3		$d_2 = D_2$	D_4	d_3	D_1
			2	17.000	18.500	15.500	16.000
	18		4	16.000	18.500	13.500	14.000
			2	19.000	20.500	17.500	18.000
20			4	18.000	20.500	15.500	16.000
			3	20.500	22.500	18.500	19.000
	22		5	19.500	22.500	16.500	17.000
			8	18.000	23.000	13.000	14.000
24			5	21.500	24.500	18.500	19.000
			8	20.000	25.000	15.000	16.000
			3	24.500	26.500	22.500	23.000
	26		5	23.500	26.500	20.500	21.000
			8	22.000	27.000	17.000	18.000
			3	26.500	28.500	24.500	25.000
28			5	25.500	28.500	22.500	23.000
			8	24.000	29.000	19.000	20.000
			3	28.500	30.500	26.500	27.000
	30		6	27.000	31.000	23.000	24.000
			10	25.000	31.000	19.000	20.000
			3	30.500	32.500	28.500	29.000
32			6	29.000	33.000	25.000	26.000
			10	27.000	33.000	21.000	22.000
			3	32.500	34.500	30.500	31.000
	34		6	31.000	35.000	27.000	28.000
			10	29.000	35.000	23.000	24.000
			3	34.500	36.500	32.500	33.000
36			6	33.000	37.000	29.000	30.000
			10	31.000	37.000	25.000	26.000
			3	36.500	38.500	34.500	35.000
	38		7	34.500	39.000	30.000	31.000
			10	33.000	39.000	27.000	28.000
			3	38.500	40.500	36.500	37.000
40			7	36.500	41.000	32.000	33.000
			10	35.000	41.000	29.000	30.000
			3	40.500	42.500	38.500	39.000
	42		7	38.500	43.000	34.000	35.000

Table 8.4 continued

Nominal Diameter d			Pitch P	Pitch Diameter	Major Diameter	Minor Diameter	
Column 1	Column 2	Column 3		$d_2 = D_2$	D_4	d_3	D_1
			10	37.000	43.000	31.000	32.000
			3	42.500	44.500	40.500	41.000
44			7	40.500	45.000	36.000	37.000
			12	38.000	45.000	31.000	32.000
			3	44.500	46.500	42.500	43.000
	46		8	42.000	47.000	37.000	38.000
			12	40.000	47.000	33.000	34.000
			3	46.500	48.500	44.500	45.000
48			8	44.000	49.000	39.000	40.000
			12	42.000	49.000	35.000	36.000
			3	48.500	50.500	46.500	47.000
	50		8	46.000	51.000	41.000	42.000
			12	44.000	51.000	37.000	38.000
			3	50.500	52.500	48.500	49.000
52			8	48.00	53.000	43.000	44.000
			12	46.000	53.000	39.000	40.000
			3	53.500	56.000	45.000	52.000
	55		9	50.500	56.000	45.000	46.000
			14	48.000	57.000	39.000	41.000
			3	58.500	60.500	56.500	57.000
60			9	55.500	61.000	50.000	51.000
			14	53.000	62.000	44.000	46.000
			4	63.000	65.500	60.500	61.000
	65		10	60.000	66.000	54.000	55.000
			16	57.000	67.000	47.000	49.000
			4	68.000	.70.500	65.500	66.000
70			10	65.000	71.000	59.000	60.000
			16	62.000	72000	52.000	54.000
			4	73.000	75.500	70.500	71.000
	75		10	70.000	75.500	70.500	71.000
			16	67.000	77.000	57.000	59.000
			4	78.000	80.500	75.500	76.000
80			10	75.000	81.000	69.000	70.000
			16	72.000	82.000	62.000	64.000
			4	83.000	85.500	80.500	81.000
	85		12	79.000	86.000	72.000	73.000
			18	76.000	87.000	65.000	67.000

8.3 FORCE ANALYSIS OF SQUARE THREADS

The major dimensions of a square threaded screw are shown in Fig. 8.2(*a*).

Pitch, p—It is defined as the distance measured parallel to the axis from a point on one thread to the corresponding point on the adjacent thread.

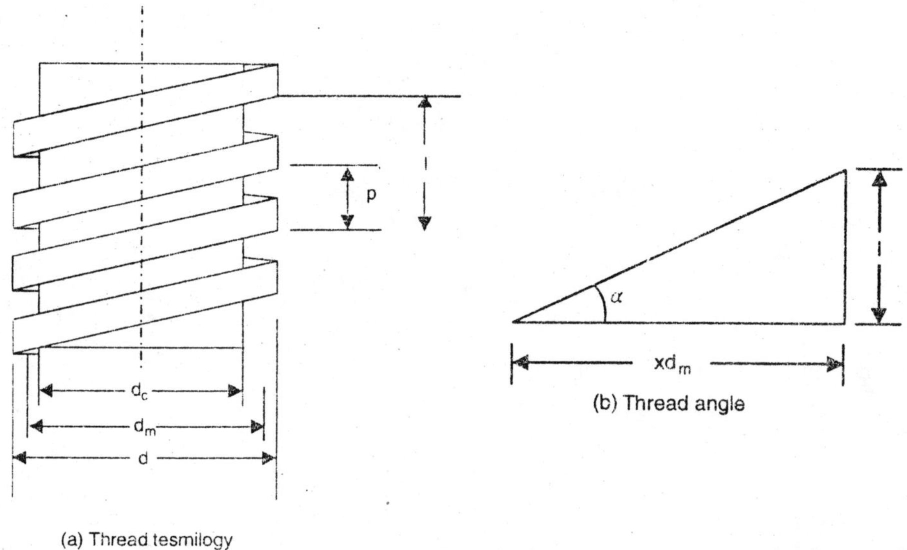

(a) Thread tesmilogy

(b) Thread angle

Fig. 8.2

Lead, *l*—It is defined as the distance measured parallel to the axis that the nut will advance in one revolution of the screw. For a single - start screw lead and pitch are the same. For a multi start screw, lead = number of starts × pitch.

The development of the thread is shown in Fig. 8.2 (*b*). The helix angle α of the thread is given by,

$$\tan \alpha = \frac{p}{\pi d_m} \text{ or } \frac{l}{\pi d_m} \qquad ...(8.1)$$

where d_m = mean diameter of screw

1. Lifting Load

Consider the body being lifted up the plane by a horizontal force P applied as shown in Fig. 8.3 (*a*). Resolving the forces perpendicular and along the plane, we have

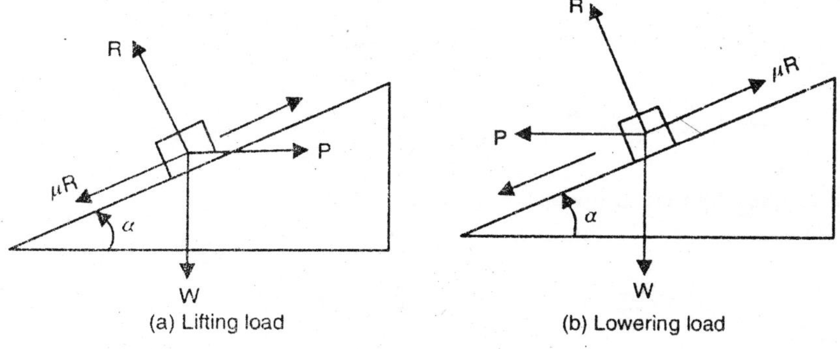

(a) Lifting load

(b) Lowering load

Fig. 8.3 Force analysis of square threads

$$P \cos \alpha = W \sin \alpha + \mu R$$

$$P \sin \alpha + W \cos \alpha = R$$

or

$$P \cos \alpha = W \sin \alpha + \tan \phi \, (P \sin \theta + W \cos \alpha)$$

$$P (\cos \alpha \cos \phi - \sin \alpha \sin \phi) = W (\sin \alpha \cos \phi + \cos \alpha \sin \phi)$$

$$P \cos (\alpha + \phi) = W \sin (\alpha + \phi)$$

$$P = W \tan (\alpha + \phi) \quad ...(8.2)$$

$$= \frac{W (\tan \alpha + \tan \phi)}{1 - \tan \alpha \tan \phi}$$

$$= \frac{W (\tan \alpha + \mu)}{1 - \mu \tan \alpha}$$

$$= \frac{W (\sin \alpha + \mu \cos \alpha)}{\cos \alpha - \mu \sin \alpha} \quad ...(8.3)$$

Where $\phi = \tan^{-1} \mu$ = angle of friction

Frictional torque required to lift the load,

$$T = \frac{Pd_m}{2} = \left(\frac{Wd_m}{2}\right) \tan (\alpha + \phi) \quad ...(8.4)$$

2. Lowering Load

Consider the body being lowered down the plane by a horizontal force P applied as shown in Fig. 8.3 (*b*). Resolving the forces perpendicular and along the plane, we have

$$P \cos \alpha + W \sin \alpha = \mu R$$

$$P \sin \alpha + R = R \cos \alpha$$

or

$$P \cos \alpha + W \sin \alpha = \mu (W \cos \alpha - P \sin \alpha)$$

$$P (\cos \alpha + \mu \sin \alpha) = W (\mu \cos \alpha - \sin \alpha)$$

$$P = \frac{W (\mu \cos \alpha - \sin \alpha)}{\cos \alpha + \mu \sin \alpha}$$

$$= \frac{W (\mu - \tan \alpha)}{1 + \mu \tan \alpha}$$

$$= \frac{W (\tan \phi - \tan \alpha)}{1 + \tan \phi \tan \alpha}$$

$$= W \tan (\phi - \alpha) \quad ...(8.5)$$

Frictional torque required to lower the load,

$$T = \frac{Pd_m}{2} = \left(\frac{Wd_m}{2}\right) \tan (\phi - \alpha) \quad ...(8.6)$$

3. Efficiency of Screw

The efficiency of the screw is defined as the ratio of the effort required without friction and with friction.

$$\eta = \frac{P_0}{P}$$

(*a*) For lifting the load,

$$\eta = \frac{\tan\alpha}{\tan(\alpha + \phi)} \quad ...(8.7)$$

(*b*) For lowering the load,

$$\eta = \frac{\tan(\alpha - \phi)}{\tan\alpha} \quad ...(8.8)$$

Torque will remain positive if $\phi > \alpha$. Such a screw is called **self locking screw.** If $\phi < \alpha$, then the torque will become negative. In other words, the screw will lower of its own. Such a screw is called **over hauling screw.**

For $\alpha = \phi$, the efficiency of the screw becomes,

$$\eta = \frac{\tan\phi}{\tan(2\phi)}$$

For a self locking screw,

$$\eta \le \frac{\tan\phi}{\tan(2\phi)} \le \left[\frac{1}{2} - \tan^2\left(\frac{\phi}{2}\right)\right]$$

$$\le 50\%$$

8.4 FORCE ANALYSIS OF TRAPEZOIDAL THREADS

For a screw having trapezoidal threads, as shown in Fig. (8.4), let 2β be the angle of the threads.

Then normal reaction on the threads,

$$R = \frac{W}{\cos\beta}$$

Frictional force,

$$F = \mu R = \left(\frac{\mu}{\cos\beta}\right).W$$

$$= \mu_e W \quad ...(8.9)$$

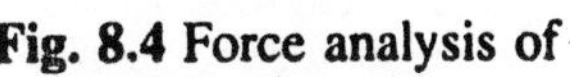

Fig. 8.4 Force analysis of trapezoidal thread

where μ_e is called the virtual or equivalent coefficient of friction.

Therefore, for a screw having trapezoidal threads, the virtual coefficient of friction should be used to calculate the torque required to lift or lower the load.

1. Lifting Load

$$P = \frac{W(\tan\alpha + \mu\sec\beta)}{1 - \mu\sec\beta\tan\alpha} \quad ...(8.10)$$

$$T_f = \left(\frac{Wd_m}{2}\right)\left[\frac{\tan\alpha + \mu\sec\beta}{1 - \mu\sec\beta\tan\alpha}\right] \qquad ...(8.11)$$

$$\eta = \frac{\tan\alpha\,(1 - \mu\sec\beta\tan\alpha)}{\mu\sec\beta + \tan\alpha} \qquad ...(8.12)$$

2. Lowering Load

$$P = \frac{W(\mu\sec\beta - \tan\alpha)}{1 + \mu\sec\beta\tan\alpha} \qquad ...(8.13)$$

$$T_f = \left(\frac{Wd_m}{2}\right)\left[\frac{\mu\sec\beta - \tan\alpha}{1 + \mu\sec\beta\tan\alpha}\right] \qquad ...(8.14)$$

8.5 CONDITION FOR EFFICIENCY TO BE MAXIMUM

Efficiency,

$$\eta = \frac{\tan\alpha}{\tan(\alpha + \phi)}, \text{ for lifting of load}$$

For efficiency to be maximum,

$$\frac{d\eta}{d\alpha} = 0$$

Now

$$\eta = \frac{\tan\alpha(1 - \tan\alpha\tan\phi)}{\tan\alpha + \tan\phi}$$

$$= \frac{\left(\frac{\sin\alpha}{\cos\alpha}\right)\left[1 - \left(\frac{\sin\alpha}{\cos\alpha}\right)\left(\frac{\sin\phi}{\cos\phi}\right)\right]}{\left[\frac{\sin\alpha}{\cos\alpha} + \frac{\sin\phi}{\cos\phi}\right]}$$

$$= \frac{\sin\alpha[\cos\alpha\cos\phi - \sin\alpha\sin\phi]}{\cos\alpha[\sin\alpha\cos\phi + \sin\phi\cos\alpha]}$$

$$= \frac{\sin\alpha\cos(\alpha + \phi)}{\cos\alpha\sin(\alpha + \phi)}$$

$$= \frac{[\sin(2\alpha + \phi) - \sin\phi]}{[\sin(2\alpha + \phi) + \sin\phi]}$$

For efficiency to be maximum, $\sin(2\alpha + \phi) = 0$

or $$2\alpha + \phi = 90^\circ$$

or $$\alpha = 45^\circ - \frac{\phi}{2} \qquad ...(8.15)$$

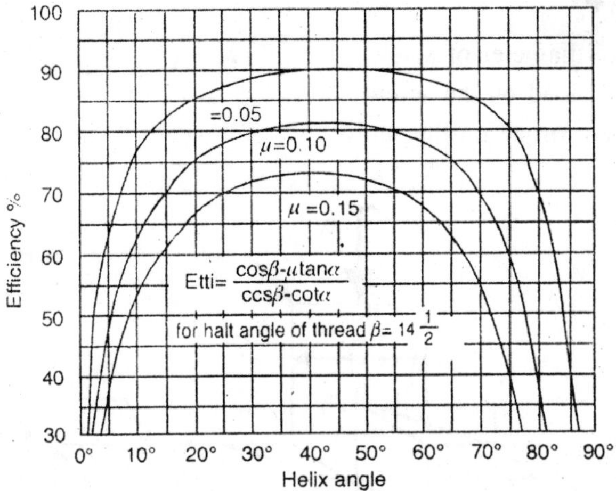

Fig. 8.5 Efficiency of power screw when collar friction is negligible

$$\eta_{\max} = \frac{1 - \sin\phi}{1 + \sin\phi} \qquad \text{....(8.16)}$$

The efficiency of power screw when collar friction is negligible is shown in Fig. 8.5

8.6. COLLAR FRICTION

Consider a collar of inside diameter D_i and outside diameter D_0, as shown in Fig. 8.6. The collar of the power screw rotates with respect to the cup which remains stationary due to load W. This results in friction from diameter D_i to D_o. According to the uniform pressure theory, the torque required to overcome friction at the collar,

$$T_c = \left(\mu_c \frac{W}{3}\right)\left[\frac{D_0^3 - D_i^3}{D_0^2 - D_i^2}\right] \qquad \text{...(8.17a)}$$

and according to uniform wear theory,

$$T_c = \left(\mu_c \frac{W}{4}\right)(D_0 + D_i) \qquad \text{...(8.17b)}$$

Where μ_c = collar coefficient of friction (Table 8.5)

Total torque,

$$T = T_f + T_c \qquad \text{....(8.18)}$$

Fig. 8.6 Collar friction

Table 8.5 Coefficient of Friction for Thrust Collars of Power Screws

Material	During starting	During running
Soft steel on cast iron	0.170	0.121
Hardened steel on cast iron	0.147	0.092
Soft steel on bronze	0.101	0.084
Hardened steel on bronze	0.081	0.063

8.7 STRESSES IN SCREW

Let d_c = core diameter of screw

W = axial load on the screw

T = Total torque on the screw

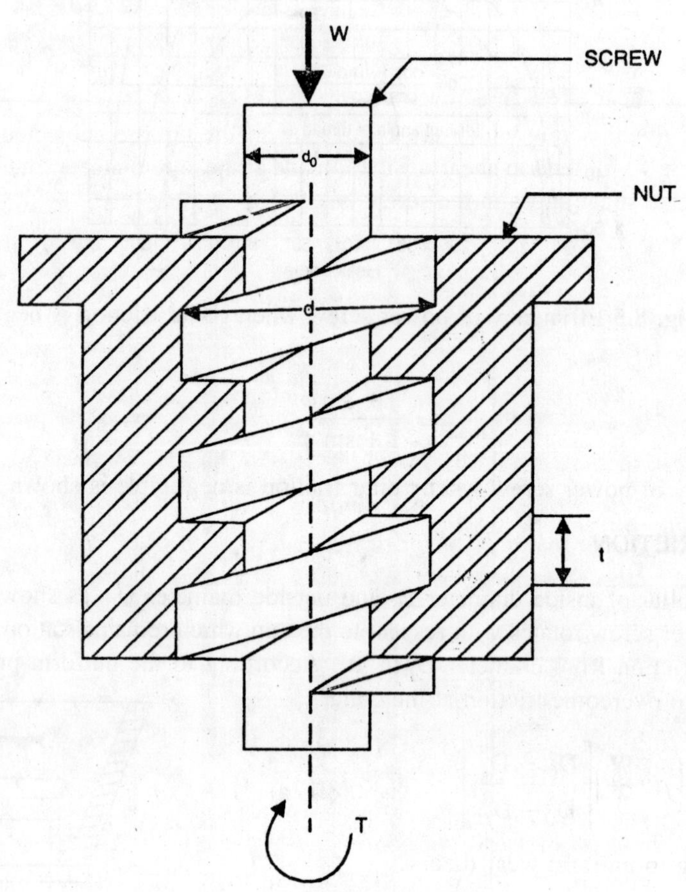

Fig. 8.7 Stresses in screw

The stresses acting on the screw shown in Fig. 8.7 are :

1. Direct compressive stress,

$$\sigma_c = \frac{4W}{\pi d_c^2} \quad ..(8.19)$$

2. Torsional shear stress,

$$\tau = \frac{16T}{\pi d_c^3} \quad ...(8.20)$$

3. Maximum normal and shear stress,

$$\sigma_1 = 0.5\left[\sigma_c + \left\{\sigma_c^2 + 4\tau^2\right\}^{1/2}\right] \quad ...(8.21a)$$

$$\tau_{max} = \left[\left(\frac{\sigma_c}{2}\right)^2 + \tau^2\right]^{1/2} \quad ...(8.21b)$$

4. Transverse shear stress in screw

The threads of the screw which are engaged with the nut are subjected to transverse shear stresses. The screw will tend to shear off the threads at the core diameter under the action of load W. The transverse shear stress in the screw is given by,

$$\tau_s = \frac{W}{\pi d_c t n} \quad ...(8.22)$$

where t = thickness of threads

n = number of threads in engagement

5. Transverse shear stress in nut,

$$\tau_n = \frac{W}{\pi d t n} \quad ...(8.23)$$

where d = outside or major diameter of nut or screw

6. Bearing pressure between screw and nut

$$p_b = \frac{4W}{\pi\left(d^2 - d_c^2\right)n} \quad ...(8.24a)$$

$$= \frac{W}{\pi d_m t n} \quad ...(8.24b)$$

7. Compressive stress in screw considering column action.

when the load has been lifted to the maximum height, the screw may be considered as a column with suitable end conditions. In such a case, the compressive stress in screw may be determined by Rankine—Gordon or Johnson's formula for buckling of columns.

Critical laod,

$$W_{cr} = \sigma_y \times A_c\left[\frac{1 - \sigma_y\left(\frac{L}{K}\right)^2}{4C\pi^2 E}\right] \quad ...(8.25)$$

Where σ_y = yield stress

L = screw length

$$K = \text{least radius of gyration} = \frac{d_c}{4}$$

$$C = \text{end condition coefficient} = 0.25$$

$$E = \text{Modulus of elasticity} \approx 20 \text{ kN/mm}^2$$

$$A_c = \text{core area of screw} = \left(\frac{\pi}{4}\right) d_c^2$$

Example 8.1 A power screw having double — start screw threads of 26 mm nominal diameter 9 (normal series) and 5 mm pitch is used to lift a load of 10 kN. The outer and inner diameters of screw collar are 5 mm and 20 mm respectively. The coefficient of thread friction and collar friction may be assumed as 0.2 and 0.15 respectively. The screw rotates at 15 rpm. Assuming uniform wear condition at the collar and allowable thread bearing pressure of 6 MPa, determine (*a*) the stress in the screw, and (*b*) number of threads of nut in engagement with screw.

Solution.

Given : $d = 26$ mm, $p = 5$ mm, W = 10 kN, $D_o = 50$ mm, $D_i = 20$ mm, $\mu = 0.2$, $\mu_c = 0.15$, N = 15 rpm, $p_b = 6$ MPa

Mean diameter of screw, $$d_m = d - \frac{p}{2} = 26 - \frac{5}{2} = 23.5 \text{ mm}$$

Core diameter, $$d_c = d - p = 26 - 5 = 21 \text{ mm}$$

Thread thickness, $$t = \frac{p}{2} = 2.5 \text{ mm}$$

Lead, $$l = 2p = 10 \text{ mm}$$

Helix angle,

$$\alpha = \tan^{-1}\left[\frac{l}{\pi d_m}\right] = \tan^{-1}\left[\frac{10}{\pi \times 23.5}\right] = 7.714°$$

Friction angle at threads,

$$\phi = \tan^{-1} \mu = \tan^{-1} 0.2 = 11.31°$$

Frictional torque,

$$T_f = \left(\frac{W d_m}{2}\right) \tan(\alpha + \phi)$$

$$= \left(\frac{10{,}000 \times 23.5}{2}\right) \tan(7.714° + 11.31°)$$

$$= 40514 \text{ N. mm}$$

Collar torque,

$$T_c = \left(\frac{\mu_c W}{4}\right)(D_o + D_i)$$

$$= \left(\frac{0.15 \times 10{,}000}{4}\right)(50 + 20)$$

$$= 26250 \text{ N.mm}$$

Total torque,

$$T = T_f + T_c = 40514 + 26250 = 66764 \text{ N.mm}$$

(*a*) Stresses in the screw

Direct compressive stress,

$$\sigma_c = \frac{4W}{\pi d_c^2} = \frac{4 \times 10{,}000}{\pi \times 21^2} = 28.87 \text{ N/mm}^2$$

Torsional shear stress,

$$\tau = \frac{16T}{\pi \times d_c^3} = \frac{16 \times 66764}{\pi \times 21^3} = 36.71 \text{ N/mm}^2$$

Maximum shear stress,

$$\tau_{max} = \left[\left(\frac{\sigma_c}{2}\right)^2 + \tau^2\right]^{1/2}$$

$$= \left[\left(\frac{28.87}{2}\right)^2 + (36.71)^2\right]^{1/2} = 39.45 \text{ MPa}$$

(*b*) Number of threads of nut in engagement.

$$n = \frac{W}{\pi d_m t p_b} = \frac{10{,}000}{\pi \times 23.5 \times 2.5 \times 6} = 9$$

Example 8.2 The lead screw of a lathe has Acme threads of 50 mm nominal diameter and 8 mm pitch. The screw must exert an axial pressure of 2 kN in order to drive the total carriage. The axial thrust is taken by a collar 100 mm outside diameter and 50 mm inside diameter. The lead screw rotates at 25 rpm. The coefficient of friction for screw and collar is 0.15 and 0.10 respectively. Determine (*a*) the power required to drive the screw, and (*b*) the efficiency of the screw.

Solution.

Given $d = 50$ mm, $p = 8$ mm, W = 2kN, $D_o = 100$ mm, $D_i = 50$ mm, $\mu = 0.15$, $\mu_c = 0.10$, N = 25 rpm, $2\beta = 29°$

Mean diameter of screw,

$$d_m = d - \frac{p}{2} = 50 - \frac{8}{2} = 46 \text{ mm}$$

Core diameter,

$$d_c = d - p = 50 - 8 = 42 \text{ mm}$$

Thread thickness,

$$t = \frac{p}{2} = 4 \text{ mm}$$

Helix angle,

$$\alpha = \tan^{-1}\left[\frac{p}{\pi d_m}\right] = \tan^{-1}\left[\frac{8}{\pi \times 46}\right] = 3.17°$$

Virtual coefficient of friction,

$$\mu_e = \frac{\mu}{\cos\beta} = \frac{0.15}{\cos 14.5°} = 0.155$$

Friction angle at threads,

$$\phi = \tan^{-1}\mu_e = \tan^{-1} 0.155 = 8.8°$$

Frictional torque, $$T_f = \left(\frac{Wd_m}{2}\right)\tan(\alpha + \phi)$$

$$= \left(2{,}000 \times \frac{46}{2}\right)\tan(3.17° + 8.8°)$$

$$= 9742.43 \text{ N.mm}$$

Collar torque, $$T_c = \left(\mu_c \frac{W}{4}\right)(D_0 + D_i)$$

$$= \left(0{,}10 \times \frac{2{,}000}{4}\right)(100 + 50)$$

$$= 7500 \text{ N.mm}$$

Total torque, $$T = T_f + T_c = 9742.43 + 7500 = 17242.43 \text{ N.mm}$$

(*a*) Power required, $$P = \frac{2\pi NT}{60 \times 10^3}$$

$$= \frac{2\pi \times 25 \times 17242.43}{60 \times 10^6} = 0.045 \text{ kW}$$

(*b*) Efficiency of screw

Torque required to drive the screw without friction,

$$T_0 = \left(\frac{Wd_m}{2}\right)\tan\alpha$$

$$= \left(2000 \times \frac{46}{2}\right)\tan 3.17° = 2547.64 \text{ N.mm}$$

Efficiency of screw, $$\eta = \frac{T_0}{T} = \frac{2547.64}{17242.43} = 0.1477 \text{ or } 14.77\%.$$

Example 8.3 A C – clamp as shown in Fig. 8.8 has trapezoidal threads of nominal diameter 10 mm and pitch 3 mm. The coefficient of friction for screw threads is 0.15 and for the collar 0.20. The mean diameter of collar is 15 mm. If the force exerted by the operator at the end of lever is 60 N, determine (*a*) the length of handle, (*b*) the maximum shear stress in screw, and (*c*) bearing pressure on the threads.

Solution.

Given : d =10 mm; p = 3 mm, μ = 0.15, μ_c = 0.20, d_c = 15 mm, F = 60 N, W = 4 *k*N, 2β = 30°; h = 25 mm, screw free length = 150 mm

Mean diameter of screw, $$d_m = \frac{d-p}{Q_2} = \frac{10-3}{Q_2} = 8.5 \text{ mm}$$

Core diameter, $$d_c = d - p = 10 - 3 = 7 \text{ mm}$$

Helix angle, $$\alpha = \tan^{-1}\left[\frac{p}{\pi d_m}\right] = \tan^{-1}\left[\frac{3}{\pi \times 8.5}\right] = 6.41°$$

Virtual coefficient of friction, $$\mu_e = \frac{\mu}{\cos\beta} = \frac{0.15}{\cos 15°} = 0.1553$$

Friction angle at threads, $$\phi = \tan^{-1}\mu_e = \tan^{-1} 0.1553 = 8.827°$$

Frictional torque,

$$T_f = \left(\frac{Wd_m}{2}\right)\tan(\alpha + \phi)$$

$$= \left(4,000 \times \frac{8.5}{2}\right)\tan(6.41° + 8.827°)$$

$$= 4630.6 \text{ N.mm}$$

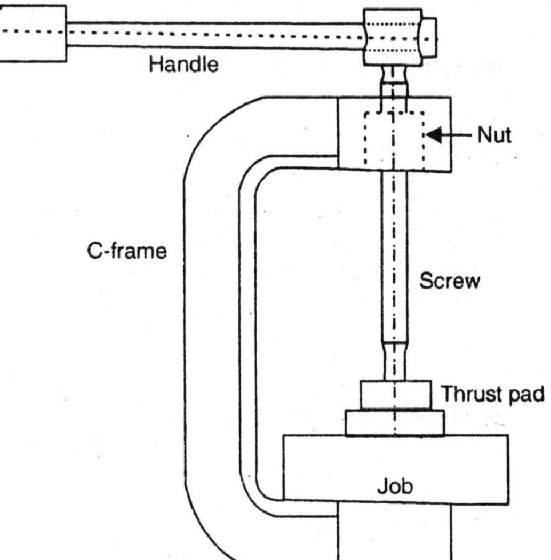

Fig. 8.8. C-clamp

Collar torque,

$$T_c = \left(\mu_c \frac{W}{2}\right)d_c$$

$$= \left(0.20 \times \frac{4,000}{2}\right) \times 15$$

$$= 6000 \text{ N.mm}$$

Total torque,

$$T = T_f + T_c = 4630.6 + 6000 = 10630.6 \text{ N.mm}$$

(*a*) Length of handle.

$$Fl = T$$

$$l = \frac{10630.6}{60} = 177.2 \text{ mm}$$

(*b*) Maximum shear stress in screw.

Section 1 – 1 :

$$\text{Shear stress, } \tau = \frac{16T}{\pi \times d_c^3} = \frac{16 \times 10630.6}{\pi \times 7^3} = 157.85 \text{ N/mm}^2$$

Bending moment, M = F × free length of screw = 60 × 150 = 9000 N.m

$$\text{Bending stress, } \sigma_b = \frac{32M}{\pi d_c^3}$$

$$= \frac{32 \times 9000}{\pi \times 7^3} = 267.27 \text{ N/mm}^2$$

$$\text{Maximum shear stress, } \tau_{max} = \left[\left(\frac{\sigma_b}{2}\right)^2 + \tau^2\right]^{1/2}$$

$$= \left[\left(\frac{267.27}{2}\right)^2 + (157.85)^2\right]^{1/2} = 206.82 \text{ MPa}$$

Section 2 – 2 :

$$\text{Direct compressive stress, } \sigma_c = \frac{4W}{\pi d_c^2} = \frac{4 \times 4{,}000}{\pi \times 7^2} = 103.94 \text{ N/mm}^2$$

$$\text{Torsional shear stress, } \tau = \frac{16T_c}{\pi \times d_c^3} = \frac{16 \times 6000}{\pi \times 7^3} = 89.1 \text{ N/mm}^2$$

$$\text{Maximum shear stress, } \tau_{max} = \left[\left(\frac{\sigma_c}{2}\right)^2 + \tau^2\right]^{1/2}$$

$$= \left[\left(\frac{103.94}{2}\right)^2 + (89.1)^2\right]^{1/2} = 103.15 \text{ MPa}$$

(*c*) Bearing pressure on threads.

$$\text{Height of nut, } h = np$$

$$n = \frac{25}{3} = 8.33 \approx 9$$

$$\text{Thickness of threads, } t = \frac{p}{2} = \frac{3}{2} = 1.5 \text{ mm}$$

Example 8.4 A machine slide weighing 25 kN is elevated by a double start Acme threaded screw at the rate of 0.96 m/min. If coefficient of friction is 0.14, calculate the power required to drive the slide. The end of the screw carries a thrust collar of 30 mm inside and 60 mm outside diameter. The coefficient of friction of collar is 0.15. Pitch of screw threads is 6 mm and outside diameter is 36 mm. If the screw is of 45C8 steel (S_{yt} = 380 MPa), calculate the factor of safety.

Solution. Given : W = 25 kN, $2\beta = 29°$, $v = 0.96$ m/min. $\mu = 0.14$, $D_i = 30$ mm, $D_0 = 60$ mm, $\mu_c = 0.15$, $p = 6$ mm, $d = 36$ mm, $S_{ut} = 630$ MPa, $S_{yt} = 380$ MPa.

Mean diameter of screw, $$d_m = d - \frac{p}{2} = 36 - \frac{6}{2} = 33 \text{ mm}$$

Core diameter, $$d_c = d - p = 36 - 6 = 30 \text{ mm}$$

Lead, $$l = 2p = 12 \text{ mm}$$

Helix angle, $$\alpha = \tan^{-1}\left[\frac{l}{\pi d_m}\right] = \tan^{-1}\left[\frac{12}{\pi \times 33}\right] = 6.6°$$

Virtual coefficient of friction, $$\mu_e = \frac{\mu}{\cos\beta} = \frac{0.14}{\cos 14.5°} = 0.1446$$

Friction angle at threads, $$\phi = \tan^{-1}\mu_e = \tan^{-1} 0.1446 = 8.23°$$

Frictional torque, $$T_f = \left(\frac{Wd_m}{2}\right)\tan(\alpha + \phi)$$

$$= \left(25{,}000 \times \frac{33}{2}\right)\tan(6.6° + 8.23°)$$

$$= 109205 \text{ N.mm}$$

Collar torque, $$T_c = \left(\frac{\mu_c W}{4}\right)(D_0 + D_i)$$

$$= \left(0.15 \times \frac{25{,}000}{4}\right)(60 + 30)$$

$$= 84375 \text{ N.mm}$$

Total torque, $$T = T_f + T_c = 109205 + 84375 = 193580 \text{ N.mm}$$

Speed of screw, $$N = \frac{v}{l} = \frac{960}{12} = 80 \text{ rpm}$$

Power required to drive the slide, $$P = \frac{2\pi NT}{60 \times 10^3}$$

$$= \frac{2\pi \times 80 \times 193580 \times 10^{-3}}{60 \times 10^3}$$

$$= 1.62 \text{ kW}$$

Direct compressive stress, $\sigma_c = \dfrac{4W}{\pi d_c^2} = \dfrac{4 \times 25{,}000}{\pi \times 30^2} = 35.37$ MPa

Torsional shear stress, $\tau = \dfrac{16T}{\pi \times d_c^3} = \dfrac{16 \times 193580}{\pi \times 30^3} = 36.51$ MPa

Maximum tensile stress, $\sigma_1 = \left(\dfrac{1}{2}\right)[\sigma_c + (\sigma_c^2 + 4\tau^2)^{1/2}]$

$$= 0.5\,[35.37 + \{(35.37)^2 + 4\,(36.51)^2\}^{1/2}$$

$$= 0.5\,[35.37 + 81.13] = 58.35 \text{ MPa}$$

Maximum shear stress, $\tau_{max} = \left[\left(\dfrac{\sigma_c}{2}\right)^2 + \tau^2\right]^{1/2}$

$$= 40.5 \text{ MPa}$$

Factor of safety in tension $= \dfrac{S_{yt}}{\sigma_1} = \dfrac{380}{58.35} = 6.51$

Factor of safety in shear $= \dfrac{S_{yt}}{2\tau_{max}} = \dfrac{380}{2 \times 40.5} = 4.69$

Example 8.5 The following data refer to a bench vice having single start square threads as shown in Fig. 8.9.

Major diameter of screw = 12 mm
Minor diameter of screw = 10 mm
Pitch = 2 mm
Collar diameter = 10 mm
Coefficient of friction for screw = 0.15
Coefficient of friction for collar = 0.18
Effort applied at the end of the lever= 100 N
Length of lever = 150 mm

Determine (*a*) the clamping force between the jaws, (*b*) the efficiency of the vice, and (*c*) the torque in the screw body at section A–A and B–B.

Solution.

Given : d = 12 mm, d_c = 10 mm, p = 2 mm, d_c = 10 mm, μ = 0.15, μ_c = 0.18, F = 100 N, l = 150 mm

Mean diameter of screw, $d_m = d_c + \dfrac{p}{2} = 10 + 1 = 11$ mm

Helix angle, $\alpha = \tan^{-1}\left[\dfrac{p}{\pi d_m}\right] = \tan^{-1}\left[\dfrac{2}{\pi \times 11}\right] = 3.31°$

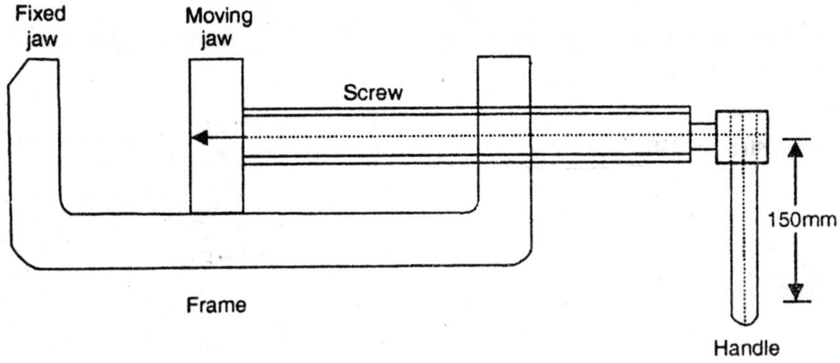

Fig. 8.9 Machine Vice

Friction angle at threads, $\phi = \tan^{-1}\mu = \tan^{-1} 0.15 = 8.53°$

Frictional torque, $T_f = \left(\frac{Wd_m}{2}\right)\tan(\alpha + \phi)$

$= \left(W \times \frac{11}{2}\right) \tan(3.31° + 8.53°)$

$= 1.153$ W N.mm

Collar torque, $T_c = \left(\frac{\mu_c W}{3}\right) d_c$, for uniform pressure

$= \left(0.18 \times \frac{W}{3}\right) \times 10$

$= 0.6$ WN.mm

Total torque at B-B, $T = T_f + T_c = (1.153 + 0.6)\,W = 1.753$ W N.mm

Torque applied by the operator, $T = Fl = 100 \times 150 = 15000$ N.mm

$W = 15000$

(*a*) Clamping force, $W = 8557$ N

(*b*) Efficiency of vice $= \frac{T_0}{T} = \left(\frac{Wd_m}{2}\right)\tan\alpha$

$= \left(8557 \times \frac{11}{2}\right) \frac{\tan 3.31}{15000} = 0.1815$ or 18.15%

(*c*) Torque at A – A $= 1.153 \times 8557 = 9866.2$ N.mm

Torque at B – B $= 15{,}000$ N.mm

8.8 DESIGN OF SCREW JACK

A screw jack for lifting load is shown in Fig. 8.10. The various parts of the screw are designed as follows :

1. Screw diameter.

Compressive stress in the screw,

$$\sigma_c = \frac{4W}{\pi d_c^2}$$

Core diameter,

$$d_c = \left[\frac{4W}{\pi \sigma_c}\right]^{1/2}$$

Where σ_c = permissible compressive stress in screw material

W = load lifted

The screw is square threaded whose standard dimensions can be determined.

Nominal or outside diameter, $d = d_c + p$

Mean diameter, $d_m = d_c + \frac{p}{2}$

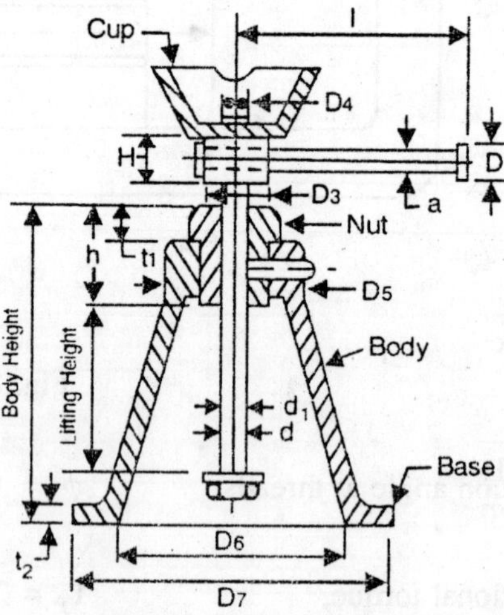

Fig. 8.10 Screw jack.

2. Maximum tensile and shear stress in screw.

Direct compressive stress, $\sigma_c = \frac{4W}{\pi d_c^2}$

Torsional shear stress, $\tau = \frac{16T_f}{\pi \times d_c^3}$

Maximum tensile stress, $\sigma_{1,2} = \left(\frac{1}{2}\right)\left[\sigma_c \pm (\sigma_c^2 + 4\tau^2)^{1/2}\right]$

Maximum tensile stress, $\tau_{max} = \left[\left(\frac{\sigma_c}{2}\right)^2 + \tau^2\right]^{1/2}$

$$\text{Where } T_f = \left(\frac{Wd_m}{2}\right)\tan(\alpha + \phi)$$

$$\alpha = \tan^{-1}\left[\frac{p}{\pi d_m}\right]$$

$$\phi = \tan^{-1}\mu$$

μ = thread friction coefficient

The maximum shear stress should be less than the permissible shear stress in screw material.

3. Height of nut.

Number of threads in contact with screw spindle, $n = \dfrac{W}{\left(\dfrac{\pi}{4}\right)(d^2 - d_c^2)\, p_b}$

Where p_b = permissible bearing pressure on the nut (Table 8.6)

Height of nut, $h = np$

Where p = pitch of threads.

Table 8.6 Allowable bearing pressure for nut

Type of machine	Material of		Allowable bearing pressure, p_b,
	Screw	Nut	MPa
1. Hand press	Steel	Bronze	17 – 22
2. Screw jack	Steel	Cast Iron	12 – 17
	Steel	Bronze	11 – 17
3. Hoisting screw	Steel	Cast Iron	4 – 6.8
	Steel	Bronze	5.5 – 9.6
4. Lead screw	Steel	Bronze	1.0 – 1.6

4. Check for stresses in Screw and Nut.

Shear stress in screw, $\tau_s = \dfrac{W}{\pi n d_c t}$

Shear stress in nut, $\tau_n = \dfrac{W}{\pi n d t}$

Where $t = \dfrac{p}{2}$ = thickness of screw

5. Nut Collar size.

Let D_2 = outside diameter of collar

D_1 = inside diameter of collar

The inside diameter of nut collar is found by considering the tearing strength of nut.

$$\left(\frac{\pi}{4}\right)[D_1^2 - d^2]\,\sigma_t = W$$

where σ_t = permissible tensile strength of nut material

The outside diameter of nut collar is found by considering the crushing strength of nut collar.

$$\left(\frac{\pi}{4}\right)[D_2^2 - D_1^2]\,\sigma_c = W$$

Where σ_c = permissible compressive strength of nut collar material.

The thickness of nut collar is found by considering the shearing strength of nut collar.

$$\pi D_1 t_c \tau_c = W$$

Where t_c = nut collar thickness

τ_c = permissible shear stress in nut collar material

6. Head dimensions.

Diameter of head on the top of screw, $D_3 = 1.75\,d$

Pin diameter in the cup, $D_4 = \dfrac{D_3}{4}$

7. Torque required between Cup and Head.

$$T_c = \left(\frac{\mu_c W}{3}\right)\left[\frac{D_3^3 - D_4^3}{D_3^2 - D_4^2}\right], \text{ according to uniform pressure theory}$$

$$= \left(\frac{\mu_c W}{4}\right)(D_4 + D_3), \text{ according to uniform wear theory}$$

Where μ_c = collar coefficient of friction

Total torque, $T = T_f + T_c$

8. Length of lever,

A normal person can apply a force F = 60 to 100 N.

Length of lever, $l = \dfrac{T}{F}$

9. Diameter of lever.

The lever is under bending.

Bending moment, $M = Fl$

Diameter of level, $d_l = \left[\dfrac{32M}{\pi\sigma_b}\right]^{1/3}$

Where σ_b = permissible bending stress in lever.

10. Height of Head

$$H = 2d_l$$

11. Check of Screw in Buckling

Unsupported length of screw, L = lift of screw + 0.5 × height of nut

Buckling or critical load for screw,

$$W_{cr} = A_c\sigma_y\left[1 - \left\{\frac{\sigma_y}{4C\pi^2 E}\right\}\left(\frac{L}{K}\right)^2\right]$$

Where σ_y = yield strength of screw material

C = end condition coefficient.

= 0.25, assuming the screw with lower end fixed and upper end free

K = least radius of gyration of screw

$$= \frac{d_c}{4}$$

For a safe design $W_{cr} > W$

12. Efficiency of screw.

$$\text{Efficiency of screw, } \eta = \frac{T_0}{T}$$

Where $$T_0 = \left(\frac{Wd_m}{2}\right)\tan\alpha$$

13. Body dimensions.

Diameter of body at top, $D_5 = 1.5\ D_2$

Thickness of base, $t_2 = 2t_1$

Thickness of body, $t_3 = 0.25\ d$

Inside diameter of bottom, $D_6 = 2.25\ D_2$

Outer diameter at the bottom, $D_7 = 1.75\ D_6$

Height of body = Maximum lift + height of nut + 100 mm

Example 8.6 Design a screw jack to lift a load of 100 kN through a height of 400 mm. The permissible stresses in screw and nut are given below :

Steel screw : Tension/compression = 100 MPa, shear = 60 MP*a*

Phosphor – bronze nut : Tension = 100 MPa, compression = 90 MPa, shear = 80 MPa

Bearing pressure between nut and screw = 15 MPa

Solution.

Given : W = 100 *k*N, lift = 400 mm, $\sigma_{ts} = \sigma_{cs}$ = 100 MPa, τ_s = 60 MPa, σ_{tn} = 50 MPa, σ_{cn} = 45 MPa, τ_n = 40 MPa, p_b = 15 MPa

Refer to Fig. 8.10.

1. Screw diameter.

$$\text{Compressive stress in the screw, } \sigma_{cs} = \frac{4W}{\left(\pi d^2{}_c\right)}$$

$$d_c = \left[\frac{4W}{\pi\sigma_{cs}}\right]^{\frac{1}{2}}$$

$$= \left[\frac{4 \times 100{,}000}{\pi \times 100}\right]^{\frac{1}{2}}$$

$= 35.68 \approx$ 40 mm, accounting for shear stress.

We select normal series square threads.

$$p = 7 \text{ mm}$$

Nominal or outside diameter, $d = d_c + p = 40 + 7 = 47$ mm

Mean diameter, $d_m = \dfrac{d_c + p}{2} = 40 + 3.5 = 43.5$ mm

Thread thickness, $t = \dfrac{p}{2} = 3.5$ mm

2. Maximum tensile and shear stresses in screw.

Direct compressive stress, $\sigma_c = \dfrac{4\,W}{\left(\pi d^2{}_c\right)} = \dfrac{4 \times 100{,}000}{\left(\pi \times 40^2\right)} = 79.6$ MPa

Helix angle, $\alpha = \tan^{-1}\left[\dfrac{p}{\pi d_m}\right] = \tan^{-1}\left[\dfrac{7}{\pi \times 43.5}\right] = 2.93^\circ$

Assume coefficient of friction between screw and nut, $\mu = 0.2$

Friction angle at threads, $\phi = \tan^{-1}\mu = \tan^{-1} 0.2 = 11.31^\circ$

Frictional torque, $T_f = \left(\dfrac{W\,d_m}{2}\right)\tan(\alpha + \phi)$

$$= \left(\frac{100{,}000 \times 43.5}{2}\right)\tan(2.93^\circ + 11.31^\circ)$$

$$= 552067 \text{ N.mm}$$

Torsional shear stress, $\tau = \dfrac{16\,T_f}{\pi \times d_c^3}$

$$= \frac{16 \times 552067}{\pi \times 40^3} = 43.93 \text{ MPa}$$

Maximum tensile stress, $\sigma_{1,\,2} = \left(\dfrac{1}{2}\right)\left[\sigma_c \pm \left(\sigma_c^2 + 4\tau^2\right)^{\frac{1}{2}}\right]$

$$= 0.5\left[79.6 \pm \left\{(79.6)^2 + 4 \times (43.93)^2\right\}^{\frac{1}{2}}\right]$$

$$= 0.5\,[79.6 \pm 118.56]$$

$= 99$ MPa (max) < 100 MPa. Hence safe.

Maximum shear stress, $\tau_{max} = \left[\left(\frac{\sigma_c}{2}\right)^2 + \tau^2\right]^{\frac{1}{2}}$

$= 59.28$ MPa < 60 MPa. Hence safe.

3. Height of nut.

Number of threads in contact with screw spindle,

$$n = \frac{W}{\left(\frac{\pi}{4}\right)\left(d^2 - d^2_c\right) p_b}$$

$$= \frac{100,000}{\left(\frac{\pi}{4}\right)\left(47^2 - 40^2\right) 15}$$

$$= 13.94 \approx 14$$

Height of nut, $h = np = 14 \times 7 = 98$ mm

4. Check for stresses in Screw and Nut.

Shear stress in screw, $\tau_s = \frac{W}{\pi n d_c t} = \frac{100,000}{\pi \times 14 \times 40 \times 3.5} = 16.24$ MPa.

Shear stress in nut, $\tau_n = \frac{W}{\pi n d t} = \frac{100,000}{\pi \times 14 \times 47 \times 3.5} = 13.82$ MPa

These stresses are within permissible limits. Hence design is safe.

5. Nut Collar size.

Let D_2 = outside diameter of collar

D_1 = inside diameter of collar

The inside diameter of nut collar is found by considering the tearing strength of nut.

$$\left(\frac{\pi}{4}\right)\left[D_1^2 - d^2\right]\sigma_{tn} = W$$

$$\left(\frac{\pi}{4}\right)\left[D_1^2 - 47^2\right] \times 50 = 100,000$$

$$D_1 = 68.96 \approx 70 \text{ mm}$$

The outside diameter of nut collar is found by considering the crushing strength of nut collar.

$$\left(\frac{\pi}{4}\right)\left[D_2^2 - D_1^2\right]\sigma_{cn} = W$$

$$\left(\frac{\pi}{4}\right)\left[D_2^2 - 70^2\right] \times 45 = 100,000$$

$$D_2 = 87.9 \approx 90 \text{ mm}$$

The thickness of nut collar is found by considering the shearing strength of nut collar.

$$\pi D_1 t_c \tau_{cn} = W$$

$$\pi \times 70 \times t_c \times 40 = 100{,}000$$

$$t_c = 11.36 \approx 12 \text{ mm}$$

6. Head dimensions.

Diameter of head on the top of screw,

$$D_3 = 1.75\, d = 1.75 \times 47 = 82.25 \approx 84 \text{ mm}$$

$$\text{Pin diameter in the cup, } D_4 = \frac{D_3}{4} = \frac{84}{4} = 21 \text{ mm}$$

7. Torque required between Cup and Head.

$$T_c = \left(\frac{\mu_c W}{3}\right)\left[\frac{D_3^3 - D_4^3}{D_3^2 - D_4^2}\right],$$

according to uniform pressure theory

$$\text{Assuming } \mu_c = 0.15$$

$$T_c = \left(\frac{0.15 \times 100{,}000}{3}\right)\left[\frac{84^3 - 21^3}{84^2 - 21^2}\right]$$

$$= 441000 \text{ N.mm}$$

$$\text{Total torque, } T = T_f + T_c = 552067 + 441000 = 993067 \text{ N.mm}$$

8. Length of lever.

A normal person can apply a force $F = 100$ to $300 \text{ N} \approx 300 \text{ N}$

$$\text{Length of lever, } l = \frac{T}{F} = \frac{993067}{300} = 3310 \approx 3300 \text{ mm}$$

9. Diameter of lever.

The lever is under bending.

$$\text{Bending moment, } M = Fl = 300 \times 3300 = 990000 \text{ N.mm}$$

$$\text{Diameter of lever, } d_1 = \left[\frac{32 M}{\pi \sigma_b}\right]^{\frac{1}{3}}$$

$$= \left[\frac{32 \times 990000}{\pi \times 100}\right]^{\frac{1}{3}} = 46.5 \approx 48 \text{ mm}$$

10. Height of Head

$$H = 2d_1 = 2 \times 48 = 96 \text{ mm}$$

11. Check of Screw in Buckling

Unsupported length of screw, $L =$ lift of screw + 0.5 × height of nut

$$= 400 + 0.5 \times 98 = 449 \text{ mm}$$

$$W_{cr} = A_c \sigma_y \left[1 - \left\{\frac{\sigma_y}{4C\pi^2 E}\right\}\left(\frac{L}{K}\right)^2\right]$$

Where $K = \frac{d_c}{4} = \frac{40}{4} = 10$ mm, $c = 0.25$, $\sigma_y = 200$ MPa

$$W_{cr} = \left(\frac{\pi}{4}\right) \times 40^2 \times 200 \left[1 - \left\{\frac{200}{4 \times 0.25 \times \pi^2 \times 200 \times 10^3}\right\}\left(\frac{449}{10}\right)^2\right]$$

$$= 199990 \approx 200 \text{ kN} > 100 \text{ kN}$$

Hence the screw is safe in buckling.

12. Efficiency of screw.

Efficiency of screw, $\eta = \frac{T_0}{T}$

Where $T_0 = \left(\frac{W d_m}{2}\right) \tan \alpha$

$$= \left(\frac{100{,}000 \times 43.5}{2}\right) \tan 2.93° = 111322 \text{ N.mm}$$

$$\eta = \frac{111322}{993067} = 0.112 \text{ or } 11.2\%$$

13. Body dimensions.

Diameter of body at top, $D_5 = 1.5D_2 = 1.5 \times 90 = 135$ mm

Thickness of base, $t_2 = 2t_c = 2 \times 12 = 24$ mm

Thickness of body, $t_3 = 0.25d = 0.25 \times 47 = 11.75 \approx 12$ mm

Inside diameter of bottom, $D_6 = 2.25D_2 = 2.25 \times 90 = 202.5 \approx 205$ mm

Outer diameter at the bottom, $D_7 = 1.75\, D_6 = 1.75 \times 205 = 358.75 \approx 360$ mm

Height of body = Maximum lift + height of nut + 100 mm

= 400 + 98 + 100 = 598 ≈ 600 mm

Example 8.7 Design a fly – press as shown in Fig. 8.11 capable of punching 50 mm diameter holes from 2 mm thick mild steel (S_{us} = 375 MPa) sheet. It can be assumed that shearing will be complete when the punch penetrates through half the thickness of sheet. The screw with square threads is made of 30C8 (S_{yt} = 420 MPa) steel, while the nut is made of bronze. The factor of safety is 3. The total working stroke consists of a one – quarter revolution, 45° in front of the press and 45° behind the press. During the return stroke, the punch is raised by 5 mm to provide clearance to insert the sheet. The working stroke is completed in one second. The balls are made of cast iron, with amass density of 7280 kg / m^3, and the radius R_m is 500 mm. Neglecting collar friction, determine (*a*) the dimensions of screw, (*b*) the length of nut, and (*c*) the size of the balls.

Solution.

Given : d = 50 mm, t_s = 2 mm, S_{us} = 375 MPa, Screw : $S_{yt} = S_{yc}$ = 420 MPa, FOS = 3, working stroke = $\frac{1}{4}$ th of revolution, clearance = 5 mm, ρ_{ci} = 7280 kg/ m^3, R_m = 500 mm.

(*a*) **Screw**

Force required to shear the sheet,

$$F = \pi d t_s S_{us} = \pi \times 50 \times 2 \times 375 = 117810 \text{ N}$$

Allowable compressive stress in screw,

$$\sigma_c = \frac{S_{yc}}{FOS} = \frac{420}{3} = 140 \text{ MPa}$$

Core diameter of screw,

$$d_c = \left[\frac{4F}{\pi \sigma_c}\right]^{\frac{1}{2}} = \left[\frac{4 \times 117810}{\pi \times 140}\right]^{\frac{1}{2}} = 32.73 \text{ mm}$$

Accounting for torsion, let us take,

$$d = 50 \text{ mm; pitch, } p = 8 \text{ mm}$$

Since the punch is to be raised by 5 mm during the return stroke of one quarter revolution, the minimum lead should be 20 mm. The screw is therefore triple threaded.

Lead, l = $3p = 3 \times 8 = 24$ mm

Mean diameter of screw,

d_m = $d - 0.5p = 50 - 4 = 46$ mm

Core diameter,

d_c = $d - p = 50 - 8 = 42$ mm

Helix angle,

$$\alpha = \tan^{-1}\left[\frac{l}{\pi d_m}\right]$$

$$= \tan^{-1}\left[\frac{24}{\pi \times 46}\right] = 9.427°$$

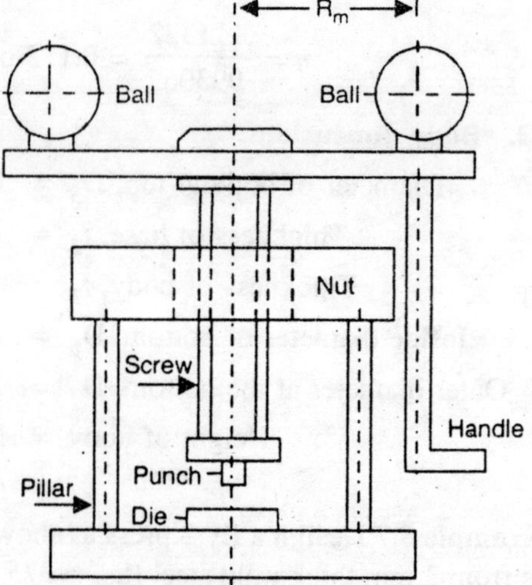

Fig. 8.11 Fly-press

Let coefficient of friction between screw threads and nut, μ = 0.15

Friction angle,

$$\phi = \tan^{-1} \mu \tan^{-1} 0.15 = 8.539°$$

$$\text{Torque, } T = F \times \left(\frac{d_m}{2}\right) \times \tan(\alpha + \phi)$$

$$= 117810 \times 23 \times \tan(9.427° + 8.539°) = 878338 \text{ N.mm}$$

Stresses in screw body :

Compressive stress, $\sigma_c = \dfrac{4F}{\pi d_c^2} = \dfrac{4 \times 117810}{\pi \times 42^2} = 85$ MPa

Shear stress, $$\tau = \frac{16T}{\pi d_c^3}$$

$$= \frac{16 \times 878338}{\pi \times 42^3}$$

$$= 60.4 \text{ MPa}$$

Maximum Shear stress, $$\tau_{max} = \left[\left(\frac{\sigma_c}{2}\right)^2 + \tau^2\right]^{\frac{1}{2}}$$

$$= \left[\left(\frac{85}{2}\right)^2 + (60.4)^2\right]^{\frac{1}{2}}$$

$$= 73.85 \text{ MPa}$$

According to distortion energy theory,

$$S_{ys} = 0.577\ S_{yt}$$

$$= 0.577 \times 420 = 242.34 \text{ MPa}$$

$$FOS = \frac{S_{ys}}{\tau_{max}}$$

$$= \frac{242.34}{73.85}$$

$= 3.28 > 3$. Hence safe.

Efficiency, $$\eta = \frac{\tan\alpha}{\tan(\alpha + \phi)}$$

$$= \frac{\tan 9.429^\circ}{\tan(9.429^\circ + 8.539)}$$

$= 0.5123$ or $51.23\ \%$

Therefore, the screw is not of the self-locking type.

(*b*) Length of nut :

Let allowable bearing pressure between bronze nut and screw,

$$p_b = 18 \text{ MPa}$$

Number of threads, $$n = \frac{4F}{\pi\left(d^2 - d_c^2\right) p_d}$$

$$= \frac{4 \times 117810}{\pi \left(50^2 - 42^2\right) \times 18}$$

$$= 11.32 \approx 12$$

Height of nut, $h = np$

$$= 12 \times 8 = 96 \text{ mm}$$

Thickness of threads, $t = \frac{p}{2} = \frac{8}{2} = 4 \text{ mm}$

Shear stress in screw threads, $\tau_s = \frac{F}{\pi d_c tn}$

$$= \frac{117810}{\pi \times 42 \times 4 \times 12}$$

$$= 18.6 \text{ MPa}$$

Shear stress in nut threads, $\tau_n = \frac{F}{\pi dtn}$

$$= \frac{117810}{\pi \times 50 \times 4 \times 12}$$

$$= 15.62 \text{ MPa}$$

(*c*) Size of balls :

Work done by punch $= F \times 0.5t$

$$= 117810 \times 0.5 \times 2 = 117810 \text{ N}$$

Work done by balls $= \frac{F}{\eta}$

$$= \frac{117810}{0.5123}$$

$$= 229963 \text{ N.mm}$$

Average angular speed, $\omega_{avg} = \frac{\left(\frac{\pi}{2}\right)}{t} = \frac{\left(\frac{\pi}{2}\right)}{1} = \frac{\pi}{2}$ rad/s

Minimum angular speed, $\omega_{min} = 0$

Now $\omega_{ang} = \frac{\omega_{max} + \omega_{min}}{2}$

$$\frac{\pi}{2} = \frac{\omega_{max}}{2}$$

$$\omega_{max} = \pi \text{ rad/s}$$

Radius of gyration of balls, $K = R_m = 0.5$ m

Kinetic energy of balls, KE $= \left(\frac{1}{2}\right) mK^2\omega^2_{max} = \left(\frac{1}{2}\right) \times m \times (0.5)^2 \times \pi^2$

$= 229963 \times 10^{-3}$

$m = 186.4$ kg

Mass of one ball $= 83.6$ kg

$$= \left(\frac{\pi d_b^3}{6}\right)\rho_{ci}$$

$$= \left(\frac{\pi d^3{}_b}{6}\right) \times 7280$$

Diameter of ball, $d_b = 0.29$ m or 290 mm

Example 8.8 A sluice gate weighing 20 kN is raised and lowered by means of square threaded screw, as shown in Fig. 8.12. The frictional resistance induced by water pressure against the gate when it is in the lowest position is 4 kN. The outside diameter of the screw is 60 mm and the pitch is 10 mm. The outside and inside diameters of washer are 150 mm and 50 mm respectively. The coefficient of friction between the screw and nut is 0.10 and for the washer an˙ eat is 0.12. Determine :

(*a*) Maximum force to be exerted at the ends of the lever raising and lowering the gate,

(*b*) efficiency of the arrangement, and

(*c*) number of threads and height of the nut, for an allowable bearing pressure of 7 MPa.

Solution.

Given : $W_g = 20$ kN, $W_f = 4$kN, $d_0 = 60$ mm, $p = 10$ mm, $D_1 = 150$ mm, $D_2 = 50$ mm, $\mu = 0.10$, $\mu_c = 0.12$, $p_b = 7$ MPa

(*a*) Core diameter of screw,

$$d_c = d_o - p$$

$$= 60 - 10 = 50 \text{ mm}$$

Mean diameter, $$d_m = \frac{d_o + d_c}{2}$$

$$= \frac{60 + 50}{2} = 55 \text{ mm}$$

Helix angle, $$\alpha = \tan^{-1}\left[\frac{p}{\pi d_m}\right]$$

$$= \tan^{-1}\left[\frac{10}{\pi \times 55}\right]$$

$$= \tan^{-1} 0.058 = 3.3123°$$

Friction angle between screw and nut,

$$\phi = \tan^{-1}\mu$$

$$\mu = \tan^{-1} 0.10$$

$$= 5.7106°$$

Raising the Gate :

Total load on screw,

$$W = W_g + W_f = 20 + 4 = 24 \text{ kN}$$

Torque required,

$$T_1 = W\frac{d_m}{2}\tan(\alpha + \phi)$$

$$= 24 \times \left(\frac{55}{2}\right) \times \tan(3.3123° + 5.7106°)$$

$$= 104.8 \text{ N.m}$$

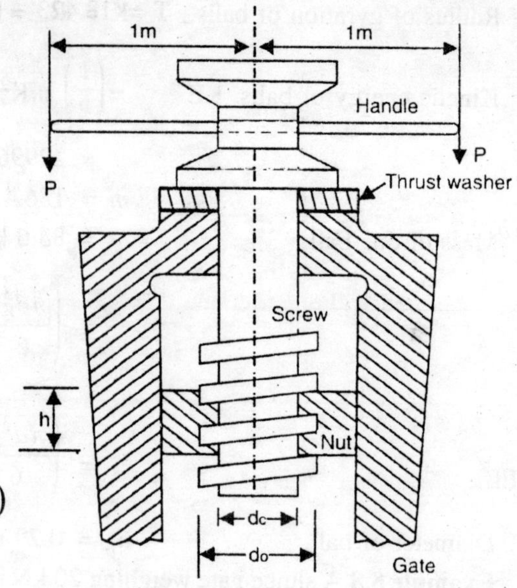

Fig. 8.12 Gate Valve

Mean radius of washer,

$$R_m = \frac{D_1 + D_2}{4} = \frac{150 + 50}{4} = 50 \text{ mm}$$

Frictional torque at the washer,

$$T_f = \mu_c W R_m$$

$$= 0.12 \times 24 \times 50$$

$$= 144 \text{ N.m}$$

$$\text{Total torque, } T = T_1 + T_f = 104.8 + 144$$

$$= 248.8 \text{ N.m}$$

Torque applied at the end of the lever

$$= 2P \times 1 = 2P \text{ N.m}$$

$$2P = 248.8$$

$$P = 124.8 \text{ N}$$

Lowering the Gate :

Total load on the screw, $W = W_g - W_f = 20 - 4 = 16 \, kN$

$$T_1 = W\frac{d_m}{2}\tan(\alpha - \phi)$$

$$16\left(\frac{55}{2}\right)\tan\left(5.7106° - 3.3123°\right) = 18.428 \text{ N.m}$$

$$T_f = \mu_c W R_m$$

$$= 0.12 \times 16 \times 50$$

$$= 96 \text{ N.m}$$

$$T = 18.423 + 96$$
$$= 114.423 \text{ N.m}$$
$$P = \frac{114.423}{2}$$
$$= 57.21 \text{ N}$$

(*b*) Efficiency :

Torque without friction, $T_o = W \frac{d_m}{2} \tan \alpha$

$$= 24 \times 27.5 \times \tan 3.3123°$$
$$= 38.28 \text{ N.m}$$

Efficiency, $\eta = \frac{T_o}{T}$

$$= \frac{38.28}{248.8}$$

= 0.1538 or 15.38%

(*c*) Number of threads :

Thickness of screw threads, $t = \frac{p}{2} = \frac{10}{2} = 5$ mm

Number of threads $n = \frac{W}{\pi d_m t p_b}$

$$= \frac{24,000}{\pi \times 55 \times 5 \times 7} = 4$$

Height of nut, $h = np$

$$= 4 \times 10 = 40 \text{ mm}$$

Example 8.9 A toggle jack as shown in Fig. 8.13 (*a*) is to be designed for lifting a load of 4 kN. When the jack is in the top position, the distance between the centre line of nuts is 50 mm and in the bottom position it is 200 mm. The 8 links of the jack are symmetrical and 110 mm long. The link pins in the base are set 30 mm apart. The links screw and pins are made of mild steel for which permissible stresses are 100 MPa in tension and 50 MPa in shear. The bearing pressure on the threads is limited to 10 MPa. Pitch of square threads is 6 mm and coefficient of frictiuon is 0.2.

Solution.

Screw :

The load on the screw is maximum when the jack is in the bottom most position. In this position, the position of link CD is shown in Fig. 8.13 (*b*).

$$\cos \theta = \frac{100 - 15}{110} = 0.77273$$
$$\theta = 39.4°$$

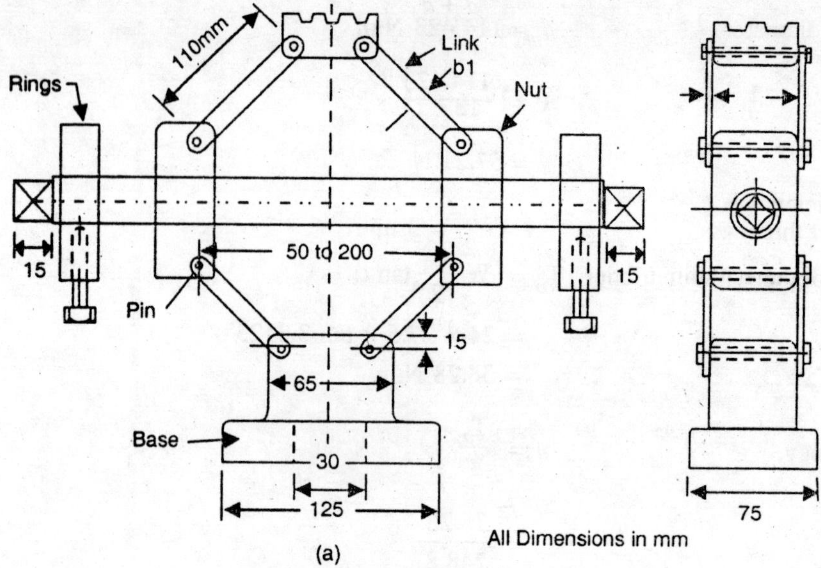

Fig. 8.13 Toggle jack

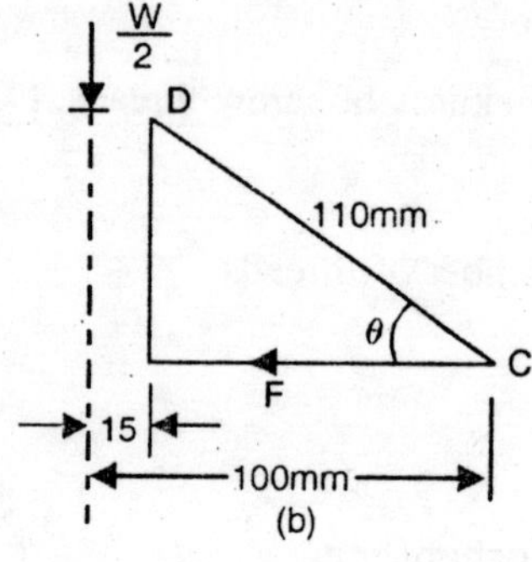

Fig. 8.13

Pull in the screw,

$$F = \frac{0.5\,W}{\tan\theta}$$

$$= \frac{0.5 \times 4}{\tan 39.4^\circ}$$

$$= 2.435 \text{ kN}$$

Total force in the screw due to both nuts,

$$W = 2F$$

$$= 4.87 \text{ kN}$$

Core diameter of screw,

$$d_c = \left[\frac{4\,W}{\pi\sigma_t}\right]^{\frac{1}{2}}$$

$$= \left[\frac{4 \times 4870}{\pi \times 100}\right]^{\frac{1}{2}}$$

$$= 7.87 \text{ mm}$$

Accounting for torsion of screw, let

$$d = 18 \text{ mm and pitch, } p = 6 \text{ mm}$$

Core diameter

$$d_c = d - p$$

$$= 18 - 6$$

$$= 12 \text{ mm}$$

Mean diameter, $d_m = \frac{d + d_c}{2}$

$= \frac{18+12}{2} = 15$ mm

Thread thickness, $t = \frac{p}{2} = 3$ mm

Helix angle, $\alpha = \tan^{-1}\left[\frac{p}{\pi d_m}\right]$

$= \tan^{-1}\left[\frac{6}{\pi \times 15}\right]$

$= 7.256°$

Friction angle, $\phi = \tan^{-1}\mu = \tan^{-1} 0.2$

$= 11.310°$

Torque required to rotate the screw,

$$T = W\left(\frac{d_m}{2}\right)\tan(\alpha + \phi)$$

$$= 4.87\left(\frac{15}{2}\right)\tan(7.256° + 11.310°)$$

$$= 12.268 \text{ N.m}$$

Torsional shear stress in screw,

$$\tau = \frac{16T}{\pi d_c^3}$$

$$= \frac{16 \times 12268}{\pi \times 12^3}$$

$$= 36.16 \text{ MPa}$$

Direct tensile stress, $\sigma_t = \frac{4W}{\pi d_c^2}$

$$= \frac{4 \times 4870}{\pi \times 12^2}$$

$$= 43 \text{ MPa}$$

Maximum principal stress,

$$\sigma_{max} = 0.5\left[\sigma_t + \left(\sigma_t^2 + 4\tau^2\right)^{\frac{1}{2}}\right]$$

$$= 0.5\left[43 + \left\{43^2 + 4 \times (36.16)^2\right\}^{\frac{1}{2}}\right]$$

$$= 63.5 \text{ MPa} < 100 \text{ MPa}$$

Hence safe.

Maximum shear stress, $\tau_{max} = \left[\left(\frac{\sigma_t}{2}\right)^2 + \tau^2\right]^{\frac{1}{2}}$

$$= \left[(21.5)^2 + (36.16)^2\right]^{\frac{1}{2}}$$

$$= 42 \text{ MPa} < 50 \text{ MPa}$$

Hence safe.

Nut :

Let n = number of threads on nut.

Considering the nut in bearing of threads,

$$\pi d_m t n p_b = W$$

$$n = \frac{4870}{\pi \times 15 \times 3 \times 10}$$

$$= 3.44 \approx 4$$

Height of nut, $h = np$

$$= 4 \times 6 = 24 \text{ mm}$$

To prevent movement of nuts beyond 200 mm, rings of 8 mm thickness are provided on the screw on both sides and fixed by set screws.

Length of screw = 200 + h + 2 × thickness of rings + 30 mm for spanner

= 200 + 24 + 16 + 30 = 270 mm

Length of spanner :

Let l = length of spanner, mm

Effort applied by operator = 50 N

Then $l = \frac{T}{50}$

$$= \frac{12268}{50}$$

$$= 245.36 \approx 250 \text{ mm}$$

Pins in the Nuts :

The pins are in double shear. If d_p = diameter of pins, then

$$2 \times \left(\frac{\pi}{4}\right) \times d_p^2 \times \tau = F$$

$$d_p = \left[\frac{2 \times 2435}{\pi \times 50}\right]^{1/2} = 5.57 \approx 6 \text{ mm}$$

Diameter of pin head $= 1.5\, d_p = 1.5 \times 6 = 9$ mm

Thickness of pin head $= \dfrac{d_p}{2} = 3$ mm

Splits pins are used to keep the pins in position.

Links :

$$\text{Load on each link} = \frac{F}{2} = 1217.5 \text{ N}$$

Let b_1 = width of link, mm

t_1 = thickness of link. mm

$b_1 = 3\, t_1.$

Area of cross-section of link, $A = 3t_1^2 \text{ mm}^2$

Least moment of inertia, $I_{yy} = \dfrac{t_1 b_1^3}{12} = 2.25\, t_1^4$

Least radius of gyration, $K_{yy} = \left(\dfrac{I_{yy}}{A}\right)^{1/2}$

$$= \left[\frac{2.25 t_1^4}{3 t_1^2}\right]^{1/2}$$

$$= 0.866\, t_1$$

For buckling of links in the vertical plane, the ends are considered hinged. Using Rankine–Gordon formula for column, we have

$$P_{cr} = \frac{\sigma_c A}{\left[1 + a\,(l_e / K)^2\right]}$$

Taking $a = \dfrac{1}{7500}$ and $\sigma_c = \sigma_t = 100$ MPa, we have

$$P_{cr} = \frac{100 \times 3t_1^2}{\left[1 + (1/7500)\,\{110/(0.866 t_1)\}^2\right]}$$

$$= \frac{300\, t_1^4}{t_1^2 + 21.5}$$

Taking a factor of safety =5, $P_{cr} = 5 \times 1217.5 = 6087.5$ N

$$6087.5 = \frac{300\, t_1^4}{t_1^2 + 21.5}$$

$$300\, t_1^4 - 6087.5 t_1^2 - 13088.125 = 0$$

$$t_1^4 - 20.29 t_1^2 - 43.63 = 0$$

$$t_1^2 = 0.5\,[20.29 \pm \{(20.29)^2 + 4 \times 43.63)^{1/2}]$$

$$= 22.25$$

$$t_1 = 4.72 \approx 5 \text{ mm}$$

$$b_1 = 15 \text{ mm}$$

The ends with the link are considered fixed for buckling in a plane perpendicular to the vertical plane.

Therefore, $$l_e = \frac{l}{2} = \frac{110}{2} = 55 \text{ mm}$$

$$P_{cr} = \frac{100 \times 5 \times 15}{\left[1 + \left(\frac{1}{7500}\right)\left\{\frac{55}{0.866 \times 5}\right\}^2\right]}$$

$$= 7342 \text{ N} > 6087.5 \text{ N. Hence safe.}$$

Example 8.10 Design a turn buckle as shown in Fig. 8.14 to take up pull of 100 kN in the tie rod. The permissible stresses are : Tension = 80 MPa, Compression = 100 M Pa, and Shear = 45 MPa.

Solution.

Let d_c = core diameter of threaded portion of turn buckle.

$$d_c = \left[\frac{4p}{\pi \sigma_t}\right]^{1/2}$$

$$= \left[\frac{4 \times 100 \times 10^3}{\pi \times 80}\right]^{1/2}$$

$$= 39.89 \text{ mm}$$

Accounting the torsion, take d = 62 mm, p = 14 mm of coarse series square threads.

Core diameter, $$d_c = d - p$$

$$= 62 - 14$$

$$= 48 \text{ mm}$$

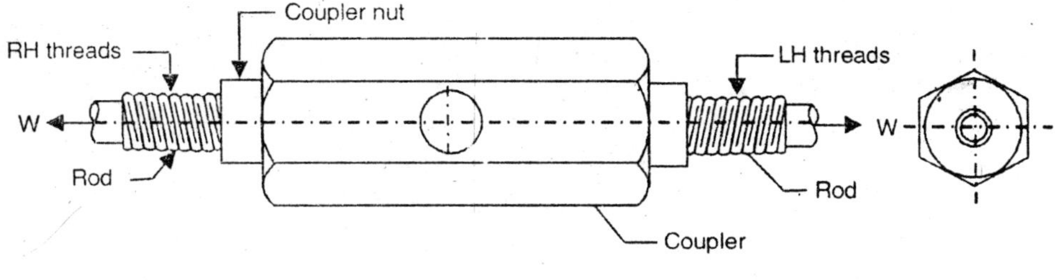

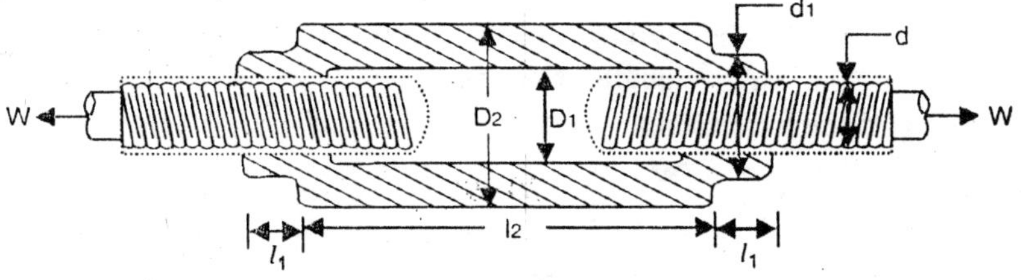

Fig. 8.14 Turn Buckle

Mean diameter, $$d_m = \frac{d + d_c}{2} = \frac{62 + 48}{2} = 55 \text{ mm}$$

Helix angle, $$\alpha = \tan^{-1}\left[\frac{p}{\pi d_m}\right] = \tan^{-1}\left[\frac{14}{\pi \times 55}\right] = 2.65°$$

Friction angle, $\phi = \tan^{-1}\mu = \tan^{-1} 0.2 = 11.310°$

Torque required to rotate the screw,

$$T = W\left(\frac{d_m}{2}\right)\tan(\alpha + \phi)$$

$$= 100\left(\frac{55}{2}\right)\tan(2.65° + 11.310°) = 683.66 \text{ N.m}$$

Torsional shear stress in screw, $$\tau = \frac{16\,T}{\pi\, d_c^3} = \frac{16 \times 683660}{\pi \times 48^3} = 31.48 \text{ MPa}$$

Direct tensile stress, $$\sigma_t = \frac{4\,W}{\pi\, d_c^2} = \frac{4 \times 100}{\pi \times 48^2} = 55.26 \text{ MPa}$$

Maximum principal, stress, $\sigma_{max} = 0.5\,[\sigma_t + (\sigma^2_t + 4\,\tau^2)^{1/2}]$

$= 0.5\,[55.26 + \{55.26^2 + 4 \times (31.48)^2\}^{1/2}$

$= 69.51 \text{ MPa} > 80 \text{ M Pa}$

Hence safe.

Maximum shear stress, $\tau_{max} = \left[\left(\frac{\sigma_t}{2}\right)^2 + \tau^2\right]^{1/2}$

$= [(27.63)^2 + (31.48)^2]^{1/2}$

$= 41.88 \text{ MPa} < 45 \text{ MPa}$

Hence safe.

Coupler Nut :

Let n = number of threads on nut.

l_1 = length of coupler

Considering shearing of the nut threads,

$$\pi d_c l_1 \tau = W$$

$$l_1 = 100 \times \frac{10^3}{\pi \times 48 \times 45}$$

$$= 14.74 \approx 15 \text{ mm}$$

Considering crushing of nut threads, we have

$$\left(\frac{\pi}{4}\right)\left(d^2 - d_c^2\right) n\sigma_c = W$$

$$n = \frac{4 \times 100 \times 10^3}{\pi (62^2 - 48^2) \times 110}$$

$$= 0.75$$

$$l_1 = np = 0.75 \times 14 = 10.5 \text{ mm}$$

Generally $l_1 > d$. We take $l_1 = 70$ mm so that there are 5 threads in contact.

Outside Diameter of Coupler Nut :

Considering tensile failure of coupler nut, we have

$$\left(\frac{\pi}{4}\right)\left(d_1^2 - d^2\right)\sigma_t = W$$

$$\left(\frac{\pi}{4}\right)\left(d_1^2 - 62^2\right) \times 80 = 100 \times 10^3$$

$$d_1 = 73.72 \approx 75 \text{ mm}$$

Coupler :

Inside diameter, $D_1 = d + 6 = 62 + 6 = 68$ mm

Considering tensile failure of coupler, we have

$$\left(\frac{\pi}{4}\right)\left(D^2 - D_1^2\right)\sigma_t = W$$

$$\left(\frac{\pi}{4}\right)\left(D^2 - 68^2\right) \times 80 = 100 \times 10^3$$

$$D = 78.83 \approx 80 \text{ mm}$$

Length of coupler, $l_2 = 6d = 6 \times 62 = 372$ mm

Thickness of coupler, $t = \dfrac{D - D_1}{2}$

$$= \frac{80-68}{2}$$

$$= 6 \text{ mm.}$$

EXERCISES

8.1 A double-threaded power screw, with metric trapezoidal thread, is used to raise a load of 400 kN. The nominal diameter is 100 mm and the pitch is 12 mm. The coefficient of friction at screw thread is 0.15. Neglecting collar friction, design the power screw and calculate the torque required to raise and to lower the load and also find the efficiency of the screw.

8.2 Show that if the power screw is to be self-locking, its efficiency can not be more than 50%.

8.3 A 70 mm screw with square threads is used in a press. It has a maximum unsupported length of 450 mm. The screw is of C-25 steel with phosphor bronze nut. Determine—

(*a*) the safe capacity of the press,

(*b*) the proper length of the nut, and

(*c*) the necessary torsional moment, assuming coefficient of friction is the threads as 0.13 and at the thrust collar as 0.15, which has an outside diameter of 90 mm and inside diameter of 26 mm.

8.4 Give the design procedure for the design of a screw jack.

8.5 In a hand vice the screw has a double start square threads of 26 mm diameter. If the lever is 0.25 m and the maximum force that can be applied at the end of the lever is 300 N, determine the force with which the job is held in the jaws of the vice. Take the coefficient of friction as 0.14.

8.6 Describe various forms of threads used for power screw, giving their merits and demerits. Discuss the procedure for the design of a screw having square threads.

8.7 A screw jack suitable to lift a maximum load of 50 kN through a height of 200 mm with ground clearance of 300 nm is to be designed. The allowable bearing pressure between nut and screw is 16 N/ mm^2. Coefficient of friction of threads and collars may be taken as 0.14. Select suitable material and stresses. Use design data hand book.

8.8 A power screw having double start square thread of 32 mm nominal diameter and 5 mm pitch, is acted upon an axial load of 12 kN. The inner and outer diameters of the screw collar are 20 mm and 50 mm respectively. The coefficient of thread friction and collar friction may be assumed as 0.15 and 0.20 respectively. The screw rotates at 24 rpm. Assume uniform wear condition. Permissible thread bearing pressure is 6N / mm^2. Allowable shear stress in nut and screw is 30 MPa. Determine :

(*i*) torque required to rotate the screw,

(*ii*) stress induced in the screw, and

(*iii*) height of nut.

8.9 A sluice gate valve used in water pipelines is raised by the spindle which is rotated by the handwheel. The spindle has single-start square threads having nominal diameter 36 mm and 6 mm pitch. The inner and outer diameters of the friction collar are 30 mm and 50 mm respectively. The coefficient of friction at the threads and the collar are 0.15 and 0.20 respectively. The weight of the gate is 8 kN and the frictional resistance to open the valve due to water pressure is 3kN. Assuming uniform wear at the collar, determine (*a*) the torque required to raise the gate, and (*b*) the overall efficiency of the mechanism.

8.10 A double-threaded power screw is used to raise a load of 6 kN. The nominal diameter of the screw having Acme threads is 60 mm and pitch is 9 mm. The coefficient of friction at the threads is 0.15. Neglecting collar friction, calculate (*a*) the torque required to raise and lower the load, and (*b*) the efficiency of the screw for lifting the load.

8.11 Design a square threaded screw jack to lift a load of 50 kN. Assume suitable data from the Data Book.

8.12 The following data apply to the C-clamp shown in Fig. 8.15 :

Outside diameter of threads = 12 mm

Core diameter = 9.853 mm

Pitch-1.75 mm

Coefficient of thread friction = 0.12

Coefficient of collar friction = 0.25

Load, W = 5 kN

Effort at lever end, F = 80 N

Effective collar radius = 6 mm

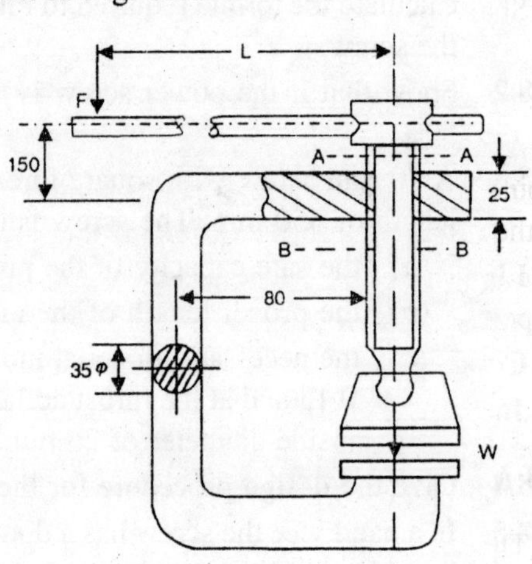

Fig. 8.15

Determine (*a*) the length of the handle, (*b*) stresses at section A–A and B–B of the screw, and (*c*) bearing pressure on the threads.

8.13 The lead screw of a lathe has 50 × 8 threads. The screw must exert an axial force of 3 kN in order to drive the tool carriage. The thrust is carried on a collar 120 mm outside diameter and 60 mm inside diameter. The screw rotates at 30 rpm. The coefficient of friction for the screw and collar is 0.15 and 0.12 respectively. Determine (*a*) the power required to drive the screw, and (*b*) the efficiency of the lead screw.

9 TRANSMISSION SHAFTS

Chapter

9.1 INTRODUCTION

A shaft is a rotating machine element which is used to transmit power from one place to another. An axle is a machine element which may or may not rotate and supports transverse loads. It does not transmit power. A transmission shaft transmits power from the prime mover to the machine. A spindle is a short rotating shaft.

9.2 CLASSIFICATION OF SHAFTS

Shafts may be classified as :

1. Rigid shaft
2. Flexible shaft

A rigid shaft does not bend appreciably about its axis. The axes of driving and driven shafts should be coaxial. Flexible shafts bend appreciably about their axis and permit the non-coaxiality of the driving and driven shafts. Rigid shafts may be further classified as :

Line shaft. It is a long and continuous shaft used in workshops and factories for the distribution of power.

Counter shaft. A shaft in between the line shaft and the machine is called the counter shaft.

Jack shaft. A shaft directly coupled to the power source is called the jack shaft.

9.3 MATERIALS FOR SHAFTS

The materials to be used for shafts should have the following properties :

1. High tensile strength.
2. Low notch sensitivity factor.
3. Good machinability.
4. High endurance strength.
5. Less prone to environment effects.

The mechanical properties of materials used for shaft are given in Table 9.1.

Table 9.1 Mechanical properties of steels used for shafts

Steel Designation	Tensile strength S_{ut}, MPa	Yield strength, S_{yt}, MPa
1. 40C 8	569–667	324
2. 45C 8	618–696	353
3. 50C 8	747–765	373
4. 50C 12	706 min	392

9.4 STANDARD SIZES OF SHAFTS

The preferred shaft diameters are given in Table 9.2.

Table 9.2 Preferred shaft diameters, mm

6, 8, 9, 10, 14, 16, 18, 19, 20, 25, 28, 30, 40, 45, 50, 56, 60, 63, 71, 75, 80, 85, 90, 95, 100, 125, 140, 150, 160, 170, 180, 190, 200, 250, 280, 300, 400, 450, 500, 530, 560, 600 and 630 mm

9.5 DESIGN AGAINST STATIC LOAD

The shaft is generally made of a ductile material, like steel, aluminium, copper, etc. Therefore, the shaft can be designed by using an appropriate theory of failure. The shaft must not only be strong enough to provide an adequate factor of safety from the static and dynamic stress view points but it must also be sufficiently rigid to prevent harmful torsional and lateral deflections. Further, the shaft must have such dimensions that its natural frequency of vibration is sufficiently remote from the operating frequency to avoid resonance with its consequent destructive stresses and displacements.

9.5.1 Theories of Failure

The theories of failure suitable for the ductile material of shafts are :

1. Maximum shear stress theory.
2. Maximum distortion energy theory.

The maximum distortion energy theory is more accurate of the two, but the maximum shear stress theory gives results on the safe side and is simple to apply. The maximum shear stress theory suffers from the drawback that the intermediate stress is ignored.

According to the maximum shear stress theory, we have

$$\tau_{max} = \frac{\sigma_1 - \sigma_2}{2}$$

$$= \frac{S_{yt}}{2n}$$

or

$$\sigma_1 - \sigma_2 = \frac{S_{yt}}{n} \qquad \text{... (9.1)}$$

where σ_1, σ_2 = maximum and minimum principal stresses respectively

S_{yt} = yield strength under simple tension

n = factor of safety

According to the maximum distortion energy theory, we have

$$(\sigma_1 - \sigma_2)^2 + (\sigma_2 - \sigma_3)^2 + (\sigma_3 - \sigma_1)^2 = 2\left(\frac{S_{yt}}{n}\right)^2$$

or

$$\left(\frac{1}{\sqrt{2}}\right) [(\sigma_1 - \sigma_2)^2 + (\sigma_2 - \sigma_3)^2 + (\sigma_3 - \sigma_1)^2]^{1/2} = \frac{S_{yt}}{n} \qquad \text{...(9.2)}$$

For the case of plane stress system, we have

$$\left[\sigma_x^2 + \sigma_y^2 - \sigma_x\sigma_y + 3\tau_{xy}^2\right]^{\frac{1}{2}} = \frac{S_{yt}}{n} \quad \text{... (9.3)}$$

9.5.2 Combined Bending and Torsion

Let M = bending moment acting on the shaft
T = applied torque to the shaft
d = diameter of shaft

(*a*) Solid shaft

$$\text{Bending stress, } \sigma_b = \frac{32M}{\pi d^3}$$

$$\text{Shear stress, } \tau = \frac{16T}{\pi d^3}$$

$$\text{Maximum shear stress in the shaft, } \tau_{max} = \left[\left(\frac{\sigma_b}{2}\right)^2 + \tau^2\right]^{\frac{1}{2}}$$

$$= \left[\frac{16}{\pi d^3}\right]\left[M^2 + T^2\right]^{\frac{1}{2}} \quad \text{...(9.4)}$$

According to maximum shear stress theory of failure,

$$S_{ys} = 0.5S_{yt}$$

where S_{yt} = yield strength of the material in tension
S_{ys} = yield strength of the material in shear

Therefore

$$\tau_{max} = \frac{S_{ys}}{n} = \frac{0.5S_{yt}}{n} \quad \text{...(9.5)}$$

where n = factor of safety.

$$\left[\frac{32}{\pi d^3}\right]\left[M^2 + T^2\right]^{\frac{1}{2}} = \frac{S_{yt}}{n} \quad \text{...(9.6)}$$

(*b*) Hollow shaft

For a hollow shaft,

$$J = \left(\frac{\pi}{32}\right)\left(d_0^4 - d_i^4\right)$$

$$= \left(\frac{\pi}{32}\right) d^4{}_0 \, (1 - k^4)$$

where

$$k = \frac{d_i}{d_0}$$

Eq. (9.6) becomes,

$$\left[\frac{32}{\pi d^3_0 (1-k^4)}\right]\left[M^2+T^2\right]^{\frac{1}{2}} = \frac{S_{yt}}{n} \quad \text{... (9.7)}$$

9.5.3 A.S.M.E. Code for the Design of Transmission Shafting

The ASME Code for the design of transmission shafting is based on the maximum shear stress theory and takes into account the effect of fatigue and shock by introducing two constants K_m and K_t into the equation of maximum shear stress theory. K_m and K_t are the combined fatigue and shock factors for bending and torsion respectively. Eq. (9.1) in the modified form can be written as :

$$t_{max} = \left[\frac{16}{\pi d^3}\right]\left[(K_m M)^2 + (K_t T)^2\right]^{\frac{1}{2}} \quad \text{....(9.8)}$$

The values of K_m and K_t are given in Table 9.3.

Table 9.3 Values of combined fatigue and shock factors.

Nature of load	K_m	K_t
Gradually applied or steady load	1.5	1.0
Suddenly applied loads with minor shocks only	1.5 – 2.0	1.0 – 1.5
Suddenly applied loads with heavy shocks.	2.0 – 3.0	1.5 – 3.0

9.5.4. Combined Axial, Bending and Torsion loading on Hollow Shaft

For a hollow shaft of outside diameter d_o and inside diameter d_i subjected to the combined action of axial force F, bending moment M and torque T, according to maximum shear stress theory of failure, we have

$$\tau_{max} = \left[\frac{16}{\pi d^3 (1-k^4)}\right]\left[\left\{K_m M + \frac{\alpha F d_0 (1+k^2)}{8}\right\}^2 + (K_t T)^2\right]^{\frac{1}{2}} \quad \text{... (9.9)}$$

where $k = \dfrac{d_i}{d_0}$

α = a constant depending upon the magnitude of column action of the unsupported shaft length between the bearings.

$$= \frac{1}{\left[1-0.0044\left(\frac{L}{K}\right)\right]} \text{ for } \frac{L}{K} < 115$$

$$= S_{yc} \frac{\left(\frac{L}{K}\right)^2}{\pi^2 n E} \quad \text{for } \frac{L}{K} > 115$$

L = Length of the shaft between the supporting bearings

K = Ladius of gyration of the shaft

$$= \frac{d_0^2 \left(1 + k^2 - k\right)}{16}$$

S_{yc} = yield strength in compression of shaft material

n = 1 for hinged ends

= 2.25 for fixed ends

9.6 FACTOR OF SAFETY

The permissible shear stress τ_d for shafts without keyway is taken as :

$$\tau_d = 0.30\ S_{yt}$$

or $= 0.18\ S_{ut}$, whichever is minimum ... (9.10)

The tensile strength of shafts is taken as :

$$\sigma_d = 0.6 S_{yt}$$

or $= 0.36 S_{ut}$, whichever is less ...(9.11)

If keyways are present, these values are to be reduced by 25%.

The static weakening effect of the keyway is given by :

$$k_k = 1.0 - 0.2\left(\frac{b}{d}\right) - 1.1\left(\frac{t}{d}\right) \approx 0.75 \quad \text{...(9.12)}$$

where $k_k = \dfrac{\text{Strength of shaft with keyway}}{\text{Strength of shaft without keyway}}$

b, t = width and depth of keyway respectively

d = shaft diameter

9.7 DESIGN FOR TORSIONAL RIGIDITY

In certain applications, the torsional deformation of the shaft should be limited as demanded by design. In such cases the rigidity or stiffness of the shaft is very important.

The angle of twist of a shaft under torsion is given by :

$$\theta = \frac{TL}{GJ}$$

Torsional rigidity, $k_t = \dfrac{T}{\theta} = \dfrac{GJ}{L}$ ...(9.13)

where T = torque applied to the shaft, N.mm

L = length of shaft, mm

G = modulus of rigidity of shaft material, N/mm^2

J = polar moment of inertia of shaft, mm^4

$$= \frac{\pi d^4}{32}$$

d = shaft diameter, mm

The torsional deflection of line shaft is limited to 1 degree per 20 diameters or 3° per metre length of shaft. For machine tools it is limited to 0.25° per metre length.

9.8 DESIGN FOR BENDING RIGIDITY

In certain applications like gears, hydrodynamic bearings, machine tools, etc., the components are designed on the basis of lateral rigidity. Permissible deflection is the criterion for design in these cases. Deflection of a shaft carrying concentrated load W at a distance 'a' from left hand bearing and of length L is given by,

$$\delta = \frac{Wa^2b^2}{3EIL} \quad .....(9.14)$$

where EI = bending stiffness of shaft

$b = L - a$

Bending rigidity, $$k_b = \frac{W}{\delta} = \frac{3EIL}{a^2b^2} \quad ....(9.15)$$

Deflection of a shaft of span L carrying concentrated load w at the middle is given by,

$$\delta = \frac{W L^3}{48 E I} \quad ...(9.16)$$

Bending rigidity, $$k_b = \frac{48EI}{L^3} \quad ...(9.17)$$

Deflection of a shaft carrying uniformly distributed load of intensity w per unit length over the whole span L, is given by,

$$\delta = \frac{5 wL^4}{384 EI} \quad ...(9.18)$$

Bending rigidity, $$k_b = \frac{384EI}{5L^3} \quad ..(9.19)$$

The permissible lateral deflection of various components are as follows :

1. Shafts carrying gears = not more than 3 mm per mm of face width ...(9.20 a)

or = not more than $0.01 \times m$ ...(9.20 b)

where m = module of gear teeth

2. Transmission shaft supported on plain bearings = not more than 0.6 times the distance from the load point to the centre of the bearing in metres. ...(9.21 *a*)

or = (0.001 – 0.003) L ...(9.21 *b*)

where L = span length between the two bearings.

The lateral rigidity depends upon the following variables :

1. Diameter of the shaft.
2. Forces acting on the shaft.
3. Modulus of elasticity of shaft material.

The lateral rigidity can be improved by the following methods.

1. Reduce the span length.
2. Increase the number of supports.
3. Reduce the number of joints.
4. Assemble the components with pre-load.
5. Lubricate the contact surfaces with high viscosity oil.

9.9 DESIGN AGAINST FATIGUE LOAD

The design method against fatigue loading under combined stresses has been explained in art. 4.12. The equivalent stresses under axial or bending loads, according to Soderberg equation are :

$$\text{Equivalent mean normal stress, } \sigma_{me} = \sigma_m + \left(\frac{S_y}{S_e}\right)\sigma_a$$

$$\text{Equivalent alternating normal stress, } \sigma_{ae} = \left(\frac{S_e}{S_y}\right)\sigma_m + \sigma_a$$

The analogous equations for torsional loadings are :

$$\text{Equivalent mean shear stress, } \tau_{me} = \tau_m + \left(\frac{S_{ys}}{S_{es}}\right)\tau_a$$

$$\text{Equivalent alternating shear stress, } \tau_{ae} = \left(\frac{S_{es}}{S_{ys}}\right)\tau_m + \tau_a$$

(*a*) Combined bending and torsion

1. Maximum shear stress theory

Maximum equivalent mean shear stress,

$$(\tau_{me})_{max} = \left[\left(\frac{\sigma_{me}}{2}\right)^2 + \tau_{me}^2\right]^{\frac{1}{2}}$$

$$= \frac{0.5S_y}{n} = \frac{S_{ys}}{n}$$

For a solid circular shaft of diameter d, we have

$$\frac{S_{ys}}{n} = \left[\frac{16}{\pi d^3}\right]\left[\left\{M_m + \left(\frac{S_y}{S_e}\right)M_a\right\}^2 + \left\{T_m + \left(\frac{S_{ys}}{S_{es}}\right)T_a\right\}^2\right]^{\frac{1}{2}} \quad ...(9.22)$$

2. Distortion energy thoery

$$\frac{S_y}{n} = \left[\sigma_{me}^2 + 3\tau_{me}^2\right]^{\frac{1}{2}}$$

$$\frac{S_y}{n} = \left[\frac{32}{\pi d^3}\right]\left[\left\{M_m + \left(\frac{S_y}{S_e}\right)M_a\right\}^2 + 0.75\left\{T_m + \left(\frac{S_{ys}}{S_{es}}\right)T_a\right\}^2\right]^{\frac{1}{2}} \quad ...(9.23)$$

(*b*) Plane stresses

$$\sigma_{mx} = \frac{(\sigma_x)_{max} + (\sigma_x)_{min}}{2}$$

$$\sigma_{ax} = \frac{(\sigma_x)_{max} - (\sigma_x)_{min}}{2}$$

$$\sigma_{my} = \frac{(\sigma_y)_{max} + (\sigma_y)_{min}}{2}$$

$$\sigma_{ay} = \frac{(\sigma_y)_{max} - (\sigma_y)_{min}}{2}$$

$$\tau_{mxy} = \frac{(\tau_{xy})_{max} + (\tau_{xy})_{min}}{2}$$

$$\tau_{axy} = \frac{(\tau_{xy})_{max} - (\tau_{xy})_{min}}{2}$$

1. Maximum shear stress theory

$$(\tau_m)_{max} = \left[\left\{\frac{\sigma_{mx} - \sigma_{my}}{2}\right\}^2 + \tau_{mxy}^2\right]^{\frac{1}{2}}$$

$$(\tau_a)_{max} = \left[\left\{\frac{\sigma_{ax} - \sigma_{ay}}{2}\right\}^2 + \tau_{axy}^2\right]^{\frac{1}{2}}$$

Using Soderberg failure criterion, we have

$$\frac{S_{ys}}{n} = (\tau_m)_{max} + \left(\frac{S_{ys}}{S_{es}}\right)(\tau_a)_{max} \quad ...(9.24)$$

2. Distortion energy theory

$$(\sigma_m)_{max} = \left[\sigma_{mx}^2 + \sigma_{my}^2 - \sigma_{mx}\,\sigma_{my} + 3\tau_{mxy}^2\right]^{\frac{1}{2}}$$

$$(\sigma_a)_{max} = \left[\sigma_{ax}^2 + \sigma_{ay}^2 - \sigma_{ax}\,\sigma_{ay} + 3\tau_{axy}^2\right]^{\frac{1}{2}}$$

Using Soderberg failure criterion, we have

$$\frac{S_y}{n} = (\sigma_m)_{max} + \left(\frac{S_y}{S_e}\right)(\sigma_a)_{max} \quad ...(9.25)$$

9.10 CRITICAL SPEED OF SHAFTS

It is essential that excessive lateral vibrations of shafts should be avoided in order to reduce dynamic stresses. Shafts operating at or near to their critical (or natural) speed develop lateral displacements of relatively high amplitudes. Therefore, critical speed of shafts should be checked.

The critical speed of a shaft of length L carrying a disc at the centre of weight W and supported on anti-friction bearings at the ends is given by,

$$\omega_{cr} = \left[\frac{48EIg}{WL^3}\right]^{\frac{1}{2}} \quad ...(9.26)$$

where I = moment of inertia of the shaft about its axis of rotation.

E = Young's modulus of elasticity of shaft material

g = acceleration due to gravity

For a central disc of weight W on a shaft rotating in sleeve (long) bearings,

$$\omega_{cr} = \left[\frac{192EIg}{WL^3}\right]^{\frac{1}{2}} \quad ...(9.27)$$

When the disc is mounted at a distance 'a' from left bearing and 'b' from right bearing, then

$$\omega_{cr} = \left[\frac{3EIg}{Wa^2b^2}\right]^{\frac{1}{2}} \text{ for short bearing} \quad ...(9.28)$$

$$= \left[\frac{3EIL^3g}{Wa^2b^2}\right]^{\frac{1}{2}} \text{ for long bearing} \quad ...(9.29)$$

Critical speed of a shaft of uniformly distributed weight w per unit length supported in short bearing is given by,

$$\omega_s = \left[\frac{384EIg}{5wL^4}\right]^{\frac{1}{2}} \quad ...(9.30)$$

For a shaft carrying n discs, the critical speed may be found by using Dunkerley's equation :

$$\frac{1}{\omega_{cr}^2} = \frac{1}{\omega_1^2} + \frac{1}{\omega_2^2} + ... + \frac{1}{\omega_n^2} + \frac{1}{\omega_s^2} \quad ....(9.31)$$

where ω_s = natural frequency of shaft itself.

Example 9.1 A line shaft carrying two pulleys A of diameter 200 mm and B of 400 mm receives power at pulley A by means of a vertical belt, which is then transmitted to pulley B carrying horizontal belt. The pulley A is located at the middle of bearings 1 m apart and pulley B overhangs on the right of right bearing by 200 mm. The ratio of tensions is 2.5 : 1 and the maximum tension in either belt is limited to 2.5 kN. For the shaft material, S_{ut} = 650 MPa and S_{yt} = 380 MPa. Determine the shaft diameter using ASME code and taking into account the keyway effect.

Solution.

Given : S_{ut} = 650 MPa, S_{yt}= 380 MPa, $\frac{F_1}{F_2} = 2.5$, F_{max} = 2.5 kN, d_a = 200 mm, d_b = 400 mm

Design shear stress, $\tau_d = 0.30S_{yt} = 0.30 \times 380 = 114$ MPa

$= 0.18Sut = 0.18 \times 650 = 117$ MPa

$= 114$ MPa (minimum value)

Accounting for keyway effect, $\tau_d = 0.75 \times 114 = 85.5$ MPa

The maximum belt tension shall occur on pulley A, being of smaller diameter.

The arrangement of pulleys is shown in Fig. 9.1.

Pulley A :

F_1 = 2500 N, F_2 = 1000 N

Torque supplied to shaft, $T = (F_1 - F_2)\,\frac{d_a}{2} = (2500 - 1000) \times \frac{200}{2}$ = 1,50,000 N.mm

Resultant pull on pulley A, $F_a = F_1 + F_2 = 2500 + 1000 = 3500$ N ↓

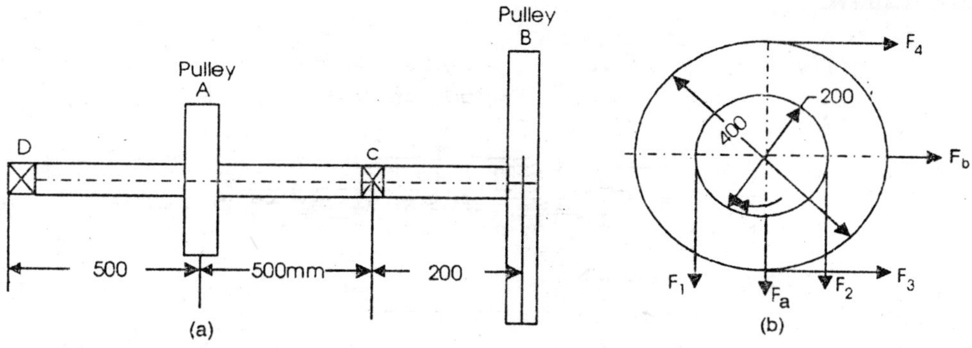

Fig. 9.1

Pulley B :

Let F_3, F_4 = tensions in the belt

$$T = (F_3 - F_4)\,\frac{d_b}{2}$$

$$F_3 - F_4 = 2 \times \frac{150,000}{400} = 750 \text{ N}$$

$$\frac{F_3}{F_4} = 2.5$$

$$F_4 = \frac{750}{1.5} = 500 \text{ N}$$

$$F_3 = 1250 \text{ N}$$

Resultant pull on pulley B, $F_b = F_3 + F_4 = 1250 + 500 = 1750 \text{ N} \rightarrow$

Bending Moments :

(*a*) **Vertical force**

$$M_{av} = \frac{F_a L}{4} = 3500 \times \frac{1000}{4} = 8,75,000 \text{ N.mm}$$

(*b*) **Horizontal force**

$$M_c = F_b \times 200$$
$$= 1750 \times 200 = 3,50,000 \text{ N.mm}$$

$$M_{ah} = \frac{350,000}{2} = 175,000 \text{ N.mm}$$

Resultant bending moment at A,

$$M = [M^2_{av} + M^2_{ah}]^{1/2}$$
$$= 1000 \times [875^2 + 175^2]^{1/2}$$
$$= 892328 \text{ N.mm}$$

Shaft diameter :

$$d^3 = \left[\frac{16}{\pi\tau_d}\right] [(K_m M)^2 + (K_t T)^2]^{1/2}$$

$$= \left[\frac{16}{(\pi \times 85.5)}\right] [(1.5 \times 892328)^2 + (1.0 \times 150{,}000)^2]^{1/2}$$

$$= 80228$$

$$d = 43.13 \text{ mm} \approx 45 \text{ mm}$$

Example 9.2 A steel shaft is subjected to a torque varying from 400 N.m to 140 N.m and a bending moment varying from 500 N.m. to 250 N.m. Determine the shaft diameter if its ultimate strength in tension is 540 MPa and tensile yield strength is 400 MPa. A factor of safety of 2 is desired. Endurance strength is 40% of the ultimate strength and $K_f = 1.25$.

Solution.

Given : $T_{max} = 400$ N.m, $T_{min} = 140$ N.m, $M_{max} = 500$ N.m, $M_{min} = 250$ N.m, $S_{ut} = 540$ MPa,

$$S_{yt} = 400 \text{ MPa}, n = 2, S'_e = 0.4\, S_{ut}$$

$$S'_e = 0.4 \times 540 = 216 \text{ MPa}$$

Corrected endurance strength, $S_e = \dfrac{S'_e}{K_f} = \dfrac{216}{1.25} = 172.8$ MPa

According to distortion energy theory,

$$S_{ys} = 0.577 S_y$$

$$= 0.577 \times 400$$

$$= 230.8 \text{ MPa}$$

$$S_{es} = 0.577\, S_e$$

$$= 0.577 \times 172.8$$

$$= 99.7 \text{ MPa}$$

$$M_m = \frac{500 + 250}{2} = 375 \text{ N.m}$$

$$M_a = \frac{500 - 250}{2} = 125 \text{ N.m}$$

$$T_m = \frac{400 + 140}{2} = 270 \text{ N.m}$$

$$T_a = \frac{400 - 140}{2} = 130 \text{ N.m}$$

Using maximum distortion energy theory,

$$\frac{S_y}{n} = \left[\frac{32}{\pi d^3}\right]\left[\left\{M_m + \left(\frac{S_y}{S_e}\right)M_a\right\}^2 + 0.75\left\{T_m + \left(\frac{S_{ys}}{S_{es}}\right)T_a\right\}^2\right]^{1/2}$$

$$\frac{400}{2} = \left[\frac{32}{\pi d^3}\right]\left[\left\{375 + \left(\frac{400}{172.8}\right)\times 125\right\}^2 + 0.75\left\{270 + \left(\frac{230.8}{99.7}\right)\times 130\right\}^2\right]^{1/2} \times 10^3$$

$$d = 34.8 \approx 35 \text{ mm}$$

Example 9.3 A shaft is supported by two bearings 2 m apart and carries two pulleys as shown in fig. 9.2. Pulley A is mounted at a distance of 0.5 m and drives a pulley directly below it with the help of a belt having maximum tension of 2.5 kN Pulley B is driven with the help of an electric motor and belt which is placed horizontally to the right. The angle of contact for both the pulleys is 180° and coefficient of friction is 0.20. Determine the diameter of solid shaft having allowable stresses in tension and shear of 60 MPa and 40 MPa respectively.

Solution.

Given : F_1 = 2.5 kN, μ = 0.20, θ = 180°, σ_t = 60 MPa, τ = 40 MPa, d_c = 0.5m, d_d = 0.3 m.

$$\frac{F_1}{F_2} = \exp(\mu\,\theta)$$

$$= \exp(0.20 \times \pi) = 1.8745$$

$$F_2 = \frac{2500}{1.8745}$$

$$= 1333.72 \text{ N}$$

Resultant load on pulley C, $F_c = F_1 + F_2$ = 2500 + 1333.72 = 3833.72 N↓

Fc
A C (c) Vertical forces D B
1437.6
383.36
(d) Vertical B.M.D
Fd
A C (e) Horizontal forces D B
2044.5
638.9
(f) Horizontal B.M.D

Fig. 9.2 (b) B.M. diagrams

Torque on pulley C, $$T_c = (F_1 - F_2)\left(\frac{d_c}{2}\right)$$

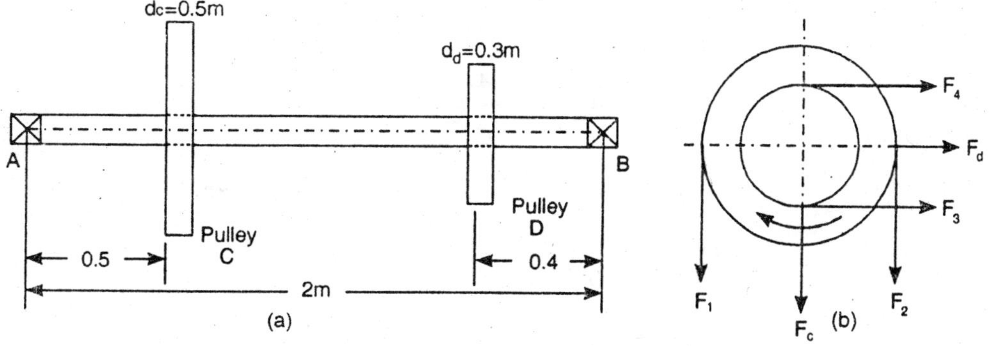

Fig. 9.2

$$= (2500 - 1333.72)\left(\frac{0.5}{2}\right)$$

$$= 291.570 \text{ N.m}$$

Now $T_c = T_d$

Therefore $$(F_3 - F_4)\left(\frac{d_2}{2}\right) = 291.570$$

$$F_3 - F_4 = 291.570 \times \frac{2}{0.3} = 1943.8 \text{ N}$$

Also $$\frac{F_3}{F_4} = 1.8745$$

$$F_4 = 2222.7 \text{ N},$$

$$F_3 = 4166.5 \text{ N}$$

Resultant load on pulley D,

$$F_d = F_3 + F_4$$

$$= 4166.5 + 2222.7$$

$$= 6389.2 \text{ N} \rightarrow$$

The vertical and horizontal forces acting on the shaft are shown in Fig. 9.2 (*c*) and (*e*) and the corresponding bending moment diagrams in Fig. 9.2 (*d*) and (*f*) respectively.

Vertical forces :

$$R_{AV} = F_c \times \frac{1.5}{2}$$

$$= 3833.72 \times \frac{1.5}{2} = 2875.3 \text{ N}$$

$$M_{CV} = 2875.3 \times 0.5 = 1437.6 \text{ N.m}$$

$$M_{DV} = 1437.6 \times \frac{0.4}{1.5} = 383.36 \text{ N.m}$$

Horizontal forces :

$$R_{AH} = F_d \times \frac{0.4}{2}$$

$$= 6389.2 \times \frac{0.4}{2}$$

$$= 1277.84 \text{ N.m}$$

$$M_{DH} = 1277.84 \times 1.6 = 2044.5 \text{ N.m}$$

$$M_{CH} = 2044.5 \times \frac{0.5}{1.6} = 638.9 \text{ N.m}$$

Resultant bending moments :

$$M_C = \left[M_{CV}^2 + M_{CH}^2\right]^{1/2}$$

$$= [(1437.6)^2 + (638.9)^2]^{1/2}$$

$$= 1573.2 \text{ N.m}$$

$$M_D = \left[M_{DV}^2 + M_{DH}^2\right]^{1/2}$$

$$= [(2044.5)^2 + (383.36)^2]^{1/2}$$

$$= 2080.13 \text{ N.}$$

$$M_{max} = 2080.13 \text{ N.m}$$

Equivalent torque,

$$T_e = \left[M_{max}^2 + T^2\right]^{1/2}$$

$$= [(2080.13)^2 + (291.57)^2]^{1/2}$$

$$= 2108.4 \text{ N.m}$$

$$\left(\frac{\pi}{16}\right) d^3 \tau = T_e$$

$$d = \left[\frac{16 \times 2108.4 \times 10^3}{\pi \times 40}\right]^{1/3}$$

$$= 64.5 \text{ mm}$$

Equivalent bending moment,

$$M_e = 0.5 \, [M_{max} + T_e]$$

$$= 0.5 \, [2080.13 + 2108.4]$$

$$= 2098.26 \text{ N.m}$$

$$\left(\frac{\pi}{32}\right) d^3 \sigma_t = M_e$$

$$d = \left[\frac{2098.26 \times 10^3 \times 32}{\pi \times 60}\right]^{1/3} = 70.88 \approx 72 \text{ mm}$$

Adopt the larger value of $d = 72$ mm

Example 9.4 A shaft is supported on two bearings 1 m apart. A straight tooth 20° pressure angle spur gear having 0.5 m pitch diameter is located at 0.2 m from the left hand bearing. A pulley of 0.6 m diameter is mounted at 0.3 m from the right hand bearing. The gear is driven by a pinion with a horizontal tangential force and the pulley drives a horizontal belt having 180° angle of wrap and weighs 2 *k*N. The maximum belt tension is 2400 N and belt tension ratio is 3. Determine the shaft diameter if allowable shear stress of its material is 45 MPa.

Solution.

Given : $\phi = 20°\ d_g = 0.5$ m, $d_p = 0.6$ m, $\theta = 180°$, $W_p = 2\ k$N, $F_1 = 2400$ N, $\frac{F_1}{F_2} = 3$, $\tau = 45$ MPa

$$F_2 = \frac{2400}{3} = 800 \text{ N}$$

Torque on pulley,
$$T = (F_1 - F_2)\frac{d_p}{2}$$
$$= (2400 - 800) \times \frac{0.6}{2}$$
$$= 480 \text{ N.m}$$

Tangential force on gear,
$$F_{tg} = \frac{2T}{d_g}$$
$$= 2 \times \frac{480}{0.5}$$
$$= 1920 \text{ N}\leftarrow$$

Radial force on gear,
$$F_{rg} = F_{tg} \tan\phi$$
$$= 1920 \tan 20°$$
$$= 698.82 \text{ N}\downarrow$$

Resultant belt force,
$$F = F_1 + F_2$$
$$= 2400 + 800$$
$$= 3200 \text{ N}\rightarrow$$

698.82 N
2000 N
A
B
(c) Vertical forces
231.8
462.2
(d) Vertical B.M.D
1920N
3200N
115.2
(e) Horizontal forces
556.8
(f) Horizontal B.M.D

Fig. 9.3 (b) B.M. diagrams

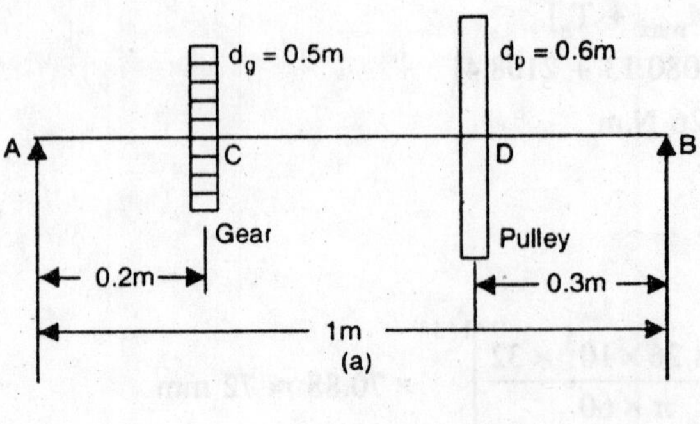

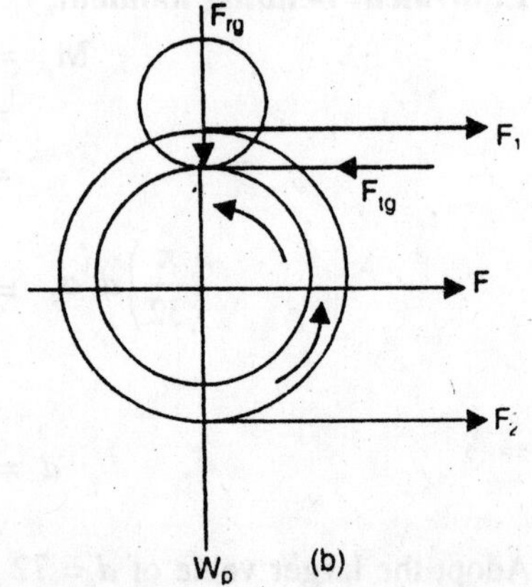

Fig. 9.3

The loading on shaft is shown in Fig. 9.3 (*a*)

Taking moments about B, we have

Vertical forces [Fig. 9.3 (*b*)] :

$$R_{AV} \times 1 = 698.82 \times 0.8 + 2000 \times 0.3$$

$$= 1159$$

$$R_{AV} = 1159 \text{ N}$$

Bending moment at C, $M_{CV} = R_{AV} \times 0.2$

$$= 1159 \times 0.2$$

$$= 231.8 \text{ N.m}$$

Bending moment at D, $M_{DV} = 1159 \times 0.7 - 698.82 \times 0.5$

$$= 461.9 \text{ N.m}$$

Horizontal forces [Fig. 9.3 (*c*)] :

$$R_{AH} \times 1 = 1920 \times 0.8 - 3200 + 0.3 = 576$$

$$R_{AH} = 576 \text{ N}$$

$$M_{CH} = 576 \times 0.2$$

$$= 115.2 \text{ N.m}$$

$$M_{DH} = 576 \times 0.7 - 1920 \times 0.5$$

$$= -556.8 \text{ N.m}$$

Resultant bending moment :

$$M_C = [(231.8)^2 + (115.2)^2]^{1/2}$$

$$= 258.85 \text{ N.m}$$

$$M_D = [(461.9)^2 + (-556.8)^2]^{1/2}$$

$$= 723.45 \text{ N.m}$$

Maximum bending moment, $M = 723.45$ N.m

Equivalent torque, $T_e = [M^2 + T^2]^{1/2}$

$$= [(723.45)^2 + (480)^2]^{1/2}$$

$$= 868.2 \text{ N.m}$$

$$\left(\frac{\pi}{16}\right) d^3 \tau = T_e$$

$$d = \left[16 \times 868.2 \times \frac{10^3}{\pi \times 45}\right]^{1/3}$$

$$= 46.14 \approx 50 \text{ mm}$$

Example 9.5 A mild steel shaft transmits 30 *k*W at 300 rpm, It carries 1 *k*N concentrated load at midspan. The distance between the bearings is 3 m. The load is applied gradually. Calculate the diameter of the shaft if permissible tensile stress is 56 MPa and shear stress is 42 MPa.

Solution.

Given : P = 30 kW, N = 300 *rpm*, W = 1 *k*N, L = 3 m, σ_t = 56 MPa, τ = 42 MPa

For gradually applied load, shock and fatigue factors are : K_m = 1.5, K_t = 1.0 [Table 9.3]

Maximum bending moment, $$M = \frac{WL}{4}$$

$$= 1 \times \frac{3}{4}$$

$$= 0.75 \text{ kN.m}$$

Power transmitted, $$P = \frac{2\pi NT}{60 \times 10^3}$$

Torque, $$T = \frac{30 \times 60 \times 10^3}{2\pi \times 300}$$

$$= 954.93 \text{ N.m}$$

Equivalent torque, $$T_e = [(K_m M)^2 + (K_t T)^2]^{1/2}$$

$$= [(1.5 \times 750)^2 + (1.0 \times 954.93)^2]^{1/2}$$

$$= 1475.64 \text{ N.m}$$

$$\frac{\pi}{16} d^3 \tau = T_e$$

$$= \left[\frac{16 \times 1475.64 \times 10^3}{\pi \times 42}\right]^{1/3}$$

$$= 56.35 \text{ mm}$$

Equivalent bending moment, $$M_e = 0.5 [K_m M + T_e]$$

$$= 0.5 [1.5 \times 750 + 1475.64]$$

$$= 1300.32 \text{ N.m}$$

$$\frac{\pi}{32} d^3 \sigma t = M_e$$

$$d = \left[\frac{32 \times 1300.32 \times 10^3}{\pi \times 56}\right]^{1/3}$$

$$= 61.84 \approx 63 \text{ mm}$$

Adopt the larger value of diameter,

$$d = 63 \text{ mm}$$

Example 9.6 A shaft 1.2 m between bearings carries a pulley (d_p = 1.2 m) and a spur gear (d_g = 0.8 m) overhanging by 0.3 m and 0.4 m from the left hand and right hand bearings respectively. The shaft transmits 25 *k*W at 200 rpm. The tangential force on the gear is vertical and the belt tensions on the pulley are downwards. The self weight of the gear and pulley is 1 kN and 3 kN respectively. The ratio of belt tensions is 2.5. Angle of wrap on the pulley is 180°. The shock and fatigue factors are 2 and 1.5 in bending and torsion respectively. Calculate the diameter of the shaft if allowable shear stress in shaft material is 63 MPa.

Solution.

Given : P = 25 *k*W, N = 200 *rpm*, $\frac{F_1}{F_2}$ = 2.5, d_p = 1.2 m, d_g = 0.8 m, W_g = 1 kN, W_p = 3 kN, K_m = 2.0, K_t = 1.5, τ = 63 MPa.

The loaded shaft is shown in Fig. 9.4 (*a*).

$$T = \frac{25 \times 60 \times 10^3}{2\pi \times 200}$$

$$= 1193.662 \text{ N.m.}$$

Torque on pulley,

$$T = (F_1 - F_2)\frac{d_p}{2}$$

$$= (F_1 - F_2) \times \frac{1.2}{2}$$

$$= 1193.662$$

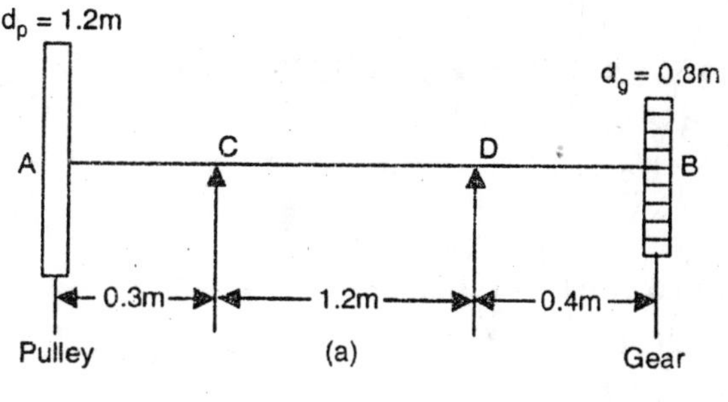

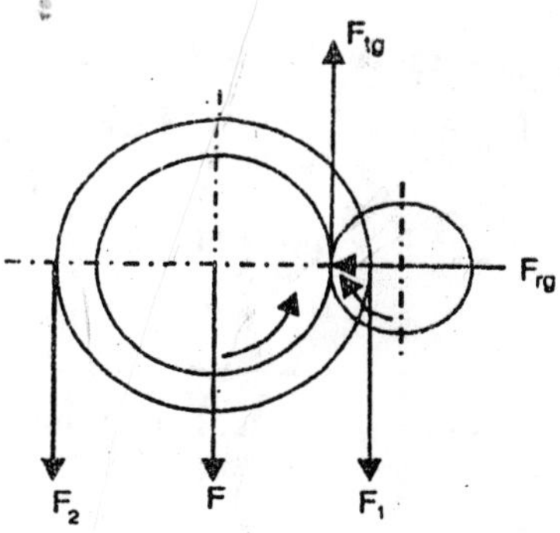

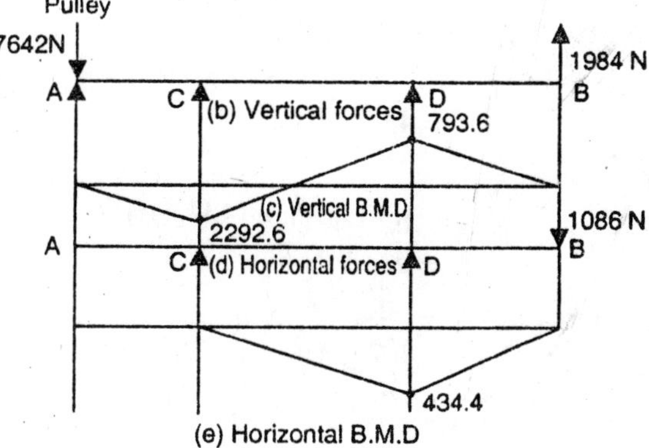

Fig. 9.4

$$F_1 - F_2 = 1989.4 \text{ N}$$

$$F_2 = \frac{1989.4}{1.5}$$

$$= 1326.3 \text{ N}$$

$$F_1 = 3315.7 \text{ N}$$

Resultant vertical force on pulley,

$$F_{AV} = F_1 + F_2 + W_{p}$$

$$= 3315.7 + 1326.3 + 3000 = 7642 \text{ N}\downarrow$$

Tangential force on gear, $F_{tg} = \dfrac{2T}{d_g}$

$$= \frac{2 \times 1193.662}{0.8}$$

$$= 2984 \text{ N}\uparrow$$

Radial force on gear, $F_{rg} = F_{tg} \tan \phi$

$$= 2984 \tan 20°$$

$$= 1086 \text{ N}\leftarrow$$

Resultant vertical force on gear,

$$F_{BV} = F_{tg} - W_g$$

$$= 2984 - 1000$$

$$= 1984 \text{ N}\uparrow$$

Vertical forces [Fig. 9.4 (*b*)] :

$\Sigma M_D = 0$; $R_{CV} \times 1.2 = 7642 \times 1.5 + 1984 \times 0.4 = 12256.6$

$$R_{CV} = 10213.8 \text{ N}$$

Bending moment at C, $M_{CV} = F_{AV} \times 0.3$

$$= 7642 \times 0.3$$

$$= -2292.6 \text{ N.m}$$

Bending moment at D, $M_{DV} = F_{BV} \times 0.4$

$$= 1984 \times 0.4$$

$$= +793.6 \text{ N.m}$$

Horizontal forces [Fig. 9.4 (*c*)] :

$$M_{DH} = 1086 \times 0.4$$

$$= 434.4 \text{ N.m}$$

Resultant bending moment at D :

$$M_D = [(739.6)^2 + (434.4)^2]^{1/2}$$

$$= 904.7 \text{ N.m}$$

Maximum bending moment, $M = -2292.6$ N.m

Equivalent torque, $T_e = [(K_m M)^2 + (K_t T)^2]^{1/2}$

$$= [(2 \times 2292.6)^2 + (1.5 \times 1193.662)^2]^{1/2}$$

$$= 4922.4 \text{ N.m}$$

$$\frac{\pi}{16} d^3 \tau = T_e$$

$$d = \left[\frac{16 \times 4922.4 \times 10^3}{\pi \times 63}\right]^{1/3}$$

$$= 73.5 \approx 75 \text{ mm.}$$

Example 9.7 A hollow steel shaft is to transmit 30 kW at 300 rpm with 20% overload. The maximum bending moment in the shaft is 1500 N.m. and axial compressive load is 10 kN. The shaft is supported on rigid bearings 2 m apart. The permissible shear stress in the shaft is 42 MPa and the diameter ratio is 0.6. The shock and fatigue factors in bending and torsion are 1.6 and 1.5 respectively. Determine the diameter of the shaft.

Solution.

Given : P = 30 kW, N = 300 *rpm*, overload = 20%, M = 1500 N.m, F = 10 kN, L = 2 m,

$k = \frac{d_i}{d_o} = 0.6$, $\tau = 42$ MPa, $K_m = 1.6$, $K_t = 1.5$

$$I = \left(\frac{\pi}{64}\right)\left(d_o^4 - d_i^4\right)$$

$$= \left(\frac{\pi}{64}\right) d_o^4 \left(1 - k^4\right)$$

$$A = \left(\frac{\pi}{4}\right)\left(d_o^2 - d_i^2\right)$$

$$= \left(\frac{\pi}{4}\right) d_o^2 \left(1 - k^2\right)$$

$$K^2 = \frac{I}{A} = \frac{d_o^2 (1 + k^2)}{16}$$

$$= d_o^2 \frac{\left[1 + (0.6)^2\right]}{16}$$

$$= 0.085 \, d_o^2$$

$$K = 0.2915 \, d_o$$

$$P = \frac{2\pi N T_m}{60 \times 10^3}$$

$$T_m = \frac{30 \times 60 \times 10^3}{2\pi \times 300}$$

$$= 954.93 \text{ N.m}$$

$$T_{max} = 1.2 \times T_m$$
$$= 1.2 \times 954.93$$
$$= 1145.9 \text{ N.m}$$

$$\tau_{max} = \left[\frac{16}{\pi d_o^3\, 1-k^4}\right]\left[\left\{K_m M + \frac{\alpha F d_o\left(1+k^2\right)}{8}\right\}^2 + (K_t T)^2\right]^{1/2}$$

α = a constant depending upon the magnitude of column action of the unsupported shaft length between the bearings.

$$= \frac{1}{\left[1-0.004\left(\frac{L}{K}\right)\right]} \text{ for } \frac{L}{K} < 115$$

$$= \left[\frac{1}{1-0.0044\left\{\frac{2000}{0.2915 d_o}\right\}}\right]$$

$$= \frac{1}{\left[\frac{1-30.189}{d_o}\right]}$$

$$= \frac{d_o}{d_o - 30.189}$$

$$42 = \left[\frac{16}{\pi d_o^3\left(1-0.6^4\right)}\right] \times [\{1.6 \times 1500 \times 10^3 + \frac{d_o}{d_o - 30.189}$$

$$\times 10{,}000 \times d_o \frac{\left(1+0.6^2\right)}{8} \}^2 + (1.5 \times 1145.9 \times 10^3)^2]^{1/2}$$

$$7.178\, d_o^3 = \left[\left\{2400 \times 10^3 + \frac{1700\, d_o^2}{\left(d_o - 30.189\right)^2}\right\}^2 + 2954445 \times 10^6\right]^{1/2}$$

When F = 0, d_o = 71 mm

By hit and trial we find that d_o = 75 mm

EXERCISES

9.1 A shaft 1.5 m long is supported on two bearings. It carries two gears A and B at 0.5 m and 1.0 m from the left hand bearing of pitch circle diameters 400 mm and 200 mm respectively. Gear A is driven at 350 *rpm* from the top by another gear and receives 5 kW power. Gear B drives another gear horizontally. If the yield tensile strength of the shaft is 480 MPa, determine the diameter of the shaft as per ASME code. Assume $K_m = 2.0$ and $K_t = 1.5$.

9.2 A steel shaft is supported on bearings 1.0 m apart. Power is transmitted to the shaft by a belt running on a 500 mm diameter pulley A which overhangs to the right of the right bearing by 300 mm. Another pulley B, 200 mm in diameter, is located midway between the bearings. 10 *k*W power is transmitted from the shaft by a belt running on pulley B. The ratio of belt tensions is 3. The shaft runs at 600 *rpm*. The material properties are : $S_{ut} = 640$ MPa, $S_{yt} = 410$ MPa. Determine the shaft diameter if the belt drives are at right angles.

9.3 A steel shaft is subjected to a torque varying from 120 N. m to 360 N.m and a bending moment varying from 220 N.m to 440 N.m. Determine the shaft diameter if its ultimate tensile strength is 540 MPa and tensile yield strength is 420 MPa. A factor of safety of 2 is desired. Fatigue stress-concentration factor is 1.25.

9.4 The shaft of an axial flow rotary compressor is subjected to a maximum torque of 2.5 *k*N.m and a maximum bending moment of 5 *k*N.m. The combined shock and fatigue factors in torsion and bending are 1.5 and 2.0 respectively. Determine the diameter of the shaft if the permissible shear stress in the shaft is 55 MPa. What percentage saving in material shall be obtained if a hollow shaft is designed ?

10 KEYS AND COUPLINGS

Chapter

10.1 KEYS

A key is a piece of metal inserted between the shaft and hub of a pulley or a gear to connect them together in order to prevent relative motion between them. A keyway is a slot or recess cut in the shaft and hub to accommodate the key. It is a temporary type of fastening.

10.1.1 Types of Keys

Keys are of the following types :

1. Sunk key 2. Saddle key 3. Tangent key. 4. Round key, and 5. Splines.

1. Sunk key. A sunk key is provided half in the keyway of the shaft and half in the keyway of the hub. This key is of the following types :

(*i*) **Rectangular sunk key.** This key is shown in Fig. 10.1 (*a*). Its usual proportions are : width, $b = \frac{d}{4}$, thickness, $t = \frac{d}{6}$, where d = diameter of the shaft. Taper on the key = 1 : 100.

(*ii*) **Square sunk key**. In this case, $b = t = \frac{d}{4}$, as shown in Fig. 10.1 (*b*).

Fig. 10.1 Rectangular and square keys

(*iii*) **Parallel sunk key.** It may be a rectangular or square key with no taper on the length of key.

(*iv*) **Gib-head key**. It is a rectangular sunk key with a head at one end known as gib head to facilitate the removal of key from the keyway, as shown in Fig. 10.2. Its usual proportions are :

Fig. 10.2 Gib-head key

$$b = \frac{d}{4} \quad \text{...(10.1 } a\text{)}$$

$$t = \frac{d}{6} \quad \text{...(10.1 } b\text{)}$$

(*v*) **Feather key.** A feather key is used where it is essential to slide a keyed gear or pulley along the shaft axis. The key is either made a tight fit in the shaft or the hub and a sliding fit in the other. When making a tight fit in the shaft, it is screwed to the shaft, as shown in Fig. 10.3 (*a*) and when making a tight fit in the hub, gib heads are provided on both sides of the key, as shown in Fig. 10.3 (*b*).

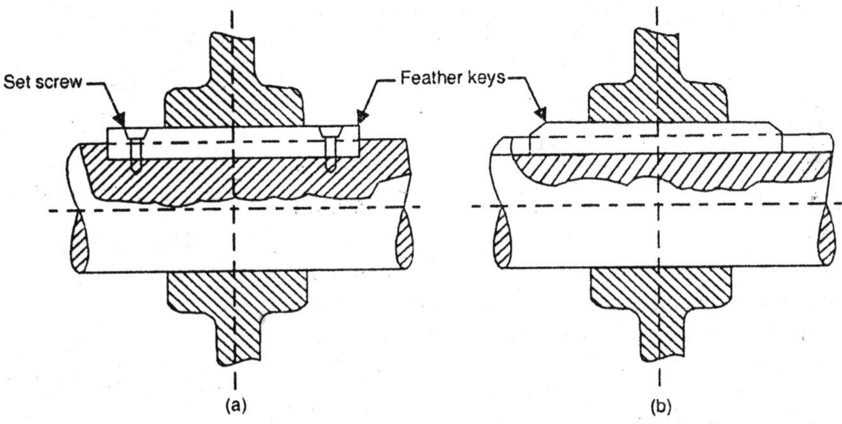

Fig. 10.3 Feather key

(*vi*) **Woodruff key.** It is an easily adjustable key cut from a cylindrical disc having segmental cross-section in front view as shown in Fig. 10.4.

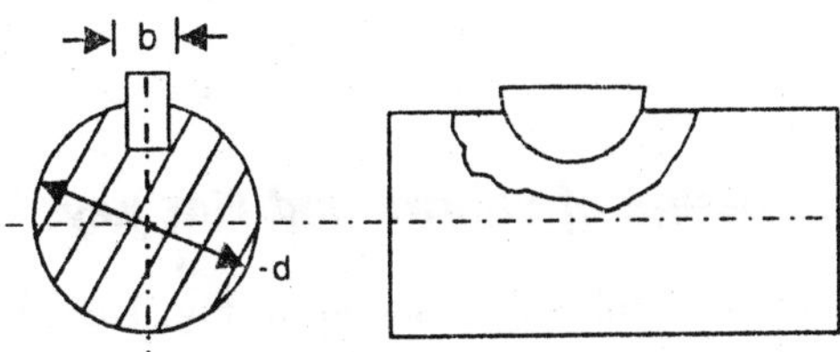

Fig. 10.4. Woodruff key

2. Saddle key. It is similar to a flat key whose bottom surface is given a radius to match the shaft radius as shown in Fig. 10.5. They are used for light loads.

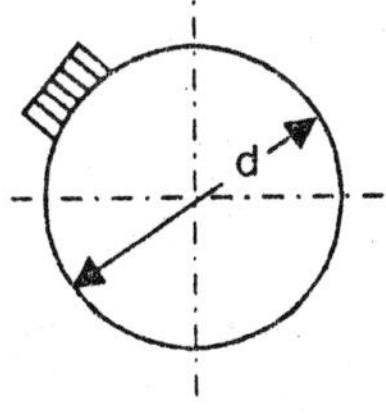

Fig. 10.5 Saddle key

3. Tangent key. The tangent keys are fitted in pairs at right angles as shown in Fig. 10.6. They remain in compression under load and are used for heavy loads.

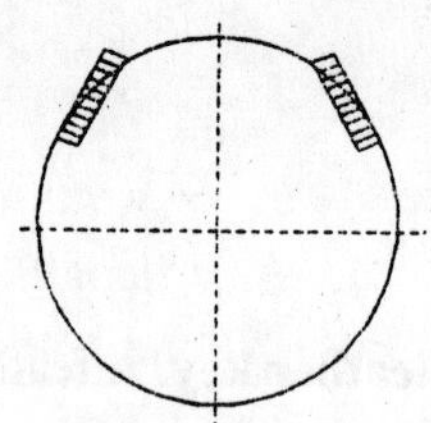

Fig. 10.6 Tangent key

4. Round key. This key is circular in section and fits into holes drilled partly in the shaft and partly in the hub as shown in Fig. 10.7. They are used for low power drives.

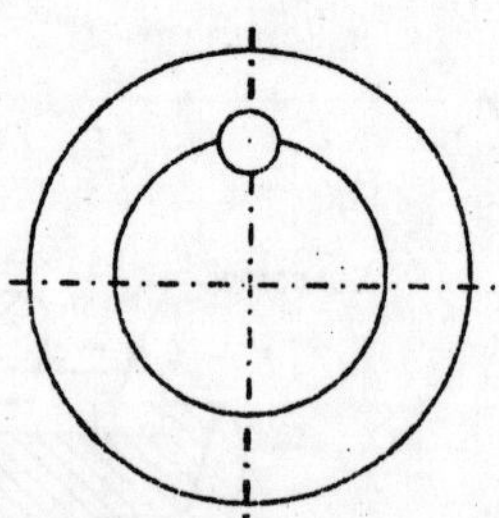

Fig. 10.7 Round key

5. Kennedy key. It consists of two square keys as shown in Fig. 10.8 and is preferred for heavy loads.

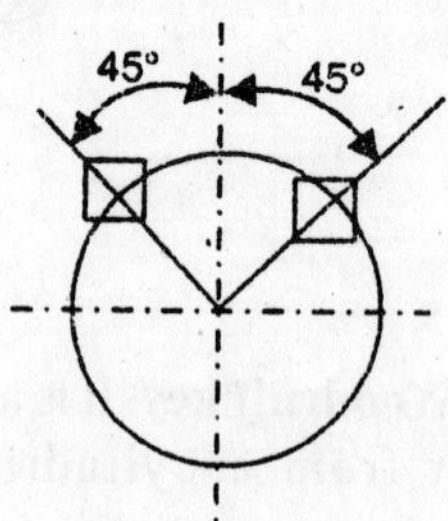

Fig. 10.8 Kennedy key

10.1.2 Design of Square and Flat Keys

A square sunk key is shown in Fig. 10.9. The design of square and flat keys is based on two criteria–failure due to shear stresses or crushing due to compressive stresses. The forces acting on the key are shown in the figure. The two equal and opposite forces F are due to the torque being transmitted from the shaft to the hub and vice versa. The forces F′ (F′ = F) act as a resisting couple preventing the key to roll in the keyway. It is assumed that the force F is tangential to the shaft. Therefore,

$$F = \frac{2T}{d} \quad ...(1)$$

where T = transmitted torque, N.mm

d = shaft diameter, mm

F = tangential force, N

The shear failure shall occur in plane AB so that the shear stress is given by :

$$\tau = \frac{F}{bl} \qquad ...(2)$$

where b = width of key

l = length of key

From Eqs. (1) and (2), we get

$$\tau = \frac{2T}{dbl} \qquad ..(10.2)$$

The failure due to compressive stresses shall occur on surfaces AC and BD. Hence,

$$AC = BD = \frac{t}{2}$$

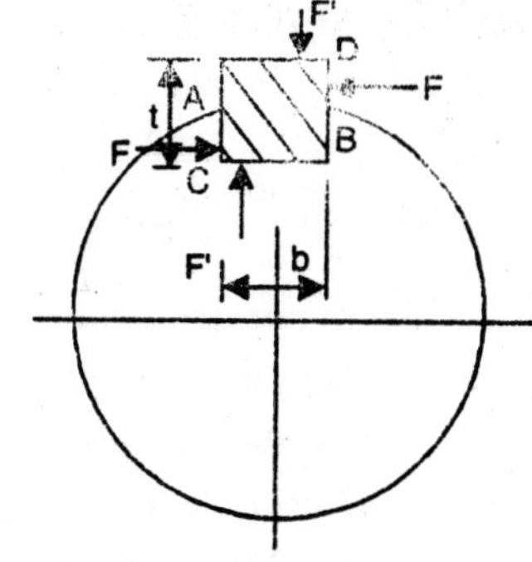

Fig. 10.9 Forces on rectangular key

The compressive stress is given by :

$$\sigma_c = \frac{F}{\frac{tl}{2}} = \frac{2F}{tl} \qquad ...(3)$$

where t = thickness of key

From Eqs. (1) and (3), we get

$$\sigma_c = \frac{4T}{dtl} \qquad ...(10.3)$$

If the key is to be equally strong in shear and crushing, then

$$\frac{2T}{dbl} = \frac{4T}{dtl}$$

$$\frac{2b}{t} = \frac{\sigma_c}{\tau} \qquad ...(10.4)$$

Generally $\sigma_c = 2\tau$

Therefore $\frac{b}{t} = 1$, i.e. a square key is the best size.

If the key is to be equally strong in shear as the shaft, then

$$\frac{2T}{dbl} = \frac{16T}{\pi d^3}$$

$$l = \frac{\pi d^2}{8b}$$

If $b = \frac{d}{4}$, then $l = \frac{\pi d}{2} \approx 1.5\, d$

10.2 DESIGN OF SPLINES

Splines are keys made integral with the shaft which fit in the keyways broached in the hub. Such shafts are called splined shafts as shown in Fig. 10.10. These shafts are stronger than shafts having a single keyway. Splined shafts are used when the force transmitted is large and the driven member is to slide over the driving member.

Let D = major diameter of splines
d = minor diameter of splines
l = length of hub
n = number of splines
p_m = permissible pressure on splines
$\approx$ 6.5 MPa
A = total area of splines
T = torque transmitted by the splines

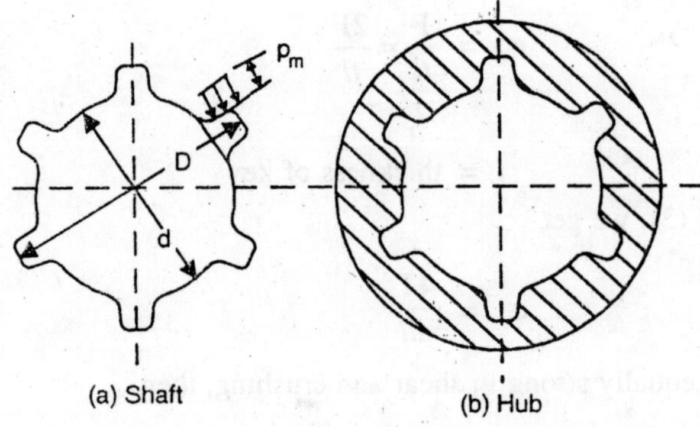

Fig. 10.10 Splines

Now $$A = (D - d)\frac{ln}{2}$$

Mean radius of splines, $$R_m = \frac{D + d}{4}$$

Torque transmitting capacity of splines, $$T = p_m A R_m$$

$$= \left(\frac{1}{8}\right) p_m ln\,(D^2 - d^2) \qquad ...(10.5)$$

Example 10.1 Design the rectangular key for a shaft of 75 mm diameter. The shearing and curshing stresses for key material are 50 MPa and 75 MPa respectively.

Solution.

Given. d = 75 mm, τ = 50 MPa, σ_c = 75 MPa

For the key to be equally strong in shear and crushing,

$$\frac{2b}{t} = \frac{\sigma_c}{\tau} = \frac{75}{50} = 1.5$$

$$b = \frac{d}{4} = \frac{75}{4} = 18.75 \approx 20 \text{ mm}$$

$$t = \frac{2 \times 20}{1.5} = 26.67 \approx 27 \text{ mm}$$

For key to be equally strong in shear as the shaft,

$$l = \frac{\pi d^2}{8b} = \frac{\pi \times 75^2}{8 \times 20} = 110.4 \approx 115 \text{ mm.}$$

Example 10.2 Design a rectangular key for a shaft of 50 mm diameter. The shearing and crushing stresses for the key and shaft are 42 MPa and 72 MPa.

Solution.

Given. d = 50 mm, τ = 42 MPa, σ_c = 72 MPa

For the key to be equally strong in shear and crushing,

$$\frac{2b}{t} = \frac{\sigma_c}{\tau} = \frac{72}{42} = 1.714$$

$$b = \frac{d}{4} = \frac{50}{4} = 12.5 \approx 15 \text{ mm}$$

$$t = \frac{2 \times 15}{1.714} = 17.5 \approx 20 \text{ mm}$$

For key to be equally strong in shear as the shaft,

$$l = \frac{\pi d^2}{8b} = \frac{\pi \times 50^2}{8 \times 15} = 65.45 \approx 70 \text{ mm.}$$

Example 10.3 Design a square key for fixing a gear on the shaft having 25 mm diameter. The gear rotates at 550 rpm and transmits 12 kW power to the meshing gear. The key is made of steel having yield stress in tension as 400 N/mm^2. The yield stress in compression and tension may be taken equal to each other. Assume factor of safety 2.5.

Solution.

Given. d = 25 mm, N = 550 rpm, P = 12 kW, σ_{yt} = 400 N/mm^2 = σ_{yc}, n = 2.5

$$P = \frac{2\pi NT}{60 \times 10^3}$$

$$T = \frac{12 \times 60 \times 10^3}{2 \times \pi \times 550}$$

$$= 208.348 \text{ N.m}$$

According to maximum shear stress theory, $\tau = 0.5\, \sigma_{yt} = 0.5 \times 400 = 200$ MPa

Design shear stress, $\tau_d = \frac{\tau}{n} = \frac{200}{2.5} = 80 \text{ N/mm}^2$

Design crushing strength, $\sigma_d = \frac{\sigma_{yc}}{n} = \frac{400}{2.5} = 160 \text{ N/mm}^2$

Width of key, $b = \frac{d}{4} = \frac{25}{4} = 6.25 \approx 7 \text{ mm}$

Thickness of key, $t = 7$ mm

Length of key based on shear failure,

$$l = \frac{2T}{db\tau_d}$$

$$= \frac{2 \times 208348}{25 \times 7 \times 80} = 29.75 \approx 30 \text{ mm}$$

Length of key based on crushing failure,

$$l = \frac{4T}{dt\sigma_d}$$

$$= \frac{4 \times 208348}{25 \times 7 \times 160} = 29.75 \approx 30 \text{ mm.}$$

10.3 COUPLINGS

A coupling is a machine element which can be used to connect two shafts. The applications of couplings are :

1. To provide a connection between two units manufactured separately.
2. To provide for easy disconnection during repair.
3. To provide as a connection between shaft lengths to facilitate easy transformation.
4. To provide mechanical flexibility for misalignment.
5. To minimize shock load transmission throughout the shaft length.

10.3.1 Types of Couplings

Shaft couplings are used to connect two shafts. The shafts may be either aligned or misaligned having lateral or angular misalignment. Accordingly we can have either rigid or flexible couplings. The couplings may be classified as follows :

1. Rigid couplings. They do not permit even minor misalignment of the axes of the connected shafts. Rigid couplings are of the following types :

(*i*) Sleeve or muff coupling
(*ii*) Clamp coupling
(*iii*) Flange coupling, and
(*iv*) Solid or forged flange coupling.

2. Flexible couplings. They permit slight misalignment between the axes of the connected shafts. Flexible couplings are of the following types :

(*i*) Oldham's coupling
(*ii*) Hooke's coupling
(*iii*) Universal flexible coupling, and
(*iv*) Slip coupling.

10.3.2. *Sleeve or Muff Coupling*

It consists of a hollow cylinder which may be keyed or pinned to the shafts as shown in Fig. 10.11. This coupling does not have any projecting part on the external surface and is very safe to use. However, this requires very close and accurate fitting and is difficult to disassemble. The sleeve may be made of cast iron or steel. The diameter of the sleeve may be determined from the torque to be transmitted by the shaft, as follows :

Torque transmitted,

$$T = \frac{\pi}{16}\left[\frac{D^4 - d^4}{D}\right] \times \tau \qquad ...(10.6)$$

where D = outside diameter of muff

d = diameter of shaft

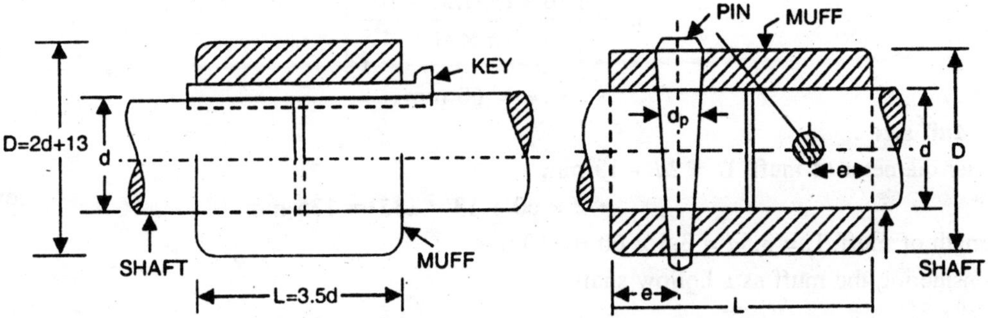

Fig. 10.11 Sleeve or muff coupling

τ = permissible shear stress for muff material.

The key can be designed as explained in art. 10.1.2. sometimes two keys are used to carry half the torque. Empirically,

$$D = 2d + 13 \text{ mm} \qquad ....(10.7a)$$

Length of muff, $L = 3.5\,d$ ...(10.7 *b*)

Edge distance, $e = 0.75\,d$ ...(10.7 *c*)

The diameter of the pins can be determined by considering its failure in shear. Thus

$$2 \times \left(\frac{\pi}{4}\right) d_p^2 \times \tau = \frac{2T}{d} \qquad ...(10.8)$$

where d_p = diameter of pin

= (0.20 – 0.5) d, empirically ...(10.9)

τ = permissible shear stress in pin material.

Example 10.4 Design a muff coupling which is used to connect two steel shafts transmitting 50 kW at 300 rpm. The material for the shafts and key is plain carbon steel for which allowable shear and crushing stresses may be taken as 40 MPa and 80 MPa respectively. The material for the muff is carbon steel for which the allowable shear stress may be assumed as 15 MPa.

Solution.

Given. P = 50 kW, N = 300 rpm, $\tau_s = 40$ MPa, $\sigma_{cs} = 80$ MPa, $\tau_c = 15$ MPa

Refer to Fig. 10.11.

1. Shaft size :

Power transmitted, $$P = \frac{2\pi NT}{60 \times 10^3}$$

$$T = \frac{50 \times 60 \times 10^3}{2\pi \times 300} = 1591.55 \text{ N.m}$$

Let d = diameter of shaft

$$\left(\frac{\pi}{16}\right) d^3 \tau_s = T$$

$$d = \left[\frac{16 \times 1591.55 \times 10^3}{\pi \times 40}\right]^{\frac{1}{3}}$$

$$= 58.74 \approx 60 \text{ mm}$$

2. Muff size :

Outer diameter of muff, D = 2d + 13 mm

$$= 2 \times 60 + 13 = 133 \approx 135 \text{ mm}$$

Length of muff, L = 3.5d = 3.5 × 60 = 210 mm

Considering the muff as a hollow shaft,

$$T = \frac{\pi}{16}\left[\frac{D^4 - d^4}{D}\right]\tau_c$$

$$1591.55 \times 10^3 = \frac{\pi}{16}\left[\frac{135^4 - 60^4}{135}\right]\tau_c$$

$$\tau_c = 3.43 \text{ MPa} < 15 \text{ MPa}$$

Hence safe.

3. Key size :

From standard tables, for shaft size of 58-65 mm, the dimensions of gib-head key are :

Width, b = 18 mm, depth, h = 18 mm, length = $\frac{L}{2}$ = 105 mm

(*a*) Check for shear stress in key.

$$T = \frac{bld\,\tau}{2}$$

Shear stress in key, $$\tau = \frac{2 \times 1591.55 \times 10^3}{18 \times 105 \times 60}$$

$$= 28 \text{ MPa} < 40 \text{ MPa}$$

(*b* Check for crushing stress in key.

$$T = \frac{h}{2} \times l \times \sigma_c \times \frac{d}{2}$$

$$\sigma_c = \frac{4T}{hld}$$

$$= \frac{4 \times 1591.55 \times 10^3}{18 \times 105 \times 60}$$

$$= 56.1 \text{ MPa} < 80 \text{ MPa}$$

Hence safe.

10.3.3 Clamp or Compression coupling

This is a modification of the muff coupling, where the muff is split into two parts for easy assembly and disassembly of the coupling. The two halves of the coupling, as shown in Fig. 10.12, are fitted on the shaft and rigidly fastened by bolts. The torque is transmitted by frictional forces set up between the muff and the shaft. A key may also be fitted to avoid slip. The ribs add strength and serve as a protection to the workers against injury from revolving bolt heads and nuts. This coupling is used for transmitting heavy torque.

Let d_b = core diameter of bolts

n = number of bolts

T = transmitted torque

d = shaft diameter

μ = coefficient of friction between shaft and sleeve materials

l = length of sleeve

σ_t = allowable tensile stress in the bolts

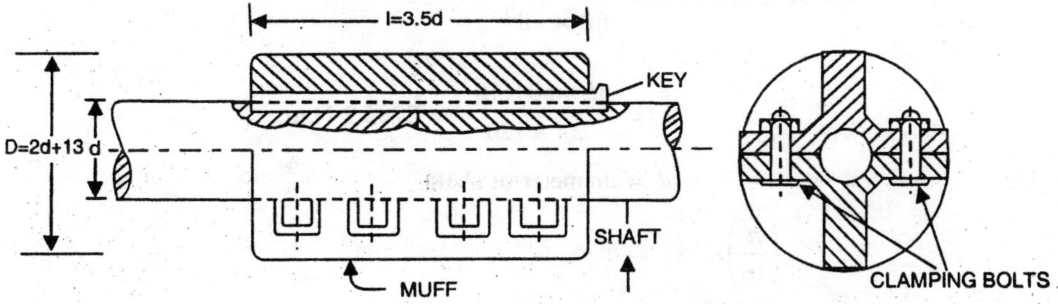

Fig. 10.12 Clamp or compression coupling

Clamping force on each shaft,

$$F_c = \left(\frac{n}{2}\right) \times \left(\frac{\pi}{4}\right) d_b^2 \times \sigma_t$$

Clamping area, $$A_c = d \times \left(\frac{1}{2}\right)$$

Intensity of clamping pressure, $p_c = \dfrac{F_c}{A_c}$

Tangential frictional force per unit area of shaft periphery, $F = \mu p_c$

Torque that can be transmitted by the coupling,

$$T = \left(\frac{d}{2}\right) \times \left(\frac{\pi d l}{2}\right) \times F$$

$$= \left(\frac{\pi^2}{16}\right) d_b^2 \, \sigma_t \mu \, n \, d \quad \text{...(10.10)}$$

Load on each bolt, $$P = \left(\frac{\pi}{4}\right) d_b^2 \, \sigma_t \quad \text{...(10.11 } a\text{)}$$

$$\therefore \quad = \frac{4T}{\pi \, d n \mu} \quad \text{...(10.11 } b\text{)}$$

Example 10.5 Design a clamp coupling to transmit 35 kW at 120 rpm. The allowable shear stress for shaft and key is 40 MPa. The number of bolts connecting the two halves are 6. The permissible tensile stress for the bolts is 70 MPa. The coefficient of friction between the muff and the shaft surfaces may be taken as 0.3.

Solution.

Given : P = 35 kW, N = 120 rpm, n = 6, τ = 40 MPa, σ_t = 70 MPa, μ = 0.3

Refer to Fig. 10.12.

1. Shaft size :

Power transmitted, $$P = \frac{2\pi NT}{60 \times 10^3}$$

$$T = \frac{35 \times 60 \times 10^3}{2\pi \times 120} = 2785.2 \text{ N.m}$$

Let d = diameter of shaft

$$\left(\frac{\pi}{16}\right) d^3 \tau_s = T$$

$$d = \left[\frac{16 \times 2785.2 \times 10^3}{\pi \times 40}\right]^{\frac{1}{3}} = 70.78 \approx 72 \text{ mm}$$

2. Muff size :

Outer diameter of muff, D = 2*d* + 13 mm

= 2 × 72 + 13 = 157 ≈ 160 mm

Length of muff, L = 3.5*d* = 3.5 × 72 = 252 mm

Considering the muff as a hollow shaft,

$$T = \left(\frac{\pi}{16}\right)\left[\frac{D^4 - d^4}{D}\right]\tau_c$$

$$2785.2 \times 10^3 = \left(\frac{\pi}{16}\right)\left[\frac{(160^4) - (72^4)}{135}\right]\tau_c$$

$$\tau_c = 3.04 \text{ MPa} < 15 \text{ MPa}$$

Hence safe.

3. Key size :

From standard tables, for shaft size of 72 mm, the dimensions of gib-head key are :

Width, b = 22 mm, depth, h = 18 mm, length = $\frac{L}{2}$ = 126 mm

(*a*) Check for shear stress in key.

$$T = \frac{bld\,\tau}{2}$$

$$\text{Shear stress in key, } \tau = \frac{2 \times 2785.2 \times 10^3}{22 \times 126 \times 72}$$

$$= 27.9 \text{ MPa} < 40 \text{ MPa}$$

(*b*) Check for crushing stress in key.

$$T = \left(\frac{h}{2}\right) \times l \times \sigma_c \times \frac{d}{2}$$

$$\sigma_c = \frac{4T}{hld}$$

$$= \frac{4 \times 2785.2 \times 10^3}{18 \times 126 \times 72}$$

$$= 68.8 \text{ MPa} < 80 \text{ MPa}$$

Hence safe.

4. Size of bolts :

Let d_b = core diameter of bolts

$$T = \left(\frac{\pi^2}{16}\right) d_b^2 \sigma_t \mu n d$$

$$d_b = \left[\frac{16 \times 2785.2 \times 10^3}{\pi^2 \times 70 \times 0.3 \times 6 \times 72}\right]^{\frac{1}{2}}$$

$$= 22.3 \approx 23.32 \text{ mm for coarse threads series.}$$

Nominal diameter = 27 mm or M 27.

10.3.4 Forged End Flange Coupling

In this coupling, the flanges are forged at the ends of the shafts as shown in Fig. 10.13. The flanges are connected together by bolts. Such type of couplings are generally used for connecting marine engine shafts.

The various dimensions of forged end coupling are :

Let d = diameter of shaft

$$d_2 = 0.2d + 5 \text{ mm}$$

$$D_2 = d + 2d_2 + 20 \text{ mm}$$

$$b = 0.3d + 5 \text{ mm}$$

$$n = \frac{d}{55} + 3$$

$$D_3 = \frac{d + D_2}{2}$$

$$b_2 = 4 \text{ to } 5 \text{ mm}$$

$$b_1 = b_2 + 1 \text{ mm}$$

$$D_1 = D_2 + d_2 + 40 \text{ mm}$$

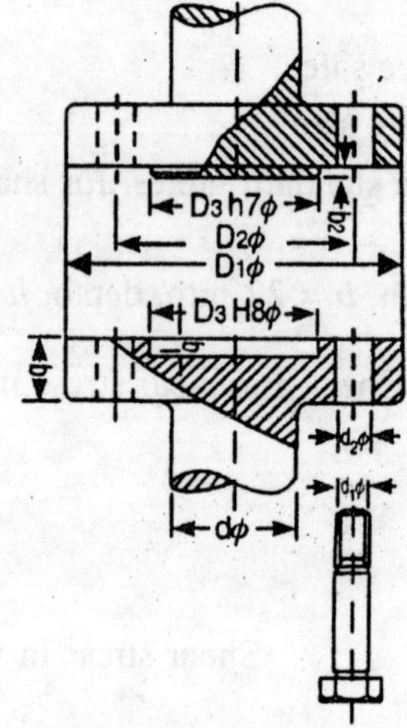

Fig. 10.13 Forged end type flange coupling

10.3.5 Rigid Protected Type Flange Coupling

It consists of two flanges keyed to the shaft ends and bolted together. To ensure proper alignment of the shafts the end of one shaft may enter into the recess provided in the flange attached to the other shaft. Protected type of flanges are often used to provide safety to the operator. These flanges project beyond the heads of the bolts and nuts. The various parts of the flange coupling are shown in Fig. 10.14. The various parts can be designed as discussed below :

1. Shaft diameter :

$$P = \frac{2\pi NT_m}{60 \times 10^3}$$

$$T_m = \frac{P \times 60 \times 10^3}{2\pi \times N}$$

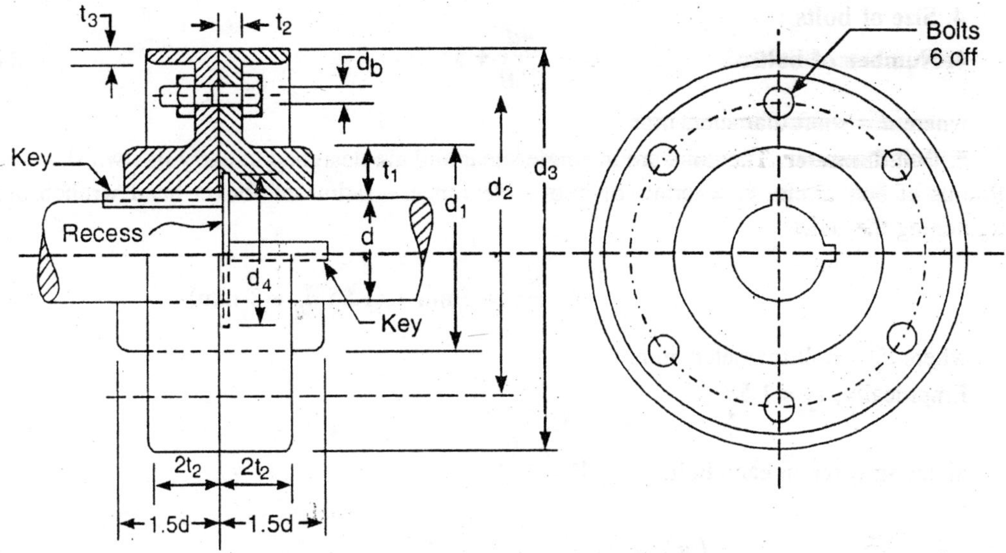

Fig. 10.14 Protected type rigid flange coupling

$$T_{max} = (1 + \% \text{ overload factor}) \times T_m$$

$$\left(\frac{\pi}{16}\right) d^3 \tau_s = T_{max}$$

2. Hub diameter. The hub is designed as a hollow shaft subjected to torsion. Thus

$$T = \left(\frac{\pi}{16}\right)\left[\frac{d_1^4 - d^4}{d_1}\right]\tau_h \qquad ...(10.12)$$

where τ_h = permissible shear stress in hub

Empirically, $d_1 = 2d$

Thickness of hub, $t_1 = \dfrac{d_1 - d}{2}$

Diameter of recess in flanges, $d_4 = d + t_1$

Outside diameter of protecting flange,

$$d_3 = 4d$$

3. Hub length. The hub length is made equal to the key length required. The key size is determined by designing the key and the shaft to be equally strong in shear.

Width of key, $b = \dfrac{d}{4}$

Length of key, $l = 1.5d$

Generally a square or a rectangular key is used. For a rectangular key

$$t = \frac{d}{6}.$$

4. Number of bolts. $n = \frac{4d}{150} + 3$ ...(10.13)

where d = shaft diameter, mm

5. Bolt diameter. The bolts are in single shear and are designed to carry its own share of load. Radius of bolt circle, r_2 = radius of hub + clearance + radius of socket head wrench used for tightening the bolts

$$= \frac{d_1}{2} + 5\text{mm} + (1.85\, d_b + 8 \text{ mm}) \quad ...(10.14)$$

where d_b = bolt diameter, mm

Empirically, $r_2 = 1.5d$

Shearing force in each bolt, $F = \frac{T}{r_2 n}$

$$\left(\frac{\pi}{4}\right) d_b^2\, \tau_b = F \quad ...(10.15)$$

where τ_b = permissible shear stress in bolts.

6. Bolt diameter based on tensile load. The necessity of reamed holes for bolts may be avoided if the coupling is designed on the basis that the torque is transmitted by friction between the flanges. In this case, the bolts will be in tension only. Assuming uniform pressure distribution between the flanges, the mean friction radius is given by,

$$r_f = \frac{2\,(r_3^3 - r_4^3)}{3\,(r_3^2 - r_4^2)}$$

Then $T = n \times \mu\, P_i \times r_f$

where P_i = initial tensile force in each bolt

r_3 = outer radius of the friction surface

$= \frac{d_3}{2}$

r_4 = inner radius of the friction surface

μ = coefficient of friction between flanges

Therefore $P_i = \left(\frac{\pi}{4}\right) d_b^2\, \sigma_t$ ...(10.16)

where σ_t = permissible tensile stress in bolts.

7. Flange thickness. Considering the bearing failure of flange.

$$F = n \times d_b \times t_2 \times \sigma_c \quad ...(10.17)$$

where σ_c = permissible bearing stress in flange.

Empirically, $t_2 = 0.5t_1 + 6$ mm

where $t_1 = \frac{d_1 - d}{2}$ = hub thickness

Considering shearing of the flange at the junction with the hub,

$$2\left(\frac{\pi d_1^2}{4}\right) \times \tau \times t_2 = \text{T} \qquad ...(10.18)$$

8. Check for crushing of bolt.

Considering crushing of the bolts,

$$n d_b t_2 \sigma_{cb} \times \frac{d_2}{2} = \text{T} \qquad ...(10.19)$$

where σ_{cb} = permissible crushing strength of bolts.

9. Thickness of protecting flange.

$$t_3 = 0.5 t_2 \text{ (min) or } 0.25\, d$$

10. Hub overlap = 3 mm (min)

Example 10.6 Design a protected type rigid cast iron flange coupling for a steel shaft transmitting 20 kW at 240 rpm and having an allowable shear stress of 45 MPa. The shearing and tensile stresses in the bolts are limited to 30 MPa and 60 MPa respectively. The material of the shaft and key is the same for which the crushing strength is twice the shear strength. The overload is 25%. The allowable shear stress for cast iron is 15 MPa and coefficient of friction is 0.15.

Solution.

Given. P = 20 kW, N = 240 rpm, τ_s = 45 MPa, τ_b = 30 MPa, σ_b = 60 MPa, $\sigma_{cs} = 2\tau_s$, τ_{ci} = 15 MPa, $\text{T}_{max} = 1.25\ \text{T}_m$, $\mu = 0.15$

Refer to Fig. 10.14

1. Shaft diameter.

$$\text{P} = \frac{2\pi \text{NT}_m}{60 \times 10^3}$$

$$\text{T}_m = \frac{20 \times 60 \times 10^3}{2\pi \times 240} = 795.775 \text{ N.m}$$

$$\text{T}_{max} = 1.25 \times 795.775 = 994.72 \text{ N.m}$$

$$\left(\frac{\pi}{16}\right) d^3 \tau_s = \text{T}_{max}$$

$$\left(\frac{\pi}{16}\right) \times d^3 \times 45 = 994.72 \times 10^3$$

$$d = 48.28 \approx 50 \text{ mm}$$

2. Hub diameter. The hub is designed as a hollow shaft subjected to torsion. Thus

$$\text{T}_{max} = \left(\frac{\pi}{16}\right)\left[\frac{d_1^4 - d^4}{d_1}\right]\tau_h$$

Where τ_h = permissible shear stress in hub made of C.I.
= 15 MPa

$$994{,}720 = \left(\frac{\pi}{16}\right)\left[\frac{d_1^4 - 50^4}{d_1}\right]\tau_h$$

Empirically, $d_1 = 2d = 100$ mm

$\tau_h = 5.4$ MPa < 15 MPa

Hence safe. Adopt $d_1 = 100$ mm

Thickness of hub, $t_1 = \dfrac{d_1 - d}{2} = \dfrac{100 - 50}{2} = 25\text{mm}$

Diameter of recess in flanges, $d_4 = d + t_1 = 50 + 25 = 75$ mm

Outside diameter of protecting flange,

$d_3 = 4d = 4 \times 50 = 200$ mm

3. Hub length. The hub length is made equal to the key length required. The key is determined by designing the key and the shaft to be equally strong in shear.

Width of key, $b = \dfrac{d}{4} = \dfrac{50}{4} = 12.5 \approx 15$ mm

Length of key, $l = 1.5d = 1.5 \times 50 = 75$ mm

We use a square key with thickness,

$t = b = 15$ mm

Hub length, $l = 75$ mm

4. Number of bolts. $n = \dfrac{4d}{150} + 3$

Where d = shaft diameter, mm

$$n = \frac{4 \times 50}{150} + 3 = 4.3 \approx 4$$

The bolts have to be even in number from balancing point of view.

5. Bolt diameter. The bolts are in single shear and are designed to carry its own share of load.

Radius of bolt circle, $r_2 = 1.5d = 1.5 \times 50 = 75$ mm

Shearing force in each bolt,

$$F = \frac{T_{max}}{r_2 n} = \frac{994720}{75 \times 4} = 3315.7 \text{ N}$$

$$\left(\frac{\pi}{4}\right) d_b^2\, \tau_b = F$$

Where τ_b = permissible shear stress in bolts.

$$\left(\frac{\pi}{4}\right) d_b^2 \times 30 = 3315.7$$

$$d_b = 11.86 \approx 12 \text{ mm}$$

6. Bolt diameter based on tensile load. The necessity of reamed holes for bolts may be avoided if the coupling is designed on the basis that the torque is transmitted by friction between the flanges. In this case, the bolts will be in tension only. Assuming uniform pressure distribution between the flanges, the mean friction radius is given by,

$$r_f = \frac{2\,(r_3^3 - r_4^3)}{3\,(r_3^2 - r_4^2)}$$

$$r_3 = \text{outer radius of the friction surface}$$

$$= \frac{d_3}{2} = \frac{200}{2} = 100 \text{ mm}$$

$$r_4 = \text{inner radius of the friction surface}$$

$$= \frac{d_4}{2} = \frac{75}{2} = 37.5 \text{ mm}$$

$$r_f = \frac{2\left[100^3 - (37.5)^3\right]}{3\{100^2 - (37.50^2\}}$$

$$= 73.48 \text{ mm}$$

Then $T_{max} = n \times \mu\, P_i \times r_f$

Where P_i = initial tensile force in each bolt

μ = coefficient of friction between flanges

$$P_i = \frac{994720}{(4 \times 0.15 \times 73.48)} = 8122.4 \text{ N}$$

Therefore $$P_i = \left(\frac{\pi}{4}\right) d_b^2\, \sigma_t$$

Where σ_t = permissible tensile stress in bolts.

$$8122.4 = \left(\frac{\pi}{4}\right) d_b^2 \times 60$$

$$d_b = 13.1 \approx 15 \text{ mm}$$

7. Flange thickness. Considering bearing failure of flange,

$$F = n \times d_b \times t_2 \times \sigma_c$$

Where σ_c = permissible bearing stress in flange.

Empiricially, $t_2 = 0.5t_1 + 6$ mm $= 0.5 \times 25 = 6 = 18.5 \approx 20$ mm

$$\sigma_{ci} = \frac{3315.7}{(4 \times 15 \times 20)} = 2.76 \text{ MPa} < 30 \text{ MPa}$$

Considering shearing of the flange at the junction with the hub,

$$2\left(\frac{\pi d_1^2}{4}\right) \times \tau \times t_2 = T_{max}$$

$$2\left(\pi \times \frac{100^2}{4}\right) \times 20 \times \tau = 994720$$

$$\tau = 3.17 \text{ MPa} < 15 \text{ MPa}$$

Hence safe.

8. Check for crushing of bolt.

Considering crushing of the bolts,

$$n d_b t_2 \sigma_{cb} \times \frac{d_2}{2} = T_{max}$$

Where σ_{cb} = permissible crushing strength of bolts.

$$4 \times 12 \times 20 \times \sigma_{cb} \times \frac{150}{2} = 994720$$

$$\sigma_{cb} = 13.8 \text{ MPa} < 60 \text{ MPa}$$

Hence safe.

9. Thickness of protecting flange.

$$t_3 = 0.5 t_2 \text{ (min) or } 0.25\, d$$

$$t_3 = 0.5 \times 20 = 10 \text{ mm}$$

10. Hub overlap = 3 mm (min)

10.4 BUSH TYPE FLEXIBLE COUPLING

In a bush type coupling, motion from one - half of flange is transmitted to the other half by means of pins or bolts. The pins are rigidly bolted to one flange and fit loosely in the corresponding holes of the other flange. This type of coupling is commonly used where the driving and driven members are mounted on a monoblock. This can take up parallel misalignment of shafts upto 0.25 mm per 10 mm outside diameter of coupling and angular misalignment upto 1.5 degree and axial displacement also. The pins are covered by leather, rubber, etc. A typical bush type flexible coupling is shown in Fig. 10.15.

The various dimensions of coupling may be determined as follows :

1. Number of bolts

$$n = 0.04d + 3$$

where d = diameter of shaft, mm

2. Outer diameter of hub

$$d_1 = 2d$$

3. Diameter of bolts

$$d_b = \frac{0.5d}{\sqrt{n}}$$

Shear force in each bolt, $F = \dfrac{2T}{nd_2}$

Where d_2 = pitch circle diameter of bolts

$$\left(\frac{\pi}{4}\right) d_b^2 \tau = F$$

Where τ = permissible shear stress in bolts

4. Outer diameter of rubber bush

d_3 = diameter of pin + 2 × thickness of rubber bush

5. Length of bush

Considering the bearing of bush,

$$F = l_2 d_3 p_b$$

Where p_b = bearing pressure for bush

$= 2$ MPa

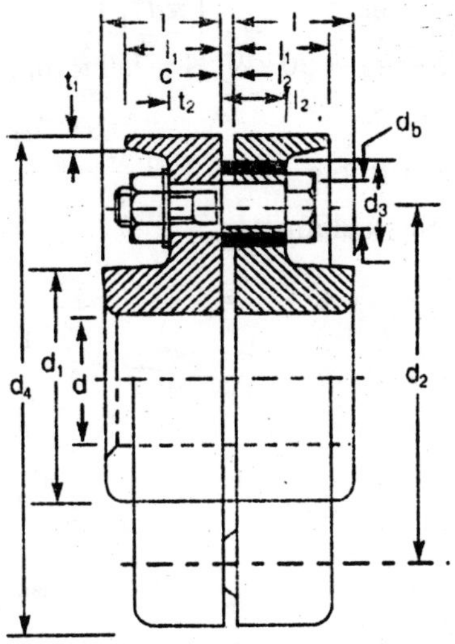

Fig. 10.15 Bush type flexible coupling

6. Check for pin in bending

Let c = clearance between two flanges. The loaded pin is shown in Fig. 10.16.

Bending moment on the pin,

$$M = Fl_3$$

$$\text{Where } l_3 = \frac{l_2 - c}{2} + c = \frac{l_2 + c}{2}$$

$$\frac{F(l_2 + c)}{2} = \left(\frac{\pi}{32}\right) d_b^3 \times \sigma_b$$

Fig. 10.16 Loading on the pin

Where σ_b = bending stress in pin

7. Maximum shear stress in pin

$$\tau_{max} = \left[\left(\frac{\sigma_b}{2}\right)^2 + \tau^2\right]^{\frac{1}{2}} < 35 \text{ MPa}$$

Example 10.7 Design a bushed - pin type flexible coupling to connect a compressor shaft to a motor shaft transmitting 30 kW at 750 rpm. The overload factor is 20%. The allowable material properties are as follows :

Shaft, key and pin :

Shear = 35 MPa

Crushing = 70 MPa

Cast iron :

Shear = 15 MPa

Rubber bush :

Bearing = 0.8 MPa

Solution.

Given : P = 30 kW, N = 750 rpm, $T_{max} = 1.2T_m$, $\tau_s = \tau_k = \tau_b = 35$ MPa, $\sigma_{cs} = \sigma_{ck} = \sigma_{cb} = 70$ MPa, $\tau_{ci} = 15$ MPa, $p_b = 0.8$ MPa

Refer to Fig. 10.16.

1. Shaft diameter.

$$P = \frac{2\pi NT_m}{60 \times 10^3}$$

$$T_m = \frac{30 \times 60 \times 10^3}{2\pi \times 750} = 381972 \text{ N.mm}$$

$$T_{max} = 1.2 \times 381972 = 458366 \text{ N. mm}$$

$$\left(\frac{\pi}{16}\right) d^3 \tau_s = T_{max}$$

$$\left(\frac{\pi}{16}\right) \times d^3 \times 35 = 458366$$

$$d = 40.55 \approx 42 \text{ mm}$$

2. Hub diameter. The hub is designed as a hollow shaft subjected to torsion. Thus

$$T_{max} = \left(\frac{\pi}{16}\right)\left[\frac{d_1^4 - d^4}{d_1}\right]\tau_h$$

Where τ_h = permissible shear stress in hub made of C.I.
= 15 MPa

$$458366 = \left(\frac{\pi}{16}\right)\left[\frac{d_1^4 - 50^4}{d_1}\right]\tau_h$$

Empirically, outer diameter of hub, $d_1 = 2d = 84$ mm

$$\tau_h = 4.2 \text{ MPa} < 15 \text{ MPa}$$

Hence safe. Adopt $d_1 = 84$ mm

Thickness of hub, $t_1 = \dfrac{d_1 - d}{2} = \dfrac{84 - 42}{2} = 21$ mm

Outside diameter of protecting flange, $d_4 = 4d = 4 \times 42 = 168 \approx 170$ mm

3. Key size and hub length. The hub length is made equal to the key length required. The key is determined by designing the key and the shaft to be equally strong in shear.

Width of key, $b = \dfrac{d}{4} = \dfrac{42}{4} = 10.5 \approx 12$ mm

Length of key, $l = 1.5\,d = 1.5 \times 42 = 63$ mm

We use a square key with thickness,

$$t = b = 12 \text{ mm}$$

Hub length, $l = 63$ mm

4. Number of bolts

$$d_b = 0.04\,d + 3$$
$$= 0.04 \times 42 + 3 = 4.68 \approx 6$$

5. Diameter of bolts

$$d_b = \frac{0.5d}{\sqrt{n}}$$

$$= \frac{0.5 \times 42}{\sqrt{6}} = 8.57 \text{ mm}$$

To account for bending of pin in the bush, let core diameter of pin, $d_b = 20$ mm so that shank diameter of bolt = 24 mm

6. Outer diameter of rubber bush

d_3 = diameter of pin + 2 × thickness of rubber bush of 2 mm + 2 × thickness of brass bush of 6 mm

$$= 24 + 2 \times 2 + 2 \times 6 = 40 \text{ mm}$$

Pitch circle diameter of bolts,

$$d_2 = d_1 + d_3 + 2 \times 6 = 84 + 40 + 12 = 136 \text{ mm}$$

7. Check for shear in bolt

Shear force in each bolt, $F = \dfrac{2T_{max}}{nd_2}$

$$F = \frac{2 \times 458366}{6 \times 136} = 1123.4 \text{ N}$$

$$\left(\frac{\pi}{4}\right) d_b^2\, \tau = F$$

Where τ = permissible shear stress in bolts

$$\left(\frac{\pi}{4}\right) \times 20^2 \times \tau = 1123.4$$

$$\tau = 3.57 \text{ MPa} < 35 \text{ MPa}$$

Hence safe.

8. Length of bush

Considering the bearing of bush,

$$F = l_2 d_3 p_b$$

Where p_b = bearing pressure for bush

$= 0.8$ MPa

l_2 = length of bush

$$l_2 = \frac{1123.4}{40 \times 0.8} = 35 \text{ mm}$$

9. Check for pin in bending

Let c = clearance between two flanges.

$= 5$ mm

The loaded pin is shown in Fig. 10.16

Bending moment on the pin,

$$M = Fl_3$$

$$\text{Where } l_3 = \frac{l_2 - c}{2} + c = \frac{l_2 + c}{2}$$

$$\frac{F(l_2 + c)}{2} = \left(\frac{\pi}{32}\right) d_b^3 \times \sigma_b$$

$$1123.4 \frac{(35 + 5)}{2} = \left(\frac{\pi}{32}\right) \times 20^3 \times \sigma_b$$

Bending stress in pin, $\sigma_b = 28.6$ MPa

10. Maximum shear stress in pin

$$\tau_{max} = \left[\left(\frac{\sigma_b}{2}\right)^2 + \tau^2\right]^{\frac{1}{2}}$$

$$= \left[\left(\frac{28.6}{2}\right)^2\right] + (3.57)^2]^{\frac{1}{2}}$$

$= 14.73 \text{ MPa} < 35 \text{ MPa}$

Hence safe.

11. Flange thickness. Considering the bearing failure of flange,

$$F = n \times d_b \times t_2 \times \sigma_c$$

Where σ_c = permissible bearing stress in flange.

Empirically, $t_2 = 0.5t_1 + 6 \text{ mm} = 0.5 \times 21 + 6 = 16.5 \approx 18 \text{ mm}$

Considering shearing of the flange at the junction with the hub,

$$2\left(\frac{\pi d^2_1}{4}\right)\times\tau\times t_2 = T_{max}$$

$$2\left(\frac{\pi\times 84^2}{4}\right)\times 15\times\tau = 458366$$

$$\tau = 8.66 \text{ MPa} < 15 \text{ MPa}$$

Hence safe.

Thickness of protecting flange, $t_1 = 0.5t_2 = 9$ mm

Length of protecting flange, $l_1 = l_2 + 1.5d_b = 35 + 1.5 \times 20 = 65$ mm

EXERCISES

10.1 What are the advantages of flexible coupling ? How do you classify flexible coupling ?

10.2 Design a flange coupling to transmit 135 kW power at 120 rpm from one shaft to another. Allowable shearing stress of the bolt material is 45 N/mm^2 while that of the shaft material is 55 N/mm^2. The flanges are made of C.I. for which the stresses in compression and shear are 75 N/mm^2 and 175 N/mm^2 respectively. Suggest suitable dimensions for the shaft, bolts and coupling. (Do not design the key).

10.3 Design a cast iron protected type flange coupling to connect two shafts of 36 mm diameter, transmitting 15 kW at 720 rpm. The overload capacity is 1.25 times the average torque. The bolts and key are made of C20 steel and the flanges are made of cast iron.

10.4 Design a square key for a 45 mm diameter shaft to transmit 10 kW at 250 rpm. The key and shaft are made of steel having $S_{yt} = S_{yc} = 380$ MPa. The factor of safety desired is 3.

10.5 Design a cast iron protected type rigid flange coupling to connect a 50 kW, 1440 rpm electric motor to a centrifugal pump. The starting torque of motor is 200 % of the rated torque. Assume suitable data from the Data Book.

10.6 Design a bushed pin type flexible coupling for connecting a motor and a pump shaft for the following data :

Power transmitted	=	20 kW
Speed	=	900 rpm
Diameter of motor and pump shafts	=	60 mm
Allowable bearing pressure in rubber bush	=	0.35 MPa

Table 10.1 Dimensions of Parallel Keys and Keyways

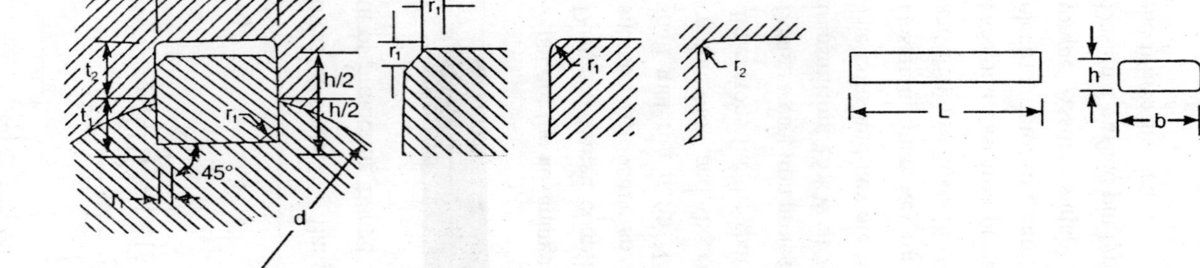

For shaft diameters	Above	6 8 10	12 17	22 30 38 44 50	58 65 75	85 95 110	130 150 170 200	230 260 290	330 380 440
	Upto	8 10 12	17 22	30 38 44 50 58	65 75 85	110 150 130	150 170 200 230	260 290 330	380 440 500
Key cross-section	Width, b	2 3 4	5 6	8 10 12 14 16	18 20 22	25 28 32	36 40 45 50	56 63 70	80 90 100
	Height, h	2 3 4	5 6	7 8 8 9 10	11 12 14	14 16 18	20 22 25 28	32 32 36	40 45 50
Keyway depth (Nominal)	in Shaft, t_1	1.2 1.8 2.5	3.0 3.5	4.0 5 5 5.5 6	7 7.5 8.5	9.0 10 11	12 13 15 17	19 20 22	25 28 31
	in Hub, t_2	1 1.4 1.8	2.3 2.8	3.3 3.3 3.3 3.8 4.3	4.4 4.9 5.4	5.9 6.4 7.4	8.4 9.4 10.4 11.4	12.4 13.4 14.4	15.4 17.4 19.5
Tolerance on	r_1	+0.05 -0.00	+0.1 -0.0			+0.15 -0.00			
Keyway depth	t_2	+0.05 -0.00	+0.1 -0.0			+0.15 -0.00			
Chamfer or radius of key	r_1 Max	0.25	0.35	0.55	0.80		1.30	2.00	2.95
	Min	0.16	0.25	0.40	0.60		1.00	1.60	2.50
Keyway radius	r_2 Max	0.16	0.25	0.40	0.60		1.00	1.60	2.50
Lingthb of key	L Min	6 6 8	10 14	18 22 28 36 45	50 56 63 71 80 90		100 110 125 140	160 180 200	220 250 280
	Max	20 36 45	50 71	90 110 140 160 180	200 220 250 280 320 360		400 400 400 400	400 400 400	400 400 400

IS : 2048 - 1962

Table 10.2 Dimensions of Parallel Keys and Keyways (All dimensions in mm)

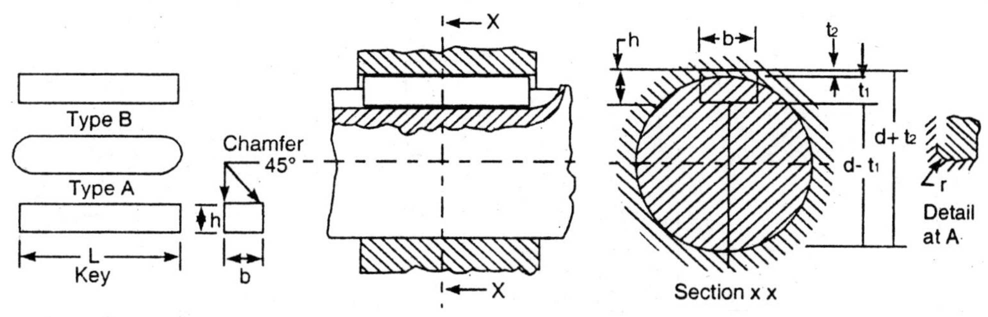

Range of Shaft dia d		Key	Keyway										Range of Key Length L			
Above	Up to	$b \times h$	b	Tol on b					t_1	Tol on t_1	t_2	Tol on t_2	t			
				Running fit		Light drive fit		Force fit								
				Shaft H9	Hub D10	Shaft N9	Hub Js9	Shaft and hub P9					Min.	Max.	Min.	Max.
6	8	2 × 2	2	+ 0.025	+ 0.060	− 0.004	+ 0.0125	− 0.006	1.0		1.2		0.08	0.16	6	20
8	10	3 × 3	3	0	+ 0.020	− 0.029	−0.0125	−0.031	1.8		1.4		0.08	0.16	6	36
10	12	4 × 4	4						2.5	+0.1	1.8	+0.1	0.08	0.16	8	45
12	17	5 × 5	5	+ 0.030	+0.078	0	+0.0150	− 0.012	3.0	0	2.3	0	0.16	0.25	10	56
17	22	6 × 6	6	0	+ 0.030	− 0.015 0	− 0.042	3.5			2.8		0.16	0.25	14	70
22	30	8 × 7	8	+0.036	+0.098	0	+0.018 0	− 0.015	4.0		3.3		0.16	0.025	18	90
30	38	10 × 8	10	0	+ 0.040	− 0.036	− 0.018 0	− 0.051	5.0		3.3		0.25	0.40	22	110
38	44	12 × 8	12						5.0		3.3		0.25	0.40	28	140
44	50	14 × 9	14	+ 0.043	+ 0.120	0	+0.215	− 0.018	5.5		3.8		0.25	0.40	36	160
50	58	16 × 10	16	0	+ 0.050	− 0.043	− 0.021	5−	0.061		6.0	4.3	0.25	0.40	45	180
58	65	18 × 11	18						7.0	+ 0.2	4.4	+ 0.2	0.25	0.40	50	200
65	75	20 × 12	20					7.5	0	4.9	0		0.40	0.60	56	220
75	85	22 × 14	22	+ 0.052	+ 0.149	0	+0.026 0	− 0.022	9.0		5.4		0.40	0.60	63	250
85	95	25 × 14	25	0	+ 0.065	− 0.052	− 0.026 0 0.074	9.0			5.4		0.40	0.60	70	280
95	110	28 × 16	28					10.0			6.4		0.40	0.60	80	320
110	130	32 × 18	32					11.0			7.4		0.40	0.60	90	360
130	150	36 × 20	36					12.0			8.4		0.70	1.00	100	400
150	170	40 × 22	40	+ 0.062	+ 0.180	0	+ 0.0310	0.026	13.0		9.4		0.70	1.00	110	400
170	200	45 × 25	45	0	+ 0.080	− 0.062	− 0.031 0	− 0.088	15.0		10.4		0.70	1.00	125	400
200	230	50 × 28	50						17.0	1	1.4		0.70	1.00	140	400
230	260	56 × 32	56					20.0	+ 0.3		12.4	+ 0.3	1.20	1.60	160	400
260	290	63 × 32	63	+ 0.074	+0.220	0	+0.037 0	− 0.032	20.0	0	12.4	0	1.20	1.60	180	400
290	330	70 × 36	70	0	+ 0.100	− 0.074	− 0.037 0	− 0.106	22.0		14.4		1.20	1.60	200	400
330	380	80 × 40	80						25.0		15.4		2.00	2.50	220	400
380	440	90 × 45	90	+0.087	+ 0.260	0	+ 0.043 5	− 0.037	28.0		17.4		2.00	2.50	250	400
440	500	100 × 50	100	0	+ 0.120	− 0.087		− 0.043 5	31.0		19.5		2.00	2.50	280	400

Table 10.3 Dimensions of Gib-head Keys and Keyways (All dimensions in mm)

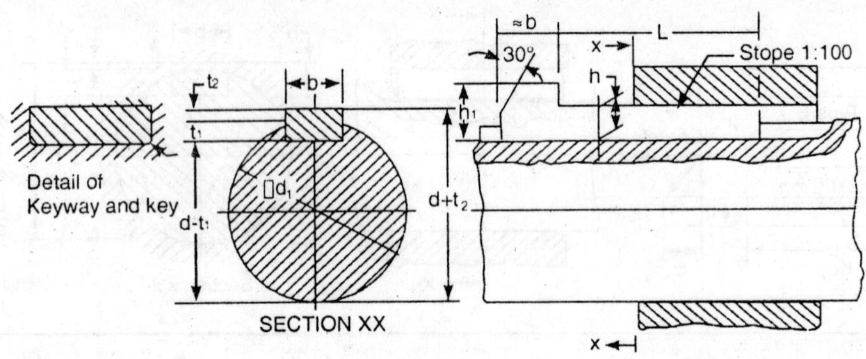

Range of Shaft Dia *d* Above	Upto	Key *b × h*	Keyway *b*	Tol on *b* D10	t_1	Tol on t_1	t_2	Tol on t_2	*r* Min.	*r* Max.	Range of Key Length *L* Min.	Max.	h_1
10	12	4 × 4	4		2.5	1.2			0.08	0.16	14	45	7
12	17	5 × 5	5	+ 0.078	3.0	+ 0.1	1.7	+ 0.1	0.16	0.25	14	56	8
17	22	6 × 6	6	+ 0.030	3.5	0	2.2	0	0.16	0.25	16	70	10
22	30	8 × 7	8		4.0		2.4		0.16	0.25	20	90	11
30	38	10 × 8	10	+ 0.098	5.0		2.4		0.25	0.40	25	110	12
38	44	12 × 8	12	+ 0.040	5.0		2.4		0.25	0.40	32	140	12
44	50	14 × 9	14		5.5		2.9		0.25	0.40	40	160	14
50	58	16 × 10	16	+ 0.120	6.0		3.4		0.25	0.40	45	180	16
58	65	18 × 11	18	+ 0.050	7.0	+ 0.2	3.4	+ 0.2	0.25	0.40	50	200	18
65	75	20 × 12	20		7.5	0	3.9	0	0.40	0.60	56	220	20
75	85	22 × 14	22		9.0		4.4		0.40	0.60	63	250	22
85	95	25 × 14	22		9.0		4.4		0.40	0.60	70	280	22
95	110	28 × 16	28	+ 0.065	10.0		5.4		0.40	0.60	80	320	25
110	130	32 × 18	32		11.0		6.4		0.40	0.60	90	360	28
130	150	36 × 20	36		12.0		7.1		0.70	1.00	100	400	32
150	170	40 × 22	40	+ 0.180	13.0		8.1		0.70	1.00	110	400	36
170	200	45 × 25	45	+ 0.080	15.0		9.1		0.70	1.00	125	400	40
200	230	50 × 28	50		17.0		10.1		0.70	1.00	140	400	45
230	260	56 × 32	56		20.0	+ 0.3	11.1	+ 0.3	1.20	1.60			50
260	290	63 × 32	63		20.0	0	11.1	0	12.0	1.60			50
290	330	70 × 36	70	+ 0.220	22.0		13.1		1.20	1.60			56
330	380	80 × 40	80	+0.100	25.0		14.1		2.00	2.50			63
380	440	90 × 45	90	+ 0.260	28.0		16.1		2.00	2.50			70
440	500	100 × 50	100	+ 0.120	31.0		18.1		2.00	2.50			80

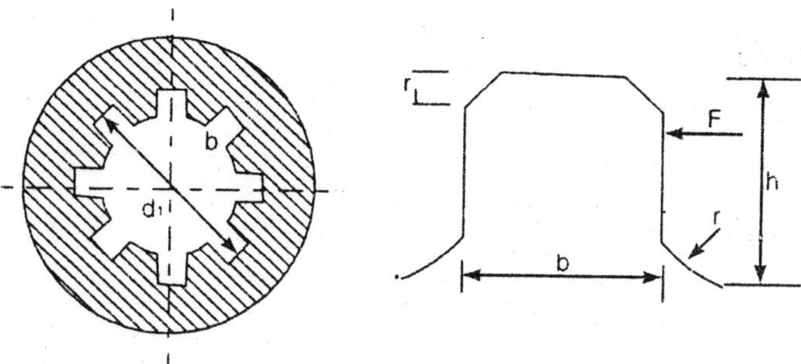

Table 10.4 Thickness of spline, mm

Number of splines	b	h (thickness)		
		Permanent fit	slides when loaded	slides when unloaded
4	0.241*d*	0.075*d*	—	0.125*d*
6	0.250*d*	0.050*d*	0.100*d*	0.075*d*
10	0.156*d*	0.045*d*	0.095*d*	0.070*d*
16	0.098*d*	0.045*d*	0.095*d*	0.070*d*

Table 10.5 Bearing pressure (steel—steel)

Fit of spline	Permanent fit	Slides when unloaded	Slides when loaded
p	21 MPa	14 MPa	7 MPa

Table 10.6 Dimensions and Tolerances for Straight-sided splines

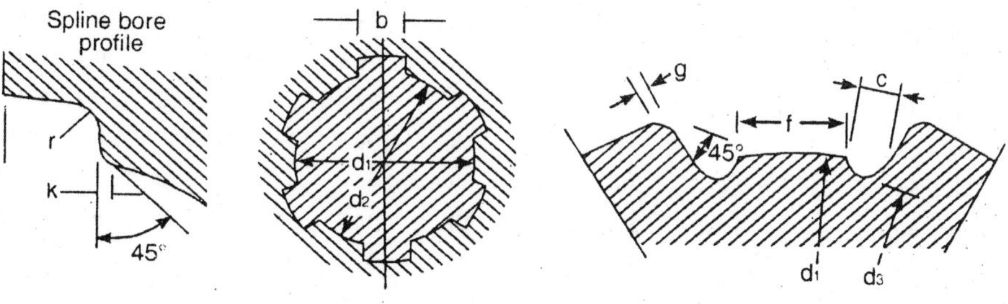

Light Duty Series

Nominal size $n \times d_1 \times d_2$	No. of spline keys n	Centering	d_1	d_2	b	d_3 Min.	e Max.	* f	* g Max.	* k Max.	r Max.
6 × 23 × 26		On	23	26	6	22.1	1.25	3.54	0.3	0.3	0.2
6 × 26 × 30	6	Inside	26	30	6	24.6	1.84	3.85	0.3	0.3	0.2
6 × 28 × 32		diameter	28	32	7	26.7	1.77	4.03	0.3	0.3	0.2
8 × 32 × 36			32	36	6	30.42	1.89	2.71	0.4	0.4	0.3
8 × 36 × 40			36	40	7	34.5	1.78	3.46	0.4	0.4	0.3
8 × 42 × 46			42	46	8	40.4	1.68	5.03	0.4	0.4	0.3
8 × 46 × 50	8	Inside	46	50	9	44.62	1.61	5.75	0.4	0.4	0.3
8 × 52 × 58		diameter	52	58	10	49.7	2.72	4.89	0.5	0.5	0.5

Light Duty Series

Nominal size $n \times d_1 \times d_2$	No. of spline keys n	Centering	d_1	d_2	b	d_3 Min.	e* Max.	f*	g* Max.	k Max.	r Max.
8 × 56 × 62		or Flanks	56	62	10	53.6	2.76	6.38	0.5	0.5	0.5
8 × 62 × 68			62	68	12	59.82	2.48	7.31	0.5	0.5	0.5
10 × 72 × 78			72	78	12	69.6	2.54	5.45	0.5	0.5	0.5
10 × 82 × 88			82	88	12	79.32	2.67	8.62	0.5	0.5	0.5
10 × 92 × 98	10		92	98	14	89.44	2.36	10.08	0.5	0.5	0.5
10 × 102 × 108			102	108	16	99.9	2.23	11.49	0.5	0.5	0.5
10 × 112 × 120			112	120	18	108.8	3.23	10.72	0.5	0.5	0.5

Medium Duty Series :

Nominal size $n \times d_1 \times d_2$	No. of spline keys n	Centering	d_1	d_2	b	d_3* Min.	e Max.	f*	g* Max.	k Max.	r Max.
6 × 11 × 14			11	14	3	9.9	1.55	—	0.3	0.3	0.2
6 × 13 × 16			13	16	3.5	12.0	1.5	0.32	0.3	0.3	0.2
6 × 16 × 20			16	20	4	14.54	2.1	0.16	0.3	0.3	0.2
6 × 18 × 22		On	18	22	5	16.7	1.95	0.45	0.3	0.3	0.2
6 × 21 × 25	6	Inside	21	25	5	19.5	1.98	1.95	0.3	0.3	0.2
6 × 23 × 28		Diameter	23	28	6	21.3	2.3	1.34	0.3	0.3	0.2
6 × 26 × 32			26	32	6	23.4	2.94	1.65	0.4	0.4	0.3
6 × 28 × 34			28	34	7	25.9	2.94	1.70	0.4	0.4	0.3
8 × 32 × 38			32	38	6	29.4	3.3	0.15	0.4	0.4	0.3
8 × 36 × 42			36	42	7	33.5	3.01	1.02	0.4	0.4	0.3
8 × 42 × 48			42	48	8	39.5	2.91	2.57	0.4	0.4	0.3
8 × 46 × 54	8	On	46	54	9	42.7	4.1	0.86	0.5	0.5	0.5
8 × 52 × 60		Inside	52	60	10	48.7	4.0	2.44	0.5	0.5	0.5
8 × 56 × 65		Diameter	56	65	10	52.2	4.74	2.5	0.5	0.5	0.5
8 × 62 × 72		or	62	72	12	57.8	5.0	2.4	0.5	0.5	0.5
10 × 72 × 82		Flanks	72	82	12	67.4	5.43	—	0.5	0.5	0.5
10 × 82 × 92			82	92	12	77.1	5.4	3.0	0.5	0.5	0.5
10 × 92 × 102	10		92	102	14	87.3	5.2	4.5	0.5	0.5	0.5
10 × 102 × 112			102	112	16	97.7	4.9	6.3	0.5	0.5	0.5
10 × 112 × 125			112	125	18	106.3	6.4	4.4	0.5	0.5	0.5

Heavy Duty Series

Nominal size $n \times d_1 \times d_2$	No. of spline keys n	Centering	d_1	d_2	b	* d_3 Min.	e Max.	f	g Max.	k Max.	r Max.
10 × 16 × 20			16	20	2.5	14	0.3		0.3	0.15	
10 × 18 × 23			18	23	3	15.6	0.3		0.3	0.15	
10 × 21 × 26		On	21	26	3	18.44	0.3		0.3	0.15	
10 × 23 × 29	10	Inside	23	29	4	20.3	0.3		0.3	0.15	
10 × 26 × 32		Diameter	26	32	4	23	0.4		0.4	0.15	
10 × 28 × 35		or Flanks	28	35	4	24.4	0.4		0.4	0.25	
10 × 32 × 40			32	40	5	28	0.4		0.4	0.25	
10 × 36 × 45			36	45	5	31.3	0.4		0.4	0.25	
10 × 42 × 52			42	52	6	36.9	0.5		0.4	0.4	
10 × 46 × 56			46	56	7	40.9	0.5		0.5	0.4	
10 × 52 × 60			52	60	5	47	0.5		0.5	0.4	
16 × 56 × 65	16		56	65	5	50.6	0.5		0.5	0.4	
16 × 62 × 72			62	72	6	56.1	0.5		0.5	0.4	
20 × 72 × 82		On	72	82	7	65.9	0.5		0.5	0.4	
16 × 82 × 92		Flanks	82	92	6	75.6	0.5		0.5	0.4	
20 × 92 × 102			92	102	7	85.5	0.5		0.5	0.4	
20 × 102 × 115	20		102	115	8	93.7	0.5		0.5	0.4	
20 × 112 × 125			112	125	9	103.7	0.5		0.5	0.4	

These values have been calculated on the basis that the spline shaft profiles are produced by the hobbing process.

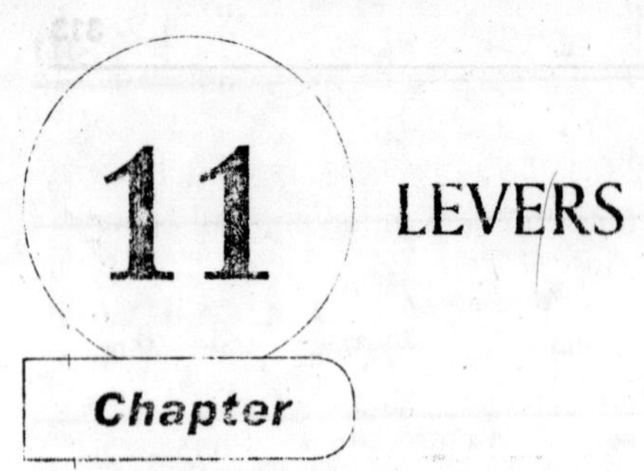

11 LEVERS

11.1 INTRODUCTION

A lever is a machine component to obtain mechanical advantage. A lever of the first kind is shown in Fig. 11.1, where P is the effort and W is the load. Taking moments about the fulcrum, we have

$$P \times l_1 = W \times l_2$$

l_2 l_1

Fulcrum

W (Load)

P (Effort)

Fig. 11.1 Lever Systerm.

Mechanical advantage, $\dfrac{W}{P} = \dfrac{l_1}{l_2}$

In the case of lever of the first kind, the fulcrum is in between the effort and load. In a lever of the second kind, the load is in between the fulcrum and effort, whereas in a lever of the third kind, the effort is in between the fulcrum and load. The ratio $\dfrac{l_1}{l_2}$ is called the lever ratio or leverage. Levers are commonly used in mechanical devices to increase the magnitude of load. Some of the examples of levers are : bell crank lever, rocker arm, balance beam, scissors, cutting pliers, handle of a hand pump, lever safety valve, nut cracker, wheel barrow, etc.

11.2 DESIGN OF LEVERS

The design of a lever can be carried out as follows :

1. Calculate the rection at the flucrum.
2. Calculate the maximum bending moment in the lever. Then,

$$M_{max} = \sigma z$$

where σ = permissible bending stress for lever

z = section modulus of lever.

The cross-section of the lever is generally tapered from the fulcrum to the ends. The thickness is kept constant and taper is given on the height. The ends of the lever should be checked for shear strength.

The fulcrum pin is designed in bending and then checked in bearing and shear. Sometimes it is also checked in deflection. The different ways in which force may be distributed on the pin are shown in Fig. 11.2

(*i*) **Pin simply supported in fork**

When the pin is supported in the fork as shown in Fig. 11.2 (*a*), the stress is assumed to be zero at the outside and maximum at the inside. The maximum bending moment in the pin will occur at the centre.

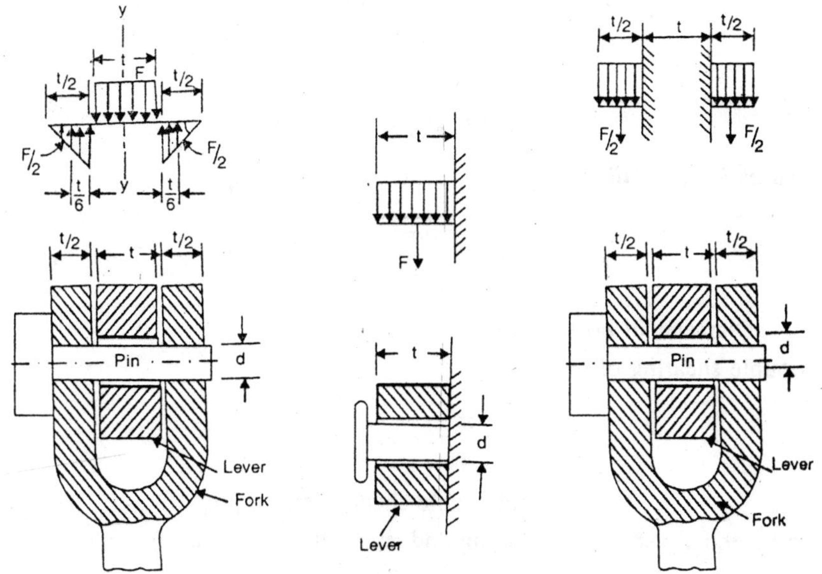

(*a*) Pin simply supported in fork.

(*b*) Cantilever type pin-bearing in lever

(*c*) Cantilever type pin-bearing in fork.

Fig. 11.2 Distribution of force on pin

$$M_{max} = \frac{1}{2}F\left(\frac{t}{2}+\frac{1}{3}\times\frac{t}{2}\right)-\frac{F}{2}\times\frac{t}{4}$$

$$= \frac{5}{24}Ft \simeq \frac{Ft}{4} \quad \text{... (1)}$$

(*ii*) **Cantilever type pin-bearing in lever**

For the cantilever type pin shown in Fig. 11.2 (*b*)

$$M_{max} = \frac{Ft}{2}$$

(*iii*) **Cantilever type pin-bearing in fork**

In this case, as shown in Fig. 11.2 (*c*)

$$M_{max} = \frac{Ft}{2}$$

Bearing pressure, $$p_b = \frac{F}{dt} \quad \text{... (2)}$$

where p_b = allowable bearing pressure

d = diameter of pin

Substituting the value of F from Eq. (2) in Eq. (1) we get

$$p_b \times dl \times \frac{l}{4} = M_{max} = \sigma \times \frac{\pi}{32} d^3$$

$$= \frac{l}{d}\sqrt{\frac{\pi}{8}} \cdot \sqrt{\frac{\sigma}{p_b}} = 0.634\sqrt{\frac{\sigma}{p_b}},$$

Similarly for cases (*ii*) and (*iii*), we get

$$\frac{l}{d} = 0.448\sqrt{\frac{\sigma}{p_b}}$$

Diameter of boss = $2d$

Considering double shearing of the pin,

$$F = 2 \times \frac{\pi}{4} d^2 \tau$$

where τ = allowable shear stress in pin.

If the pin is to be equally strong in bending and shear, then for case (*i*), we have

$$\frac{\pi}{4} d^3 \sigma \times \frac{4}{t} = 2 \times \frac{\pi}{4} d^2 \tau$$

or

$$\frac{\tau}{\sigma} = \frac{d}{4t}$$

For case (*ii*) and (*iii*), the pin is in single shear, then,

$$\frac{\pi}{4} d^2 \tau = \frac{\pi}{32} d^3 \sigma \times \frac{2}{t}$$

or

$$\frac{\tau}{\sigma} = \frac{d}{4t}$$

If $d = t$. then

$$\frac{\tau}{\sigma} = 0.25$$

Therefore, the shearing stress is one-fourth of the bending stress, and it may not be necessary to check the pin under shearing loading.

If l is the length of the pin, then

$$\frac{l}{d} \simeq 1 \text{ to } 1.3$$

A phosphor bronze bush of thickness 3 mm may be used to decrease wear of pin and lever. Proper lubrication should be provided at the fulcrum to decrease friction.

Example. 11.1 *Design the handle as shown in Fig. 11.3. The operator can apply an effort of 500 N at the end of the handle. The handle is attached to the spindle by means of a tapered round pin. Assume allowable stresses in tension and shear to be 105 MPa and 55 MPa respectively.*

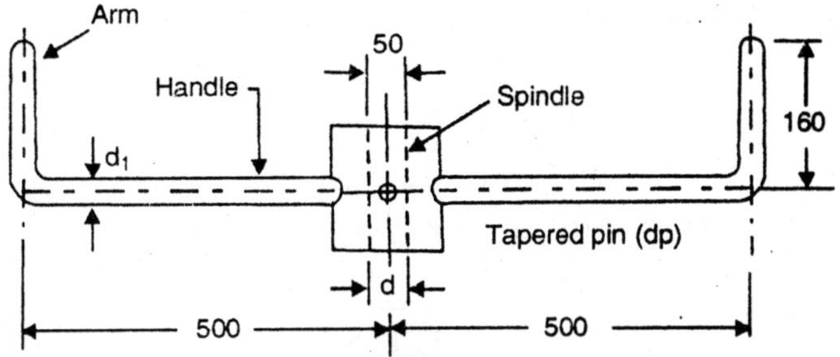

Fig. 11.3 Handle for spindle.

Sol. Torsional moment on the taper pin,

$$T = 2 \times 500 \times 500 \times 10^{-3} = 500 \text{ N.m}$$

The pin is in double shear. Let d_p be diameter of the pin and d the diameter of spindle. Then considering the shear failure of pin,

$$T = 2 \times \frac{\pi}{4} d_p^2 \times \tau \times \frac{d}{2}$$

$$500 \times 10^3 = 2 \times \frac{\pi}{4} d_p^2 \times 50 \times \frac{50}{2}$$

$$d_p = \sqrt{\frac{500 \times 10^3 \times 8}{2 \times \pi \times 55 \times 50}}$$

$$= 15.2 \simeq 16 \text{ mm (say)}$$

Maximum bending moment on handle,

$$M = 500 \times 500 \times 10^{-3} = 250 \text{ N.m}$$

Turning moment on handle, assuming effort to act in the middle of arm,

$$T = 500 \times \frac{160}{2} \times 10^{-3} = 40 \text{ N.m}$$

Equivalent torque in handle

$$T_e = \sqrt{M^2 + T^2}$$

$$= \sqrt{(250)^2 + (40)^2} = 253.18 \text{ N.m}$$

Equivalent bending moment,

$$M_e = \frac{1}{2}\left(M + \sqrt{M^2 + T^2}\right)$$

$$= \frac{1}{2}\left(250 + \sqrt{(250)^2 + (40)^2}\right)$$

$$= 251.59 \text{ N.m}$$

Considering shearing failure of handle,

$$T_e = \frac{\pi}{16} d_1^3 \tau$$

$$253.18 \times 10^3 = \frac{\pi}{16} \times d_1^3 \times 55$$

$$d_1 = \sqrt[3]{\frac{253.18 \times 16 \times 10^3}{\pi \times 55}} = 28.6 \simeq = 30 \text{ mm}$$

Checking the handle under bending, we have

$$M_e = \frac{\pi}{32} d_1^3 \sigma$$

$$\sigma = \frac{32 \times 251.59 \times 10^3}{\pi \times (30)^3} = 94.9 \text{ MPa}$$

which is less than the allowable stress of 105 MPa, hence satisfactory.

Diameter of boss $= 2d = 2 \times 50 = 100$ mm

Height of boss $= 2d_1 = 2 \times 30 = 60$ mm

11.3 TYPES OF LEVERS

1. Hand Lever

A simple hand lever is shown in Fig. 11.4

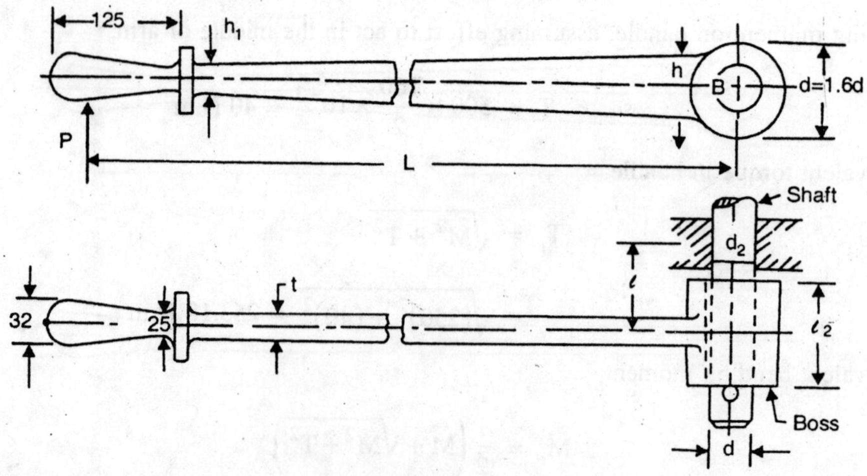

Fig. 11.4 Hand Lever.

Let h = height of the lever

t = thickness of the lever

L = length of the lever.

Maximum bending moment in the lever,

$$M = PL$$

where P = effor applied by hand = 350 to 400 N

Section modulus, $$z = \frac{th^2}{6}$$

Considering bending stress in the lever,

$$PL = \sigma \times \frac{th^2}{6}$$

where σ = permissible bending stress in the lever.

Generally, near the boss, $h = 5t$. Then

$$t = \sqrt[3]{\frac{6PL}{25\sigma}}$$

The thickness is kept constant and the height is tapereds towards the handle.

Height near the handle, $h_1 = 0.5\,h$

The handle is attached to the shaft by a square key.

Twisting moment on shaft, $T = Pl$

Equivalent torque in shaft, $T_e = \sqrt{M^2 + T^2}$

Considering torsional shear stress in shaft, we have

$$T_e = \frac{\pi}{16} d_2^3 \tau$$

where τ is the permissible shear stress in shaft.

$$\frac{\pi}{16} d_2^3 \tau = P\sqrt{L^2 + l^2}$$

$$d_2 = \left[\frac{16P}{\pi\tau}\sqrt{L^2 + l^2}\right]^{\frac{1}{3}}$$

Equivalent bending moment in shaft,

$$M_e = \frac{1}{2}\left(M + \sqrt{M^2 + T^2}\right)$$

$$= \frac{P}{2}\left(L + \sqrt{L^2 + l^2}\right)$$

Checking the shaft in bending, we have

$$\frac{\pi}{32} d_2^3 \sigma = M_e$$

$$= \frac{P}{2}\left(L + \sqrt{L^2 + l^2}\right)$$

$$\sigma = \frac{16P}{\pi d_2^3}\left(L + \sqrt{L^2 + l^2}\right)$$

Length of the boss, $l_2 = (1 \text{ to } 1.25)\, d$

Thickness of boss, $t_2 = 0.3\, d$

Diameter of boss, $d_1 = 1.6\, d$

The thickness of the boss can be checked against tearing parallel to the axis. In that case,

$$PL = \sigma \times l_2 t_2 \times \left(\frac{d + t_2}{2}\right)$$

$$l_2 = \frac{2PL}{\sigma l_2 (d + t_2)}$$

2. Foot Lever

A foot lever is shown in Fig. 11.5. A man can apply foot effort of 700 to 800 N. The design of the foot lever can be carried out in the same way as the hand lever.

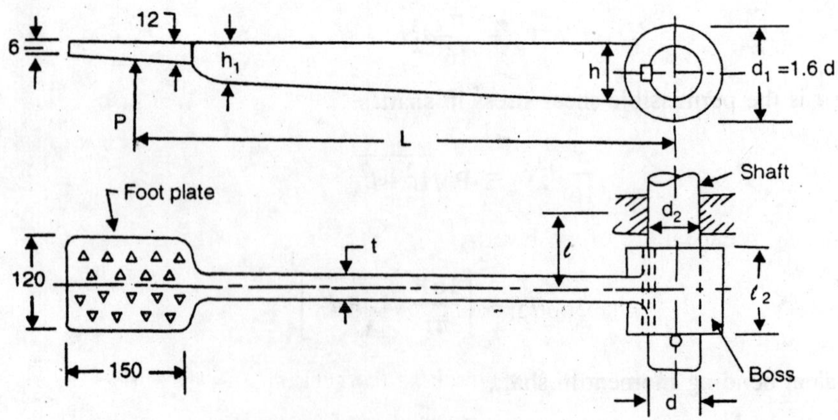

Fig. 11.5 Foot Lever.

3. Cranked Lever

A cranked lever as shown in Fig. 11.6, is used for hoisting winches. The lever may be operated either by a single person or two persons. A person can apply a force of 150 to 200 N continuously with a velocity of 1 m/s approximately. The maximum force which a person can apply to operate the lever may be taken as 400 N. The size of lever may be estimated as follows :

Length of handle,

l = 300 mm, when operated by one person.

= 500 mm, when operated by two persons.

Length of lever, L = 350 to 400 mm

Size of square shaft end in the handle =30 to 40 mm

Diameter of handle 40 to 60 mm

Considering the effort P to act at a distance of $\frac{2l}{3}$ the bending moment in the handle is,

$$M = 2\frac{Pl}{3}$$

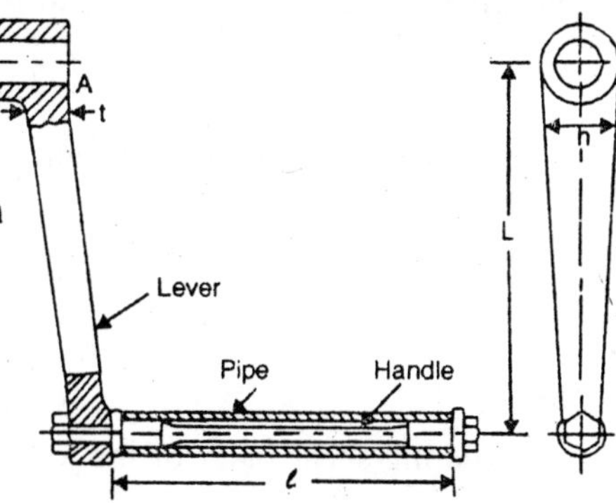

Fig. 11.6 Cranked Lever

Let d = diameter of handle

σ = permissible bending stress in handle.

Then
$$\frac{2}{3}Pl = \frac{\pi}{32}d^3\sigma$$

or
$$d = \sqrt[3]{\frac{64Pl}{3\pi\sigma}}$$

Maximum bending moment in lever,

$$M = PL$$

Twisting moment in lever,

$$T = \frac{2}{3}Pl$$

Equivalent torque in lever,

$$T_e = \sqrt{M^2 + T^2}$$

$$= P\sqrt{L^2 + \frac{4}{9}l^2} \quad \text{... (1)}$$

Considering shear failure of the lever,

$$T_e = \tau z_p \quad \text{... (2)}$$

where z_p = section modulus in torsion

$= \frac{2}{9}ht^2$ for a rectangular section

$= \frac{\pi}{16}t^2h$ for an elliptical section of major axis h and minor axis t.

Eqs. (1) and (2) will enable to find the size of the lever.

Shaft Journal :

Twisting moment, $T = PL$

Bending moment, $M = P\left(\frac{2}{3}l + x\right)$

x = distance from side A of boss to the centre of journal.

Equivalent twisting moment,

$$T_e = P\sqrt{L^2 + \left(\frac{2}{3}l + x\right)^2}$$

Let d_2 = diameter of the journal.

Considering shear failur of the journal, we have

$$\frac{\pi}{16}d_2^3\tau = P\sqrt{L^2 + \left(\frac{2}{3}l + x\right)^2} \quad \text{... (3)}$$

d_2 can be determined from eq. (3)

4. Overhung crank

An overhung crank lever is shown in Fig. 11.7

Let P = force acting on the crank pin

d = diameter of crank pin

l = length of crank pin.

p_b = permissible bearing pressure for pin

σ = permissible bending stress for pin.

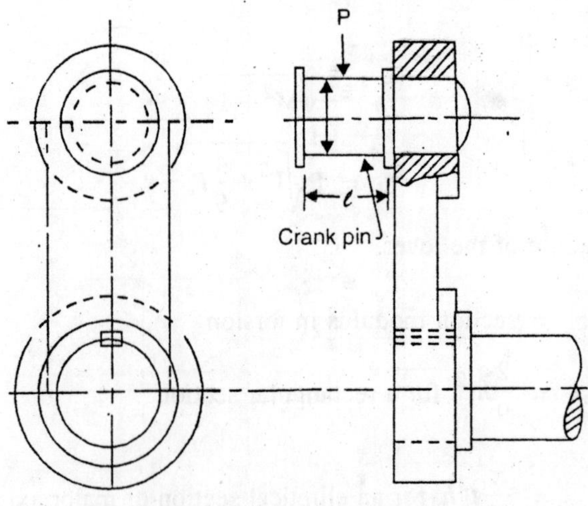

Fig. 11.7 Overhung crank

Maximum bending moment on crank pin,

$$M = \frac{1}{2}Pl$$

Also $$P = p_b \times dl$$

$\therefore$ $$M = \frac{1}{2}\frac{P^2}{p_b d} = \frac{\pi}{32}d^2\sigma$$

or $$d = \sqrt[3]{\frac{16P^2}{\pi P_b \sigma}}$$

$$\frac{1}{d} = \sqrt{\frac{\pi}{16}\frac{\sigma}{p_b}}$$

5. Lever of a safety valve

The lever of a safety valve is shown in Fig. 11.8. Taking moments about the fulcrum, we have

$$Pl_1 = Wl_2$$

or $$W = \frac{Pl_1}{l_2}$$

where $$P = \frac{\pi}{4}D^2 p$$

D = valve diameter,

p = steam pressure when the valve blows off.

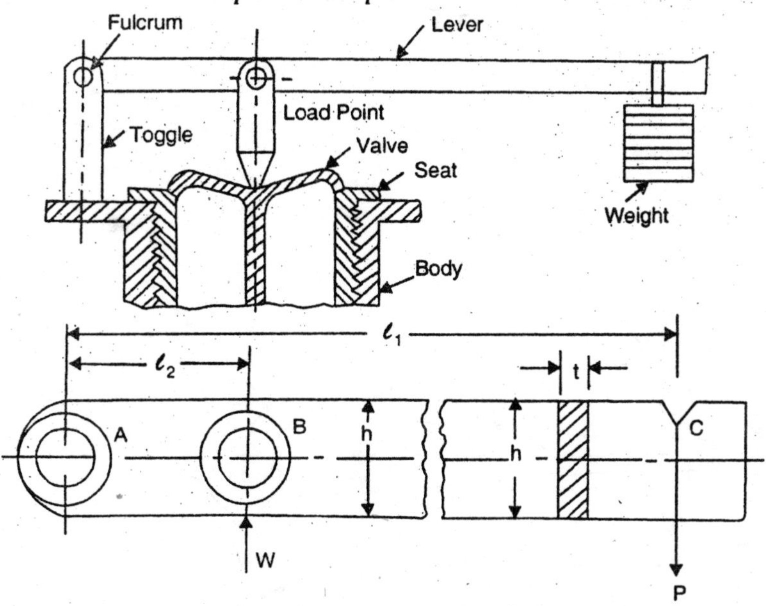

Fig. 11.8 Lever of a safety valve.

Maximum bending moment in the lever,

$$M = W(l_2 - l_1) = \frac{1}{6} t h^2 \sigma$$

Generally, $$t = \frac{h}{4}$$

$\therefore$ $$W(l_2 - l_1) = \frac{h^3 \sigma}{24}$$

or $$h = \sqrt[3]{\frac{24 W(l_2 - l_1)}{\sigma}}$$

The pins at the fulcrum and load point are in double shear.

$$\frac{t}{d} = 0.634 \sqrt{\frac{\sigma}{p_b}}$$

Considering bearing of the pin,

$$p_b \times dt = P$$

Thus, we can determine d and t.

Thickness of phosphor bronze bush $= 1.5$ mm

Diameter of hole in the lever, d_0 $= d + 3$ mm

Diameter of boss $= 2d_0$

6. Rocker Arm for Exhaust Valve

The rocker arm for an exhaust valve is shown in Fig. 11.9. Taking moments about the fulcrum, we get

$$(P \cos 45°) \times L = F \times L$$

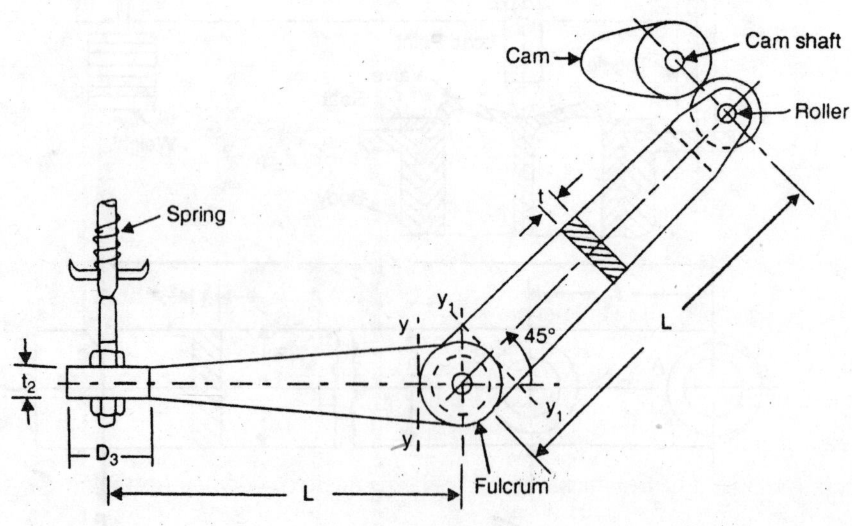

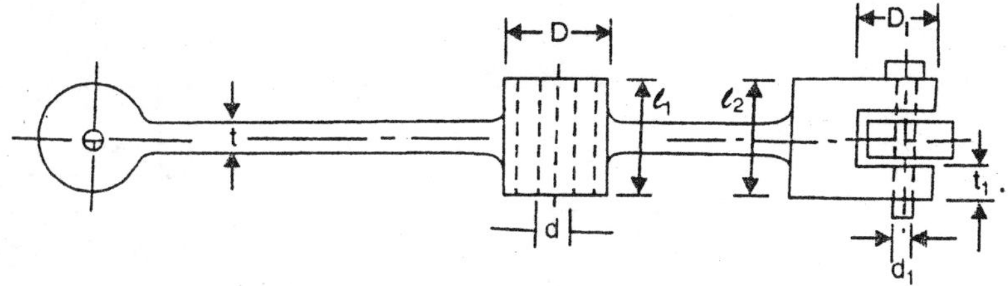

Fig. 11.9 Exhaust valve gear rocker arm.

or $$P = \sqrt{2}\,F$$

Reaction at the fulcrum pin, $R = F + P$

Considering bearing of the fulcrum pin, we have

$$p_b \times l_1 d = R \qquad \text{... (1)}$$

where l_1 = length of fulcrum pin

d = diameter of fulcrum pin.

Considering the fulcrum pin to be simply supported at the ends and carrying concentrated load R at the centre, the maximum bending moment in the pin is,

$$M = \frac{R l_1}{4}$$

Resisting moment of fulcrum pin $= \frac{\pi}{32} d^3 \sigma$

$$\therefore \qquad \frac{R l_1}{4} = \frac{\pi}{32} d^3 \sigma \qquad \text{... (2)}$$

where σ = allowable bending stress for fulcrum pin.

Eqs. (1) and (2) may be solved to find l_1 and d.

Empricially, $$\frac{l_1}{d} = 1.25$$

Considering double shearing of the pin, we have

$$2 \times \frac{\pi}{4} d^2 \tau = R$$

τ should be less than allowable shear stress.

External diameter of boss, $D = 2d$

Thickness of brass bush = 3 mm

Diameter of hole in the lever, $d_0 = d + 6$ mm

Neglecting moment of resistance of the bush, the section modulus for the boss is (Fig. 11.10)

$$z = \frac{\frac{1}{12} l_1 \left[D^3 - d_o^3 \right]}{\frac{D}{2}}$$

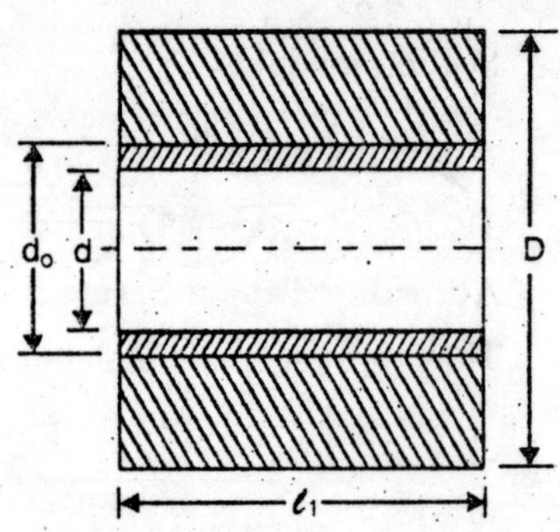

Fig. 11.10 Bearing bush

$$M = FL$$

Bending stress in boss, $\sigma_b = \frac{M}{z}$

which should be less than the permissible bending stress.

Roller Pin :

Considering bearing of the pin in the roller, we have

$$P = p_b \times d_1 l'_2$$

Assuming $\frac{l'_2}{d_1} = 1.5$, we can find l'_2 and d_1.

Maximum bending moment in the pin,

$$M = \frac{P l'_2}{4}$$

Bending stress in pin, $\sigma = \frac{M}{z}$

where $z = \frac{\pi}{32} d_1^3$

The pin is in double shear. Hence

$$2 \times \frac{\pi}{4} d_1^2 \tau = P$$

Thikness of eye, $t_1 = \frac{l_2'}{2}$

Overall diameter of eye, $D_1 = 2d_1$

Outside diameter of roller $= D_1 + (3 \text{ to } 5)$ mm

Outer size of fork, $l_2 = l_2' + 2t_1 + 2 \times 1.5$ mm.

Where clearance between the roller and fork has been taken as 1.5 mm.

Cross-section of Lever :

Maximum bending moment in lever,

$$M = F \left(L - \frac{D}{2} \right)$$

Section modulus, $z = \frac{1}{6}th^2$

$$M = \sigma z$$

Taking $h = 1.5\ D_1$, t can be determined.

Adjusting Tappet Screw :

The tappet screw is under compression. Hence

$$\frac{\pi}{4} d_c^2 \sigma_c = F$$

From which d_c can be determined by knowing permissible compressive stress s_c in the screw.

Outside diameter of screw, $d_o = \frac{d_c}{0.84}$

$$D_3 = 2d_o$$

Depth at screw end, $t_2 = 2d_o$

7. Bell Crank Lever

In a bell crank lever (Fig. 11.11), the two arms are inclined at right angle. Taking moments about the fulcrum B, we have

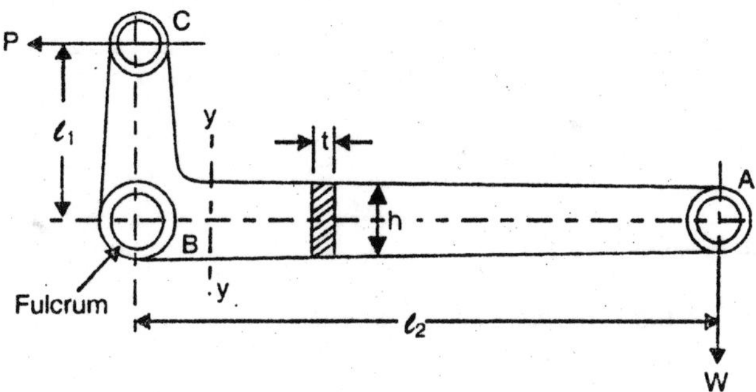

Fig. 11.11 Bell Crank Lever.

$$Pl_1 = Wl_2$$

$$P = W\frac{l_2}{l_1}$$

Reaction at the fulcrum, R = $\sqrt{P^2 + W^2}$

Fulcrum Pin :

Considering bearing of the pin, we have $p_b \times ld = R$

Where pb = allowable bearing pressure for the pin.

l, d = length and diameter of the pin, respectively.

Taking $l/d = 1.25$, we can determine l and d.

Considering double shearing of the pin, we have

$$2 \times \frac{\pi}{4} d^2 \tau = R$$

The shear stress τ in the pin should be less than the allowable shear stress.

Considering the pin to be simply supported at the ends and the load to be between concentrated and uniformly distributed type, we have

$$M = \frac{Rl}{6}$$

$$= \frac{\pi}{32} d^3 \sigma$$

or $$\sigma = \frac{16Rl}{3\pi d^3}$$

which should be below the permissible bending stress for the pin material.

Thinkness of brass bush $= 3$ mm

Diameter of hole in the lever, $d_o = d + 6$ mm

Diameter of boss of fulcrum, $D = 2d$

Section modulus of boss, $$z = \frac{1}{12} \times l\,[D^3 - d_0^3] \times \frac{2}{D}$$

Bending moment at the fulcrum, $M = W_2 l_2$

Bending stress in boss, $$\sigma = \frac{M}{z}$$

which should be less than the permissible stress.

Lever :

Bending moment in the lever near the fulcrum,

$$M = W\left(l_2 - \frac{D}{2}\right)$$

Ler $$h = 3t.$$

Then $$M = \frac{th^2}{6} \times \sigma$$

$$= 1.5t^3\sigma$$

or $$t = \sqrt[3]{\frac{M}{1.5\sigma}}$$

Pin at C :

The load at C is much below the load at the fulcrum. We adopt the same size of the pin and the boss as at the fulcrum.

Pin at A :

The end A of the lever is of the fork type. Therefore, the pin is in double shear. Let $l_1/d_1 = 1.25$. Then

$$2 \times \frac{\pi}{4} d_1^2 \times \tau = W$$

$$pb \times l_1 d_1 = W$$

$$\text{Width of eye} = 0.5\, l_1$$

Thickness of phosphor bronze bush = 3 mm

Inner diameter of eye $= d_1 + 6$ mm

Maximum bending moment in the pin,

$$M = \frac{Wl_1}{4}$$

Section modulus, $$z = \frac{\pi}{32} d_1^3$$

$$M = \sigma z$$

The bending stress should be less than the permissible stress.

Outside diameter of fork = $2d_1$.

8. Cross Lever

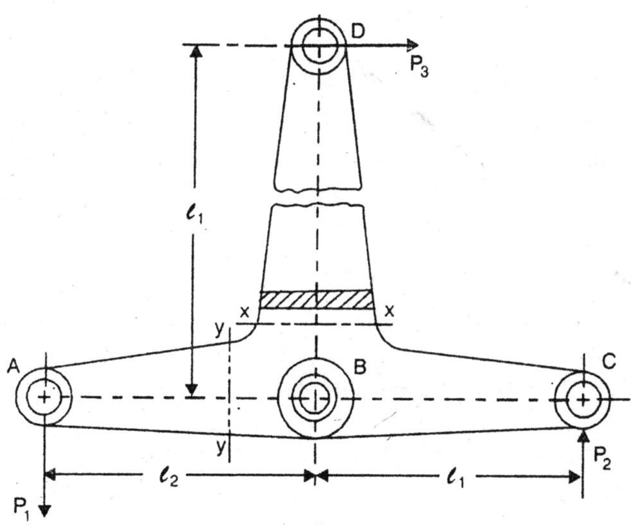

Fig. 11.12 Cross lever.

A cross lever is shown in Fig. 11.12. Taking moments about the fulcrum B, we have

$$P_3 l_1 = (P_1 + P_2)\, l_2$$

$$P_3 = (P_1 + P_2)\, \frac{l_2}{l_1}$$

When $P_2 = 0$, then

$$P_3' = P_1 \frac{l_2}{l_1}$$

and the reaction at the fulcrum is,

$$R_1 = \sqrt{P_1^2 + P_3'^2}$$

when $P_2 \neq 0$, then

$$R_2 = \sqrt{P_3^2 + (P_2 - P_1)^2}$$

Pins :

Design the pins for maximum loads.

$$\frac{l}{d} = 0.634\sqrt{\frac{\sigma}{p_b}}$$

Considering the bearing of the pin, we have

$$p_b \times ld = F$$

where $F = \max(P_1, P_2, P_3, R_1, R_2)$

From these two equations we can determine l and d.

The pin is in double shear. Hence

$$2 \times \frac{\pi}{4} d^2 \times \tau = F$$

The shear stress τ should be less than the permissible value.

Thickness of phosphor bronze bush = 3 mm

Diameter of hole in the lever, d_o = $d + 6$ mm

Diameter of boss, $D = 2d_o$

Vertical Arm :

Maximum bending moment in vertical arm,

$$M = P_3 l_1$$

$$M = \frac{1}{6} th^2 \times \sigma$$

Taking $h = 3t$, we can determine h and t.

Horizontal Arms :

Maximum bending moment in horizontal arm,

$$M = \max(P_1, P_2) \times l_2$$

Let h_1 = depth of horizontal arm.

$$M = \frac{1}{6} th_1^2 \times \sigma$$

or

$$h_1 = \sqrt{\frac{6M}{t\sigma}}$$

Example 11.2 *Design the lever as shown in Fig 11.13, assuming that a man can apply a maximum force of 450 N. The permissible stresses for pin material are 50 MPa in bending and 10 MPa in bearing pressure. The allowable bending stress in lever is 75 MPa.*

Sol. Let W = load lifted by the lever.

P = effort applied by the man.

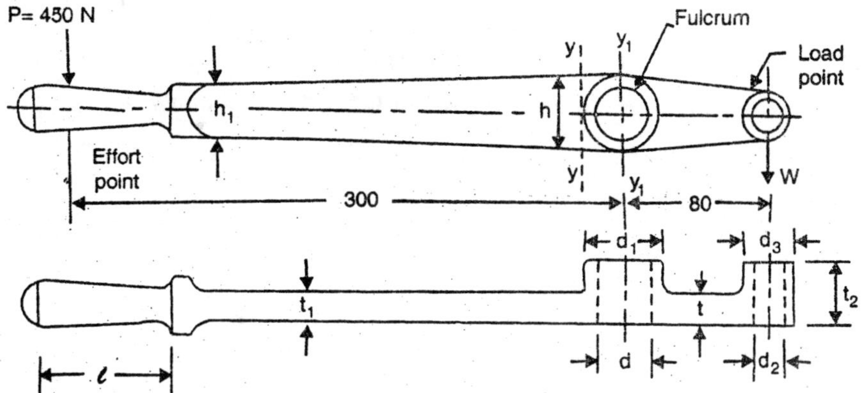

Fig. 11.13 Hand lever.

Taking moments about the fulcrum, we have

$$P \times 300 = W \times 80$$

$$W = \frac{450 \times 300}{80} = 1687.5 \text{ N}$$

Reaction at fulcrum,

$$R = P + W$$

$$= 450 + 1687.5 = 2137.5 \text{ N}$$

Fulcrum Pin :

Considering the bearing and bending of the pin in the fork, we have

$$\frac{t}{d} = 0.448\sqrt{\frac{\sigma}{p_b}}$$

$$= 0.448\sqrt{\frac{50}{10}} = 1.0$$

Also

$$Pb \times td = R$$

$$td = \frac{2137.5}{10} = 213.75$$

$$\therefore \quad t = d = 14.62 \simeq 15 \text{ mm}$$

Lever :

Outer diameter of boss, $d_1 = 2d = 2 \times 15 = 30$ mm

Bending moment in the lever at section $y - y$,

$$M = P\left(300 - \frac{d_1}{2}\right)$$

$$= 450\ (300 - 15) = 128250 \text{ N.mm}$$

Let t_1 = thickness of the lever.

h = height of the lever at section $y - y$.

Taking $h = d_1 = 30$ mm,

$$M = \frac{1}{6} t_1 h^2 \sigma$$

$$128250 = \frac{1}{6} \times t_1 \times 900 \times 75$$

$$t_1 = \frac{128250 \times 6}{900 \times 75}$$

$$= 11.4 \simeq 12 \text{ mm}$$

Now consider the section $y_1 - y_1$ at the fulcrum.

Section modulus at $y_1 - y_1$,

$$z = \frac{\frac{1}{12}(td_1^3 - td^3)}{d_1/2}$$

$$= \frac{15\,(30^3 - 15^3)}{6 \times 30} = 1968.75 \text{mm}^3$$

Maximum bending moment at $y_1 - y_1$,

$$M = 450 \times 300 = 135000 \text{ N.mm}$$

Bending stress, $\sigma = \frac{M}{z}$

$$= \frac{135000}{1968.75} = 68.57 \text{ N/mm}^2$$

which is less than the permissible value of 75 MPa. Hence safe.

Pin at the Load point :

Considering bearing of the pin, we have

$$P_b \times d_2 t_2 = W$$

$$d_2 t_2 = \frac{1687.5}{10} = 168.75 \text{ mm}^2$$

Also $\frac{t_2}{d_2} = 1$

$\therefore$ $d_2 = t_2 = 13$ mm

Diameter of boss, $d_3 = 2d_2 = 26$ mm

Height of lever near the handle, $h_1 = 0.5\,h = 15$ mm

Handle :

Length of handle = 100 mm

Minimum diameter = 20 mm

Maximum diameter = 25 mm.

Example 11.3 *The lever for a self energizing block brake is shown in Fig. 11.14. Determine the size of the lever and fulcrum. Assume the following permissible stresses :*

Lever–Tension and compression = 80 MPa

Pin–Tension = 80 MPa

Bearing pressure = 20 MPa

Shear stress = 20 MPa

Maximum breaking force = 5 kN.

Sol. Let P_1 = force applied at the lever end.

Taking moment about the fulcrum, we get

$$P_1 \times 500 = P \times 250$$

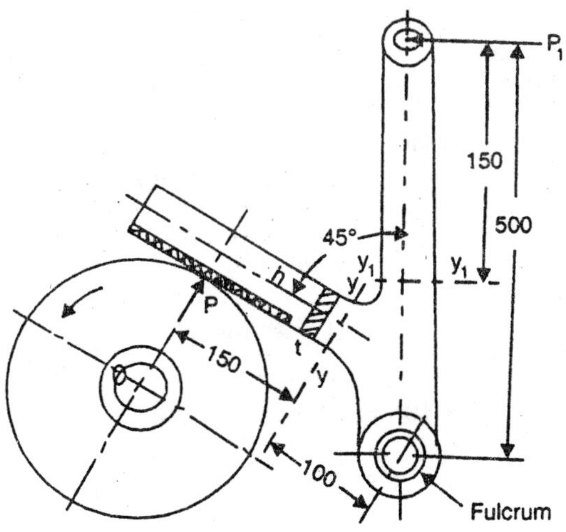

Fig. 11.14 Block Brake.

$$P_1 = \frac{5 \times 250}{500} = 2.5\ kN$$

Reaction at the fulcrum,

$$R = \sqrt{P^2 + P_1^2 + 2PP_1 \cos\theta}$$

$$= \sqrt{25 + 6.25 + 2 \times 5 \times 2.5 \cos 45^\circ}$$

$$= 6.995 \text{ kN}$$

Pin at the Fulcrum :

Let l, d = length and diameter of pin respectively. If the pin is to be equally strong in bearing pressure and bending, then

$$\frac{l}{d} = 0.634 \sqrt{\frac{\sigma}{p_b}}$$

$$= 0.634 \sqrt{\frac{80}{20}} = 1.268$$

Considering bearing of the pin, we get

$$pb \times ld = R$$

$$ld = \frac{6.995 \times 10^3}{20} = 349.75$$

$$1.268\, d^2 = 349.75$$

$$d = 16.6 \simeq 18 \text{ mm},\ l = 23 \text{ mm}$$

Bush thickness = 3 mm

Hole size, $d_0 = 18 + 2 \times 3 = 24$ mm

Boss diameter, $D = 2d_0 = 48$ mm

Lever :

Maximum bending moment in the lever at $y - y$,

$$M = P \times 150$$

$$= 5 \times 150 = 750 \text{ kN.mm}$$

$$= \frac{1}{6} th^2 \times \sigma$$

Taking $h = D = 48$ mm

$$750 \times 10^3 = \frac{1}{6} t \times (48)^2 = 80$$

$$t = 24.4 \simeq 25 \text{ mm}$$

Bending moment at $y_1 - y_1$ is,

$$M = 2.5 \times 150 = 375 \text{ kN.mm}$$

$$= \frac{1}{6} th^2 \sigma$$

Keeping $t = 25$ mm,

$$375 \times 10^3 = \frac{1}{6} \times 25h^2 \times 80$$

$$h = 33.54 \simeq 35 \text{ mm.}$$

Example 11.4 *The line diagram of a bench shearing machine is shown in Fig. 11.15. Determine the cross-section of the lever. The size of the work piece is 6 mm × 3 mm. Maximum operating force is 180 N. The ultimate shearing strength of work piece is 400 MPa. Assume permissible stress in tension for pins, link and lever to be 85 MPa and allowable bearing pressure as 20 MPa.*

Sol. Area of work piece to be sheared, $A_s = 6 \times 3 = 18 \text{ mm}^2$

Required shearing force, $F = \tau_u \times A_s$

$$= 400 \times 18 = 7200 \text{ N}$$

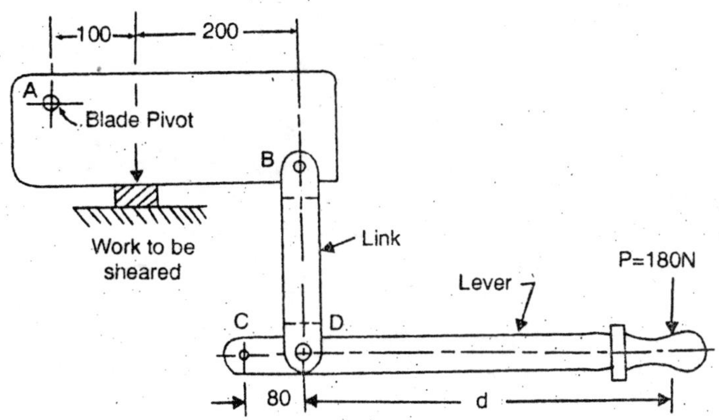

Fig. 11.15 Bench Shearing Machine.

Let P_1 = force in the link.

Taking moments about the blade, pivot, we get

$$P_1 \times 300 = F \times 100$$

$$P_1 = \frac{7200 \times 100}{300} = 2400 \text{ N}$$

Force in pin at blade pinot $= F - P_1$

$$= 7200 - 2400 = 4800 \text{ N}$$

Force in pins at B and D $= P_1 = 2400$ N

Force in pin at C $= P_1 + P$

$$= 2400 + 180 = 2580 \text{ N}$$

Taking moments about D, we have

$$2580 \times 80 = P \times d$$

$$d = \frac{2580 \times 80}{180} = 1146.67 \text{ mm}$$

Pins at B, C, D :

Maximum force in pin = max (2400, 2580) N

= 2580 N

If the pin is to be equally strong in bearing pressure and bending stresses, then

$$\frac{l}{d} = 0.634\sqrt{\frac{\sigma}{p_b}}$$

$$= 0.634\sqrt{\frac{85}{20}} = 1.307$$

Considering bearing of the pin, we have

$$p_b \times ld = 2580$$

$$ld = \frac{2580}{20} = 129 \text{ mm}^2$$

$$d = \sqrt{\frac{129}{1.307}} = 9.93 \simeq 10 \text{ mm}$$

$$l = 13 \text{ mm}$$

Considering double shearing of the pin, we have

$$2 \times \frac{\pi}{4} d^2 \tau = 2580$$

$$\tau = \frac{2580 \times 2}{\pi \times 100} = 16.42 \text{ MP}a$$

whch is quite safe.

Using 2 mm thick brass bush, the size of hole will be 14 mm and diameter of boss will be 28 mm.

Link :

Let the link be circular in cross-section of dia d_1 and having forked ends. As the link is in tension, therefore,

$$\frac{\pi}{4} d_1^2 \sigma_t = 2580$$

$$d_1 = \sqrt{\frac{2580 \times 4}{\pi \times 85}} = 6.21 \simeq 10 \text{ mm}$$

Lever :

Thickness of lever, $t = l = 13$ mm

Let h = height of the lever.

Maximum bending moment in lever,

$$M = 2580 \times 80$$

$$= 206400 \text{ N. mm}$$

$$M = \sigma_b z$$

$$206400 = 85 \times \frac{1}{6} \times 13h^2$$

Using 2 mm thick brass bush, the size of hole will be 14 mm and diameter of boss will be 28 mm.

Link :

Let the link be circular in cross-section of dia d_1 and having forked ends. As the link is in tension, therefore,

$$\frac{\pi}{4} d_1^2 \sigma_t = 2580$$

$$d_1 = \sqrt{\frac{2580 \times 4}{\pi \times 85}} = 6.21 \simeq 10 \text{ mm}$$

Lever :

Thickness of lever, $t = l = 13$ mm

Let h = height of the lever.

Maximum bending moment in lever,

$$M = 2580 \times 80$$
$$= 206400 \text{ N. mm}$$

$$M = \sigma_b z$$

$$206400 = 85 \times \frac{1}{6} \times 13h^2$$

$$h = \sqrt{\frac{206400 \times 6}{85 \times 13}} = 33.47 \simeq 34 \text{ mm}$$

Section modulus of the lever at D,

$$z = \frac{\frac{1}{12}[13 \times (34)^3 - 13 \times (14)^3]}{\frac{34}{2}} = 2329.8 \text{ mm}^3$$

$$\text{Bending stress} = \frac{206400}{2329.8} = 85.6 \text{ MPa}$$

which is nearly equal to the safe value of 85 MP*a*.

The handle may be made 25 mm to 35 mm in diameter and 125 mm long.

Pin A :

$$\frac{l_2}{d_2} = 1.307$$

$$l_2 d_2 = \frac{4800}{20} = 240 \text{ mm}^2$$

$$d_2 = \sqrt{\frac{240}{1.307}} = 13.55 \simeq 14 \text{ mm}$$

11.3 A hand lever for a brake is 0.75 m long from the centre of gravity of the spindle to the point of application of pull of 350 N. The effective overhang from the nearest bearing is 100 mm. If the permissible stresses in tension, a shear and crushing are not to exceed 60 MPa, 45 MPa and 80 MPa respectively, design the spindle, key and lever. Assume the arm of the lever to be rectangular having width twice its thickness.

11.4 Design a foot brake lever from the following data :
Length of lever from centre of gravity of spindle to the point of load application = 1.25 m
Maximum load on foot plate = 750 N
Overhang from the nearest bearing = 100 mm
Permissible tensile stress = 75 MP*a*
Permissable shear stress = 60 MP*a*

11.5 Design a cranked lever for the following dimensions :
Length of handle = 350 mm
Length of lever arm = 400 mm
Overhang of journal = 100 mm
The lever is operater by a single person who can apply a maximum force of 450 N at a distance of $\frac{1}{4}$th length of handle from its free end. The permissible shear stress in lever material is 50 MP*a* and for shaft material is 40 MPa

11.6 A lever safety valve is 60 mm in diameter. It is required to blow off at 1.25 MP*a*. Design a mild steel lever of rectangular cross-section if the permissible stresses are 70 MP*a* in tension, 50 MP*a* in shear and 20 MPa in bearing. The material of pin is same as that of lever. The distance from the fulcrum to the centre of gravity of lever is 800 mm and the distance between the fulcrum pin and valve spindle link pin is 100 mm.

11.7 Design a right angled bell crank lever having one arm 500 mm and the other 200 mm long. A load of 4.5 kN is to be raised acting on a pin at the end of 500 mm arm and the effort is applied at the end of 200 m arm. The permissible stresses for pin and lever are 85 MPa in tension and compression and 65 MP*a* in shear. The bearing pressure on the pin should not exceed 10 MP*a*.

12 COTTER AND KNUCKLE JOINTS

Chapter

12.1 INTRODUCTION

A cotter is a flat wedge-shaped piece of rectangular cross-section having a tapering width. A cotter is used to make a temporary joint and is used to connect two axial rods which may be either in tension or compression. The taper on the width varies from 1:48 to 1:8. If the taper is more than 1:8. It is called a ***self-locking*** cotter. A taper of 1:24 is generally provided. The tendency of the cotter to slacked back can be decreased by providing a gib either on one side or both sides of the cotter.

A knuckle joint is used to connect two rods which are in tension only. The two rods should lie in the same plane but may not be necessarily coaxial.

12.2. SPIGOT AND SOCKET COTTER-JOINT

A spigot and socket cotter joint is shown in Fig. 12.1. The end of the rod which goes into the socket is called the spigot. A rectangular hole is made in the socket and spigot through which the cotter is inserted. The load applied is usually axial but it may be tensile or compressive. The compressive load is taken by the collar on the spigot. The various parts may be designed as follows :

1. Diameter of Rods

Let F = force in the rods in tension or compression

d = diameter of the rods

$\sigma_t = \sigma_c$ = permissible tensile or compressive stress in the rods

Considering failure of the rods in tension or compression, we have

$$\left(\frac{\pi}{4}\right) d^2 \sigma_t = F$$

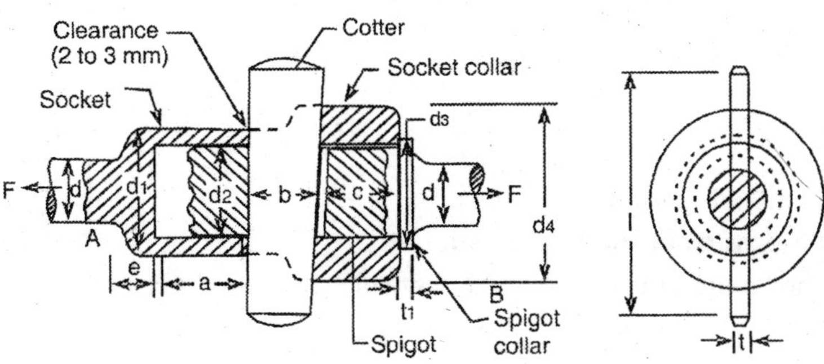

Fig. 12.1 Socket and spigot cotter joint.

$$d = \left[\frac{4F}{\pi\sigma_t}\right]^{1/2} \qquad ...(1)$$

2. Diameter of Spigot

The weakest section of the spigot is where the cotter passes through it, as shown in Fig. 12.2. Considering failure of the spigot in tension, we have

$$\left[\left(\frac{\pi}{4}\right)d_2^2 - d_2 t\right]\sigma_t = F \qquad ...(2)$$

Where d_2 = diameter of spigot

t = thickness of cotter

σ_t = permissible tensile stress in spigot.

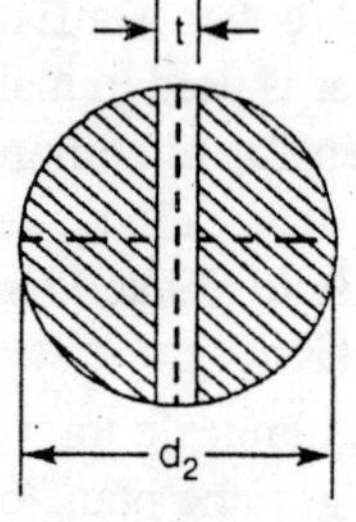

Fig. 12.2

Generally $t = \dfrac{d_2}{4}$

Thus, d_2 can be determined.

3. Check for Crushing failure of Rod or Cotter

Let σ_c = allowable compressive stress for rod or cotter.

Then $d_2 t \sigma_c = F$

$$\sigma_c = \frac{F}{d_2 t} \qquad ...(3)$$

4. Outside Diameter of Socket

Considering the failure of socket across the slot in tension or compression, as shown in Fig.12.3, we have

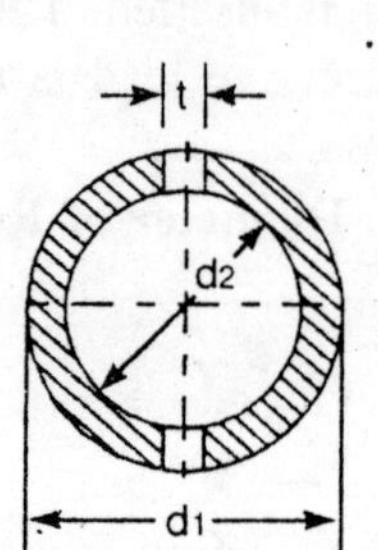

Fig. 12.3

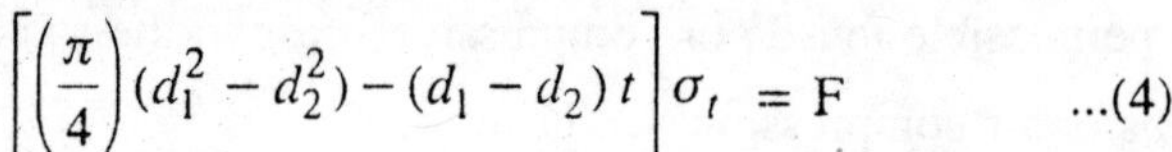

$$\left[\left(\frac{\pi}{4}\right)(d_1^2 - d_2^2) - (d_1 - d_2)\,t\right]\sigma_t = F \qquad ...(4)$$

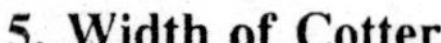

where d_1 = outside diameter of socket.

5. Width of Cotter

Considering double shear of cotter, as showin in Fig. 12.4, we have

$$2bt\tau = F$$

Width of cotter, $b = \dfrac{F}{2t\tau}$ ...(5)

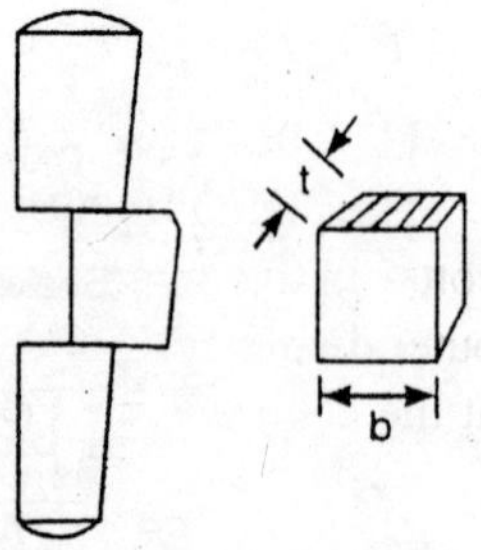

Fig. 12.4

Where τ = permissible shear stress in cotter.

6. Length of Spigot from slot to the end.

Considering the failure of the socket end in double shear, we have

$$2ad_2\tau = F$$

Length of spigot from cotter slot to end,

$$a = \frac{F}{2d_2\tau} \qquad ...(6)$$

7. Socket Collar Diameter

Considering failure of the socket collar in crushing, as showin in Fig. 12.5, we have

$$(d_4 - d_2)t\sigma_c = F \qquad ....(7)$$

Where d_4 = diameter of socket collar.

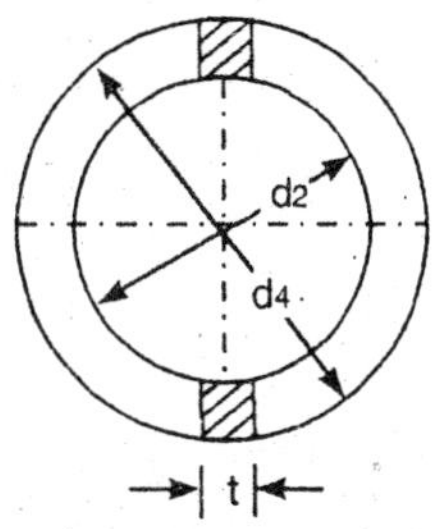

Fig. 12.5

8. Socket Collar Thickness

Considering the failure of socket collar in double shear, we have

$$2(d_4 - d_2)c\tau = F \qquad ...(8)$$

Where c = thickness of socket collar.

9. Diameter of Spigot Collar

Considering the failure of spigot collar in crushing, as shown in Fig. 12.6, we have

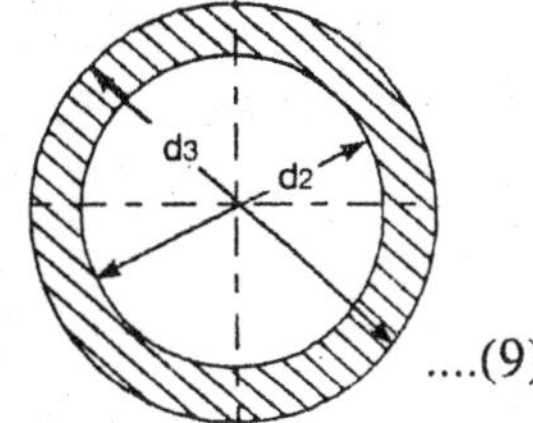

$$\left(\frac{\pi}{4}\right)(d_3^2 - d_2^2)\sigma_c = F \qquad ....(9)$$

Fig. 12.6

Where d_3 = diameter of spigot collar.

10. Thickness of Spigot Collar

Considering the failure of spigot collar in shear, as shown in Fig. 12.7, we have

$$\pi d_2 t_1 \tau = F \qquad ...(10)$$

Where t_1 = thickness of spigot collar.

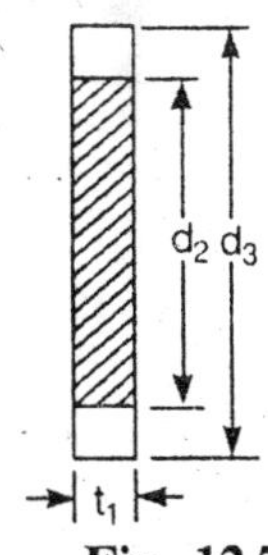

Fig. 12.7

11. Check for Cotter in Bending

The distribution of load on the cotter is shown in Fig. 12.8. It is assumed that the load on the cotter in the spigot end is uniformly distributed while in the socket end it varies from zero at the outer diameter d_4 and maximum at the inner diameter d_2. The maximum bending moment occurs at the centre of the cotter and is given by,

$$M_{max} = \left(\frac{F}{2}\right)\left[\frac{(d_4 - d_2)}{6} + \frac{d_2}{2}\right] - \left(\frac{F}{2}\right) \times \left(\frac{d_2}{4}\right)$$

$$= \left(\frac{F}{24}\right)[2d_4 + d_2]$$

Section modulus of cotter,

$$z = \frac{tb^2}{6}$$

Bending stress in cotter,

$$\sigma_b = \frac{M_{max}}{z}$$

$$= \left[\frac{F}{4tb^2}\right](2d_4 + d_2) \quad ...(11)$$

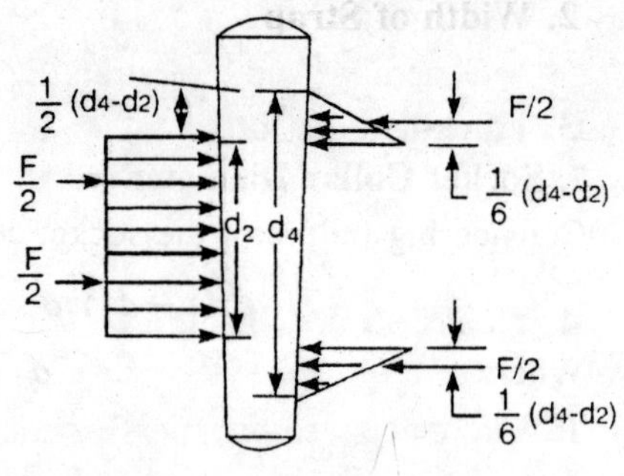

Fig. 12.8 Load distribution on the cotter

The bending strees should be less than the permissible bending stress of cotter material.

12.3 GIB AND COTTER JOINT

12.3.1. *Strap End of Connecting Rod*

The gib and cotter joint for the strap end (or big end) of a connecting rod is shown in Fig. 12.9. The various parts may be designed as follows :

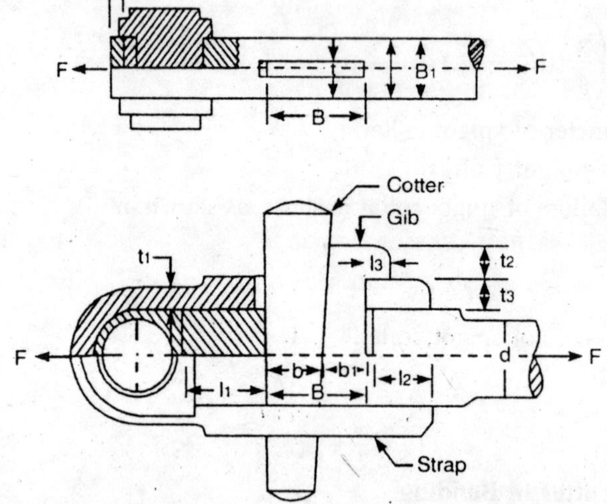

Fig. 12.9 Gib and cotter joint for strap end of a connecting rod.

1. Diameter of Rod

The connecting rod is subjected to tensile and compressive load. Considering the failure of rod of diameter *d* under tensile load F, we have

$$\left(\frac{\pi}{4}\right)d^2\sigma_t = F$$

$$d = \left[\frac{4F}{\pi\sigma_t}\right]^{1/2} \quad ....(1)$$

2. Width of Strap

Generally, width of strap, $B_1 = d$ ....(2)

3. Thickness of Cotter

$$t = \frac{B_1}{4} \qquad ...(3)$$

4. Thickness of Gib

It is kept equal to thickness of cotter $= t = \dfrac{B_1}{4}$ ...(4)

5. Height and Lenght of Gib Head

Height of gib head, $t_2 = t$ ...(5)

Length of gib head, $l_3 = t$ ...(6)

6. Thickness of Strap

Considering tearing of the strap at the thinnest section, we have

$$2B_1 t_1 \sigma_t = F$$

$$\text{Thickness of Strap, } t_1 = \frac{F}{2B_1\sigma_t} \qquad ...(7)$$

Considering tearing strength of strap at the cotter shoulder to be equal to that at the thinnest section,

we have

$$2t_3(B_1 - t) = 2t_1 B_1$$

$$t_3 = \frac{B_1 t_1}{B_1 - t} \qquad ...(8)$$

7. Total Width of Gib and Cotter

Width of gib, $b_1 = 0.55$ B ....(9)

Width of cotter, $b = 0.45$ B ....(10)

8. Other Dimensions

$l_1 = 2t_1$ and $l_2 = 2.5t_1$ ...(11)

$t_4 = 91.15$ to $1.5)\ t_1$ ...(12)

12.3.2 For Square Rods

The gib and cotter joint for a square rod is shown in Fig. 12.10. The various parts can be deigned as follows :

1. Size of Rod

Considering the failure of square rod of cross-sectional area x^2 under tensile force F, we have

$$\sigma_t \times x^2 = F$$

$$x = \left(\frac{F}{\sigma_t}\right)^{1/2} \qquad ...(1)$$

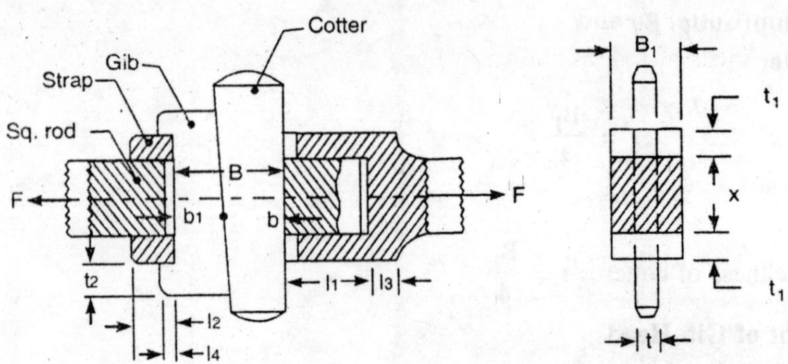

Fig. 12.10 Gib and cotter joint for square rods.

where σ_t = permissible tensile stress in rod.

The other dimensions are fixed as follows :

$$\text{Width of strap, } B_1 = x$$

$$\text{Thickness of cotter, } t = \frac{B_1}{4}$$

$$\text{Thickness of gib} = t$$

$$\text{Height of gib, } t_2 = t$$

$$\text{Length of gib head, } l_4 = t$$

2. Width of Gib and Cotter

Considering the double shear of gib and cotter, we have

$$2Bt\tau = F$$

$$B = \frac{F}{2t\tau} \qquad ...(2)$$

Width of gib, $b_1 = 0.55B$, for a single gib ...(3*a*)

$= 0.30B$, for a double gib ...(3*b*)

Width of cotter, $b = 0.45B$, for a single gib ...(4*a*)

$= 0.40b$, for a double gib. ...(4*b*)

3. Thickness of Strap

Considering the failure of strap end in tension at the location of gib and cotter, as shown in fig. 12.11, we have

$$2(B_1 t_1 - t t_1)\,\sigma_t = F$$

Now $B_1 = x$.

Thus $2\,(x - t)\,t_1\,\sigma_t = F$

Fig. 12.11

$$\text{Thickness of strap, } t_1 = \frac{F}{2\,(x - t)\,\sigma_t} \qquad ...(5)$$

4. Check for Strap or Gib in crushing

$$2tt_1\sigma_c = F$$

$$\sigma_c = \frac{F}{2tt_1} \qquad ...(6)$$

Which should be less than the allowable compressive stress in strap or gib.

5. Length of Rod from Cotter to end

Considering the double shear of rod, as shown in Fig. 12.12, we have

$$2l_1 \times \tau = F$$

$$l_1 = \frac{F}{2 \times \tau} \quad ...(7)$$

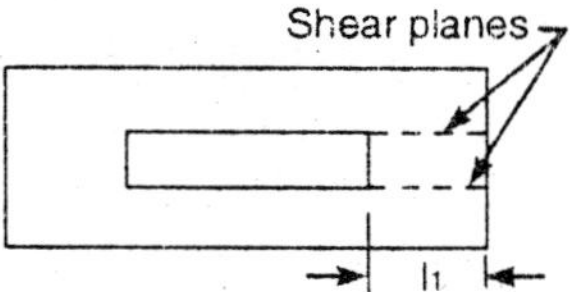

Fig. 12.12

6. Length of Rod from Gib to the end of Strap

Considering the double shear of the strap. As shown in Fig. 12.13, we have

$$2 \times 2l_2 t_1 \tau = F$$

$$l_2 = \frac{F}{4t_1\tau} \quad ...(8)$$

Fig. 12.13

7. Other Dimensions

$$l_3 = \frac{2x}{3} \quad ...(9)$$

12.4 KNUCKLE JOINT

A knuckle joint is used to conect two rods subjected to tensile loads. An eye is forged at the end of one rod and a fork at the end of other rod. The eye and fork are connected by means of a pin. This joint provides flexibility in one plane and the joint can be connected and disconnected quickly. The knuckle pin is secured by means of a collar and split pin. Its rotation in the fork is prevented by means of a small stop, pin, peg or snug. The knuckle joint is made of steel or wrought iron.

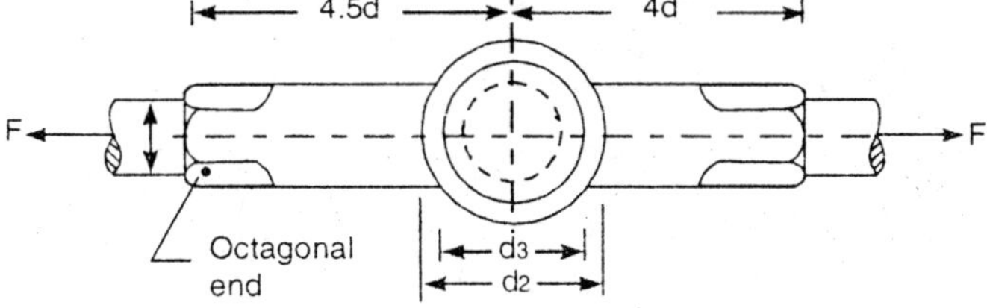

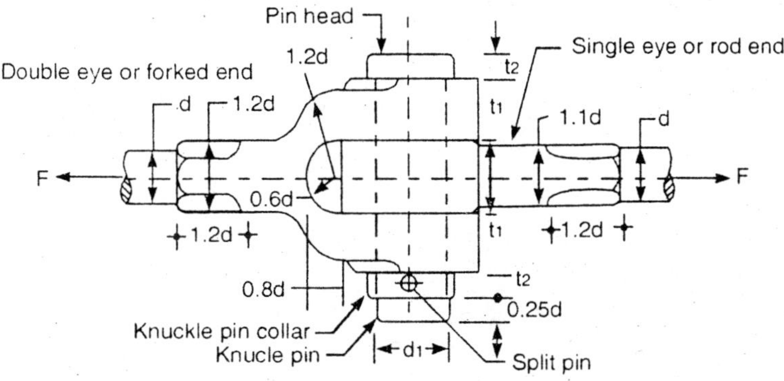

Fig. 12.14 Kunckle joint

Two views of a knuckle joint are shown in Fig. 12.14. The various parts of the joint may be designed as follows :

1. Diameter of Rod.

Let d = diameter of the rod, and F = tensile load. Considering the failure of the rod in tension, we have

$$\left(\frac{\pi}{4}\right)d^2\sigma_t = \text{F}$$

$$d = \left[\frac{4\text{F}}{\pi\sigma_t}\right]^{1/2} \quad \text{...(1)}$$

2. Diameter of the Knuckle Pin.

Considering the double shear of the pin, we have

$$2\left(\frac{\pi}{4}\right)d_1^2\tau = \text{F}$$

$$d_1 = \left[\frac{2\text{F}}{\pi\tau}\right]^{1/2} \quad \text{...(2)}$$

3. Check for the Knuckle Pin in Bending.

The pin fits in the eye and the fork in order to permit angular movment of eye in the fork. This introduces bending in the pin in addition to shear. To determine the bending stress in the pin, it is assumed that the load is uniformly distributed in the eye and varies uniformly over the fork, as shown in Fig. 12.15. The bending moment will be maximum at the centre of pin.

$$\text{M}_{max} = \left(\frac{\text{F}}{2}\right)\left(\frac{t_1}{3}+\frac{t}{2}\right)-\left(\frac{\text{F}}{2}\right)\times\left(\frac{t}{4}\right)$$

$$= \left(\frac{\text{F}}{2}\right)\left(\frac{t_1}{3}+\frac{t}{4}\right)$$

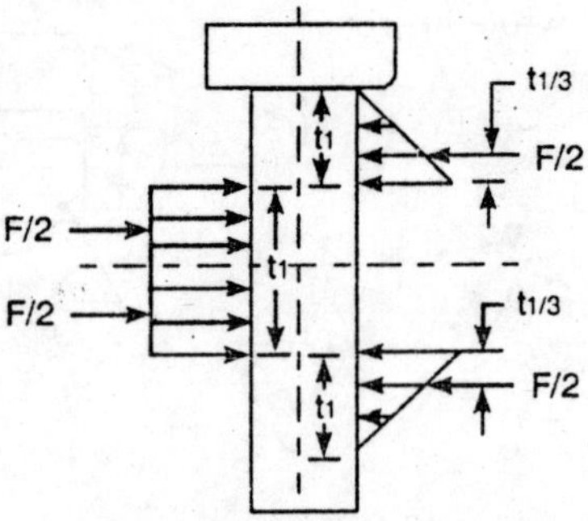

Fig. 12.15 Load distribution on the pin

Section modulus of pin, $z = \frac{\pi d_1^3}{32}$

Bending stress in pin, $$\sigma_b = \frac{\text{M}_{max}}{z} = \frac{16\text{F}\left(\frac{t_1}{3}+\frac{t}{4}\right)}{\pi d_1^3} \quad \text{....(3)}$$

Which should be less than the permissible bending stress for pin material

4. Outer Diameter of Eye.

Considering shearing of the eye, we have

$$(d_2 - d_1)\,t\tau = \text{F}$$

$$d_2 = d_1 + \frac{\text{F}}{t\tau} \quad \text{...(4)}$$

5. Check for crushing of Eye.

The single eye or pin may fail in crushing due to tensile load.

$$d_1 t \sigma_c = F$$

$$\sigma_c = \frac{F}{d_1 t} \qquad ...(5)$$

Which should be less than the permissible compressive stress in eye material.

6. Thickness of Fork end.

Considering the tensile failure of fork end, we have

$$2(d_2 - d_1) t_1 \sigma_c = F$$

$$t_1 = \frac{F}{2(d_2 - d_1)\sigma_t} \qquad ....(6)$$

7. Check for the Fork end in shear.

The for end may fail in shear due to tensile load

$$2(d_2 - d_1) t_1 \tau = F$$

$$\tau = \frac{F}{2(d_2 - d_1) t_1} \qquad ...(7)$$

Which should be less than the permissible shear stress in fork.

8. Check for Fork end in crushing.

The fork end or pin may fail in crushing due to tensile load.

$$d_1 \times 2t_1 \times \sigma_c = F$$

$$\sigma_c = \frac{F}{2d_1 t_1} \qquad ...(8)$$

Which should be less than the permissible compressive stress in fork.

9. Empirical Dimensions.

$$d_1 = d$$
$$d_2 = 2d$$
$$d_3 = 1.5d$$
$$t = 1.25d$$
$$t_1 = 0.75d$$
$$t_2 = 0.5d$$

Example 12.1 Design a spigot and socket cotter joint to support a load of 25kN in tension and compression. The material used is carbon steel having following allowable stresses :

Tension and compression = 50 MP*a*, shear = 40 MP*a*, crushing = 80 MP*a*.

Solution.

Given; $F = 25$ kN, $\sigma_t = 50$ MP*a*, $\tau = 40$ MP*a*, $\sigma_c = 80$ MP*a*

The spigot and socket cotter joint is shown in Fig. 12.1.

1. Diameter of Rods, d

Let d = diameter of the rods

Considering failure of the rods in tension or compression, we have

$$\left(\frac{\pi}{4}\right) d^2 \sigma_t = F$$

$$d = \left[\frac{4F}{\pi\sigma_t}\right]^{\frac{1}{2}} = \left[\frac{4 \times 25 \times 10^3}{\pi \times 50}\right]^{\frac{1}{2}} = 25.23 \approx 26 \text{ mm}$$

2. Diameter of spigot, d_2

Considering failure of the spigot in tension, we have

$$\left[\left(\frac{\pi}{4}\right) d_2^2 - d_2 t\right] \sigma_t = F$$

where d_2 = diameter of spigot

t = thickness of cotter = $\frac{d_2}{4}$

σ_t = permissible tensile stress in spigot.

$$\left[\left(\frac{\pi}{4}\right) d_2^2 - \frac{d_2 \times d_2}{4}\right] 50 = 25000$$

$$d_2 = 30.56 \approx 32 \text{ mm}$$

$$t = \frac{32}{4} = 8 \text{ mm}$$

3. Check for Crushing failure of Rod or Cotter

Let σ_c = allowable compressive stress for rod or cotter.

Then $d_2 t \sigma_c = F$

$$\sigma_c = \frac{F}{d_2 t} = \frac{25000}{32 \times 8} = 97.65 \text{ MPa} > 80 \text{ MPa}$$

Hence not safe. We determine the value of d_2 from the safe crushing stress.

$$d_2 \times \left(\frac{d_2}{4}\right) \times 80 = 25000$$

$$d_2 = 35.35 \approx 36 \text{ mm}$$

$$t = 9 \text{ mm}$$

4. Outside Diamter of Socket, d_1

Considering the failure of socket across the slot in tension or compression,

$$\left[\left(\frac{\pi}{4}\right)\left(d_1^2 - d_2^2\right) - \left(d_1 - d_2\right) t\right] \sigma_t = F$$

$$\left[\left(\frac{\pi}{4}\right)\left(d^2{}_1 - 36^2\right) - \left(d_1 - 36\right) \times 9\right] \times 50 = 25000$$

$$\left(\frac{\pi}{4}\right)\left(d^2{}_1 - 1296\right) - \left(d_1 - 36\right) \times 9 = 500$$

$$d^2{}_1 - 1296 - 11.46d_1 + 412.53 = 636.62$$

$$d^2{}_1 - 11.46d_1 - 1920.1 = 0$$

$$d_1 = 0.5\left(11.46 \pm \left\{(11.46)^2 + 4 \times 1920.1\right\}^{\frac{1}{2}}\right)$$

$$= 49.92 \approx 50 \text{ mm}$$

5. Width of Cotter, b

Considering double shear of cotter,

$$2bt\tau = \text{F}$$

$$\text{Width of cotter, } b = \frac{\text{F}}{2t\tau} = \frac{25000}{2 \times 9 \times 40} = 34.7 \approx 35 \text{ mm}$$

6. Length of Spigot from slot to the end, a

Considering the failure of the socket end in double shear, we have

$$2ad_2\tau = \text{F}$$

Length of spigot from cotter slot to end,

$$a = \frac{\text{F}}{2d_2\tau}$$

$$= \frac{25000}{2 \times 36 \times 40} = 8.68 \approx 10 \text{ mm}$$

7. Socket Collar Diamter, d_4

Considering failure of the socket collar in crushing, as shown in Fig. 12.5, we have

$$(d_4 - d_2)\ t\sigma_c = \text{F}$$

$$(d_4 - 36) \times 9 \times 80 = 25000$$

$$d_4 = 70.7 \approx 72 \text{ mm}$$

8. Socket Collar Thickness, c

Considering the failure of socket collar in double shear, we have

$$2\ (d_4 - d_2)\ c\ \tau = \text{F}$$

$$2\ (72 - 36) \times c \times 40 = 25000$$

$$c = 10.64 \approx 12 \text{ mm}$$

9. Diameter of Spigot Collar, d_3

Considering the failure of spigot collar in crushing,

$$\left(\frac{\pi}{4}\right)\left(d^2{}_3 - d^2{}_2\right)\sigma_c = \text{F}$$

$$\left(\frac{\pi}{4}\right)\left(d^2{}_3 - 36^2\right)\times 80 = 25000$$

$$d_3 = 41.16 \approx 42 \text{ mm}$$

10. Thickness of Spigot Collar, t_1

Considering the failure of spigot collar in shear,

$$\pi\, d_2 t_1\, \tau = \text{F}$$

$$t_1 = \frac{25000}{\pi \times 36 \times 40} = 5.53 \approx 6 \text{ mm}$$

11. Check for Cotter in Bending

$$\text{Bending stress in cotter, } \sigma_b = \left[\frac{\text{F}}{4td^2}\right](2d_4 + d_2)$$

$$= \left[\frac{25000}{4\times 9\times 35^2}\right](2\times 72 + 36)$$

$$= 10.2 \text{ MPa} < 80 \text{ MPa.}$$

Hence safe.

12. Length of cotter, $l = 4d = 4 \times 26 = 104$ mm

13. Dimension $e = 1.2d = 1.2 \times 36 = 31.2 \approx 32$ mm

Example. 12.2 The big end of a connecting rod, as shown in Fig. 12.9, is subjected to a maximum load of 40 kN. The diameter of the circular part of rod adjacent to the strap end is 70 mm. Design the joint assuming the following permissible stresses :

Strap : Tension = 25 MPa

Gib and cotter : Shear = 20 MPa

Solution.

Given : F = 40 KN, d = 70 mm, σ_t = 20 MPa

The gib and cotter joint for the strap end (or big end) of a connecting rod is shown in Fig. 12.9.

1. Width of Strap, B_1

Generally, width of Strap, $B_1 = d = 70$ mm

2. Thickness of Cotter, t

$$t = \frac{B_1}{4} = \frac{70}{4} = 17.5 \approx 18 \text{ mm}$$

3. Thickness of Gib

It is kept equal to thickness of cotter =

$$t = \frac{B_1}{4}$$

4. Height and Length of Gib Head

Height of gib head, $t_2 = t = 18$ mm

Length of gib head, $l_3 = t = 18$ mm

5. Thickness of strap, t_1

Considering tearing of the strap at the thinnest section, we have

$$2B_1 t_1 \sigma_t = F$$

$$\text{Thickness of strap, } t_1 = \frac{F}{2B_1 \sigma_t} = \frac{40,000}{2 \times 70 \times 25} = 11.42 \approx 12 \text{ mm}$$

Considering thearing strength of strap at the cotter shoulder to be equal to that at the thinnest section, we have

$$2t_3 (B_1 - t) = 2t_1 B_1$$

$$t_3 = \frac{B_1 t_1}{B_1 - t} = \frac{70 \times 12}{70 - 18} = 16.15 \approx 18 \text{ mm}$$

6. Total width of Gib and Cotter, B

Considering double shear failure of gib and cotter,

$$2Bt\tau = F$$

$$B = \frac{40,000}{2 \times 18 \times 20} = 55.5 \approx 56 \text{ mm}$$

Width of gib, $b_1 = 0.55 \text{ B} = 0.55 \times 56 = 30.8 \approx 32$ mm

Width of cotter, $b = 0.45 \text{ B} = 24$ mm

7. Other Dimensions

$$l_1 = 2t_1 = 2 \times 12 = 24 \text{ mm}$$

$$l_2 = 2.5t_1 = 2.5 \times 12 = 30 \text{ mm}$$

$$t_4 = (1.15 \text{ to } 1.5)\ t_1 = 1.25 \times 12 = 15 \text{ mm}$$

Example 12.3 Design a gib and cotter joint for a square rod to carry a maximum load of 30 kN. Assume that the gib, cotter and rod are made of the same material having the following allowable stresses :

Tension = 20 MPa, shear = 15 MPa, and crushing 50 MPa.

Solution.

Given : F = 30 *k*N, $\sigma_t = 20$ MPa, $\tau = 15$ MPa, $\sigma_c = 50$ MPa

The gib and cotter joint for a square rod is shown in Fig. 12.10.

1. Size of Rod, x

Considering failure of square rod under tensile force F,

$$\sigma_t \times x^2 = F$$

$$x = \left(\frac{F}{\sigma_t}\right)^{\frac{1}{2}} = \left(\frac{30,000}{20}\right)^{\frac{1}{2}} = 38.72 \approx 40 \text{ mm}$$

The other dimensions are fixed as follows :

Width of strap, $B_1 = x = 40$ mm

$$\text{Thickness of cotter, } t = \frac{B_1}{4} = \frac{40}{4} = 10 \text{ mm}$$

$$\text{Thickness of gib} = t = 10 \text{ mm}$$

$$\text{Height of gib, } t_2 = t = 10 \text{ mm}$$

$$\text{Length of gib head, } l_4 = t = 10 \text{ mm}$$

2. Width of Gib and Cotter

Considering the double shear of gib and cotter,

$$2Bt\tau = F$$

$$B = \frac{F}{2t\tau} = \left(\frac{30{,}000}{2 \times 10 \times 15}\right) = 100 \text{ mm}$$

$$\text{Width of gib, } b_1 = 0.55B, \text{ for a single gib}$$

$$= 0.55 \times 100 = 55 \text{ mm}$$

$$\text{Width of cotter, } b = 0.45B, \text{ for a single gib}$$

$$= 0.45 \times 100 = 45 \text{ mm}$$

3. Thickness of Strap

Considering the failure of strap end in tension at the location of gib and cotter,

$$2\,(B_1 t_1 - t t_1)\,\sigma_t = F$$

Now $B_1 = x$

Thus $2\,(x - t)\,t_1\,\sigma_t = F$

$$\text{Thickness of strap, } t_1 = \frac{F}{2(x-t)\sigma_t}$$

$$= \frac{30{,}000}{2(40-10) \times 20} = 25 \text{ mm}$$

4. Check for Strap or Gib in crushing

$$2tt_1\sigma_c = F$$

$$\sigma_c = \frac{F}{2tt_1}$$

$$= \frac{30{,}000}{2 \times 10 \times 25} = 60 \text{ MPa} > 50 \text{ MPa}$$

Therefore, we calculate t_1 from crushing failure,

$$t_1 = \frac{30{,}000}{2 \times 10 \times 50} = 30 \text{ mm}$$

5. Length of Rod from Cotter to end

Considering the double shear of rod,

$$2l_1 \times \tau = F$$

$$l_1 = \frac{F}{2x\tau} = \frac{30,000}{2 \times 40 \times 15} = 25 \text{ mm}$$

6. Length of Rod from Gib to the end of Strap

Considering the double shear of the strap.

$$2 \times 2l_2 t_1 \tau = F$$

$$l_1 = \frac{F}{4t_1\tau} = \frac{30,000}{4 \times 30 \times 15} = 16.67 \approx 18 \text{ mm}$$

7. Other Dimensions

$$l_3 = \frac{2x}{3} = \frac{2 \times 40}{3} = 26.67 \approx 27 \text{ mm}$$

Example 12.4 Design a knuckle joint to transmit 100 kN. The design stresses are : Tension = 80 MPa, shear = 60 MPa, and crushing = 120 MPa.

Solution.

Given : F=100 kN, σ_t = 80 MPa, τ = 60 MPa, σ_c = 120 MPa

The knuckle joint is shown in Fig. 12.14.

1. Diameter of Rod, d

Considering the failure of the rod in tension,

$$\left(\frac{\pi}{4}\right) d^2 \sigma_t = F$$

$$d = \left[\frac{4F}{\pi\sigma_t}\right]^{\frac{1}{2}}$$

$$= \left[\frac{4 \times 100,000}{\pi \times 80}\right]^{\frac{1}{2}} = 39.89 \approx 40 \text{ mm}$$

2. Diameter of the Knuckle Pin, d_1

Empirically, $d_1 = d = 40$ mm

Considering the double shear of the pin,

$$2\left(\frac{\pi}{4}\right) d^2_1 \tau = F$$

$$\tau = \frac{2F}{\pi d^2_1}$$

$$= \frac{2 \times 100,000}{\pi \times 40^2} = 39.8 \text{ MPa} < 60 \text{ MPa}$$

Hence safe.

3. Thickness of single eye or rod end, t

$$t = 1.25d = 1.25 \times 40 = 50 \text{ mm}$$

4. Thickness of fork, t_1

$$t_1 = 0.75d = 0.75 \times 40 = 30 \text{ mm}$$

5. Thickness of pin head, t_2

$$t_2 = 0.5d = 0.5 \times 40 = 20 \text{ mm}$$

6. Check for the Knuckle Pin in Bending.

$$\text{Bending stress in pin, } \sigma_b = 16\text{F}\frac{\left(\frac{t_1}{3}+\frac{t}{4}\right)}{\pi d^3{}_1}$$

$$= 16 \times 100{,}000 \frac{\left(\frac{30}{3}+\frac{50}{4}\right)}{\pi \times 40^3} = 179 \text{ MPa}$$

Which gives a lower value of factor or safety.

7. Outer Diamter of Eye.

Considering shearing of the eye, we have

$$(d_2 - d_1)\, t\, \tau = \text{F}$$

$$d_2 = d_1 + \frac{\text{F}}{t\tau} = 40 + \frac{100{,}000}{50 \times 60} = 73.3 \approx 75 \text{ mm}$$

8. Check for Thickness of Fork eye.

The single eye or pin may fail in crushing due to tensile load.

$$d_1 t \sigma_c = \text{F}$$

$$\sigma_c = \frac{\text{F}}{d_1 t} = \frac{100{,}000}{40 \times 50} = 50 \text{ MPa} < 120 \text{ MPa}$$

Hence safe.

9. Check for Thickness of Fork end.

Considering the tensile failure of fork end,

$$2\,(d_2 - d_1)\, t_1\, \sigma_c = \text{F}$$

$$\sigma_t = \frac{\text{F}}{2(d_2 - d_1)t_1} = \frac{100{,}000}{2(75-40) \times 30}$$

$$= 47.6 \text{ MPa} < 80 \text{ MPa}$$

Hence safe.

10. Check for the Fork end in shear.

The fork end may fail in shear due to tensile load.

$$2\,(d_2 - d_1)\, t_1 \tau = \text{F}$$

$$\tau = \frac{F}{2(d_2 - d_1)t_1}$$

$$= \frac{100,000}{2(75-40)\times 30} = 47.6 \text{ MPa} < 60 \text{ MPa}$$

Hence safe.

11.Check for Fork end in crushing.

The fork end or pin may fail in crushing due to tensile load.

$$d_1 \times 2t_1 \times \sigma_c = F$$

$$\sigma_c = \frac{F}{2d_1t_1} = \frac{100,000}{2\times 40\times 30} = 41.67 \text{ MPa} < 120 \text{ MPa}$$

Hence safe.

EXERCISES

12.1 Design a spigot and socket joint to connect two rods to take a pull of 60 kN. The maximum permissible stresses are : 50 MPa in tension, 40 MPa in shear and 100 MPa in crushing.

12.2 Desing a gib and cotter joint to take an axial load of 100 kN which alternately changes from tension to compression. The allowable stresses are : 60 MPa in tension, 40 MPa in shear and 75 MPa in crushing.

12.3 The big end of a connecting rod is subjected to a load of 45 kN. The diameter of the circular part adjacent to the strap is 50 mm. Design the joint assuming that permissible tensile stress in strap is 35 MPa and shear stress in gib and cotter is 20 MPa.

12.4 Design a knuckle joint to withstand a tensile load of 30 kN. The permissible stresses are : Tension = 65 MPa, shear = 50 MPa, and crushing = 85 MPa.

13 PIPE JOINTS

Chapter

13.1 INTRODUCTION

The pipes are used for transporting various fluids like water, steam, gases, oil and chemicals from one place to another. In engineering practice, cast iron, wrought iron, steel, and brass are the materials generally used for the pipes. Cast iron pipes are generally used for water and sewage systems. Wrought iron and steel pipes are used for conveying steam, air and oil. Brass pipes are used for pressure lubrication systems for prime movers. Pipe joints are essential to obtain the desired length of pipe.

13.2 STRESSES IN PIPES

A pipe may be considered as thin if its diameter to thickness ratio is less than 20, otherwise it shall be a thick pipe. For a thin pipe of diameter d and thickness t under internal pressure p, hoop stress is given by,

$$\sigma_\theta = \frac{pd}{2t} + c \quad \text{... (13.1)}$$

where c = corrosion allowance.

For a thick pipe, the maximum hoop stress occurs at the inner surface, and is given by Lame's equation,

$$(\sigma_\theta)_{max} = p\left[\frac{k^2+1}{k^2-1}\right] \quad \text{... (13.2)}$$

where k = outside radius / inside radius = $\frac{r_o}{r_i}$

Wall thickness of thick pipe,

$$t = r_i\left[\left\{\frac{\sigma_\theta + p}{\sigma_\theta - p}\right\}^{1/2} - 1\right] \quad \text{... (13.3)}$$

If Q = volume flow rate in the pipe

v = velocity of fluid flow

Then

$$Q = \left(\frac{\pi}{4}\right) d^2 v$$

Or

$$d = \left[\frac{4Q}{\pi v}\right]^{1/2} \quad \text{... (13.4)}$$

13.3 DESIGN OF PIPE JOINTS

13.3.1 Circular flanged joint.

A circular flanged pipe joint is shown in Fig. 12.1. Such a joint is designed on the basis that the fluid pressure acts in between the flanges and tends to separate them with a pressure existing at the point of leaking. The flanges are kept in position by bolts, which are subjected to tensile stress.

The effective diameter on which the fluid pressure acts is taken as the diameter of a circle touching the bolt holes. Let

D = pitch circle diameter of bolts.

d_b = outside or nominal diameter of bolts.

D_1 = diameter of the circle touching the bolt holes

$= D - d_b$

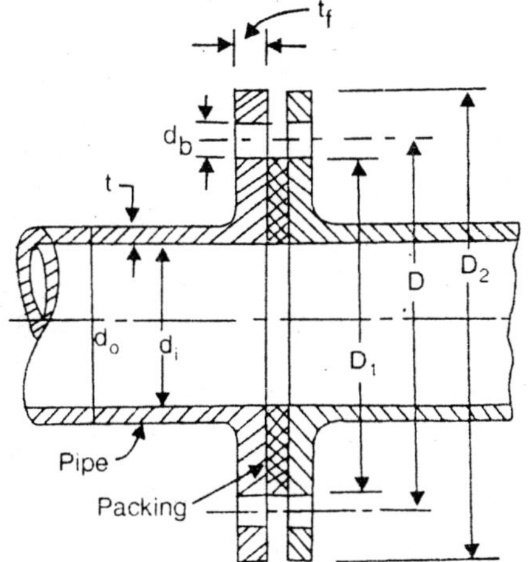

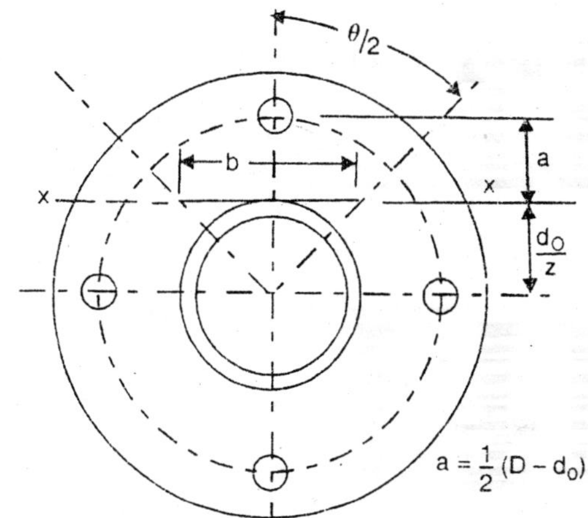

Fig. 13.1 Circular flanged joint.

n = number of bolts

p_i = fluid pressure in the pipe.

d_c = core diameter of bolts.

σ_t = permissible tensile stress in bolts.

Separating force acting on the flanges $= \frac{\pi}{4} D_1^2 p_i$

Resisting force of bolts $= n \times \frac{\pi}{4} d_c^2 \sigma_t$

For equilibrium of flanges,

$$n \times \frac{\pi}{4} d_c^2 \sigma_t = \frac{\pi}{4} D_1^2 p_i$$

$$d_c = D_1 \cdot \sqrt{\frac{p_i}{n.\sigma_t}} \quad \text{...(1)}$$

Thickness of pipe, $t = \frac{p_i d_i}{2\sigma} + c$ ...(2)

where σ = permissible tensile stress in pipe material.

c = corrosion allowance $\approx$ 9 mm

d_i = inside diameter of pipe.

Outside diameter of pipe, $d_o = d_i + 2t$

At the initial stage D_1 may be estimated as,

$D_1 \approx d_o$ + estimated clearance between bolt hole edge and pipe. ...(3)

Having calculated d_c from Eq. (1), the bolt hole diameter d_b may be known from standard tables.

Then

$D = d_o + 2d_b + 2 \times$ clearance between bolt head and pipe ($\approx$ 20 mm) ... (4)

and exact value of D_1 is,

$$D_1 = D - d_b \quad \text{...(5)}$$

The exact value of D_1 may be compared with the estimated value, which should not differ too much.

Tensile force on each bolt, $F = \frac{1}{n}\left(\frac{\pi}{4} D_1^2 p_i\right)$

Angle between two bolts, $\theta = \frac{360°}{n}$

Pitch of bolts, $p = \frac{\pi D}{n}$ ...(6a)

$\approx 20\sqrt{d_b}$ to $30\sqrt{d_b}$ for a leak proof joint. ...(6b)

Outside diameter of flangle, $D_2 = D + 2d_b + 2 \times$ clearance between bolt head and flange edge.

Thickness of flange :

The flange is assumed to be fixed at the outside of pipe and subjected to bending with a force acting at the bolt centre.

$$M_{max} = Fa$$

where $a = \frac{1}{2}(D - d_o)$

Let t_f = thickness of flange.

The effective depth of flange is taken symmetrically between bolt centres at a cross-section $x - x$ tangential to the pipe as shown in Fig. 12.1.

Depth of flange, $b = 2 \times \frac{d_o}{2} \tan \frac{\theta}{2}$

Moment of resistance of flange, $z = \frac{b t_f^2}{6}$.

If σ_b = permissible tensile stress for flange, then $M_{max} = \sigma_b z$. ...(7)

From where, t_f can be calculated.

Width of flange, $$B = \frac{1}{2}(D_2 - d_o)$$

Empirically, the dimesntions of flange may be estimated from the following :

$$d_b = 0.75\,t + 1 \text{ mm} \nless 16 \text{ mm} \quad (a)$$

$$n = 0.275\,d_i + 1.6 \quad (b)$$

$$t_f = 1.5\,t + 3 \text{ mm} \quad (c)$$

Width of flange, $$B = 2.3\,d_b \quad (d)$$

$$D_2 = d_i + 2t + 2B \quad (e)$$

$$D = d_i + 2t + 2d_b + 12 \text{ mm.} \quad (f)$$

whac c = corrosion allowance ≃ 9 mm.

Example 13.1 *A pipe of 300 mm internal diameter is subjected to an internal pressure of 1.25 MPa and is to be joined by a circular flange joint. Calculate the pipe thickness by taking the permissible tensile stress in the pipe to be 20 MPa. Also calculate the size of flanges and number and size of bolts required. Take permissible bending stress in flanges to be 20 MPa, and for bolts in tension to be 25 MPa.*

Sol. Pipe thickness, $$t = \frac{p_i d_i}{2\sigma_t} + c$$

where c = corrosion allowance = 9 mm.

$$t = \frac{1.25 \times 300}{2 \times 20} + 9 = 18.375 \text{ mm} \simeq 9 \text{ mm}$$

Outside diameter of pipe, $d_o = 300 + 2 \times 19 = 338$ mm

The clearance between outside of pipe and edge of holes may be assumed 50 mm (Fig. 13.2).

$$D_1 = 338 + 2 \times 50 = 438 \text{ mm}$$

D_1 is taken as the effective diameter on which the fluid pressure acts.

Let number of bolts, $n = 8$.

$$n \times \frac{\pi}{4} d_c^2 \sigma_t = \frac{\pi}{4} D_1^2 p_i$$

$$d_c = 438 \times \sqrt{\frac{1.25}{8 \times 25}}$$

$$= 34.62 \text{ mm} \simeq 34.903 \text{ mm}$$

Therefore, bolt size we select is M 39 × 4.

Bolt head diameter = 2 × 39 = 78 mm

Keeping a clearance of 20 mm between bolt head and pipe,

Bolt circle diameter,

D = 338 + 2 × 20 + 2 × 39 = 456 mm

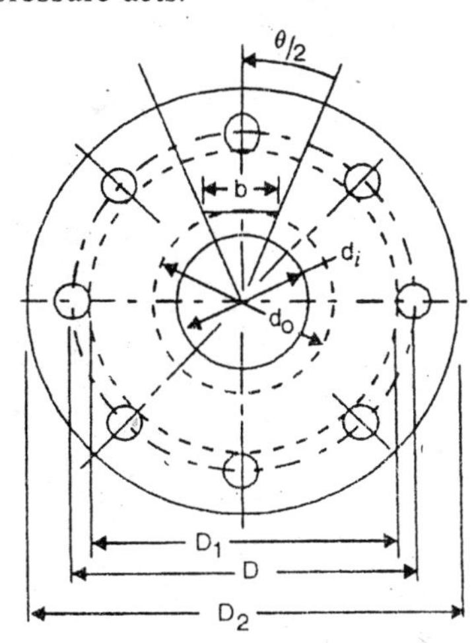

Fig. 13.2

Actual $D_1 = 456 - 39 = 417$ mm

Actual force on each both séparating the flanges.

$$= \frac{1}{8} \times \frac{\pi}{4} (417)^2 \times 1.25 = 21339.4 \text{ N}$$

Outside diameter of flange,

$D_2 = 456 + 2 \times 39 + 2 \times 10 = 554$ mm

The flange may be assumed to be fixed at the outside of pipe, where maximum bending moment due to force in the bolts is,

$$M_{max} = 21339.4 \times \left(\frac{D - d_o}{2}\right)$$

$$= 21339.4 \times \left(\frac{456 - 338}{2}\right)$$

$$= 1259024.6 \text{ N-mm}$$

Let t_f = thickness of flange, mm

Effective depth of flange at outside of pipe taking bending moment is,

$$b = 2 \times \frac{338}{2} \times \tan\left(\frac{360}{2 \times 8}\right)^\circ = 140 \text{ mm}$$

$$\text{Moment of resistance of flange} = \frac{b\,t_f^2}{6}$$

$$M_{max} = \sigma_b z$$

$$1259024.6 = 20 \times \frac{140 \times t_f^2}{6}$$

$$t_f = 51.94 \text{ mm} \simeq 52 \text{ mm}$$

13.3.2 Oval flanged joint

An oval flanged joint is shown in Fig. 12.3. In order to locate the pipes coaxially, a spigot and socket is provided on the flanges. To make the joint leak proof, packing is used according to the profile of the flange.

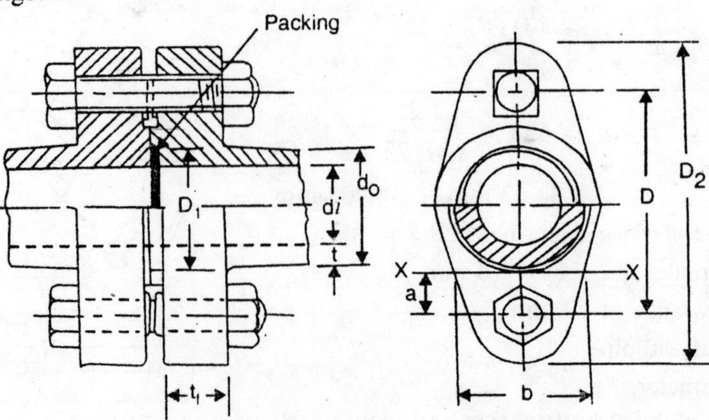

Fig. 13.3 Oval flanged pipe joint.

Assuming that the packing material is compressed to the same pressure as that of inside the pipe, the total force separating the two flanges is,

$$F = \frac{\pi}{4} d_i^2 \, p_i + \frac{\pi}{4}(D_1^2 - d_i^2) p_i$$

$$= \frac{\pi}{4} D_1^2 p_i \qquad \text{...(1)}$$

Here number of bolts, $n = 2$

$$\text{Force on each bolt} = \frac{F}{2}$$

Hence $$\frac{\pi}{4} d_c^2 \sigma_t = \frac{F}{2} \qquad \text{...(2)}$$

Nominal diameter of bolts, d_b should not be less than 12 mm.

The thickness of flange is obtained by considering it in bending due to the force in the bolt.

$$M_{max} = \frac{F}{2} \times a$$

where $$a = \frac{D - d_o}{2}.$$

Section modulus, $$z = \frac{1}{2} b t_f^2$$

where b = width of flange at section $x - x$, tangential to pipe.

$$M_{max} = \sigma z$$

$$\frac{F}{2} \times a = \sigma \times \frac{1}{6} b t_f^2$$

$\therefore$ $$t_f = \sqrt{\frac{3Fa}{\sigma b}} \qquad \text{...(3)}$$

where σ = permissible bending stress in flange.

Empirically,

$$d_b = 0.75t + 10 \text{ mm} \qquad (a)$$

$$t_f = 1.5t + 3 \text{ mm} \qquad (b)$$

$$D_2 = d_i + 2t + 4.6\, d_b \qquad (c)$$

$$D = D_2 - (3t + 20 \text{ mm}) \qquad (d)$$

....(4)

Example 13.2 Design an oval flanged pipe joint for a pipe having 50 mm bore and subjected to internal fluid pressure of 7.5 MPa. The permissible stresses in tension in pipe material and bolts are 20 MPa and 60 MPa respecitvely.

Sol. Wall thickness of pipe,

$$t = \frac{d_i}{2}\left[\sqrt{\frac{\sigma_t + p_i}{\sigma_t - p_i}} - 1\right]$$

$$= \frac{50}{2}\left[\sqrt{\frac{20+75}{20-7.5}} - 1\right] = 12 \text{ mm}$$

Let us assume that width of packing = 10 mm

Outside diameter of packing, $D_1 = d_i + 2 \times \text{width of packing}$

$= 50 + 2 \times 10 = 70$ mm

Outside diameter of pipe, $d_o = d_i + 2t = 50 + 2 \times 12 = 74$ mm

Separating force on flanges,

$$F = \frac{\pi}{4} D_1^2 p_i$$

$$= \frac{\pi}{4} \times (70)^2 \times 7.5 = 28863.4 \text{ N}$$

Force on each bolt $= \frac{F}{2} = 14431.7$ N

If d_c = core diameter of bolts, then

$$\frac{\pi}{2} d_c^2 \sigma = \frac{F}{2}$$

$$d_c = \sqrt{\frac{4 \times 14431.7}{\pi \times 60}} = 17.5 \text{ mm.} \simeq 18.933 \text{ mm}$$

We select M22 × 2.5 size bolt.

Outside diameter of flange,

$$D_2 = d_i + 2t + 4.6\, d_b$$

$$= 50 + 2 \times 12 + 4.6 \times 22$$

$$= 50 + 24 + 101.2 = 175.2 \simeq 176 \text{ mm}$$

Pitch circle diameter of bolts,

$$D = D_2 - (3t + 20 \text{ mm})$$

$$= 176 - (3 \times 12 + 20) = 120 \text{ mm}$$

Let $D_3 = 100$ mm

The flange profile is shown in Fig. 12.4, which is taken as elliptical.

Semi major axis, $a_1 = \frac{D_2}{2} = 88$ mm

Semi minor axis, $b_1 = \frac{D_3}{2} = 50$ mm

Equation of ellipse is :

$$\frac{x^2}{a_1^2} + \frac{y^2}{b_1^2} = 1$$

At $\quad x = \frac{d_o}{2} = 37$ mm

$$\frac{37^2}{88^2} + \frac{y^2}{50^2} = 1$$

$$y = 45.36 \text{ mm}$$

$\therefore \quad b = 2y = 90.72$ mm

$$a = \frac{1}{2}(\text{D} - d_o)$$

$$= \frac{1}{2}(120 - 74) = 23 \text{ mm}$$

Fig. 13.4 Elliptical flange profile.

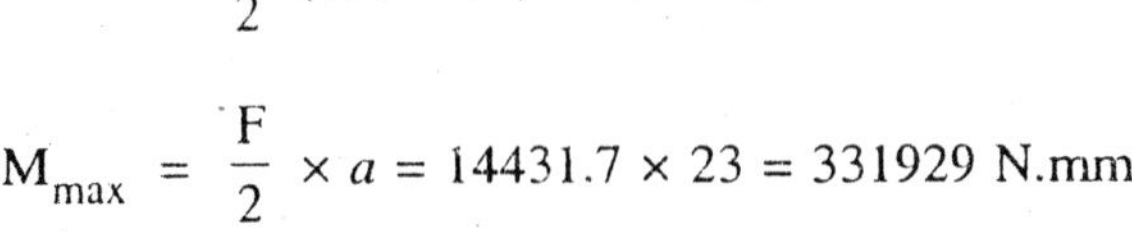

$$\text{M}_{max} = \frac{\text{F}}{2} \times a = 14431.7 \times 23 = 331929 \text{ N.mm}$$

Section modulus of flange $= \frac{1}{6} b t_f^2$

$$\frac{1}{6} \times 90.72 \times t_f^2 \times 20 = 331929$$

$$t_f = 33 \text{ mm} \simeq 35 \text{ mm}$$

13.3.3 Square flanged joint

A square flange is shown in Fig. 13.5

Let $\quad \text{D}_1$ = diameter of packing.

$\quad d_c$ = core diameter of bolts.

Then $$4 \times \frac{\pi}{4} d_c^2 \sigma = \frac{\pi}{4} \text{D}_1^2 p_i$$

$$\therefore \quad d_c = \text{D}_1 \sqrt{\frac{p_i}{4\sigma}} \qquad ...(1)$$

Nominal bolt diameter, $\quad d_b = \frac{d_c}{0.84}$

Thickness of pipe, $$t = \frac{d_i}{2}\left[\sqrt{\frac{\sigma_t + p_i}{\delta_t - p_i}} - 1\right] \qquad ...(2)$$

Outside diameter of pipe, $$d_o = d_i + 2t \qquad ...(3)$$

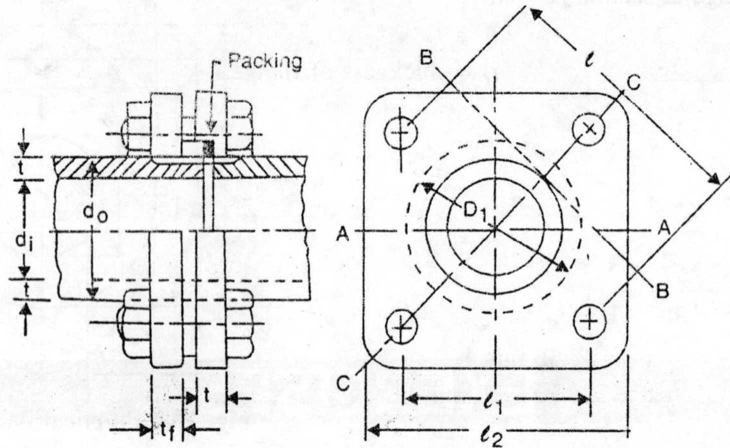

Fig. 13.5 Square flanged pipe joint.

Minimum diagonal length of bolt centres,

$$l = d_o + 2 \times d_b + 2 \times \text{clearance between pipe and bolt head.} \quad ...(4)$$

Distance between bolt centres,

$$l_1 = \frac{l}{\sqrt{2}} \quad ...(5)$$

Width of flange, $l_2 = l_1 + 2d_b + 2 \times$ clearance between nuts and flange edge. ...(6)

The flange is weakest in bending about section A – A. Bending moment due to force in two bolts about section A – A.

$$M_1 = 2F \times \frac{l_1}{2}$$

where $$F = \frac{1}{4}\left(\frac{\pi}{4} D_1^2 p_i\right)$$

The flanges may be screwed with metric pipe threads of pitch 0.907, 1.337, 1.814 and 2.309 mm.

Depth of threads, $h = 0.640327 \times \text{pitch}$

Core diameter of threads, $d_1 = d_o - h$

Mean diameter of threads, $d_m = \frac{1}{2}(d_1 + d_o)$

Centroid of the radius of the arc from A – A over which the load due to fluid pressure may be taken to be concentrated.

$$r = 0.6366 \times \frac{1}{2} \times d_m$$

Bending moment due to fluid pressure

$$M_2 = 2F \times r$$

Resultant bending moment on the flange about section A – A,

$$M = M_1 - M_2$$

Width of flange at section A – A,

$$b = l_2 - d_o$$

Let t_f = thickness of flange.

$$M = \sigma_b z = \sigma_b \times \frac{1}{6} b t_f^2$$

or

$$t_f = \sqrt{\frac{6M}{b\sigma_b}} \quad ...(7)$$

EXERCISES

13.1 Determine the dimensions of a flanged joint for a cast iron pipe 300 mm diameter to carry a pressure of 0.8 MP*a*. The allowable tensile stress for cast iron is 14 MP*a*. The corrosion allowance may be taken as 9 mm.

13.2 A flanged cast iron pipe having 250 mm internal diameter is subjected to a fluid pressure of 0.4 MP*a*. The flange is connected by a circular flange joint. Calculate the pipe thickness, size of flanges and number and size of bolts required. Take the following permissible stresses :

Tensile stress in pipe	= 15 MP*a*
Bending stress in flanges	= 15 MP*a*
Tensile stress in bolts	= 25 MP*a*

13.3 Design an oval flanged pipe joint for a pipe having 60 mm bore and subjected to an internal fluid pressure of 6 MP*a*. The maximum tensile stress in pipe is limited to 20 MP*a* and in the bolts 60 MP*a*.

13.4 Design a square flanged pipe joint for a pipe of internal diameter 70 mm subjected to an internal fluid pressure of 8 MP*a*. The maximum tensile stress in the pipe material is limited to 20 MP*a* and in the bolts 30 MP*a*.

13.5 A scamless steel pipe carries 2500 m^3 of steam per hour at a pressure of 1.25 MP*a*. The velocity of flow is 30 m/s. Find the inside diameter of pipe. Also calculate its wall thickness if permissible tensile stress is 40 MP*a*.

14 MECHANICAL SPRINGS

Chapter

14.1 INTRODUCTION

A spring may be defined as an elastic body whose function is to absorb energy in the form of strain energy when loaded and release that energy on unloading. The main applications of a spring are :

1. To absorb energy due to either shock or vibration as in the case of vehicle springs, railway buffers, shock absorbers and vibration dampers.
2. To apply force as in the case of brakes, clutches and spring loaded safety valves.
3. To control motion by maintaining contact between two elements as in a cam and follower.
4. To measure forces as in a spring balance and engine indicators.
5. To store energy as in watches and toys.

14.2 TYPES OF SPRINGS

The various types of springs are :

1. Helical springs. They are made up of a wire coiled in the form of a helix. They are used for tensile or compressive loading. In the case of close-coiled helical spring, the helix angle is generally less than 10°. A close-coiled helical spring is shown in Fig. 14.1. Such springs are subjected to torsional shear stresses only. If the helix angle is more than 10° then the spring is called open-coiled helical spring. Such springs are subjected to both bending and shear stresses.

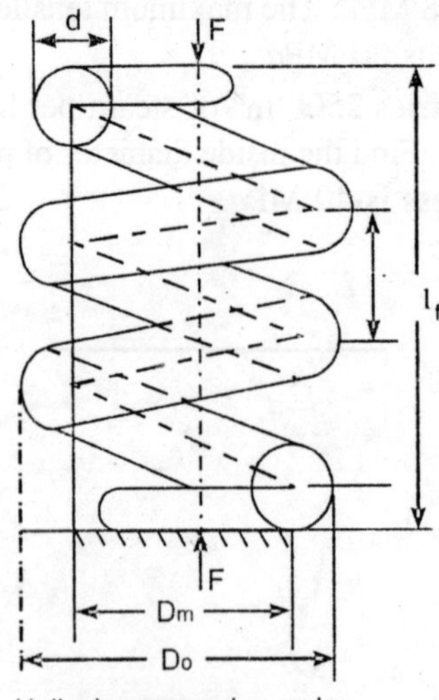

Fig. 14.1 Helical compression spring

2. Flat spiral springs. These springs are made of flat wire and wound in the form of a spiral. They are mainly used in clocks and are subjected to bending stresses.

3. Torsion springs. It is a device used to transmit the torque to a particular component of a machine or mechanism. The spring can be loaded by a torque about the axis of the coils. They resist bending moment.

4. Laminated (or carriage or leaf) springs. These springs consist of graduated leaves assembled by U-clips. They are subjected to bending stresses and are widely used in vehicles.

14.3 NOMENCLATURE

The nomenclature of compression spring is shown in Fig. 14.1.

Solid length, l_s – It is the length of the spring when all the coils touch each other.

$$l_s = id \qquad \text{....(14.1)}$$

where i = number of active coils

d = wire diameter.

Free length, l_f — It is the length of the spring in the unloaded condition.

l_f = solid length + maximum deflction + clearance between adjacent coils (clash allowance)

$$= id + \delta_{max} + 0.15\,\delta_{max} \qquad \text{...(14.2)}$$

Spring index, C — It is the ratio of the mean coil diameter to the wire diameter.

$$C = \frac{D_m}{d} \qquad \text{...(14.3)}$$

where D_m = mean coil diameter

Spring stiffness (or rate), k — It is defined as the load F per unit deflection δ of the spring.

$$k = \frac{F}{\delta} \qquad \text{...(14.4)}$$

Coil pitch, p — It is the axial distance between adjacent coils in the uncompressed state.

$$p = \frac{l_f}{(i-1)}$$

$$= \frac{(l_f - l_s)}{i} + d \qquad \text{...(14.5)}$$

14.4 DESIGN OF HELICAL COMPRESSION SPRINGS AGAINST STATIC LOAD

1. Stresses in the spring.

A close—coiled helical spring is subjected to the following stresses under axial load :

(*i*) Direct shear stress.

$$\text{Direct shear stress in spring wire, } \tau_d = \frac{4F}{\pi d^2} \qquad \text{...(14.6)}$$

The direct shear stress distribution is shown in Fig. 14.2 (*a*).

(*ii*) Torsional shear stress in the spring wire is given by,

$$\tau_s = \frac{8FD_m}{\pi d^3} = \frac{8FC}{\pi d^2} \qquad ...(14.7)$$

The torsional shear stress distribution is shown in Fig. 14.2 (*b*).

(*iii*) Maximum shear stress in spring wire,

$$\tau_{max} = \tau_s + \tau_d$$

$$= \left[\frac{8FD_m}{\pi d^3}\right]\left[1 + \frac{0.5}{C}\right] \qquad ...(14.8a)$$

$$= K_s \times \left[\frac{8FD_m}{\pi d^3}\right] \qquad ...(14.8b)$$

$$= K_s\left[\frac{8FC}{\pi d^2}\right] \qquad ...(14.8c)$$

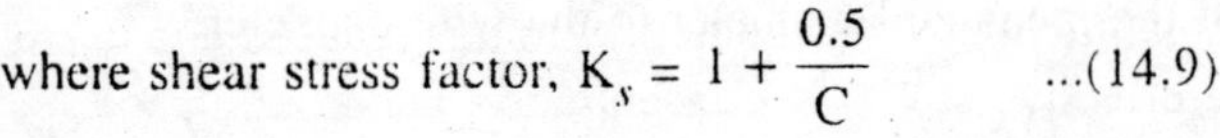

where shear stress factor, $K_s = 1 + \frac{0.5}{C}$ ...(14.9)

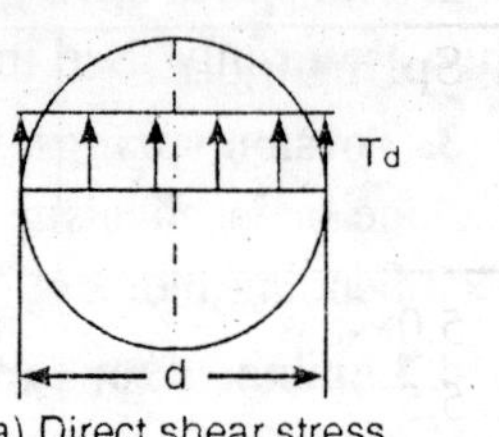

(a) Direct shear stress

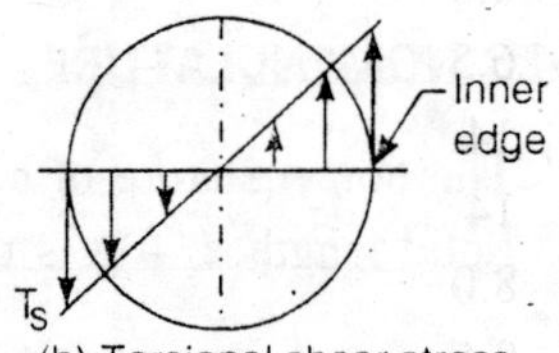

(b) Torsional shear stress

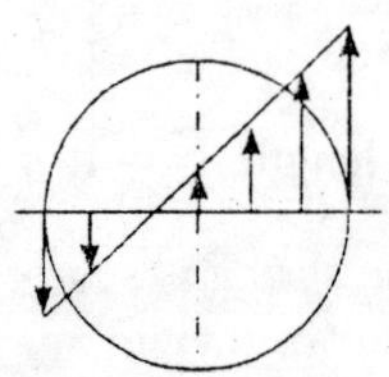

(c) Total shear stress

Fig. 14.2 Stresses in spring wire

The total shear stress distribution is shown in Fig. 14.2 (*c*)

(*iv*) Stresses due to curvature of the wire.

The effect of direct shear stress and curvature was introduced by A.M. Wahl, and is called the Wahl factor, which is given by,

$$K_w = \frac{4C - 1}{4C - 4} + \frac{0.615}{C} \qquad ...(14.10)$$

The Wahl correction factor K_w consists of two factors, Correction factor K_s for direct shear stress and correction factor K_c for the curvature effect, Therefore,

$$K_w = K_s K_c$$

$$K_c = \frac{K_w}{K_s} \qquad ...(14.11)$$

Shear stress considering direct shear stress and curvature of wire,

$$\tau_{max} = K_w\left[\frac{8FC}{(\pi d^2)}\right] \qquad ...(14.12)$$

The value of these factors are given in Table 14.1.

Table 14.1 Spring stress factors

Spring index	Wahl factor K_w	Direct shear stress factor K_s	Coil curvature factor K_c
5.0	1.311	1.100	1.192
5.5	1.278	1.091	1.171
6.0	1.253	1.083	1.157
6.5	1.231	1.077	1.143
14.0	1.213	1.071	1.133
14.5	1.197	1.067	1.122
8.0	1.184	1.063	1.114
8.5	1.172	1.059	1.107
9.0	1.162	1.056	1.100
10.0	1.145	1.050	1.090

The stress — concentration due to curvature is localized at the inner side of the spring coil. When the spring is subjected to a static load, the effect of stress — concentration K_c is neglected due to localized yielding. Therefore, for static loading the maximum shear stress in the spring wire is given by Eq. (14.8).

When the spring is subjected to fluctuating loads, the endurance limit is reduced due to stress–concentration factor K_c. Therefore, the effective stress — concentration factor is K_w.

2. Deflection of spring.

$$\text{Deflection of spring, } \delta = \frac{8FD_m^3 i}{Gd^4} \quad ...(14.13a)$$

$$= \frac{8FC^3 i}{Gd} \quad ...(14.13b)$$

3. Stiffness of spring

$$\text{Stiffness of spring, } k = \frac{Gd^4}{8D_m^3 i} \quad ...(14.14a)$$

$$= \frac{Gd}{8C^3 i} \quad ...(14.14b)$$

4. Buckling of compression springs.

If the free length of the spring is more than four times the pitch diameter of the spring, then the spring behaves like a column and may fail by buckling.

Critical buckling laod, $F_{cr} = kk_l l_f$ ....(14.15)

where k = spring stiffness

k_l = buckling factor depending upon the ratio $\frac{l_f}{D_m}$ (see Fig. 14.3)

l_f = free length of spring

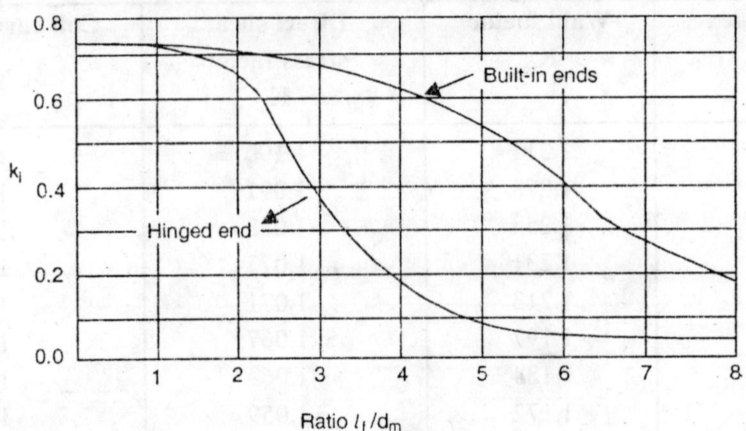

Fig. 14.3 Curve for buckling load correction factor

For safe in buckling, $F_{cr} > F_{max}$

If $\dfrac{l_f}{D_m} \leq 2.6$, guide is not necessary ...(14.16*a*)

> 2.6, guide is necessary ...(14.16*b*)

5. Surge in springs

If the spring is subjected to dynamic load which fluctuates between two limits then resonance may occur if the forcing frequency coincides with the natural frequency of the spring. This results in very large deflection of the coils and correspondingly very high stresses. This may result into failure of the spring. This phenomenon is called **surge.**

The fundamental frequency of a compression spring is given by..,

$$f_n = \left[\frac{d}{\pi i D_m^2}\right]\left[\frac{G}{8\rho}\right]^{\frac{1}{2}} \text{ Hz, when the spring is guided at both ends} \quad ....(14.17a)$$

$$= \left[\frac{d}{\pi i D_m^2}\right]\left[\frac{G}{32\rho}\right]^{\frac{1}{2}} \text{ Hz, when the spring has one end fixed and the other free.}$$

...(14.17*b*)

where ρ = density of spring material, kg/m^3.

The natural frequency of the spring should be at least 20 times the forcing frequency to avoid resonance.

The surge in springs may be eliminated by using the following methods :

1. By using friction dampers on the centre coils so that wave propagation dies out.

2. By using springs of high natural frequency.

3. By using springs of different pitch of coils near the ends than at the centre to have different natural frequencies.

14.5 SPRING WIRE MATERIALS

There are four basic varieties of steel wire which are used in springs.

(*i*) Patented and cold – drawn steel wires.

(*ii*) Oil – hardened and tempered spring steel wires and valve spring wires (unalloyed).

(*iii*) Oil – hardened and tempered steel wires (alloyed), and

(*iv*) Stainless steel spring wires for normal corrosion resistance

The applications of these steel wires are given in Table 14.2.

Table 14.2 Applications of steel wires for helical springs

Type of wire	Applications			
(*i*) Patented and cold drawn steel	Grade 1	Grade 2	Grade 3	Grade 4
	Springs subjected to static or low-load cycles	Springs subjected to moderate-load cycles	Highly stressed static springs or subjected to moderate dynamic loads	Springs subjected to severe stresses
(*ii*) Unalloyed, oil – hardened and tempered spring steel,	Grade SW		Grade VW	
	Springs subjected to moderate fluctating stresses		Springs subjected to high amplitude of fluctuating stresses	
(*iii*) Alloyed, oil – hardened and tempered steel	1S and 2S		1D and 2D	
	For static working springs		For dynamic working springs	
(*iv*) Stainless steel for normal corrosion resistance	For corrosion resistance environment applications			

For plain carbon steels spring wires,

$$S_{yt} = 0.75\ S_{ut} \qquad ...(14.18a)$$

$$S_{ys} = 0.577\ S_{yt} \qquad ...(14.18b)$$

Design shear stress,

$$\tau_d = \frac{S_{ys}}{n} \qquad ...(14.19)$$

where n = factor of safety

$$\tau_d = 0.55S_{ut} \text{ for alloy steels} \qquad ...(14.20a)$$

$$= 0.8S_{ys} \text{ for light service} < 10^4 \text{ cycles} \qquad ...(14.20b)$$

$$= 0.6S_{ys} \text{ for average service} \geq 10^4 \text{ and} < 10^6 \text{ cycles} \qquad ....(14.20c)$$

$$= 0.4S_{ys} \text{ for} \geq 10^6 \text{ cycles} \qquad ....14.20d)$$

The shear strength of spring steel wire decreases as diameter increases. The size factor is given by,

$$B = 0.86 + \frac{1.75}{d} \text{ for } d < 12.5 \text{ mm} \qquad ...(14.21a)$$

$$= \frac{1.885}{d^{0.25}} \text{ for } d \geq 12.5 \text{ mm} \quad ...(14.21b)$$

Design stress, $\tau'_d = B\tau_d$ ...(14.22)

The modulus of rigidity for spring wires is given in Table 14.3. The physical properties of spring steel wires are given in Tables 14.4 to 14.7.

Table 14.3 Modulus of Rigidity G for Spring Wires

Material	G N/mm^2
Patented and cold drawn steel wires—unalloyed, Grades 1, 2, 3 and 4 as per IS : 4454 (Part I)—1981.	81,370
Oil hardened and tempered spring steel wire and valve spring wire-ualloyed, Grades SW and VW as per IS : 4454 (Part II)—1975.	81,370
Alloyed, oil hardened and tempered valve spring wire and spring steel wire for use under moderately elevated temperatures, Grades IS, 2S, ID and 2D as per IS : 4454 (Part III)—1975.	81,370
Hot rolled steel for hot formed springs, Grades 1, 2 as per IS : 3431—1965.	80,000
Hot rolled steel for hot formed springs, Grade 98C6, 113 C6, 55 Si 7, 60 Si 7, 50 Cr 4V2 and 60 Cr 4V2 as per IS : 3195—1975.	80,000
Stainless spring steel wire, as per IS : 4454 (Part IV)—1975.	
Grade 1 : Tempered	
Untempered	73,530
Grade 2 : Tempered	69,610
Untempered	78,430
Hard drawn brass wires as per IS : 4076—1967	73,530
Phosphor bronze wires as per IS : 1385—1968	35,000
	42,000

Table 14.4 Physical Properties of Patented and Cold Drawn, Steel Wires Unalloyed for Cold formed Springs

Uncoated wire	Tensile, strength, N/mm^2, Min.			
dia d (nominal), mm	Grade 1	Grade 2	Grade 3	Grade 4
0.07	—	—	2550	—
0.08	—	—	2450	—
0.09	—	—	2530	—
0.10	—	—	2530	—
0.11	—	—	2520	—
0.12	—	—	2520	—
0.14	—	—	2510	—
0.16	—	—	2500	—
0.18	—	—	2500	—

Uncoated wire dia d (nominal), mm	Tensile, strength, N/mm², Min.			
	Grade 1	Grade 2	Grade 3	Grade 4
0.20	—	—	2490	2700
0.22	—	—	2480	2680
0.25	—	—	2470	2670
0.28	—	—	2460	2660
0.30	1720	2060	2460	2660
0.32	1710	2050	2450	2650
0.34	1710	2050	2440	2640
0.36	1700	2040	2430	2630
0.38	1700	2040	2430	2620
0.40	1700	2040	2430	2620
0.43	1690	2030	2420	2610
0.45	1680	2020	2410	2600
0.48	1680	2020	2400	2590
0.50	1670	2010	2390	2580
0.53	1660	2000	2380	2570
0.56	1660	2000	2370	2560
0.60	1650	1990	2360	2550
0.63	1640	1980	2340	2540
0.65	1640	1980	2330	2540
0.70	1630	1970	2320	2530
0.75	1620	1960	2300	2500
0.80	1610	1950	2280	2480
0.85	1600	1930	2260	2460
0.90	1590	1920	2250	2440
0.95	1580	1910	2250	2420
1.00	1570	1900	2240	2400
1.05	1560	1890	2210	2380
1.10	1550	1880	2190	2370
1.20	1540	1860	2170	2340
1.25	1530	1850	2140	2320
1.30	1520	1840	2130	2300
1.40	1500	1800	2110	2290
1.50	1490	1800	2100	2260
1.60	1470	1780	2080	2250
1.70	1460	1760	2050	2220
1.80	1440	1750	2030	2190
1.90	1430	1730	2010	2180
2.00	1420	1720	1990	2160
2.10	1410	1700	1960	2130
2.25	1400	1680	1940	2100
2.40	1380	1660	1910	2070
2.50	1370	1640	1890	2050
2.60	1360	1620	1860	2030
2.80	1340	1600	1840	2000
3.00	1320	1570	1830	1980
3.20	1310	1550	1790	1960
3.40	1290	1530	1760	1920

Uncoated wire	Tensile, strength, N/mm^2, Min.			
dia d (nominal), mm	Grade 1	Grade 2	Grade 3	Grade 4
3.60	1270	1510	1750	1920
3.80	1260	1490	1720	1860
4.00	1250	1480	1700	1840
4.25	1250	1460	1680	1820
4.50	1230	1440	1660	1800
4.75	1210	1420	1620	1770
5.00	1190	1390	1600	1750
5.30	1170	1370	1570	1720
5.60	1150	1350	1550	1690
6.00	1130	1320	1530	1670
6.20	1120	1310	1500	1640
6.50	1110	1290	1480	1620
7.00	1090	1260	1460	1610
7.50	1070	1250	1430	1570
8.00	1050	1220	1400	1540
8.50	1050	1200	1370	1500
9.00	1000	1180	1350	1480
9.50	990	1150	1310	—
10.0	980	1130	1290	—
10.5	—	1100	—	—
11.0	—	1080	—	—
12.0	—	1040	—	—
12.5	—	1030	—	—
13.0	—	1020	—	—
14.0	—	990	—	—
15.0	—	970	—	—
16.0	—	960	—	—
17.0	—	950	—	—

IS : 4454 (Part I)—1981

Table 14.5 Physical Properties of Oil Hardened and Tempered Spring Steel Wire and Valve Spring Wire Unalloyed

Wire Dia (Nominal) mm	Tolerance mm	Tensile* sgrength N/mm^2, Min For grades	
		SW**	VW†
1.00 1.05 1.10	±0.015	1760—1960	1670—1810
1.20 1.25 1.30 1.40		1720—1960	1620—1760

Wire Dia (Nominal) mm	Tolerance mm	Tensile* sgrength N/mm², Min For grades	
		SW**	VW†
1.50 1.60 1.70 1.80 1.90	±0.020	1670—1860	1570—1690
2.00 2.10 2.25 2.40	±0.030	1620—1760	1520—1620
2.50 2.50 2.80		1570—1720	1470—1620
3.00 3.20		1520—1670	1430—1530
3.40	±0.04		
3.60 3.80 4.00		1480—1630	1400—1500
4.25 4.50 4.75 5.00		1440—1590	1370—1470
5.30 5.60		1400—1500	1340—1440
6.00	±0.06		
6.30 6.50 7.00 7.50		1360—1510	1300—1400
8.00 8.50		1290—1440	
9.00 9.50 10.00	±0.08		
10.5 11.0 12.0 12.5 13.0 14.0	±0.12	1250—1400	

IS : 4454 (Part II)—1975

*Yield strength = 0.9 × Ultimate tensile strength. *For moderately stressed springs.

Table 14.6 Physical Properties of Oil Hardened and Tempered Steel Wire-alloyed

Wire Dia (Nominal) mm	Tolerance mm	Tensile strength, N/mm², Min For Grades				Reduction in area, percent, Min.
		IS*	ID †	2S*	2D †	
1.00		—	—	2060	2060	
1.05						
1.10	±0.015	1860	1760	1960	1960	
1.20						
1.25						
1.30						
1.40						
1.50						
1.60						
1.70	±0.020					50
1.80						
1.90						
2.00						
2.10						
2.25						
2.40						
2.50	±0.030	1760	1670	1910	1910	
2.60						
2.80						
3.00						
3.20						
3.40						
3.60						
3.80		1670	1570	1860	1860	
4.00						
4.25						
4.50	±0.040					45
4.75						
5.00						
5.30						
5.60		1520	1470	1760	1760	
6.00						
6.30						
6.50	±0.06	1470	1420	1720	1720	42
7.00						
7.50				1670	1670	
8.00						

*For static working springs. † For dynamic working springs IS : 4454 (Part III)—1975

Table 14.7 Specification for Steel Wires for Cold Formed Stainless Spring Steel Wire for Normal Corrosion Resistance

Wire dia (Normal) mm	Tolerance mm	Tensile Strength N/mm^2, Min
0.10		2060
0.11	±0.005	
0.125		2010
0.14		
0.16		
0.18		
0.20		
0.22		
0.25		1960
0.28		
0.30		
0.32		
0.34		
0.36		
0.38		
0.40		
0.43	±0.010	1910
0.45		
0.48		
0.50		
0.53		
0.56		1860
0.60		
0.63		
0.65		
0.70		
0.75		
0.80		
0.85		
0.90		
0.95		
1.00		
1.05		
1.10	±0.015	1760
1.15		
1.20		
1.25		
1.30		
1.40		

Wire dia (Normal) mm	Tolerance mm	Tensile Strength N/mm^2, Min
1.50		
1.60		
1.70	±0.020	1670
1.80		
1.90		
2.00		
2.10		
2.25		
2.40		
2.50	±0.030	1579
2.60		
2.80		
3.00		
3.15		
3.20		
3.40		
3.60		
3.80		
4.00		
4.25	±0.040	1470
4.50		
4.75		
5.00		
5.30		
5.60		
6.00		
6.30		
6.50		
7.00	±0.060	1320
8.00		
8.50		
9.00		
9.50		
10.00	±0.080	1270

IS : 4454 (Part IV)—1975

† For higher dynamic stressed and moderate impact loaded springs.

14.6 SPRING ENDS

For transmitting the axial load the spring ends are so formd that for any position of the spring, the spring action is axial as far as possible. This is generally achieved by decreasing the pitch at the runout coil. The spring end is then ground so that a flat seating surface is obtained as shown in Fig. 14.4. The total number of coils and block lengths for helical compression springs are given in Table 14.8.

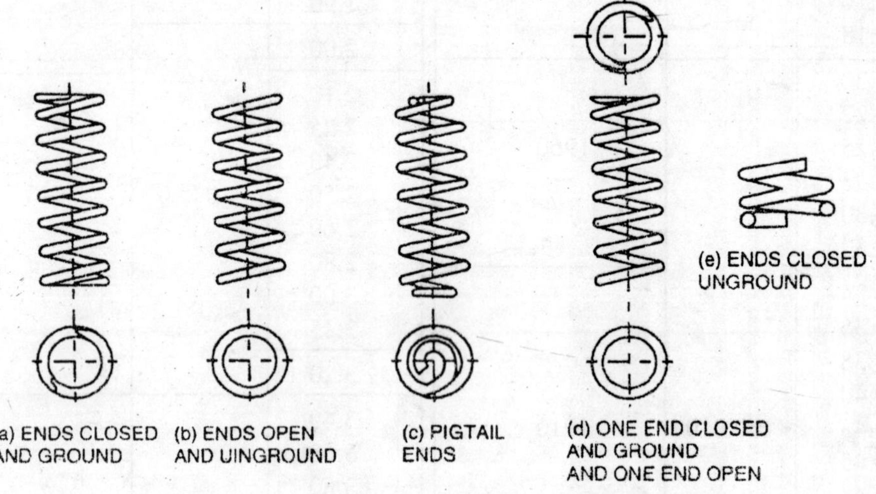

Fig. 14.4 Spring ends.

Table 14.8 Total number of coils for helical compression springs.

Type of ends	Total number of coils, i_t
1. Hot coiled springs	
Closed (squared) and ground	$i_f + 1.50$
Open and unground	$i_f + 1.00$
Pigtail	$i_f + 1.50$
One end closed and ground and one end open	$i_f + 1.67$
2. Cold coiled springs	
Squared	$i_f + 2$
Squared and ground	$i_f + 2$
Plain	i_f
Plain and ground	i_f

14.7 DESIGN OF HELICAL COMPRESSION SPRINGS AGAINST FATIGUE LOAD

The Soderberg plot of the $\tau_m - \tau_a$ diagram for two – way shear is shown in Fig. 14.5. The factor of safety n is given by,

$$\frac{\tau_m}{S_{ys}} + \frac{\tau_a}{S'_{es}} = \frac{1}{n} \quad ...(14.23)$$

where S'_{es} = endurance limit for given number of cycles.

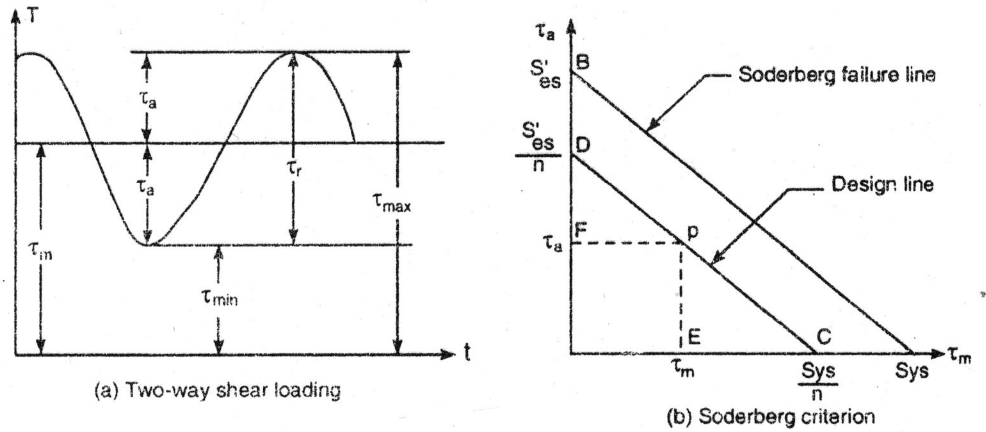

Fig. 14.5 Two way shear design of spring wire

The endurance limit of the spring wire is given for one – way shear and Eq. (14.23) is not applicable. In that case a modified Soderberg procedure based on one – way shear is used. The one – way shear test is based on a shear stress varying continually from zero to a maximum value S'_{es} as shown in Fig. 14.6 (*a*). It is assumed that S'_{es} is the value of shear stress for which a part is on the verge of failure after an infinite number of cycles. In Fig. 14.6 (*a*), $\tau_m = \tau_a = \frac{S'_{es}}{2}$. The modified Soderberg plot is shown in Fig. 14.6 (*b*) for factor of safety n. AB is the Soderberg failure line. CD is the design line with factor of safety n which is parallel to line AB. Considering similar traingles PCF and AEB, we have

$$\frac{PF}{FC} = \frac{AE}{EB}$$

$$\frac{\tau_a}{\left(\frac{S_{ys}}{n} - \tau_m\right)} = \frac{\left(\frac{S'_{es}}{2}\right)}{\left(S_{ys} - \frac{S'_{es}}{2}\right)} \quad ...(14.24)$$

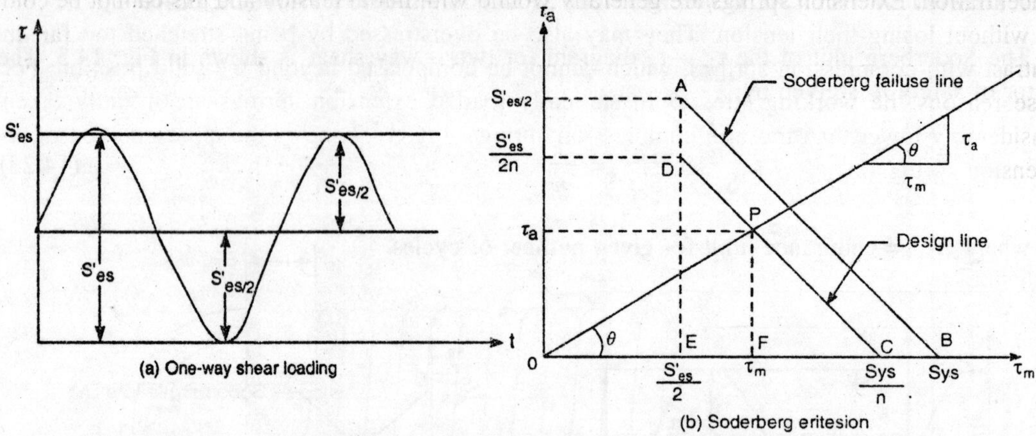

Fig. 14.6 One-way shear design of spring wire

Or $$\frac{\tau_m - \tau_a}{S_{ys}} + \frac{2\tau_a}{S'_{es}} = \frac{1}{n} \qquad \text{...(14.25)}$$

Where $$\tau_m = K_s\left[\frac{8F_m D_m}{\pi d^3}\right]$$

$$\tau_a = K_w\left[\frac{8F_a D_m}{\pi d^3}\right]$$

$$F_m = \frac{F_{max} + F_{min}}{2}, \text{ and } F_a = \frac{F_{max} - F_{min}}{2}$$

Eq. (14.24) or (14.25) is used in the design of springs against fatigue loads.

The wrie diameter then becomes,

$$d^3 = K_w\left[\frac{8F_m D_m}{\pi \tau_m}\right] \qquad \text{...(14.26)}$$

The endurance limit of steel spring wires is given by,

$$S'_{es} = 0.21S_{ut} \text{ for cold drawn steel wires} \qquad \text{...(14.27}a\text{)}$$

$$= 0.22\ S_{ut} \text{ for oil-hardened and tempered steel wires} \qquad \text{...(14.27}b\text{)}$$

The yield shear strength of steel spring wires is given by,

$$S_{ys} = 0.42S_{ut} \text{ for cold drawn steel wires} \qquad \text{...(14.28}a\text{)}$$

$$= 0.45\ S_{ut} \text{ for oil-hardened and tempered steel wires} \qquad \text{...(14.28}b\text{)}$$

14.8 DESIGN OF HELICAL EXTENSION SPRINGS

(*a*) Static Load

The design of helical extension springs follows that of compression springs. However, consideration must be given to the effects of the turns which generally involve additional stress concentration. Extension springs are generally wound with initial tension and this cannot be cold set without losing their tension. They may also be overstressed by being stretched too far, in contrast with compression springs, which cannot be compressed beyond the solid position. For these reasons the working stresses in statically loaded extension springs are initially taken considerably lower than those in compression springs. Fig. 14.7 shows the representation of an extension spring.

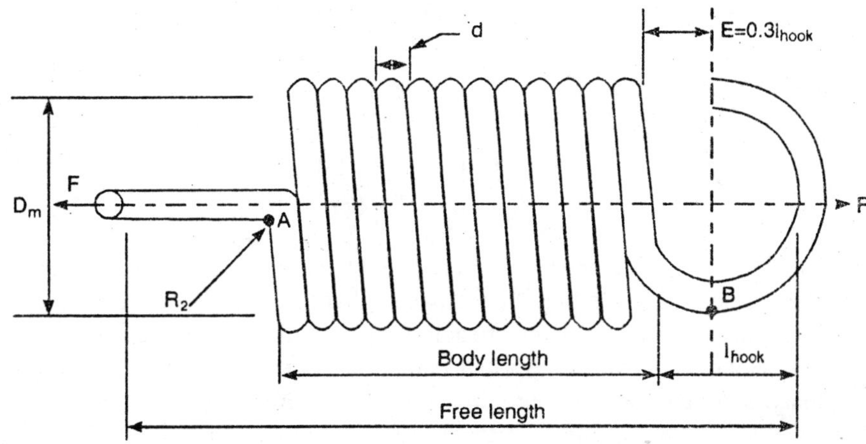

Fig. 14.7 Helical extension spring

All the design formulae given for compression springs apply to the extension spring. However, with initial tension F_i the difference $F - F_i$ is substituted for F.

Initial tension, $$F_i = F - \delta K$$

$$= \frac{F - Gd^4\delta}{8D_m^3 i_f} \qquad ...(14.29)$$

The total number of coils in an extension spring with initial tension is approximately given by :

$$i_t = \frac{L_k}{d} - 1 \qquad ...(14.30)$$

where L_k = body length of coils when not loaded but subjected to initial tension.

The exact number of coils depend on the shape of the end loop. The end coils for the spring shown in Fig. 14.7 is 0.25. The length L_k is given by,

$$L_k = (i_t + 1)d_{max} \qquad ...(14.31)$$

The unloaded length of the spring L_i for equal loops measured between the inside edges of the hooks is given by,

$$L_i = L_k + 2L_h \qquad ....(14.32)$$

where L_h = distance of inside edge of hook from body of spring

$$\approx D_i$$

Spring stiffness,

$$k = \frac{F - F_i}{\delta} = \frac{Gd^4}{8D_m^3\, i} \quad ...(14.33)$$

Spring deflection,

$$\delta = \frac{8(F - F_i)d_m^3\, i}{Gd^4} \quad ...(14.34)$$

Surging in extension spring,

$$f = \left[\frac{d}{\pi D_i^2}\right]\left[\frac{G}{8\rho}\right]^{\frac{1}{2}} \quad ....(14.35)$$

The initial tension is the load needed to overcome the force which presses against one another. Initial tension is introduced by coiling the springs so that the coils exert certain pressure against each other. The initial tension is only practicable for cold coiled extension springs which are not finally hardened. Hot coiled extension springs cannot be made with initial tension. For hot coiled springs the space between coils is approximately 0.5 to 5 mm corresponding to a permissible shear stress of 400 – 600 MPa. The permissible stresses are given in Table 14.9.

Table 14.9 Permissible stresses in extension springs

Type of steel wire	$\frac{S_{yt}}{S_{ut}}$	$\frac{S_{ys}}{S_{ut}}$
1. Cold drawn plain carbon steel	0.75	0.45
2. Hardened and tempered steel	0.75	0.50
3. Stainless steel	0.55	0.35

For hot coiled extension springs, τ = 600 MPa.

(*b*) Fatigue Load

The design of extension springs subjected to fatigue load may be carried out in the same way as for compression springs but with around 20–25% less stress. The life of extension springs is affected by the form of the loops or end fittings. It is advisable to avoid using extension springs under conditions of alternating load. When the use of an extension spring under alternating load is unavoidable the spring specified should be of the cold coiled type with tapered-in or screwed-in end fittings. Hot coiled extension springs should not be used under alternating loads.

Example 14.1. Design a close – coiled helical compression spring to sustain a load ranging from 250 N to 300 N. The axial deflection of the spring is limited to 8 mm. Assume a spring index of 8. The permissible shear stress in the spring wire is 420 MPa and modulus of rigidity is 84 GPa.

Solution.

Given : F_{min} = 250 N, F_{max} = 300 N, δ = 8mm, C = 8, τ_d = 420 MPa, G = 84 GPa

1. Wahl's correction factor,

$$K_w = \frac{4C - 1}{4C - 4} + \frac{0.615}{C}$$

$$= \frac{4\times 8-1}{4\times 8-4}+\frac{0.615}{8}=1.184$$

2. Wire diameter,

$$\tau_d = K_w\left[\frac{8F_{max}C}{\pi d^2}\right]$$

$$= 1.184\left[\frac{8\times 300\times 8}{\pi d^2}\right]$$

$$d = 4.15 \approx 4.25 \text{ mm}$$

3. Mean coil diameter,

$$D_m = 8\times 4.25 = 34 \text{ mm}$$

4. Stiffness of spring,

$$k = \frac{F_{max}-F_{min}}{\delta} = \frac{300-250}{8} = 6.25 \text{ N/mm}$$

5. Number of active turns,

$$i = \frac{Gd}{8C^3 k}$$

$$= \frac{84\times 10^3\times 4.25}{8\times 8^3\times 6.25}$$

$$= 13.945 \approx 14$$

6. Total number of turns for squared and ground ends,

$$i_t = i+2 = 14+2 = 16$$

7. Free length of spring,

$$l_f = l_s + \delta_{max} + \text{ clash allowance } (= 0.15\,\delta_{max})$$

$$= i_t d + \delta_{max} + 0.15\,\delta_{max}$$

$$= 16\times 4.25 + 1.15\times 8\times\frac{300}{50}$$

$$= 123.2 \approx 124 \text{ mm}$$

8. Pitch of coils,

$$p = \frac{l_f}{i_t-1} = \frac{124}{15} = 8.27 \approx 8.3\text{mm}$$

9. Check for buckling,

$$m = \frac{l_f}{D_m} = \frac{124}{34} = 3.647$$

> 2.6, provide guide.

Critical load for buckling, $F_{cr} = kK_l l_f$

$$K_l = 0.22 \text{ for hinged ends}$$

$$= 0.62 \text{ for fixed ends}$$

$$F_{cr} = 6.25 \times 0.22 \times 124 = 170.5 \text{ N for hinged ends} < F_{max}$$

$$= 6.25 \times 0.62 \times 124 = 480.5 \text{ N for fixed ends} > F_{max}$$

Therefore, the spring is safe in buckling for fixed ends.

10. Lowest natural frequency for both ends fixed,

$$f_n = \left[\frac{d}{\pi i D_m^2}\right]\left[\frac{G}{8\rho}\right]^{\frac{1}{2}} \text{ Hz}$$

$$= \left[\frac{4.25}{\pi \times 14 \times 34^2}\right]\left[\frac{84 \times 10^3}{8 \times 7800 \times 10^{-9}}\right]^{\frac{1}{2}}$$

$$= 3.067 \text{ Hz}$$

Example 14.2 A helical compression spring is subjected to an initial preload of 60 N and the maximum operating load during the load – cycle is 140 N. The wire diameter is 3 mm and the mean coil diameter is 18 mm. The spring is made of oil – hardened and tempered valve spring wire of grade VW having S_{ut} = 1430 MPa. Determine the factor of safety used in design.

Solution.

Given : F_{min} = 60 N, F_{max} = 140 N, d = 3 mm, D_m = 18 mm, S_{ut} = 1430 MPa

$$C = \frac{D_m}{d} = \frac{18}{3} = 6$$

$$K_w = \frac{4C-1}{4C-4} + \frac{0.615}{C}$$

$$= \frac{4 \times 6 - 1}{4 \times 6 - 4} + \frac{0.615}{6} = 1.2525$$

$$K_s = 1 + \frac{0.5}{C} = 1 + \frac{0.5}{6} = 1.083$$

$$F_m = \frac{F_{max} + F_{min}}{2} = \frac{140 + 60}{2} = 100 \text{ N}$$

$$F_a = \frac{F_{max} - F_{min}}{2} = \frac{140 - 60}{2} = 40 \text{ N}$$

$$\tau_m = K_s\left[\frac{8F_m C}{\pi d^2}\right] = 1.083\left[8 \times 100 \times \frac{6}{\pi \times 3^2}\right] = 183.86 \text{ MPa}$$

$$\tau_a = K_w\left[\frac{8F_a C}{\pi d^2}\right] = 1.2525\left[8 \times 40 \times \frac{6}{\pi \times 3^2}\right] = 85.05 \text{ MPa}$$

For oil – hardened and tempered steel,

$$S'_{es} = 0.22S_{ut} = 0.22 \times 1430 = 314.6 \text{ MPa}$$

$$S_{ys} = 0.45S_{ut} = 0.45 \times 1430 = 643.5 \text{ MPa}$$

Using Eq. (14.24), we have

$$\frac{\tau_m}{S_{ys}} + \left(\frac{\tau_a}{S'_{es}}\right)\left[2 - \frac{S'_{es}}{S_{ys}}\right] = \frac{1}{n}$$

$$\frac{183.86}{643.5} + \left(\frac{85.05}{314.6}\right)\left[2 - \frac{314.6}{643.5}\right] = \frac{1}{n}$$

$$n = 1.44$$

Example 14.3 An extension spring is assembled with a preload of 40 N. The wire diameter of spring is 3 mm with a spring index of 6. The spring has 15 active coils. The properties of spring wire are :

Allowable shear stress = 650 MPa

Modulus of rigidity = 84 kN/mm^2

Determine (*a*) initial torsional shear stress in wire; (*b*) spring stiffness; and (*c*) spring load to cause yielding.

Solution.

Given : F_i = 40 N, d = 3 mm, C = 6, i = 15, τ = 650 N/mm^2 , G = 84 kN/mm^2

(*a*) Wahl's stress facor,

$$K_w = \frac{4C-1}{4C-4} + \frac{0.615}{C}$$

$$= \frac{4 \times 6 - 1}{4 \times 6 - 4} + \frac{0.615}{6} = 1.2525$$

Initial torsional shear stress,

$$\tau_i = K_w\left[\frac{8F_iC}{\pi d^2}\right]$$

$$= 1.2525\left[8 \times 40 \times \frac{6}{\pi \times 3^2}\right] = 85.05 \text{ MPa}$$

(*b*) Spring stiffness,

$$k = \frac{Gd}{8C^3 i} = \frac{84 \times 10^3 \times 3}{8 \times 6^3 \times 15} = 9.72 \text{ N/mm}$$

(*c*) Spring load to cause yielding,

$$\tau = K_w\left[\frac{8FC}{\pi d^2}\right]$$

$$650 = 1.2525\left[8 \times F \times \frac{6}{\pi \times 3^2}\right]$$

$$F = 305.7 \text{ N}$$

Example 14.4 A helical compression spring is subjected to a varying load between 500 N to 1200 N. The spring index is 6 and a factor of safety of 1.5 is desired. The yield stress in shear is 760 MPa and ebdurance limit in shear is 350 MPa. Calculate (*a*) diameter of spring wire; (*b*) mean coil diameter; (*c*) number of active coils; and (*d*) free length of spring.

The compression of spring at the maximum load is 25 mm and modulus of rigidity is 82 KN/mm^2.

Solution.

Given : F_{min} = 500 N, F_{max} = 1200 N, C = 6, n = 1.5, S_{ys} = 760 MPa, S'_{es} = 350 MPa, δ = 25 mm,

$$G = 82 \text{ kN/mm}^2$$

$$K_w = \frac{4C - 1}{4C - 4} + \frac{0.615}{C}$$

$$= \frac{4 \times 6 - 1}{4 \times 6 - 4} + \frac{0.615}{6} = 1.2525$$

$$K_s = 1 + \frac{0.5}{C} = 1 + \frac{0.5}{6} = 1.083$$

$$F_m = \frac{F_{max} + F_{min}}{2} = \frac{1200 + 500}{2} = 850\text{N}$$

$$F_a = \frac{F_{max} - F_{min}}{2} = \frac{1200 - 500}{2} = 350\text{ N}$$

$$\tau_m = K_s\left[\frac{8F_m C}{\pi d^2}\right] = 1.083\left[8 \times 850 \times \frac{6}{\pi \times d^2}\right]$$

$$= \frac{14065}{d^2}\text{MPa}$$

$$\tau_a = K_w\left[\frac{8F_a C}{\pi d^2}\right] = 1.2525\left[8 \times 350 \times \frac{6}{\pi \times d^2}\right]$$

$$= \frac{6698}{d^2}\text{MPa}$$

(*a*) $$\frac{\tau_m - \tau_a}{S_{ys}} + \frac{2\tau_a}{S_{es}} = \frac{1}{n}$$

$$\left(\frac{1}{d^2}\right)\left[\frac{14065-6698}{760}+2\times\frac{6698}{350}\right] = \frac{1}{1.5}$$

$$d^2 = 71.95$$

$$d = 8.48 \approx 9 \text{ mm}$$

(*b*) $$D_m = Cd = 6 \times 9 = 54 \text{ mm}$$

(*c*) $$\delta = \frac{8F_{max}C^3 i}{Gd}$$

$$25 = \frac{8 \times 1200 \times 6^3 i}{82 \times 10^3 \times 9}$$

$$i = 8.89 \approx 9$$

(*d*) For squared and ground ends,

$$i_t = i + 2 = 9 + 2 = 11$$

Free length, $$l_f = i_t d + 1.15\delta = 11 \times 9 + 1.15 \times 25 = 1214.75 \text{ mm}$$

14.9 LEAF SPRINGS

These springs are made of a number of graduated overlapping flat plates. The thickness and width of the plates is kept constant. This gives a beam of uniform strength. The plates are assembled with a clamp. The springs can be semi-elliptical or quarter-elliptical. A semi-elliptical spring is shown in Fig. 14.8. The longest leaf is fastened to the supports. It is called the main or master leaf. To take up the shear at the ends, two or more leaves can be made of full length. Their ends are bent to form an eye. The centre bolt holds the leaves together. The U-clips (Fig. 14.9) reduces the bending stresses in the central part of the spring. The rebound clips serve to distribute some of the load to the shorter leaves which otherwise would be taken by the master leaf alone. Half the difference in length between two neighbouring leaves is called the overlap.

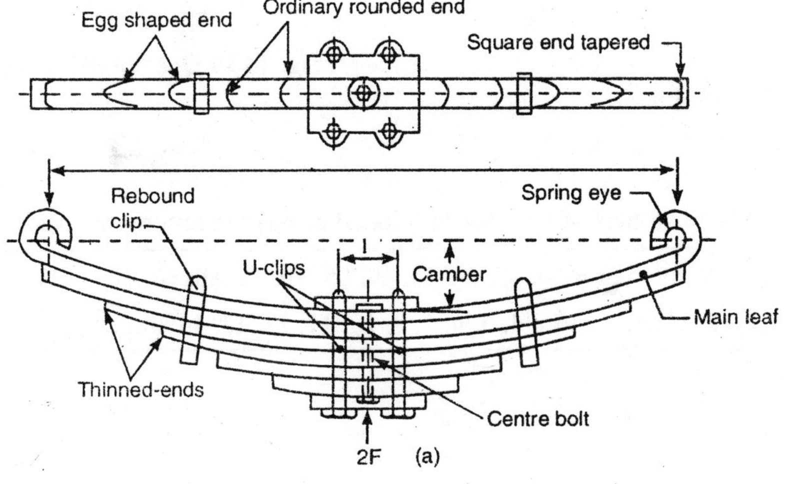

Fig. 14.8 Semi-elliptical leaf spring

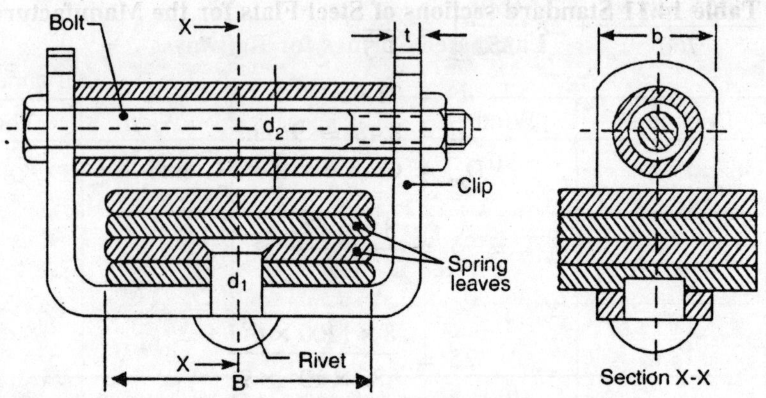

Fig. 14.9 U-clip

14.9.1 Quarter-Elliptic Spring

If a triangular plate is cut into strips of uniform width and placed one below the other, it becomes a quarter–elliptic spring as shown in Fig. 14.10. The bending stress in the spring is given by,

$$\sigma = \frac{6FL}{(nbh^2)} \quad ...(14.36)$$

And deflection,

$$\delta = \frac{6FL^3}{(nEbh^3)} \quad ...(14.37)$$

Fig. 14.10 Quarter-elliptic leaf spring

Table 14.10 Size of flats for laminated srings for automotive suspension.

Width, mm = 40, 45, 50, 55, 60, 65, 70, 75, 80, 90, 100, 120 and 150
Thickness, mm = 4, 5, 6, 7, 8, 10, 11, 12, 13, 14 and 16

Table 14.11 Standard sections of Steel Flats for the Manufacture of Laminated Springs for Railways

All dimensions in mm

Width	Thickness	Width	Thickness	Width	Thickness
50	10	90	6	115	10
	13		8		11
63	3		10		13
	8		11		16
	10		13		19
	11		16	120	16
	13		19		19
75	6	100	8		10
	8		10	130	13
	10		11		16
	11		13	140	11
	13		16		13
	16		19	150	11
					13
					16

IS : 3885 (Part 1)—1977

Table 14.12 Steel for manufacture of laminated springs (Railway rolling stock) – Rib and groove sections

Thickness, mm = 6, 8, 10, 11, 13, 14, 16 and 19		
Width, mm = 50, 63, 75, 90, 100, 115, 120, 125, 140 and 150		
Grade	Designation	Tensile strength, MPa
1	55C6	1180 – 1420 H & T
2	75C6	1180 – 1570 H & T
4	55Si7	1275 – 1471 H & T

Table 14.13 Physical Properties of Cold-rolled steel Strips for Springs

Material Designation grades		Annealed			Hardened and Tempered		
		Vickers Hardness HV max	Yield stress N/mm^2 min	Tensile strength N/mm^2 max	Vickers Hardness HV	Yield stress N/mm^2 min	Tensile strength N/mm^2
1	45C8	185					
2	55C6	195					
3	65C6	205	270	590	350—425	1030—1420	1180—1420
4	70C6	215					
5	75C6	220					
6	80C6	220	290	640	350—475	1180	1180—1570
7	85C6	220					
8	98C6	225	310	690	475—540	1470	1570—1760
9	55Si7	240	340	780	475—570	1470	1570—1960
10	50Cr4	240	340	780	505—650	1570	1670—2960
11	50Cr4V2	240	340	780	555—670	1760	1880—2960
Standar Thickness (mm) : 010, 0.15, 0.20, 0.25, 0.30, 0.40, 0.50, 0.60, 0.80, 1.00, 1.25, 1.50, 1.80, 2.00, 2.50, 3.00, 3.55, 4.00 and 5.00							

IS : 2507—1975

Table 14.14 Dimensions for centre bolts for leaf springs (All dimensions in mm)

Width of leaves	≤60	≤80	≤80	≥80≤100	≥90	≥90	≥90	≥90	≥90
Diameter of bolt	M8	M10	M12	M14 × 1.5	M16	M16 × 1.5	M_{18} × 1.5	M20 × 2	M24 × 2
Diameter of bolt head	12	15	17	20	20	22	25	28	34
Length of bolt head	10	10	10	11	11	11	12	14	18

Table 14.15 Recommended bore diameter of eye, mm

Eye bore diameter, mm = 18, 20, 22, 23, 25, 27, 28, 30, 32, 35, 38, 50, 55, 60

Table 14.16 Dimensions for U-clips, rivets and bolts (All dimensions in mm)

Spring width, B	Clip section, $b \times t$	Rivet dimaeter, d_1	Bolt diameter, d_2
Upto 50	20 × 4	6	6
50, 55 and 60	25 × 5	8	8
76, 70, 75 and 80	25 × 6	10	8
90, 100 and 125	32 × 6	12	10

Table 14.17 Factor of safety for laminated springs

Type of load/service	Factor of safety
Static load	1.5
Railroad service	2 – 2.25
Automobile suspension :	
Rear	2.25 – 3.5
Front	3.25 – 3.5
Dynamic load	1.25 – 1.5

14.9.2 Semi - Elliptic Leaf Spring

A semi – elliptic spring is shown in Fig. 14.8

1. Bending stress. The bending stress in a leaf spring is given by,

$$\sigma = \frac{3FL}{2nbh^2} \quad ...(14.38)$$

Where F = load on the spring

L = length of the longest leaf

b = width of the leaf

h = thickness of leaf

n = number of graduated leaves

2. Deflection. The deflection of the spring is given by,

$$\delta = \frac{3FL^3}{8nEbh^3} \quad ...(14.39)$$

where E = modulus of elasticity of spring material.

14.9.3 Graduated and Full Length Leaves

The bending moment at the ends of the semi – elliptical spring is zero. This shall require zero width of leaves at the ends. But there is maximum shear force at the ends. Therefore, to take up the shear force at the ends and to stiffen the leaves at the ends, two or more leaves are made of full length, as shown in Fig. 14.11.

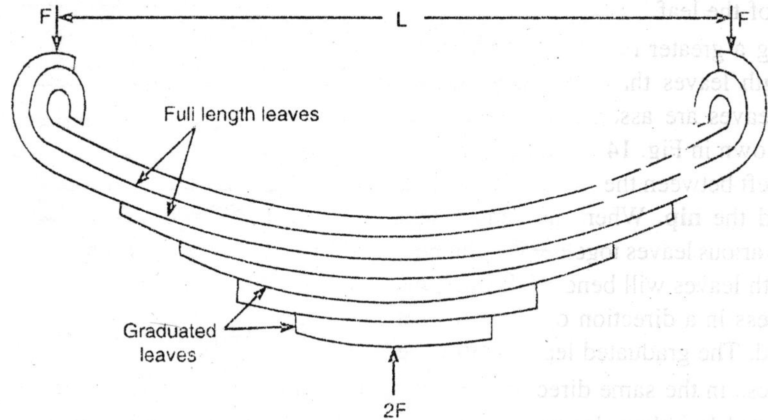

Fig. 14.11 Semi-elliptical spring having full length and graduated leaves

1. Load sharing between the leaves.

Let n_f = number of full length leaves

n_g = number of graduated leaves

Then load shared by the full length leaves,

$$F_f = \frac{3n_f F}{2n_g + 3n_f} \quad \text{....(14.40}a\text{)}$$

Load shared by the graudated leaves,

$$F_g = \frac{2n_g F}{2n_g + 3n_f} \quad \text{...(14.40}b\text{)}$$

2. Stresses in the leaves.

$$\sigma_g = \frac{12FL}{(2n_g + 3n_f)\, bh^2} \quad \text{...(14.41}a\text{)}$$

$$\sigma_f = \frac{18FL}{(2n_g + 3n_f)\, bh^2} = 1.5\sigma_g \quad \text{...(14.41}b\text{)}$$

3. Deflection.

$$\delta_g = \delta_f = \frac{12FL^3}{(2n_g + 3n_f)\, Ebh^3} \quad \text{...(14.42)}$$

4. Initial gap or nip in the leaves.

To equalize the stresses in the full length and graduated leaves for better utilization of the material, the following two ways may be adopted :

Full length leaves may be made of smaller thickness than the graduated leaves. In this way, the full length leaves will induce smaller bending stress due to smaller distance from the neutral axis to the edge of the leaf.

By giving a greater radius of curvature to the full length leaves than graduated leaves, before the leaves are assembled to form the spring, as shown in Fig. 14.12. By doing so, a gap will be left between the leaves. This initial gap is called the **nip.** When the central bolt holding the various leaves together is tightened, the full length leaves will bend back and have an initial stress in a direction opposite to that of under load. The graduated leaves will have an initial stress in the same direction as that of under load. Consequently, the full length leaves will be stressed less than the graduated leaves. The initial gap can be adjusted to equalize the stresses in full length and graduated leaves.

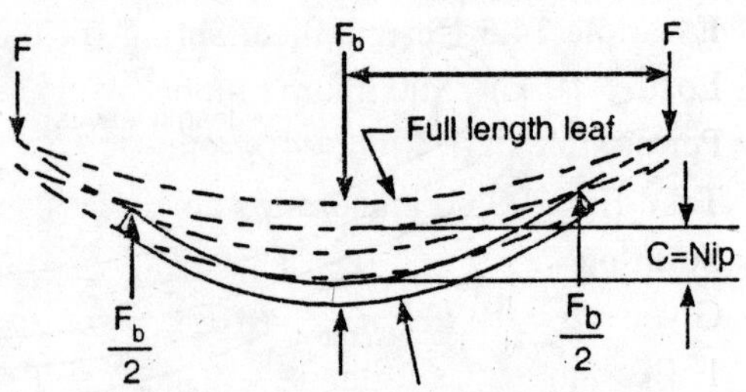

Fig. 14.12 Nip in a leaf spring

$$\text{Initial gap,} \quad c = \frac{2FL^3}{nEbh^3} \qquad ...(14.43)$$

5. Bolt load.

Initial bolt load required to close the gap,

$$F_b = \frac{2n_g n_f F}{n(2n_g + 3n_f)} \qquad ...(14.44)$$

6. Maximum stress in spring.

$$\sigma_f = \sigma_g = \frac{6FL}{nbh^2} \qquad ...(14.45)$$

7. Camber = maximum deflection

14.9.4 Length of leaves

Let L = span or overall length of the spring

l = width of bend or distance between centres of U-bolts.

Effective length of the spring, $L_e = L - l$, when band is used

$= L - \frac{2l}{3}$, when U-bolts are used

If a leaf spring has two full length leaves, then the length of leaves may be obtained as follows :

$$\text{Length of smallest leaf} = \frac{L_e}{n-1} + 1$$

where $n = n_f + n_g$

$$\text{Length of next leaf} = \frac{2L_e}{n-1} + 1$$

$$\text{Length of } (n-1)\text{th leaf} = \frac{n - 1L_e}{n-1} + 1$$

Length of master leaf = $L + 2\pi (d + h)$

where d = inside diameter of eye

h = thickness of master leaf

Example 14.5 Design a leaf spring for the following specifications :

Load = 10 kN, Maximum number of leaves = 10, Span of spring = 0.6 m

Permissible deflection = 100 mm.

Take E = 200 GPa and allowable stress for spring material = 600 MPa.

Solution.

Given : F = 10 kN, n_{max} = 10, L = 0.6 m, δ = 100 mm, E = 200 GPa, σ = 600 MPa

1. Bending stress in the spring,

$$\sigma = \frac{6FL}{nbh^2}$$

$$600 = 6 \times 10,000 \times \frac{600}{n \times b \times h^2}$$

$$nbh^2 = 60,000 \quad \text{...(1)}$$

2. Deflection of spring,

$$\delta = \frac{6FL^3}{nEbh^3}$$

$$100 = \frac{6 \times 10,000 \times 600^3}{n \times 200 \times 10^3 \times bh^3}$$

$$nbh^3 = 648000 \quad \text{...(2)}$$

Dividing Eq. (2) by (1), we get

$$h = 10.8 \approx 11 \text{ mm}$$

From Eq. (1), we have

$$b = \frac{60,000}{10 \times 11^2} = 49.6 \text{ mm}$$

From Eq. (2), we have

$$b = 48.68 \text{ mm}$$

We adopt b = 63 mm

Example 14.6 A laminated spring of an automobile is to carry a load of 5 kN. The spring is to be made 1000 mm between the supports. Design and sketch the spring. The number of leaves may be taken as 8 with two of them of full length.

Solution.

Given : 2F = 5 kN, F = 2500 N, $2L_1$ = 1000 mm, n = 8, n_f = 2, n_g = 6

1. Preliminary decisions.

We select hardened and tempered leaf spring steel, giving

$$\sigma_{ut} = 1820 - 2060 \text{ MPa} = 1940 \text{ MPa, and E} = 195 \text{ GPa}$$

Taking a FOS = 3,

Endurance limit,

$$\sigma_e = 0.5 \times 1940 = 970 \text{ MPa}$$

So that

$$\sigma_d = \frac{970}{3} \approx 325 \text{ MPa}$$

2. Effective length of spring.

Let distance between U-bolts = 60 mm

$$2L = 2L_1 - \left(\frac{2}{3}\right) \times 60$$

$$= 1000 - 40 = 960$$

$$L = 480 \text{ mm}$$

3. Final stress,
$$\sigma = \frac{6FL}{ibh^2}$$

$$= \frac{6 \times 2500 \times 480}{8 \times bh^2}$$

$$= 325$$

$$bh^2 = 2769.23 \text{ mm}^3$$

Let h = 6 mm, then b = 76.92 ≈ 75 mm

Actual stress,
$$\sigma_a = \frac{6 \times 2500 \times 480}{8 \times 75 \times 6^2} = 333.3 \text{ MPa}$$

4. Bolt load,

$$F_b = \frac{2i_g i_f F}{i(2n_g + 3n_f)}$$

$$= \frac{2 \times 6 \times 2 \times 2500}{8\,(2 \times 6 + 3 \times 2)}$$

$$= 416.7 \text{ N}$$

5. Length of leaves.

To determine the length of individual leaves, one of the full length leaves is treated as part of graduated leaves. The leaf spring is shown in Fig. 14.13.

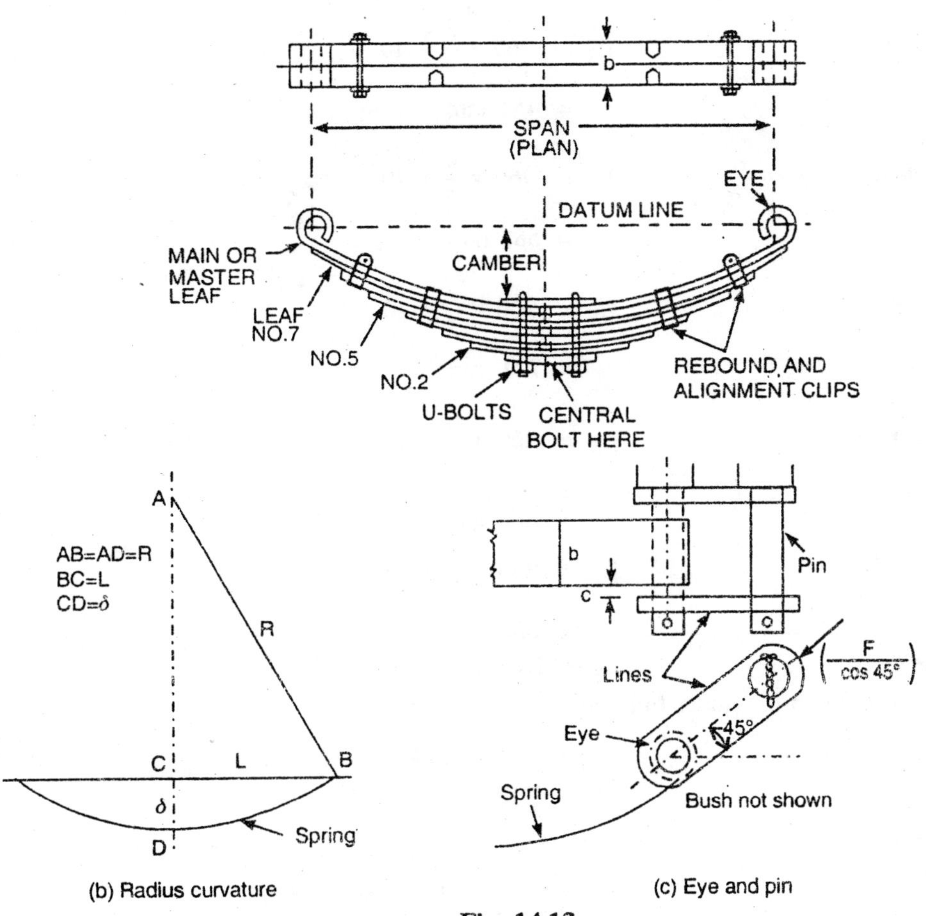

(b) Radius curvature

(c) Eye and pin

Fig. 14.13

Ineffective length,

$$l = \left(\frac{2}{3}\right) \times \text{distance between bolt centres}$$

$$= 40 \text{ mm}$$

Length of smallest leaf,

$$L_1 = \frac{\text{Effective length}}{i - 1} + l$$

$$= \frac{960}{8 - 1} + 40$$

$$= 177 \text{ mm}$$

Second leaf, $L_2 = 960 \times \frac{2}{7} + 40$

$= 314$ mm

Third leaf, $L_3 = 960 \times \frac{3}{7} + 40$

$= 451$ mm

Fourth leaf, $L_4 = 960 \times \frac{4}{7} + 40$

$= 588$ mm

Fifth leaf, $L_5 = 960 \times \frac{5}{7} + 40$

$= 721$ mm

Sixth leaf, $L_6 = 960 \times \frac{6}{7} + 40$

$= 863$ mm

Seventh leaf $L_7 = 960 \times \frac{7}{7} + 40$

$= 1000$ mm

Eight leaf, $L_8 = L_7 = 1000$ mm

6. Deflection of spring under full load,

$$\delta = \frac{12FL^3}{Ebh^3\left(2n_g + 3n_f\right)}$$

$$= \frac{12 \times 2500 \times 480^3}{195 \times 10^3 \times 75 \times 6^3 (2 \times 6 + 3 \times 2}$$

$= 58.35$ mm

14. Radius of curvature

Taking camber = maximum deflection

$$R^2 = (R - \delta)^2 + L_1^2$$

$$= (R - 58.35)^2 + 500^2$$

$$= R^2 - 116.7R + 250116.7$$

$$R = 2143.24 \text{ mm}$$

8. Eye and pin.

Load on the pin, $F_p = \frac{F}{\cos 45^\circ}$

$$= \frac{2500}{\cos 45^\circ}$$

$$= 3674 \text{ N}$$

Considering bearing of the pin, assuming bearing pressure, $p_b = 10$ MPa, we have

$$b d_p p_b = F_b$$

$$d_p = \frac{3674}{75 \times 10}$$

$$= 4.89 \approx 6 \text{ mm}$$

Length of pin, $l_p = b + 2 \times \text{clearance}$

$$= 75 + 2 \times 2 = 79 \text{ mm}$$

Check of pin in bending :

Maximum bending moment on pin

$$M = \frac{F_p b}{4} = \frac{3674 \times 75}{4}$$

$$= 688814.5 \text{ N. mm}$$

$$= \left(\frac{\pi}{32}\right) d_p^3 \sigma_p$$

$$d_p = \left[\frac{688814.5 \times 32}{\pi \times 100}\right]^{1/3}$$

$$= 19.15 \approx 20 \text{ mm} > 6 \text{ mm}$$

We adopt, $d_p = 20$mm

9. Eye diameter,

Let bush thickness = 3 mm

Eye diameter = $d_p + 2 \times 3$

$= 26$ mm ≈ 27 mm

10. Central bolt,

Diameter of central bolt for $b > 65$ mm, $d_b = 16$ mm

$$\text{Stress in the central bolt} = \frac{F_b}{A_b}$$

$$= \frac{4 \times 416.7}{\pi \times 16^2}$$

$= 2.07$ MPa, which is quite safe.

11. Size of clips, rivets and bolt

For $b = 75$ mm, clip section = 25 mm × 6 mm

Rivet diameter = 10 mm

Diameter of bolt = 8 mm

EXERCISES

14.1 Design a spring for a boiler safety valve of 60 mm diameter required to blow off at a stream pressure of 1.25 MPa. The initial compression of spring is 40 mm and lift is limited to 20 mm. The maximum shear stress in the material of wire is limited to 600 MPa. The modulus of rigidity for spring material is 85 kN/mm^2. The spring index = 6.

14.2 Give the design procedure for the design of compressive spring of an I.C. Engine. Indicate the procedure for checking of this spring for buckling.

14.3 Give, in detail, the design procedure for the design of a leaf spring of an automobile.

14.4 Figure 14.14 given below shows a combined spring and dead weight loaded valve. The valve diameter is 30 mm and it weights 10 N. The weight of the lever is 25 N. The spring (free length = 100 mm) is stretched up to 120 mm when loaded. Calculate the dead load P needed for the valve to just open at a pressure 3.5 N/mm^2 on the valve. A 2 mm lift of the valve is needed for escape of gases. Assume spring index 6, permissible shear stress for spring wire 500N/mm^2. G= 0.84 × 10^5 N/mm^2 . Design the spring.

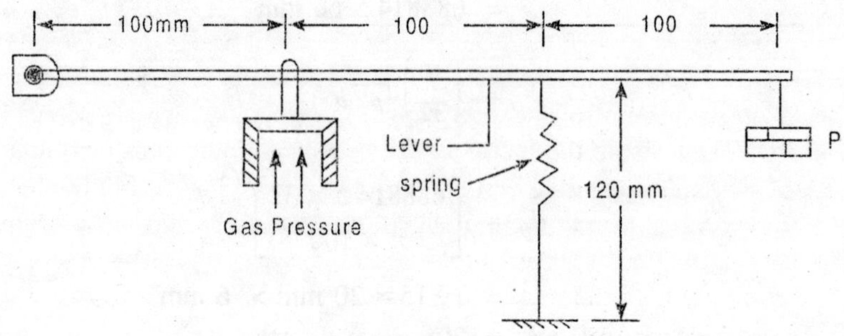

Fig. 14.14

14.5 Design a spring for a balance to measure 0 to 1000 N over a scale of length 80 mm. The spring is to be enclosed in a casing of 25 mm diameter. The approximate number of turns is 30. The modulus of rigidity is 86 kN/mm^2. Calculate the maximum shear stress induced.

14.6 (*i*) What do you understand by A.M. Wahl's factor and state its importance in design of Springs.

(*ii*) What is nipping in leaf springs. Discuss its role.

14.7 (*a*) A helical compression spring is made of oil-tempered steel and is subjected to a load varying from 600 N to 1000 N. The spring index is 6 and the factor of safety is 1.5. If the yield shear stress is 700 N/mm^2 and endurance stress is 350 N/mm^2, find the wire diameter of the spring.

(*b*) Differentiate between full length and gaduated leaves of a spring, mentioning the advantage of heaving graduated leaves. What is nipping in a leaf spring ? What is its role ?

14.8 (*i*) Explain surging in springs.

(*ii*) Design a helical spring for spring loaded safety valve for the following conditions

Operating pressure = 1 N/mm^2

Maximum pressure when the valve blows off freely = 1.075 N/mm^2

Maximum lift of the valve when the pressure is 1.075 N/mm^2 = 6 mm

Diameter of the valve seat = 100 mm
maximum shear stress = 400 N/mm^2
Modulus of rigidity = 80 × 10^3 N/mm^2
Spring index = 5.5

14.9. A helical spring is made from a wire of 6 mm diameter and has outside diameter of 75 mm. If the permissible shear stress is 350 N/mm^2 and modulus of rigidity is 84 *k*N/mm^2, find the axial load which the spring can carry and defection per active turn;

(*i*) neglecting the effect of curvature, and

(*ii*) considering the effect of curvature.

14.10 A semi-elliptical laminated vehicle spring is to be designed to carry a load of 3 *k*N and consists of two full length leaves and five graduated leaves of width 65 mm. The spring is to be 1100 mm in length and attached to the axle by two U-bolts 80 mm apart. These bolts hold the central position of the spring so rigidity that they may be considered equivalent to a band having a width equal to the distance between the bolts. If allowable stress is 350 MPa and modulus of elasticity is 210 GPa, determine :

(*i*) thickness of leaves

(*ii*) defection of spring

(*iii*) length of leaves.

14.11 A helical compression spring is subjected to an initial preload of 60 N and the maximum load during the load cycle is 320 N. The wire diameter is 6 mm and the spring index is 5. The spring is made of oil-hardened and tempered steel wire of Grade SW having S_{ut} = 1440 MPa. Determine the factor of safety from fatigue considerations.

14.12 A semi-elliptic spring consists of two extra full-length leaves and five graduated leaves. Each leaf is 6 mm thick and 50 mm wide. The centre-to-centre distance between the two eyes is 1.0 m. The leaves are prestressed in such a way that when the load is maximum, stresses induced in all the leaves are equal to 350 MPa. Determine the maximum load that the spring can withstand.

14.13 A safety valve of 50 mm diamter is to blow off at a pressure of 1.2 MPa. It is held on its seat by a close-coiled helical spring. The maximum lift of the valve is 10 mm. Determine the main diamensions of a compression spring of spring index 5.5. The initial compression of the spring is 30 mm. The maximum shear stress in the spring wire is limited to 500 MPa.

14.14 A compression spring is required to exert minimum force of 300 N and a maximum force of 650 N with a deflection of 20 mm. The spring is to fit in a hole of 40 mm diameter. Design the spring if the load is to be static.

14.15 A cylinder relief valve is to operate under varying load from 250 N to 600 N. The spring index is to be 6. Using cold drawn spring steel wire, calculate the diameter of wire for infinite life. A factor of safety of 1.5 is desired.

14.16 Design a leaf spring for a truck with the following specifications :

Maximum load on springs = 150 *k*N
Number of springs = 4
Permissible bending stress in springs = 600 MPa
Maximum number of leaves = 10
Spring span = 1 m
Permissible defection = 80 mm

Assume missing date suitably.

15 FLAT AND V-BELTS

Chapter

15.1 INTRODUCTION

Power can be transmitted either by flexible or non – flexible type of machine elements. In **flexible** elements, the centre distance can be changed conveniently and is **non-flexible** type of machine elements, the centre distance cannot be changed. The flexible type of elements are : belts, chains, and ropes. The non – flexible type of elements are : gears, clutches, couplings, and power screws. Flexible elements are used for power transmission over long distances and non flexible type over short distances.

15.2 FACTORS FOR SELECTION OF BELT DRIVE

The following factors must be considered while selecting a belt drive.

1. Speed of of driving and driven shaft.
2. Speed reduction ratio.
3. Power to be transmitted.
4. Centre distance of shafts.
5. Space available.
6. Layout of shaft.

15.3 *Factors on which Coefficient of Friction Depends*

The coefficient of friction between belt and pulley depends upon the following factors :

1. Belt material.
2. Pully material.
3. Belt slip.
4. Belt speed.

15.4 DESIGN OF FLAT BELTS

Flat belts can be used for horizontal, vertical, quarter – turn, right – angled, crossed and reversed drive. A horizontal open type flat belt drive is shown in Fig. 15.1. Flat belt drives are classified as :

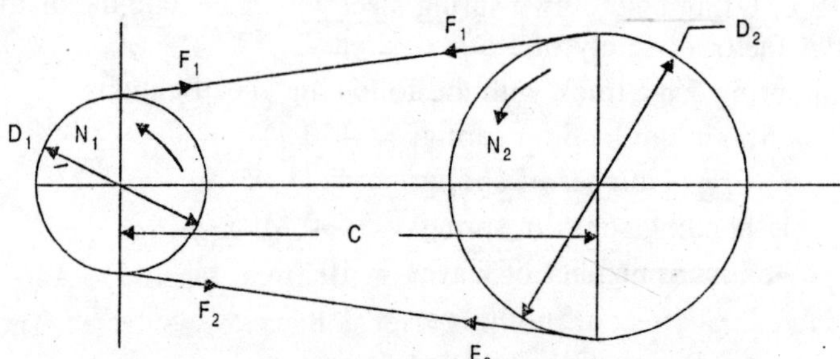

Fig. 15.1 Open flat belt drive

Light : $v \leq 12$ m/s and upto 7.5 kW
Medium : $v = 12 - 24$ m/s and 7.5 – 15 kW
Heavy : $v > 24$ m/s and above 15 kW

1. Speed ratio :

Neglecting slip and thickness of belt, the speed ratio.

$$i = \frac{N_2}{N_1} = \frac{D_1}{D_2} \quad \text{... (15.1}a\text{)}$$

where N_1, N_2 = rpm of smaller and bigger pulley respectively.
D_1, D_2 = diameters of smaller and bigger pulley respectively, mm.

Considering slip and thickness of belt,

Speed ratio,
$$i = \left(1 - \frac{s}{100}\right)\left[\frac{D_1 + t}{D_2 + t}\right] \quad \text{... (15.1}b\text{)}$$

where t = thickness of belt.
s = total percentage slip on the pulleys
$= s_1 + s_2 - 0.01 s_1 s_2$
s_1, s_2 = % slip on smaller and bigger pulley respectively

2. Angle of arc of contact

For open belt drive,

angle of arc of contact,
$$\theta = \pi \pm 2 \sin^{-1}\left[\frac{D_2 - D_1}{2C}\right] \text{rad} \quad \text{... (15.2}a\text{)}$$

where C = centre distance

Use + ve sign for the bigger pulley and – sign for the smaller pulley.

For crossed belft drive (Fig. 15.2),

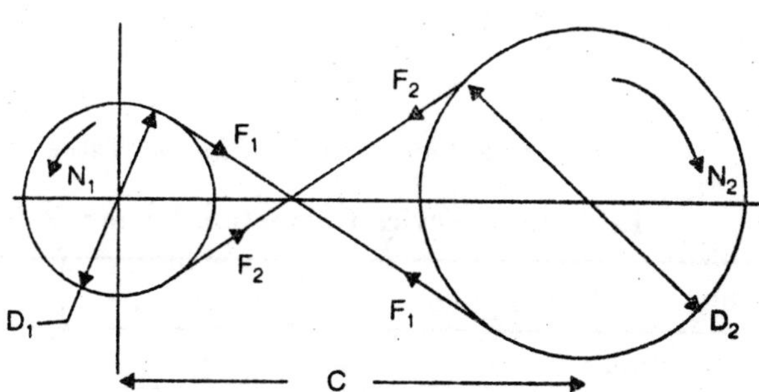

Fig. 15.2 Crossed flat belt drive

$$\theta = \pi \pm 2 \sin^{-1}\left[\frac{D_2 + D_1}{2C}\right] \text{rad} \quad \text{... (15.2}b\text{)}$$

3. Belt tension and power transmitted neglecting centrifugal tension

At low speeds ($v < 12$ m/s), the effect of centrifugal tension may be neglected. The ratio of belt tensions is,

$$\frac{F_1}{F_2} = \exp(\mu\theta) \quad \text{... (15.3)}$$

where F_1, F_2 = belt tension on the tight and slack side respectively

μ = coefficient of friction between belt and pulley (Table 15.1)

Table 15.1 Coefficient of Friction for Belts

Belt material	Pulley material						
	Cast iron, steel			Wood	Compressed paper	Leather face	Rubber face
	Dry	Wet	Greasy				
Leather, Oak tanned	0.25	0.20	0.15	0.30	0.33	0.38	0.40
Leather, chrome tanned	0.35	0.32	0.22	0.40	0.45	0.48	0.50
Canvas, stitched	0.20	0.15	0.12	0.23	0.25	0.27	0.30
Cotton, woven	0.22	0.15	0.12	0.25	0.28	0.27	030
Camel hair, woven	0.35	0.25	0.20	0.40	0.45	0.45	0.45
rubber	0.30	0.18	—	0.32	0.35	0.40	0.42
Balata	0.32	0.20	—	0.35	0.38	0.40	0.42

For tanned leather belts on C.I. pulleys,

$$u = 0.54 - \frac{0.712}{v + 2.542} \quad v \text{ in m/s} \quad \text{... (15.4)}$$

Peripheral speed of belt, $v = \dfrac{\pi D_1 N_1}{60 \times 10^3}$ m/s

Corrected power, $P_c = \dfrac{P \times K_s}{K_j}$ ... (15.5)

where K_s = service correction factor (Table 15.2)

Table 15.2 Service Correction Factor, K_s

Service condition	K_s
(*a*) Atmospheric condition :	
Clean, scheduled maintenance on large drives	1.2
Normal	1.0
Oily, wet or dusty	0.7
(*b*) Angle of centre line :	
Horizontal to 60° from horizontal	1.0
60° to 75° from horizontal	0.9
75° to 90° from horizontal	0.8

Service condition	K_s
(*c*) Pulley material :	
Fibre on motor and small pulleys	1.2
Cast iron on steel	1.0
(*d*) Service :	
Temporary or infrequent	1.2
Normal	1.0
Important or continuous	0.8
(*e*) Peak loads :	
Steady belt loads as obtained with steam engines, turbines, diesel and multi-cylinder gas engines, fans, centrifugal pumps, and steady line shaft loads.	1.0
Jerky belt loads, as obtained with large induction motors ; compensator started, shunt wound, direct current motors ; single cylinder gas engines ; reciprocating machines ; and machines developing series of peak loads, such as compressors, rock crushers and punch presses etc.	0.8
Shock and reversing belt loads on all motors 10 hp. and under, all cross-the-line start motors, wound-rotor (slip-ring) motors, synchronous motors, and reversing loads such as printing presses, elevator service and laundary washers etc.	0.6

K_j = correction factor for belt joint (Table 15.3)

Table 15.3 Correction Factors K_j for Belt Joints

Type of Joint	K_j
Cemented by belt maker	1.00
Cemented	0.98
Wire laced by machine	0.90
Wire laced by hand	0.82
Rawhide faced	0.60
Metal belt hoops	0.35

Power transmitted,

$$P_c = \frac{Fv}{10^3} \text{ kW} \quad \text{... (15.6)}$$

$F = F_1 - F_2$, tangential force in N

Power transmitted per square mm of belt cross – section,

$$P_c = \left[\frac{\sigma_1 v}{10^3}\right]\left[1-\exp(-\mu\theta)\right] \quad \text{... (15.7)}$$

where $\sigma_1 = \frac{F_1}{bt}$ is the allowable stress in the belt material

b = belt width, mm

4. Belt tensions and power transmitted considering centrifugal tension

At high speeds ($v > 12$ m/s), the effect of centrifugal tension must be considered. The ratio of belt tension is,

$$\frac{F_1 - F_c}{F_2 - F_c} = \exp(\mu\theta) \quad \text{... (15.8)}$$

where F_c = centrifugal tension per unit belt length

$= \rho btv^2$

ρ = density of belt material, kg/m³ (Table 15.4)

Power transmitted,

$$P_c = \left[\frac{(F_1 - F_c)v}{10^3}\right]\left[1 - \exp(-\mu\theta)\right] \text{ kW} \quad \text{... (15.9)}$$

Table 15.4 Properties of belt Materials

Material	*Design strength N/mm²*	*Endurance strength N/mm²*	*Elastic modulus N/mm²*	*Density kg/m³*	*Maximum velocity*
Rubber	1.6	6.0	10.0	$1.1 - 1.2 \times 10^{-3}$	30
Leather	2.0	6.0	30.0	$1.0 - 1.1 \times 10^{-3}$	30 – 50
Fabric	1.5	3.0	15.0	$0.9 - 1.0 \times 10^{-3}$	20 – 25
Plastic	4.0	6.0	60.0	$1.0 - 1.1 \times 10^{-3}$	60

5. Condition for maximum power transmission

$$v = \left[\frac{\sigma_d}{3\rho}\right]^{\frac{1}{2}} \quad \text{... (15.10)}$$

where σ_d = design stress for the belt material per unit width of belt (Table 15.4, 15.6, 15.8 and 15.10)

6. Initial tension

$$F_i = \frac{F_1 + F_2}{2} \quad \text{... (15.11)}$$

7. Wdith of belt

Neglecting centrifugal tension,

Width of belt,

$$b = \frac{F_1}{\sigma_d} \quad \text{... (15.12)}$$

Considering centrifugal tension,

$$bt = \left[\frac{10^3 P_c}{v(\sigma_d - \rho v^2)}\right]\left[\frac{1}{1 - \exp(-\mu\theta)}\right] \quad \text{... (15.13)}$$

For standard width of belt, refer to Table 15.5, 15.7, 15.8 and 15.9)

Table 15.5 Widths of Friction Surface Rubber Transmission Belting

Nominal belt width, mm	*Tolerance, mm*
25,32, 40, 50, 63	± 2.0
75,80, 90, 100, 112, 125	± 3.0
140, 160, 180, 200, 224, 250	± 4.0
280, 315, 355, 400, 450, 500	± 5.0

Table 15.6 Tensile Strength of Solid Woven Impregnated Cotton* and Hair Belting**

Type of belt	*Minimum tensile strength, N/mm²*
cotton	51.68
Hair	
Grade 1	37.04
Grade 2	28.86

* IS : 529 – 1976 ** IS : 530–1976

Table 15.7 Dimensions for Solid Woven Impregnated Cotton* and Hair Belting for Power Transmission**

Pulley width mm	*Nominal belt width mm*	*Tolerance on width mm*	*Nominal belt thickness, mm*				
			5 + 0.5 – 0.0	*6.5 + 0.5 – 0.3*	*8 + 0.6 – 0.3*	*10 + 0.7 – 0.5*	*12 + 0.8 – 0.6*
32	25		*x*	—	—	—	—
40	32		*y*	—	—	—	—
50	40		*x*	—	—	—	—
63	50	± 1.6	*x*	*x*	—	—	—
71	63		*x*	*x*	—	—	—
80	71		—	*y*	—	—	—
90	80		—	*y*	—	—	—
100	90		—	*y*	*y*	—	—
112	100		—	*x*	*x*	—	—
125	112		—	*x*	*x*	—	—
140	125		—	*x*	*x*	*x*	—
160	140		—	*y*	*y*	*y*	—
180	160		—	—	*y*	*y*	—
200	180	± 2.4	—	—	*y*	*y*	—
224	200		—	—	*x*	*x*	—
250	224		—	—	*x*	*x*	—

Pulley width	*Nominal belt*	*Tolerance*	*Nominal belt thickness, mm*				
mm	*width mm*	*on width mm*	*5 + 0.5 − 0.0*	*6.5 + 0.5 − 0.3*	*8 + 0.6 − 0.3*	*10 + 0.7 − 0.5*	*12 + 0.8 − 0.6*
280	250		—	—	–	*x*	*y*
315	280		—	—	—	*y*	*y*
355	315	± 3.2	—	—	—	*y*	*y*
400	355		—	—	—	*x*	*x*
450	400		—	—	—	*x*	*x*
500	450		—	—	—	—	*x*
560	500		—	—	—	—	*x*
675	600		—	—	—	—	*z*

x = preferred size, *y* = permissible size, *z* = discourage size.

* IS : 529 – 1976, **IS : 530 – 1976

Table 15.8 Specification for Vegetable Tanned Leather Belting

Type of belt	*Width of belt mm*	*Length of splice mm*	*Tensile strength N/mm²*
Single	Upto and including 25	110	24.05
	Over 25 upto and including 40	115	
	" 40 " " " 50	125	
	" 50 " " " 71	140	
	" 71 " " " 80	150	
	" 90 " " " 140	180	
	" 160 and above 200		
Double	Upto and including 125	100	
	Over 125	125	24.05
Splices (single and double) —			20.20

IS : 2240 – 1976

Table 15.9 Leather Belt Data

Thickness in mm

Grade of belt	Single ply pulley dia upto 200 mm	Double ply pulley dia upto 300 mm	Triple ply pulley dia upto 600 mm	Quadruple pulley dia above 600 mm	Width mm	Increment mm
Light	3	6	—	—	12 to 24	3
					24 to 102	6
					102 to 198	12
Medium	4	8	12.5	17.5	200 to 800	25
					800 to 1400	50
Heavy	5	10	15	20	1500 to 2100	100

Table 15.10 Weight and Tensile Strength of Rubber Transmission Belting

Type of duck		Nominal mass g/m^2	Tensile strength, N/mm of width	
			Warp	Weft
Soft	28	810	60.11	28.83
	32	915	67.37	31.77
Hard	31	910	60.11	33.73
	34	970	67.37	43.25

IS : 1370 – 1976

Table 15.11 Rating in Kilowatts per mm Width of Rubber Transmission Belting at 180° arc of Contact

Type of duck		Mass of fabric g/m^2	Pulley dia mm	Belt speed m/s	Rating per mm width per ply in kW
Soft	28	810	250	5	0.012
	32	915	250	5	0.015
Hard	31	910	250	5	0.012
	34	970	250	5	0.015

IS : 1370 – 1976

Table 15.12 Thickness of Friction Surface : Rubber Transmission Belting

Ply construction	Nominal belt thickness, mm	Tolerance, mm
3 ply	3.9	± 0.5
4 ply	5.1	± 0.7
5 ply	6.4	± 0.8
6 ply	7.7	± 0.9
7 ply	9.1	± 1.0
8 ply	10.4	± 1.1
9 ply	11.7	± 1.2
10 ply	13.0	± 1.3

IS : 1370-1976

8. Maximum stress in the belt

$$\sigma_{max} = \text{stress due to maximum belt tension + centrifugal stress + bending stress while turning over the pylleys}$$

$$= \frac{F_1}{bt} + \rho v^2 + \frac{Et}{D_1} \qquad \text{... (15.14)}$$

where E = modulus of elasticity of belt material

9. Length of belt

The Length of open belt is,

$$L_o = \left(\frac{\pi}{2}\right)(D_1 + D_2) + 2C + \frac{(D_2 - D_1)^2}{4C} \qquad \text{... (15.15)}$$

The length of crossed – belt is,

$$L_c = \left(\frac{\pi}{2}\right)(D_1 + D_2) + 2C + \frac{(D_2 + D_1)^2}{4C} \qquad \text{... (15.16)}$$

10. Centre distance

$$C = \frac{L_o}{4} - 0.3925(D_1 + D_2)$$

$$+\left[\left\{\frac{L_o}{4} - 0.3925(D_2 - D_1)\right\}^2 - \frac{(D_2 - D_1)^2}{8}\right]^{\frac{1}{2}} \qquad \text{... (15.17)}$$

15.5 PULLEYS FOR FLAT BELTS

(*a*) Cast Iron Pulley

A cast iron pulley is shown in Fig. 15.3

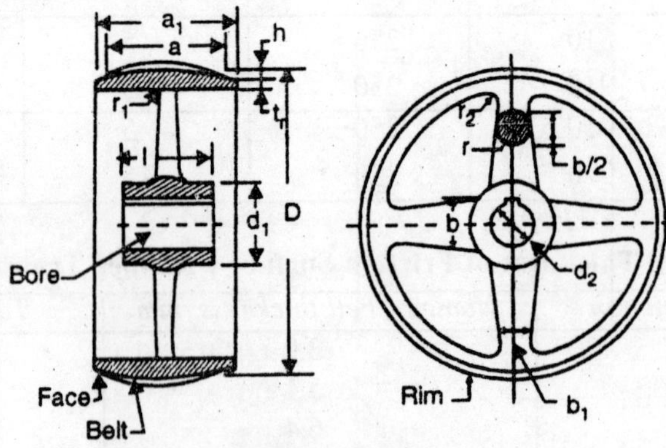

Fig.15.3 Cast iron flat belt pulley

Material : Cast iron of Grade FG 200 of IS : 210 – 1978

For nominal diameter of pulley, refer to Table 15.13, 15.14.

Table 15.13 Crown of cast iron and mild steel flat pulleys of diameters from 400 mm to 2000 mm

Nominal dia D mm	*Crown h of pulleys of width*						
	125 and smaller	*140 and 160*	*180 and 200*	*224 and 250*	*280 and 315*	*355*	*400 and larger*
400	1	1.2	1.2	1.2	1.2	1.2	1.2
450	1	1.2	1.2	1.2	1.2	1.2	1.2
500	1	1.5	1.5	1.5	1.5	1.5	1.5
560	1	1.5.	1.5.	1.5	1.5	1.5.	1.5
630	1	1.5	2	2	2	2	2
710	1	1.5	2	2	2	2	2
800	1	1.5	2	2.5	2.5	2.5	2.5
900	1	1.5	2.	2.5	2.5	2.5	2.5
1000	1	1.5	2	2.5	3	3	3
1120	1.2	1.5	2	2.5	3	3	3.5
1250	1.2	1.5	2	2.5	3	3.5	4
1400	1.5	2	2.5	3	3.5	4	4
1600	1.5	2	2.5	3	3.5	4	5
1800	2	2.5	3	3.5	4	4.5	5
2000	2	2.5	3	3.5	4	4.5	6

All dimension in mm **IS : 1691—1980**

Table 15.14 Width and diameters of flat cast iron and mild steel pulleys

Width, mm	*Tolerance, mm*	*Nominal dia, mm*	*Tolerance, mm*
20, 25, 32, 40, 50, 63, 71	± 1	40	± 0.5
		45, 50	± 0.6
80, 90, 100 112, 121, 140	± 1.5	56, 63	± 0.8
		71,80	± 1.0
160, 180, 200 224, 250, 280, 315	± 2	90, 100, 112	± 1.2
		125, 140	± 1.9
355, 490, 450 500, 560, 630	± 3	160, 180, 200	± 2.0
		224, 250	± 2.5
Material of C.I. Pulley		280, 315, 355	± 3.0
Grade FG 200 of IS : 210—1978		400, 450, 500	± 4.0
		560, 630, 710	± 5.0
		800, 900, 1000	± 6.3
Material for M.S. pulley :		1120, 1250, 1400	± 8.0
IS : 226–1975		1600, 1800, 2000	± 10.0

IS : 1961—1980

Table 15.15 Width of flat pulleys

Belt width, mm	*Pulley to be wider than the belt width by, mm*
Upto and including 125	13
From 125 upto and including 250	25
From 250 upto and including 375	38
From 375 upto and including 500	50

IS : 2122—1973

1. Rim size.

Bartho's formula for width of rim,

$$B = 1.19b + 10 \text{ mm of single belt} \quad \text{... (15.18a)}$$

$$= 1.10b + 5 \text{ mm for double belt} \quad \text{... (15.18b)}$$

Bartho's formula for thickness of pulley rim,

$$t_r = 0.25\sqrt{D} + 15.5 \text{ mm for light duty} \quad \text{... (15.19 a)}$$

$$= 0.375\sqrt{D} + 3.2 \text{ mm for heavy duty} \quad \text{... (15.19b)}$$

For cast iron flat pulleys, rim thickness,

$$t_r = \frac{D}{200} + 3 \text{ mm for single belt} \quad \text{... (15.20a)}$$

$$= \frac{D}{200} + 6 \text{ mm for double belt} \quad \text{... (15.20b)}$$

Refer to Table 15.15 for width of flat pulleys.

2. Crowning of pulley rim.

A convex curvature (called crowning) is given to the pulley rim to make the belt run in the middle of the pulley width. The crowning depends upon the diameter and width of the pulley rim. For crown height. refer to Table 15.13.

3. Number of arms. (Table 15.16)

Provide web for pulleys upto 200 mm diameter. The number of arms for other diameters are as follows :

$$\text{For } D = 200 - 450 \text{ mm, number of arms, } n = 4 \quad \text{... (15.21a)}$$

$$\text{For } D > 450 \text{ mm, } n = 6 \quad \text{... (15.21b)}$$

4. Cross – section of arms.

The cross – section of arms is elliptical.

$$\text{Major axis of arm near boss, } b_1 = 2.94\left[\frac{BD}{4n}\right]^{\frac{1}{3}} \text{ for single belt} \quad \text{... (15.22a)}$$

$$= 2.94\left[\frac{BD}{2n}\right]^{\frac{1}{3}} \text{ for double belt} \quad \text{... (15.22b)}$$

$$\text{Minor axis near boss} = \frac{b_1}{2} \quad \text{... (15.23)}$$

$$\text{Cross – section near rim} = \frac{2}{3} \text{ rd of that near boss} \quad \text{... (15.24)}$$

Radius of cross – section of arms,

$$r = \frac{3b_1}{4} \quad \text{... (15.25)}$$

$$\text{Radius, } r_1 = r_2 = \frac{b_1}{2} \text{ near rim} \quad \text{... (15.26)}$$

Analytical method for arm size :

Maximum bending moment on arms,

$$M_{max} = (F_1 - F_2)\frac{D}{2} \quad \text{... (15.27)}$$

Bending moment in each arm, assuming that only half the number of arms take the bending moment,

$$M = \frac{2\,M_{max}}{n}$$

$$= \left(\frac{\pi}{64}\right) b_1^3 \sigma_t \quad \text{... (15.28)}$$

where σ_t = permissible tensile strength for material of arms.

Table 15.16 Number of arms in mild steel pulley

Diameter of pulley, mm	*No. of arms*	*Dia of arms, mm*
280 to 500	6	19
560 to 710	8	19
800 to 1000	10	22
1120	12	22
1250	14	22
1400	16	22
1600	18	22
1800	18	22
2000	22	22

IS : 1691—1980

Table 15.17 Minimum Pulley Diameter for Given Belt Speeds and Belt Plies in mm

No. of plies	*Maximum belt speed, m/s*				
	10	*15*	*20*	*25*	*30*
3	90	100	112	140	180
4	140	160	180	200	250
5	200	224	250	315	355
6	250	315	355	400	405
7	355	400	450	500	560
8	450	500	560	630	710
9	560	630	710	800	900
10	630	710	800	900	1000

IS : 1370—1976

5. Size of boss.

Minimum length of bore, $l = \frac{2B}{3}$ ... (15.29)

The diameter d_1 of boss should be such that,

$$\frac{d_1 - d_2}{2} = 0.412\,(BD)^{1/3} + 6 \text{ mm for single belt} \quad \text{... (15.30a)}$$

$$= 0.529\,(BD)^{1/3} + 6 \text{ mm for double belt} \quad \text{... (15.30b)}$$

(*b*) Steel Pulley.

A steel pulley is shown in Fig. 15.4.

Material : Mild steel conforming to IS : 226 – 1975

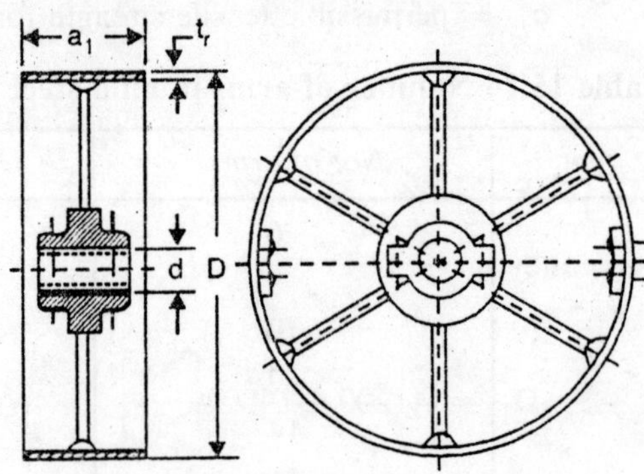

Fig. 15.4 Mild steel flat belt pulley

Thickenss of rim, $t_r = 5$ mm ... (15.31)

Length of boss $= \frac{1}{2} \times$ width of face

= 100 mm minimum for pulleys with 19 mm diameter spokes

= 138 mm minimum for pulleys with 22 mm diameter spokes

Cross section of spokes = circular

Rows of arms = single for pulleys upto 300 mm width ... (15.33*a*)

= two for width more than 300 mm ... (15.33*b*)

15.6 SHAFT AND KEY SIZE

Resultant pull on pulley due to belt tensions,

$$P = \left[\frac{F}{K-1}\right]\left[K^2+1+2K\cos(180^\circ-\theta)\right]^{\frac{1}{2}} \quad \text{... (15.34)}$$

where $K = \dfrac{F_1 - F_c}{F_2 - F_c}$; $F = F_1 - F_2$

Let W = weight of the pulley, and α = angle between P and W

Resultant force on the pulley,

$$S = [P^2 + W^2 + 2PW\cos\alpha]^{1/2} \quad \text{... (15.35)}$$

Let x = overhang of pulley

M = S*x*, bending moment due to S

$$\text{Torque in the shaft, } T = (F_1 - F_2)\frac{D}{2} \quad \text{... (15.36)}$$

$$\text{Equivalent torque, } T_e = \left[(K_m M)^2 + (K_t T)^2\right]^{\frac{1}{2}} \quad \text{... (15.37)}$$

where K_m and K_t are shock and fatigue factors for bending and torsion respectively.

Let d_2 = diameter of shaft

$$T_e = \left(\frac{\pi}{16}\right) d_2^3 \tau \quad \text{... (15.38)}$$

Where τ = permissible shear stress for the shaft material.

A suitable key for the shaft and pulley assembly is a square key with a gib head.

15.7 DIAMETER OF MOTOR PULLEY

Diameter of motor pulley,

$$D_1 = (1200 \text{ to } 1400)\left[\frac{P}{N_1}\right]^{\frac{1}{3}} \quad \text{... (15.39)}$$

15.8 SELECTION OF FLAT-BELTS FROM MANUFACTURER'S CATALOGUE

The following information is required to select the flat – belt from manufacturer's catalogue

1. Power to be transmitted
2. Input and output speeds.
3. Centre distance available.

The basic procedure for belt selection from the catalogue of The Dunlop Rubber Co. (India)

1. Calculate the maximum power transmitted by the belt from the rated power.

$$(kW)_{max} = F_a\ (kW) \quad \text{... (15.40)}$$

Where F_a = load correction factor (Table 15.18)

Table 15.18 Load Correction Factors, F_a

	Type of load	F_a
Normal load	Exact peak load is known	1.0
Steady load	Screens, centrifugal pumps and fans, agitators, belt conveyor, evaporators, light machine tools, printing and textile machinery	1.2
Intermittent load	Heavy duty fans and blowers, reciprocating pumps and compressors, saw mill and paper mill machinery, brick work machinery, heavy machine tools, elevators and line shafts	1.3
shock load	vacuum pumps, rolling mills, tube and ball mills, grinders, hammers, automatic machinery	1.5

2. The rated power is based on 180° angle of arc of contact. For this angle less than 180°, the belt is required to carry additional tension. For this purpose a factor called 'arc of contact factor' F_d is used to calculate the corrected power to be transmitted by the belt.

$$(kW)_{corrected} = F_a F_d\ (kW) \quad \text{... (15.41)}$$

where F_d = arc of contact factor (Table 15.19)

3. Calculate the corrected power rating for the belt.

The power transmitting capacities of belts are :

Type of belt	Power rating in kW per mm per ply	Corrected kW rating	Optimum belt speed, m/s
HI – SPEED	0.0118	$\frac{0.0118 \times v}{5.08}$	17.8 – 22.0
FORT	0.0147	$\frac{0.0147 \times v}{5.08}$	17.8 – 22.0

$$\text{Corrected kW rating} = \frac{0.0118 \times v}{5.08} \text{ for HI-SPEED belt} \quad \text{... (15.42}a\text{)}$$

$$= \frac{0.0147 \times v}{5.08} \text{ for FORT belt} \quad \text{... (15.42}b\text{)}$$

where v = belt velocity, m/s

The rating in kW of rubber transmission belting is given in Table 15.11.

4. Calculate the product of (width × number of plies),

$$\text{Width} \times \text{number of plies} = \frac{\text{corrected power}}{\text{corrected kW rating}} \quad \text{... (15.43)}$$

5. Select the standard belt width and number of plies from Table 15.12 and 15.20.

The nominal belt thickness for number of plies of rubber transmission belting is given in Table 15.12. The standard width of Dunlop HIGH SPEED and Dunlop FORD Flat Belt is given in Table 15.20

Table 15.19 Arc of contact correction factor, F_d

Arc of contact	F_d	Arc of contact	F_d
90°	1.68	190°	0.97
120°	1.33	200°	0.94
130°	1.26	210°	0.91
140°	1.19	220°	0.88
150°	1.13	230°	0.86
160°	1.08	240°	084
170°	1.04	250°	0.82
180°	1.00		

Table 15.20 Standard Width of Dunlop HIGH SPEED and Dunlop FORD Flat Belt

3 *ply* *mm*	4 *ply* *mm*	5 *ply* *mm*	6 *ply* *mm*	8 *ply* *mm*
25	25	–	–	–
32	32	–	–	–
40	40	–	–	–
44	44	–	–	–
50	50	–	–	–
63	63	–	–	–
76	76	76	–	–
90	90	90	–	–
100	100	100	100	–
–	112	112	112	–
–	125	125	125	–
–	140	–	–	–
–	152	152	152	–
–	–	180	180	–
–	200	200	200	200
–	–	224	–	–
–	–	250	250	250
–	–	–	–	305
–	–	–	–	355
–	–	–	–	400

15.9 CREEP IN BELTS

The tension on the tight side of a belt drive is more than the tension on the slack side. As a result of this, the belt is stretched more on the tight side as compared to the slack side. Therefore, the driver pulley receives more length of the belt and delivers less. Hence, the belt creeps (jumps) forward. The reverse occurs on the driven pulley, which receives less length of the belt and delivers more. Thus the belt jumps backward. This phenomenon is called creeping of belt.

$$\text{Creep} = \frac{F_1 - F_2}{bt\,E} \quad \text{... (15.44)}$$

where F_1, F_2 = tensions on the tight and slack sides respectively

b, t = width and thickness of belt respectively

E = Young's modulus of elasticity of belt material.

Example 15.1 It is required to select a flat – belt drive for a compressor running at 720 rpm, which is driven by a 25 kW, 1440 rpm motor. Space is available for a centre distance of 3 m. The belt is open type.

Solution.

Given : $N_1 = 1440$ rpm, $N_2 = 720$ rpm, P = 25 kW, C = 3 m

We take, $D_1 = 280$ mm

$$D_2 = \frac{280 \times 1440}{720} = 560 \text{ mm}$$

Angle of arc of contact,

$$\theta = \pi - 2\sin^{-1}\left[\frac{D_2 - D_1}{2C}\right]$$

$$= \pi - 2\sin^{-1}\left[\frac{560 - 280}{2 \times 3000}\right]$$

$$= 174.65° \text{ or } 3.048 \text{ rad}$$

Length of open belt,

$$L_o = 2C + \left(\frac{\pi}{2}\right)(D_1 + D_2) + \frac{(D_2 - D_1)^2}{4C}$$

$$= 2 \times 3 + \left(\frac{\pi}{2}\right)(0.280 + 0.560) + \frac{(0.560 - 0.280)^2}{12}$$

$$= 7.326 \text{ m}$$

Pitch line velocity,

$$v = \pi D_1 N_1 / (60 \times 10^3)$$

$$= \frac{\pi \times 280 \times 1440}{60 \times 10^3} = 21.11 \text{ m/s}$$

Load correction factor for compressor, (Table15.18)

$$F_a = 1.3$$

$$F_a = 1.3$$

Arc of contact factor, $F_d = 1.04 - \frac{(1.04-1.00)(174.65-170)}{180-170}$ (Table 15.19)

$$= 1.0214$$

Corrected power, $P_c = P \times F_a \times F_d = 25 \times 1.3 \times 1.0214 = 33.1955$ kW

We select a HI – SPEED Dunlop transmission duck belting of power rating 0.012 kW per mm width per ply. The optimum range of belt velocity for this belt is 17.8 – 22.9 m/s.

$$\text{Corrected belt rating} = \frac{0.0118v}{5.08} = \frac{0.012 \times 21.11}{5.08} = 0.049 \text{ kW}$$

$$\text{Width} \times \text{No. of plies} = \frac{\text{Corrected power}}{\text{Corrected belt rating}}$$

$$= \frac{33.1955}{0.049} = 677.459$$

Belt widths :

$$4 \text{ plies, } b = \frac{677.459}{4} = 169.36 \text{ mm}$$

$$5 \text{ plies, } b = \frac{677.459}{5} = 135.\ 49 \text{ mm}$$

$$6 \text{ plies, } b = \frac{677.459}{6} = 112.91 \text{ mm}$$

We select a belt of 152 mm preferred width and 5 plies from Table 15.20

Example 15.2 A leather belt is used in a flat-belt drive to transmit 15 kW from a motor pulley running at 1440 rpm to another pulley running at 480 rpm. The centre distance between the pulleys is twice the diameter of the bigger pulley. The belt should operate at a velocity of 20 m/s approximately. The maximum stress in the belt should not exceed 2.25 MPa. The density of leather is 950 kg/m^3 ad the coefficient of friction is 0.35. The belt thickness is 5 mm. Calculate the length and width of belt.

Solution.

Given : P =15 kW, N_1 = 1440 rpm, N_2 = 480 rpm. C = $2D_2$, $v \approx 20$ m/s, $\sigma_w = 2.25$ MPa,

$$\rho = 950 \text{ kg/m}^3, \mu = 0.35, t = 5 \text{ mm}$$

1. Diameter of motor pulley,

$$D_1 = \frac{60 \times 10^3 \times 20}{\pi \times 1440} = 265.26 \approx 250 \text{ mm}$$

$$v = \frac{\pi \times 250 \times 1440}{60 \times 10^3} = 18.85 \text{ m/s}$$

$$D_2 = \frac{250 \times 1440}{480} = 750 \text{ mm}$$

2. Centre distance, $C = 2D_2 = 2 \times 750 = 1500$ mm

3. Length of open belt, $L_o = 2C + \dfrac{\pi(D_1+D_2)}{2} + \dfrac{(D_2-D_1)^2}{4C}$

$$= 2\times 15.5 + \frac{\pi(0.25+0.75)}{2} + \frac{(0.75-0.25)^2}{4\times 15.5}$$

$$= 4.61246 \text{ m}$$

4. Mass of belt per metre length,

$$m = \rho bt = 950 \times b \times 5 \times 10^{-6} \times 1$$

$$= 4.75 \times 10^{-3} \times b \text{ kg/m}$$

5. Centrifugal tension, $F_c = \rho v^2 = 4.75 \times 10^{-3} \times 18.85^2 \times b = 15.6878b$ N

6. Angle of arc of contact, $\theta = 180° - 2\sin^{-1}\left[\dfrac{D_2-D_1}{2C}\right]$

$$= 180° - 2\sin^{-1}\left[\frac{750-250}{2\times 1500}\right]$$

$$= 160.81° \text{ or } 2.8067 \text{ rad}$$

7. $\text{Exp}(\mu\theta) = \exp(0.35 \times 2.8067) = 2.67$

8. $\dfrac{F_1 - F_c}{F_2 - F_c} = \exp(\mu\theta)$

$$\frac{F_1 - 15.6878b}{F_2 - 15.6878b} = 2.67 \quad \text{... (1)}$$

9. Maximum belt tension, $F_1 = \sigma_w bt = 2.25 \times b \times 5 = 115.25b$ ... (2)

10. Tangential force, $F_1 - F_2 = \dfrac{10^3 P}{v} = \dfrac{10^3 \times 15}{18.85} = 795.756$ N ... (3)

11. Solving Eqs. (1) to (3), we have

$$\frac{115.25b - 15.6878b}{115.25b - 795.756 - 15.6878b} = 2.67$$

$$\frac{9.5622b}{9.5622b - 795.756} = 2.67$$

$$3.5183b = 9.5622b - 795.756$$

$$b = 133 \text{ mm or 4 ply 140 mm}$$

We select belt width, $b = 140$ mm

15.10 V - BELTS

V – belts give more positive drive and are used for short centre distance between the driving and driven pulleys.

15.10.1 Characteristics of V-Belt Drive

1. It requires a grooved pulley.
2. The belt comes into contact with the sides of the pulley groove only.
3. The belts are endless.
4. It gives more positive drive than flat belts.
5. The slip of the belt is less as compared to flat belts.
6. The wedging action of the belt in the pulley groove offset the decrease in the arc of contact.
7. V–belts are used as a matched set.
8. The pulley groove angle is 40°.
9. They are suitable for speed ratio upto 7 and belt speed upto 25 m/s.
10. The size of a V-belt is specified by the cross-section followed by inside length.

15.10.2 Design of V-Belt Drive

1. Equivalent coefficent of friction.

Fig. 15.5 shows a V – belt in a grooved pulley. The normal force on the belt face is,

$$F_n = \frac{F}{2 \sin \beta}$$

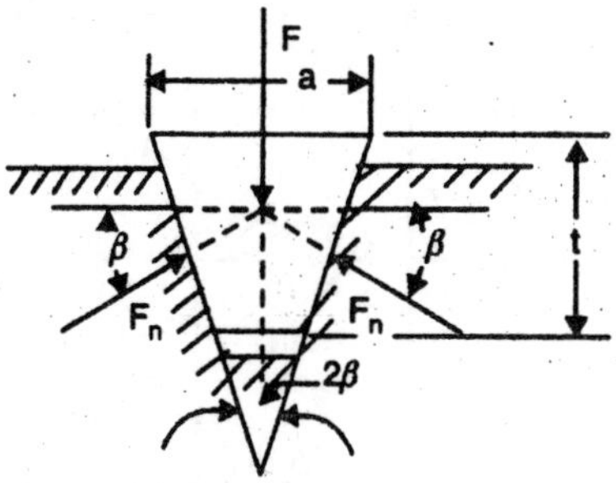

where 2β = angle of the groove = 40°

Tractive force $= 2\mu F_n$

$$= \left(\frac{\mu}{\sin \beta}\right) F = \mu_e F \quad \dots (15.45)$$

Fig. 15.5

where $\mu_e = \frac{\mu}{\sin \beta}$, is the equivalent of effective coefficient of friction

2. Selection of belt section

The cross – section suitable for a given power to be transmitted can be selected from Table 15.21 and 15.23.

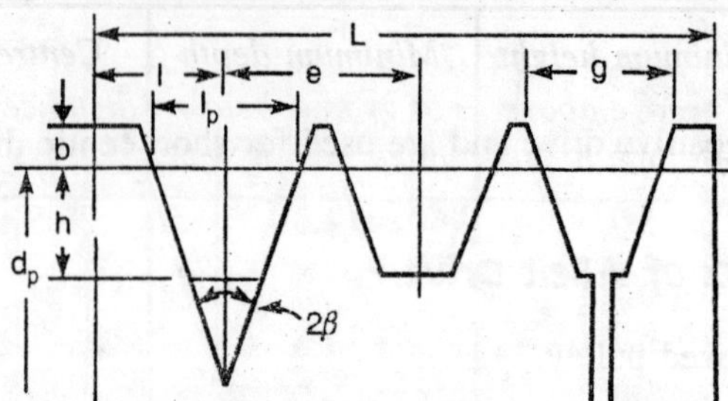

Fig. 15.6 Dimensions for Standard V-Grooved Pulley

Table 5.21 Classification of V-belts

Type of Belt	*IS Code No.*	*Cross-section symbol*	*Nominal Top width, a, mm*	*Nominal thickness, t, mm*	*Angle of V-Belt, 2β degree*
Industrial purposes	IS : 2494—1974	A	13	8	
		B	17	11	40
		C	22	14	
		D	32	19	
		E	38	23	
Automotive fan	IS : 5635—1970	FB 10	10	8	
	FB 13	13	11	40	
Agricultural machinery	IS : 8777—1978	HI	25.4	12.7	
		HJ	31.8	15.1	
		HK	38.1	17.5	40
		HL	44.5	19.8	
		HM	50.8	22.2	

Table 15.22 Properties of V-belts for industrial purposes

Cross-section of belt symbol	*Area of cross-section mm^2*	*Max. working load F_l,N*	*Power* range kW*	*Min. groove pulley dia, mm*	*Weight per m. N*	*Density of belt material kg/m^3*	*Max. number of pulley grooves*
A	80.7	200	upto 3.42	75	0.882	1115	6
B	143.0	300	0.5-6.61	137	1.628	1160.8	9
C	236.7	750	0.80-13	229	2.962	1275.8	14
D	476.6	1400	1.65-32.2	330	5.923	1267.3	14
E	681.4	1800	3.48-51.8	533	8.885	1329.6	20

IS : 9515—1980

Table 15.23 Dimensions for standard V-grooved pulleys

Groove section	*Pitch width l_p, mm*	*Minimum height of groove above pitch line, b, mm*	*Minimum depth of groove below pitch line, h, mm*	*Centre to centre distance of grooves, e, mm*	*Edge of pulley to first groove centre, f, mm*
A	11	3.3	8.7	15 ± 0.3	10 + 2 − 1
B	14	4.2	10.1	19 ± 0.4	12.5 + 2 − 1

Groove section	*Pitch width l_p, mm*	*Minimum height of groove above pitch line, b, mm*	*Minimum depth of groove below pitch line, h, mm*	*Centre to centre distance of grooves, e, mm*	*Edge of pulley to first groove centre, f, mm*
C	19	5.7	14.3	25.5 ± 0.5	17^{+2}_{-1}
D	27	8.1	19.9	37 ± 0.6	24^{+3}_{-1}
E	32	9.6	23.4	44.5 ± 0.7	29^{+4}_{-1}

IS : 3142—1965

3. Number of belts.

The number of belts n, may be determined from the following formula :

$$n\left(F_1 - F_c\right)\left[1 - \frac{1}{\exp\left(\mu_e \theta\right)}\right] = \frac{10^3 \times P \times K_s}{v} \quad \text{... (15.46)}$$

where $v = \dfrac{\pi D_1 N_1}{60 \times 10^3}$ m/s

D_1 = diameter of smaller pulley, mm

N_1 = rpm of smaller pulley

K_s = correction factor according to service

$$\theta = 2 \cos^{-1}\left[\frac{D_2 - D_1}{2C}\right] \quad \text{... (15.47)}$$

According to IS : 2494 – 1974, the number of belts are given by,

$$n = \frac{P \times K_a}{kW \times K_c \times K_d} \quad \text{... (15.48)}$$

where P = drive power, kW

kW = rating of V – belts at 180° arc of contact (Table 15.25)

K_a = correction factor for industrial service (Table 15.26)

K_c = correction factor for belt length (Table, 15.28)

K_d = correction factor for arc of contact (Table 15.29)

Equivalent pitch diameter, $d_e = d_p \times K_b$

K_b = small diameter factor

Table 15.24 Selection of Centre Distance

Speed ratio (D/b)	*Recommended C/D ratio*
1	1.50
2	1.20
3	1.00
4	0.95
5	0.90
6	0.85
7	0.85
8	0.85
9	0.85

Table 15.25 Power rating of V–belts in kW

Belt cross-section symbol	*Formula*	*Maximum value of d_c in formula, mm*
A	$P\ (kW) = \left(0.45v^{-0.09} - \dfrac{19.62}{d_e} - 0.765 \times 10^{-4}\ v^2\right) v$	125
B	$P\ (kW) = \left(0.79v^{-0.09} - \dfrac{50.8}{d_e} - 1.32 \times 10^{-4}\ v^2\right) v$	175
C	$P\ (kW) = \left(1.47v^{-0.09} - \dfrac{142.7}{d_e} - 2.34 \times 10^{-4}\ v^2\right) v$	300
D	$P\ (kW) = \left(3.22v^{-0.09} - \dfrac{506.7}{d_e} - 4.78 \times 10^{-4}\ v^2\right) v$	425
E	$P\ (kW) = \left(4.58v^{-0.09} - \dfrac{952}{d_e} - 7.05 \times 10^{-4}\ v^2\right) v$	700
where	P = power at 180° arc of contact for a belt of average length. v = belt speed in m/s. d_e = equivalent pitch diameter = $d_p \times K_b$ d_p = pitch diameter of the smaller pulley K_b = small diameter factor (Table 15.27)	

IS : 2494–1974

Table 15.26 Correction factors for industrial serivce, K_a

Severity of service	*Type of driven machines*	*A.C. motors ; normal torque, squirrel cage, synchronous and split phase.* *D.C. motors, shunt wound, multiple cylinder internal combustion engines over 600 rpm.*			*Type of driving unit. A.C. motors, high torque, high slip repulsion, induction, single phase, series wound and slipring.* *D.C. motors, series wound and compound wound, single and multi-cylinder I.C. engines under 600 rpm, line shaft, clutches, brakes.*		
		Operational hr. per day			*Operational hr. per day*		
		upto 10 hrs.	*over 10 to 16 hr.*	*to 16 hr and continuous service*	*upto 10 hrs.*	*over 10 to 16 hr.*	*over 16 hr and continuous service*
Light duty	Agitators for liquids, blowers and exhausters, centrifugal pumps and compressors, fans upto 7.5 kW and light duty conveyors.	1.0	1.1	1.2	1.1	1.2	1.3
Medium duty	Belt conveyors for sand grains etc., dough mixers, fans over 7.5 kW, generators, lines shafts, laundry machinery, machine tools, punches, presses and shears, printing machinery, positive displacement rotary pumps, and revolving and vibrating screens.	1.1	1.2	1.3	1.1	1.2	1.3
Heavy duty	Brick machinery, bucket elevators, conveyors (dragpan-screw), hammer mills, paper mill beaters, piston pumps, positive displacement blowers, pulverizers, saw mill and textile machinery.	1.2	1.3	1.4	1.4	1.5	1.6
Extra heavy duty	Crushers (gyratory jaw-roll), mills (ball-rod-tube), hoists	1.3	1.4	1.5	1.5	1.6	1.8

IS : 2494–1974

Table 15.27 Small diameter factor, K_b

Speed ratio range	K_b
1.000 to 1.019	1.00
1.020 to 1.032	1.01
1.033 to 1.055	1.02
1.056 to 1.081	1.03
1.082 to 1.109	1.04
1.110 to 1.142	1.05
1.143 to 1.178	1.06
1.179 to 1.222	1.07
1.223 to 1.274	1.08
1.275 to 1.340	1.09
1.341 to 1.429	1.10
1.430 to 1.562	1.11
1.563 to 1.814	1.12
1.815 to 2.948	1.13
2.949 and over	1.14

IS : 2494—1974

Table 15.28 Correction factors for belt length, K_c

Nominal inside length mm	*Belt cross-section symbol*				
	A	*B*	*C*	*D*	*E*
(1)	*(2)*	*(3)*	*(4)*	*(5)*	*(6)*
610	0.80				
660	0.81				
711	0.82				
787	0.84				
813	0.85				
889	0.87	0.81			
914	0.87				
965	0.88	0.83			
991	0.88				
1016	0.89	0.84			
1067	0.90	0.85			
1092	0.90				
1168	0.92	0.87			
1219	0.93	0.88			
1295	0.94	0.89	0.80		
1372		0.90			
1397	0.96	0.90			
1422	0.96	0.90			

Nominal inside length mm	*Belt cross-section symbol*				
	A	*B*	*C*	*D*	*E*
(1)	*(2)*	*(3)*	*(4)*	*(5)*	*(6)*
1473	0.97				
1524	0.98	0.92	0.82		
1600	0.99				
1651	1.00	0.94			
1727	1.00	0.95	0.85		
1778	1.01	0.95			
1905	1.02	0.97	0.87		
1981	1.03	0.98			
2032	1.04				
2057	1.04	0.98	0.89		
2159	1.05	0.99	0.90		
2286	1.06	1.00	0.91		
2438	1.08		0.92		
2464		1.02			
2540		1.03			
2667	1.10	1.04	0.94		
2845	1.11	1.05	0.95		
3048	1.13	1.07	0.97	0.86	
3150			0.97		
3251	1.14	1.08	0.98	0.87	
3404			0.99		
3658		1.11	1.00	0.90	
4013		1.13	1.02	0.92	
4115		1.14	1.03	0.92	
4394		1.15	1.04	0.93	
4572		1.16	1.05	0.94	
4953		1.18	1.07	0.96	
5334		1.19	1.08	0.96	0.94
6045			1.11	1.00	0.96
6807			1.14	1.03	1.99
7569			1.16	1.05	1.01
8331			1.19	1.07	1.03
9093			1.21	1.09	1.05
9855			1.23	1.11	1.07
10617			1.24	1.12	1.09
12141				1.16	1.12
13665				1.18	1.14
15189				1.20	1.17
16713				1.23	1.19

IS : 2494–1974

Table 15.29 Correction Factors for arc of contact factor, K_d

Arc of contact on small pulley, degree		*Correction factor, K_d*	
		VV	*V-flat*
180		1.00	0.75
177		0.99	0.76
174		0.99	0.76
171		0.98	0.77
169		0.97	0.78
166		0.97	0.79
163		0.96	0.79
160		0.95	0.80
157		0.94	0.80
154		0.93	0.81
151		0.93	0.81
148		0.92	8.83
145		0.91	0.83
142		0.90	0.84
139		0.89	0.85
136		0.88	0.85
133		0.87	0.86
130		0.86	0.86
127		0.85	0.85
123		0.83	0.83
120		0.82	0.82
117	Discouraged range	0.81	0.81
113		0.80	0.80
110		0.78	0.78
106		0.77	0.77
103		0.75	0.75
99		0.73	0.73
95		0.72	0.72
91		0.70	0.70
87		0.68	0.68
83		0.65	0.65

IS : 2494—1974

4. Power rating of V – belts,.

$$\text{kW} = \left(\frac{\text{nav}}{10^3}\right)\left[\sigma_d - \frac{\rho v^2}{10^6}\right]\left[1 - \frac{1}{\exp(\mu_e \theta)}\right] \quad \text{... (15.49)}$$

where σ_d = design stress = 2.26 N/mm²

ρ = mass density of belt material = 1080 kg/m³

a = area of cross – section of belt, mm²

5. Initial tension.

$$F_i = \frac{F_1 + F_2}{2} \quad \text{... (15.50)}$$

6. Optimum velocity of belt for maximum power transmission.

$$v = \left(\frac{F_i}{3m}\right)^{\frac{1}{2}} \quad \text{... (15.51)}$$

where m = mass per meter length of belt

7. Ratio of belt tensions.

$$\frac{F_1 - mv^2}{F_2 - mv^2} = \exp(\mu_e \theta) \quad \text{... (15.52)}$$

8. Pitch length of belt.

$$L_p = 2C + 15.57\left(D_1 + D_2\right) + \frac{\left(D_2 - D_1\right)^2}{4C} \quad \text{... (15.53)}$$

Refer to Table 15.30 for standard nominal inside length and nominal pitch length.

9. Centre distance.

$$C = A + (A^2 - B)^{1/2} \quad \text{... (15.54)}$$

where A $= \frac{L_p}{4} - \frac{\pi\left(D_1 + D_2\right)}{8}$

$$B = \frac{\left(D_2 - D_1\right)^2}{8}$$

$$C_{max} = 2\,(D_1 + D_2) \quad \text{... (15.55)}$$

$$C_{min} = 0.55\,(D_1 + D_2) + t$$

Refer to Table 15.24 for selection of centre distance.

10. Pulley size.

Refer to Table 15.23 for dimesions of Standard V – grooved pulleys and Table 15.31 for standard pulley pitch diameters.

Table 15.30 Nominal inside length, nominal pitch length and permissible length variations for V-belts

Nominal inside length, mm	*Nominal pitch length, mm, cross-section symbol*					*Pitch length*	*Variation*
	A	*B*	*C*	*D*	*E*	*PLL**	*MVL***
(1)	*(2)*	*(3)*	*(4)*	*(5)*	*(6)*	*(7)*	*(8)*
610		645					
660		696				+ 11.4	
711		747				– 6.4	
787		823					
813		848				+ 12.5	
889		925	932			– 7.5	
914		950					2.5
965		1001	1008				
991		1026					
1016		1051	1059			+ 14.0	
1067		1102	1110			– 8.9	
1092		1128					
1168		1204	1212				
1219		1255	1262				
1295		1331	1339	1351		+ 16.0	
1372			1415				
1397		1433	1440				
1422		1451	1466				
1473		1509					
1524		1560	1567	1580			5.0
1600		1636				+ 17.8	
1626		1661				– 12.5	
1651		1687	1694				
1727		1763	1770	1783			
1778		1814	1821				
1905		1941	1948	1991			
1981		2017	2024				
2032		2068					
2057		2093	2101	2113		+ 30	
2159		2195	2202	2215		– 16	7.5
2286		2322	2329	2342			
2438		2474		2494			

Nominal inside length, mm	*Nominal pitch length, mm, cross-section symbol*					*Pitch length*	*Variation*
	A	*B*	*C*	*D*	*E*	*PLL**	*MVL***
(1)	*(2)*	*(3)*	*(4)*	*(5)*	*(6)*	*(7)*	*(8)*
	2464	....	2507	....	....	....	+ 34
	2540	....	2583	....	....	....	– 18
	2667	2703	2710	2723	....	....	
	2845	2880	2888	2901	....	....	
	3048	3084	3091	3104	3127	....	
3150				3205			
3251		3287	3294	3307	3330	+ 38	
3404				3459		– 21	12.5
3658		3693	3701	3736			
4013		4056	4069	4092		+ 43	
4115		4158	4171	4194		– 24	
4394		4437	4450	4473			
4572		4615	4628	4651			12.5
4953		4996	5009	5032	....	+ 49	
5334		5377	5090	5413	5426	– 28	
6045			6101	6124	6137		
6807			6863	6886	6889	+ 56	
7569			7625	7648	7661	– 32	
8331			8387	8410	8423	+ 65	
9093			9149	9172	9185		15
9855				9934	9947	– 37	
10617			10696	10709	+ 76		
12141			12220	12233	– 43	17.5	
13665			13744	13757	+ 89		
15189			15268	15281	– 50		
16713			16792	16805	+ 105		
					– 59		

IS : 2494–1974

PLL Pitch length limits, MVL*** maximum variation in length within a matched set.

Table 15.31 Recommended standard pulley pitch diameters

Series of pitch diameters			*Degree of preferece* for pitch diameters, according to groove section*				
Nominal value	*Pitch dia limits, mm*						
mm	*Max.*	*Min.*	*A*	*B*	*C*	*D*	*E*
(1)	*(2)*	*(3)*	*(4)*	*(5)*	*(6)*	*(7)*	*(8)*
75	75	76.5	3				
80	80	81.3	3				
85	85	86.4	3				
90	90	91.4	1				
95	95	96.5	2				
100	100	101.6	1				
106	106	107.7	2				
112	112	113.8	1				
118	118	119.9	2				
125	125	127.0	1	2			
132	132	134.1	2	2			
140	140	142.2	1	1			
150	150	152.4	2	2			
160	160	162.6	1	1			
170	170	172.7	3	2			
180	180	182.9	1	1			
190	190	193.0	3	3			
200	200	203.2	1	1	1		
212	212	215.4			2		
224	224	227.6	2	2	1		
236	236	239.8			2		
250	250	254.0	1	1	1		
265	265	269.2			2		
280	280	284.5	2	2	1		
300	300	304.8	2	2	2		
315	315	320.0	1	1	1		
355	355	360.7	2	2	2	1	
375	375	381.0		2	2	2	
400	400	406.4	1	1	1	1	
425	425	431.8					

Series of pitch diameters			Degree of preferece* for pitch diameters, according to groove section				
Nominal value	Pitch dia limits, mm						
mm	Max.	Min.	A	B	C	D	E
(1)	(2)	(3)	(4)	(5)	(6)	(7)	(8)
450	450	457.2	2	2	2	1	
475	475	482.6			2		
500	500	508.8	1	1	1	1	1
530	530	538.8		3	3	3	2
560	560	569.0	2	2	2	2	1
600	600	609.6		2	2	2	2
630	630	640.0	1	1	1	1	1
670	670	680.7					2
710	710	721.4	2	2	2	2	1
750	750	762.0		2	2	2	3
800	800	812.8	3	1	1	1	1
900	900	914.4		2	2	2	2
1000	1000	1016.0		1	1	1	1
1060	1060	1077.0				2	
1120	1120	1137.9		3	2	2	2
1250	1250	1270.0			1	1	1
1400	1400	1422.4			2	2	2
1500	1500	1524.0				2	2
1600	1600	1625.6			1	1	1
1800	1800	1828.4				2	2
1900	1900	1930.4					2
2000	2000	2032.0				1	1
2240	2240	2275.8					2
2500	2500	2540.0					1

*1 = Recommended, 2 = Second preference, 3 = Not recommended, IS : 3142–1965

Example 15.3 Design a V–belt drive to transmit 7.5 kW from an induction motor running at 1440 rpm to a fan running approximately at 480 rpm. The service is for 24 hours per day. Centre distance is 1.25 m.

Solution.

Given : $P = 7.5$ kW, $N_1 = 1440$ rpm, $N_2 \approx 480$ rpm, service = 24 hours, $C = 1.25$ m

1.Select cross – section of belt. (Table 15.22)

For P = 0.8 – 13 kW, Cross – section C of V – belt is suitable.

Top width of belt, w = 22 mm, Nominal thickness,

t = 14 mm

Weight per metre = 2.962 N/m,

Minimum pully diameter, D_1 = 229 mm, v_{max} = 25 m/s

Let D_1 = 240 mm

2. Diameter of fan pulley, $D_2 = \frac{240 \times 1440}{480} = 920$ mm

3. Belt speed, $v = \frac{\pi \times 0.24 \times 1440}{60} = 18$ m/s

4. Picth length of belt, $L = 2C + \frac{\pi(D_1 + D_2)}{2} + \frac{(D_2 - D_1)^2}{4C}$

$$= 2 \times 1.25 + \frac{\pi(0.24 + 0.72)}{2} + \frac{(0.72 - 0.24)^2}{(4 \times 15.25)}$$

= 4.016 m or 4016 mm

The standard nominal pitch length of C – belt nearest to above length is 4069 mm. [Table 15.30]

5. Exact centre distance, $C = A + (A^2 - B)^{\frac{1}{2}}$

Where $A = \frac{L_p}{4} - \frac{\pi(D_1 + D_2)}{8}$

$$= \frac{4069}{4} - \frac{\pi(240 + 720)}{8} = 640.26 \text{ mm}$$

$$B = \frac{(D_2 - D_1)^2}{8} = \frac{(720 - 240)^2}{8} = 28800$$

$C = 640.26 + [(640.26)^2 - 28800]^{½}$

= 1257.6 mm or 1.2576 m

6. Inside length of belt, L_i = 4013 mm

The belt size is C – 4013 : V – belt.

7. Angle of arc of contact, $\theta = 180° - 2\sin^{-1}\left[\frac{D_2 - D_1}{2C}\right]$

$$= 180° - 2\sin^{-1}\left[\frac{0.72 - 0.24}{2 \times 1.2576}\right]$$

= 158°

8. Small diameter factor, $K_b = 1.14$ for speed ratio = 3 [Table 15.27]

Correction factor for industrial service,

$$K_a = 1.3 \quad \text{[Table 15.26]}$$

Correction factor for belt length,

$$K_c = 1.02 \quad \text{[Table 15.28]}$$

Correction factor for arc of contact,

$$K_d = 0.80 \quad \text{[Table 15.29]}$$

9. Equivalent pitch diameter, $d_e = d_1 \times K_b = 210 \times 1.14 = 273.6$ mm

Maximum value of d_e in the C – belt power rating formula,

$$d_e = 300 \text{ mm}$$

$$\text{Power rating for C – belt} = \left(1.47v^{-0.09} - \frac{142.7}{d_e} - 2.34 \times 10^{-4} v^2\right) v$$

$$= \left(1.47 \times 18^{-0.09} - \frac{142.7}{273.6} - 2.34 \times 10^{-4} \times 18^2\right) \times 18$$

$$= 9.646 \text{ kW}$$

10. $$\text{Number of belts} = \frac{(\text{Transmitted power}) \times K_a}{(\text{kW rating} \times K_c \times K_d)}$$

$$= \frac{7.5 \times 1.3}{(9.646 \times 1.02 \times 0.8)}$$

$$= 1.239 \approx 2 \text{ belts.}$$

EXERCISES

15.1 A flat belt is required to transmit 20 kW from a pulley 1.2 m diameter running at 240 rpm. The angle of contact is 165° and the coefficient of friction between belt and pulley rim is 0.25. Belt thickness is 6 mm and density of its material is 1050 kg/m^3. Determine the belt width if permissible stress in belt is 2.5 MPa.

15.2 An electric motor drives a compressor to transmit 25 kW at 750 rpm of motor pulley. Calculate the width of belt of thickness 6 mm having permissible stress of 2.5 MPa. account for centrifugal tension. The following data is given :

Item	Motor pulley	Compressor pulley
Diameter	350	1440
Angle of contact	160°	170°
Coefficient of friction	0.30	0.25

15.3 Design a belt drive to transmit 100 kW at 20% overload between two pulleys of diameters 1 m and 1.5 m having centre distance of 3.5 m. The belt speed is limited to 20 m/s and coefficient of friction is 0.35. A slip of 1.25% and 6% friction loss at each pulley is allowed.

15.4 In a horizontal flat belt drive, an electric motor of 15 kW power running at 1500 rpm drives a blower at 600 rpm. The centre distance is twice the diameter of the bigger pulley. The density of belt material is 1200 kg/m^3 and allowable stress is 4 MPa. Coefficient of friction on motor pulley is 0.5 and blower pulley 0.4 Peripheral speed of belt is limited to 20 m/s. Calculate the following :

(*a*) Pulley diameters, (*b*) width of belt

(*c*) cross – section area of belt, and (*d*) initial tension.

15.5 Design a flat belt drive to transmit 120 kW at belt speed of 25 m/s between two pulleys of diameters 300 mm and 600 mm having centre distance of 1.5 m. The allowable belt stress is 8 MPa. Belts are available having thickness to width ratio of 0.1 and a material density of 1100 kg/m^3. Determine the width and thickness of belt required if coefficient of friction between belt and pulley is 0.3.

15.6 Design cast iron pulley to transmit 20 kW at 300 rpm. The distance between the pulleys is 1 m and the angle of lap is 165°. The coefficient of friction between the belt and the pulley is 0.3. The allowable tension per metre width of belt is 2.5 N. The allowable stresses are :

Shear stress in shaft = 45 MPa, bending stress in pulley arms = 15 MPa.

15.7 A compressor requiring 100 kW is to run at about 250 rpm. It is driven by an electric motor running at 750 rpm. The diameter of compressor pulley should not be larger than 1.2 m while the centre distance between the pulley is limited to 1.6 m. The belt speed should not exceed 25 m/s. Determine the number of C – type V – belts required and its inside length. Assume coefficient of friction between belt and pulley to be 0.25.

15.8 An electric motor transmits 50 kW running at 750 rpm to a compressor at 300 rpm by means of V–belts. The approximate larger pulley diameter is 1.5 m and centre distance is 1.75 m. The overload factor is 15%. The coefficient of friction between belt and pulley is 0.27. The driven pulley is overhung to the extent of 300 mm from the nearest bearing. The permissible shear stress for the shaft material is 42 MPa while accounting for keyway. Determine (*a*) the number of C – type V – belts required, (*b*) diameter of shaft, and (*c*) pulley dimensions.

16 CHAINS AND ROPES

Chapter

16.1 INTRODUCTION

In order to avoid slipping, steel chains are used. The chains are made up of number of rigid links which are hinged together by pin joints in order to provide the necessity flexibility for wraping round the driving and driven sprockets. The sprockets have projecting teeth of special profile and fit into the corresponding recesses in the links of the chain as shown in Fig. 16.1. The sprockets and chain are constrained to move together without slipping and ensure perfect velocity ratio.

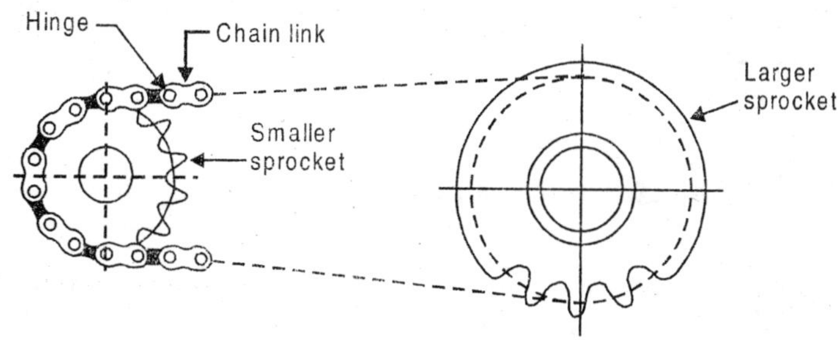

Fig. 16.1 Sprockets and chain

16.2 ADVANTAGES AND DISADVANTAGES OF CHAIN DRIVE OVER BELT AND ROPE DRIVES.

The advantages and disadvantages of chain drive over belts or ropes are given below :

Advantages

1. Can be used for short as well as long centre distances.
2. A number of shafts can be driven in the same or opposite directions from a single driving sprocket.
3. More compact drive as compared to belt and rope drives.
4. High efficiency.
5. More positive drive, *i.e.* less slip.
6. No fire hazards and temperature effects.
7. Can be easily replaced.

Disadvantages

1. Unsuitable for precise motion due to polygonal effect and wear of joints.
2. Require precise alignment of shafts.
3. Require proper maintenance, lubrication and slack adjustment.
4. Generates noise.

16.3 TYPES OF CHAINS

The chains are classified as :

(*a*) Hoisting and hauling chains.

(*b*) Transmission chains

(*a*) Hoisting and Hauling Chains

The hoisting and hauling chains are used for lifting of load. These chains are of the following two types as shown in Fig. 16.2.

1. Chain with oval links. The links of this chain are oval in shape. The cross – section of the link is generally circular—The joint should be perpendicular to the line of action of the load where the stress is minimum. The joint is generally welded. The sprockets for these chains have receptacles to receive the links. These chains are used at low speeds such as chain hoists.

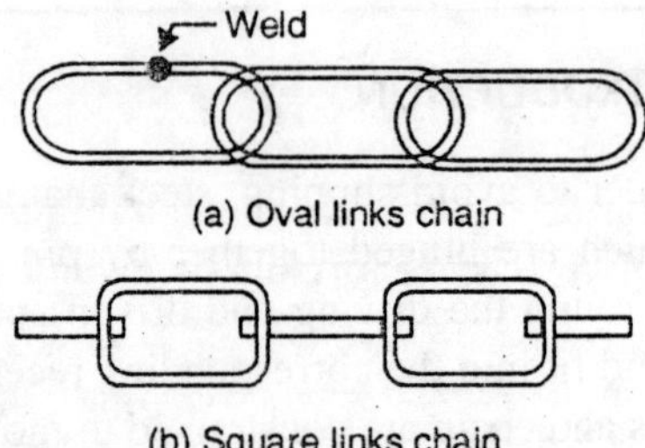

Fig. 16.2 Hoisting and Hauling chains

2. Chains with square links – The links of this chain are of square shape. Such type of chains are used in hoists, and cranes, etc.

(*b*) Transmission chains

These chain are used for power transmission.

16.4 DESIGN OF TRANSMISSION ROLLER CHAIN

The chain on the sprocket of a transmission chain is shown in Fig. 16.3. The pitch (*p*) of the chain is the distance between the hinge centre of a link and the corresponding hinge centre of the adjacent link. The pitch circle diameter (*d*) of chain sprocket is the diameter of the circle on which the hinge centres of the chain lie, when the chain is wraped round a sprocket. The points A, B, C and D are the hinge centres of the chain and the circle drawn through these centres is called the pitch circle and its diameter is called the pitch circle diameter.

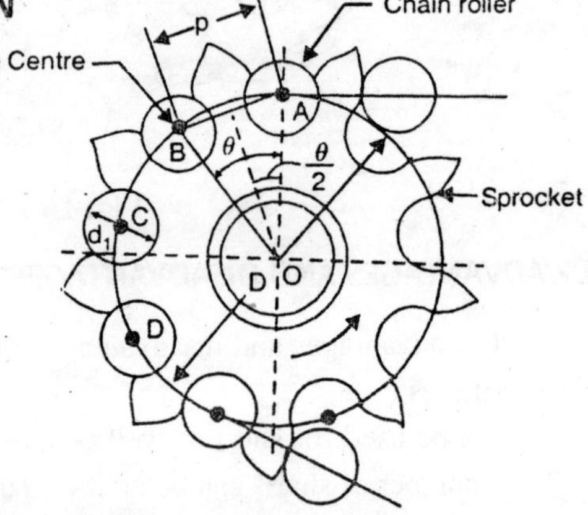

Fig. 16.3 Terms used in chain drive.

1. Transmission ratio.

$$\text{Transmission ratio, } i = \frac{n_1}{n_2} = \frac{d_2}{d_1} = \frac{z_2}{z_1} \quad ...(16.1)$$

where n = *rpm*, d = pitch diameter of sprocket, z = number of teeth on sprocket

Refer to Table 16.1 for preferred transmission ratios.

2. Average chain speed.

$$v = \frac{p z_1 n_1}{60 \times 10^3} \text{ m/s} \quad ....(16.2)$$

where p = pitch of chain, mm

Refer to Table 16.2 for maximum velocity of various types of chains.

3. Minimum number of teeth on sprocket.

The minimum number of teeth on smaller sprocket are 17 and the maximum number of teeth on the bigger sprocket are 114.

$$z_{min} = \frac{4d_r}{p} + 5 \text{ for pitches of 16.54 mm} \quad ....(16.3a)$$

$$= 4d_r + 4 \text{ for pitches of 3.175 to 5.785 mm} \quad ....(16.3b)$$

Where d_r = roller pin diameter, mm

For recommended number of teeth on sprockets, refer to Table 16.3 and 16.4.

4. Maximum sprocket speed.

(*a*) Based on allowable amount of impact between a roller and a sprocket, the speed of sprocket, according to Bartlett formula, is given by :

$$n \leq 1170\left[\frac{A}{(W_f p)}\right]^{1/2} \text{ rpm} \quad ...(16.4a)$$

where W_f = weight of chain, N/m

$A = b_1 \times d_r$ = projected area of roller, m^2

b_1 = length of roller, m

p = pitch, m

$$v_{max} \leq 19.48\, z_1 \left[\frac{A}{W_f p}\right]^{1/2} \text{ m/s} \quad ...(16.4b)$$

(*b*) Based on energy of impact per tooth per minute,

$$n \leq \left(\frac{1437}{p}\right)\left[\frac{A}{W_f}\right]^{1/2} \text{ rpm} \quad ....(16.5a)$$

$$v_{max} \leq 24z_1 \left[\frac{A}{W_f}\right]^{1/2} \text{ m/s} \quad ...(16.5b)$$

(*c*) Based on the effect of centrifugal force,

$$n \leq \left(\frac{36350}{p}\right) W_f z_1]^{1/2} \text{ rpm} \quad ...(16.6a)$$

$$v_{max} \leq 600\left[\frac{Az_1}{W_f}\right]^{1/2} \text{ m/s} \quad ....(16.6b)$$

Refer to Table 16.2 for maximum speed of sprocket pinion.

5. Chain pitch.

$$p \leq 250\left(\frac{90}{n_1}\right)^{2/3} \text{ mm} \quad \text{...(16.7)}$$

Standard pitches for transmission steel roller chains are given in Table 16.5.

6. Sprocket pitch diameter.

$$d = \frac{p}{\sin\left(\frac{180°}{z}\right)} \quad \text{...(16.8a)}$$

$$\text{Root diameter, } d_f = d - d_r \quad \text{..(16.8b)}$$

7. Centre distance.

The centre distance in a transmission chain is shown in Fig. 16.4.

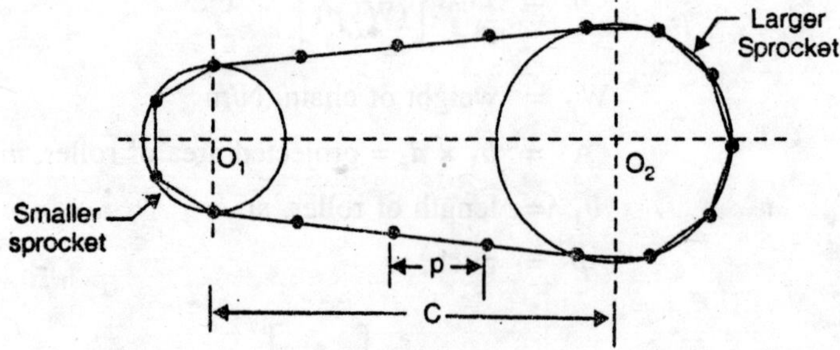

Fig. 16.4 Length of chain

$$C_{min} = K_{min} C' \quad \text{....(16.9)}$$

Where

$$C' = \frac{d_{a1} + d_{a2}}{2}$$

$$d_a = \text{tip diameter of sprocket}$$

$$= \frac{p}{\tan\left(\frac{180°}{z}\right)} + 0.6p$$

$$K_{min} = \text{a constant (Table 16.6)}$$

$$C_{max} = 80p \quad \text{...(16.10)}$$

Centre distance in pitches, C_p = (20 – 30) p for highly pulsating loads ....(16.11*a*)

= (30 – 50) p for uniform loads ...(16.11*b*)

Centre distance in chain drive is given in Table 16.6.

8. Chain length.

Chain length in pitches,

$$L_p = 2C_p \cos\alpha + \frac{(z_1 + z_2)}{2} + \frac{\alpha\,(z_1 - z_2)}{180°} \quad \text{...(16.12a)}$$

$$= 2C_p + \frac{(z_1 + z_2)}{2} + \frac{k(z_1 - z_2)^2}{C_p} \qquad ...(16.12b)$$

Where $\alpha = \sin^{-1}\left[\frac{d_2 - d_1}{C_p}\right]$

k = a variable dependent on $\frac{(z_1 - z_2)}{C_p}$ (Table 16.7)

Chain length, $L = 2C\cos\alpha + z_1 p\left[\frac{180° + 2\alpha}{360°}\right] + z_2 p\left[\frac{180° - 2\alpha}{360°}\right]$ ..(16.13)

$$L = mp \qquad ....(16.14)$$

Where m = number of chain links

$$= \frac{2C}{p} + \frac{(z_1 + z_2)}{2} + \frac{p(z_2 - z_1)^2}{4\pi^2 C}$$

Exact centre distance,

$$C = \left(\frac{P}{4}\right)\left[m - \frac{(z_1 + z_2)}{2}\right] + \left[\left\{m - \frac{(z_1 + z_2)}{2}\right\}^2 - 8\left\{\frac{z_2 - z_1}{(2\pi)}\right\}^2\right]^{1/2} \qquad ....(16.15)$$

9. Allowable chain pull.

$$F_a = \frac{F_u}{K_f} \qquad ...(16.16)$$

Where F_u = ultimate strength of chain, N

K_f = working factor of safety (Table 16.8)

= 5 for z upto 40 and v upto 0.5 m/s

= 18 for z upto 10 and v upto 6 m/s

10. Power transmitted.

$$P = \frac{Fv}{10^3 \times K_s} kW \qquad ....(16.17)$$

Where K_s = service factor (Table 16.9)

11. Design power. Design power = rated power × K_s

Rated power per strand, kW = $p^2\left[0.98v - (0.1988v)\left\{1 + 50\sin^2\left(\frac{90°}{z_1}\right)\right\}\right]$ ...(16.18)

12. Actual factor of safety.

$$K_{fa} = \frac{F_u}{F + F_c + F_s} \qquad ...(16.19)$$

Where $F = \dfrac{10^3 \times P \times K_s}{v}$, is the tangential force in N

$F_c = \dfrac{W_f v^2}{g}$, is the centrifugal force in N

$F_s = K_{sg}\ W_f C$, is the sag force in N

W_f = weight of chain per meter, N/m

K_{sg} = coefficent of sag (Table 16.10)

For a safe design, $K_{fa} > K_f$

13. Number of strands.

If $F > F_a$, ten number of strands,

$$i = \frac{F}{F_a} \qquad ....(16.20)$$

14. Bearing pressure on roller

$$p_b = \frac{F}{A} \qquad ....(16.21)$$

where $A = b_1 d_r$ = area of roller.

Power transmitted on the basis of allowable stress,

$$P = \frac{p_b A v}{K_s} \qquad ...(16.22)$$

Where p_b = allowable bearing pressure on rollers, N/mm^2 (Table 16.11)

15. Load on shaft.

$$F_{shaft} = K_L F \qquad ...(16.23)$$

Where K_L = load factor (Table 16.12)

16. Diametral dimensions of wheel rims and profile.

The diametral dimensions of wheel rims are shown in Fig. 16.5 and wheel rim profile in Fig. 16.6.

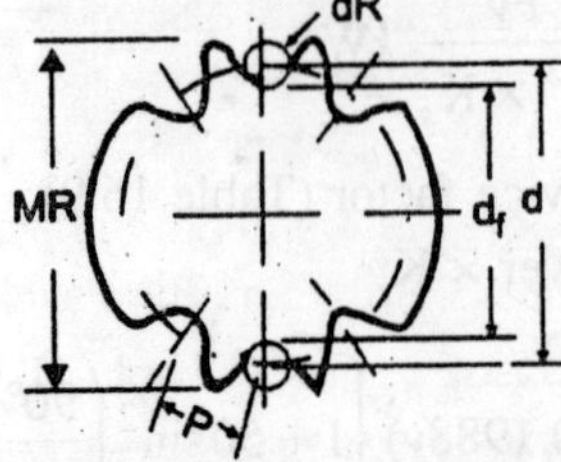

dR
MR
d d_r
P

Even number of teeth

p = chordal pitch = chain pitch

dR = measuring pin diameter

z = number of teeth

Odd number of teeth

d = pitch circle diameter = p (sin 180°/z)

dt = root diameter

MR = Measurement over pins

Fig. 16.5 Diametral dimensions of wheel rims.

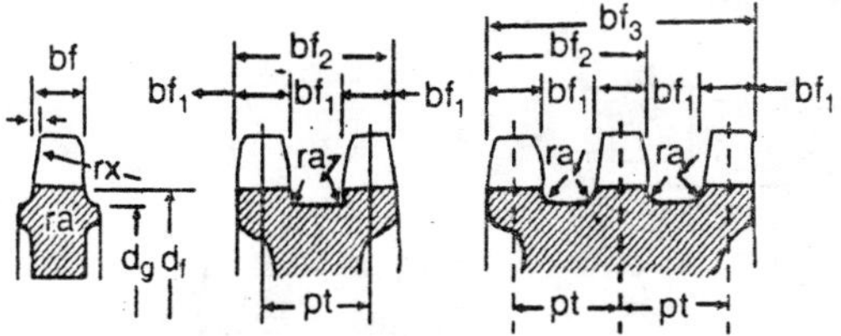

Fig. 16.6 Wheel of rim profile

Refer to Table 16.13 for tooth width of chain wheels. Pitch circle diameters for chain wheels are given in Table 16.16 and 16.17.

17. Wheel tooth gap forms.

The wheel tooth gap forms are shown in Fig. 16.7.

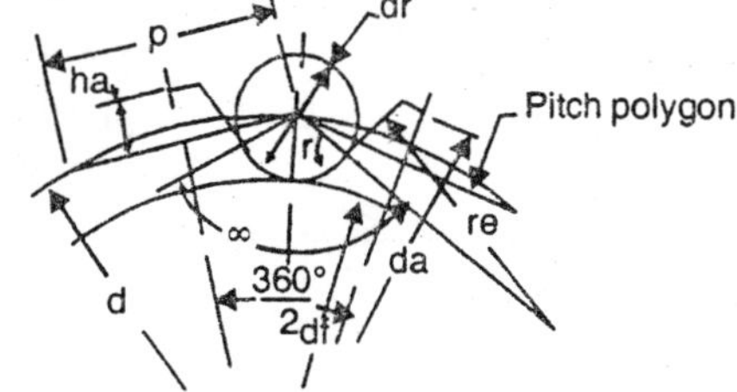

Fig. 16.7 Wheel tooth gap forms for steel roller chain

18. Sprocket proportions.

Refer to Figs. 16.8, 16.9 and Tables 16.14 and 16.15.

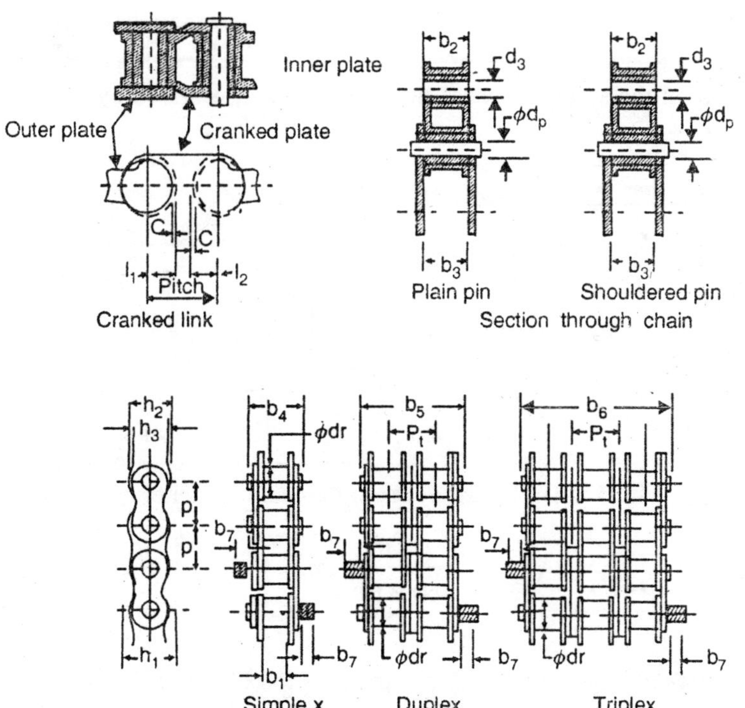

Fig. 16.8 Dimensions for transmission steel roller chains

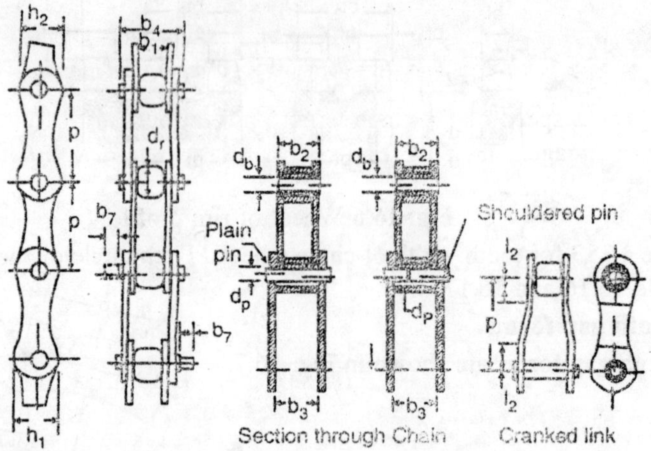

Fig. 16.9 Extended pitch precision transmission roller chain

16.5 CHORDAL (OR POLYGONAL) EFFECT IN CHAIN

The chain passes around the sprocket as a series of chordal links. This action is similar to that of a non-slipping belt wrapped around a rotating polygon. The chordal action on a sprocket having only four teeth is shown in Fig. 16.10. In Fig. 16.10 (*a*), the chain link AB is at a distance of $\left(\frac{D}{2}\right)$ from the centre of the sprocket wheel. Its linear velocity is given by,

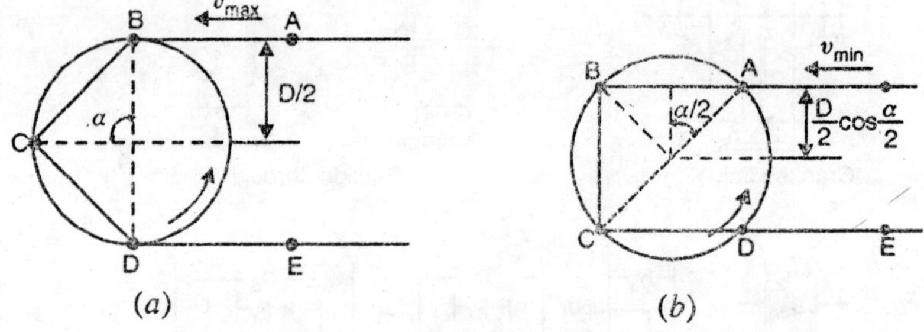

Fig. 16.10 Chordal action on a chain sprocket

$$v_{max} = \frac{\pi DN}{60 \times 10^3} \text{ m/s} \qquad ...(1)$$

As the sprocket rotates through an angle $\left(\frac{\alpha}{2}\right)$, the position of the chain link AB is shown in Fig. 16.10 (*b*). In this case, the link is at a distance of $\left(\frac{D}{2}\right)\cos\left(\frac{\alpha}{2}\right)$ from the centre of the sprocket and its linear velocity is given by,

$$v_{min} = \frac{\pi DN \cos\left(\frac{\alpha}{2}\right)}{60 \times 10^3} \text{ m/s} \quad ...(2)$$

Therefore, the linear speed of the chain is not uniform but varies from v_{max} to v_{min} during every cycle of tooth engagement. This results in a pulsating and jerky motion. The variation in velocity is given by,

$$(v_{max} - v_{min}) \propto \left[1 - \cos\left(\frac{\alpha}{2}\right)\right]$$

Or

$$\propto \left[1 - \cos\left(\frac{180°}{z}\right)\right]$$

In order to decrease the velocity variation, the number of teeth on the sprocket should be increased. For smooth operation at moderate and high speeds, it is considered a good practice to use a driving sprocket with at least 17 teeth. From durability and noise considerations, the minimum number of teeth on the driving sprocket is 19 or 21. It is better to use odd number of teeth on the driving sprocket with an even number of chain links for more uniform wear. This also ensures that the same link does not come into contact with the same sprocket teeth in every rotation. The maximum number of teeth recommended on the sprocket wheel are 100 to 120.

Example 16.1 Design a chain drive to connect a 7.5 kW, 1440 rpm electric motor to a centrifugal pump running at 720 rpm. The service conditions involve moderate shocks.

Solution.

Given : P = 7.5 kW, N_1 = 1440 rpm, N_2 = 720 rpm, $i = 2$

1. Preliminary decisions.

In order to reduce polygonal effect, the number of teeth on the driving sprocket for transmission ratio 2 are,

$$z_1 = 27 \text{(Table 16.4)}$$

It is further assumed that the chain is a simplex roller chain with a single strand.

2. Minimum factor of safety, $n = 11.7 + \frac{(13.2 - 11.7)(1440 - 1200)}{400} = 12.6$ (Table 16.8)

3. We select ISO 10B simplex roller chain having following specifications (Table 16.14): Pitch, p = 15.875 mm, weight per meter length, w = 8.92 N/m, breaking load, F_u = 22.26 kN

4. Pitch diameter of driving sprocket, $d_1 = \frac{p}{\sin\left(\frac{180°}{z_1}\right)}$

$$= \frac{15.875}{\sin\left(\frac{180°}{27}\right)} = 136.74 \text{ mm}$$

5. Number of teeth on driven sprocket, $z_2 = i \times z_1 = 2 \times 27 = 54$

6. Pitch line velocity, $v = \dfrac{\pi d_1 N_1}{60 \times 10^3} = \dfrac{\pi \times 136.74 \times 1440}{60 \times 10^3} = 10.31$ m/s

7. Centre distance, C = 40*p* = 40 × 15.875 = 635 mm

Length of chain in pitches,
$$L_p = \frac{2C}{p} + \frac{z_1 + z_2}{2} + \frac{p(z_2 - z_1)^2}{4C\pi^2}$$

$$= 2 \times 40 + \frac{(27 + 54)}{2} + \frac{15.875(54 - 27)^2}{4 \times 635 \times \pi^2}$$

$$= 120.96 \text{ or } 121 \text{ links}$$

8. Service factor :

$$K_s = K_1K_2K_3K_4K_5K_6$$
K_1 = 1.25 for load with minor shocks
K_2 = 1.0 for adjustable supports
K_3 = 1.0 for C = (30 – 50) *p*
K_4 = 1.0 for horizontal drive
K_5 = 1.0 for drop lubrication
K_6 = 1.0 for single shift of 8 hours per day
K_s = 1.25 × 1 × 1 × 1 × 1 × 1 = 1.25

9. Power transmitted on the basis of F_u.

$$P = \frac{F_u v}{nK_s} = \frac{22700 \times 10.31}{12.6 \times 1.25} = 14.859 \text{ kW}$$

10. Tangential force,
$$F = \frac{10^3 P}{v} = \frac{10^3 \times 7.5}{10.31} = 727.45 \text{ N}$$

Centrifugal force, $F_c = \dfrac{wv^2}{g} = \dfrac{8.92 \times (10.31)^2}{9.81} = 96.65$ N

Sag tension, $F_s = K_{sag}\, wC$

K_{sag} = 6 for horizontal position of drive (Table 16.10)

F_s = 6 × 8.92 × 0.635 = 33.985 N

11. Actual factor of safety, $n' = \dfrac{F_u}{F + F_c + F_s}$

$$= \frac{22260}{727.45 + 96.65 + 33.985}$$

$$= 25.94 > 12.6$$

Hence the design is safe.

16.6 ROPES

Ropes are used for long distance transmission and lifting of loads in mines, drilling rigs, multistorey building construction, etc. Basically there are two types of ropes : fibre ropes and metallic ropes. Fibre ropes are made of hemp, cotton, sisal, manila, jute, nylon, poly-propylene and coir. Metallic ropes are made of steel, aluminium alloys, copper, bronze and stainless steel. Fibre ropes are generally used where flexibility is of utmost importance, low power transmission, as in instrument devices. Metallic ropes are generally made of steel wire with either fibre core or metallic core.

16.6.1 Selecting Fibre Ropes

Fibre ropes are selected based on the load to be lifted. A factor of safety of about 35 is acceptable. The velocity should be from 15 – 30 m/s. The diameter of rope sheave should be (36 – 40) d, where d = rope diameter. The number of grooves on the pulley should not be more than 24. Fibre ropes in pulley groove are show in in Fig. 16.11.

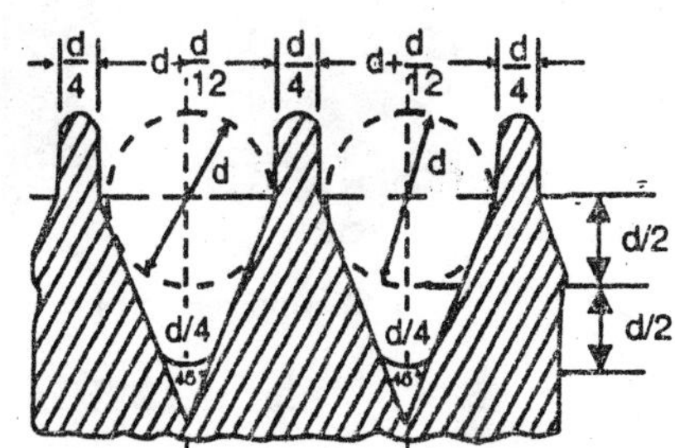

Fig. 16.11 Fibre rope pully grooves

$$\text{Length of rope, } L = \frac{\pi(D_1 + D_2)}{2} + (D_2 - D_1)\sin^{-1} + \left[\frac{D_2 - D_1}{2C}\right] + 2C\cos\beta \quad ...(16.24)$$

Where D_1, D_2 = diameters of smaller and bigger pulley respectively

C = centre distance

β = semi-groove angle of pulley

Minimum centre distance, $C_{min} = (D_2 - D_1) + 1.5D_2$ ...(16.25)

Considering centrifugal tension,

$$\frac{F_1 - F_c}{F_2 - F_c} = \exp(\mu_e \theta) \quad ...(16.26)$$

Where $\mu_e = \dfrac{\mu}{\sin\beta}$

$$\text{Total power transmitted, } P = \frac{n(F_1 - F_2)v}{10^3} kW \quad ...(16.27)$$

Where n = number of ropes

v = rope velocity, m/s

Fibre ropes are joined together either by splicing or with the help of eyelet thimbles.

16.6.2 Steel Wire Ropes

A wire rope is made up of strands and a strand is made up of one or more layers of wires. The number of strands in a rope denotes the number of groups of wires that are laid over the central core. For example a 6 × 19 wire rope construction means that the rope has 6 strands and each

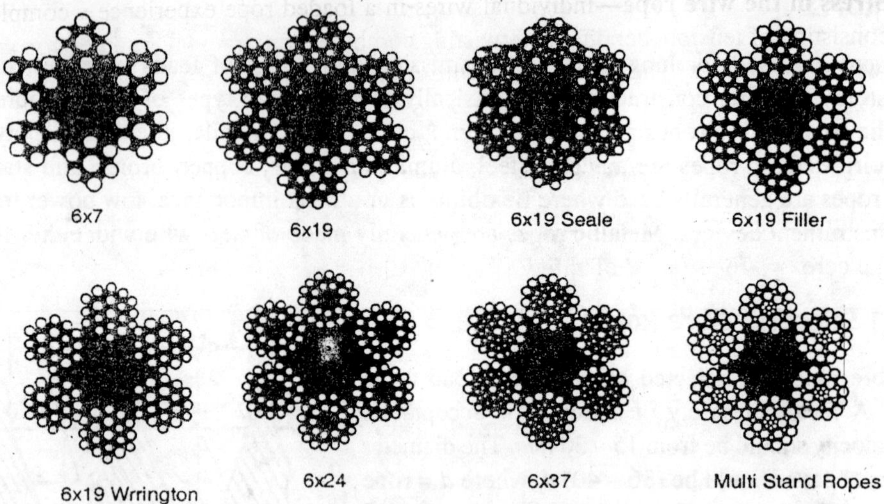

Fig. 16.12 Examples of rope construction, designation

strand is composed of 19(12/6/1) wires. It means that there are 114 wires in the rope. The tensile designation of the wire rope corresponds to the minimum tensile strength of the wire in N/mm^2.

Some of the steel wire ropes are shown in Fig. 16.12.

Common wire rope applications are given in Table 16.18.

16.6.3 Selection of a Wire Rope

The following points must be considered while making a choice for a wire rope :

1. Strength
2. Abrasion resistance
3. Flexibility
4. Resistance to crushing
5. Fatigue strength, and
6. Corrosion resistance.

1. Number of bends—When a wire rope passes over drums and pulleys, it bends. It has been found that rope life is approximately inversely proportional to the number of bends. One bend means the transition of the rope either from its straight position to a bent one or from a bent position into a staight one. Reverse bending is equal to two single bends towards the same side.

The ratio of $\frac{D_{min}}{d}$ (D_{min} = minimum diameter of pulley, d = rope diameter) for a given number of bends are given in Table 16.19.

2. Diameter of wire rope—The diameter of wire rope is approximately given by,

$$d = 1.5\, d_w \sqrt{i} \qquad \text{...(16.28)}$$

where i = number of wires in the rope

d_w = wire diameter

3. Stress in the wire rope—Individual wires in a loaded rope experience a complex state of stress consisting of tension, bending and twisting combined with mutual compression and rubbing of the wires and strands. The stress in the wire rope is given by,

$$\sigma = \frac{\sigma_{br}}{K_f} = \frac{W}{A} + \frac{d_w E_r}{D_{min}} \qquad ...(16.29)$$

Where σ_{br} =ultimate breaking strength of the rope wire material, N/mm^2

K_f = rope factor of safety (Table 16.20)

A = area of useful cross – section of the rope, mm^2 (Table 16.21)

W = tension in the rope, N

E_r = $\frac{3E}{8}$, corrected elastic modulus of rope accounting for torsion effects in the wires, N/mm^2 (Table 16.22)

4.Maximum allowable tension in the rope :

$$W_a = \frac{F_{br}}{K_f} \text{ N} \qquad ...(16.30)$$

Where F_{br} = breaking strength of the rope, N (Tables 16.23 and 16.24)

5.Steel rope weight and breaking load

The approximate steel rope weight is given by,

$$W_r = Kd^2 \text{ N/m length} \qquad ...(16.31)$$

Where K =empirical factor for weight of steel cored ropes (Table 16.22)

d = diameter of rope, mm

Minimum breaking load,

$$F_{br} = \frac{K'd^2 R_0}{10^3} \text{ N}$$

where R_0 = tensile grade of wire, N/mm^2 ...(16.32)

K = empirical factor for the minimum breaking load for a given rope construction Refer to Table 16.22 for values of K'

16.6.4 Stresses in Hoisting Ropes

The following stresses are induced in hoisting ropes :

1. Direct tensile stress due to load hoisted and the weight of the rope.
2. Stresses due to bending of the rope over the sheave.
3. Starting stresses, and
4. Streses due to change of rope speed including stops.

1. Direct tensile stress—The direct tensile stress is given by,

Effective load, $W_e = W + W_r$

$$\sigma_d = \frac{W + W_r}{A} \qquad ....(16.33)$$

Where W = load lifted

W_r = self weight of the rope

A = area of cross – section of the rope

$$= \frac{i\pi d_w^2}{4}$$

2. Bending stresses—The bending stresses are induced in the rope when the rope passes over the sheaves and drums. The bending stress is given by,

$$\sigma_b = \frac{d_w E_r}{D} \qquad ...(16.34)$$

Where D = sheave diameter

Bending load, $W_b = \sigma_b \times A$

Effective load, $W_e = W + W_r + W_b$

3. Starting stresses—If the rope starts from rest without any slackness and acquires a velocity Tm/s in t seconds, then acceleration of rope is,

$$a = \frac{v}{t}$$

Acceleration load, $W_a = \frac{(W + W_r)a}{g}$

Effective load, $W_e = W + W_r + W_b + W_a$ ...(16.35)

If there is *a* slack in the rope of amount *h* before starting, then impact load in the rope during starting is given by,

$$W_{st} = (W + W_r)\left[1 + \left\{1 + \frac{v_{st}^2 E_r}{\sigma L g}\right\}^{1/2}\right] \qquad ...(16.36)$$

Where v_{st} = velocity of rope at the instant when the rope is taut = $(2ah)^{1/2}$

$\sigma = \frac{W + W_r}{A}$, is the static stress in the rope

L = length of the rope

Effective load during starting, $W_e = W_{st} + W_b$ ...(16.37)

4. Stresses due to change in speed or sudden stopping of hoisting drum.

(*a*)Let the speed of rope change from v_1 to v_2 in time *t*. Then acceleration of rope.

$$a = \frac{v_2 - v_1}{t}$$

Acceleration load, $W_a = \frac{(W + W_r)\,a}{g}$

Effective load, $W_e = W + W_r + W_a + W_b$ ...(16.38)

(*b*) Let x = distance in which the load being lifted is stopped.

L = length of the rope

Then
$$x_s = \frac{(W + W_r)\,L}{AE_r}$$

$$\frac{x}{x_s} = \frac{\sigma_{sud}}{\sigma}$$

where σ_{sud} = sudden stress in the rope

Sudden load, $W_{sud} = \sigma_{sud} \times A$

Effective load, $W_e = W + W_r + W_b + W_{sud}$ ....(16.40)

5. Factor of safety.

$$\text{Factor of safety} = \frac{\text{Breaking load}}{\text{Maximum effective load on the rope}} \quad \text{....(16.41)}$$

Example 16.2 Select a suitable steel wire rope for a mine hoist carrying a load of 2 tonnes to be lifted from a depth of 100 m. A rope speed of 10 m/s must be attained in 10 seconds.

Solution.

1 Preliminary decisions :

Minimum factor of safety for mine ropes = 10 (Table 16.20.

The rope is also subjected to sudden and bending loads, each of which double the load. Therefore, a nominal factor of safety of about 40 may be assumed.

2. Breaking load for selection of rope, $W_{br} = 2 \times 40 = 80$ tonnes $\approx$ 785 kN

3. We find that 6 × 19(9/9/1) steel wire rope with steel core and 1570 tensile designation of wire gives minimum breaking load of 898 kN with nominal diameter of 40 mm.
Approximate mass = 656 kg/100 m [Table 16.22 (*d*)]

4. Average diameter of sheave, D = 45*d* = 45 × 40 = 1800 mm (Table 16.18)

5. Bending load,
$$W_b = \frac{AE_r d_w}{D}$$

$A = 0.38d^2$ (Table 16.21)

$E_r = 64700$ N/mm² (Table 16.22)

$d_w = 0.063d$ (Table 16.21)

$$W_b = \frac{(0.38 \times 40^2) \times 64700 \times 0.063 \times 40}{1800}$$

$= 55073$ N

6. Acceleration, a
$$= \frac{v}{t} = \frac{10}{10} = 1 \text{ m/s}^2$$

7. Weight of the rope,
$$W_r = \frac{656 \times 9.81 \times 100}{100} = 6.435 \text{ kN}$$

Weight to be hoisted, $W = 2 \times 9.81 = 19.62$ kN

8. Acceleration load, $W_a = \frac{(W + W_r)\, a}{g} = \frac{(19.62 + 6.435) \times 1}{9.81} = 16.656$ kN

9. Effective load, W_e

(*a*) Starting load without slack, $W_{st} = 2\,(W + W_r) = 2\,(19.62 + 6.435) = 52.11$ kN

$$W_e = W_{st} + W_b = 52.11 + 55.073 = 107.183 \text{ kN}$$

(*b*) During the first 10 seconds, after starting.

$$W_e = W + W_r + W_b + W_a$$
$$= 19.62 + 6.435 + 55.073 + 16.656 = 97.784 \text{ kN}$$

(*c*) Maximum effective load, $(W_e)_{max} = 107.183$ kN

10. Operating factor of safety $= \frac{\text{Breaking strength}}{(W_e)_{max}}$

$$= \frac{898}{107.183} = 8.378 < 10$$

We observe that the operating factor of safety is less than the minimum desired. If we increase the sheave diameter to 3000 mm, then bending load, $W_b = 33.044$ kN

$$W_e = 75.755 \text{ kN giving FOS} = 11.85 > 10.$$

16.7 WIRE ROPE SHEAVES

Small sheavas are usually cast in one piece without ribs. Large sheaves are provided with ribs and holes or with cross-shaped spokes. The arms of the sheave may be elliptical in cross-section. Sheaves are freely mounted on fixed axles on anti-friction bearings or bronze bushings. The ratio of pulley hub length to axle diameter should be 1.5 to 1.8. The unit bearing pressure is :

$$p = \frac{W}{l_a d_a} \qquad \text{....(16.42)}$$

Where

l_a = length of axle bushing

d_a = diameter of sheave axle

Ratio of drum and sheave diameter to rope diameter is given in Table 16.25 and dimensions of sheave in Table 16.26.

16.8 STEEL WIRE ROPE DRUMS

With a power drive the drum should always be provided with helical grooves so that the rope winds up uniformly and is less subject to wear. The radius of the helical grooves should be selected so as to prevent jamming of the rope. Dimensions of drum grooves are given in Table 16.29. Annular or concentric grooves should not be greater in depth than 10 per cent of the rope diameter. The clearance between adjacent rope is specified in Table 16.30.

Drums with one coiling rope have only one helix—right hand; drums designed for two rope members are provided with two helices—right hand and left hand (Fig. 16.15). The number of turns on the drum for one rope member is :

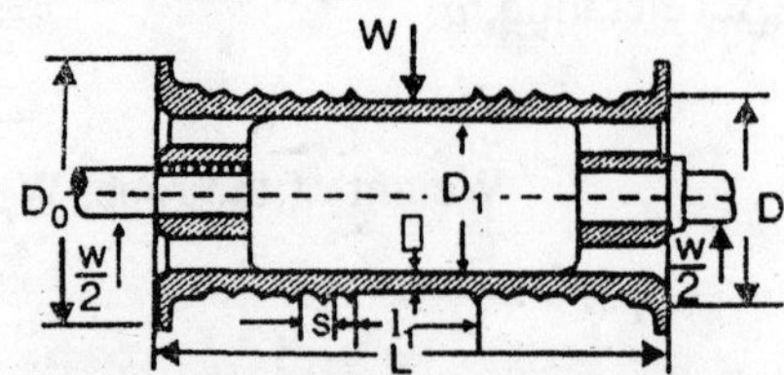

Fig. 16.15 Wire rope drum

$$z = \frac{Hi}{\pi D} + 2 \qquad \text{...(16.43)}$$

where i = ratio of pulley system

D = drum diameter

H = height to which the load is raised

The factor 2 accounts for the holding turns.

$$\text{Length of drum, } L = \left(\frac{Hi}{\pi D} + 7\right) s \text{ for one rope.} \qquad \text{...(16.44a)}$$

$$= \left(\frac{2Hi}{\pi D} + 12\right) s + l_1 \text{ for two ropes.} \qquad \text{...(16.44b)}$$

where s = pitch of the helix

l_1 = space in the middle of the drum between the right hand and left hand helices, dictated by design.

Wall thickness of a cast iron drum,

$$w = 0.02D + 6 \text{ to } 10 \text{ mm} \qquad \text{...(16.45)}$$

$$\text{Outside diameter of drum, } D_0 = D + 6d \qquad \text{....(16.46)}$$

16.8.1 Stresses in Rope Drums

$$\text{Compressive stress in drum, } \sigma_c = \frac{W}{w \times s} \qquad \text{....(16.47)}$$

$$\text{Maximum bending stress, } \sigma_b = \frac{8WLD}{\pi (D^4 - D_i^4)} \qquad \text{...(16.48)}$$

$$\text{Maximum torque on drum, } T_{max} = \frac{W(D+d)}{2} \qquad \text{...(16.49)}$$

$$\text{Maximum shear stress, } \tau_{max} = \frac{16T_{max.} \times D}{\pi (D^4 - D_i^4)} \qquad \text{....(16.50)}$$

Rope drums are generally made of grey cast iron for which maximum principal stress theory is applicable. Thus maximum stress in the drum becomes :

$$\sigma = \sqrt{\sigma_b^2 + 4\tau_{max}^2 + \sigma_c^2} \le \sigma_{allowable} \qquad \text{....(16.51)}$$

16.8.2 Holding Capacity of Wire Rope Reel

The holding capacity of the reel (L) in metres in any size length may be calculated from the following formula :

$$L = \frac{\pi (A + D_r) \times A \times C}{1000d} \qquad \text{...(16.52)}$$

where A = depth of rope layer in mm = $\frac{1}{2}(D_d - D_r)$

D_d = diameter of drum in mm as measured over the outermost layer filling the drum.

D_r = diameter of reel barrel in mm.

C = width between reel flanges, mm

d = rope diameter in mm.

The dimensions for fastening ropes on drums are given in Table 16.31.

EXERCISES

16.1 Design a roller chain to transmit power from a 25 kW, 750 rpm electric motor to a reciprocating pump running at 250 rpm. The pump is to operate for 24 hours continuously per day and centre distance is 0.5 m. Calculate the number of teeth on each sprocket, chain pitch and size.

16.2 Design a chain drive to run a blower at 600 rpm from a 10 kW, 1440 rpm electric motor. The centre distance is 1 m.

16.3 A rope drive is required to transmit 750 kW from a pulley of diameter 1.25 m running at 500 rpm. The safe pull in each rope is 2250 N and the mass of the rope is 1 kg/m length. The angle of lap and groove angle is 150° and 45° respectively. Find the number of ropes required for the drive if the coefficient of friction between the rope and the pulley is 0.3.

16.4 Select a suitable steel wire rope to lift a load of 15 kN from a well 100 m deep. The rope should have a minimum factor of safety of 8. The weight of the bucket is 5 kN. The load is lifted up with a maximum speed of 2.5 m/s which is attained in one second.

Table 16.1 Preferred transmission ratios for chain drive, R 20 series

i = 1, 1.12, 1.25. 1.4, 1.6, 1.8, 2, 2.25, 3.15, 4, 4.5, 5, 5.6, 6.3 and 7.1.

Table 16.2 Maximum velocity for various types of chains, rpm

Type of chain	*Number of sprocket teeth*	*Chain pitch p, mm*				
	z	*12*	*15*	*20*	*25*	*30*
Roller chain	15	2300	1900	1350	1150	1000
	19	2400	2000	1450	1200	1050
	23	2500	2100	1500	1250	1100
	27	2550	2150	1550	1300	1100
	30	2600	2200	1550	1300	1100
Silent chain	17 to 35	12.7	15.87	19.05	25.4	31.75
		3300	2650	2200	1650	1300

Table 16.3 Preferred range of teeth on roller chain sprockets

Sprockets for steel roller chains z = 17, 19, 21, 23, 25, 38, 57, 76, 95 & 114
Sprockets for extended pitch precision chains z = 7, 9, 10, 11, 13, 19, 27, 38 and 57

Table 16.4 Recommended number of teeth on smaller sprocket

Type of chain	*Number of teeth at speed ratio*				
	1—2	*2—3*	*3—4*	*4—5*	*5—7*
Roller	31–27	27–25	25–23	23–21	21–17
Silent	40–35	35–31	31–27	27–23	23–19

Table 16.5 Standard pitches for transmission steel roller chains

Preferred pitches (mm)— 8.0, 9.325, 12.70, 15.875, 19.05, 25.40, 31.75 38.10, 44.45, 50.80, 63.50, 76.20, 88.90, 101.60 and 114.30

IS : 2403—1975

Table 16.6 Centre distance in chain drive

Transmission ratio *min*	*K* *mm*	*Minimum centre distance,* C_{min} *mm*
3	1 + (30 to 50/C*)	$(d_{a1} + d_{a2})$ + (30 to 50 mm)
3 to 4	1.2	$0.3\ (d_{a1} + d_{a2})$
4 to 5	1.3	$0.65\ (d_{a1} + d_{a2})$
5 to 6	1.4	$0.7\ (d_{a1} + d_{a2})$
6 to 7	1.5	$0.75\ (d_{a1} + d_{a2})$

$c^* = \frac{1}{2}(d_{a1} + d_{a2})$, d_a = tip dia. of sprocket.

Table 16.7 Condensed values of k in chain length, pitches

$\dfrac{z_1 - z_2}{C_p}$	0.1	1.0	2.0	3.0	4.0	5.0	6.0
k	0.02533	0.02538	0.02525	0.02584	0.02631	0.02704	0.02828

Table 16.8 Factor of safety K_f for chain drives

Type of chain	*RPM of smaller sprocket*								
	50	*200*	*400*	*600*	*800*	*1000*	*1200*	*1600*	*2000*
Bush roller chain									
p = 12 to 15 mm	7	7.8	8.55	9.35	10.2	11.0	11.7	13.2	14.8
= 20 to 25 mm	7	8.2	9.35	10.3	11.7	12.9	14.0	16.3	—
= 30 to 35 mm	7	8.55	10.2	13.2	14.8	16.3	19.5	—	—
Silent chain									
p = 12.7 to 15.87 mm	20	22.2	24.2	28.7	29.0	31.0	33.4	37.8	42.0
= 19.05 to 25.4 mm	20	23.4	26.7	30.0	33.4	36.8	40.0	46.5	53.5

Table 16.9 Service Factor K_s

Service factor, K_s	= $K_1, K_2, K_3, K_4, K_5, K_6$
Load factor K_1	= 1.0 for constant load = 1.25 for variable load with mild shock = 1.50 for heavy shock loads.
Factor for distance regulation K_2	= 1.0 for adjustable supports = 1.1 for chains with idlers = 1.25 for fixed supports
Factor for centre distance K_3	= 1.25 for $c_p < 25$ = 1.0 for c_p = 30 to 50 = 0.8 for c_p = 60 to 80
Factor for position of shaft K_4	= 1.0 for inclination upto 60° to horizontal. = 1.25 for inclination over 60°
Lubrication factor K_5	= 0.8 for continuous lubrication = 1.0 for drop lubrication = 1.5 for periodic lubrication.
Rating factor K_6	= 1.0 for 8 hours/day = 1.25 for 16 hours/day = 1.50 for continuous service

Design power = Rated power × K_s

Table 16.10 Coefficient for sag K_{sg}

Correction Factor	*Position of chain drive*			
	Horizontal	*Upto 40°*	*More than 40°*	*Vertical*
K_{sg}	6	4	2	1

Table 16.11 Allowable bearing stress p_b on rollers, N/mm²

(for $K_s = 1$ and $z_1 = 15$ to 30

Pitch	*speed of small sprocket, rpm*										
mm	*<50*	*200*	*400*	*600*	*800*	*1000*	*1200*	*1600*	*2000*	*2400*	*2800*
9.525 12.70 15.875 19.05	34.32	30.9	28.14	25.7	23.73	21.97	20.6	18.14	16.18	14.7	13.43

Table 16.12 Load factor, K_L

Load factor	*Position of chain drive*			
	Horizontal or upto 40°		*Vertical or more than 40°*	
	Contant load	*Shock load*	*Constant load*	*Shock load*
K_L	1.15	1.3	1.05	1.15

Table 16.13 Tooth width of simple chain wheels and multiplex chain wheels

Dimensions	*p ≤ 12.7 mm*	*Tol*	*p > 12.7 mm*	*Tol*
bf_1 for simple chain wheels	0.93 b_1		0.95 b_1	
bf_1 for duplex and triplex chain wheels	0.91 b_1	h14	0.93b_1	h14
bf_1 for quadruplex chain wheels and above	0.88 b_1		0.93 b_1	

Table 16.14 Chain Dimensions, Measuring Loads and Breaking Loads of Base Chains (Transmission steel roller chains) (Refer Fig. 16.8)

All dimensions in millimetres

ISO Chain Number	Pitch	Roller Dia	Width between Inner Plates	Bearing Pin Body Diameter	Chain Path Depth	Inner Plate Depth	Outer Intermediate Plate Depth	Cranked Link Dimensions*			Transverse Pitch	Width Over Inner Link	Width between Outer Plates
	p	d_r Max	b_1 Min	d_p Max	h_1 Min	h_2 Max	h_3 Max.	l_1 Min	l_2 Min	c	p_t	b_2 Max	b_3 Min
05B	8.00	5.00	3.00	2.31	7.37	7.11	7.11	3.71	3.71	0.08	5.64	4.77	4.90
06B	9.525	9.53	5.72	3.28	8.52	8.26	8.26	4.37	4.32	0.08	10.24	8.53	8.66
08B	12.70	8.51	7.75	4.45	12.07	11.81	10.92	5.66	6.12	0.08	13.92	11.30	11.43
10B	15.875	10.16	9.65	5.08	14.99	14.73	13.72	7.11	7.62	0.10	16.59	13.28	13.41
12B	10.05	12.07	11.63	5.72	16.39	16.13	16.13	8.33	8.33	0.10	19.46	15.62	15.75
16B	25.40	15.88	17.02	8.28	21.34	21.08	21.08	11.15	11.15	0.13	31.88	25.45	25.58
20B	31.75	19.05	19.56	10.19	26.28	26.42	26.42	26.42	13.89	0.15	36.45	29.01	29.14
24B	38.10	25.40	25.40	14.63	33.73	33.40	33.40	17.25	17.55	0.18	48.36	37.92	38.05
28B	44.45	27.94	30.99	15.90	37.46	37.08	37.08	19.51	19.51	0.20	59.56	46.58	46.71
32B	50.80	29.21	30.98	17.81	42.72	42.29	42.29	22.20	22.50	0.50	58.55	45.57	45.70
40B	63.50	39.76	38.10	22.89	54.49	52.96	52,96	27.76	27.76	0.20	72.29	55.75	55.88
48B	76.20	48.26	45.72	29.24	64.52	63.88	63.88	33.45	33.45	0.20	91.21	70.56	70.69
56B	88.90	53.98	53.34	34.32	78.64	77.85	77.85	40.61	40.61	0.20	106.60	71.33	81.46
64B	101.60	63.50	60.96	39.40	91.08	90.17	90.17	47.07	47.07	0.20	119.89	90.02	92.15
72B	114.30	72.39	68.51	44.48	104.67	102.63	103.63	53.37	53.37	0.20	136.27	103.81	103.94

Table 16.14 continued.

ISO Chain Number	Width Over Bearing Pins			Additional Width for Joint Fastener+	Measuring Load, N			Breaking Load, kN			Average weight Wf, N/m		
	Simple	Duplex	Triplex		Simple	Duplex	Triplex	Simplex	Duplex	Triplex	Simplex	Duplex	Triplex
	b_4 Max	b_5 Max	b_6 Max	b_7 Max				Min	Min	Min			
05B	8.6	14.3	19.9	3.1	49	98	147	4.51	7.84	11.18	1.76	3.24	5.10
06B	13.5	23.8	34.0	3.3	78.5	147	216	8.92	16.96	24.96	4.02	7.55	10.69
08B	17.0	31.0	44.9	3.9	127.5	255	382	17.85	31.18	44.52	6.86	12.94	19.12
10B	19.6	36.2	52.8	4.1	196.1	392	588	22.26	44.52	66.78	8.92	17.85	26.77
12B	22.7	42.2	61.7	4.6	284.4	559	843	28.93	57.86	88.79	11.47	23.14	34.71
16B	36.1	68.0	99.9	5.4	500.1	1002	1491	42.27	84.53	126.80	26.48	51.97	76.98
20B	43.2	79.7	116.1	6.1	774.7	1559	2334	64.53	129.05	193.58	35.79	70.91	116.70
24B	53.4	101.8	150.2	6.6	1108	2226	3334	97.87	195.74	293.62	67.17	142.20	196.13
28B	65.1	124.7	184.3	7.4	1510	3020	4440	129.05	258.11	387.17	—	—	—
32B	67.4	126.0	184.5	7.9	2000	4001	6001	169.07	338.13	507.20	—	—	—
40B	82.6	154.9	227.2	10.2	3118	6227	9346	262.52	525.05	787.57	—	—	—
48B	99.1	190.4	281.6	10.5	4452	8895	13347	400.40	800.71	1201.02	—	—	—
56B	114.6	221.2	—	11.7	6090	12,189	—	542.70	1085.40	—	—	—	—
64B	130.9	250.8	—	13.9	7963	15,926	—	711.77	1423.43	—	—	—	—
72B	147.4	283.7	—	14.3	10,101	20,192	—	898.58	1797.71	—	—	—	—

IS : 2403—1975

Table 16.15 Chain Dimensions, Measuring Loads and Breaking Loads for extended pitch transmission precision roller chains

All dimensions in millimetres

Chain Number	Pitch	Roller dia	Width between Inner Plates ner	Bearing pin body dia	Bush bore	Chain path depth	Plate depth	Cranked Link Dimension	Width over Inner Link	Width between outer plates	Width Over Bearing pin	Addl. Width for Joint* Fast-	Measuring load	Breaking load	Weight per m
	P	d_3 Max	b_1 Min	d_p Max	d_b Min	h_1 Min	h_1 Max	l_2 Min	b_2 Max	b_3 Max	b_4 Max	b_7 Max.	N	F_u kN	N
208A	25.40	7.92	7.95	3.96	4.01	12.33	12.07	6.9	11.18	11.31	17.8	3.9	127.5	13.83	4.71
208B	25.40	8.51	7.75	4.45	4.50	12.07	11.81	6.9	11.30	11.43	17.0	3.0	127.4	17.85	4.31
210A	31.75	10.16	9.53	5.08	5.13	15.35	15.09	8.4	13.84	13.97	21.8	4.1	196.1	21.77	—
210B	31.75	10.16	9.65	5.08	5.13	14.99	14.73	8.4	13.28	13.41	19.6	4.1	196.1	22.26	5.88
212A	38.10	11.91	12.70	5.94	5.99	18.34	18.08	9.9	17.75	17.88	26.9	4.6	284.4	31.18	14.22
212B	38.10	12.07	11.68	5.72	5.77	16.39	16.13	9.9	15.62	15.75	22.7	4.6	284.4	28.93	7.35
216A	50.80	15.88	15.88	7.92	7.79	24.39	24.13	13.0	22.61	22.74	33.5	5.4	500.1	55.60	—
216B	50.80	15.88	17.02	8.28	8.33	21.34	21.08	13.0	25.25	25.58	36.1	5.4	500.1	42.27	—
220A	63.50	19.05	19.05	9.53	9.58	30.48	30.18	16.0	27.46	27.59	41.1	6.1	774.7	86.79	—
220B	63.50	19.05	19.56	10.19	10.24	26.68	26.42	16.0	29.41	29.14	43.2	6.1	774.7	64.53	—
224A	76.20	22.23	25.40	11.10	11.25	36.55	36.20	19.1	34.46	35.59	50.8	6.6	1108.1	124.54	—
224B	76.20	25.40	25.40	14.63	14.68	33.73	33.40	36.20	37.92	38.05	53.4	6.6	1108.1	97.87	—
228B	88.90	27.94	30.99	15.90	15.95	37.46	37.08	21.3	46.58	46.71	65.1	7.4	1510.2	129.05	—
232B	101.60	29.21	30.99	17.81	17.86	42.72	42.29	24.4	45.57	45.70	67.4	7.9	2000.5	169.07	—

IS : 3542—1980

Table 16.16 Pitch Circle Diameters for Chain Wheels. For Transmission Steel Roller Chains

The following table gives correct pitch circle diameters for wheels to suit a chain of unit pitch (1 mm). For chain of any other pitch, the pitch circle diameter of the wheel is obtained by multiplying the figures given below by the pitch of that chain : IS : 2403—1975

Number of teeth diameter	Pitch circle diameter	Number of teeth diameter	Pitch circle diameter	Number of teeth	Pitch circle diameter	Number of teeth	Pitch circle diameter
9	2.923 8	45	14.335 6	81	25.789 5	117	37.245 7
10	3.236 1	46	14.653 7	82	26.107 8	118	37.565 0
11	3.549 4	47	14.971 7	83	26.426 0	119	37.883 3
12	3.863 7	48	15.289 8	84	26.744 3	120	38.201 6
13	4.178 6	49	15.607 9	86	27.062 5	121	38.519 8
14	4.494 0	50	15.926 0	86	27.380 7	122	38.838 1
15	4.809 7	51	16.244 1	87	27.699 0	123	39.156 4
16	5.125 8	52	16.562 2	88	28.017 2	124	39.474 6
17	5.442 2	53	16.880 3	89	28.335 5	125	39.792 9
18	5.758 8	54	17.198 4	90	28.653 7	126	40.111 2
19	6.075 5	55	17.516 6	91	28.971 9	127	40.429 5
20	6.392 5	56	17.834 7	92	29.290 2	128	40.747 8
21	6.709 5	57	18.152 9	93	29.608 4	129	41.066 0

Number of teeth diameter	*Pitch circle*	*Number of teeth diameter*	*Pitch circle*	*Number of teeth diameter*	*Pitch circle diameter*	*Number of teeth*	*Pitch circle*
22	7.026 6	58	18.4710	94	29.926 7	130	41.384 3
23	7.343 9	59	18.789 0	95	30.244 9	131	41.702 6
24	7.661 3	60	19.107 3	96	30.563 2	132	42.020 9
25	7.978 7	61	19.425 5	97	30.881 5	133	42.339 1
26	8.296 2	62	19.743 7	98	31.199 7	134	42.657 4
27	8.613 8	63	20.061 9	99	31.518 0	135	42.975 7
28	8.931 4	64	20.380 0	100	31.836 2	136	43.294 0
29	9.249 1	65	20.698 2	101	32.154 5	137	43.612 3
30	9.566 8	66	21.016 4	102	32.472 7	138	43.930 6
31	9.884 5	67	21.334 5	103	32.791 0	139	44.248 8
32	10.202 3	68	21.652 8	104	33.109 3	140	44.567 1
33	10.520 1	69	21.971 0	105	33.427 5	141	44.885 4
34	10.838 0	70	22.289 2	106	33.745 8	142	45.203 7
35	11.155 8	71	22.607 4	107	34.064 0	143	45.522 0
36	11.473 7	72	22.925 6	108	34.382 3	144	45.840 3
37	11.791 6	73	23.243 8	109	34.700 6	145	46.158 5
38	12.109 6	74	23.562 0	110	35.018 8	146	46.476 8
39	12.427 5	75	23.880 2	111	35.337 1	147	46.795 1
40	12.745 5	76	24.198 5	112	35.655 4	148	47.113 4
41	13.063 5	77	24.516 7	113	35.973 7	149	47.401 7
42	13.381 5	78	24.834 0	114	36.291 9	150	47.750 0
43	13.699 5	79	25.153 1	115	36.610 2		
44	14.017 6	80	25.471 3	116	36.928 5		

Table 16.17 Pitch Circle Diameter for Extended Pitch Precision Roller Chains

The following table gives correct pitch circle diameter, for wheels to suit a chain of unit pitch (for example, 1 mm). The pitch circle diameters for wheels to suit a chain of any other pitch are directly proportional to the pitch of the chain :

Number of teeth	*Pitch circle diameter*	*Number of teeth*	*Pitch circle diameter*	*Number of teeth*	*Pitch circle diameter*	*Number of teeth*	*Pitch circle diameter*
5	1.701 3	23	7.343 9	41	13.063 5	58½	18.630 1
5½	1.849 6	23½	7.502 6	41½	13.222 5	59	18.789 2
6	2.000 0	24	7.661 3	42	13.381 5	59½	18.948 2
6½	2.151 9	24½	7.820 0	42½	13.540 5	60	19.107 3
7	2.304 8	25	7.978 7	43	13.699 5	60½	19.266 5
7½	2.458 6	25½	8.137 5	43½	13.858 5	61	19.425 5

Number of teeth	Pitch circle diameter	Number of teeth	Pitch circle diameter	Number of teeth	Pitch circle diameter	Number of teeth	Pitch circle diameter
8	2.613 1	26	8.296 2	44	14.017 6	62½	19.584 7
8½	2.768 2	26½	8.455 0	44½	14.176 5	62	19.743 7
9	2.923 8	27	8.613 8	45	14.335 6	62½	19.902 9
9½	3.079 8	27½	8.772 6	45½	14.494 6	63	20.061 9
10	3.236 1	28	8.931 4	46	14.653 7	63½	20.221 0
10½	3.392 7	28½	9.090 2	46½	14.812 7	64	20.380 0
11	3.549 4	29	9.249 1	47	14.971 7	64½	20.539 3
11½	3.706 5	29½	9.408 0	47½	15.130 8	65	20.098 2
12	3.863 7	30	9.566 8	48	15.289 8	65½	20.857 5
12½	4.021 1	30½	9.725 6	48½	15.448 8	66	21.016 4
13	4.178 6	31	9.884 5	49	15.607 9	66½	21.175 7
13½	4.336 2	31½	10.043 4	49½	15.766 9	67	21.334 6
14	4.494 0	32	10.202 3	50	15.926 0	67½	21.493 9
14½	4.651 8	32½	10.361 2	50½	16.085 0	68	21.652 8
15	4.809 7	33	10.520 1	51	16.244 1	68½	21.812 1
15½	4.967 7	33½	10.679 0	51½	16.403 1	69	21.971 0
16	5.125 8	34	10.838 0	52	16.562 2	69½	22.130 3
16½	5.284 0	34½	10.996 9	52½	16.721 2	70	22.289 2
17	5.442 2	35	11.155 8	53	16.880 3	70½	22.448 5
17½	5.600 5	35½	11.314 8	53½	17.039 3	71	22.607 4
18	5.758 8	36	11.473 7	54	17.198 4	71½	22.766 7
18½	5.917 1	36½	11.632 7	54½	17.357 5	72	22.925 6
19	6.075 5	37	11.791 6	55	17.516 6	72½	23.084 9
19½	6.234 0	37½	11.950 6	55½	17.675 6	73	23.243 8
20	6.392 5	38	12.109 5	56	17.834 7	73½	23.403 1
20½	6.550 9	38½	12.268 5	56½	17.993 8	74	23.562 0
21	6.709 5	39	12.427 5	57	18.152 9	74½	23.721 3
21½	6.868 1	39½	12.586 5	57½	18.311 9	75	23.880 2
22	7.026 6	40	12.745 5	58	18.471 0		
22½	7.185 3	40½	12.904 5				

IS : 3842—1966

Table 16.18 Common wire rope applications

Type of service	*Type of rope*	*Sheave diameter*	
		Recommended	*Minimum*
Haulage rope Mine haulage Factor yard haulage Inclined planes Tramways Power transmission Guy wires	6 × 7	72 d	42 d
Standard hoisting rope (most commonly used rope) Mine hoists Quarries Ore docks Cargo hoists Car pullers Cranes Derricks Drodges Tram ways Well drilling Elevators	6 × 19	45 d 60–100 d 20–30 d	30 d
Extra-flexible hoisting rope	8 × 19	31 d	21 d
Special flexible hoisting rope : Steel mill ladles Cranes High speed elevators Service where sheave Diameters are limited	6 × 37	27 d	18 d

Table 16.19 D_{min}/d as a functions of the number of bends in the systems

No. of bends	1	2	3	4	5	6	7	8	9	10	11	12	13	14	15	16
D_{min}/d	16	20	23	25	26.5	28.0	30	31	32	33	34	35	36	37	37.5	38

Table 16.20 Recommended factors of safety for wire ropes, K_f

Rope application	*Factor of safety*		
Mining ropes	*10 or other figure laid down by the statutory authority*		
Wire ropes used on the cranes and other hoisting equipment. Fixed guys	Class 1	Class 2 & 3	Class 4
Unreeved rope bridles of jib cranes or ancillary appliances, such as lifting beams. Ropes which are straight between terminal fittings	3.5	4.0	4.5
Hoisting, lifting and reeved bridle systems of inherently flexible cranes (for example, mobile crawler tower, guy derrick, stiff leg derrick, where jibs are supported by ropes or where equivalent shock absorbing devices are incorporated in jib supports.	4.0	4.5	5.5
Cranes and hoists in general hoist blocks.	4.5	5.0	6.0

IS : 3973—1967

Table 16.21 Approximate Wire Rope and Sheave Data

Rope Construction	*Weight N/m*	*Wire dia (d_w) (mm)*	*Area mm^2*	*Recommended sheave dia*	
				Average	*Minimum*
6 × 7	0.0333 d^2	0.106d	0.38d^2	75d	42d
6 × 19	0.0363 d^2	0.063d	0.38d^2	45d	30d
6 × 37	0.0343 d^2	0.045d	0.38d^2	27d	18d
8 × 19	0.0339 d^2	0.050d	0.38d^2	31d	21d

Table 16.22 Numerical values of factors K and K′

Rope designation	*Ropes mass factor*			$\frac{K_2}{K_{1n}}$	$\frac{K_2}{K_{1p}}$ cored ropes	*Min. breaking load factor*		$\frac{K_2'}{K_1'}$	E_{k2} N/mm²
	Natural fibre cored ropes K_{1n}	*Polypropylene fibre cored ropes,* K_{1p}	*Steel cored ropes* K_2			*Fibre cored ropes* K_1'	*Steel croed ropes* K_2'		
	kg (100 m. mm²)								
6 × 7	.3574	0.3485	0.3485	1.10	1.135	1.3322	0.3588	1.08	68600
6 × 19	0.3460	0.3374	0.3906	1.10	1.135	0.3073	0.3319	1.08	64700
6 × 37	0.3460	0.3374	—	—	—	0.2948	—	—	
6 × 19 Seale	0.3726	0.3633	0.4099	1.10	1.135	0.3309	0.3574	1.08	64700
6 × 19 Filler	0.3802	0.3707	0.4182	1.10	1.135	0.3377	0.3647	1.08	
6 × 26 Warrington—Seale									
6 × 31 Warrington—Seale	0.3802	0.3707	0.4182	1.10	1.135	0.3299	0.3563	1.08	
6 × 36 Warrington Seale									
6 × 41 Warrington Seale									
6 × 12	0.2506	0.2306	—	—	—	0.2086	—	—	
6 × 24	0.3184	0.3041	—	—	—	0.2802	—	—	
8 × 19 Seale	0.3484	0.3362	0.4252	1.22	1.28	0.2870	0.3386	1.18	68600
8 × 19 Filler	0.3655	0.3440	0.4349	1.22	1.28	0.2936	0.3464	1.18	
17 × 7 & 18 × 7	0.3828	0.3790	0.4019	1.05	1.06	0.3186	0.3281	1.03	53900
34 × 7 & 36 × 7	0.3902	0.3863	0.4019	1.03	1.035	0.3122	0.3184	1.02	

IS : 6494—1977

Table 16.23 Wire ropes for general engineering purposes

(a) 6 × 7 (6/1) construction

Nominal dia, mm	*Approximate mass Kg/100 m*		*Minimum breaking load in kN corresponding to tensile designation of wires of*					
	Fibre core	*Steel core*	*1570*		*1770*		*1960*	
			Fibre core	*Steel core*	*Fibre core*	*Steel core*	*Fibre core*	*Steel core*
8	22.9	25.2	33	36	38	41	42	45
9	28.9	31.8	42	46	48	51	53	57
10	35.7	39.1	52	56	59	64	65	70
11	43.2	47.6	63	68	71	77	79	85
12	51.5	56.6	75	81	85	91	94	101

(*b*) 6 × 19 (12/6/1) construction (aggregate breaking load = 1.163 × tabulated load)

8	22.1	31	35	39
9	38.0	39	44	49
10	34.6	48	54	60
11	41.9	58	66	73
12	49.8	69	78	82
13	58.5	82	92	102
14	67.8	95	107	118
16	88.6	124	139	154
18	112	156	176	195
19	125	174	199	217
20	138	193	218	241
22	167	234	263	291
24	199	278	313	347
26	234	326	368	407
28	271	378	426	472
32	354	494	557	617
36	448	625	705	781
40	554	772	870	964
44	670	934	1053	1166
48	797	1112	1253	1388
52	936	1305	1471	1629

Table 16.23 continued

(*c*) 6 × 37 (18/12/6/1) construction

8	22.1	24.4	30	32	33	36	37	40
9	28.0	30.8	37	41	42	46	47	51
10	34.6	38.1	46	50	52	56	58	62
11	41.9	46.1	56	60	63	68	70	76
12	49.8	54.8	67	72	75	81	83	90
13	58.5	64.3	78	84	88	95	98	106
14	67.8	74.6	91	98	102	110	113	122
16	88.6	97.4	118	128	133	144	148	160
18	112	123	150	162	169	183	187	202
19	125	137	167	180	188	203	209	224
20	138	152	185	200	209	225	231	250
22	167	184	224	242	252	273	280	302
24	199	219	267	289	300	225	333	359
26	234	257	313	338	333	381	390	422
28*	271	–	363	–	409	–	453	–
32*	354	–	474	–	531		592	–
36*	448	–	600	–	676	–	749	–
40*	554	–	740	–	835	–	924	–
44*	670	–	896	–	1010	–	1119	–
48*	797	–	1070	–	1200		1331	–
52*	936	–	1250	–	1410	–	1562	–
56*	1085	–	1450	–	1640	–	1812	–

Aggregate load for fibre core =1.21 × tabulated load
Aggregate load for steel core =1.316 × tabulated load
*Sizes not preferred.

(*d*) 6 × 17 (8/8/1) and 6 × 19 × (9/9/1) construction

Nominal dia. mm	Approximate mass kg/100 m		Minimum breaking load in kN corresponding to tensile designation of wires of					
			1570		1770		1960	
	Fibre core	Steel core	Fibre core	Steel core	Fibre core	Steel core	Fibre core	Steel core
8	23.8	26.2	33	36	37	40	42	45
9	30.2	33.2	42	45	47	51	53	57
10	37.3	41.0	52	56	59	69	65	70
11	45.1	49.6	63	68	71	77	78	85
12	53.7	49.0	75	81	84	91	93	101

Nomimal dia. mm	Approximate mass kg/100 m		Minimum breaking load in kN corresponding to tensile designation of wires of					
			1570		1770		1960	
	Fibre core	Steel core	Fibre core	Steel core	Fibre core	Steel core	Fibre core	Steel core
13	63.0	69.3	88	95	99	107	110	118
14	73.0	80.3	102	110	115	124	127	137’
16	95.4	105	133	144	150	162	166	179
18	121	133	168	182	190	205	210	227
19	135	148	188	203	211	228	234	253
20	149	164	208	224	234	253	259	280
22	180	198	251	272	283	306	314	339
24	215	236	299	323	337	364	374	403
26	252	277	351	379	396	428	438	479
28	292	321	407	440	459	496	508	549
32	382	420	532	575	600	648	664	717
36	483	531	673	727	759	820	841	908
40	596	656	831	898	937	1012	1038	1121
44	721	794	1006	1086	1134	1225	1226	1356
48	858	944	1197	1293	1349	1458	1494	1614
52	1008	1108	1405	1517	1584	1711	1754	1894

Table 16.24 Specifications for steel wire suspension ropes for lifts elevators and hoists (*a*) 6 × 19 (12/6/1) construction

Nominal dia, mm	*Approximate mass kg/100*	*Minimum breaking load in kN corresponding to tensile designation of wires of*		
		1230	*1420*	*1570*
6	12.5	13.6	15.7	17.4
7	17.0	18.5	21	24
8	22.1	24	28	31
9	28.0	31	35	39
10	34.6	38	44	48
11	41.9	46	53	58
12	49.8	54	63	69

(*b*) **6 × 19 (9/9/1) construction**

6	13.4	14.7	16.9	18.7
7	18.3	19.9	23	25
8	23.8	26	30	33
9	30.2	33	38	42
10	37.3	41	47	52
11	45.1	49	57	63
12	53.7	59	68	75
13	63.0	69	79	88
14	73.0	80	92	102
16	95.4	104	120	133
18	121	132	152	168
19	135	147	170	189
20	149	162	188	208

(*c*) **6 × 19 (12/6/6F/1) construction**

6	13.7	15.0	17.3	19.1
7	18.6	20.4	23	26
8	24.3	27	31	34
9	30.8	34	39	43
10	38.0	42	48	53
11	46.0	50	58	64
12	54.7	60	69	76
13	64.3	70	81	90
14	74.5	81	94	104
16	97.3	106	123	136
18	123	135	155	172
19	137	150	173	191
20	152	166	192	212

The nominal diameter 19 mm is non-preferred.

Table 16.25 Ratio of drum and sheave diameter to rope diameter

Purpose	*Construction*	*Minimum ratio* upto v = 50 m/min.*		
Mining installations	*All*	*100*		
		Class 1	Class 2 & 3	Class 4
Cranes and allied hoisting equipment	6 × 37 8 × 19 Filler wire	15	17	22
	8 × 19 8 × 19 Warrington 8 × 19 Seale 37 × 7 Non-rotating	17	18	24
	6 × 24	18	19	25
	6 × 19 Filler wire	18	20	23
	6 × 19 6 × 19 Warrington 17 × 9 Non-rotating 18 × 7 Non-rotating	19	23	27
	6 × 19 Seale	24	28	35

*Increase this ratio by 8% for each additional speed of 50 m/min. IS : 3973–1967

Table 16.26 Dimensions for sheave for steel wire ropes.

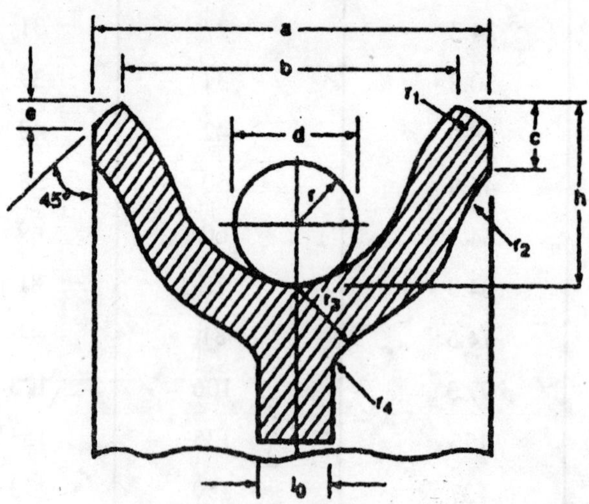

Fig. 16.13 Dimensions of sheave for steel wire ropes.

All dimensions in mm

Rope dia, d	*a*	*b*	*c*	*e*	*h*	l_0	*r*	r_1	r_2	r_3	r_4
4.8	22	15	5	0.5	12.5	8	4	2.5	2	8	6
6.2	22	15	5	0.5	12.5	8	4	2.5	2	8	6
8.7	28	20	6	1	15	8	5	3	2.5	9	6
11.0	40	30	7	1	25	10	8.5	4	3	12	8
13.0	40	30	7	1	25	10	8.5	4	3	12	8
15.0	40	30	7	1	25	10	8.5	4	3	12	8
19.5	55	40	10	1.5	30	15	12	5	5	17	10
24.0	65	50	10	1.5	37.5	18	14.5	5	5	20	15
28.0	80	60	12	3	45	20	17	6	7	25	15
34.5	90	70	15	2	55	22	20	7	8	28	20
39.0	110	85	18	2.2	65	22	25	9	10	40	30

Rudenko, N 'Materials Handling Equipment', Mir Publishers, Moscow (1969), p. 86, Table 16.

Table 16.27 Tolerance on pulley or sheave groove diameters

Nominal dia of	6 to 8	10 to 20	22 to 29	32 to 38	31 to 57	Above 57
Minimum clearance	0.80	1.60	2.40	3.20	4.00	4.80

IS : 3973–1967

The unit pressure should not exceed the values given in table 16.26.

Table 16.28 Unit pressure for sheave bearings

v, m/s	0.1	0.2	0.3	0.4	0.5	0.6	0.7	0.8	0.9	1.0	1.1	1.2	1.3	1.4	1.5
p, N/mm²	7.35	6.86	6.47	6.08	5.88	5.59	5.39	5.29	5.19	5.09	5.00	4.90	4.80	4.71	4.61

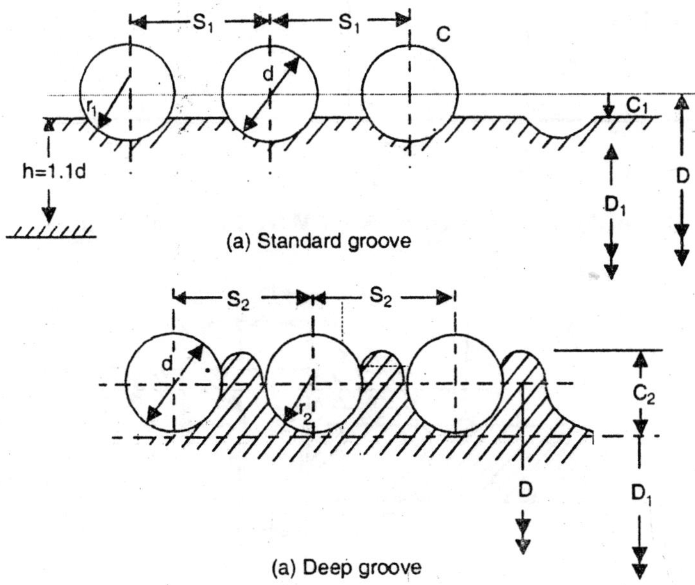

Fig. 16.14 For Table 16.29 Dimensions of drum groove.

Table 16.29 Dimensions of drum grooves

All dimensions in mm

Rope dia. d	$r_1 = 0.53\ d$	Standard groove $s_1 = 1.15\ d$	Standard groove $c_1 = 0.25\ d$	r_2	Deep groove s_2	Deep groove c_2
4.8	3.5	7	2	1.0	9	4.5
6.2	4	8	2	1.5	11	4.5
8.7	5	11	3	1.5	13	6.5
11.0	7	13	3	1.5	17	8.5
13.0	8	15	4	1.5	19	9.5
15.0	9	17	5	2.0	22	11.0
19.5	11.5	22	5	2.0	27	13.5
24.0	13.5	27	6	2.5	31	16
28.0	15,5	31	8	2.5	36	18
34.5	19	38	10	3.0	41	22
39	21	42	12	3.5	50	24.5

Rudenko, N., "Materials Handling Equipment," Mir Publishers, Moscow (1969), p. 90, table 17.

Table 16.30 Clearance between adjacent ropes

Type of groove	Rope diameter,mm From	Rope diameter,mm Upto and including	Clearance
Concentric groove, CG	–	12	1.5
	12	28	2.5
	29	36	3.0
Helical groove, HG	–	32	1.5
	32	44	2.5
	44	51	3.0

IS : 3973–1967

Table 16.31 Dimensions for Fastening Steel Wire Ropes on Drums

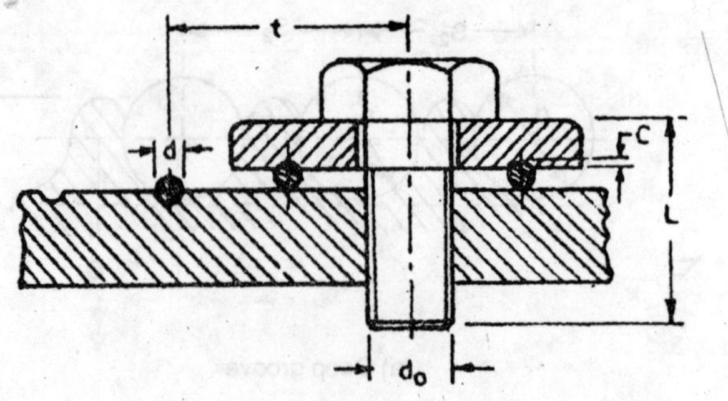

Fig. 16.16 Dimensions for fastening steel wire ropes on drums.

All dimensions in mm

Rope dia d	Locating dimensions		Screw size		Plate	
	Pitch of screws	t	d_0	l	c	Number of fastenings
(1)	(2)	(3)	(4)	(5)	(6)	(7)
4.8 6.2	28	19	8	20	2	1
7.7 8.7	34	25	10	30	4	1
11 13	52	35	16	45	5	1
15 17.5 18	53	43	18	50	7	1
19.5 20.5 21.5	58	53	22	60	8	1
23 24 26	78	60	22	60	10	1
28 28.5 30 31	78	70	27	75	12	1
32.5 34.5 36	100	85	30	100	14	2

Rudenko, N., ''Materials Handling Equipment'', Mir Publishers, Moscow, (1969), p. 94, Table 18.

17 SPUR GEARS

Chapter

17.1 INTRODUCTION

A gear may be defined as any toothed member designed to transmit or receive motion from another member by successively engaging tooth. The smaller gear is called the pinion and the bigger one the gear wheel. Gears are used in automobiles, tractors, metal cutting machine tools, hoisting and transporting machinery, rolling mills etc. Gears provide many advantages over other modes of power transmission like belts, ropes and chains, etc.

Advantages :

Gears offer the following advantages :

1. Compact drive on account of small centre distance.
2. Positive drive due to negligible slip between contacting surfaces.
3. Give higher speed ratio and can transmit higher power.
4. It is possible to transmit power between parallel, non-parallel, intersecting, and non-intersecting shafts.
5. Power can be transmitted at higher speeds.

However, the error in tooth meshing may cause undesirable vibrations and noise during operation.

Limitations :

The limitations of gear drive are :

1. Costlier than belts and chain drives.
2. Power cannot be transmitted over long distances.
3. Precise alignment of shafts is required.
4. Require continuous lubrication.

17.2 CLASSIFICATION OF GEARS

The gears may be classified as given below :

1. Based upon the shape of the gears.

   (*i*) Spur gears (*ii*) Helical and Herringbone gears (*iii*) Bevel gears
   (*iv*) Spiral gears (*v*) Hypoid gears, and (*vi*) Worm gears.

2. Based upon the orientation of the shafts.

(*i*) Gears for parallel shafts, e.g. spur gears, single helical and double helical gears.

(*ii*) Gears for intersecting shafts e.g. straight bevel, spiral bevel, zerol bevel, and face gears.

(*iii*) Gears for non-parallel and non-intersecting shafts, e.g. spiral gears, hypoid gears, and worm gears.

The various types of gears are shown in Fig. 17.1

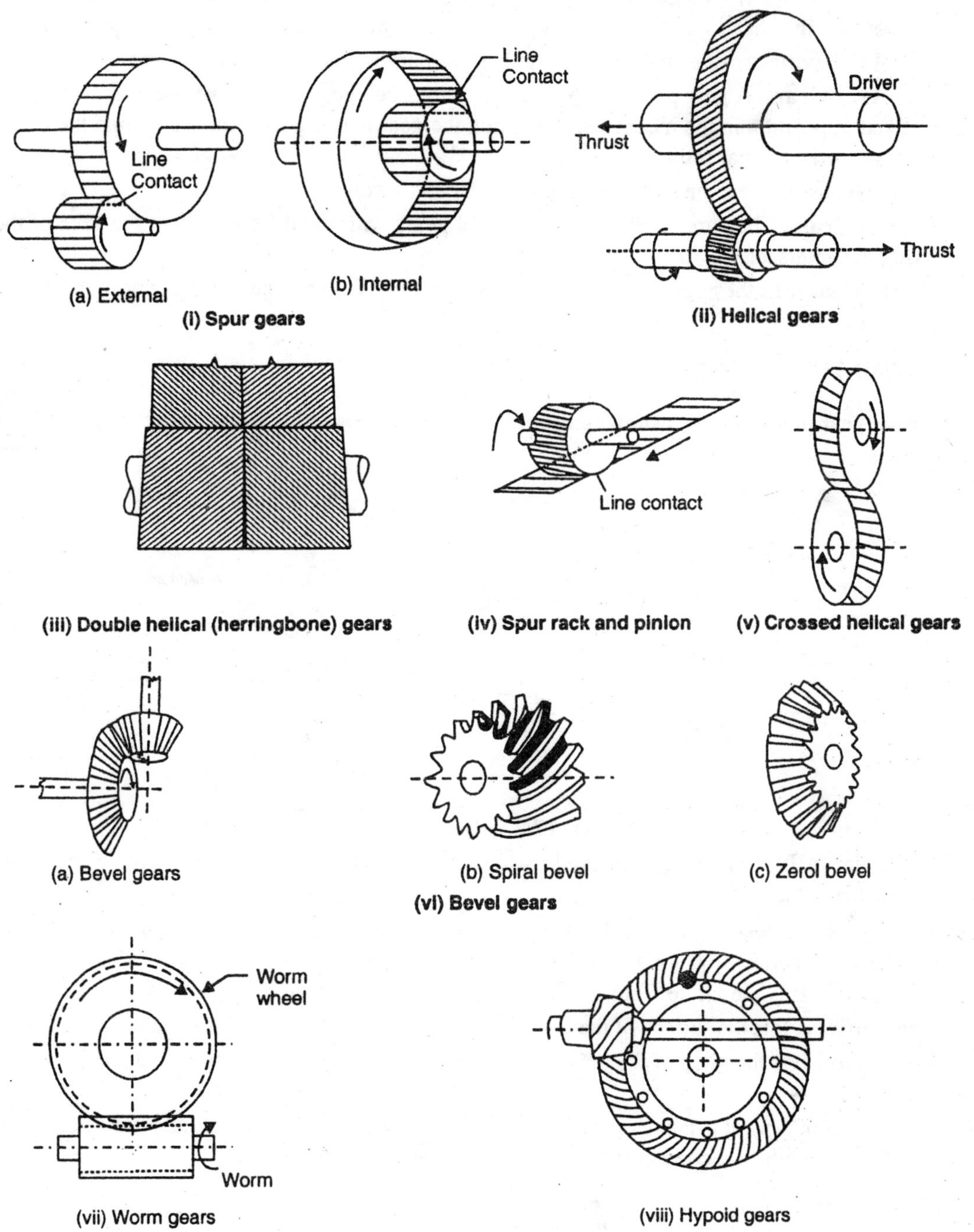

Fig. 17.1 Types of Gears

Spur Gear. It is a cylindrical gear whose tooth traces are straight line generators of the reference cylinder. The teeth are parallel to the shaft axis.

Helical Gear. In these gears, the teeth are cut in the form of a helix on the pitch cylinder between meshing gears. The teeth are inclined to the shaft axis.

Double helical or Herringbone Gears. It is a cylindrical gear in which a part of the face width of the tooth is right hand and the other part left hand, with or without a gap between them.

Spiral Gears. In these gears, the tooth traces are curved lines other than helices.

Planetary Gears. A gear pair (or a train comprising such gears) one of whose axes, instead of being fixed in position in the mechanism of which the gear pair is a part, moves around the other, is called a planetary gear train.

Bevel Gears. In these gears the reference surface is a cone.

Hypoid Gears. They are similar to spiral bevel gears with the difference that the axes of the shafts do not intersect.

Worm Gears. In these gears, one gear has screw threads. The axes of the worm and worm wheel are at right angles.

17.3 GEAR TERMINOLOGY

A spur gear pair in mesh is shown in Fig. 17.2. The various terms related to gears are defined as follows :

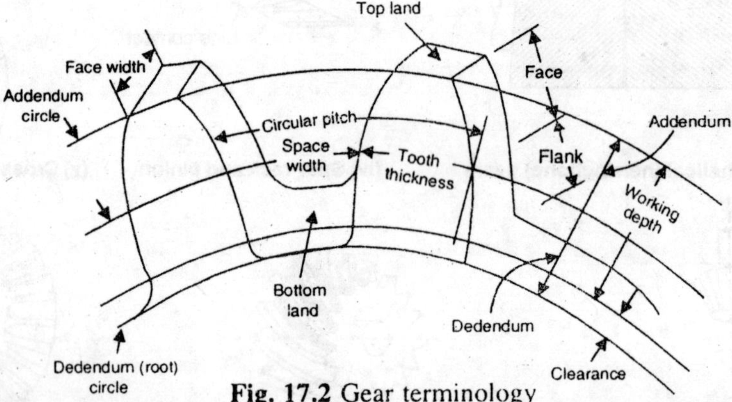

Fig. 17.2 Gear terminology

Pitch circle. It is an imaginary circle which by pure rolling action, would give the same motion as the actual gear.

Pitch circle diameter (*d*). It is the diameter of a circle which by pure rolling action would produce the same motion as the toothed gear wheel.

Base circle. It is the circle from which involute form is generated.

Pitch surface. It is the surface of the disc which the toothed gear has replaced at the pitch circle.

Pitch point. It is the pitch of the tangency or the point of contact of the two pitch circles of the mating gears.

Circular pitch (p). It is the distance measured along the circumference of the pitch circle from a point on one tooth to a corresponding point on the adjacent tooth.

$$p = \pi \frac{d}{z} \quad ...(17.1)$$

where z = number of teeth.

Base pitch (p_b). It is the distance measured along the circumference of the base circle from a point on one tooth to a corresponding point on the adjacent tooth.

Base pitch, $$p_b = p \times \cos \phi \quad ...(17.2)$$

where ϕ = pressure angle of gear tooth profile.

Diametral pitch (P). It is expressed as the number of teeth per unit pitch circle diameter.

$$P = \frac{z}{d} \quad \text{...(17.3)}$$

$$P_p = \pi \quad \text{....(17.4)}$$

Module (m). It is expressed as the length of the pitch circle diameter per unit number of teeth.

$$m = \frac{d}{z} = \frac{1}{P} \quad \text{...(17.5)}$$

Addendum (h_a). The radial height of the tooth above pitch circle.

Addendum circle. A circle bounding the top of the teeth.

Dedendum (h_f). The radial depth of a tooth below the pitch circle.

Dedendum circle. A circle passing through the roots of all the teeth.

Clearance (c). The radial height difference between addendum and dedendum of a teeth.

Working depth. It is the radial distance of tooth from addendum circle to clearance circle.

Total depth. It is the sum of addendum and dedendum or the radial distance from dedendum circle to addendum circle.

Face. The part of the tooth surface lying below the pitch surface.

Backlash. The minimum distance between the non driving side of a tooth and adjacent side of the mating tooth at the pitch circle.

Profile. The curve forming face and flank.

Tooth thickness (t). This is the arc distance measured along the pitch circle from its intercept with one flank to its intercept with the other flank of the same tooth.

Face width (b). The width of the gear tooth measured axially along the pitch surface.

Top land. It is the surface of the top of the tooth.

Tooth fillet. The radius that connects the root circle to the profile of the tooth.

Tooth Space. It is the width of the space between two teeth measured on the pitch circle.

Pressure angle (ϕ). The angle between the common normal at the point of contact and the common tangent at the pitch point. The pressure angle is either 14.5° or 20°.

Path of contact. It is the locus of the point of contact of two mating teeth from the beginning of engagement to the end of engagement. It is a straight line.

Path of approach. It is the portion of the path of contact from the beginning of engagement to the pitch point.

Angle of approach. The angle turned by gears during path of approach.

Path of recess. It is the portion of the path of contact from the pitch point to the end of engagement of the two mating teeth.

Angle of recess. The angle turned through during path of recess.

Arc of contact. It is the locus of a point on the pitch circle, from the beginning of engagement to the end of engagement of pair of teeth in mesh.

Involute. The involute of a circle is the curve traced by the end of a thread as it is unwound from a stationary cylinder.

Base Circle Diameter (d_b). It is the diameter of the base circle.

$$d_b = d \cos \phi \quad \text{...(17.6)}$$

Cycloid. It is the locus of a point on the circumference of a circle which rolls without slipping on a fixed straight line.

Centre distance (C). It is the distance between the centres of rotation of the two gears in mesh.

$$C = \frac{d_1 + d_2}{2} = \frac{m(z_1 + z_2)}{2} \quad \text{.... (17.7)}$$

17.4 SELECTION OF GEARS

The following factors are considered while deciding to select a type of gear :

1. General layout of shafts.
2. Speed reduction.
3. Power transmission.
4. Input speed.
5. Cost.
6. Quiet operation.

The selection factors for gears are given in Table 17.1.

Table 17.1 Selection factors for gears.

Selection criteria	Type of gears
1. Layout of shafts :.	
(*i*) Parallel	Spur, helical and herringbone gears.
(*ii*) Intersecting	Straight bevel, spiral bevel, zerol bevel, and face gears.
(*iii*) Non-parallel and non-intersecting	Spiral, hypoid, and worm gears.
2. Speed reduction :	
(*i*) Upto 3 : 1	Bevel gears.
(*ii*) Upto 6 :1	Single pair of spur or helical gears.
(*iii*) Upto 10 : 1	Two stage or three stage reduction spur or helical gears.
(*iv*) 60 : 1 to 100 : 1	Worm gears.
3. Tooth speed :	
(*i*) Parallel shafts :	
Upto 5 m/s	Spur or helical gears
25 to 200 m/s	Helical gears
(*ii*) Intersecting shafts	
Upto 2.5 m/s	Straight or spiral bevel gears
Upto 60 m/s	Spiral bevel gears
(*iii*) Non-intersecting shafts	
Upto 50 m/s	Cross helical, worm and worm wheel
4. Power transmitted :	
(*i*) Low speed — upto 3 m/s	Spur, bevel and worm gears
(*ii*) Medium speed—3 to 15 m/s	Helical gears
(*iii*) High speed—more than 15m/s	Helical gears
5. Cost :	Spur gears are the cheapest.
6. Quiet operation	Helical gears.

17.5 CONDITION FOR CORRECT GEARING

17.5.1 Conjugate Action

The profiles of the teeth of two mating gears are designed so as to provide constant speed ratio throughout the length of contact. When the profiles of the gears satisfy this condition, they are said to have conjugate action. For the conjugate action to take place, the profiles must conform to the fundamental law of gearing.

17.5.2 Fundamental Law of Gearing

The fundamental law of gearing may be stated as follows :

The shape of the teeth of a gear must be such that the common normal at the point of contact between two teeth must always pass through a fixed point on the line of centres.

The fixed point is called the pitch point. In order to obtain conjugate action, the gear profiles are cut to conform to an involute curve. The possibility of altering the centre distance without destroying the conjugate action is an important property of the involute profile.

If ω_1, ω_2 = angular speeds of smaller (pinion) and bigger wheel (gear) respectively

r_1, r_2 = pitch radii of pinion and gear respectively

z_1, z_2 = number of teeth on pinion and gear respectively

n_1, n_2 = *rpm* of pinion and gear respectively

Then, according to the fundamental law of gearing

$$\text{Speed ratio} = i = \frac{\omega_1}{\omega_2} = \frac{r_2}{r_1} = \frac{z_2}{z_1} = \frac{n_1}{n_2} \qquad ...(17.8)$$

17.6 PROFILE OF GEARS

The two commonly used profiles for gears are : involute and cycloidal. Involute profile is commonly used with 14.5° or 20° pressure angle. The involute profile has the following properties :

1. The involute of a circle is not defined inside the base circle.
2. The generating line is tangent to the base circle and normal to the involute it generates.
3. All involutes drawn from the same circle are alike.
4. The involute of a circle of larger radius shall be flatter as compared to that of smaller circle.

20° pressure angle has the following advantages :

1. Reduces the risk of undercutting.
2. Gives stronger tooth with a higher load carrying capacity.
3. Provides greater length of contact.
4. Ensures larger radius of curvature.

17.7 INTERFERENCE IN GEARS

The portion of the gear tooth below the base circle is cut as a radial line and not as an involute curve. Therefore, if contact occurs below the base circle, non-conjugate action would result. In Fig. 17.3, points A and B which are the points of tangency of the base circles with the pressure line, are known as the interference points. If contact occurs outside these points, interference will occur. Therefore, the interference points A and B should limit the length of path of action. Point C shows the beginning of contact and point D the end of contact. The beginning of contact occurs before the interference point A, therefore interference will occurs.

Radius of addendum circle of pinion $r_{a1} = r_1 + h_{a1}$

Base circle radius of pinion, $r_{b1} = r_1 \cos\phi$

Radius of addendum circle of gear, $r_{a2} = r_2 + h_{a2}$

Base circle radius of gear, $r_{b2} = r_2 \cos\phi$

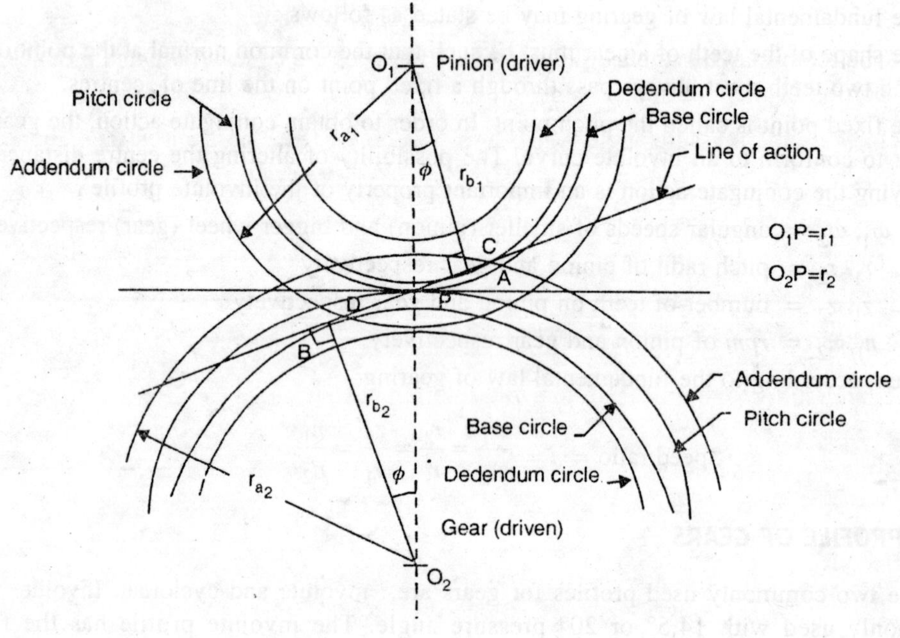

Fig. 17.3 Interference in gears.

Where r_1, r_2 = pitch circle radii of pinion and gear respectively

h_{a1}, h_{a2} = addenda of pinion and gear respectively

Length of path of approach, $L_a = \left[r_{a2}^2 - r_{b2}^2\right]^{\frac{1}{2}} - r_2 \sin\phi$ ...(17.9*a*)

Length of path of recess, $L_r = \left[r_{a1}^2 - r_{b1}^2\right]^{\frac{1}{2}} - r_1 \sin\phi$ ....(17.9*b*)

Maximum possible length of path of approach, $(L_a)_{max} = r_1 \sin\phi$ ...((17.10*a*)

Maximum possible length of path of recess, $(L_r)_{max} = r_2 \sin\phi$ .....(17.10*b*)

Interference will occur, when

$L_a < (L_a)_{max}$ ....(17.11*a*)

and $L_r < (L_r)_{max}$ ....(17.11*b*)

17.7.1 Methods to Prevent Interference

The following methods given in Table 17.2 may be used to prevent interference :

Table 17.2 Methods to prevent interference.

Method adopted to prevent interference	Consequences
1. Undercutting the portion of the tooth below the base circle causing interference.	Reduces the contact ratio resulting in noisier and rougher gear action. Reduces the section modulus at the base of the tooth where stress is already maximum.
2. The tooth may be stubbed, i.e. remove portion of the tooth tip.	Reduces the contact ratio and produces noisier and rougher gear action.
3. Increase the pressure angle.	Decreases the base circle and increases the separating force
4. Cut gears with long and short addendum gear teeth.	This gives non-standard gears which are non-interchangeable and more expensive.

17.7.2 Minimum Number of Teeth

The minimum number of teeth to avoid interference are :

Gear :

$$z_2 = \frac{2a_w}{\left[\left\{1+\left(\frac{1}{i^2}\right)\sin^2\phi+\left(\frac{2}{i}\right)\sin^2\phi\right\}^{\frac{1}{2}}-1\right]} \qquad \text{...(17.12)}$$

Pinion :

$$z_1 = \frac{2a_p}{\left[\left\{1+i^2\sin^2\phi+2i\sin^2\phi\right\}^{\frac{1}{2}}-1\right]} \qquad \text{...(17.13)}$$

Rack :

$$z_{min} = \frac{2a_r}{\sin^2\phi} \qquad \text{....(17.14}$$

where $i = \frac{z_2}{z_1}, a_w = \frac{h_{a2}}{m}, a_p = \frac{h_{a1}}{m}, a_r = \frac{h_r}{m}$

h_{a2}, h_{a1}, h_r = addenda of gear, pinion and rack respectively.

ϕ = pressure angle

17.8 GEAR MATERIALS

The desirable properties of gear materials are :

1. The gear material should have sufficient static strength (ultimate or yield) and endurance strength against fluctuating loads to resist failure due to the breakage of tooth.
2. It should have sufficient surface endurance strength to avoid failure due to destructive pitting caused by excessive wear.
3. The material should have a low coefficient of friction to avoid failure due to scoring caused by high sliding velocities during high speed power transmission.
4. The coefficient of thermal expansion of material should be low to limit the thermal stresses causing distortion and warping.

The Bureau of Indian Standards (BIS) have recommended numerous materials for gears as per IS : 4460 – 1967. These materials are :

Malleable cast iron, cast iron, cast steel, forged steel, surface hardened steels, case hardened steels, and phosphor bronze.

A summary of the gear materials and their properties and applications are given in Table 17.3.

Table 17.3 Selection of gear materials

	Material	Properties	Applications
1.	Cast iron	Low wear, excellent machinability, ease of casting, low noise, shock absorbent, low tensile strength.	Suitable for complicated shapes.
2.	Plain carbon Steels	High toughness, tensile strength and endurance strength.	Used for industrial gears.
3.	Cast steels	Can with stand high stresses.	Used for gears difficult to fabricate.
4.	Alloy steels	High tooth strength and low wear.	Used for industrial and automobile gears.
5.	Heat treated Steels	High toughness and tensile strength, long life.	Used for gears subjected to excessive wear and high loads.
6.	Case hardened Steels	Hard shell and soft core.	Used for gears subjected to varying loads and shocks.
7.	Surface hardened steels	Gives hard surface and soft core.	Suitable for transmitting shock loads.
8.	Non-ferrous Metals	High corrosion resistance, low inertia of rotating parts.	Used in instruments and appliances.
9.	Non-metallics	Quiet operation, internal lubrication, damping of shock and vibrations, manufacturing economy, low load carrying capacity. The mechanical properties of Cast Irons and Steels as gear materials are given in Table 17.4 and 17.5 respectively.	Used for toys, instruments and appliances.

Table 17.4 Mechanical properties of cast irons used for gears.

Material	Condition (Finished Gear)	Ultimate tensile strength, S_{ut} MPa	Brinell Hardness Number, BHN MPa
Grey Cast Iron			
FG 200	As cast	200	179 (min)
FG 220	As cast	220	197 (min)
FG 350	As cast	350	207 (min)
FG 350	Heat treated	350	300 (min)
Malleable Cast Iron			
White heart Grade B	-	275	217 (max)
Black heart Grade B	-	314	149 (max)
Nodular (Ductile) Cast Iron			
SG 800/2	As cast	800	245 – 350
SG 700/2	As cast	700	225 – 300
SG 600/3	As cast	600	190 – 270
SG 500/7	As cast	500	170 – 240
SG 400/12	As cast	400	200 (max)

Table 17.5 Mechanical properties of steels used for gears.

Material	Heat treatment	Ultimate tensile strength, MPa	Brinell Hardness Number, BHN
Cast Steel :			
Grade 27–54	-	540	145
Grade 1	Annealed	620	182 (min)
	H + T	790	229 (min)
	Surface	620	165 (core, min)
	Hardened	460	(case, min)
High Tensile Steel Casting :			
Grade 1 CS 640	-	640	190 (min)
Grade 2 CS 700	-	700	207 (min)
Grade 3 CS 840	-	840	248 (min)
Grade 4 CS 1030	-	1030	305 (min)
Grade 5 CS 1230	-	1230	355 (min)
Forged Steel :			
30C8	N	490	143
30C8	H + T	590	152

Material	Heat treatment	Ultimate tensile strength, MP*a*	Brinell Hardness Number, BHN
40C8	N	570	152
40C8	H + T	590	179
55C8	N	706	201
55C8	H + T	686	223
55C8		784	248
Carbon Chromium Steels :			
55 Cr 3	N	883	225 (min)
55 Cr 3	H + T	981	285 (min)
40 Cr 4	H + T	785	229 (min)
Carbon Manganese Steels :			
27 Mn 6	N	540	
27 Mn 6	H + T	590	170 min
27 Mn 6	H + T	687	201 min
37 Mn 6	-	687	201 min
37 Mn 6	-	785	229 min
Manganese Molybdenum Steels :			
35 Mn 6 Mo 3	N	687	201 min
35 Mn 6 Mo 3	H + T	785	229 min
35 Mn 6 Mo 4	N	785	229 min
35 Mn 6 Mo 4	H + T	883	255 min
35 Mn 2 Mo 28	H + T	700	200 min
35 Ni 1Cr60	H + T	700	200 min
Chromium Molybdenum Steels :			
40 Cr 4 Mo 3	H + T	687	201 min
40 Cr 5 Mo 6	H + T	883	248 min
15 Cr 13 Mo 6	H + T	883	255 min
25 Cr 13 Mo 6	H + T	883	255 min
Nickel steels :			
40 Ni 14	H + T	785	229 min
Nickel Chromium Steels :			
30 Ni 6 Cr 5	H + T	1510	444 min
Nickel Chromium Molybdenum Steels :			
40 Ni 6 Cr 4 Mo 3	H + T	883	255 min

Material	Heat treatment	Ultimate tensile strength, MPa	Brinell Hardness Number, BHN
31 Ni 10 Cr 3 Mo6	H + T	883	255 min
40 Ni 10 Cr 3 Mo6	H + T	980	285 min
Surface Hardened Steels :			
40 C 8	-	540	145 core 460 case
55 C 6	-	695	200 core 520 case
55 C 3	-	850	250 core 500 case
40 Cr 4	-	695	200 core 300 case
3.5 % Ni Steels	-	695	200 core 300 case
3.5 % Ni Cr Steels	-	850	250 core 500 case
1.5 % Ni 1 % Cr Steels	-	850	250 core 500 case
Case Hardened Steels :			
0.12 – 0.22 % C steels	-	495	650 case
0.15 % C steels	-	495	140 core
0.20 % C steels	-	495	140 core 640 case
3% Ni steels	-	695	200 core 300 case
3.5 % Ni steels	-	695	200 core 620 case
5% Ni steels	-	850	600 case

H = Hardened, T = Tempered, N = Normalised

The recommended combination of materials is given in Table 17.6.

Table 17.6 Recommended combination of steels for pinion and gear.

Wheel	Surface hardness, BHN	Gear ratio, i	Materials					
Pinion	≤ 350		45C8	50C4	50C4	40Cr1	35Ni1Cr60	40Ni1Cr1Mo15
Gear	≤ 350		35C8	35C8	45C8	50C4	40Cr1	35NiCr60
		i < 8	St 50	Cast	St63	55C	Cast steel	
			Cast	steel	Cast	Cast steel		
			steel		steel			
Pinion	> 350		45C8	55C8	40Cr1	35Ni1Cr60	40Ni1Cr1Mo15	40ni2Cr1Mo28
		i < 4	50C4					
Gear	> 350		35C8	45C8	50C4	40Cr1	35Ni1Cr60	15Ni2Cr1Mo15
				St63	55C8			
Pinion	> 350		45C8	50C4	40Cr1	15Ni2Cr1Mo15	40NiCr1Mo28	
Gear	≤350		Cast	35Ni1	Cast	45C8	55C8	
		i ≥ 4	steel	Cr60	steel			
					35Ni1			
					Cr60			

17.9 GEAR MANUFACTURE

The various methods used for gear manufacture are :

1. Casting
2. Forming, and 3. Metal removal

The applications of various methods and their limitations are given in Table 17.7

Table 17.7 Methods for gear manufacture.

Method	Properties and Applications	Limitations
1. Casting		
(*i*) Sand casting	Large size gears, low operating speed, farm implements, hand operated devices, small batches	Rough surface, inaccurate dimensions.
(*ii*) Die casting	Fair accuracy, requires little machining, light duty applications, used for gears in toys, cameras and counters.	Suitable for low melting pint metals like alloys of aluminium and copper
(*iii*) Investment Casting	Accurate with good surface finish, can withstand heavy loads.	For low melting alloys, costly method.
(*iv*) Injection Moulding	Used for large volume production of light gears of thermoplastics, used in instruments and household appliances.	Low tooth accuracy, high initial tooling cost.
(*v*) Centrifugal Casting	Used for phosphor bronze worm and worm wheel rim, steel gears	Suitable for small gears
(*vi*) Shell Moulding	Used for small gears	Low accuracy.

Method	Properties and Applications	Limitations
2. Forming : (*i*) Roll forming	Excellent accuracy, mirror polish, chipless method, lot of material saving, suitable for spur and helical gears.	Very costly machinery, suitable for mass production only.
(*ii*) Drawing and Extrusion	Any shape of tooth form can be made, gears for watches, clocks, type writers.	Suitable for high volume production of gears made of non-ferrous metals.
(*iii*) Stamping	For small and thin gears used in toys, watches and instruments.	Thick gears cannot be made.
3. Form cutter methods :	In this method, the spaces between the teeth are cut conforming to the gear tooth form. Rotary form relieved gear cutters are used for spur and helical gears.	
(*i*) Milling	Simple, economical and flexible method, suitable for spur, helical, bevel and rack gears. Gears suitable for slow speed machinery.	Low quality of gears produced, very slow process, separate cutter is required for every pitch and number of teeth.
(*ii*) Shaper or Planer Machining	Suitable for small batch production. Spur, helical, bevel and rack gears can be cut.	Very slow process
(*iii*) Broaching	Highly accurate gears with excellent surface finish, suitable for **internal gears,** racks and gear sectors of high volumes.	Tool is not universal for each size, parts with blind end cannot be machined.
4. Gear generation	The gear generation is based on the relative rolling motion of the cutter and the blank. This process in which the cutter is similar to a mating pinion or a rack is called ''gear shaping'. If the generating tool (cutter) is threaded and gashed, it is called a ''hob', and the process is called gear hobbing.	
(*i*) Gear shaping	One cutter cannot be used for cutting all spur gears of the same module irrespective of the number of teeth on the gear. It can be used to cut spur, helical, herringbone, internal and cluster gears, racks, splines, pawls, etc. Gears find applications in automobiles, machine tools, instruments, machinery, clocks, and other appliances.	The cutting action is reciprocating and takes place only during one half of the stroke. Separate helical guide is required for cutting helical gears.
(*ii*) Gear hobbing	The method is versatile and can generate spur, helical, **worm and worm wheels** The method is rapid, economical and highly productive, The cutter can cut all gears of same module irrespective of the number of teeth on the gear. The produced gears are used in automobiles, machine tools, instruments, clocks, etc.	It cannot cut internal gears.

17.10 GEAR TOOTH FAILURES

The two basic modes for gear tooth failure are :

1. Breakage of the tooth due to static and dynamic loads.
2. Surface destruction

1. Breakage of Tooth

The complete breakage of the tooth can be avoided by adjusting the module and face width so that the beam strength of the gear tooth is more than the sum of static and dynamic loads. The static beam strength of a gear tooth was suggested by Wilfred Lewis. The dynamic load is caused due to small machining errors resulting into inertia and impact loads on the gear tooth. The calculations for dynamic load have been suggested by Prof. Spotts and Earle Buckingham. Considering these aspects the breakage of the gear tooth can be ascertained.

2. Surface Destruction

The wear of gear tooth takes place due to the combined action of rolling and sliding.

Rolling causes contact stresses and sliding causes rubbing action. Pinion is subjected to more rubbing action as it rotates faster than the gear. The calculations for wear load have been suggested by Earle Buckingham based on the Hertz's theory of contact stresses.

The principal types of gear tooth wear are : abrasive wear, corrosive wear, pitting, and scoring. We shall discuss them briefly.

(i) **Abrasive Wear.** The tooth surface is scratched by foreign particles in the lubricant , such as dirt, rust, weld spatter or metallic debris. This can be reduced by oil filter, using high viscosity lubricants, and surface hardness.

(ii) **Corrosive Wear.** The corrosion of the tooth surface is caused by corrosive elements, such as extreme pressure (EP) additives present in the lubricating oils and foreign materials due to external contamination. These elements attack the tooth surface, resulting in fine wear uniformly distributed over the entire surface. The corrosive wear can be controlled by complete enclosure of the gears, selecting proper additives and replacing the lubricant at regular intervals of use.

(iii) **Pitting.** It is a type of fatigue failure caused by repeated applications of stress cycles. Pitting phenomenon is of two types : initial pitting and destructive pitting. The initial (or corrective) pitting is a localized phenomenon, characterized by small pits at high spots. Such high spots are progressively worn out and the load is redistributed. Initial pitting is caused by the errors in tooth profile, surface irregularities and mis-alignment. This can be controlled by precise machining of gears and their correct alignment so that the load is uniformly distributed across the full face width and the dynamic load is reduced.

Destructive pitting is a surface fatigue failure which occurs when the load on the gear tooth exceeds the surface endurance strength of the gear material. This type of failure is characterized by pits, which continue to grow resulting in complete destruction of the tooth surface. In some cases, this may even break the tooth permanently. This type of failure can be avoided by ensuring that the wear strength of the gear tooth is more than the sum of static and dynamic loads. The surface endurance strength is a function of the hardness and can be increased by improving the surface hardness of the gear tooth surface by using an appropriate heat treatment process.

(iv) **Scoring.** The oil film between the gear teeth may breakdown under excessive surface pressure, high sliding velocity and inadequate supply of lubricant. This results in generation of excessive frictional heat and overheating of the contacting surfaces of gear teeth. This may lead to metal – to – metal contact. Scoring is a stick – slip phenomenon, in which alternate welding and

shearing takes place rapidly at the high spots. This increases the wear rate faster. Scoring can be controlled by selecting proper surface speed, surface pressure and flow rate of lubricant to keep the temperature of contacting surfaces within permissible limits. The bulk temperature of lubricant can be reduced by providing fins on the gear box, air cooling by a fan or circulating cold water.

17.11 GEAR LUBRICATION AND LUBRICANTS

For the satisfactory performance and durability of the gears, proper lubrication of the gear teeth is essential. The main function of the lubricant is to separate the tooth surfaces and remove heat from the gears. The relative motion between two teeth in contact is a combination of rolling and sliding motion. This combination of motion increases the lubrication problems. Splash and pumped lubrication are the two methods by which lubricant is usually applied to a gear system. In splash lubrication, the gears are enclosed in a box and dipped in a bath of lubricating oil. In the pumped lubrication system, a jet of oil is directed towards the meshing teeth. For low velocities, under 15 m/s, the splash system is used, whereas for velocities above 25 m/s, pump fed lubrication is preferred.

Gears are lubricated by grease, straight mineral oils or extreme pressure lubricants. Grease is used as a lubricant for hand – operated gear mechanisms or drives working at low speeds and low loads working intermittently. For enclosed gear drives operating under high loads and normal speeds, straight mineral oils provide satisfactory lubrication. Small gears operating at low speeds (below 1200 rpm) and lightly loaded should be lubricated by SAE 80 grade mineral oils. Gears operating at low speeds and heavily loaded should be lubricated with SAE 90 grade mineral oil at normal ambient temperatures and SAE 140 grade mineral oil at high temperatures. For heavy duty applications, EP lubricants are used. EP lubricants are mineral oils with additives. They are used in the automobile gear box and heavy industrial gear boxes. Synthetic oils should be used for aircraft turbine engines.

17.12 GEAR PROFILE STANDARDS

A set of gears are interchangeable when any two gears selected from the set will mesh and satisfy the fundamental law of gearing. For interchangeability all gears of the set must have the same modules, circular pitch, pressure angle, addendum and dedendum. The tooth thickness must be half of the circular pitch. A standard tooth system specifies the pressure angle and the relations for tooth dimensions in terms of module.

17.12.1 *Larger and Smaller Pressure Angles*

(*a*) Larger Pressure Angle

The advantages of larger pressure angle are :

1. Lesser number of teeth on pinion to avoid interference.
2. Smaller bending stress due to large tooth thickness at the base of tooth.
3. Smaller contact stresses due to larger radii of curvature at the pitch line.

The limitations of larger pressure angle are :

1. Larger radial force.
2. Low contact ratio giving rise to more noise.

(*b*) Smaller Pressure Angle

The advantages of smaller pressure angle are :

1. Smaller radial force.
2. High contact ratio giving quiet operation.

The limitations of smaller pressure angle are :

1. Requires larger number of teeth on pinion to avoid interference.
2. High bending stresses.
3. High contact stresses.

17.13 STATIC BEAM STRENGTH OF SPUR GEARS

The analysis for the determination of bending stresses in a spur gear tooth was presented by Wilfred Lewis in 1892. In this analysis, the gear tooth is treated as a cantilever beam, as shown in Fig. 17.3. A spur gear tooth with the force acting at the tip is shown in Fig. 17.4 for the analysis. The normal force F_n is resolved into its radial component F_r and tangential component F_t acting at A, the intersection of the line of action of F_n and the centre line of the tooth.

The static strength is determined by making the following assumptions :

1. The effect of the radial force component F_r which produces compresses stresses, is neglected.
2. The tangential component F_t which produces bending stresses, is uniformly distributed over the face width of the gear tooth.
3. The effect of stress-concentration is neglected.
4. At any time only one pair of teeth is in contact and takes the total load.

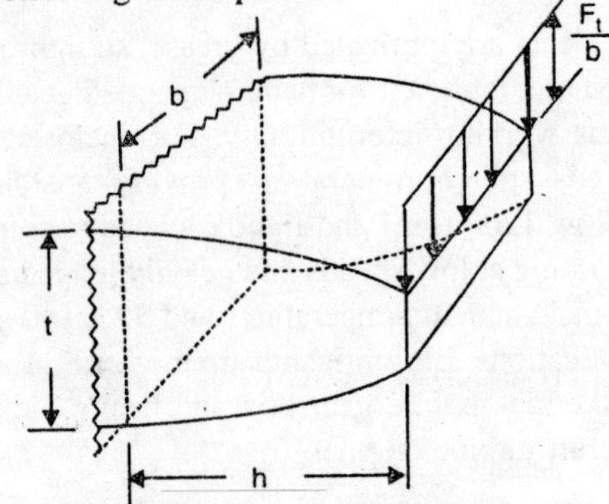

Fig. 17.4 Gear tooth as a cantilever

The maximum bending stress is located and computed as follows :

Through the point A in Fig. 17.5, draw a parabola (shown in dashed lines) tangent to the tooth curves at B and D. This parabola represents the outline of a beam of uniform strength. The maximum bending stress in the actual tooth will be at the point of tangency B or D. This stress is calculated as follows :

At section B – D, bending moment,

$$M = F_t \times h$$

Moment of inertia of tooth,

$$I = \frac{bt^3}{12}$$

Where h = height of the tooth from A to plane B – D

b = face width

t = tooth thickness at BD

Bending stress,

$$\sigma = \frac{My}{I}$$

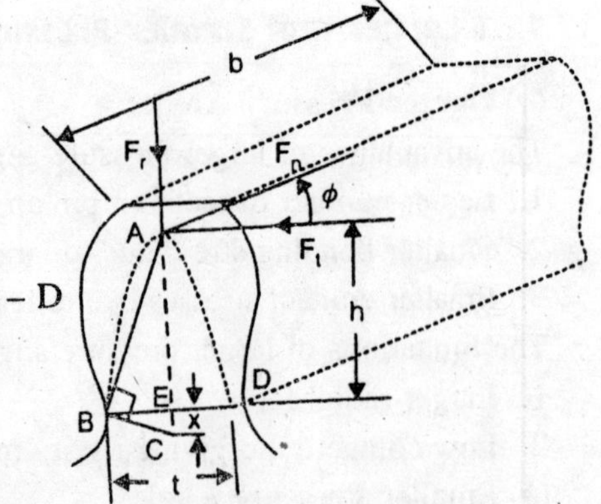

Fig. 17.5 Gear tooth as parabolic beam

Where $$Y = \frac{t}{2}$$

$$\therefore \quad \sigma = \frac{6F_t h}{bt^2}$$

Or $$h = \left[\frac{\sigma b}{6F_t^2}\right] t^2 = \text{const.} \times t^2$$

This equation represents a parabola, as assumed above.

Now ∆s ABE and BCE are similar. Thus

$$\frac{x}{\left(\frac{t}{2}\right)} = \frac{\left(\frac{t}{2}\right)}{h}$$

Or $$h = \frac{t^2}{4x}$$

$$\therefore \quad F_t = \sigma b \times \left(\frac{2x}{3}\right) \qquad \text{...(a)}$$

Define $y = \frac{2x}{(3p)}$ = Lewis form factor based upon circular pitch

Then Eq. (*a*) becomes,

$$F_t = \sigma b p y \qquad \text{...(17.15a)}$$

$$= \frac{\sigma b Y}{P} \qquad \text{...(17.15b)}$$

$$= \sigma b m Y \qquad \text{...(17.15c)}$$

The standard pressure angles are : 14.5°, 20° full depth (FD), and 20° stub. The proportions of standard tooth forms are given in Table 17.8.

Table 17.8 Standard tooth forms of involute profile.

Particulars	**14.5° full-depth**	**20° full-depth**	**20° stub**
Addendum, h_a	m	m	0.8m
Dedendum, h_f	1.157 m	1.250 m	m
Total depth, h	2.157 m	2.125 m	1.8 m
Working depth, h_w	2 m	2 m	1.6 m
Tooth thickness, t	1.5708 m	1.5708 m	1.5708 m
Clearance, c	0.157 m	0.250 m	0.200 m
Fillet radius, r	0.209 m	0.300 m	0.304 m
Minimum number of teeth	32	18	14

Standard gears can be corrected in following two ways :

1. Without alteration of shafts centre distance.
2. With alteration of centre distance.

The pressure angle recommended by the BIS is 20° full depth. In addition to the other proportions given in Table 17.5 are :

Addendum (or tip) diameter, $d_a = m\ (z + 2)$

Dedendum (or root) diameter, $d_f = m\ (z - 2.5)$

Tooth space, $e = t$

The standard modules are given in Table 17.9, gear ratios in Table 17.10 and number of teeth and sums of number of teeth in Table 17.11.

Table 17.9 Standard modules.

First choice	1.0, 1.25, 1.5, 2, 2.5, 3, 4, 5, 6, 8, 10, 12, 16, 20, 25, 32, 40 and 50
Second Choice	1.125, 1.375, 1.75, 2.25, 2.75, 3.5, 4.5, 5.5, 7, 9, 11, 14, 18, 22, 28, 36 and 45

Table 17.10 Standard Gear Ratios i for Spur Gears

Single	1.25, 1.4, 1.6, 1.8, 2.0, 2.24, 2.5, 2.8, 3.15, 3.55, 4.0, 4.5, 5.0, 5.6, 7.1, 8.0, 9.0 and 10.
Double	8, 9, 10, 11, 12, 12.5, 14, 16, 18, 20, 22.4, 25, 28, 31.5, 35.5, 40, 45 and 50.
Triple	40, 45, 50, 56, 63, 71, 80, 90, 100, 112, 125, 140, 160, 180, 200, 250, 280, 315, 355 and 400

Table 17.11 Number of Teeth and Sums of numbers of Teeth for Standard Transmission Ratios

Transmission Ratio	1	1.12	1.25	1.4	1.6	1.8	2.0	2.24	2.5	2.8	3.15	3.55	4.0
Sum of number of teeth, $z_1 + z_2$	*Number of teeth-pinion,* z_1 / *gear,* z_2												
100	50/50	41/53	44/56	–	39/61	36/64	–	31/69	–	26/79	24/76	22/78	20/80
101	15/50	–	45/56	42/54	39/62	36/65	34/67	31/70	29/72	–	24/77	22/79	20/81
102	52/51	48/54	45/57	42/60	–	37/65	34/68	–	29/73	27/55	–	–	–
103	52/51	–	–	43/60	40/63	37/66	–	32/71	–	27/76	25/78	–	21/82
104	52/52	49/55	46/58	43/61	40/69	–	35/69	32/72	–	–	25/79	23/81	21/83
105	53/52	–	–	–	41/64	38/67	35/70	–	30/75	–	25/80	23/82	21/834
106	53/53	50/56	47/59	44/62	41/65	38/68	–	33/73	30/76	28/78	–	23/83	21/85
107	54/53	–	47/60	44/63	41/66	–	36/71	33/74	–	28/79	26/81	–	–
108	54/55	51/57	44/60	45/63	42/66	39/69	36/72	33/75	31/77	28/80	26/82	24/84	22/86
109	50/54	51/58	48/61	45/64	42/67	39/70	–	24/75	31/78	–	26/85	24/85	22/87

$$F_t = \sigma b \pi m y \qquad ...(17.15d)$$

Where Y = Lewis form factor based upon diametral pitch.

Eq. (17.15) gives the beam strength of a gear tooth. Thus

Beam strength, $$F_b = \sigma_d b \pi m y \qquad ...(17.16)$$

Where σ_d = allowable static bending stress of grar material

$$= \frac{S_{ut}}{FOS} = \frac{S_{ut}}{3} \quad ...(17.17)$$

b = face width, which varies from 3 πm to 4πm.

m = module

Eq. (17.16) is known as the Lewis equation for beam strength of spur gears tooth. The allowable bending stress is based on ultimate tensile strength with a factor of safety of 3.

17.14 LEWIS FORM FACTORS

Lewis form factors based upon circular pitch are given by,

$$y = 0.124 - \frac{0.684}{z} \text{ for } 14.5° \text{ involute FD} \quad ...(17.18a)$$

$$= 0.154 - \frac{0.912}{z} \text{ for involute FD} \quad ...(17.18b)$$

$$= 0.170 - \frac{0.950}{z} \text{ for } 20° \text{ involute stub} \quad ...(17.18c)$$

where FD = full depth

In the design process, it is required to decide which one is the weaker out of the pinion and gear. In Eq. (17.16), we observe that b and m are same for the pinion and the gear. When different materials are used for the pinion and the gear, the product $\sigma \times y$ decides the weaker out of the two. The weaker one should be selected for design.

17.15 TANGENTIAL LOAD ON GEAR TOOTH

The tangential load on the tooth at the pitch line is given by,

$$F_t = \frac{10^3 \times P}{v} \text{ N} \quad ...(17.19)$$

Where P = power transmitted in kW

v = mean pitch line velocity = $\frac{\pi d n}{60 \times 10^3}$ m/s

d = pitch circle diameter, mm

n = speed in rpm

17.16 SERVICE FACTOR, C_s

The service factor accounts for increase in the tangential force due to fluctuation of the torque developed by the prime mover and the torque required to run the machine. It depends upon the prime mover and the driven machine.

$$C_s = \frac{\text{maximum torque}}{\text{rated torque}} \quad ...(17.20)$$

The values of service factor are given in Table 17.12.

Table 17.12 Service factor, C_s

Type of load	Type of service		
	Intermittent 3 hours per day	8-10 hours per day	Continuous 24 hours per day
Steady	0.8	1.00	1.25
Light shock	1.00	0.25	1.50
Medium shock	1.25	0.50	1.80
Heavy shock	1.5	1.8	2.00

The type of loads for different applications are given in table 17.13.

Table 17.13 Type of loads for different machinery.

S. No.	Type of load	Driven machinery
1.	Steady	Generators, liquid mixing agitators, centrifugal blowers, compressor and pumps, medium duty conveyors, fans, etc.
2.	Moderate shock	Liquid and solid missing agitators, multi-cylinder reciprocating compressors and pumps, heavy duty conveyors, sugar cane crushers, hoists, concrete mixers, textile machinery, etc.
3.	Heavy shock	Single-cylinder reciprocating compressors and pumps, stone crushers, passenger lifts, heavy duty hoists, hammer mills, presses, rolling mills, tumbling barrels, printing presses, etc.

17.17 LOAD DISTRIBUTION FACTOR, C_m FOR SPUR GEARS

The load distribution factor accounts for the non-uniform distribution of load across the face width of the gear. It depends upon the following factors :

1. Accuracy of alignment of gears and bearing mountings.
2. Bearing clearances.
3. Elastic rigidity of shafts, gears, bearings and housings, etc.
4. Manufacturing accuracy of gear tooth.

The values of load distribution factor are given in Table 17.14.

Table 17.14 Load distribution factor for spur gears, C_m

S. No.	Characteristics of support	Face width upto, mm			
		50	150	225	400
1.	Accurate mountings, small bearing clearances, minimum deflection, precision gears	1.3	1.4	1.5	1.8
2.	Less rigid mountings and less accurate gears, contact across the full face	1.6	1.7	1.8	2.0
3.	Accuracy and mounting such that less full face contact exists.	More than 2.0			

17.18 MAXIMUM LOAD, $(F_t)_{MAX}$

$$(F_t)_{max} = C_s C_m F_t \quad ...(17.21)$$

17.19 DESIGN FOR DYNAMIC LOAD

The dynamic force is introduced in the gear teeth due to the following factors.

1. In accuracies of the tooth profile.
2. Errors in tooth spacing.
3. Misalignment between bearings.
4. Elasticity of parts, and
5. Inertia of rotating masses.

These forces introduce impact and fatigue stresses in the gear tooth. These stresses become more severe as the pitch line velocity increases.

The following methods have been suggested for the design of gear teeth under dynamic load :

1. Velocity factor method.
2. Earle Buckingham's increment load method.
3. Prof. Spotts adjacent pitch error method

1. Velocity Factor method

To account for the additional stresses due to dynamic load, a velocity factor, C_v, is introduced into the Lewis equation. The velocity factor is an empirical relationship developed by past experience. It given by :

$$C_v = \frac{3.05}{3.05+v} \text{ for ordinary industrial gears operating up to 8 m/s.} \quad ...(17.22a)$$

$$= \frac{6.1}{6.1+v} \text{ for accurately cut gears operating between 6 to 20 m/s.} \quad ...(17.22b)$$

$$= \frac{5.55}{5.55+\sqrt{v}} \text{ for precision gears cut with a high degree of accuracy and} \quad ...(17.22c)$$

operating at $v \geq 20$ m/s.

The effective load between two meshing gears becomes,

$$F_{eff} = \frac{(F_t)_{max}}{C_v} \quad ...(17.23)$$

Eq. (17.23) may be used to determine the face width of the gear at the initial stages when the dimensions of the gear teeth are unknown.

Calculate static beam strength of gear tooth,

$$F_b = \sigma_d b \pi y m$$

Then $$F_b = F_{eff}$$

Face width, $$b = \frac{C_s C_m F_t}{C_v \sigma_d \pi y m} \quad \text{... (17.24)}$$

2. Earle Buckingham's Increment Load Method

The dynamic load minaly depends uon the tooth error and the pitch line velocity of the gear. Based on the accuracy of the gear teeth, BIS has classified the gears into 12 different grades from Grade 1 to Grade 12 in the decreasing order of the precision. Table 17.15 specifies the manufacturing methods, typical applications, and permissible pitch line velocities for different grades of gears.

Table 17.15 BIS Gear Grades.

IS Grades	Accuracy level	Manufacturing methods	Typical applications	Permissible pitch line velocity, *v* m/s
1, 2 or 3	Ultra – high (AA)	Finish grinding followed by honing.	Used for master gears.	$40 \leq v < 200$
4 or 5	High (A)	Shaving and finish grinding with first rate machine tools and skilled operators.	Used for turbine gearing, aerospace gearing, and sometimes critical industrial gears.	$v \geq 30$
6 or 7	Medium high (B)	Shaving and grinding with emphasis on production rate rather than highest quality.	Used for medium speed industrial gears and critical vehicle gears.	$25 \leq v < 30$
8 or 9	Medium (C)	Fine hobbing or hobbing followed by shaving with first rate machine tools and skilled operators.	Used for vehicle gears, geared motors, and low speed industrial gears.	$20 \leq v < 25$
10 or 11	Low (D)	Hobbing, shaping or milling	Used for low speed gears.	$v < 20$
12	Very low (E)	Casting or moulding	Used for low speeds and light loads.	$v < 10$

Earle Buckingham pointed out that the dynamic load is an incremental load that should be added to the transmitted load. The dynamic load is given by,

$$F_d = (F_t)_{max} + F_i \quad \text{.... (17.25)}$$

Where $(F_t)_{max} = C_s C_m F_t$

F_i = increment load

$$= \frac{20.67v\left[Cb + (F_t)_{max}\right]}{\left[20.67v + \left\{Cb + (F_t)_{max}\right\}^{\frac{1}{2}}\right]} \quad \text{... (17.26)}$$

C = load stress (or dynamic) factor, N/mm

$$= \frac{e}{\left[K_1\left(\frac{1}{E_p} + \frac{1}{E_g}\right)\right]}$$

e = sum of errors between two meshing teeth, mm

$= e_p + e_g$

e_p = error for pinion

e_g = error for gear

The error depends upon the quality of the gear and the method of manufacture. The expected error on gear tooth for the 12 IS Grades is considered to be equal to tolerance. The tolerances for adjacent pitch error (e) are given in Table 17.16. These tolerances are calculated by using the following basic equation.

$$\phi_p = m + 0.25\sqrt{d} \quad \text{... (17.27)}$$

Where ϕ_d = tolerance factor

m = module, mm

d = pitch circle diameter, mm

Table 17.16 Tolerances on the adjacent pitch error.

IS Grade	Adjacent pitch error *e*, microns
1	$0.80 + 0.06\phi_p$
2	$1.25 + 0.10\phi_p$
3	$2.00 + 0.16\phi_p$
4	$3.20 + 0.25\phi_p$
5	$5.00 + 0.40\phi_p$
6	$8.00 + 0.63\phi_p$
7	$11.00 + 0.90\phi_p$
8	$16.00 + 1.25\phi_p$
9	$22.00 + 1.80\phi_p$
10	$32.00 + 2.50\phi_p$
11	$45.00 + 3.55\phi_p$
12	$63.00 + 5.00\phi_p$

K_1 = *a* constant

= 9.345 for 14.5° involute FD ... (17.28*a*)

= 9.000 for 20° involute FD ... (17.28*b*)

= 8.700 for 20° involute stub ... (17.28*c*)

E_p, E_g = modulii of elasticity of pinion and gear materials

Values of load stress factor are given in table 17.17

Table 17.17 Values of load stress factor, C in N/mm for 20° pressure angle FD involute teeth.

Material		Poisson's ratio, ν		Modulus of elasticity, E GPa		Load stress factor, C
Pinion	Gear	Pinion	Gear	Pinion	Gear	N/mm
Steel	Steel	0.292	0.292	207	207	11500*e*
Cast iron	Cast iron	0.211	0.211	100	100	8900e
Steel	Cast iron	0.292	0.211	207	100	7500*e*
Steel	Cast iron	0.292	0.211	207	160	10000*e*

3. Prof. Spotts Adjacent Pitch Error Method

According to Prof. Spotts, the dynamic load on gear tooth is given by,

$$F_d = (F_t)_{max} + F_i \quad \text{... (17.29)}$$

Depending upon the materials of the pinion and the gear, there are three equations for the increment Load.

$$F_i = \frac{e n_p z_p b r_p r_g}{\left[K_2\left(r_p^2 + r_g^2\right)^{\frac{1}{2}}\right]} \quad \text{... (17.30)}$$

Where K_2 = 2530 for steel pinion with steel gear ... (17.31*a*)

= 3785 for C.I. pinion with C.I. gear ... (17.31*b*)

= 3260 for steel pinion with C.I. gear. ... (17.31*c*)

Where n_p = *rpm* of pinion

r_p, r_g = pitch circle radii of pinion and gear respectively, mm

e = sum of errors between two meshing teeth, mm

= $e_p + e_g$

e_p = error for pinion

e_g = error for gear

Multiply r_g^2 in the bracket by 0.92 for steel pinion with C.I. gear.

The error depends upon the quality of the gear and the method of manufacture. The expected error on gear both for the 12 IS Grades is considered to be equal to tolerance. The tolerances for adjacent pitch error (*e*) are given in Table 17.14.

17.20 ENDURANCE STRENGTH OF GEAR

The endurance limit for steel gears is given by,

$$S_e = k_a k_b k_c k_d k_e k_f S'_e \quad \text{... (17.32a)}$$

Where S'_e = uncorrected endurance strength ... (17.32*b*)

= 0.5 S_{ut} for steel

k_a = surface finish factor (Fig. 4.7 or 4.8)

k_b = size factor (Table 4.3)

k_c = reliability factor (Table 4.4)

k_d = temperature factor (Eq. 4.15)

k_e = mean stress factor

= 1.0 for idler gears

= 1.4 for input and output gears

k_f = fatigue stress reduction factor (Table 17.16)

Table 17.18 Fatigue strength reduction factor for gears, k_f

z	20	24	30	40	50	80	125	250
k_f	0.944	0.941	0.924	0.915	0.901	0.896	0.886	0.861

$$S_e = 1.75 \times (\text{BHN})_{\text{core}} \text{ for steel} \qquad \dots (17.33)$$

The endurance strength of gear teeth is given by

$$F_e = S_e\, b\pi y m \qquad \dots (17.34)$$

Where S_e = corrected endurance limit of gear material

For a safe design,

$$F_e \geq k\, F_d \qquad \dots (17.35)$$

Where k = factor of safety

= 1.25 for steady loads

= 1.35 for pulsating loads

= 1.5 for shock loads.

17.21 DESIGN FOR WEAR

The failure of the gear tooth due to pitting occurs when the Hertz's contact stresses between two meshing teeth exceed the surface endurance strength of the material. Pitting is a surface fatigue failure which is characterized by small pits on the surface of the gear tooth. In order to avoid this type of failure, the proportions of gear tooth and surface hardness should be selected in such a way that the wear strength of the gear tooth is more than the effective load between the meshing teeth. The analysis of wear strength was done by Earle Buckingham which gives the wear strength of gear tooth. This equation is based on Hertz's theory of contact stresses.

The wear load is determined by the surface endurance limit of the material, curvature of the surface, and relative hardness of the surfaces. The pinion should always be harder to allow for work hardening of the gear to preserve the involute profile and to allow for greater abrasive wear on the pinion, and to decrease the possibility of seizing.

Consider two contacting cylinders under load F, as shown in Fig. 17.6. The Hertz's contact stresses between two cylinders is given by,

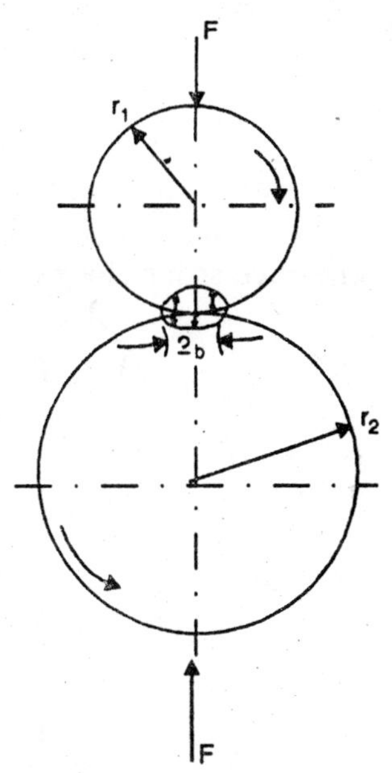

Fig. 17.6 Contact stresses

$$\text{Surface stress,} \quad \sigma = \left[\frac{F\left(\frac{1}{r_1} - \frac{1}{r_2}\right)}{\left\{\pi L\left[\frac{(1-\nu_1^2)}{E_1} + \frac{(1-\nu_2^2)}{E_2}\right]\right\}} \right]^{\frac{1}{2}} \qquad \text{... (17.36)}$$

Where L = length of contact of cylinder

r_1, r_2 = radii of curvature of cylinders

ν_1, ν_2 = Poisson's ratio of cylinders

E_1, E_2 = modulii of elasticity of cylinders

For applying this equation to spur gears, the contact between the two teeth is taken at the pitch point as shown in Fig. 17.7. The force F is replaced by the allowable wear load F_w and σ is replaced by σ_{es}, the allowable surface endurance strength. r_1 and r_2 are replaced by the radii of curvature of the pinion and gear respectively, and L by face width b.

Now $r_1 = r_p \sin\phi,\ r_2 = r_g \sin\phi$

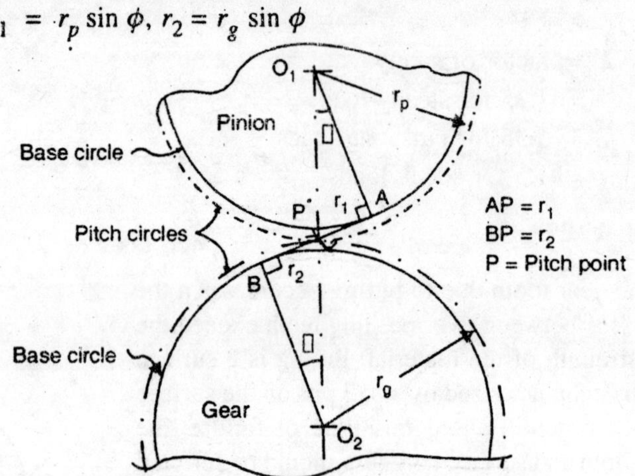

Fig. 17.7 Contact between meshing teeth

The force acting along the pitch line in Fig. 17.7 is F_n. Therefore,

$$F = F_n = \frac{F_t}{\cos\phi}$$

$$\sigma_{es} = \left[F_w \times \left[\frac{1}{(\sin\phi\cos\phi)}\right] \frac{\left(\frac{1}{r_p} - \frac{1}{r_g}\right)}{\left\{\pi b\left[\frac{(1-\nu_p^2)}{E_p} + \frac{(1-\nu_g^2)}{E_g}\right]\right\}} \right]^{\frac{1}{2}}$$

$$F_w = \left[\frac{d_p d_g}{d_p + d_g}\right]\left[\frac{1-v_p^2}{E_p} + \frac{1-v_g^2}{E_g}\right]\left[\frac{d_p d_g}{d_p + d_g}\right]\left[\frac{1-v_p^2}{E_p} + \frac{1-v_g^2}{E_g}\right] \quad \text{... (17.37)}$$

Now $v_p \approx v_g = 0.3$ so that $1 - v_p^2 = 1 - v_g^2 = 1 - (0.3)^2 = 0.91$

and $\dfrac{4}{(\pi \times 0.91)} = 1.4$

$$F_w = \left(\frac{\sigma_{es}^2\, b \sin\phi \cos\phi}{1.4}\right)\left[\frac{2d_p d_g}{d_p + d_g}\right]\left[\frac{1}{E_p} + \frac{1}{E_g}\right] \quad \text{... (17.38)}$$

Let $$Q = \frac{2d_g}{d_p + d_g} = \frac{2z_g}{z_p + z_p} \quad \text{... (17.39)}$$

and $$K = \left(\frac{\pi S_{es}^2 \sin\phi \cos\phi}{4}\right)\left[\frac{1-v_p^2}{E_p} + \frac{1-v_g^2}{Eg}\right] \quad \text{... (17.40)}$$

Then $F_w = d_p b Q K$ ... (17.41)

Where K = load – stress factor.

$$= c_1 \left(\frac{\text{BHN}}{100}\right)^2 \quad \text{... (17.42a)}$$

Refer to Table 17.17 for values of c_1. ... (17.42n)

According to G. Niemann, the approximate value of S_{es} for steel pinion is given by

$$S_{es} = 2.65\ (\text{BHN})_{\text{case}}\ \text{MPa} \quad \text{... (17.43)}$$

Where BHN = Brinell hardness number in MPa

For various combinations of pinion and gear materials, the values of K are given in Table 17.19

Table 17.19 Values of load – stress factor, K for ϕ = 20° FD involute.

$$K = c_1 \left(\frac{\text{BHN}}{100}\right)^2 \text{ MPa}$$

Material		Poisson's ratio		Modulus of elasticity, GPa		c_1
Pinion	Gear	v_p	v_g	E_p	E_g	
Steel	Steel	0.30	0.30	210	210	0.1536
Cast Iron	Cast Iron	0.28	0.28	100	100	0.3267
Steel	Cast Iron	0.30	0.28	210	100	0.2402
Steel	Steel	0.30	0.30	207	207	0.15585
Steel	Cast Iron	0.30	0.28	207	100	0.2413
Steel	Steel	0.292	0.292	207	207	0.16
Cast Iron	Cast Iron	0.211	0.211	100	100	0.21
Steel	Cast Iron	0.292	0.211	207	160	0.18

For a safe design $F_w \geq F_d$ ... (17.44)

The pinion should invariably be designed for wear load. If $F_w < F_d$, then the hardness of the pinion should be increased. In that case put $F_w = F_d$, calculate the value of K from Eq. (17.41) and substitute that value in Eq. (17.42) to determine the hardness of pinion required.

17.22 FORCE ANALYSIS

Consider a pair of spur gears having their teeth in contact at the pitch point P, as shown in Fig. 17.8. The force that the tooth of one gear exerts on the meshing tooth of the other gear is normal to the tooth surface, and therefore, acts along the pressure angle. The normal force F_n is the force exerted by the follower gear tooth on the driver gear tooth. The driver gear tooth would exert a force equal to F_n in magnitude but opposite in direction. The normal force can be resolved into two components, F_t, the tangential force, and F_r, the radial force.

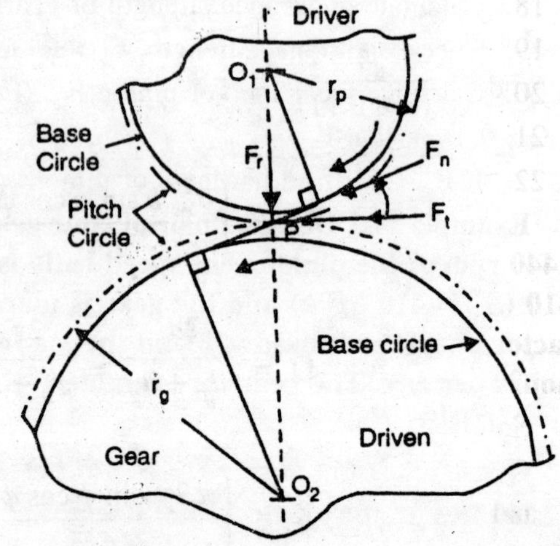

Fig. 17.8 Forces on a spur gear tooth

$$F_t = F_n \cos \phi \qquad \text{... (17.45)}$$

$$F_r = F_n \sin \phi = F_t \tan \phi$$

Where ϕ = pressure angle

Now $$F_t = \frac{2T}{d_1} \qquad \text{... (17.46)}$$

where d_1 =pitch circle diameter of pinion.

17.23 DESIGN PROCEDURE FOR SPUR GEARS

1. Assume module for the gears and minimum number of teeth on pinion if not given.
2. Select materials for the pinion and gear and their allowable bending stresses.
3. Calculate pitch circle diameters of pinion and gear.
4. Calculate Lewis form factors y for pinion and gear.
5. Decide which one is weaker out of pinion and gear by the product $\sigma_d y$. Design the weaker one.
6. Calculate pitch line velocity, v
7. Calculate tangential force, F_t. (Eq. 17.19)
8. Select service factor, C_s from tables.Table (17.12)
9. Determine load distribution factor, C_m from Table (17.14)
10. Calculate the velocity factor C_v from eq. (17.24).
11. Calculate maximum tangential load, $(F_t)_{max} = C_s C_m F_t$. (Eq: 17.21)
12. Calculate the effective load, $F_{eff} = \dfrac{(F_t)_{max}}{C_v}$ (Eq. 17.23)

13. Calculate beam strength of gear tooth, $F_b = \sigma_d \pi ymb$.
14. Calculate face width from. $F_b = F_{eff}$
15. Calculate increment load, F_i by Earle Buckingham's method or Prof. Spotts method.
16. Calculate dynamic load, F_d. [Eq. 17.25) or Eq. (17.27]
17. Calculate endurance limit of tooth, S_e. [Eq. 17.32 (*a*)]
18. Calculate endurance strength of gear, $F_e = S_e \pi ymb$ (Eq. 17.34)
19. Check endurance strength. $F_e > F_d$ to give a desirable FOS. (Eq. 17.35)
20. Calculate wear load of pinion F_w. (Eq. 17.41)
21. Check that $F_w > F_d$.
22. If $F_w < F_d$, find hardness of pinion.

Example 17.1 Design a pair of spur gears with 20°FD involute teeth to transmit 10 kW at 1440 rpm of the pinion. The speed ratio is 4 : 1. The pinion is made of plain carbon steel Fe 410 (S_{ut} = 410 MPa) and the gear is made of grey cast iron FG 200 (S_{ut} = 200 MPa). The factor of safety desired is 3 and the load is steady with medium shocks running for 8 to 10 hours per day. The gears are commercial gears.

Solution.

Given : Pressure angle, $\phi = 20°$ FD involute, P = 10 kW, $n_1 = 1440$ rpm, $i = 4 : 1$

(*a*) Design for Static Beam Strength

1. Preliminary decisions :

Minimum number of teeth on pinion to avoid interference,

$$z_1 = 18, \text{ module, } m = 5 \text{ mm}$$

$$\sigma_1 = 410 \text{ MPa}, \sigma_2 = 200 \text{ MPa}.$$

2. Calculate number teeth on the gear,

$$z_2 = iz_1 = 4 \times 18 = 72$$

3. Calculate pitch diameters of pinion and gear.

$$d_1 = mz_1 = 5 \times 18 = 90 \text{ mm}, d_2 = 4 \times 90 = 360 \text{ mm}$$

4. Calculate Lewis form factors

$$y = 0.154 - \frac{0.912}{z}$$

$$y_1 = 0.154 - \frac{0.912}{18} = 0.1033$$

$$y_2 = 0.154 - \frac{0.912}{72} = 0.1431$$

5.Check the weaker wheel.

$$\sigma_1 y_1 = 410 \times 0.1033 = 42.35$$

$$\sigma_2 y_2 = 200 \times 0.1431 = 28.26$$

$$\sigma_2 y_2 < \sigma_1 y_1$$

Hence the gear is the weaker and shall be considered for design.

6.Calculate pitch line velocity.

$$v = \frac{\pi d_1 n_1}{60 \times 10^3} = \frac{\pi \times 90 \times 1440}{60 \times 10^3} = 6.79 \text{ m/s}$$

7. Calculate tangential load.

$$F_t = \frac{10^3 P}{v} = \frac{10^3 \times 10}{6.79} = 1472.75 \text{ N}$$

8. Find service factor from Table 17.10

Find the given conditions,

$$C_s = 1.5$$

9. Find load distribution factor for accurate mountings and face width upto 50 mm from Table 17.14

$$C_m = 1.3$$

10. Calculate maximum load.

$$(F_t)_{max} = C_s C_m F_t$$

$$= 1.5 \times 1.3 \times 1472.75 = 2871.87 \text{ N}$$

11. Calculate velocity factor for $v < 8$ m/s,

$$C_v = \frac{3.05}{3.05 + v}$$

$$= \frac{3.05}{3.05 + 6.79} = 0.31$$

12. Calculate effect load

$$F_{eff} = \frac{(F_t)_{max}}{C_v}$$

$$= \frac{2871.87}{0.31} = 9264 \text{ N}$$

13. Calculate beam strength of gear tooth,

$$F_b = \sigma_{d2} \pi y_2 m b$$

$$= \left(\frac{200}{3}\right) \times \pi \times 0.1413 \times 5 \times b = 147.96\, b \text{ N}$$

14. Calculate face width of gear.

$$F_b = F_{eff}$$

$$147.96b = 9264$$

$$b = 62.6 \text{ mm}$$

The limits for face width are :

$$9.5 \text{ m} = 9.5 \times 5 = 47.5 \text{ mm and } 12.5 \times 5 = 62 \text{ mm}$$

Adopt b = **62 mm and m = 5 mm**

(b) Design for Dynamic Load

15. Calculate dynamic load according to Earle Buckingham's analysis,

$$F_d = (F_t)_{max} + F_i$$

Where incremental load is given by

$$F_i = 20.67v\left[\frac{Cb+(F_t)_{max}}{20.67\,v+\left\{Cb+(F_t)_{max}\right\}^{\frac{1}{2}}}\right]$$

Let us assume Grade 6 gears, for which

$$e = 8.0 + 0.63\,(m + 0.25\sqrt{d})$$

For pinion, $e_p = 8.0 + 0.63\,(5 + 0.25\sqrt{90}) = 12.644\ \mu m$

For pinion, $e_g = 8.0 + 0.63\,(5 + 0.25\sqrt{360}) = 14.138\ \mu m$

$$e = e_p + e_g = 12.644 + 14.138 = 26.782\ \mu m$$

$$E_p = 207 \text{ GPa for steel,}$$

$$E_g = 114 \text{ GPa for FG 200}$$

$$\nu_p = 0.29,\ \nu_g = 0.26$$

$$K_1 = 9.00 \text{ for } 20° \text{ FD teeth}$$

$$(F_t)_{max} = 2871.87 \text{ N}$$

Dynamic factor based on machining error.

$$C = \frac{e}{\left[K_1\left(\frac{1}{E_p}+\frac{1}{E_g}\right)\right]}$$

$$= \frac{26.782\times10^3\times10^3}{\left[9\left(\frac{1}{207}+\frac{1}{114}\right)\right]}$$

$$= 218.76 \text{ N/mm}$$

$$F_i = \frac{20.67\times6.79\,(218.76\times62+2871.87)}{\left[20.67\times6.79+(218.76\times62+2871.87)^{\frac{1}{2}}\right]}$$

$$= 8589.3 \text{ N}$$

$$F_d = 2871.87 + 8589.3 = 11461.2 \text{ N}$$

16. Calculate endurance strength of the gear.

$$F_e = S_e b\pi y m$$

Where S_e = endurance strength of gear material, grey C.I. having BHN = 179 – 233 (core)

$$\approx 0.4\,S_{ut} = 0.4 \times 200 = 80 \text{ MPa}$$

$$F_e = 80 \times 62 \times \pi \times 0.1413 \times 5 = 11009 \text{ N}$$

17.Check for safe design.

As $F_e < F_d$, therefor hardness of grey cast iron has to be increased.

18.Calculate dynamic load according to Prof. Spotts method.

For steel pinion and cast iron gear, we have

$$F_i = \frac{e n_p z_p b r_p r_g}{3260\,(r_p^2 + 0.92 r_g^2)^{1/2}}$$

$$= \frac{26.782 \times 10^{-3} \times 1440 \times 18 \times 62 \times 45 \times 180}{3260\,(45^2 + 0.92 \times 180^2)^{1/2}}$$

$$= 599.4 \text{ N}$$

$$F_d = (F_t)_{max} + F_i = 2871.87 + 599.4 = 3471.24 \text{ N}$$

$F_e > F_d$. Hence safe.

(c) Design for Wear

19.Calculate wear load. The limiting load for wear.

$$F_w = d_p b Q K$$

Here $d_p = 90$ mm, $b = 62$ mm

$$Q = \frac{2z_g}{z_g + z_p} = \frac{2 \times 72}{72 + 18} = 1.6$$

K = load stress factor

$$= 0.2413 \left(\frac{\text{BHN}}{100}\right)^2$$

Let surface hardness of pinion made of Fe 410 = 300 BHN

$$K = 0.2413 \left(\frac{300}{100}\right)^2 = 2.172$$

$$F_w = 90 \times 62 \times 1.6 \times 2.172 = 19391.6 \text{ N}$$

$$\text{FOS} = \frac{F_w}{F_d} = \frac{19391.6}{11461.2} = 1.69. \text{ Hence safe}$$

Example 17.2 A compressor running at 250 rpm is driven by a 15 kW, 750 electric motor through a pair of 20° full depth spur gears. The pinion is made of 45C8 steel (heat treated) and gear of 40C8 steel (untreated), Estimate the module, face width, and number of teeth on each gear. Check for dynamic and wear loads.

Solution.

Given : $n_1 = 750$ *rpm*, $n_2 = 750$ *rpm*, $\phi = 20°$, P = 15 kW

1.Preliminary decisions :

Allowable stresses: $\sigma_{d1} = \dfrac{630}{3} = 210$ MP*a*, $\sigma_{d2} = \dfrac{580}{3} = 193.3$ MP*a*

$z_{min} = z_1 = 18$ for $\phi = 20°$ to avoid interference, $m = 4$ mm for preliminary calculations.

2. Speed ratio, $$i = n_1 / n_2 = \frac{750}{250} = 3$$

3. Number of teeth, $$z_2 = iz_1 = 3 \times 18 = 54$$

4. Pitch diameters, $$d_1 = mz_1 = 4 \times 18 = 72 \text{ mm}$$
$$d_2 = mz_2 = 4 \times 54 = 216 \text{ mm}$$

5. Lewis form factor based on circular pitch :

$$y = 0.154 - \frac{0.912}{z}$$

$$y_1 = 0.154 - \frac{0.912}{18} = 0.1033$$

$$y_2 = 0.154 - \frac{0.912}{54} = 0.1371$$

6. Decide the weaker gear.

$$\sigma_{d1} y_1 = 210 \times 0.1033 = 21.69$$
$$\sigma_{d2} y_2 = 193.3 \times 0.1371 = 26.50$$

Since $\sigma_{d1} y_1 < \sigma_{d2} y_2$, therefore, pinion is the weaker and shall be considered for design.

7. Pitch line velocity,

$$v = \frac{\pi d_1 n_1}{60 \times 10^3}$$

$$= \frac{\pi \times 72 \times 750}{60 \times 10^3} = 2.83 \text{ m/s}$$

8. Velocity factor

$$C_v = \frac{3.05}{3.05 + v} \text{ for } v < 8 \; m/s \text{ and ordinary cut gears.}$$

$$= \frac{3.05}{3.05 + 2.83} = 0.519$$

9. Service factor assuming medium shock conditions and 8 to 10 hours working daily

$$C_s = 1.5$$

10. Tangential load, $$F_t = \frac{10^3 P}{v}$$

$$= \frac{10^3 \times 15}{2.83} = 5300 \text{ N}$$

11. Maximum tangential load, $$(F_t)_{max} = F_t C_s$$
$$= 5300 \times 1.5 = 7950 \text{ N}$$

Effective load, $F_{eff} = \frac{(F_t)_{max}}{C_v} = \frac{7950}{0.519} = 15318 \text{ N}$

12. Static beam strength of pinion teeth,

$$F_b = \sigma_{d1} \pi\, y_1\, mb$$
$$= 210 \times \pi \times 0.1033 \times 4 \times b$$
$$= 272.6\, b \text{ N}$$

13. Calculate face width of teeth.

$$F_b = F_{eff}$$
$$272.6\, b = 15318$$
$$b = 56 \text{ mm}$$

The limits for face width are :

$9.5m \leq b \leq 12.5$ m

$38mm \leq b \leq 50$ mm

The estimated face width is nearly equal to the maximum limit. We adopt $b = 50$ mm, $m = 4$ mm, $d_1 = 72$ mm, $d_2 = 216$ mm

14. Dynamic load according to Earle Buckingham's analysis,

$$F_d = F_{max} + F_i$$

$$F_i = \frac{20.67\, v\, \{Cb + (F_t)_{max}\}}{20.67\, v + (Cb + (F_t)_{max})^{1/2}}$$

Let us assume Grade 10 gears, for which

$$e = 32.0 + 2.50\, (m + 0.25 \sqrt{d})$$

For pinion, $e_p = 32.0 + 2.50\, (4 + 0.25 \sqrt{72}] = 47.30\ \mu m$

For gear, $e_g = 32.0 + 2.50\, (4 + 0.25 \sqrt{216}) = 51.18\ \mu m$

$$e = e_p + e_g = 47.30 + 51.18 = 98.48\ \mu m$$

Let $E_p = 207$ GPa, $E_g = 100$ GPa

$$K_1 = 9.00 \text{ for } 20° \text{ FD teeth}$$

$$(F_t)_{max} = 7950 \text{ N}$$

Dynamic factor based on machining error,

$$C = \frac{e}{K_1\, (1/E_p + 1/E_g)}$$

$$= \frac{98.48 \times 10^{-3} \times 10^3}{9\, (1/207 + 1/100)}$$

$$= 737.8 \text{ N/mm}$$

$$F_i = \frac{20.67 \times 2.83\, (737.8 \times 50 + 7950)}{20.67 \times 2.83 + (737.8 \times 50 + 7950)^{1/2}}$$

$$= 9706 \text{ N}$$

$$F_d = 7950 + 9706 = 17656 \text{ N}$$

15. Calculate endurance strength of the gear.

$$F_e = S_e b\pi y m$$

Where S'_e = endurance strength of pinion material, steel, having BHN = 229 (core)

$\approx 0.5S_{ut} = 0.5 \times 630 = 315$ MPa

$$S'_e = k_a k_b k_c k_d k_e k_f S'_e$$

For machined surface having

$S_{ut} = 630$ MPa, $k_a = 0.74$

$k_b = 0.85$, $k_c = 0.897$ for R = 90 %, $k_c = 1.0$, $k_d = 1.4$, $k_f = 0.944$

$S_e = 0.74 \times 0.85 \times 0.897 \times 1.0 \times 1.4 \times 0.944 \times 315 = 235$ MPa

$F_e = 235 \times 50 \times \pi \times 0.1033 \times 4 = 15252.7$ N

16. Check for safe design.

$F_e \geq 1.5\ F_d$ for shock loads

As $F_e < F_d$, therefore hardness of pinion steel has to be increased.

(c) Design for Wear

17. Calculate wear load. The limiting load for wear.

$$F_w = d_p b Q K$$

Here $d_p = 72$ mm, $b = 50$ mm

$$Q = \frac{2z_g}{(z_g + z_p)} = \frac{2 \times 54}{(54 + 18)} = 1.5$$

K = load stress factor

$$= 0.1536 \left(\frac{BHN}{100}\right)^2$$

Let hardness of pinion made of 45C8 = 500 BHN (case)

$$K = 0.1536 \left(\frac{500}{100}\right)^2 = 3.84$$

$$F_w = 72 \times 50 \times 1.5 \times 3.84 = 20736 \text{ N}$$

$F_w > F_d$, Hence safe

Example 17.3 A pair of spur gears with 20° full depth involute teeth consists of a 20 teeth pinion meshing with a 40 teeth gear. The pinion is mounted on a crankshaft of 7.5 kW single – cylinder diesel engine running at 1500 *rpm*. The driven shaft is connected to a two stage compressor. The pinion and gear are made of 40C8 steel ($S_{ut} = 600$ MPa). The module and face width of the gears are 4 mm and 40 mm respectively. (*a*) Determine the factor of safety using the velocity factor method. (*b*) If the factor of safety is 2 for wear failure, suggest surface hardness of gears. (*c*) If the gears are manufactured to Grade 8, determine the factor of safety using Spott's equation.

Solution.

Given : $\phi = 20°$, $z_p = 20$, $z_g = 40$, P = 7.5 kW, $n_p = 1500$ *rpm*, $\sigma_p = \sigma_g = 600$ MPa, $m = 4$ mm, $b = 40$ mm

(a) Form factors :

$$y_p = 0.154 - \frac{0.912}{z_p}$$

$$= 0.154 - \frac{0.912}{20} = 0.1084$$

$$y_g = 0.154 - \frac{0.912}{40} = 0.1312$$

Since $\sigma_p = \sigma_g$ and $y_p < y_g$, therefore, pinion is the weaker

Design stresses :

$$\sigma_{dp} = \sigma_{dg} = \frac{S_{ut}}{FOS} = \frac{600}{3} = 200 \text{ MPa}$$

Beam strength :

$$F_b = mb\pi y_p \sigma_{dp} = 4 \times 40 \times \pi \times 0.1084 \times 200 = 10897.6 \text{ N}$$

Static load :

$$d_p = mz_p = 4 \times 20 = 80 \text{ mm}, d_g = 4 \times 40 = 160 \text{ mm}$$

$$v_p = \frac{\pi d_p n_p}{60 \times 10^3} = \frac{\pi \times 80 \times 1500}{60 \times 10^3} = 6.283 \; m/s$$

$$F_t = \frac{10^3 P}{v_p} = \frac{10^3 \times 7.5}{6.283} = 1193.7 \text{ N}$$

Velocity factor :

$$C_v = \frac{3.05}{3.05 + v_p} = \frac{3.05}{3.05 + 6.283} = 0.327$$

Service factror, $C_s = 1.5$ from Table 17.10.

Effective load, $F_{eff} = \frac{C_s F_t}{C_v} = \frac{1.5 \times 1193.7}{0.327} = 5475.7 \text{ N}$

Factor of safety, $FOS = \frac{F_b}{F_{eff}} = \frac{10897.6}{5475.7} = 1.99$

(b) Wear load, $F_w = F_{eff} \times FOS = 5475.7 \times 2 = 10951.14 \text{ N}$

$$Q = \frac{2z_g}{z_g + z_p} = \frac{2 \times 40}{40 + 20} = 1.33$$

For steel on steel with E = 207 GPa, $K = 0.15585 \left(\frac{BHN}{100}\right)^2$ from table 17.17.

$$F_w = bQd_pK$$

$$10951.14 = 40 \times 1.33 \times 80 \times 0.15585 \left(\frac{\text{BHN}}{100}\right)^2$$

$$\text{BHN} = 406.32 \approx 410$$

(*c*) Dynamic load :

For Grade 8, $e = 16 + 1.25\,(m + 0.25\sqrt{d})$

For pinion, $e_p = 16 + 1.25\,(4 + 0.25\sqrt{80}) = 23.79\ \mu\text{m}$

For gear, $e_g = 16 + 1.25\,(4 + 0.25\sqrt{160}) = 24.95\ \mu\text{m}$

$$e = e_p + e_g = 23.79 + 24.95 = 48.74\ \mu\text{m}$$

Increment load,

$$F_i = \frac{e n_p z_p b r_p r_g}{2530\,(r_p^2 + r_g^2)^{1/2}}$$

$$= \frac{48.74 \times 10^{-3} \times 1500 \times 20 \times 40 \times 40 \times 80}{2530\,(40^2 + 80^2)^{1/2}}$$

$$= 827 \text{ N}$$

Dynamic load, $F_d = (F_t)_{max} + F_i = C_s F_t + F_i = 1.5 \times 1193.7 + 827 = 2617.6 \text{ N}$

$$\text{FOS} = F_b/F_d = \frac{10897.6}{2617.6} = 4.16.$$

Example 17.4 Design a spur gear speed ruducer for a compressor running at 300 *rpm* driven by a 7.5 kW, 900 *rpm* electric motor. The centre distance between the axes of the gear shafts should be exactly 300 mm. The starting torgue of the motor is 150% of the rated torque. The gears are made of carbon steel 50C4 (S_{ut} = 700 MP*a*). The pressure angle is 20°. The factor of safety is 2 and the gears accuracy Grade is 10. Specify the surface hardness of gears.

Solution.

Given : $n_1 = 900$ *rpm*, P = 7.5 kW, $n_2 = 300$ *rpm*, C = 300 mm, $\sigma_1 = \sigma_2 = 700$ MP*a*, FOS = 2, $\phi = 20°$

$$\frac{d_1 + d_2}{2} = 300$$

$$d_1 + d_2 = 600 \text{ mm}$$

$$\frac{d_1}{d_2} = \frac{n_1}{n_2} = \frac{300}{900} = \frac{1}{3}$$

$$d_1 + 3d_1 = 600$$

$$d_1 = 150 \text{ mm}$$

$$d_2 = 450 \text{ mm}$$

Pitch line velocity,

$$v = \frac{\pi d_1 n_1}{60 \times 10^3} = \frac{\pi \times 150 \times 900}{60 \times 10^3} = 7.06 \text{ m/s}$$

Velocity factor, $C_v = \frac{3.05}{3.05 + v} = \frac{3.05}{3.05 + 7.06} = 0.3017$

Tangential force, $F_t = \frac{10^3 P}{v} = \frac{10^3 \times 7.5}{7.06} = 1062.3$ N

Effective force, $F_{eff} = \frac{C_s F_t}{C_v} = \frac{1.5 \times 1062.3}{0.3017} = 5281.7$ N

Let $z_1 = 30$ so that $z_2 = 90$. Then $m = \frac{d_1}{z_1} = \frac{150}{30} = 5$ mm

Let face width, $b = 10\,m = 10 \times 5 = 50$ mm

Form factor, $y_1 = 0.154 - \frac{0.912}{30} = 0.1236$

$y_2 = 0.154 - \frac{0.912}{90} = 0.1438$

Pinion is the weaker, as $y_1 < y_2$

Beam strength, $F_b = mb\sigma_{1d}\pi y_1 = 5 \times 50 \times \frac{700}{3} \times \pi \times 0.1236 = 22651$ N

Dynamic load :

For Grade 10, $e = 32 + 2.5\,(m + 0.25\sqrt{d})$

For pinion, $e_p = 32 + 2.5\,(5 + 0.25\sqrt{150}) = 52.1546\mu m$

For gear, $e_g = 32 + 2.5\,(5 + 0.25\sqrt{450}) = 57.7582\mu m$

$e = e_p + e_g = 52.1546 + 57.7582 = 109.9128\ \mu m$

Increment load, $F_i = \frac{e n_p z_p b r_p r_g}{2530\,(r_p^2 + r_g^2)^{1/2}}$

$= 109.9128 \times 10^{-3} \times 900 \times 30 \times 50 \times 75 \times \left[\frac{225}{2530\,(75^2 + 225^2)^{1/2}}\right]$

$= 4172.96$ N

Dynamic load, $F_d = (F_t)_{max} + F_i = C_s F_t + F_i = 1.5 \times 1062.3 + 4172.96 = 5766.4$ N

FOS $= \frac{F_b}{F_d} = \frac{22651}{5766.4} = 3.92$

Surface hardness :

$Q = \frac{2 z_2}{z_2 + z_1} = \frac{2 \times 90}{90 + 30} = 1.5$

$F_w = F_{eff} \times FOS = 5281.7 \times 2 = 10563.4$ N

$$= bQd_1K = 50 \times 1.5 \times 150 \times 0.1536 \left(\frac{\text{BHN}}{100}\right)^2$$

$$\text{BHN} = 247.2 \approx 250$$

Example 17.5 Design a spur gear drive to tansmit 30 kW at a pinion speed of 900 rpm. The speed ratio is 3 : 1. The teeth are 20° FD involute with 20 teeth on pinion. Both the gears are made of steel and machined to Grade 6.

Solution.

Given : P =30 kW, $n_p = 900$, $i = 3$, $\phi = 20°$ FD involute, $z_p = 20$

Let us select the following materials for the pinion and gear :

Pinion : Plain carbon steel, 55C8, $S_{up} = 720$ MP*a*

Gear : Plain carbon steel, 40C8, $S_{ug} = 630$ MP*a*

$$z_g = iz_p = 3 \times 20 = 60$$

Let module, $m = 5$ mm for the preliminary design.

$$d_p = mz_p = 5 \times 20 = 100 \text{ mm}$$

$$d_g = mz_g = 5 \times 60 = 300 \text{ mm}$$

Design stresses :

$$\sigma_{dp} = \frac{S_{up}}{3} = \frac{720}{3} = 240 \text{ MPa}$$

$$\sigma_{dg} = \frac{S_{ug}}{3} = \frac{630}{3} = 210 \text{ MPa}$$

Lewis form factors :

$$y = 0.154 - \frac{0.912}{z}$$

$$y_p = 0.154 - \frac{0.912}{20} = 0.1084$$

$$y_g = 0.154 - \frac{0.912}{60} = 0.1388$$

$$\sigma_{dg}y_p = 240 \times 0.1084 = 26.016$$

$$\sigma_{dg}y_g = 210 \times 0.1388 = 29.148$$

Since $\sigma_{dp}y_p < \sigma_{dg}y_g$, therefore pinion is the weaker and shall be considered for design.

Pitch line velocity, $$v_p = \frac{\pi d_p n_p}{60 \times 10^3} = \frac{\pi \times 100 \times 900}{60 \times 10^3} = 4.71 \ m/s$$

Let service factor, $$C_s = 1.5$$

Velocity factor, $$C_v = \frac{3.05}{3.05 + v_p} = \frac{3.05}{3.05 + 4.71} = 0.393$$

Tangential force, $F_t = \frac{10^3 \times P}{v_p} = \frac{10^3 \times 30}{4.71} = 6369.4\text{N}$

$$(F_t)_{max} = C_s F_t$$
$$= 1.5 \times 6369.4 = 9554 \text{ N}$$

Static beam strength of pinion teeth,

$$F_b = \sigma_{dp} b m \pi y_p = 240 \times b \times 5 \times \pi \times 0.1084 = 408.66b \text{ N}$$

Effective load, $F_{eff} = \frac{(F_t)_{max}}{C_v} = \frac{9554}{0.393} = 24310 \text{ N}$

Face width :

$$F_b = F_{eff}$$
$$408.66b = 24310$$
$$b = 59.5 \approx 60 \text{ mm}$$

The specified limits for face width are :

$9.5\,m = 9.5 \times 5 = 47.5$ mm and $12.5\,m = 12.5 \times 5 = 62.5$ mm

The calculated face width lies between the specified limits.

Adopt b = 60mm, m = 5 mm, d_p = 100 mm, d_g = 300 mm, Centre distance, C = 0.5 (100 + 300) = 200 mm, addendum, h_a = 1m = 5 mm, dedendum, h_f = 1.25m = 6.25 mm

Dynamic Load :

For Grade 6 gears, $e = 8.0 + 0.63\,(m + 0.25\sqrt{d})$

For pinion, $e_p = 8.0 + 0.63\,(5 + 0.25\sqrt{100}] = 12.725\ \mu m$

For gear, $e_g = 8.0 + 0.63\,(5 + 0.25\sqrt{300}) = 13.878\ \mu m$

$$e = e_p + e_g = 12.725 + 13.878 = 26.603\ \mu m$$

Let $E_p = E_g = 207$ GPa

$$K_1 = 9.00 \text{ for } 20° \text{ FD teeth}$$
$$(F_t)_{max} = 9554 \text{ N}$$

Dynamic factor based on machining error,

$$C = \frac{e}{K_1\,(1/E_p + 1/E_g}$$

$$= \frac{26.603 \times 10^{-3} \times 10^3}{9\,(1/207 + 1/207)}$$

$$= 305.93 \text{ N/mm.}$$

Increment load, $F_i = \frac{20.67\,v\,\{Cb + (F_t)_{max}\}}{20.67\,v + \{Cb + (F_t)_{max}\}^{1/2}}$

$$F_i = \frac{20.67 \times 4.71\,(305.93 \times 60 + 9554)}{20.67 \times 4.71 + (305.93 \times 60 + 9554)^{1/2}}$$

$$= 10276 \text{ N}$$

$$F_d = (F_t)_{max} + F_i = 9554 + 10276 = 19830 \text{ N}$$

16.Calculate endurance strength of the gear.

$$F_e = S_e b\pi y m$$

Where S'_e = uncorrected endurance strength of pinion material, plain carbon steel, 55C8, having

$$BHN = 265$$

$$\approx 0.5S_{up} = 0.5 \times 720 = 360 \text{ MPa}$$

Accounting for various factors affecting endurance strength,

$$S_e \approx 0.8\ S'_e = 288 \text{ MPa}$$

$$F_e = 288 \times 60 \times \pi \times 0.1084 \times 5 = 29424 \text{ N}$$

17. Check for safe design.

$$\frac{F_e}{F_d} = \frac{29424}{19830} = 1.48$$

As $F_e > F_d$. Hense safe.

18.Calculate dynamic load according to Prof. Spotts method.

For steel pinion and steel gear, we have

$$F_i = \frac{e n_p z_p b r_p r_g}{2530\,(r_p^2 + r_g^2)^{1/2}}$$

$$= \frac{26.603 \times 10^{-3} \times 900 \times 20 \times 60 \times 50 \times 150}{2530\,(50^2 + 150^2)^{1/2}}$$

$$= 538.7 \text{ N}$$

$$F_d = (F_t)_{max} + F_i = 9554 + 538.7 = 10092.7 \text{ N}$$

$$\frac{F_e}{F_d} = \frac{29424}{10092} = 2.91. \text{ Hence safe.}$$

(*c*) Design for Wear

19.Calculate wear load. The limiting load for wear.

$$F_w = d_p b Q K$$

$$\text{Here } d_p = 100 \text{ mm}, b = 60 \text{ mm}$$

$$Q = \frac{2z_g}{z_g + z_p} = \frac{2 \times 60}{60 + 20} = 1.5$$

K = load stress factor

$$= 0.156 \left(\frac{BHN}{100}\right)^2$$

Surface hardness of case hardened pinion made of 55C8 = 520 BHN

$$K = 0.156\left(\frac{520}{100}\right)^2 = 4.218$$

$$F_w = 100 \times 60 \times 1.5 \times 4.218 = 37962 \text{ N}$$

$F_w > F_d$, Hence safe.

17.24 CONSTRUCTION OF PINION AND GEAR WHEEL

A gear pinion and wheel may be constructed as discussed below :

1. **Integral Pinion :** When the dedendum circle diameter of the pinion is very close to the shaft diameter, it is made integral with the shaft, as shown in Fig 17.9 (*a*). This type of construction eliminates the key and hence increases the rigidity of the system and reduces the manufacturing cost.

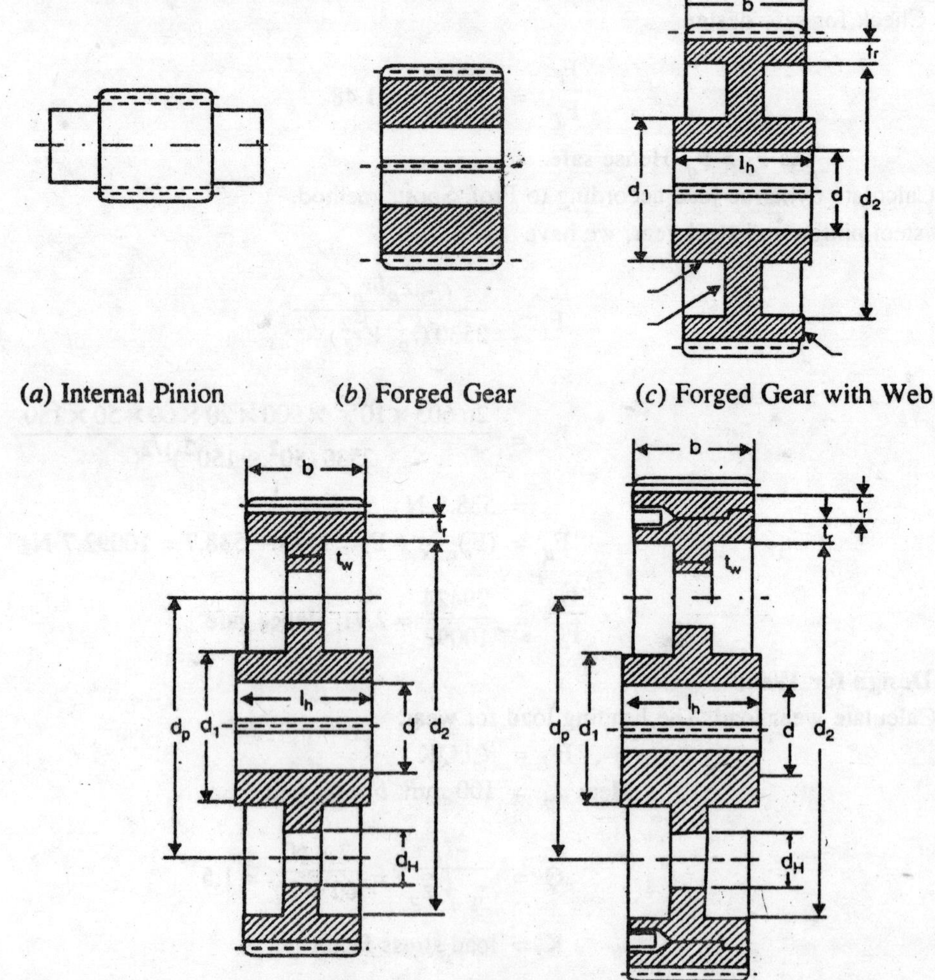

(*a*) Internal Pinion (*b*) Forged Gear (*c*) Forged Gear with Web

(*d*) Forged Gear with Web (*d*) Rimmed Gear with Composite Construction

Fig. 17.9

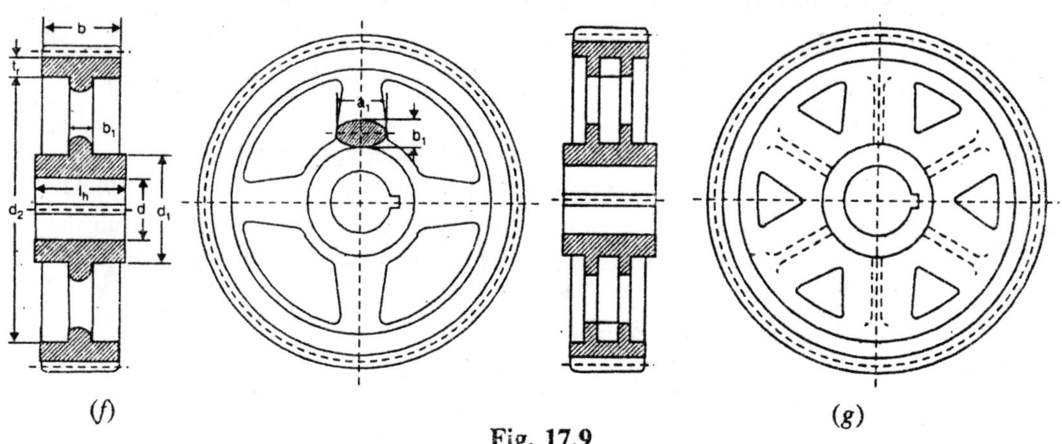

Fig. 17.9

2. **Forged or Solid Gear :** Small steel gears upto 250 mm diameter are made either from cylindrical discs cut from rolled bars or from forged solid blanks without web, as shown in Fig. 17.9 (*b*). Cast iron gears upto 250 mm diameter may be solid cast.
3. **Forged Gear with Web :** Forged gears upto 500 mm diameter are generally made from forged steel blanks with web, as shown in Fig. 17.9 (*c*) and (*d*).
4. **Rimmed Gear with Composite Construction :** Very large gears are either welded or made of composite construction. In composite construction, the steel gear rim is either press – fitted or shrink – fitted on the cost iron hub, as shown in Fig. 17.9 (*e*). Set screws are fitted between the rim and the hub to eliminate the possibility of any displacement between them. Holes are drilled in the web to reduce the weight of gear.
5. **Rim and Arm type Gear :** Cost iron gears upto 500 mm rim diameter are cast with rim connected to hub by number of arms, as shown in Fig. 17.9 (*f*). These type of gears are much lighter in weight as compared to solid gears.
6. **Two – walled Construction :** Very large and wide cast iron gears are generally made with two walls, as shown in Fig. 17.9 (*g*). Holes are drilled in the rims to reduce the weight of the gear.

17.26 DIMENSIONS OF PINION AND GEAR WHEEL

The shaft diameter is first determined from the power to be transmitted and the location of the gear. If the pinion teeth are cut integral with the shaft, the root diameter should be slightly larger than the required shaft diameter. When a solid pinion is keyed to the shaft, the minimum pitch diameter is given by :

$$(d_p)_{min} = 2 \times \text{bore} + 0.25 \times m \qquad \text{... (17.46)}$$

The minimum thickness of metal between the keyway and the root circle is :

$$t_{min} = m\sqrt{\left(\frac{z}{5}\right)} \qquad \text{... (17.47)}$$

Outside diameter of hub $= 1.8 \times$ bore for steel gear ... (17.48*a*)

$= 2 \times$ bore for cast iron gear ... (17.48*b*)

$= 1.65 \times$ bore for forged steel ... (17.48*c*)

Hub length $= 1.75 \times$ bore for light service ... (17.49*a*)

> face width of gear teeth ... (17.49*b*)

Web thickness for small pinions = (1.6 – 1.9)m ... (17.50)

Number of arms = 4 for split gears upto 1 metre diameter ... (17.51*a*)

= 6 for gears upto 3 meter diameter ... (17.51*b*)

= 8 for larger gears ... (17.51*c*)

The arms should be designed as cantilever beams loaded at the pitch line, with the load equally distributed to all the arms. The design load is the stalling load that shall develop in the arms at zero velocity.

Stalling load, $F_o = \sigma b \pi y m$

Section modulus of arms, $z = \dfrac{I}{c}$

Where I = moment of inertia of arms about the neutral axis

c = distance of the outermost fibres of the arm cross – section

Bending moment on the arms= $\dfrac{F_o d}{2}$

Bending moment in one arm, $M = \dfrac{F_o d}{2n_a}$

Where d = pitch cricle diameter of gear

n_a = numbefr of arms

$$\frac{I}{c} = \frac{F_o d}{2n_a \sigma} \qquad \text{... (17.52)}$$

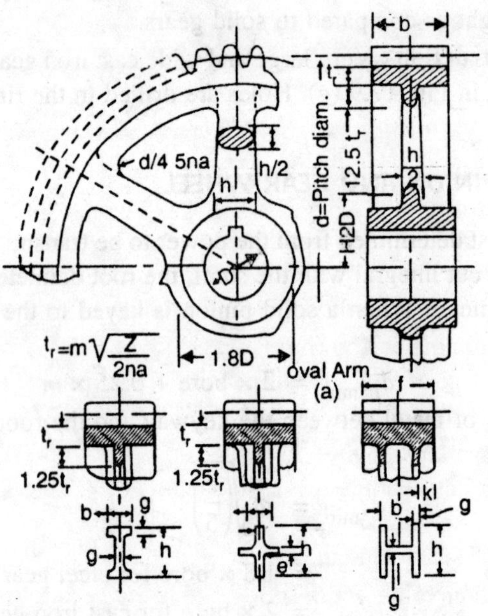

Fig. 17.10 Proportions of spur gear wheels.

The cross – section of the arms is generally elliptical with the major axis of arms lying in the plane of rotation, as shown in Fig. 17.10. The major axis is made twice the minor axis. The major axis at the outside of the hub is given by :

$$h = 4\left[\frac{I}{\pi c}\right]^{1/3} = \left[\frac{20.4 \times I}{c}\right]^{1/3} \quad \text{... (17.53)}$$

The arms are usually tapered towards the rim about 1 : 16.

Rim thickness,

$$t_r = m\left[\frac{z}{2n_a}\right]^{1/3} \quad \text{... (17.54)}$$

Very large and wide gears are made with arms cross – section of I, cross or H – section, as shown in Fig. 17.9. The key size may be determined by knowing the shaft diameter.

EXERCISES

17.1 A pair of carefully cut gears with 20° involute teeth is to transmit 25 kW at 300 rpm of the gear at a speed reduction of 5 : 1. The pinion should not be smaller than 76 mm in pitch diameter and is made of forged 30C8 steel with a hardness of 250 BHN, and drives cast iron gear. Select a suitable spur gear pair and determine the module, number of teeth and face width of the gears.

17.2 A pair of gears is to be designed to transmit 4 kW at 600 rpm of the pinion to gear rotating at 150 rpm. 20° FD involute teeth are to be used. The centre distance should be as small as possible. Only from strength considerations, determine the module, face width and number of teeth on spur gears.

17.3 The pinion in spur gear set having a speed ratio 3 rotates at 900 rpm and has 20 teeth FD of 20° pressure angle. The face width is 30 mm and the module is 3 mm. Determine the power rating of the gear set based on bending strength of teeth only. Both the pinion and the gear are made of same material having ultimate tensile strength of 400 MP*a* and yield strength of 210 MP*a*. A factor of safety of 3 is desired.

17.4 A gear drive is required to transmit a maximum power of 22.5 kW. The velocity ratio is 1 : 2 and rpm of pinion is 200. The approximate centre distance between the shafts may be taken as 600 mm. The teeth have 20° stub involute profiles. The allowable static stress for the gear / pinion material (which is cast iron) may be taken as 60 MP*a* and face width as 10 times the module. Find module of gears.

17.5 A compressor running at 250 rpm is driven by a 15 kW, 750 rpm motor through 14.5° FD involute spur gears. Estimate the module, tooth width and number of teeth on each gear. Assume suitable data.

17.6 A pair of spur gears is required to reduce speed from 600 rpm to 120 rpm for 12 hours running time per day continuously. The pinion is 40C8 steel normalised and has 20 teeth. The gear wheel is of cast iron of Grade FG 200 and has 100 teeth. The gears are 8 mm module, 100 mm face width and 20° pressure angle involute. Calculate the power rating.

17.7 A pair of 20° FD involute tooth spur gears is required to transmit 30 kW at 240 rpm of the pinion. The speed ratio is 4 : 1. Select suitable material and other data for the pinion and gear and design the gear pair.

18 HELICAL GEARS

Chapter

18.1 INTRODUCTION

Helical gears are used to transmit power between parallel shafts. A comparison of spur and helical gears is given in table 18.1.

Table 18.1 Comparison of spur and helical gears.

Spur gears	Helical gears
1. Teeth are cut parallel to the axis of the shaft.	1. Teeth are cut in the form of a helix on the pitch cylinder between meshing gears.
2. Contact between meshing teeth occurs along the entire face width of the tooth.	2. Contact between meshing gears begins with a point on the leading edge of the tooth and gradually extends along the diagonal line across the tooth.
3. Load application is sudden resulting into impact conditions and generating noise in high speed applications.	3. Pick up of load by the tooth is gradual, resulting in smooth engagement and quiet operation even at high speeds.
4. Used for parallel shafts only.	4. Crossed helical gears are used on shafts with crossed axes.
5. Speed is limited to about 20 m/s.	5. Used in automobiles, turbines and high speed applications upto 50 m/s.
6. Imposes radial load only.	6. Imposes radial and axial thrust loads.
7. Contact ratio is low.	7. Contact ratio is high.

The helical gears may be of single helical type of double helical type (herringbone gears). The axial thrust present in single helical type is automatically cancelled in double helical type gears.

18.2 TERMINOLOGY

The important terms of a helical gear are shown in Fig. 18.1.

Helix angle. It is the angle between a line drawn through one of the teeth and the centre line of the shaft on which the gear is mounted. It is denoted by β. It varies from 15° to 30°.

Normal circular pitch, p_n. It is the distance between corresponding points of adjacent teeth as measured in a plane perpendicular to the tooth elements. This is shown by plane y–y in Fig. 18.2, which is perpendicular to teeth elements. AC represents normal circular pitch.

Transverse circular pitch, p_t. It is measured in a plane perpendicular to the shaft axis, as shown by plane $x - x$ in Fig. 18.1. AB represents the transverse circular pitch.

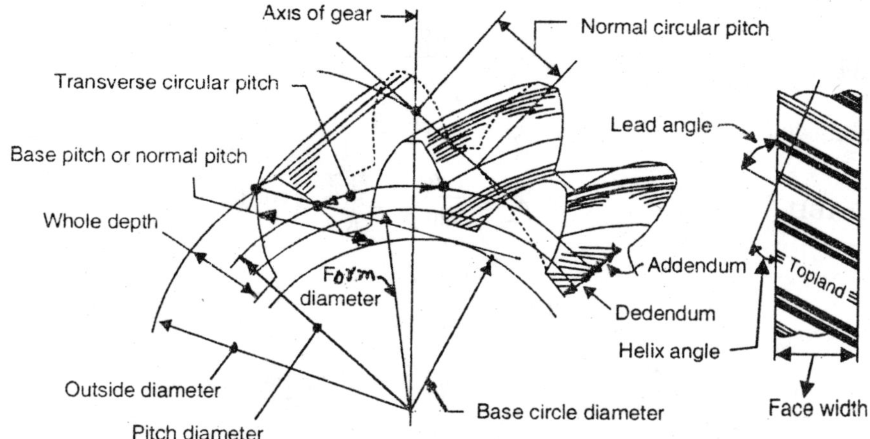

Fig. 18.1 Helical gear terminology

Transverse diameteral pitch, P_t. It is the diameteral pitch measured in the plane of ratation, *i.e.* transverse to the axis of rotation.

Normal diameteral pitch, P_n. It is the diameteral pitch measured in the plane normal to the helix. It is equal to diameteral pitch of the hob.

Transverse presure angle, ϕ_t. It is the pressure angle measured in the transverse plane or plane of rotation.

Normal presure angle, ϕ_n. It is the pressure angle measured in the normal plane or plane perpendicular to the teeth.

Axial pitch, p_x. It is the distance measured parallel to the shaft axis, AD represents the axial pitch.

In ΔABC, we have

$$\frac{AC}{AB} = \frac{p_n}{p_t} = \cos\beta$$

Normal circular pitch,

$$p_n = p_t \cos\beta \qquad ...(18.1)$$

Transverse circular pitch,

$$p_t = \frac{\pi d}{z} \qquad ...(18.2)$$

Transverse diametral pitch,

$$P_t = \frac{z}{d}$$

Transverse module,

$$m_t = \frac{1}{P_t} = \frac{d}{z}$$

$$p_t P_t = \pi = p_n P_n$$

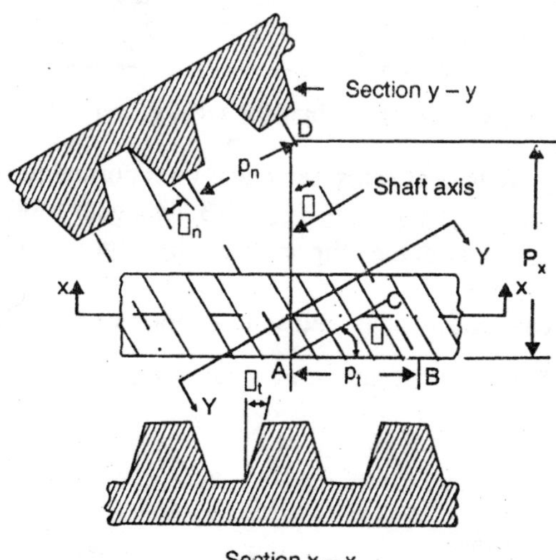

Fig. 18.2 Tooth relationships

Normal diametral pitch, $P_n = \dfrac{P_t}{\cos\beta} = \dfrac{z}{d\cos\beta}$ ...(18.3)

Normal module, $m_n = m_t \cos\beta$ ... (18.4)

From ΔABD, $p_x = \dfrac{p_t}{\tan\beta} = \left(\dfrac{\pi d}{z}\right)\cot\beta$ ... (18.5)

Also $\cos\beta = \dfrac{\tan\phi_n}{\tan\phi_t}$ ... (18.6)

$$d = \frac{zp_t}{\pi} = zm_t = \frac{zm_n}{\cos\beta} \quad \text{... (18.7)}$$

Centre distance, $C = \left(\dfrac{1}{2}\right)(d_1 + d_2)$

$$= \left[\frac{m_n}{2\cos\beta}\right] z_1 + z_2 \quad \text{... (18.8)}$$

Speed ratio, $i = \dfrac{z_2}{z_1} = \dfrac{n_1}{n_2}$ ... (18.9)

The standard normal modules are given in Table 18.2.

Table 18.2 Standard normal modules for helical gears in mm

First choice	1.0, 1.25, 1.5, 2, 2.5, 3, 4, 5, 6, 8, 10, 12, 16, 20, 25, 32, 40 and 50
Second choice	1.125, 1.375, 1.75, 2.25, 2.75,3.5, 4.5, 5.5, 7, 9, 11, 14, 18, 22, 28, 36 and 45

18.3 VIRTUAL NUMBER OF TEETH

In helical gears, the plane $x - x$ normal to the gear teeth intersects the pitch cylinder to form an ellipse, as shown in Fig. 18.3. The gear tooth profile generated in this plane, using the radius of curvature of the ellipse, would be a spur gear having the same properties as the actual helical gear.

The semi – major and semi – minor axes of this ellipse are : $a = \dfrac{d}{2\cos\beta}$ and $b = \dfrac{d}{2}$ respectively.

The radius of curvature r_c at point A is,

$$r_c = \frac{a^2}{b} = \frac{d}{2\cos^2\beta}$$

where d = pitch circle diamter

In the design of helical gears, an imaginary spur gear is considered in the plane $x - x$ with a pitch circle radius r_c and module m_n. It is called a "formative" or 'virtual" spur gear.

Pitch circle diameter of virtual gear, $d_c = \dfrac{d}{\cos^2\beta}$

The number of teeth of the equivalent (or virtual) spur gear in the normal plane are called the virtual number of teeth, z_v.

$$z_v = \frac{\pi d_c}{p_n}$$

$$= \left(\frac{\pi d}{\cos^2 \beta}\right) \times \left[\frac{1}{(\pi m_n)}\right]$$

$$= \frac{d}{(m_n \cos^2 \beta)}$$

$$= \frac{z}{\cos^3 \beta} \qquad \text{... (18.10)}$$

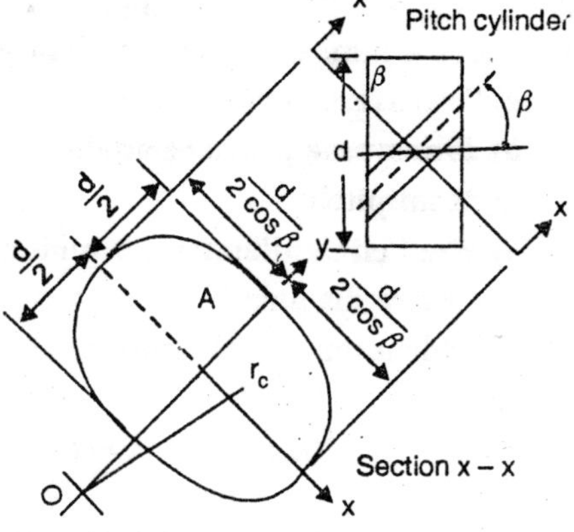

Fig. 18.3 Concept of virtual number of teeth.

The design of helical gears is based on virtual number of teeth. The cutter is also selected based upon the virtual number of teeth.

18.4 TOOTH PROPORTIONS

The proportions of a helical gear tooth are given in Table 18.3.

Table 18.3 Proportions of helical gear tooth.

Item		Proportions
Addendum, h_a		m_n
Dedendum, h_f		$1.25\ m_n$
Total depth, h		$18.25\ m_n$
Working depth, h_w		$2\ m_n$
Clearance, c		$0.25\ m_n$
Addendum circle diamter, d_a		$d + 2h_a$
		$m_n\left[\frac{z}{\cos \beta} + 2\right]$
Dedendum circle diamter, d_f		$d - 2h_f$
		$m_n\left[\frac{z}{\cos \beta} - 2.5\right]$
Base circle diamter, d_b		$d \cos \phi_t$
Lead, L		$\pi d \cot \beta$
Pressure angle, ϕ_t	max.	25°
	min.	15°
Helix angle, β	max.	45°
	min.	20°
Face width, b	min.	$\geq \frac{p_t}{\tan \beta}$
		$\geq \frac{\pi m_n}{\sin \beta}$

Example 18.1 A pair of parallel helical gears consists of a 20 teeth pinion meshing with a 60 teeth gear. The helix angle is 25° and the normal pressure angle is 20°. The normal module is 4 mm. Calculate the following :

(*a*) Transverse module

(*b*) Transverse pressure angle

(*c*) Axial pitch

(*d*) Pitch circle diamters of pinion and gear

(*e*) Centre distance

(*f*) Addendum and dedendum circle diameters of pinion.

Solution.

Given : $z_p = 20$, $z_g = 60$, $\beta = 25°$, $\phi_n = 20°$, $m_n = 4$ mm

(*a*) Transverse module, $m_t = \dfrac{m_n}{\cos\beta} = \dfrac{4}{\cos 25°} = 4.41$ mm

(*b*) Transverse presure angle, $\tan\phi_t = \dfrac{\tan\phi_n}{\cos\beta} = \dfrac{\tan 20°}{\cos 25°} = 0.4016$ mm

$$\phi_t = 21.88°$$

(*c*) Axial pitch, $p_x = \dfrac{p_t}{\tan\beta} = \dfrac{\pi m_t}{\tan\beta} = \dfrac{\pi \times 4.41}{\tan 25°} = 29.71$ mm

(*d*)

$$d_p = z_p m_t = 20 \times 4.41 = 88.2 \text{ mm}$$

$$d_g = z_g m_t = 60 \times 4.41 = 264.6 \text{ mm}$$

(*e*) Centre distance, $C = \dfrac{d_p + d_g}{2} = \dfrac{88.2 + 264.6}{2} = 176.4$ mm

(*f*)

$$d_a = m_n\left[\frac{z_p}{\cos\beta} + 2\right] = 4\left[\frac{20}{\cos 25°} + 2\right] = 96.27$$

$$d_f = m_n\left[\frac{z_p}{\cos\beta} - 2.5\right] = 78.27 \text{ mm.}$$

18.5 FORCE ANALYSIS

The forces acting on a helical gear tooth are shown in Fig. 18.4.

$$\text{Tangential force, } F_t = \frac{10^3 \times P}{v}$$

Where P = power transmitted, kW

v = pitch line velocity, m/s

$$= \frac{\pi dn}{60 \times 10^3}$$

where $n = rpm$, d = pitch diamter, mm

In ΔABC, we have

Radial force, $F_r = AB = EC = F_n \sin \phi_n$ ... (1)

$BC = F_n \cos \phi_n$ ... (2)

In ΔBDC, we have

Axial force,

$F_a = BD = FC = BC \sin \beta = F_n \cos \phi_n \sin \beta$ ... (3)

$F_t = DC = BC \cos \beta = F_n \cos \phi_n \cos \beta$... (4)

From Eqs. (3) and (4), we get

$$F_a = F_t \tan \beta \qquad \text{... (18.11)}$$

From Eqs. (1) and (4), we get

$$F_r = F_t \left(\frac{\tan \phi_n}{\cos \beta} \right) \qquad \text{... (18.12)}$$

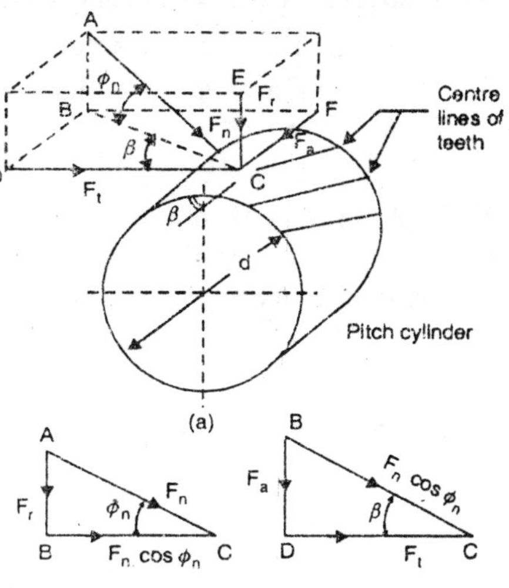

Fig. 18.4 Tooth forces in a heilcal gear

18.6 THRUST DIRECTIONS

The direction of the thrust component depends upon the hand of the helix, direction of rotation and whether the gear is the driver or the driven. The following guide lines may be used to determine the direction of the thrust component :

1. Select the driving gear.
2. Use right hand for right hand helix and left hand for the left hand helix.

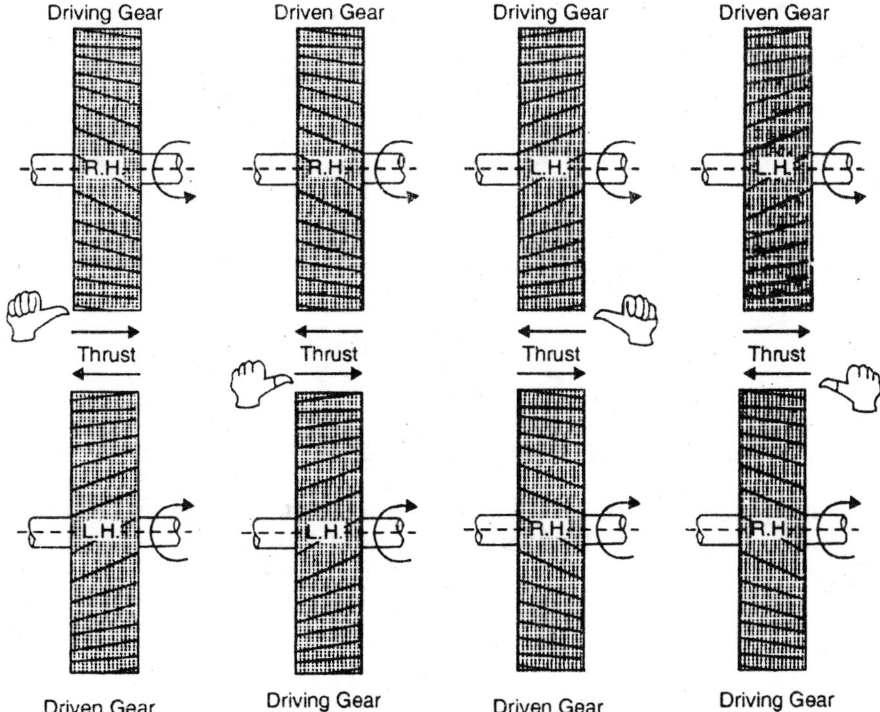

Fig. 18.5 Thrust derections in helical gear

3. Keep the fingers in the direction of rotation of the gear then the thumb will indicate the direction of the thrust component on the driving gear.
4. The direction of thrust component for the driven gear will be opposite to that for the driving gear.

The thrust directions are shown in Fig. 18.5.

Example 18.2 A pair of parallel helical gears are required to transmit 7.5 kW at 900 rpm. The number of teeth on the pinion are 20 and on the gear 40. The normal module is 5 mm and normal pressure angle 20°. The helix angle is 25°. Determine the coponents of the tooth force.

Solution.

Given : P = 7.5 kW, $N_p = 900$ rpm, $z_p = 20$, $z_g = 40$, $m_n = 5$ mm, $\beta = 25°$, $\phi_n = 20°$

Power transmitted,
$$P = \frac{2\pi NT}{60 \times 10^3}$$

$$7.5 = \frac{2\pi \times 900 \times T}{60 \times 10^3}$$

$$T = 79577 \text{ N.mm}$$

Pitch diamter of pinion,
$$d_p = \frac{z_p m_n}{\cos\beta} = \frac{20 \times 5}{\cos 25°} = 110.34 \text{ mm}$$

Tangential force,
$$F_t = \frac{2T}{d_p}$$

$$= \frac{2 \times 79577}{110.34} = 1079.9 \text{ N}$$

Axial thrust,
$$F_a = F_t \tan\beta$$
$$= 1079.9 \times \tan 25° = 503.56 \text{ N}$$

Radial force,
$$F_r = \frac{F_t \tan\phi_n}{\tan\beta}$$

$$= \frac{1079.9 \times \tan 20°}{\tan 25°} = 842.9 \text{ N.}$$

18.7 MINIMUM NUMBER OF TEETH ON HELICAL GEARS

In order to avoid interference in helical gears, it must be ensured that interference does not occur in an equivalent spur gear pair. The minimum number of teeth on a pinion to avoid interference is given by,

$$(z_v)_{min} = \frac{2h_a}{m_n \sin^2\phi_n}$$

$$\frac{z_p}{\cos^3 \beta} = \frac{2h_a}{m_n \sin^2 \phi_n}$$

$$z_p = \frac{2h_a \cos^3 \beta}{m_n \sin^2 \phi_n} \quad \text{... (18.13)}$$

18.8 STATIC BEAM STRENGTH

The helical gear is considered to be equivalent to a virtual spur gear. The virtual gear is an imaginary spur gear in a plane perpendicular to the tooth element. Its pitch circle diamter is d_c, number of teeth z_v and module m_n.

For a spur gear, beam strength, $F_b = mb\pi y\sigma$

in Fig. 18.6, we have

$F_b = (F_b)_n$ = beam strength perpendicular to tooth element

$$m = m_n$$

$$b = \frac{b}{\cos \beta}$$

$$z = z_v$$

$$\therefore (F_b)_n = \frac{m_n b\pi y\sigma C_v}{\cos \beta}$$

$$F_b = (F_b)_n \cos \beta$$

$$= m_n b\pi y\sigma_d \quad \text{... (18.14)}$$

Where σ_d = allowable static bending strength of tooth material.

Eq. (18.14) is the Lewis equation for helical gears.

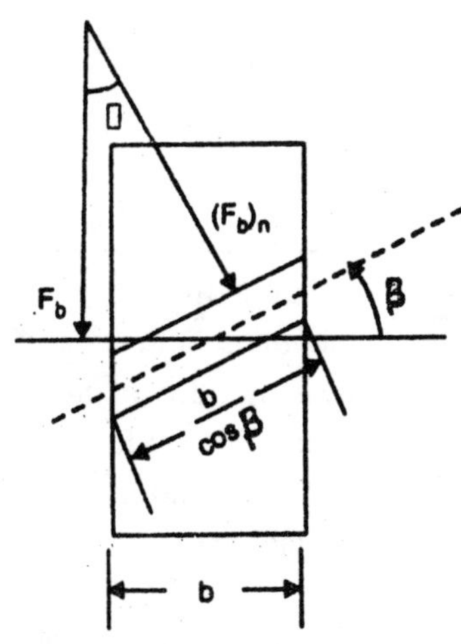

Fig. 18.6 Beam strength of helical gear tooth

18.9 LEWIS FORM FACTOR

Lewis form factors based upon circular pitch are given by,

$$y = 0.124 - \frac{0.684}{z_u} \text{ for 14.5° involute FD} \quad \text{... (18.15a)}$$

$$= 0.154 - \frac{0.912}{z_u} \text{ for 20° involute FD} \quad \text{... (18.15b)}$$

$$= 0.170 - \frac{0.950}{z_u} \text{ for 20° involute stub} \quad \text{... (18.15c)}$$

18.10 WEAR AND LUBRICATION FACTOR, C_W

The wear and lubrication factor depends upon the nature of lubrication. Its value are given in Table 18.4.

Table 18.4 Wear and lubrication factor, C_w

Nature of lubrication	C_w
Enclosed gears continuously lubricated with oil of the proper viscosity and character	1.15
Scant lubrication but regular with frequent inspection	1.25
Indifferent lubrication	1.35

18.11 LOAD DISTRIBUTION FACTOR, C_m

To account for the non-uniform distribution of load across the face width of the helical gear teeth, load distribution (or stress – concentration) factor is introduced in the analysis. Its values are given in Table 18.5

Table 18.5 Load distribution factor for helical gears, C_m

Characteristics of support	Face width upto, mm			
	50	150	225	400
1. Accurate mountings, small bearing clearances, minimum deflection, precision gears.	1.2	1.3	1.4	1.7
2. Less rigid mountings, less accurate gears, contact across the full face.	1.5	1.6	1.7	2.0
3. Accuracy and mounting such that less than full face contact exists.	Over 2.0			

18.12 MAXIMUM TANGENTIAL LOAD

$$\text{Tangential force, } F_t = \frac{10^3 P}{v}$$

$$P = \text{power transmitted, kW}$$

$$\text{Where } v = \frac{\pi d n}{60 \times 10^3}, \; m/s$$

$$n = rpm, \; d = \text{pitch circle diameter, mm}$$

$$\text{Maximum load, } (F_t)_{max} = C_w C_m F_t \quad \text{... (18.16)}$$

Where C_w = wear and lubrication factor

C_m = load distribution factor

18.13 VELOCITY FACTOR, C_v

The velocity factors are given by,

$$C_v = \frac{4.58}{4.58 + v} \quad \text{for } v \leq 5 \text{ m/s} \quad \text{... (18.17}a\text{)}$$

$$= \frac{6.1}{6.1 + v} \text{ for } 5 \le v \le 10 \text{ m/s} \quad \text{... (18.17}b\text{)}$$

$$= \frac{15.25}{15.25 + v} \text{ for } 10 \le v \le 20 \text{ m/s} \quad \text{... (18.17}c\text{)}$$

$$= \frac{5.55}{5.55 + \sqrt{v}} \text{ for } v \ge 20 \text{ m/s} \quad \text{... (18.17}d\text{)}$$

$$= \frac{0.7625}{1.0167 + v} \text{ for non - metallic gears} \quad \text{... (18.17}e\text{)}$$

18.14 EFFECTIVE TANGENTIAL LOAD

Effective load,
$$F_{eff} = \frac{(F_t)_{max}}{C_v} \quad \text{... (18.18)}$$

18.15 FACE WIDTH

Equating Eqs. (18.14) and (18.18), we get

$$\sigma_d m_n b \pi y = C_w C_m \frac{F_t}{C_v}$$

$$b = \frac{C_w C_m F_t}{C_v m_n \pi y \sigma_d} \quad \text{... (18.19)}$$

$$b_{min} = \frac{1.15 \pi m_n}{\sin \beta} \quad \text{... (18.20)}$$

For helical gears : $12.5 m_n \le b \le 20 m_n$ ... (18.21)
For herringbone gears : $20 m_n \le b \le 30 m_n$ ... (18.22)

18.16 DYNAMIC LOAD

(*a*) Earle Buckingham's Method

The dynamic load is given by,

$$F_d = (F_t)_{max} + F_i \quad \text{...(18.23)}$$

where F_i = increment load

$$= \frac{20.67v\left[Cb\cos^2\beta + (F_t)_{max}\right]\cos\beta}{\left[20.67v + \left\{Cb\cos^2\beta + (F_t)_{max}\right\}^{\frac{1}{2}}\right]} \quad \text{... (18.24)}$$

C = load stress (or dynamic) factor, N/mm

$$= \frac{e}{\left[K_1\left(\frac{1}{E_p}+\frac{1}{E_g}\right)\right]} \quad \text{... (18.25)}$$

e = probable error in action between gears, mm (Table 18.6)

$= e_p + e_g$ ... (18.26)

where e_p = pitch error for pinion

e_g = pitch error for gear

Tolerance factor, $\phi = m_n + 0.25\sqrt{d}$ ... (18.27)

Where m_n = normal module, mm

d = pitch circle diameter, mm

Table 18.6 Tolerances on the adjacent pitch error.

IS Grade	Adjacent pitch error e, microns
1	$0.80 + 0.6\phi_p$
2	$1.25 + 0.10\phi_p$
3	$2.00 + 0.16\phi_p$
4	$3.20 + 0.25\phi_p$
5	$5.00 + 0.40\phi_p$
6	$8.00 + 0.63\phi_p$
7	$11.00 + 0.90\phi_p$
8	$16.00 + 1.25\phi_p$
9	$22.00 + 1.80\phi_p$
10	$32.00 + 2.50\phi_p$
11	$45.00 + 3.55\phi_p$
12	$63.00 + 5.00\phi_p$

K_1 = *a* constant

= 9.345 for 14.5° involute FD ... (18.28*a*)

= 9.000 for 20° involute FD ... (18.28*b*)

= 8.700 for 20° involute stub ... (18.28*c*)

E_p, E_g = modulii of elasticity of pinion and gear materials

C = 11,500*e* N/ mm, for steel pinion and steel gear ... (18.29*a*)

= 8900*e* N/ mm, for cast iron pinion and cast iron gear ... (18.29*b*)

= 10,000*e* N/ mm, for steel pinion and cast iron gear (18.29*c*)

(*b*) **Prof. Spotts Method**

Increment load,
$$F_i = \frac{e n_p z_p b r_1 r_2}{2530\left(r_1^2 + r_2^2\right)^{\frac{1}{2}}} \quad \text{... (18.30)}$$

The increment load is assmued to be inclined at an angle ϕ_n to the tangent plane. The tangential component of F_i is $F_i \cos \phi_n \cos \beta$. Effective load is given by,

$$F_{eff} = C_s C_w F_t + F_i \cos \phi_n \cos \beta \quad \text{... (18.31)}$$

18.17 ENDURACE STRENGTH OF GEAR

Endurance strength of gear is given by,

$$F_e = S_e b \pi y m_n$$

Where S_e = endurance limit of gear material

For gears made of steel,

$$S_e = 1.75\ (\text{BHN})_{\text{core}} \quad \text{... (18.33)}$$

18.18 CHECK FOR SAFE DESIGN

For a safe design,

$$F_e \geq k\, F_d \quad \text{... (18.34)}$$

where k = factor of safety

= 1.25 for steady loads ... (18.35*a*)

= 1.35 for pulsating loads ... (18.35*b*)

= 1.5 for shock loads ... (18.35*c*)

18.19 WEAR LOAD

As shown in Fig. 18.7, we have

Face width of virtual pinion,

$$b_v = \frac{b}{\cos \beta}$$

Pitch circle diameter of virtual pinion,

$$d_{pv} = \frac{d_p}{\cos^2 \beta}$$

$$F_w = d_{pv}\, b_v\, \frac{QK}{\cos^2 \beta} \quad \text{... (18.36)}$$

Load Stress factor,

$$K = \left[\frac{\pi S_{es}^2 \sin \phi \cos \phi}{4}\right]\left[\frac{1 - v_p^2}{E_p} + \frac{1 - v_g^2}{E g}\right] \quad \text{... (18.37a)}$$

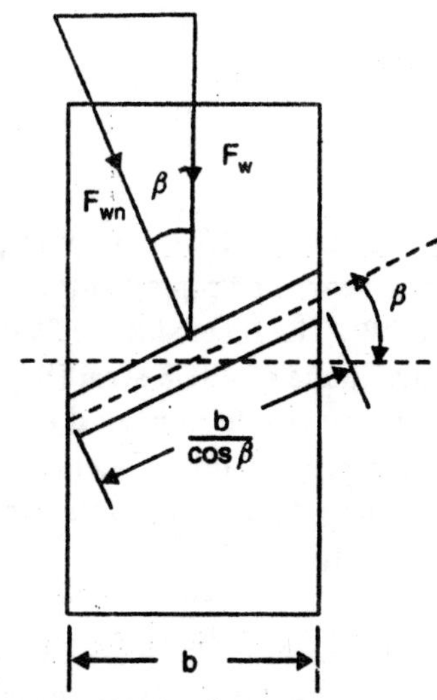

Fig. 18.7 Wear strength of helical gear tooth

$$= \left[\frac{S_{es}^2 \sin\phi \cos\phi}{1.4}\right]\left[\frac{1}{E_p} + \frac{1}{E_g}\right] \text{ for } v_p = v_g = 0.30 \qquad \text{... (18.37b)}$$

$$= c\left(\frac{BHN}{100}\right)^2 \text{ N/mm}^2$$

c = 0.16 for steel ponion and steel gear ... (18.38a)

= 0.21 for cast iron pinion and cast iron gear ... (18.38b)

= 0.18 for steel pinion and cast iron gear ... (18.38c)

For more values of c, refer to Table 17.17.

S_{es} = surface endurance limit

= 2.75 (BHN) – 70 MPa for steel

= 2.65 BHN ... (18.39a)

$\tan\phi_n = \tan\phi_t \cos\beta$

$$Q = \frac{2d_g}{d_p + d_g} = \frac{2z_g}{z_p + z_g} \qquad \text{... (18.40)}$$

For safe design $F_w \geq F_d$ ... (18.41)

18.20 Factor of Safety

$$\text{Factor of safety, } n = \frac{F_b}{F_{eff}}, \text{ against bending failure} \qquad \text{... (18.42a)}$$

$$= \frac{F_d}{F_{eff}}, \text{ against dynamic load} \qquad \text{... (18.42b)}$$

$$= \frac{F_w}{F_{eff}}, \text{ against wear failure} \qquad \text{... (18.42c)}$$

Example 18.3 A pair of parallel helical gears has 20 teeth on the pinion and 80 teeth on the gear. The pinion rotates at 750 rpm. The normal pressure angle is 20° and the helix angle is 30°. The face width is 50 mm and the normal module is 5 mm. Both the pinion and the gear are made of 40C8 steel (S_{ut} = 600 MPa) and heat treated to a surface hardness of 300 BHN. The service factor, load distribution factor, and wear and lubrication factor are 1.5, 1.2 and 1.15 respectively. Calculate the power transmitting capacity of the gear pair assuming that velocity factor accounts for the dynamic load. Factor of safety is 2.0.

Solution.

Given : $z_p = 20$, $z_g = 80$, $N_p = 750$ rpm, $\phi_n = 20°$, $\beta = 30°$, $b = 50$ mm, $m_n = 5$ mm, $S_{ut} = 600$ MPa, BHN = 300 (case), $C_s = 1.5$, $C_m = 1.2$, FOS = 2.0

1. Virtual number of teeth :

$$z_{pv} = \frac{z_p}{\cos^3\beta} = \frac{20}{\cos^3 30°} = 30.792$$

$$z_{gv} = \frac{z_g}{\cos^3 \beta} = \frac{80}{\cos^3 30°} = 123.168$$

2. Lewis form factors :

$$y_p = 0.154 - \frac{0.912}{z_{pv}}$$

$$= 0.154 - \frac{0.912}{30.792} = 0.1157$$

$$y_g = 0.154 - \frac{0.912}{123.168} = 0.1466$$

3. Decide the weaker wheel :

Pinion and gear are made of the same material and $y_p < y_g$. Therefore, pinion is the weaker and shall be considered for design.

4. Design stress :

Taking a factor of safety of 3, the design stress, $\sigma_{dp} = \frac{S_{ut}}{FOS} = \frac{600}{3} = 200$ MPa.

5. Static Beam strength :

$F_b = m_n b \pi y_p \sigma_{dp} = 5 \times 50 \times \pi \times 0.1157 \times 200 = 18174$ N

6. Pitch line velocity :

$$d_p = \frac{z_p\, m_n}{\cos \beta} = \frac{20 \times 5}{\cos 30°} = 115.47 \text{ mm}$$

$$v_p = \frac{\pi d_p N_p}{(60 \times 10^3)} = \frac{\pi \times 115.47 \times 750}{(60 \times 10^3)} = 4.53 \text{ m/s}$$

7. Velocity factor :

$$C_v = \frac{4.58}{(4.58 + V)} \text{ for } V \leq 5 \text{ m/s}$$

$$= \frac{4.58}{(4.58 + 4.53)} = 0.503$$

8. Wear load

$$F_w = bQ\, d_p\, K / \cos^2 \beta$$

$$b = 50 \text{ mm}$$

$$Q = \frac{2 z_g}{(z_g + z_p)} = \frac{2 \times 80}{(80 + 20)} = 1.6$$

$$d_p = 115.47 \text{ mm}$$

$$K = 0.16\left(\frac{BHN}{100}\right)^2 = 0.16\left(\frac{300}{100}\right)^2 = 1.44 \text{ N/mm}^2$$

$$F_w = \frac{50 \times 1.6 \times 115.47 \times 1.44}{\cos^2 30°} = 17736 \text{ N}$$

9. Effective load :

$$F_{eff} = \frac{C_s C_m C_w F_t}{C_r}$$

10. Safe tangential load :

$$= \frac{1.5 \times 1.2 \times 1.15 \times F_t}{0.503} = 4.1153\, F_t \text{N}$$

$$F_{eff} \times FOS = F_w$$

$$4.1153\, F_t \times 2 = 17736$$

$$F_t = 2154.9 \text{ N}$$

11. Power transmitting capacity :

$$\text{Torque transmitted, } T = \frac{F_t\, d_p}{2} = \frac{2154.9 \times 115.47}{2} = 124414 \text{ N. mm}$$

$$\text{Power transmitting capacity, } P = \frac{2\pi N_p T}{(60 \times 10^6)}$$

$$= \frac{2\pi \times 750 \times 214414}{(60 \times 10^6)} = 9.77 \text{ kW}$$

Example 18.4 A pair of parallel helical gears consists of 30 teeth pinion rotating at 1200 rpm and transmitting 3 kW power to a gear. The speed reduction is 3 : 1. The normal pressure angle and helix angles are 20° and 25° respectively. Both gears are made of hardened steel (S_{ut} = 600 MPa) with a hardness of 300 BHN core and 520 BHN case. The service is with light shocks from 8–10 hours per day and the factor of safety is 2. Lubrication is scant but regular with proper inspection. The load distribution factor may be taken as 1.2. Gears are finished to meet the accuracy of Grade 8. Determine the module and face width of the gears. Check for dynamic and wear loads.

Solution.

Given : z_p =30, n_p = 1200 rpm, P = 3 kW, i = 3, ϕ_n = 20°, β = 25°, S_{ut} = 600 MPa, BHN = 300 core and 520 case.

Torque transmitted,

$$T = \frac{60 \times 10^6 \times 3}{(2\pi \times 1200)} = 23873 \text{ N. mm}$$

Let m_n = 2 mm for the preliminary design.

$$d_p = \frac{z_p\, m_n}{\cos\beta} = \frac{30 \times 2}{\cos 25°} = 66.2 \text{ mm}$$

$$d_g = 3 \times 66.2 = 198.6 \text{ mm}$$

$$F_t = \frac{2T}{d_p} = \frac{2 \times 23873}{66.2} = 721.24 \text{ N}$$

$$V = \frac{\pi d_p N_p}{(60 \times 10^3)} = \frac{\pi \times 66.2 \times 1200}{(60 \times 10^3)} = 4.16 \text{ m/s}$$

$$C_v = \frac{4.58}{(4.58 + V)} = \frac{4.58}{(4.58 + 4.16)} = 0.524$$

Service factor, $C_s = 1.25$

Lubrication and wear factor, $C_w = 1.25$

Load distribution factor, $C_m = 1.2$

Maximum tangential load, $(F_t)_{max} = C_s\, C_w\, C_m\, F_t = 1.25 \times 1.25 \times 1.2 \times 721.24$
$= 1352.325$ N

Effective load, $$F_{eff} = \frac{(F_t)_{max}}{C_v} = \frac{1362.325}{0.524} = 2580.8 \text{ N}$$

$$z_{pv} = \frac{z_p}{\cos^3 \beta} = \frac{30}{\cos^3 25°} = 40.3$$

$$z_{gv} = 3 \times 40.3 = 120.9$$

$$y = 0.154 - \frac{0.912}{z_v}$$

$$y_p = 0.154 - \frac{0.912}{40.3} = 0.1314$$

$$y_g = 0.154 - \frac{0.912}{120.9} = 0.1465$$

$$\sigma d_p = \sigma d_g = \frac{600}{3} \text{ MPa}$$

As $y_p < y_g$, therefore, pinion is the weaker and shall be considered for design.

Static bending strength of pinion tooth, $F_b = m_n b \pi\, y_p\, \sigma_{dp} = 2 \times b \times \pi \times 0.1314 \times 200 = 165.12$ b N

$$F_{eff} \times \text{FOS} \times 2 = F_b$$
$$2580.8 \times 2 = 165.12\, b$$
$$b = 31.26 \text{ mm}$$

$12.5\, m_n \le b \le 20\, m_n$

$$25 \text{ mm} \le b \le 40 \text{ mm}$$

Also $$b_{min} = \frac{1.15\, \pi\, m_n}{\sin \beta} = \frac{1.15 \times \pi \times 2}{\sin 25°} = 17 \text{ mm}$$

Hence the calculated value of face width is within the limits. We adopt $b = 32$ mm

Dynamic load :

$$F_d = (F_t)_{max} + F_i$$

For Grade 8, the error is, $e = 16.00 + 1.25\,(m + 0.25\,\sqrt{d}\,)$

$$e_p = 16.00 + 1.25\,(2 + 0.25 + \sqrt{66.2}) = 21.04\ \mu m$$

$$e_g = 16.00 + 1.25\,(2 + 0.25 + \sqrt{198.6}) = 22.9\ \mu m$$

$$e = e_p + e_g = 21.04 + 22.9 = 43.94\ \mu m$$

$$r_p = \frac{d_p}{2} = \frac{66.2}{2} = 33.1 \text{ mm}$$

$$r_p = \frac{198.6}{2} = 99.3 \text{ mm}$$

$$F_i = \frac{e\, n_p\, z_p\, r_p\, r_g}{[2530\,(r_p^2 + r_g^2)^{1/2}]}$$

$$= \frac{43.94 \times 10^{-3} \times 1200 \times 30 \times 32 \times 33.1 \times 99.3}{\left[2530\left\{(33.1)^2 + (99.3)^2\right\}^{1/2}\right]}$$

$$= 628.26 \text{ N}$$

$$F_d = (F_t)_{max} + F_i \cos\phi_n \cos\beta$$

$$= 1352.325 + 628.26 \times \cos 20° \times \cos 25° = 1887.38 \text{ N}$$

$$FOS = \frac{F_b}{F_d} = \frac{165.12 \times 32}{1887.38} = 2.8$$

Wear load :

$$F_w = \frac{b\,Q\,d_p\,K}{\cos^2\beta}$$

$$b = 32 \text{ mm}$$

$$Q = \frac{2\,z_g}{(z_g + z_p)} = \frac{2 \times 90}{(90 + 30)} = 1.5$$

$$d_p = 66.2 \text{ mm}$$

$$K = 0.16\left(\frac{BHN}{100}\right)^2 = 0.16\left(\frac{300}{100}\right)^2 = 1.44 \text{ N/mm}^2$$

$$F_w = \frac{32 \times 1.5 \times 66.2 \times 1.44}{\cos^2 25°} = 5570 \text{ N}$$

$$F_w = F_d \times FOS$$

$$FOS = \frac{5570}{1887.38} \doteq 2.95.$$

Example 18.5 A pair of parallel helical gears consists of 20 teeth pinion rotating at 740 rpm and transmitting 10 kW to a 100 teeth gear. The normal pressure angle is 20° and the helix angle is 25°. The pinion and the gear are made of steel 40C8 (S_{ut} = 600 MPa) and heat treated to a surface hardness of 500 BHN and 250 BHN core. The gears are lubricated continuously in an enclosure and are accurately cut. The service factor is 1.5. Calculate the face width and the normal module.

Solution.

Given : $z_1 = 20$, $z_2 = 100$, P = 10 kW, $n_1 = 740$ rpm, $\phi_n = 20°$, $\beta = 25°$, $\sigma_{ut} = 600$ MPa

1. Preliminary decisions :

$$\text{Allowable static strength, } \sigma_d = \frac{S_{ut}}{3} = \frac{600}{3} = 200 \text{ MPa,}$$

$m_n = 4$ mm for the initial calculations

2. Calculate virtual number of teeth.

$$z_v = \frac{z}{\cos^3 \beta}$$

$$z_{1v} = \frac{20}{\cos^3 25°} = 26.866$$

$$z_{2v} = \frac{100}{\cos^3 25°} = 134.33$$

3. Calculate Lewis form factors

$$y = 0.154 - \frac{0.912}{z_y}$$

$$y_1 = 0.154 - \frac{0.912}{26.866} = 0.12$$

$$y_2 = 0.154 - \frac{0.912}{134.33} = 0.147$$

Since σ_d is same and $y_1 < y_2$, therefore, pinion is the weaker and shall be considered for design.

4. Calculate pitch diameters of gears,

$$d = \frac{m_n z}{\cos \beta}$$

$$d_p = \frac{4 \times 20}{\cos 25°} = 88.27 \text{ mm}$$

$$d_g = 441.35 \text{ mm}$$

5. Calculate pitch line velocity,

$$V = \frac{\pi\, d_p\, n_1}{(60 \times 10^3)} = \frac{\pi \times 88.27 \times 740}{(60 \times 10^3)} = 3.42 \text{ m/s}$$

6. Calculate velocity factor,

$$C_v = \frac{4.58}{(4.58 + v)} \text{ for } v < 5 \text{ m/s}$$

$$= \frac{4.58}{(4.58 + 3.42)} = 0.5725$$

7. Find lubrication and wear factor

For enclosed gears continuously lubricated, $C_w = 1.15$

8.Calculate tangential load on the pinion.

$$F_t = \frac{10^3 \times P}{v} = \frac{10^3 \times 10}{3.42} = 2924 \text{ N}$$

9. Calculate maximum tangential load.

$$(F_t)_{max} = C_s\, C_w\, F_t = 1.5 \times 1.15 \times 2924 = 5043.9 \text{ N}$$

10. Effective load, $F_{eff} = \frac{(F_t)_{max}}{C_v} = \frac{5043.9}{0.5725} = 8810.3 \text{ N}$

11.Calculate static beam strength of tooth,

$$F_b = \sigma_d\, b\pi\, y_1\, m_n$$
$$= 200 \times b \times \pi \times 0.12 \times 4 = 301.6\, b \text{ N}$$

12. Calculate face width.

$$F_b = F_{eff}$$
$$301.6\, b = 8810.3$$
$$b = 29.2 \text{ mm}$$

$$b_{min} = \frac{1.15 \times \pi\, m_n}{\sin \beta} = \frac{1.15 \times \pi \times 4}{\sin 25°} = 34.2 \text{ mm}$$

$b < b_{min}$. Thus we repeat the calculations with $m_n = 3$ mm

13.

$$d_p = 66.2 \text{ mm}$$
$$V = 2.565 \text{ m/s}$$
$$F_t = 3899 \text{ N}$$
$$C_v = 0.641$$
$$F_t = 3898.6 \text{ N}$$
$$(F_t)_{max} = 6725 \text{ N}$$
$$F_{eff} = 10492 \text{ N}$$
$$F_b = 226.2\, b \text{ N}$$
$$b = 46.4 \text{ mm}$$
$$b_{min} = 25.6 \text{ mm}$$
$$12.5\, m_n = 37.5 \text{ mm and } 20\, m_n = 60 \text{ mm}$$

We adopt $b = 50$ mm; $m_n = 3$ mm

14.Calculate dynamic load

$$F_d = (F_t)_{max} + F_i$$

where F_i = increment load

$$= \frac{20.67\,V\,[Cb\cos^2\beta + (F_t)_{max}]\cos\beta}{[20.67\,V + \{Cb\cos^2\beta + (F_t)_{max}\}^{1/2}]}$$

C = load stress (or dynamic) factor

For Grade 8, the error is,

$$e = 16.00 + 1.25\,(m_n + 0.25\,\sqrt{d})$$

$$e_p = 16.00 + 1.25\,(2 + 0.25\,\sqrt{88.27}) = 21.443\ \mu m$$

$$e_g = 16.00 + 1.25\,(2 + 0.25\,\sqrt{88.27}) = 25.065\ \mu m$$

$$e = e_p + e_g = 21.443 + 25.065 = 46.508\ \mu m$$

$$C = \frac{e}{\left[K_1\left(1/E_p + 1/E_g\right)\right]}$$

Let $E_p = E_g = 207$ GPa

$K_1 = 9.0$ for 20° FD teeth

$$C = \frac{46.508\times10^{-3}}{[9.0\times 2/207000]} = 534.84\ \text{N/mm]}$$

$$F_i = 20.67\times 2.565\,\frac{\left[534.84\times 50\times\cos^2 25° + 6725\right]\cos 25°}{\left[20.67\times 2.565 + \{534.84\times 50\times\cos^2 25° + 6725\}^{1/2}\right]}$$

$$= 6198.8\ \text{N}$$

$$F_d = 6725 + 6198.8 = 12923.78\ \text{N}$$

15.Calculate endurance strength of the gear,

$$F_e = S_e\,b\pi\,y m_n$$

where S_e = endurance limit of pinion material

$= 1.75\,(BHN)_{core} = 1.75 \times 250 = 437.5$ MPa

$$F_e = 437.5\times 50\times\pi\times 0.12\times 3 = 24740\ \text{N}$$

16.Check for safe design

$$\text{Factor of safety} = \frac{F_e}{F_d} = \frac{24740}{12923.78} = 1.91 > 1.5 \text{ for shock loads.}$$

Hence safe.

17.Calculate wear load

$$F_w = \frac{d_p\,b\,Q\,K}{\cos^2\beta}$$

$$Q = \frac{2z_g}{(z_p + z_g)}$$

$$= \frac{2 \times 100}{(100 + 20)} = 1.67$$

$$K = 0.16\left(\frac{BHN}{100}\right)^2 = 0.16\left(\frac{500}{100}\right)^2 = 4$$

$$F_w = \frac{66.2 \times 50 \times 1.67 \times 4}{\cos^2 25°}$$

$$= 26918.6 \text{ N}$$

$$\text{FOS} = \frac{F_w}{F_d} = \frac{26918.6}{12923.78} = 2.08$$

Hence safe.

Example 18.6 Design a pair of helical gears to transmit 50 kW. The pinion revolves at 4800 rpm and the speed ratio is 3. Select suitable missing data.

Solution.

Given : P = 50 kW, n_1 = 4800 rpm, i = 3

1. Preliminary decisions :

We select 45C8 (S_{ut} = 630 MPa, BHN = 229 core and 520 case) steel for the pinion 30C8 (S_{ut} = 500 MPa, BHN = 179) steel for the gear, normal pressure angle, $\phi_n = 20°$, helix angle $\beta = 25°$, $z_1 = 20$, normal module, m_n = 3 mm. FOS = 3 so that $\sigma_{d1} = \frac{630}{3} = 210$ MPa, $\sigma_{d2} = \frac{500}{3} = 167$ MPa.

Let the accuracy Grade be 8.

2. Number of teeth on the gear, $z_2 = i\, z_1 = 3 \times 20 = 60$

3. Virtual number of teeth

$$z_v = \frac{z}{\cos^3 \beta}$$

$$z_{1v} = \frac{20}{\cos^3 25°} = 26.86$$

$$z_{2v} = \frac{60}{\cos^3 25°} = 80.6$$

4. Lewis form factors

$$y = 0.154 - \frac{0.912}{z_v}$$

$$y_1 = 0.154 - \frac{0.912}{26.86} = 0.120$$

$$y_2 = 0.154 - \frac{0.912}{80.6} = 0.142$$

5.Decide the weaker wheel.

$$\sigma_{1d}\, y_1 = 210 \times 0.120 = 25.2$$
$$\sigma_{2d}\, y_2 = 167 \times 0.142 = 23.7$$

Since $\sigma_{2d} y_2 < \sigma_{1d} y_1$, therefore, gear is the weeker and shall be considered for design.

6.Pitch diameters of gears,

$$d = \frac{m_n z}{\cos \beta}$$

$$d_1 = \frac{3 \times 20}{\cos 25°} = 66.2 \text{ mm}$$

$$d_2 = 198.6 \text{ mm}$$

7. Calculate pitch line velocity.

$$V = \frac{\pi\, d_1\, n_1}{(60 \times 10^3)} = \frac{\pi \times 66.2 \times 4800}{(60 \times 10^3)} = 16.64 \text{ m/s}$$

8. Velocity factor,

$$C_v = \frac{15.25}{(15.25 + v)} \quad \text{for } 10 \le v \le 20 \text{ m/s}$$

$$= \frac{15.25}{(15.25 + 16.64)} = 0.478$$

9. Lubrication and wear factor.

For enclosed gears continuously lubricated, $C_w = 1.15$

10.Service factor, $C_s = 1.5$

11. Tangential load on the pinion.

$$F_t = \frac{10^3 \times P}{v} = \frac{10^3 \times 50}{16.64} = 3004.8 \text{ N}$$

12. Calculate maximum load,

$$(F_t)_{max} = F_t C_w C_s$$
$$= 3004.8 \times 1.15 \times 1.5 = 5183.3 \text{ N}$$

13. Effective load, $E_{ff} = \frac{(F_t)_{max}}{C_v} = \frac{5183.3}{0.478} = 10843.7$

14. Static beam strength of pinion teeth,

$$F_b = \sigma_{2d}\, b\pi\, y_2\, m_n$$

$$= 167 \times b \times \pi \times 0.142 \times 3$$
$$= 223.5\, b \text{ N}$$

15. Calculate face width.

$$F_b = F_{eff}$$
$$223.5\, b = 10843.7$$
$$b = 48.5 \text{ mm}$$

$$b_{min} = \frac{1.15 \times \pi \text{ M}n}{\sin \beta}$$

$$= \frac{1.15 \times \pi \times 3}{\sin 25°} = 25.64 \text{ mm}$$

Also $12.5\, m_n \leq b \leq 20\, m_n$

$$37.5 \text{ mm} \leq b \leq 60 \text{ mm}$$

We observe that the estimated value of face width is within the specified limits.

We adopt b = 50 mm and m_n = 3 mm

16. Dynamic load,

$$F_d = (F_t)_{max} + F_i$$

where F_i = increment load

$$= \frac{20.67\, v\,[\text{Cb} \cos^2 \beta + (F_t)_{max}] \cos \beta}{[20.67\, v + \{\text{Cb} \cos^2 \beta + (F_t)_{max}\}^{1/2}]}$$

$$C = \text{load stress (or dynamic) factor, N/mm}$$

For Grade B, the error is,

$$e = 16.00 + 1.25\,(m_n + 0.25\sqrt{d})$$

$$e_p = 16.00 + 1.25\,(2 + 0.25\sqrt{66.2}) = 21.04\ \mu\text{m}$$

$$e_g = 16.00 + 1.25\,(2 + 0.25\sqrt{198.6}) = 22.9\ \mu\text{m}$$

$$e = e_p + e_g = 21.04 + 22.9 = 43.94\ \mu\text{m}$$

$$C = \frac{e}{[K_1\,(1/E_p + 1/E_g)]}$$

Let $E_p = E_g$ = 207 GPa

K_1 = 9.0 for 20° FD teeth

$$C = \frac{43.94 \times 10^{-3}}{[9.0 \times 2/207000]} = 505.31 \text{ N/mm]}$$

$$F_i = \frac{20.67 \times 16.64\,(505.31 \times 50 \times \cos^2 25° + 5183.3) \cos 25°}{[20.67 \times 16.64 + (505.31 \times 50 \times \cos^2 25° + 5183.3)^{1/2}]}$$

$$= 16009.9 \text{ N}$$

$$F_d = 3751 + 9589.5 = 13340.5 \text{ N}$$

17. Endurance strength of the gear,

$$F_e = S_e b\pi y_1 m_n$$

where S_e = endurance limit of gear material

$$=1.75 \text{ BHN} = 1.75 \times 179 = 313.25 \text{ MPa}$$

$$F_e = 313.25 \times 50 \times \pi \times 0.142 \times 3 = 20961.4 \text{ N}$$

18. Check for safe design

$$\text{FOS} = F_e / F_d = 20961.4 / 13340.5 = 1.57$$

19. Wear load

$$F_w = \frac{d_1 \, b\, QK}{\cos^2 \beta}$$

$$d_1 = 66.2 \text{ mm}, b = 50 \text{ mm}$$

$$Q = \frac{2d_2}{(d_2 + d_1)}$$

$$= \frac{2 \times 60}{(60 + 20)} = 1.5$$

$$K = 0.16\left(\frac{\text{BHN}}{100}\right)^2 = 0.16\left(\frac{520}{100}\right)^2 = 4.3264$$

$$F_w = \frac{66.2 \times 50 \times 1.5 \times 4.3264}{\cos^2 25°}$$

$$= 26151 \text{ N}$$

$$F_w < F_d$$

$$\text{FOS} = \frac{F_w}{F_d} = \frac{26151}{13340.5} = 1.96.$$

Example 18.7. A parallell helical pinion having 24 teeth is to be made of plain carbon steel 55C8 (S_{ut} = 720 MPa). It is required to mesh with a plain carbon steel 40C8 (S_{ut} = 580 MPa). It is required to transmit 15 kW from an electric motor running at 1000 rpm to a machine running at 250 rpm. The starting torque of the motor is 125% of the rated torque. The factor is safety required is 2. The normal pressure angle is 20° FD involute and helix angle is 25°. The gears are to be machined to meet the specifications of Grade 6. The pinion and gear are to be case hardened to 300 BHN and 250 BHN respectively. Design the gear pair and check for Buckingham's dynamic load.

Solution.

Given. z_p =24, $(S_{ut})_p$ = 720 MPa, $(S_{ut})_g$ = 580 MPa, n_p = 1000 rpm, P = 15 kW, n_g = 250 rpm, torque overload = 125%, FOS = 2, ϕ_n = 20°, β = 25°, surface accuracy = Grade 6, $(\text{BHN})_p$ = 300, $(\text{BHN})_g$ = 250.

Design stresses :

$$\sigma_{dp} = \frac{720}{3} = 240 \text{ MPa}, \sigma_{dg} = \frac{580}{3} = 193.3 \text{ MPa}$$

Speed ratio,
$$i = \frac{1000}{250} = 4$$

$$z_g = 24 \times 4 = 96$$

$$z_{vp} = \frac{z_p}{\cos^3 \beta} = \frac{24}{\cos^3 25°} = 32.24$$

$$z_{vg} = \frac{z_g}{\cos^3 \beta} = \frac{96}{\cos^3 25°} = 128.96$$

Lewis form factors :

$$y = 0.154. - \frac{0.912}{z_v}$$

$$y_p = 0.154 - \frac{0.912}{32.24} = 0.1257$$

$$y_g = 0.154 - \frac{0.912}{128.96} = 0.1469$$

Decide the weaker wheel :

$$\sigma_{dp} \times y_p = 240 \times 0.1257 = 30.168$$
$$\sigma_{dg} \times y_g = 193.3 \times 0.1469 = 28.396$$

Since $(\sigma_{dg} \times y_g) < (\sigma_{dp} \times y_p)$, therefore, gear is the weaker and shall be considered for design.

Design Factors :

Let the service be from 8–10 hours per day with medium shocks giving service factor, C_s = 1.25

For enclosed gears and continuously lubricated, wear and lubrication factor, C_w = 1.15

Also assume that the face width is less than 50 mm and the support is rigid so that load distribution factor, C_m = 1.2.

Initial normal module :

Let us assume m_n = 3 mm

Pitch diameters :

$$d_p = \frac{m_n z_p}{\cos \beta} = \frac{3 \times 24}{\cos 25°} = 79.44 \text{ mm}$$

$$d_g = 79.44 \times 4 = 317.76 \text{ mm}$$

Tangential force :

$$v_g = \frac{\pi d_g n_g}{60 \times 10^3} = \frac{\pi \times 317.76 \times 250}{60 \times 10^3} = 4.16 \text{ m/s}$$

Velocity factor, $$C_v = \frac{4.58}{4.58 + v} = \frac{4.58}{4.58 + 4.16} = 0.524$$

Tangential load, $$F_t = \frac{10^3 \text{ P}}{v} = \frac{10^3 \times 15}{4.16} = 3605.77 \text{ N}$$

$$(F_t)_{max} = C_s \, C_w \, C_m \, F_t$$

$$= 1.25 \times 1.15 \times 1.2 \times 3605.77 = 6220 \text{ N}$$

$$F_{eff} = \frac{(F_t)_{max}}{C_v} = \frac{6220}{0.524} = 11870 \text{ N}$$

Beam strength :

$$F_b = \sigma_{dg} b \pi y_g \, m_n$$

$$= 193.3 \times b \times \pi \times 0.1469 \times 3 = 267.62 \, b$$

Face width :

$$F_b = F_{eff}$$

$$267.62 \, b = 11870$$

$$b = 44.35$$

$$b_{min} = \frac{1.15 \times \pi \, m_n}{\sin \beta}$$

$$= \frac{1.15 \times \pi \times 3}{\sin 25°} = 25.64 \text{ mm}$$

Also $$12.5 \, m_n \leq b \leq 20 \, m_n$$

$$37.5 \text{ mm} \leq b \leq 60 \text{ mm}$$

We observe that the estimated value of face width is within the specified limits.

We adopt $b = 45$ mm and $m_n = 3$ mm

Dynamic load,

$$F_d = (F_t)_{max} + F_i$$

Where F_i = increment load

$$= \frac{20.67 \, v \, [Cb \cos^2 \beta + (F_t)_{max}] \cos \beta}{[20.67 \, v + \{Cb \cos^2 \beta + (F_t)_{max}\}^{1/2}]}$$

$$C = \text{load stress (or dynamic) factor, N/mm}$$

For Grade 6, the error is, $e = 8.00 + 0.63\,(m_n + 0.25\sqrt{d})$

$$e_p = 8.00 + 0.63\,(3 + 0.25\sqrt{79.44}) = 11.294\ \mu m$$

$$e_g = 8.00 + 0.63\,(3 + 0.25\sqrt{317.76}) = 12.697\ \mu m$$

$$e = e_p + e_g = 11.294 + 12.697 = 23.991\ \mu m$$

$$C = \frac{e}{[K_1\,(1/E_p + 1/E_g)]}$$

Let $E_p = E_g = 207$ GPa

$K_1 = 9.0$ for 20° FD teeth

$$C = \frac{23.991 \times 10^{-3}}{[9.0 \times 2/207000]} = 275.896\ \text{N/mm}]$$

$$F_i = \frac{20.67 \times 4.16\,(275.896 \times 45 \times \cos^2 25° + 6220)\cos 25°}{[20.67 \times 4.16 + (275.896 \times 45 \times \cos^2 25° + 6220)^{1/2}]}$$

$$= 5975.44\text{N}$$

$$F_d = 6220 + 5975.44 = 12195.44\ \text{N}$$

Check for dynamic load :

$$\text{FOS} = \frac{F_d}{F_{eff}} = \frac{12195.44}{11870} = 1.023$$

EXERCISES

18.1 A parallel helical gear drive is required to transmit a maximum power of 22.5 kW. The velocity ratio is 1 : 2 and rpm of pinion is 200. The approximate centre distance between the shafts may be taken as 600 mm. The normal pressure angle is 20° full depth involute and the helix angle is 25°. The allowable static stress for the pinion and gear material, which is cast iron may be taken as 60 MPa. The face width is 10 times the module. Assuming suitable missing data, determine the normal module.

18.2 A pair of straight teeth helical gears is to transmit 20 kW when the pinion rotates at 300 rpm. The velocity ratio is 1 : 3. The allowable static stresses for pinion and gear materials are 120 MPa and 100 MPa respectively. The pinion has 18 teeth and its face width is 12 times the module. The normal pressure angle is 20° full depth involute and helix angle is 30°. Determine (*a*) normal module (*b*) face width (*c*) pitch circle diameters of pinion and gear from the stand point of strength only, taking into consideration the effect of dynamic loading. The gears have been manufactured with accuracy Grade 6.

18.3 A pair of parallel helical gears with 30° helix angle is used to transmit 10 kW at 900 rpm of the pinion. The speed ratio is 4 : 1. The normal pressure angle is 20° and the helix angle is 25°. Select suitable data and design the teeth based on static strength. Check for dynamic and wear loads.

18.4 A pair of carefully cut helical gears with 20° FD involute normal pressure angle and 25° helix angle is to transmit 25 kW at 300 rpm of the gear at a speed reduction of 5 : 1. The pinion should not be smaller than 76 mm in pitch diameter. It is made of forged steel 30C8 steel with a hardness of 250 BHN. The gear is made of cast iron FG 200. Determine the module, number of teeth and face width of gears.

18.5 The pinion in a helical gear set having a speed ratio 3 rotates at 900 rpm and has 20 full depth teeth of normal pressure anglee 20° and helix angle 25°. The face width is 30 mm and the normal module is 3 mm. What is the power rating of the gear set based only on tooth bending strength if both the pinion and gear are made of the same material having an ultimate tensile strength of 400 MPa and yield strength of 210 MPa ? Assume a factor of safety of 2.

18.6 A pair of helical gears are used to transmit 20 kW at 800 rpm of the pinion. The teeth are 20° stub in diametral plane and the helix angle is 25°. The gear and the pinion have a pitch diameter of 320 and 80 mm respectively. Both gear and pinion are made of cast steel with allowable static strength of 100 MPa. Suggest a suitable module and face width of the gear pair and check the design in wear. Take modulus of elasticity for cast steel as 200 GPa and S_{es} = 620 MPa.

BEVEL GEARS

Chapter

19.1 INTRODUCTION

Bevel gears are used to transmit power between two shafts intersecting generally at right angles. Two common types of bevel gears are with straight and spiral teeth. In case of straight bevel gears, the elements of the teeth are straight lines which converge into a common apex point of the pitch cone. Spiral bevel gears have curved teeth. We shall study the design procedure for straight bevel gears only. Involute profile is used for these gears with pressure angle of 20°.

19.2 STRAIGHT BEVEL GEAR TERMINOLOGY

The terminology of a straight bevel gear is shown in Fig. 19.1 Some of the important terms are

Pitch Cone. It is the pitch surface of the bevel gear as represented by OB.

Back Cone. It is an imaginary cone the elements of which are perpendicular to the elements of the pitch cone at the larger end of the tooth.

Cone centre. It is the apex of the pitch cone where the axes of the two mating gears intersect each other.

Pitch circle diameter, d. It is the diameter of the pitch circle at the large end of the teeth.

Pitch cone angle, δ It is the angle made by the pitch cone element with the axis of the gear.

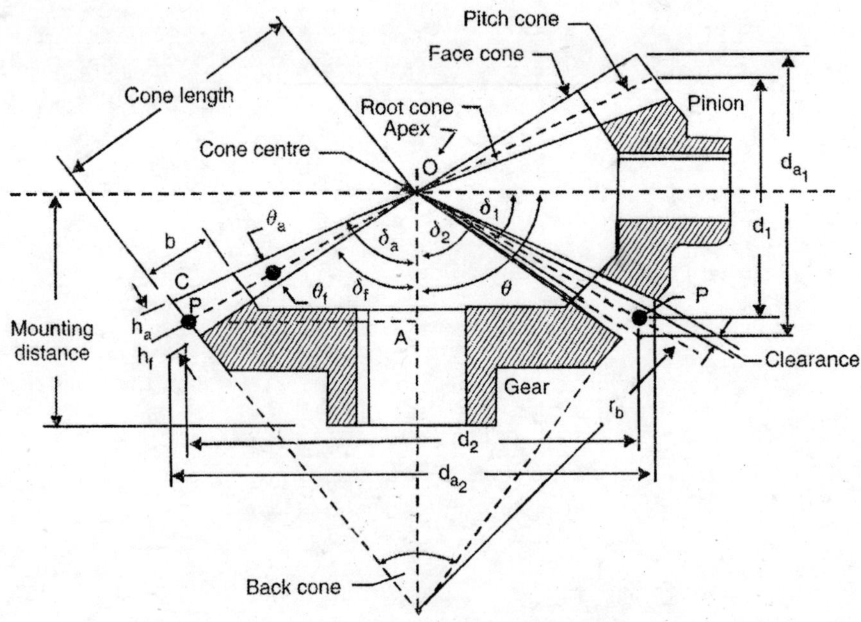

Fig. 19.1 Bevel gear terminology

δ = pitch cone angle

δ_a = face angle

δ_f = root angle

θ_a = addendum angle

θ_f = dependum angle

d_a = addendum diameter

d_f = dedendum diameter

r_b = back cone radius

L = OP = cone length or pitch cone distance

OP = pitch cone

Pitch cone distance – It is the length of the pitch cone element.

$$L = \left[r_1^2 + r_2^2\right]^{\frac{1}{2}} \quad \text{... (19.1)}$$

where r_1, r_2 = pitch circle radii of pinion and gear respectively.

Module, $$m = \frac{d}{z} \quad \text{... (19.2)}$$

where d = pitch circle diameter at the larger end of the teeth

z = number of teeth

Addenedum, $h_a = 1.0$ m ... (19.3*a*)

Dedendum, $h_f = 1.2$ m ... (19.3*b*)

Speed ratio, $$i = \frac{n_1}{n_2} = \frac{z_2}{z_1} = \frac{d_2}{d_1} \quad \text{... (19.4)}$$

Back cone radius. It is the length of the back cone element.

$$r_{b1} = \frac{r_1}{\cos \delta_1} \quad \text{... (19.5a)}$$

$$r_{b2} = \frac{r_2}{\cos \delta_2} \quad \text{... (19.5b)}$$

Mean radius (r_m). It is the pitch radius at the midpoint along the face width of the teeth.

$$r_{m1} = r_1 - \frac{b \sin \delta_1}{2} \quad \text{... (19.6a)}$$

$$r_{m2} = r_2 - \frac{b \sin \delta_2}{2} \quad \text{... (19.6b)}$$

where b =face width

Table 19.1 Standard normal modules for helical gears in mm

First choice	1.0, 1.25, 1.5, 2, 2.5, 3, 4, 5, 6, 8, 10, 12, 16, 20, 25, 32, 40, and 50
Second choice	1.125, 1.375, 1.75, 2.25, 2.75, 3.5, 4.5, 5.5, 7, 9, 11, 14, 18, 22, 28, 36 and 45

19.3 ANGLE RELATIONSHIP

Let θ = angle between the axes of the pinion and gear

(i) For $\theta < 90°$, we have

$$\tan \delta_1 = \frac{d_1 \sin \theta}{d_2 + d_1 \cos \theta} = \frac{\sin \theta}{\frac{z_2}{z_1} + \cos \theta} \quad \text{... (19.7a)}$$

$$\delta_2 = \theta - \delta_1 \tan \theta_a = \frac{2h_a \sin \delta_1}{d_1} = \frac{2h_a \sin \delta_2}{d_2} \quad \text{... (19.7b)}$$

$$\tan \theta_f = \frac{2h_f \sin \delta_1}{d_1} = \frac{2h_f \sin \delta_2}{d_2} \quad \text{... (19.8)}$$

$$d_{a1} = d_1 + 2h_a \cos \delta_1 \quad \text{... (19.9a)}$$

$$d_{a2} = d_2 + 2h_a \cos \delta_2 \quad \text{... (19.9b)}$$

(ii) For $\theta > 90°$

$$\tan \delta_1 = \frac{d_1 \sin (180° - \theta)}{d_2 + d_1 \cos (180° - \theta)} \quad \text{... (19.10a)}$$

$$= \frac{\sin (180° - \theta)}{\frac{z_2}{z_1} + \cos (180° - \theta)} \quad \text{... (19.10b)}$$

Radius of imaginary spur gear,

$$r_b = BC = \frac{D}{2 \cos \delta}$$

(iii) For $\theta = 90°$

$$\tan \delta_1 = \frac{z_1}{z_2} = \frac{1}{i} \quad \text{... (19.11a)}$$

$$\tan \delta_2 \frac{z_2}{z_1} = i \quad \text{... (19.11b)}$$

19.4 VIRTUAL NUMBER OF TEETH

The profile of the teeth of a bevel gear at the large end is the same as the profile of a spur gear laid out on a pitch radius equal to the back cone radius, r_b. The number of teeth on this imaginary gear is called the virtual or formative number of teeth.

$$z_v = \frac{2r_b}{m}$$

where m = module at the large end of teeth.

Actual number of teeth on bevel gear,

$$z = \frac{d}{m}$$

where d = pitch circle diameter at the large end.

$$\frac{z_v}{z} = \frac{2r_b}{d}$$

In ΔABP, sin ∠ABP $= \dfrac{AP}{BP}$

Or

$$\sin(90° - \delta) = \frac{d}{2r_d}$$

$$r_b = \frac{d}{2\cos\delta} \qquad \text{... (19.12)}$$

$$z_v = \frac{z}{\cos\delta} \qquad \text{... (19.13)}$$

19.5 MINIMUM NUMBER OF TEETH

In order to avoid interference in a bevel gear pair, it must be ensured that interference does not occur in an equivalent spur gear pair.

Thus

$$z_v \geq \frac{2h_a}{m\sin^2\phi}$$

Or

$$\frac{z_1}{\cos\delta_1} \geq \frac{2h_a}{m\sin^2\phi}$$

$$z_1 \geq \frac{2h_a\cos\delta_1}{m\sin^2\phi} \qquad \text{... (19.14)}$$

19.6 LEWIS FORM FACTOR

The Lewis form factor is based upon virtual number of teeth. Lewis form factors based upon circular pitch are given by,

$$y = 0.125 - \frac{0.684}{z_v} \quad \text{for 14.5° involute FD} \qquad \text{... (19.15a)}$$

$$= 0.154 - \frac{0.912}{z_v} \quad \text{for 20° involute FD} \qquad \text{... (19.15b)}$$

$$= 0.170 - \frac{0.950}{z_v} \quad \text{for 20° involute stub} \qquad \text{... (19.15c)}$$

19.7 STATIC BEAM STRENGTH

The load acting on the tooth varies linearly along the face of a bevel gear. The tooth thickness also varies linearly along the face of the gear. To determine the beam strength, consider an element *dx* along the face of the tooth, as shown in Fig. 19.2.

Let m_x = module at a distance x from the apex O of the pitch cone

y = Lewis form factor based on circular pitch

$d\,F_x$ = incremental allowable bending force acting on dx

Then, the beam strength of element dx is given by

$$dF_x = \sigma\, m_x\, \pi y\, dx$$

Multiply both sides by r_x and integrating, we have

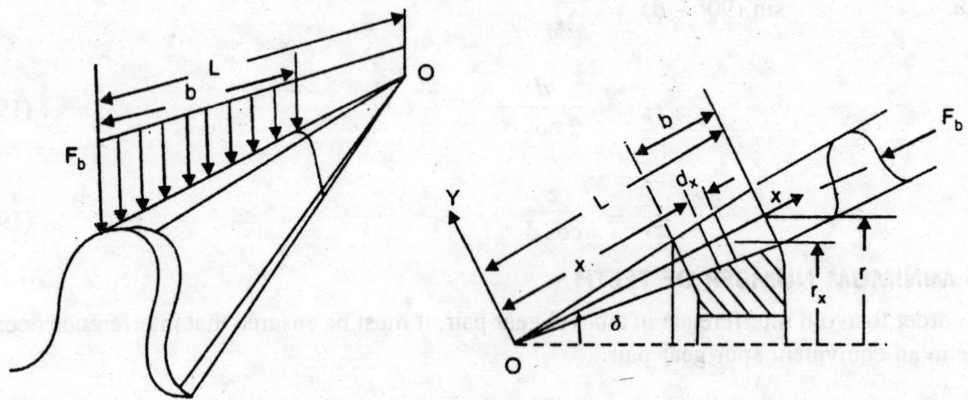

Fig. 19.2

$$\int r_x\, d\,F_x = \int \sigma\, m_x r_x\, \pi y\, .\, dx$$

Now the module varies linearly along the length. Therefore,

$$\frac{m}{m_x} = \frac{L}{x}$$

or

$$\frac{r_x}{r} = \frac{x}{L}$$

Torque,

$$T = \int_{L-b}^{L} r_x\, d\,F_x = \int_{L-b}^{L} \sigma \left(\frac{mx}{L}\right)\left(\frac{rx}{L}\right) \pi y\, .\, dx$$

$$= \left(\frac{\sigma m r \pi y}{L^2}\right) \int_{L-b}^{L} x^2\, dx$$

$$= \left(\frac{\sigma m r \pi y}{L^2}\right)\left|\frac{x^3}{3}\right|_{L-b}^{L}$$

$$F_t \times r = (\sigma m r \pi y b)\left[1 - \frac{b}{L} + \frac{b^2}{3L^2}\right]$$

Static beam strength, $$F_b = F_t = (\sigma_d \, b m \pi y)\left[1 - \frac{b}{L}\right] \text{ for } b \le \frac{L}{3} \quad \text{... (19.16)}$$

where $$L = \left[r_1^2 + r_2^2\right]^{\frac{1}{2}}$$

$$r = \frac{d}{2}$$

$$\left[1 - \frac{b}{L}\right] = \text{bevel factor.}$$

$$\sigma_d = \text{design stress}$$

Eq. (19.16) is the Lewis equation for beam strength of straight bevel gears.

19.8 VELOCITY FACTOR

The velocity factor for straight bevel gears is given by,

$$C_v = \frac{3.05}{3.05 + v}, \text{ for teeth cut by form cutter} \quad \text{... (19.17a)}$$

$$= \frac{6.1}{6.1 + v}, \text{ for generated tooth} \quad \text{... (19.17b)}$$

19.9 MAXIMUM LOAD

Tangential load, $$F_t = 10^3 \times \frac{P}{v}$$

Maximum tangential load;

$$(F_t)_{max} = C_s C_m F_t \quad \text{... (19.18)}$$

where C_s = service factor (Table 19.2)

C_m = load distribution factor (Table 19.4)

P = power transmitted

Table 19.2 Service Factor, C_s

Type of load	Type of service		
	Intermittent 3 hours per day	8 - 10 hours per day	Continuous 24 hours per day
Steady	0.80	1.00	1.25
Light shock	1.00	1.25	1.50
Medium shock	0.25	1.50	1.80
Heavy shock	1.5	1.80	2.00

The type of loads for different applications are given in Table 19.3

Table 19.3 Type of loads for different machinery.

S. No.	Type of load	Driven machinery
1.	Steady	Generators, liquids mixing agitators, centrifugal blowers, compressor and pumps, medium duty conveyors, fans, etc.
2.	Moderate shock	Liquid and solid mixing agitators, multi – cylinder reciprocating compressors and pumps, heavy duty conveyors, sugar cane crushers, hoists, concrete mixers, textile machinery, etc.
3.	Heavy shock	Single – cylinder reciprocating compressors and pumps, stone crushers, passenger lift, heavy duty hoists, hammer mills, presses, rolling mills, tumbling barrels, printing presses, etc.

Table 19.4 Load distribution factor, C_m for bevel gears

Type of application Type of mounting	General commerical gears	High quality gears
1. Both gears straddle mounted	1.25	1.00
2. One gear straddle mounted and other overhung mounted	1.4	1.1
3. Both gears overhung mounted	1.5	1.25

19.10 EFFECTIVE LOAD

$$F_{eff} = \frac{(F_t)_{max}}{C_v} \quad \text{... (19.20)}$$

$$FOS = \frac{F_b}{F_{eff}} \quad \text{... (19.21)}$$

19.11 FACE WIDTH

The face width is calculated by equating Eq. (19.16) with (19.20).

$$F_{eff} = (\sigma_d\, bm\pi y)\left[1-\frac{b}{L}\right] \quad \text{... (19.22)}$$

The face width lies between 6m to 10m or $b \le \frac{L}{3}$, whichever is less.

19.12 DYNAMIC LOAD

(*a*) Earle Buckingham's Method

Dynamic load, $F_d = (F_d)_{max} + F_i$

$$F_i = \frac{20.67v\left[Cb+(F_t)_{max}\right]}{\left[20.67v\left\{Cb+(F_t)_{max}\right\}^{\frac{1}{2}}\right]} \quad \text{... (19.23)}$$

where C = dynamic factor

$$= K_1 . e\left[\frac{E_p E_g}{E_p + E_g}\right] \quad \text{... (19.24)}$$

e = sum of errors on the meshing teeth (Table 19.5)

$= e_p + e_g$

Tolerance factor, $\phi_p = m + 0.25\sqrt{d_m}$ ... (19.25)

d_m = mean diameter at the larger end of the teeth, mm

m = module, mm

Table 19.5 Tolerance on the adjacent pitch error.

IS Grade	Adjacent pitch error e, microns
1	$0.80 + 0.06\phi_p$
2	$1.25 + 0.10\phi_p$
3	$2.00 + 0.16\phi_p$
4	$3.20 + 0.25\phi_p$
5	$5.00 + 0.40\phi_p$
6	$8.00 + 0.63\phi_p$
7	$11.00 + 0.90\phi_p$
8	$16.00 + 1.25\phi_p$
9	$22.00 + 1.80\phi_p$
10	$32.00 + 2.50\phi_p$
11	$45.00 + 3.55\phi_p$
12	$63.00 + 5.00\phi_p$

K_1 = tooth form factor

= 0.111 for 20° FD involute

= 0.107 for 14.5° FD involute

E_p, E_g = modulii of elasticity for pinion and gear respectively

The values of C are as follows :

$$C = 11500\, e \text{ N/mm for steel pinion and steel gear} \quad \text{... (19.26a)}$$

$$= 8900\, e \text{ N/mm for cast iron pinion and cast iron gear} \quad \text{... (19.26b)}$$

$$= 10000\, e \text{ N/mm for steel pinion and cast iron gear} \quad \text{... (19.26c)}$$

(*b*) Prof. Spotts Method

For bevel gears made of steel the incremental load is given by,

$$F_i = \frac{e n_p z_p b_v r_{mp} r_{mg}}{2530\left(r_{mp}^2 + r_{mg}^2\right)^{\frac{1}{2}}} \quad \text{... (19.27)}$$

where

$$b_v = \frac{b}{\cos \delta_p}$$

$$r_{mp} = \frac{d_p - b \sin \delta_p}{2}$$

$$r_{mg} = \frac{d_g - b \sin \delta_g}{2}$$

$$F_d = (F_t)_{\max} + F_i \quad \text{... (19.28)}$$

19.13 ENDURANCE STRENGTH

$$F_e = S_e\, b\pi y m \left(1 - \frac{b}{L}\right) \quad \text{... (19.29)}$$

$$FOS = \frac{F_e}{F_d} \quad \text{... (19.30)}$$

19.14 WEAR LOAD

$$F_w = bQd_{1v}K$$

Where d_{1v} = pitch circle diameter of formative pinion

$$= 2r_b = \frac{d_1}{\cos \delta_1}$$

Hence

$$F_w = bQd_1 \frac{K}{\cos \delta_1} \quad \text{... (19.31)}$$

Ratio factor $$Q = \frac{2z_{2v}}{z_{1v} + z_{2v}}$$

where $$z_v = \frac{z}{\cos\delta}$$

Load stress factor, $$K = \left(\frac{\pi S_{es}^2 \sin\phi \cos\phi}{4}\right)\left[\frac{1-\nu_1^2}{E_1} + \frac{1-\nu_2^2}{E_2}\right]$$

where S_{es} = surface endurance strength of pinion teeth material

The values of k are as given below :

$$K = 0.16\left(\frac{BHN}{100}\right)^2 \text{ N/mm}^2 \text{ for steel pinion and steel gear} \quad \text{... (19.32a)}$$

$$= 0.21\left(\frac{BHN}{100}\right)^2 \text{ N/mm}^2 \text{ for cast iron pinion and gear} \quad \text{... (19.32b)}$$

$$= 0.18\left(\frac{BHN}{100}\right)^2 \text{ N/mm}^2 \text{ for steel pinion and cast iron gear} \quad \text{... (19.32c)}$$

$$FOS = \frac{F_w}{F_{eff}} \quad \text{... (19.33)}$$

The recommended factor of safety for bevel gears is 1.5 – 2.0.

19.15 FORCE ANALYSIS

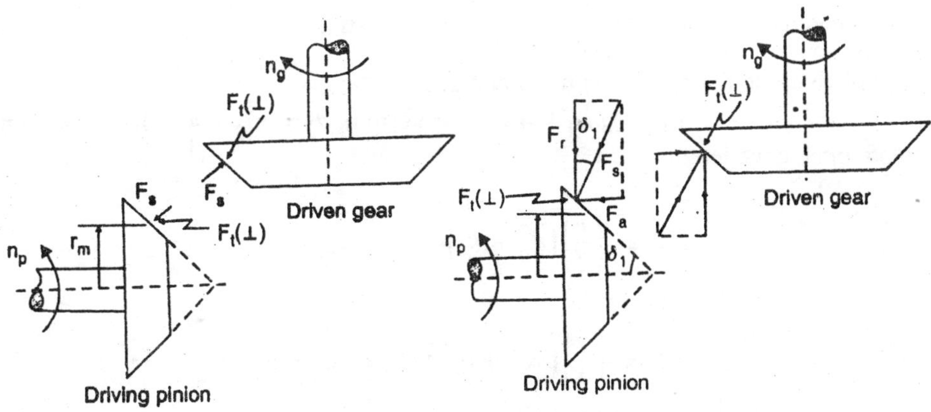

Fig. 19.3 Force analysis in Bevel gears.

The force acting on the gear tooth are shown in Fig. 19.3.

Mean radius, $r_m = \dfrac{d_1}{2} - \dfrac{b \sin \delta_1}{2}$

Tangential force, $F_t = \dfrac{T}{r_m}$ ... (19.34)

Normal force, $F_n = \dfrac{F_t}{\cos \phi}$

Separating force, $F_s = F_n \sin \phi = F_t \tan \phi$

Pinion :

Radial force, $F_r = F_s \cos \delta_1 = F_t \tan \phi \cos \delta_1$ ... (19.35)

Axial force, $F_a = F_s \sin \delta_1 = F_t \tan \phi \sin \delta_1$ ... (19.36)

Gear :

Radial force, $F_r = F_s \cos \delta_2 = F_t \tan \phi \cos \delta_2$ ... (19.37)

Axial force, $F_a = F_s \sin \delta_2 = F_t \tan \phi \sin \delta_2$ ... (19.38)

Example 19.1 A pair of straight bevel gears is mounted on shafts intersecting at right angles. The number of teeth on pinion and gear are 24 and 40 respectively. The pressure angle is 20°. The input power to the pinion is 15 kW at 1440 rpm. The pinion and gear are made of 50C4 (S_{ut} = 660 MPa) steel and heat treated to a surface hardness of 300 BHN and core hardness of 250 BHN. The gears are machined accurately by a hobbing process to Grade 6. Determine the module and the face width of gears. The service is with light shocks working daily from 8 to 10 hours.

Solution.

Given : $\theta = 90°$, $z_1 = 24$, $z_2 = 40$, $\phi = 20°$, P = 15 kW, $n_1 = 1440$ rpm, $S_{ut} = 660$ MPa, $(BHN)_{surface} = 300$, $(BHN)_{core} = 250$

1. Preliminary decisions :

Assume m = 4 mm for the initial calculations

Allowable bending stress, $\sigma_d = \dfrac{S_{ut}}{3} = \dfrac{660}{3} = 220$ MPa

2.Calculate pitch circle diameters of pinion and gear.

$$d_1 = mz_1 = 4 \times 24 = 96 \text{ mm}, \; d_2 = mz_2 = 4 \times 40 = 160 \text{ mm}$$

3. Calculate pitch cone length

$$L = \left(\frac{1}{2}\right)\left(d_1^2 + d_2^2\right)^{\frac{1}{2}}$$

$$= \left(\frac{1}{2}\right)\left(96^2 + 160^2\right)^{\frac{1}{2}} = 98.3 \text{ mm}$$

4.Calculate pitch angles of pinion and gear.

$$\tan \delta_1 = \frac{z_1}{z_2}$$

$$= \frac{24}{40} = 0.6$$

$$\delta_1 = 30.96° = 90° - 30.96° = 59.04°$$

5.Calculate virtual number of teeth.

$$z_v = \frac{z}{\cos\delta}$$

$$z_{1v} = \frac{24}{\cos 30.96°} = 28$$

$$z_{2v} = \frac{40}{\cos 59.04°} = 77.75$$

6.Calculate Lewis form factors based on circular pitch.

$$y = 0.154 - \frac{0.912}{z_v}$$

$$y_1 = 0.154 - \frac{0.912}{28} = 0.1214$$

$$y_2 = 0.154 - \frac{0.912}{77.75} = 0.1423$$

7. Decide the weaker wheel.

The material of pinion and gear is same and $y_1 < y_2$. Therefore, pinion is the weaker and shall be considered for design.

8. Mean pitch live velociy, $v = \frac{\pi d_1 n_1}{60 \times 10^3} = \frac{\pi \times 96 \times 1440}{60 \times 10^3} = 7.24$ m/s

9. Tangential load, $F_t = \frac{10^3 \text{P}}{v} = \frac{10^3 \times 15}{7.24} = 2072.33$ N

10.For generated teeth by hobbing process, the velocity factor,

$$C_v = \frac{6.1}{6.1 + v} \quad \text{for } v \leq 8 \text{ m/s}$$

$$= \frac{6.1}{6.1 + 7.24} = 0.457$$

11.Service factor for light shocks working 8 to 10 hours daily, $C_s = 1.25$

12.Load distribution factor for commerical gears straddle mounted, $C_m = 1.25$

13.Calculate maximum tangential load

$$(F_t)_{max} = C_s C_m F_t = 1.25 \times 1.25 \times 2072.33 = 3238 \text{ N}$$

14. Calculate effective load

$$F_{eff} = \frac{(F_t)_{max}}{C_v} = \frac{3238}{0.457} = 7085 \text{ N}$$

15/ Calculate static beam strength.

$$F_b = \sigma_d b \pi y b m \left(1 - \frac{b}{L}\right)$$

Where $\sigma_d = 220$ MPa

16. Calculate face width.

$$\sigma_d b \pi y_1 b m \left(1 - \frac{b}{L}\right) = F_{eff}$$

$$220 \times \pi \times 0.1214 \times 4 \times b \left[1 - \frac{b}{98.3}\right] = 7085$$

$$b^2 - 98.3\, b + 2075 = 0$$

$$b = \frac{\left[98.3 \pm \left\{(98.3)^2 - 4 \times 2075\right\}^{\frac{1}{2}}\right]}{2}$$

$$= \frac{98.3 \pm 36.9}{2}$$

$$= 30.7 \text{ mm } 67.6 \text{ mm}$$

Now $L = 98.3 \text{ mm} < 30 \text{ m} = 120 \text{ mm}$

Therefore $b = 6\text{m to } 7\text{m or } 24 \text{ mm to } 28 \text{ mm}$

Also $b \leq \frac{L}{3} \approx 33 \text{ mm}$

We adopt $b = 26 \text{ mm}, \text{m} = 4 \text{ mm}$

17. Calculate dynamic load

$$F_d = (F_t)_{max} + F_i$$

$$= (F_t)_{max} + \frac{20.67v\left[Cb + (F_t)_{max}\right]}{\left[20.67v\left\{Cb + (F_t)_{max}\right\}^{\frac{1}{2}}\right]}$$

For accuracy Grade 6, tolerance for adjacent pitch error.

$$e = 8.0 + 0.63\,(m + 0.25\sqrt{d_m})$$

$$d_m = d - b \sin \delta$$

$$d_{1m} = 96 - 26 \sin 30.96° = 82.62 \text{ mm}$$

$$d_{2m} = 160 - 26 \sin 59.04° = 137.70 \text{ mm}$$

$$e_1 = 8 + 0.63\,(4 + 0.25\sqrt{82.62}) = 11.950\ \mu\text{m}$$

$$e_2 = 8 + 0.63\,(4 + 0.25\,\sqrt{137.70}) = 12.368\ \mu m$$

$$e = e_1 + e_2 = 11.950 + 12.368 = 24.318\ \mu m$$

For steel pinion and gear, C $=11500e = 11500 \times 24.318 \times 10^{-3} = 279.66$ N/mm

$$F_d = 3238 + \frac{20.67 \times 7.24\,[279.66 \times 26 + 3238]}{20.67 \times 7.24 + \{279.66 \times 26 + 3238\}^{\frac{1}{2}}}$$

$$= 3238 + 6236.8 = 9474.8\ \text{N}$$

18.Calculate the endurance strength of pinion teeth

$$F_e = \sigma_e b \pi y_1 m \left(1 - \frac{b}{L}\right)\ \text{N}$$

For core BHN = 250 of steel

$$S_e = 1.75\ \text{BHN} = 1.75 \times 250 = 437.5\ \text{MPa}$$

$$F_e = 437.5 \times 26 \times \pi \times 0.1214 \times 4 \left(1 - \frac{26}{98.3}\right) = 12763.4\ \text{N}$$

19. Factor of safety $= \dfrac{F_e}{F_d} = \dfrac{12763.4}{9474.8} = 1.347$

Which is alright for pulsating loads.

20.Calculate limiting wear load.

$$F_w = \frac{d_1 b Q K}{\cos \delta_1}$$

For surface hardness of BHN=300 for steel on steel and $\phi = 20°$,

For steel pinion and gear,

$$K = 0.16\left(\frac{\text{BHN}}{100}\right)^2 = 0.16\left(\frac{300}{100}\right)^2 = 1.44$$

$$Q = \frac{2z_{2v}}{z_{2v} + z_{1v}} = \frac{2 \times 77.75}{77.75 + 28} = 1.47$$

$$F_w = \frac{96 \times 26 \times 1.47 \times 1.44}{\cos 30.96°} = 6161.4\ \text{N}$$

$F_w < F_{eff}$, therefore hardness of pinion has to be increased.

For $F_w = 1.5\ F_{eff}$

$$= 1.5 \times 7085 = 10627.5$$

$$\frac{96 \times 26 \times 1.47 \times K}{\cos 30.96°} = 10627.5$$

$$K = 2.484$$

$$= 0.16\left(\frac{\text{BHN}}{100}\right)^2$$

$$\text{BHN} = 394 \approx 400$$

Therefore, hardness of pinion should be raised to 400 BHN.

Example 19.2 A straight bevel pinion having 20 teeth is to mesh with a straight bevel gear having 60 teeth. The axes of the pinion and gear are at right angles. The pinion and gear are made of case hardened steel (S_{ut} = 720 MPa and 580 MPa respectively). The gear pair is manufactured by generation. The pinion shaft receives 15 kW at 1440 rpm from an electric motor. The surface hardness of the gear pair is to be 350 BHN. Design the gear pair with a factor of safety of 1.5 assuming velocity factor accounting for dynamic load.

Solution.

Given : z_p =20, z_g = 60, θ = 90°, S_{up} = 720 MPa, S_{ug} = 580 MPa, P = 15kW, n_p = 1440 rpm.

$$\text{BHN} = 350, \text{FOS} = 1.5$$

1. Preliminary decisions :

$$\text{Allowable bending stress, } \sigma_{dp} = \frac{S_{up}}{3} = \frac{720}{3} = 240 \text{ MPa}$$

$$\sigma_{dg} = \frac{580}{3} = 193.33 \text{ MPa}$$

2.Calculate pitch angles of pinion and gear.

$$\tan \delta_p = \frac{z_p}{z_g}$$

$$= \frac{20}{60} = 0.333$$

$$\delta_p = 18.435°$$

$$\delta_g = 90° - 18.345° = 71.155°$$

3.Calcualte virtual number of teeth.

$$z_v = \frac{z}{\cos \delta}$$

$$z_{pv} = \frac{20}{\cos 18.345°} = 21.07$$

$$z_{gv} = \frac{60}{\cos 71.655°} = 190.63$$

4.Calculate Lewis form factors based on circular pitch.

$$y = 0.154 - \frac{0.912}{z_v}$$

$$y_p = 0.154 - \frac{0.912}{21.07} = 0.1107$$

$$y_g = 0.154 - \frac{0.912}{190.63} = 0.149$$

5. Decide the weaker wheel.

$$\sigma_{dp} \times y_p = 240 \times 0.1107 = 26.568$$
$$\sigma_{dg} \times y_g = 193.33 \times 0.149 = 28.8$$
$$\sigma_{dp} \cdot y_p < \sigma_{dg} \cdot y_g.$$

Therefore pinion is the weaker and shall be considered for design.

6. Pitch diameters :

$$d_p = mz_p = 20\text{m}$$
$$d_g = mz_g = 60\text{m}$$

7. Pitch cone distance :

$$L = \left[\left(\frac{d_p}{2}\right)^2 + \left(\frac{d_g}{2}\right)^2\right]^{\frac{1}{2}}$$

$$= \left[(10m)^2 + (30m)^2\right]^{\frac{1}{2}} = 31.623 \text{ m}$$

8. Face width :

$$b = \frac{L}{3} \text{ or } 10 \text{ m whichever is smaller}$$

$$= \frac{31.623m}{3} = 10.54\text{m or } 10 \text{ m}$$

$$= 10\text{m}$$

9. Beam strength :

$$F_b = \sigma_{dp} bm\pi y_p \left[1 - \frac{b}{L}\right]$$

$$= 40 \times 10m \times m \times \pi \times 0.1107 \left[1 - \frac{10m}{31.623m}\right]$$

$$= 570.72 \text{ m}^2 \text{ N}$$

10. Wear strength :

$$F_w = \frac{d_p bQK}{\cos \delta_1}$$

$$Q = \frac{2z_{vg}}{z_{vg} + z_{vp}}$$

$$= \frac{2 \times 190.63}{190.63 + 21.07} = 1.8$$

$$K = 0.16\left(\frac{BHN}{100}\right)^2 = 0.16\left(\frac{350}{100}\right)^2 = 1.96 \text{ N/mm}^2$$

$$F_w = \frac{20m \times 10m \times 1.8 \times 1.96}{\cos 18.435°}$$

$$= 743.77m^2 \text{ N}$$

$$F_w > F_b.$$

Therefore, the gear pair should be designed for safety against static beam strength.

11. Mean pitch line velocity, $v = \frac{\pi d_p n_p}{60 \times 10^3} = \frac{\pi \times 20m \times 1440}{60 \times 10^3} = 1.5m$ m/s

12. Tangential load, F_t. $= \frac{10^3 P}{v} = \frac{10^3 \times 15}{1.5m} = \frac{10000}{m}$ N

13. For generated teeth by hobbing process, the velocity factor,

$$C_v = \frac{6.1}{6.1 + v} \text{ of } v \leq 8 \text{ m/s}$$

$$= \frac{6.1}{6.1 + 1.5m}$$

14. Service factor for light shocks working 8 to 10 hours daily, $C_s = 1.25$

15. Load distribution factor for commerical gears straddle mounted, $C_m = 1.25$

16. Calculate maximum tangential load,

$$(F_t)_{max} = C_s C_m Ft = \frac{1.25 \times 1.25 \times 10000}{m} = \frac{15625}{m} \text{ N}$$

17. Calculate effective load,

$$F_{eff} = \frac{(F_t)_{max}}{C_v} = \frac{\left(\frac{15625}{m}\right)}{\frac{6.1}{6.1 + 1.5m}}$$

$$= \frac{2561.5(6.1 + 1.5m)}{m}$$

18. Calculate module.

$$F_b = FOS \times F_{eff}$$

$$570.72m^2 = 1.5 \times 2561.5\,(6.1 + 1.5m)/m$$

$$570.72m^3 = 3842.25\,(6.1 + 1.5m)$$

$$= 23438 + 5763.4\text{ m}$$

$$570.72m^3 - 5763.4m - 23438 = 0$$

Let $f(m) = 570.72m^3 - 5763.4m - 23438$

By hit and trial, we have

m, mm	4	4.5
$f(m)$	–9965.5	2633.56

Adopt $m = 4.5$ mm

19.Calculate dimensions of pinion tooth

$$b = 10\text{ m} = 45\text{ mm}$$

$$d_p = mz_p = 4.5 \times 20 = 90\text{ mm}$$

$$d_g = mz_g = 4.5 \times 60 = 270\text{ mm}$$

$$L = 31.623\text{m} = 31.623 \times 4.5 = 142.3\text{ mm}$$

$$h_a = 1\text{m} = 4.5\text{ mm}$$

$$h_f = 1.2\text{ m} = 1.2 \times 4.5 = 5.4\text{ mm.}$$

Example 19.3 A 20 teeth straight bevel pinion rotating at 750 rpm transmits 10 kW to a 40 teeth bevel gear. The module is 5 mm and the pressure angle is 20°. The face width is 50 mm. If the shaft angle is 90°, determine the components of force acting on meshing teeth.

Solution.

Given : $z_p = 20$, $z_g = 40$, $n_p = 750$ rpm, $P = 10$kW, $m = 5$ mm, $b = 50$ mm, $\theta = 90°$, $\phi = 20°$

1.Calculate pitch angles of pinion and gear.

$$\tan\delta_p = \frac{z_p}{z_g}$$

$$= \frac{20}{40} = 0.5$$

$$\delta_p = 26.565°$$

$$\delta_g = 90° - 26.565° = 63.435°$$

2. Calculate pitch diameters :

$$d_p = mz_p = 5 \times 20 = 100\text{ mm}$$

$$d_g = mz_g = 5 \times 40 = 200\text{ mm}$$

3.Calculate mean radius at larger end of teeth.

$$r_{mp} = \frac{d_p}{2} - \frac{b\sin\delta_p}{2}$$

$$= \frac{100}{2} - \frac{50\sin 26.565°}{2} = 38.82\text{ mm}$$

4.Calculate mean pitch line velocity.

$$v_m = \frac{2\pi r_{mp} n_p}{60\times10^3} = \frac{2\pi\times38.82\times750}{60\times10^3} = 2.05 \text{ m/s}$$

5. Tangential force, $F_t = \frac{10^3 P}{v_m} = \frac{10\times10}{2.05} = 3278.7$ N

6. Radial force on pinion, $F_{rp} = F_t \tan\phi \cos\delta_p = 3278.7 \times \tan 20° \times \tan 26.565°$

$= 596.67$ N

7. Axial force on pinion, $F_{ap} = F_t \tan\phi \sin\delta_p = 3278.7 \times \tan 20° \times \sin 26.565°$

$= 533.68$ N

8. Radial force on gear, $F_{rg} = F_{ap} = 533.68$ N

9. Axial force on gear, $F_{ag} = F_{rp} = 596.67$ N.

Example 19.4 A pair of bevel gears with 20° pressure angle consists of a 24 teeth pinion meshing with 36 teeth gear. The module is 4 mm and the face width is 20 mm. The material for the pinion and gear is steel 50C4 (S_{ut} = 660 MPa). The gears are machined to meet the specifications of Grade 8 and the surface hardness is 400 BHN. The pinion rotates at 600 rpm and receives power from an electric motor whose starting torque is 150% of the rated torque, Assuming a factor of safety of 2, determine the rated power that the gears can transmit.

Solution.

Given : $\phi = 20°$, $z_p = 24$, $z_g = 36$, $m = 4$ mm, $b = 20$ mm, $\sigma_{dp} = \sigma_{dg} \frac{660}{3} = 220$ MPa, BHN $= 400$, $C_s = 1.5$, $C_m = 1.0$, FOS $= 2$

1.Calculate pitch angles of pinion and gear.

$$\tan\delta_p = \frac{z_p}{z_g}$$

$$= \frac{24}{36} = 0.66667$$

$$\delta_p = 333.69°$$

$$\delta_2 = 90° - 33.69° = 56.31°$$

2.Calculate virtual number of teeth.

$$z_v = \frac{z}{\cos\delta}$$

$$z_{pv} = \frac{24}{\cos 33.69°} = 28.84$$

$$z_{gv} = \frac{36}{\cos 56.31°} = 64.90$$

3.Calculate Lewis form factors based on circular pitch.

$$y = 0.154 - 0912$$

$$y_p = 0.154 - \frac{0.912}{28.84} = 0.1224$$

$$y_g = 0.154 - \frac{0.912}{64.90} = 0.140$$

4. Decide the weaker wheel.

The material of pinion and gear is same and $y_1 < y_2$. Therefore, pinion is the weaker and shall be considered for design.

5.Calculate pitch circle diameters.

$$d_p = mz_p = 4 \times 24 = 96 \text{ mm}$$
$$d_g = mz_g = 4 \times 36 = 144 \text{ mm}$$

6.Calculate pitch cone distance.

$$L = \left[\left(\frac{d_p}{2}\right)^2 + \left(\frac{d_g}{2}\right)^2\right]^{\frac{1}{2}}$$

$$= \left[48^2 + 72^2\right]^{\frac{1}{2}} = 86.53 \text{ mm}$$

7.Calculate static beam strength

$$F_b = \sigma_{dp} b \pi y m \left(1 - \frac{b}{L}\right)$$

$$= 220 \times 20 \times \pi \times 0.1224 \times 4 \left(1 - \frac{20}{86.53}\right) = 5203.5 \text{ N}$$

8.Calculate limiting wear load.

$$F_w = \frac{d_p b Q K}{\cos \delta_p}$$

For surface hardness of BHN = 400 for steel on steel and $\phi = 20°$,

For steel pinion and gear,

$$K = 0.16\left(\frac{\text{BHN}}{100}\right)^2 = 0.16\left(\frac{400}{100}\right)^2 = 2.56 \text{ N/mm}^2$$

$$Q = \frac{2z_{vg}}{z_{gv} + z_{pv}} = \frac{2 \times 64.9}{64.9 + 28.84} = 1.385$$

$$F_w = \frac{96 \times 20 \times 1.385 \times 2.56}{\cos 33.69°} = 8181.65 \text{ N}$$

9.We find that $F_b < F_w$. Therefore, the pinion shall fail under bending.

10. Calculate dynamic load

$$F_d = (F_t)_{max} + F_i$$

$$= (F_t)_{max} + \frac{20.67v\left[Cb + (F_t)_{max}\right]}{\left[20.67v + \left\{Cb + (F_t)_{max}\right\}^{\frac{1}{2}}\right]}$$

For accruacy Grade 8, tolerance for adjacent pitch error.

$$e = 16.0 + 1.25\,(m + 0.25\,\sqrt{d_m})$$

$$d_m = d - (b - \sin\delta)\, d - b\sin\delta$$

$$d_{pm} = 96 - 20\sin 33.69° = 84.9 \text{ mm}$$

$$d_{gm} = 160 - 20\sin 56.31° = 143.36 \text{ mm}$$

$$e_p = 16 + 1.25\,(4 + 0.25\,\sqrt{84.9}) = 23.879\ \mu\text{m}$$

$$e_g = 16 + 1.25\,(4 + 0.25\,\sqrt{143.36}) = 24.741\ \mu\text{m}$$

$$e = e_p + e_g = 23.879 + 24.741 = 48.620\ \mu\text{m}$$

According to Prof. Spotts.

$$F_i = \frac{e n_p z_p b_v r_{pm} r_{gm}}{2530\left(r_{pm}^2 + r_{gm}^2\right)^{\frac{1}{2}}}$$

where $$b_v = \frac{b}{\cos\delta_p} = \frac{20}{\cos 33.69°} = 24.04 \text{ mm}$$

$$F_i = \frac{48.620 \times 10^{-3} \times 600 \times 24 \times 24.04 \times 42.45 \times 71.68}{2530\left\{(42.45)^2 + (71.68)^2\right\}^{\frac{1}{2}}}$$

$$= 243 \text{ N}$$

$$F_b = (C_s C_m F_t + F_i) \times FOS$$

$$5203.5 = (1.5 \times 1.0 \times F_t + 243) \times 2$$

$$F_t = 1572.5\text{N}$$

Torque transmitted, $T = \frac{F_t d_p}{2} = \frac{1572.5 \times 96}{2} = 75480$ N.mm

Power transmitted, $P = \frac{2\pi n_p T}{60 \times 10^3}$

$$= \frac{2\pi \times 600 \times 75480}{60 \times 10^6} = 4.74 \text{ kW}$$

EXERCISES

19.1 A pair of bevel gears is required to transmit 10kW at 600 rpm from an electric motor shaft to a machine shaft. The speed reduction is 3 : 1 and the shafts are inclined at 90°. The pinion is to have 24 teeth with 20° pressure angle and is to be made of cast iron having allowable tensile strength of 75 MPa. The gear is to be made of cast iron having allowable tensile strength of 55 MPa. The pinion is mounted midway on the shaft which is supported between two bearings having span of 200 mm. Design the gear.

19.2 A pair of straight bevel gears are required to transmit 10 kW at 1200 rpm of the pinion having 18 teeth. The speed reduction ratio is 3 : 1. Use 14.5° FD teeth. Select suitable material and detemine module, face width, pitch diameter and pitch cone angles for both pinion and gear.

19.3 A pair of straight bevel gears 20° FD involute mounted on shafts intersecting at 90° consists of 24 teeth pinion meshing with 48 teeth gear. The pinion shaft is connected to 7.5 kW, 950 rpm electric motor. The starting torque of the motor is 130% of the rated torque. The load distribution factor is 1.2. The gear is to be made of grew cast iron FG 350 (S_{ut} = 350 MPa) while the pinion is to be made of case hardened steel 40C8 (S_{ut} = 580 MPa). The gears are to be generated and finished to meet the specifications of Grade 8. Design the gear pair for the following criterion :

(*a*) FOS required is 3.0 against bending failure by using velocity factor.

(*b*) FOS required is 2.0 against wear failure. specify the hardness required on pinion.

19.4 A straight bevel pinion having 24 teeth is made of alloy steel (S_{ut} = 800 MPa) is to mesh with bevel gear to be made of plain carbon steel (S_{ut} = 720 MPa). The axes of pinion and gear intersect at right angles. The gear pair is required to transmit 15 kW from a spindle running at 600 rpm to a machine running at 300 rpm. The starting torque of the motor is 120% of the rated torque. The factor of safety required is 2.0. The tooth system is 20° FD involute. The gears are to be cut to meet the specifications of Grade 6. The pinion and gear are to be case hardened to 400 BHN and 350 BHN respectively. Design the gear pair by using Buckingham's equation for dynamic load.

19.5 A pair of stragith bevel gears, manufactured by cutting, consists of a 20 teeth pinion rotating at 1500 rpm meshing with a bevel gear rotating at 500 rpm. The axes of pinion and gear intersect at 90°. The module is 4 mm and the face width is 40 mm. The pressure angle is 20° FD involute. The pinion is made of plain carbon steel 45C8 (S_{ut} = 650 MPa, E = 207 GPa) and the gear is made of grey cast iron FG 260 (E = 120 GPa). The Poisson's ratios for steel and grey cast iron and 0.30 and 0.26 respectively. The surface endurance strength of gear pair is 600 MPa. The service factor and factor of safety are 1.5 and 2.0 respectively. Assuming that the velocity factor accounts for the dynamic load, determine the rated power the gear pair can transmit.

19.6 A pair of bevel gears consists of a 24 teeth pinion meshing with a 48 teeth gear. The gears are mounted on shafts which are interesting at right angles. The module at the large end of the teeth is 5 mm. Calculate pitch circle diameters, pitch cone angles and the cone distance of the pinion and gear.

19.7 A pair of straight bevel gears consists of a 36 teeth pinion meshing with a 60 teeth gear. The module and face width are 6 mm and 60 mm respectively. The pinion and the gear are made of steel (S_{ut} = 650 MPa). Calculate the beam strength and check for wear load.

---

20 WORM GEARS

Chapter

20.1 INTRODUCTION

Worm gears are used to transmit power between two non-intersecting shafts. The shafts are generally at right angles to each other. The worm gear drive consists of a worm and a worm wheel. The worm is a threaded screw which can have single, double, triple or multi–start threads. Worm is generally the driver. Worm gears are used in mechanical handling equipment, machine tools and automobiles.

20.2 ADVANTAGES AND DISADVANTAGES OF WORM GEARS

The advantages and disadvantages of worm gears are given in Table 20.1.

Table 20.1 Advantages and disadvantages of worm gears.

Advantages	Disadvantages
1. High speed ratio from 60 : 1 to 100 : 1 with a single pair is possible.	1. Low efficiency due to excessive heat generation.
2. Compact drive with small overall dimensions.	2. Costly worm made of phosphor bronze.
3. Smooth and silent operation.	3. Excessive heat generation requires large amount of lubricating oil to dissipate heat.
4. Self – locking operation provision possible.	4. Low power transmission capacity upto 100 kW only.

20.3 TERMINOLOGY

The single and multi-start threaded worms and gear are shown in Fig. 20.1. The terminology of worm and worm gear is shown in Fig. 20.2.

Axial pitch, p_x. It is the distance measured parallel to the axis of worm between two consecutive corresponding profiles,

$$p_x = \pi\, m_x, \text{ where } m_x = \text{axial module} \qquad \text{... (20.1)}$$

Diameter quotient, $q = \dfrac{d_w}{m_x}$

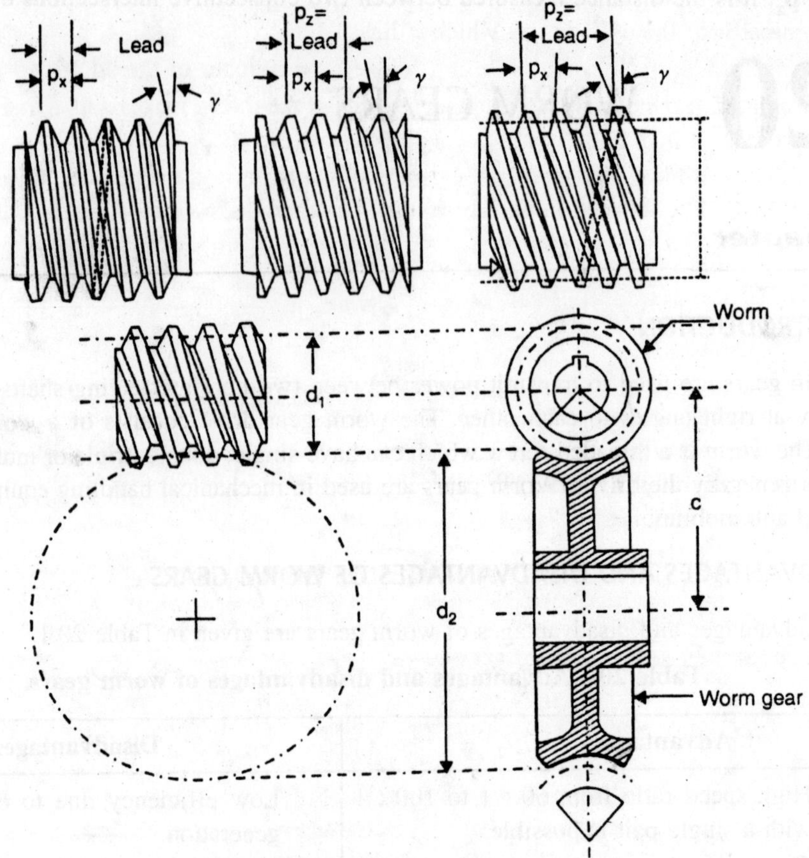

Fig. 20.1 Worms and worm wheel

· **Lead p_z,** It is the distance measured betweent wo consecutive intersections of a helix and a straight generator of the cylinder on which it lies.

$$p_z = Z_w\, p_x, \text{ where } z_w = \text{number of threads on worm} \quad ... (20.2)$$

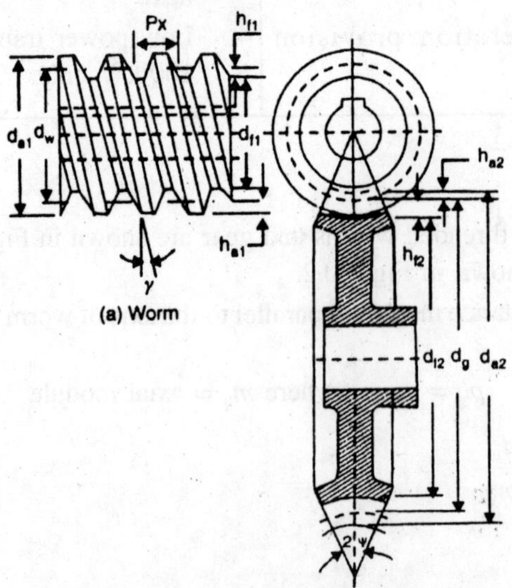

Fig. 20.2 Dimensions of Worm gears

Lead, p_x. It is the distance measured between two consecutive intersections of a helix and a straight generator of the cylinder on which it lies.

$$p_z = z_w p_x, \text{ where } z_w = \text{number of thread on worm} \quad ...(20.2)$$

Helix angle, α. It is the angle which the centre line of the thread makes with a line percpendicular to the worm axis. It is defined in case of a single start threaded worm.

Lead angle, γ. It is the angle which the centre line of the thread makes with a line perpendicular to the worm axis. It is defined in case of multi–start threaded worm.

$$\tan \gamma = \frac{p_z}{\pi d_w} = \frac{\pi m_x z_w}{\pi q m_x} = \frac{z_w}{q} = \frac{m_x z_w}{d_w} \quad ... (20.3a)$$

$$= \left(\frac{n_g}{n_w}\right)^{1/3}, \text{ for a compact drive} \quad ... (20.3b)$$

Speed ratio, *i*. It is the ratio of the speed of the worm to that of the gear.

$$i = \frac{n_w}{n_g} = \frac{z_g}{z_w} \quad ... (20.4)$$

Centre distance, *a*. It is the distance between the axis of the worm and the axis of the gear.

$$a = \left(\frac{1}{2}\right)(d_w + d_g) = \left(\frac{m_x}{2}\right)(q + z_g) \quad ... (20.5a)$$

$$a \approx 8\,[(i + 5)\,k\text{W}]^{0.588} \text{ mm} \quad ... (20.5b)$$

Normal circular pitch, p_n. It is the circular pitch of the worm in the normal plane to the thread elements.

$$p_n = p_x \cos \gamma \quad ... (20.6)$$

Axial module, m_x. It is the module measured parallel to the worm axis.

$$m_x = \frac{m_n}{\tan \gamma} \quad ... (20.7)$$

Width (or face) angle, 2ψ. It is the angle at the centre included between the points of intersection of the generating circle with the lateral faces of the teeth.

$$\tan \psi = \frac{\tan \phi_n}{\tan \gamma} \quad ... (20.8)$$

Normal pressure angle, ϕ_n – $\tan \phi_n = \tan \phi_t \cos \gamma$ ... (20.9)

Where ϕ_t = transverse pressure angle

Normal module, m_n. $m_n = m_x \cos \gamma$ ... (20.10)

Designation of worm drive. $z_1 / z_2 / q / m_x$ ... (20.10)

20.4 NUMBER OF STARTS

The number of starts on the worm should not normally exceed 6. The recommended values of number of starts are given in Table 20.2.

Table 20.2 Number of starts on worm.

Speed ratio, i	Number of starts on worm, z_w
20 and over	1
12-36	2
8-12	3
6-12	4
4-10	6

20.5 STANDARD AXIAL MODULES ON WORM

The standard axial modules for the worm are given in Table 20.3.

Table 20.3 Standard axial modules for worm, m_x (mm)

Fist choice	1.0, 1.25, 1.5, 2, 2.5, 3, 4, 5, 6, 8, 10, 12, 16, 20, 25, 32, 40 & 50.
Second choice	1.125, 1.375, 1.75, 2.25, 2.75, 3.5 4.5, 5.5, 7, 9, 11, 14, 18, 22, 28 & 36 45.

20.6 NORMAL PRESSURE ANGLE, ϕ_n

The normal pressure angle in relation to lead angle is given in Table 20.4.

Table 20.4 recommended normal pressure angle in relation to lead angle

Lead angle, γ, deg	0–15	15–30	30–35	35–40	40–50
Normal pressure angle, ϕ_n	14.5°	20°	25°	25°	30°

20.7 PROPORTIONS OF WORM AND WORM GEAR

The proportions of worm and worm gear are given in Table 20.5.

Table 20.5 Proportions of worm and worm gear.

Worm	**Worm wheel**
Pressure angle, $\phi = 20°$ Addendum, $h_{a1} = m_x$ Debendum, $h_{f1} = 1.25\ m_x$ Tooth depth, $h = 2.25\ m_x$ Bottom clearance, $c = (0.2 \text{ to } 0.3)\ m_x$ Tooth thickness : Axial $= 1.57\ m_x$ Normal $= 1.57\ m_x \cos\beta$ Reference (or pitch) diameter, $d_1 + qm_x$ Tip diameter, $d_{a1} = d_1 + 2m_x$ Root diameter, $d_{f1} = d_1 - 2m_x - 20$ Addendum modification coefficient, $x = \dfrac{a}{m_x} - 0.5\ (q + z_2)$ Base diameter, $d_{b1} = m_x\ (q + 2x)$	Pitch diameter, $d_2 = m_x z_2$ Base circle diameter, $d_{b2} = d_2$ Tip diameter, $d_{a2} = d_2 + 2m_x = d_2 + 2h_{a2}$ Throat diameter, $d_{f2} = d_2 - 2m_x - 2c$ $= d_2 - 2h_{f2}$

20.8 DIMENSIONS OF WORM

The dimensions of worm are given in Table 20.6.

Table 20.6 Dimension of Worms
All dimensions in mm.

Module m	Axial pitch p_x	No. of starts z_1	Diametral Quotient, q	Reference dia, d_1	Tip dia d_{a1}	Root dia d_{f1}	Lead angle γ
(1)	(2)	(3)	(4)	(5)	(6)	(7)	(8)
1	3.142	1	16	16.00	18	13.60	3°35′
1.25	3.927	1	(18)	22.50	25	19.50	3°11′
1.50	4.712	1	16	24	27	20.40	3°35′
			(18)	27	30	23.40	3°11′
2.0	6.283	1	(11)	22	26	17.50	5°12′
2.0	6.283	1	(18)	36	40	31.20	3°11′
		2	(11)	22	26	17.20	10°18′
		4	(11)	22	26	17.20	19°59′
		1	(11)	27.50	32.50	21.50	5°12′
		1	16	40.00	45.00	34.00	3°35′
2.5	7.854	2	(11)	33	39	25.80	10°18′
		4	(11)	33	39	25.80	19°59′
		6	(11)	44	52	36.55	28°37′
		1	(11)	33	39	25.80	5°12′
		1	(18)	54	60	46.80	3°11′
3	9.425	2	(11)	33	39	25.80	10°18′
		4	(11)	33	39	25.80	19°59′
		6	(11)	33	39	27.41	28°27′
		1	10	40	48	30.40	5°43′
		1	(18)	79	80	62.40	3°11′
4.	12.566	2	10	40	48	30.40	11°19′
		4	10	40	48	30.40	21°48′
		6	(11)	44	52	36.55	28°37′
		1	10	50	60	38.00	5°43′
		1	(18)	90	100	78.00	3°11′
5	15.708	2	10	50	60	38.00	11°19′
		4	10	50	60	38.00	21°48′
		6	(11)	55	65	45.69	28°37′

Module *m*	*Axial pitch* p_x	*No. of starts* z_1	*Diametral Quotient, q*	*Reference dia,* d_1	*Tip dia* d_{a1}	*Root dia* d_{f1}	*Lead angle* γ
(1)	(2)	(3)	(4)	(5)	(6)	(7)	(8)
		1	10	60	72	45.60	5°43′
		1	(18)	108	120	93.60	3°11′
6	18.860	2	10	60	72	45.60	11°19′
		4	10	60	72	45.60	21°48′
		6	10	60	72	49.36	30°58′
		1	10	80	96	60.80	5°43′
		1	(18)	144	160	124.80	3°11′
8	25.133	2	10	80	96	60.80	11°19′
		4	10	80	96	60.80	21°48
		4	8	80	80	48.52	26°34′
		6	(9)	72	88	58.71	33°41′
		8	(11)	88	104	75.33	36°02′
		1	10	100	120	76.00	5°43′
		1	(18)	180	200	156.00	3°11′
10	31.416	2	10	100	120	76.00	11°19′
		4	10	100	120	76.00	21°48′
		8	(11)	110	130	94.42	36°02′
		1	(9)	008	132	79.20	6°20′
		1	(18)	216	240	187.20	3°11′
12	37.699	2	(9)	108	132	79.20	12°32′
		4	(9)	108	132	79.20	23°58′
		8	(11)	132	156	113.30	36°02′
		1	(9)	144	176	105.60	6°20′
		2	(9)	144	176	105.60	12°32′
16	50.266	4	(9)	144	176	105.60	23°58′
		8	(11)	176	208	151.06	36°02′
		1	(9)	180	220	132.00	6°20′
		2	(9)	180	220	132.00	12°32′
20	62.832	4	(9)	180	220	132.00	23°58′
		8	(11)	220	260	188.83	36°02′

Note. The values given within brackets are non-preferred. IS : 3734—1966

Pressure angles = 20 deg.

20.9 RECOMMENDED VALUES OF TRANSMISSION RATIOS, CENTRE DISTANCE FOR WORM GEARS

The recommended values of transmission ratio and centre distances are given in Table 20.7.

Table 20.7 Recommended values of transmission ratios, centre, distane *$z_1/z_2/q/m$ for general purpose worm gears
All dimensions in millimetres

Centre distance	*Transmission Ratio*					
	5	*7.5*	*10*	*12.5*	*15*	*20*
25	2/10/(11)2.5	1/7(18)2	—	—	1/15/(18)1.5	1/21(18)1.25
31.5	4/21(11)/2	2/15/(11)2.5	2/21(11)/2	1/13(18)/2.5	1/15/(11)/2.5	1/10(11)2
40	4/21(11)/2.5	—	2/21(11)/2/5	—	1/16/16/2.5	1/21(11)/2.5
50	9/29(11)/2.5	4/29(11)/2.5	4/38/(11)/2	—	2/29/(11)/2.5	2/38/(11)/2
63	6/31/(11)3	4/31/(11)/3	4/39/(11)/2.5	4/51/(11)/2	2/31(11)/3	2/39(11)/2.5
80	6/29/(11)/4	4/30/(10)/4	4/42(11)/3	4/53/(11)/2.5	2/30/10/4	2/42/(11)/3
100	6/29/(11)/5	4/30/10/5	4/40(10)/4	4/49/(18)/3	2/30/10/5	2/40/(10)/4
125	6/32/(10)/6	4/31/10/6	4/40/10/5	4/52/10/4	2/31/10/6	2/40/10/5
160	6/31/(9)/8	4/30/10/8	4/43/10/6	4/54/10/5	2/30/10/8	—
200	8/39/(11)/8	4/30/10/10	4/40/10/8	—	2/30/10/10	2/40/10/8
250	8/39/(11)/10	—	4/40/10/10	4/52/8/8	—	2/40/10/10
315	8/41/(11)/12	4/30/(9)/16	4/43/(9)/12	4/53/10/10	7/30/(9)/16	—
400	8/39/(11)/16	4/31/(9)/20	4/41/(9)/16	—	2/31/(9)/20	2/41/(9)/16
500	9/39/(11)/20	—	4/41/(9)/20	4/53/(9)/16	—	2/41/(9)/20

Centre distance	*Transmission Ratio*				
	25	*30*	*40*	*50*	*60*
25	—	1/33/16/1	—	— •	—
31.5	1/26/16/1.5	1/32/(18)/1.25	—	1/48/16/1	—
40	—	1/29/(11)/2	—	—	1/62/61/1
50	—	1/29(11)/2.5	1/38/(11)/2	—	1/62/(18)/1.25
63	2/51/(11)/2	1/31/(11)/3	1/39/(11)/2.5	1/51/(11)/2	—
80	2/53/(11)/2.5	1/30/10/4	1/42(11)/3	1/53/(11)/2.5	1/62/(18)/2
100	2/54/(11)/3	1/30/10/5	1/40/10/4	1/55/(11)/3	1/64/16/2.5
125	2/52/(10)/4	1/31/10/6	1/40/10/5	1/52/10/4	—
160	2/52/10/5	1/30/10/8	—	1/54/10/5	1/62/(18)/4
200	—	1/30/10/10	1/40/10/8	—	—
250	2/52/10/8	—	1/40/10/10	1/52/10/8	—
315	2/53/10/10	1/30/(9)/16	—	1/53/13/10	1/60/(18)/8
400	—	1/31/(9)/20	1/41/(9)/16	1/49/(49)/16	1/58/(9)/12
500	2/53/(9)/16	—	1/41/(9)/20	1/53/(9)/16	—

Note. The value within brackets are non-preferred. IS : 3734 – 1966

*6/32/10/6 denotes, 6 starts on the worm, 32 teeth on the wheel, q = 10 and m = 6.

20.10 SELECTION OF NUMBER OF TEETH ON WORM WHEEL

For the given axial module, diametral quotient and centre distance, the number of teeth on worm wheel may be selected from Table 20.8.

Table 20.8 Selection of Number of Teeth, z_2 on the Worm Wheel *

m_x, mm	4	5	5	6	6	8	10	10	10	12	12	16	20
q →	11	10	12	9	11	8	11	8	11	8	9	11	8
a, mm ↓													
100	39	30	28	—	—	—	—	—	—	—	—	—	—
122	49	38	36	31	29	—	—	—	—	—	—	—	—
150	64	50	48	41	39	29*	—	—	—	—	—	—	—
180	79	62	60	51	49	37	34	28	—	—	—	—	—
210	—	74	72	61	59	44*	41*	34	31	—	—	—	—
240	—	—	—	71	69	52	49	40	37	32	29	—	—
270	—	—	—	81	79	59*	56*	46	43	37	34	—	—
300	—	—	—	—	—	67	64	52	49	42	39	28*	—
360	—	—	—	—	—	82	79	64	61	52	49	36	28
420	—	—	—	—	—	—	—	76	73	62	59	43*	34
480	—	—	—	—	—	—	—	—	—	72	69	51	40
540	—	—	—	—	—	—	—	—	—	82	79	58*	46
600	—	—	—	—	—	—	—	—	—	—	—	66	52

* Could be ± 2 teeth from values found from the table above except numbers marked with * where the deviation permissible is – 1 or + 2 teeth.

20.11 EFFECTIVE FACE WIDTH OF WORM WHEEL

In Fig. 20.3, we have

$$AC^2 = OA^2 - OC^2$$

$$\frac{b^2}{4} = \frac{d_{a_1}^2}{4} - \frac{d_1^2}{4} = \left[\frac{m_x\,(q+2)}{2}\right]^2 - \left(\frac{qm_x}{2}\right)^2$$

$$\sin\psi = \frac{b}{(d_{a_1} + 2c)}$$

$$\tan\psi = \frac{\tan\phi_n}{\tan\gamma}$$

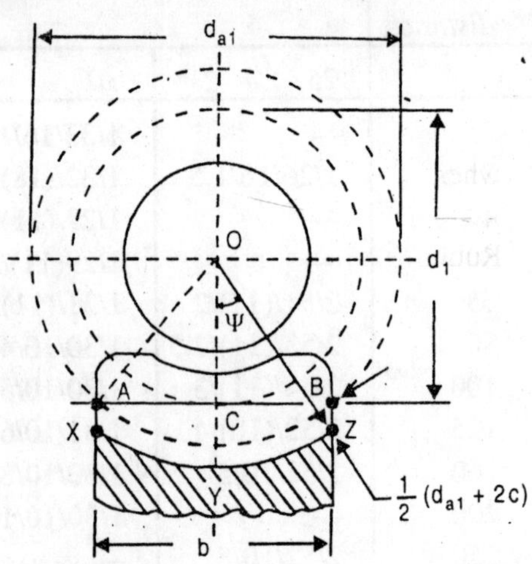

Fig. 20.3 Face width of worm wheel

where c = clearance = $0.2\, m_x \cos\gamma$

Length of root of worm wheel teeth,

$$l_r = \left[\frac{2\psi}{2\pi}\right]\pi\,(d_{a_1} + 2c)$$

$$\Rightarrow \quad = (d_{a_1} + 2c)\,\psi = (d_{a_1} + 2c)\sin^{-1}\left[\frac{b}{d_{a_1} + 2c}\right] \quad \text{.... (20.11a)}$$

Length of worm,
$$L_w = \pi m_x \left(4.5 + \frac{z_g}{50}\right) \quad \text{... (20.11b)}$$

20.12 FRICTION IN WORM GEARS

The coefficient of friction between the mating teeth of worm and worm gear depends upon the following factors ;

1. Materials of worm and worm gear.
2. Surface finish of worm and worm gear teeth.
3. Sliding velocity between the teeth.

The variation of coefficient of friction μ between the worm and worm gear v's rubbing speed is shown in Fig. 20.4

Velocity of worm,
$$v_w = \frac{\pi d_w n_w}{60 \times 10^3} \text{ m/s}$$

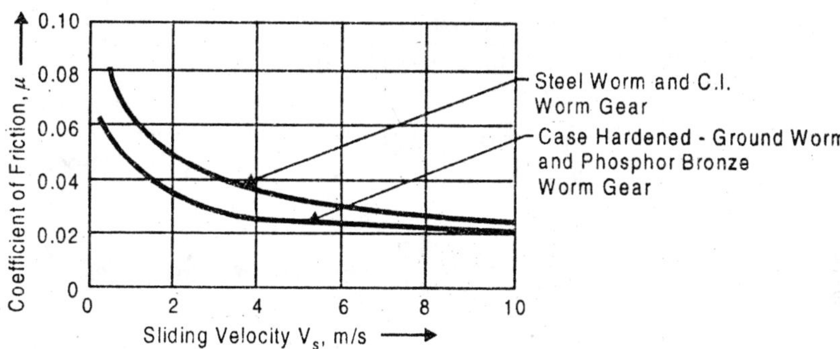

Fig. 20.4 Plot of Coefficient of Friction Versus Sliding Velocity

where d_w = pitch diameter of worm, mm

n_w = *rpm* of worm

Rubbing (or sliding) velocity,

$$v_s = \frac{v_w}{\cos\gamma} = \frac{v_g}{\cos\gamma}$$

$$= \frac{\pi d_w n_w}{60 \times 10^3 \times \cos\gamma} \text{ m/s} \quad \text{... (20.12)}$$

Coefficient of friction,
$$\mu = \frac{0.0422}{(v_s)^{0.28}} \text{ for } 0.2 \text{ m/s} < V_r < 2.8 \text{ m/s} \quad \text{... (20.13a)}$$

$$= 0.025 + 3.281\frac{v_s}{10^3} \text{ for } v_r > 2.8 \text{ m/s} \quad \text{... (20.13b)}$$

Also $$\mu = 0.051 \times \frac{S}{(v_s + 0.4)^{1/2}} \quad \text{... (20.14)}$$

where S = a factor which depends upon the material combination and surface finish of mating teeth of worm and worm gear. The values of factor S are given in Table 20.9.

Table 20.9 Values of factor S

Material of worm	Material of worm gear	S
1. Steel—Hardened and Ground	Cu-Sn Bronze	1.0
	Al—Alloy	1.15
	Cast iron	1.25
2. Steel—Heat Treated but not Ground	Cu—Sc Bronze	1.5
	Al—Alloy	1.73
	Cast Iron	1.83

20.13 EFFICIENCY OF WORM AND WORM GEAR DRIVE

Efficiency, $$\eta = \frac{\text{Power output to the gear}}{\text{Power input to the worm}}$$

$$= \frac{P_g}{P_w} = \frac{F_{aw} v_g}{F_{tw}\, v_w} \quad (\text{Because} = F_{tg} = F_{aw})$$

$$= \frac{\cos \phi_n - \mu \tan \gamma}{\cos \phi_n + \mu \cot \gamma} \text{ when worm is the driver} \quad \text{... (20.15a)}$$

$$= \frac{\cos \phi_n - \mu \tan \gamma}{\cos \phi_n + \mu \cot \gamma} \text{ when worm gear is the driver} \quad \text{... (20.15b)}$$

If $\eta < 50\%$ then the drive is self—locking type.

For maximum efficiency, $\tan\gamma = (1 + \mu^2)^{1/2} - \mu$ ... (20.16)

For a compact design,

$$\gamma = 45° - 0.5 \tan^{-1}\left(\frac{\mu}{\cos \phi_n}\right) \quad \text{... (20.17)}$$

The efficiency of the worm gear pair depends upon the following factors :

1. Lead angle, γ.
2. Coefficient of friction between the worm and worm gear.
3. Normal pressure angle, ϕ_n.

The efficiency of worm gear drive is low due to sliding velocity between worm and worm gear teeth. It generally ranges between 50% to 95%. The variation

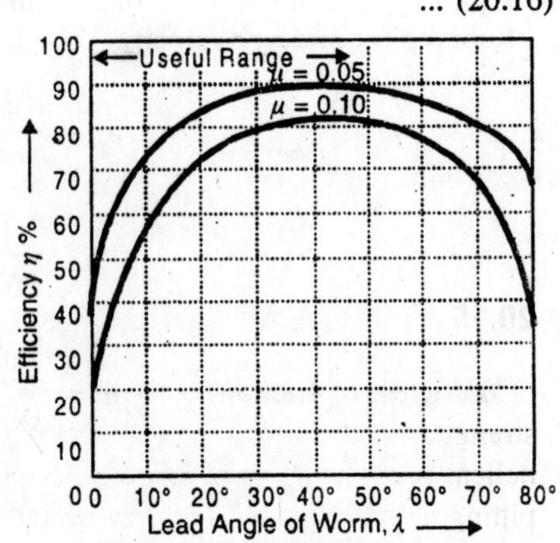

Fig. 20.5 Efficiency of worm Gear Pair

of efficiency v's lead angle is shown in Fig. 20.5. The efficiency increases rapidly upto lead angle of about 20° after which the increase in efficiency is slow. Better efficiency can be obtained by using multi—start threads. Approximate efficiencies of worm gear drive are given in Table 20.10.

Table 20.10 Approximate efficiencies of worm gear drives

Number of starts	Efficiency, %
1	70—75
2	70—85
3 and 4	80—92

20.14 THFRMAL CAPACITY OF WORM GEARS

Heat generated, $$H_g = (1 - \eta)\, P\ kW \quad \text{... (20.18a)}$$

$$= (1 - \eta)\frac{F_{tw}\, v_w}{10^3\ kw} \quad \text{... (20.18b)}$$

Heat dissipated, $$H_d = h_{cr}\, A_c \cdot \frac{\Delta t}{10^3}\ kW \quad \text{... (20.19)}$$

where
P = input power, kW
η = efficiency of worm gear drive
h_{cr} = combined heat transfer coefficient, $W/m^2 . °C$
= 10 to 18 $W/m^{2}°C$ with natural air circulation
= 20 to 28 $W/m^{2}°C$ with forced circulation with a far.
A_c = effective surface area of housing, m^2
= $1.14 \times 10^{-4} \times a^{1.7}\ m^2$
a = centre distance, mm
Δt = rise in temperature of lubricant, $°C \leq 50°C$

For a safe design, $H_d > H_g$
For $H_g = H_d$, we have

$$(1 - \eta)\, P = h_{cr}\, A_c \cdot \frac{\Delta t}{10^3}$$

$$P = h_{cr}\, A_c \cdot \frac{\Delta t}{[(1 - \eta) \times 10^3]}\ kW \quad \text{... (20.20)}$$

20.15 STATIC BEAM STRENGTH OF WORM GEARS

The worm threads are made stronger than the gear teeth. Therefore, analysis for bending strength is carried out for the gear only. For this purpose the worm gear can be regarded as a helical gear. The gear is unlikely to fall under bending stresses. The most likely failure is under pitting abrasive wear, and wear due to seizure.

Select normal module and calculate the module,

$$m_x = \frac{m_n}{\cos \gamma}$$

Approximate diameter of worm,

$$d_w = \frac{a^{0.875}}{2.2} \approx 3 \times \text{circular pitch of worm gear} \qquad \text{... (20.21}a\text{)}$$

$$d_w = \frac{m_x z_w}{\tan \gamma} \qquad \text{... (20.21}b\text{)}$$

$$d_g = m_x z_g$$

$$v_g = \frac{\pi d_g n_g}{60 \times 10^3} \text{ m/s}$$

Velocity factor,

$$C_v = \frac{3.05}{3.05 + v} \text{ for teeth cut using a form cutter}$$

$$= \frac{6.1}{6.1 + v} \text{ for generated teeth by hobbing}$$

Virtual number of teeth, $z_{vg} = z_g \cos^3 \gamma$

Lewis form factor, $y = 0.154 - \dfrac{0.912}{z_v}$

Values of Lewis form factor for worm gears are given in Table 20.11.

Table 20.11 Values of Lewis form factor for worm gears

Normal pressure angle, deg	14.5	20	25	30
Lewis form factor, y	0.100	0.125	0.150	0.175

Tangential load, $F_t = \dfrac{10^3}{v_g}$ ... (20.22)

Maximum load, $(F_t)_{max} = C_s F_t$ ... (20.23)

Effective load, $F_{eff} = \dfrac{(F_t)_{max}}{C_v} = \dfrac{C_s F_t}{C_v}$ ... (20.23)

where C_s = Service factor, which depends on type of load and machinery

= 1.25 to 2.0

Lewis equation for beam strength of gear teeth,

$$F_b = \sigma_{dg} \pi m_n b y \qquad \text{... (20.24)}$$

Where all the terms have their usual meanings.

σ_{dg} = allowable static bending strength of gear tooth material

$$= \frac{S_{ut}}{3}$$

Calculate face width from,

$$F_b = F_{eff} \quad \text{... (20.25)}$$

The face width of worm gear is taken as the larger of the following two values :

$$b = 2m\,(q + 1)^{1/2} \quad \text{... (20.26a)}$$

or

$$= 0.73\, d_w \quad \text{... (20.26a)}$$

whichever is less.

Factor of safety against bending,

$$\text{FOS} = \frac{F_b}{F_{eff}} \quad \text{... (20.27)}$$

Face width of the wheel is given in Table 20.12.

Table 20.12 Face width of worm wheel, *b*

Number of starts, z_w	Face width, *b*
1	$0.75d_w$
2 or 3	$0.75d_w$
4	$0.67d_w$

20.16 MATERIALS FOR WORM AND WORM GEAR

The threads of the worm are subjected to repetitive bending stresses and contact stresses.The body of the worm also acts as a shaft.Therefore the surface of the worm should be hard and the core should be soft. Accordingly the material for the worm is case hardened. The materials used for worm are given in Table 20.13.

Table 20.13 Mechanical Properties of Materials used for Worm

Material	Heat treatment	Ultimate tensile strength, MPa	Brinell hardness Number, BHN S_{ut} Core	MPa Case
40C8	N	580	217	217
55C8	N	720	265	265
10C4	CH	340	140	400
14C6	CH	370	140	400
16Ni80Cr60	CH	700	250	500
20Ni2Mo25	CH	750	250	500
15Ni2Cr1Mo15	CH	800	250	500
40Ni2Cr1Mo28	CH	800	250	600
13Ni3Cr18	CH	900	250	600
15Ni4Cr1	CH	1000	250	600

N = Normalized, CH = Case Hardened

Phosphor bronze is widely used for worm gear. The worm gears are either sand cast or sand cast and chilled. The mechanical properies of worm gear materials are given in Table 20.14.

Table 20.14 Mechanical Properties of Materials for Worm Gear

SI. No.	Material	Ultimate tensile strength, S_{ut} MPa	Brinell Hardness Number, BHN MPa	Applications
1.	Grey Cast Iron	200 – 350	179 – 241	Used for light duty and low speed drive, $v_s < 2$ m/s
2.	Gun Metal	220	80	Used for light to medium duty and low speed drive, $v_s < 5$ m/s
3.	Phosphor Bronze	245	95	Used for medium to heavy duty and high speed drive, $5 < v_s < 25$ m/s
4.	Nickel Bronze	340	95	Used for medium to heavy duty and high speed drive, $5 < v_s < 25$ m/s
5.	Aluminium Bronze	550	120	Used for heavy duty and low speed drive, $v_s < 5$ m/s

20.17 WEAR LOAD

$$F_w = d_g\, bK \qquad \text{... (20.28)}$$

where K = load stress factor for wear of worm gear.

The value of K depends upon the combination of materials for worm and worm gear and lead angle. Table 20.15 givesthe values of K.

Table 20.15 Worm gear load stress factor, K

Material		Load stress factor, K (MPa)		
Worm	Worm Gear	$\gamma < 10°$	$10° \leq \gamma \leq 25°$	$\gamma > 25°$
Hardened Streel	Cast Iron	0.345	0.430	0.515
Steel, 250 BHN	Phosphor Bronze	0.415	0.515	0.620
Hardened Steel	Phosphor Bronze	0.550	0.685	0.825
Hardened Steel	Chilled Phosphor Bronze	0.825	1.030	1.235
Cast Iron	Phosphor Bronze	1.030	1.285	1.545

Also $$F_w = A \cos \gamma \times \frac{C_v \sigma_c}{C_s} \quad \text{... (20.29)}$$

where $$A = \text{projected area of tooth, mm}^2 = \frac{h_w d_w \psi}{57.3} \quad \text{... (20.30)}$$

h_w = 2.16 m

ψ = half the face angle, deg

σ_c = compressive strength, MPa (Table 20.16)

C_s = Service factor for worm gears (Table 20.17)

For a safe design, $F_w > F_d$

Factor of safety againt wear,

$$FOS = \frac{F_w}{F_{eff}} \quad \text{... (20.31)}$$

Table 20.16 Permissible Compressive Stress σ_c*, N/mm²

Material of Worm	*Material of Gear*	*Number of teeth in the gear*							
		10	*20*	*30*	*40*	*50*	*60*	*70*	*80 and over*
20 *C* steel, untreated	Cast iron	0.49	1.85	3.23	5.20	6.18	7.46	8.63	9.32
40 *C* steel, untreated	Bronze, sand cast	0.79	2.35	4.61	7.36	9.32	12.47	31.05	13.83
40 *C* steel, heat treated ground	Bronze, sand cost	1.18	3.53	6.47	10.99	13.83	18.54	19.62	29.70
10 *C* alloy steel carburized, Hardened ground	Bronze, sand cost	1.86	5.00	8.63	14.81	18.54	22.37	26.19	27.57
	Bronze, chill cast	2.16	6.38	11.87	20.30	25.60	31.00	36.20	37.97
	Nickel Bronze, sand cast	2.94	7.75	20.50	23.05	31.00	37.57	43.75	46.20
	Nickel Bronze, chill cast	3.14	9.32	12.27	29.63	37.18	44.83	53.39	55.13

* Values tabulated are for 14.5° pressure angles. Multiply by 1.05 for 20° and 1.10 for 30° pressure angles.

Table 20.17 Service factor for worm gears, C_s

Type of services	Service factor, C_s
1. Intermittent with light shock	1 – 1.5
2. Continuous with medium shock, line shafts, crushers, etc.	1.5 – 2
3. Continuous with heavy shock, reciprocating pumps, paper and rubber mills, etc.	2 – 2.5
4. Continuous with frequent and heavy shocks, main line drives, steel mills, etc.	2.5 – 3

20.18 FORCE ANALYSIS

The forces on the worm and worm wheel are shown in Fig. 20.6.

$$F_{tg} = -F_{aw},\ F_{ag} = -F_{tw},\ F_{rg} = -F_{rw}$$

Let F = resultant force on the worm.

Then in the plane ABCD (Fig. 20.6*b*), we have

$$F_n = F \cos \phi_n$$

$$F_{rw} = F \sin \phi_n$$

In the plane AEBF (Fig. 20.6*c*), we have

$$F_{aw} = F_n \cos \gamma = F \cos \phi_n \cos \gamma$$

$$F_{tw} = F_n \sin \gamma = F \cos \phi_n \sin \gamma$$

$$F_{rw} = F \sin \phi_n$$

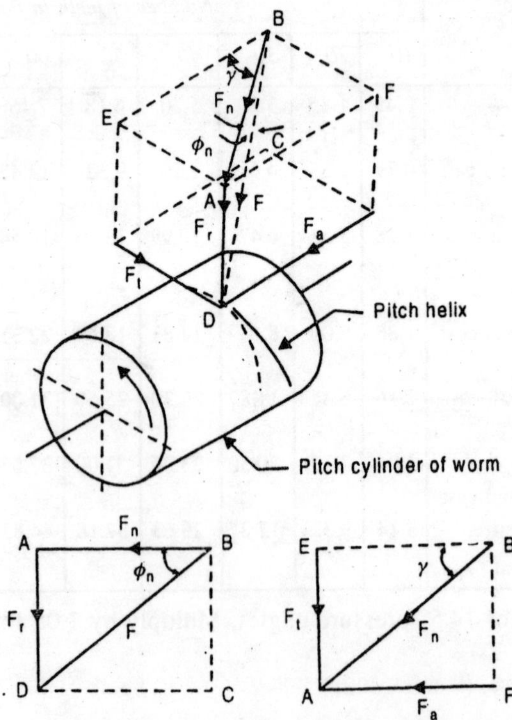

Fig. 20.6 Components of normal force on worm

The frictional forces are shown in Fig. 20.7.

Resultant forces becomes,

$$F_{tw} = F(\cos\phi_n \sin\gamma + \mu\cos\gamma)$$
$$F_{aw} = F(\cos\phi_n \cos\gamma + \mu\sin\gamma)$$
$$F_{rw} = F\sin\phi_n$$

$$\text{Now}F_{tw} = \frac{2T}{d_w} \quad ...(20.32)$$

$$F_{aw} = F_{tw}\left[\frac{\cos\phi_n\cos\gamma - \mu\sin\gamma}{\cos\phi_n\sin\gamma + \mu\cos\gamma}\right] ...(20.33)$$

$$F_{rw} = F_{tw}\left[\frac{\sin\phi_n}{\cos\phi_n\sin\gamma + \mu\cos\gamma}\right] (20..34)$$

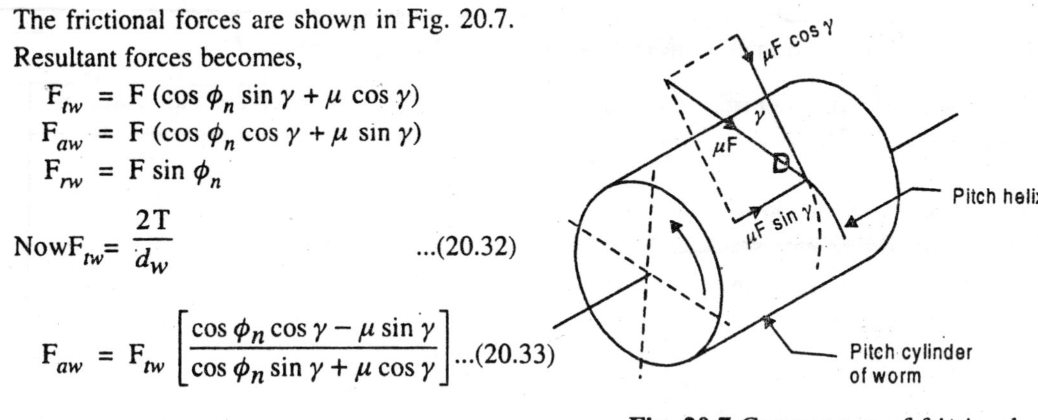

Fig. 20.7 Components of frictional force on worm

20.19 WORM GEAR REDUCERS

There are four possible arrangements of worm gear reducer as shown in Fig. 20.8

1.Worm – Down Type : In this type of reducer, the worm is down the worm gear. It is the most simplest and most widely used arrangement.

2.Worm – Up Type : In this case, the worm is above the worm gear.

3.Worm – Level Type : The worm and the worm gear are in same horizontal plane.

4.Worm – Vertical Type : The axis of the worm is vertical.

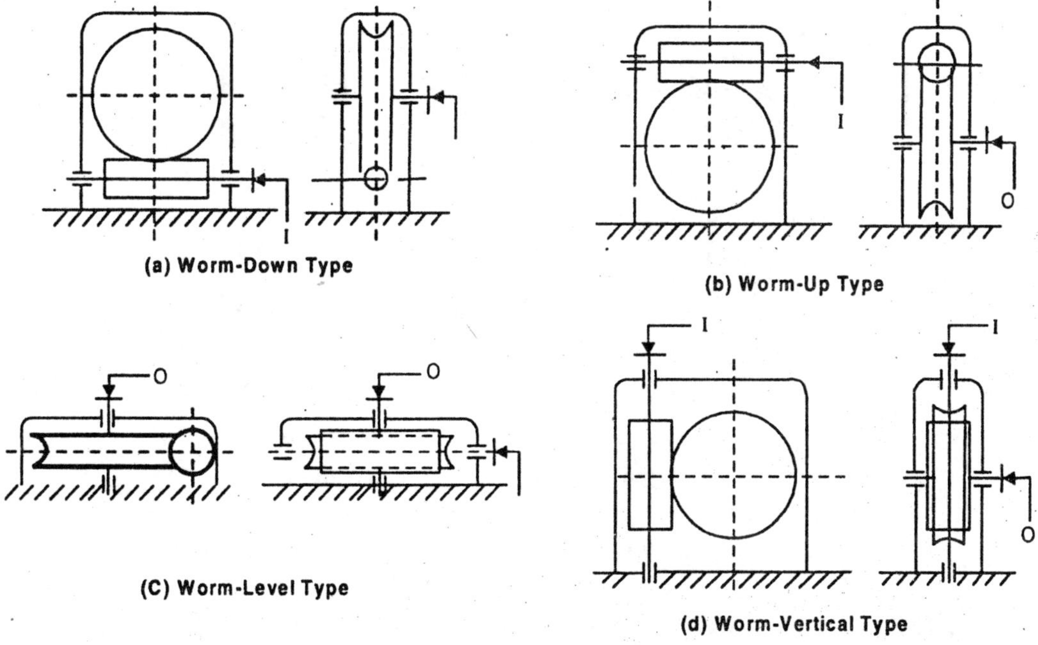

Fig. 20.8 Arrangements of Worm Gear Reducers.

20.20 CONSTRUCTION OF WORM GEARS

The hub and arms are made of cast iron and the outer rim, on which teeth are cut, is made of phosphor bronze. The gear rim is secured to the hub by one of the following methods shown in Fig. 20.9 :

1. Bolted construction.
2. Press—fitting with screw.
3. Bimetallic construction.
4. Braze welding.

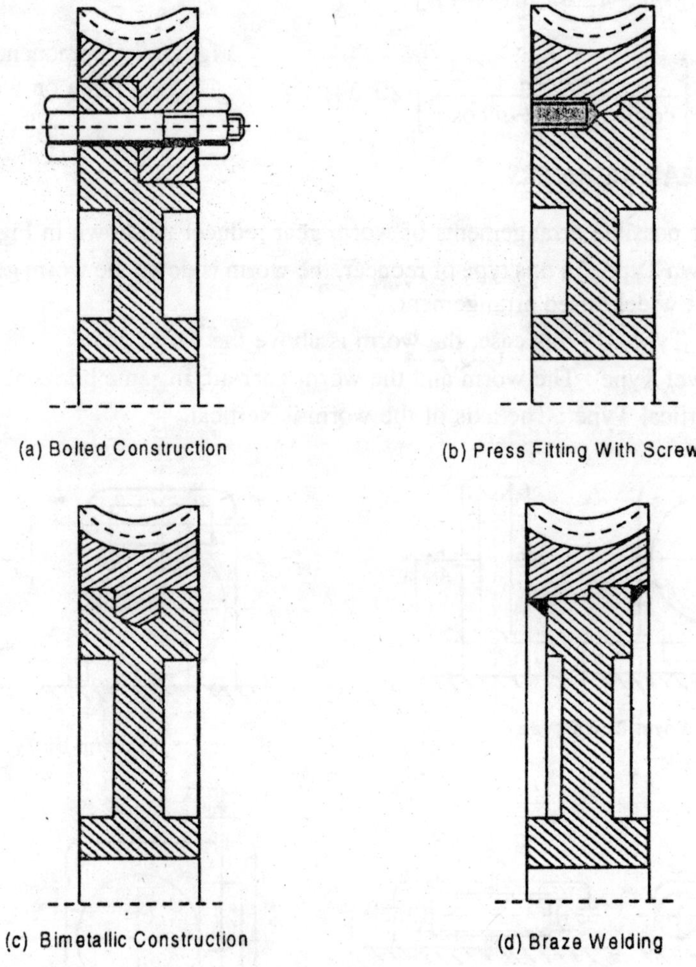

Fig. 20.9 Constructions of Worm Gears

Example 20.1 Design a worm and worm gear to transmit 2 kW from an electric motor revolving at 1440 rpm to a machine running at 72 rpm. Load is intermittent and steady. Gear is made of phosphor bronze and the worm of hardened steel. The gears are manufactured by hobbing process.

Solution.

Given : P =2 *k*W, n_w = 1440 *rpm*, n_g = 72 *rpm*, $i = \dfrac{1440}{72} = 20$

1. For a compact design, $\tan\gamma = \left(\dfrac{n_g}{n_w}\right)^{\frac{1}{3}} = \left(\dfrac{72}{1440}\right)^{\frac{1}{3}} = 0.368$

   Lead angle, $\gamma = 20°$

2. For i = 15 to 30, $z_w = 2$
3. For γ = 15° to 30°, normal pressure angle, $\phi_n = 20°$
4. Number of teeth on the gear, $z_g = iz_w = 20 \times 2 = 40$
5. Select m_n = 5 mm for the initial design.

$$m_x = \frac{m_n}{\cos\gamma} = \frac{5}{\cos 20°} = 5.32 \text{ mm}$$

$$d_g = m_x\, z_g = 5.32 \times 40 = 212.8 \text{ mm}$$

$$d_w = \frac{m_x z_w}{\tan\gamma}$$

$$= \frac{5.32 \times 2}{\tan 20°} = 29.23 \text{ mm}$$

6.
$$v_w = \frac{\pi d_w n_w}{10^3 \times 60} = \frac{\pi \times 29.23 \times 1440}{10^3 \times 60} = 2.2 \text{ m/s}$$

$$v_g = \frac{\pi d_g n_g}{10^3 \times 60} = \frac{\pi \times 212.8 \times 72}{10^3 \times 60} = 0.8 \text{ m/s}$$

7. Velocity factor for gear, $C_v = \dfrac{6.1}{6.1 + v_g}$, for gears manufactured by hobbing process

$$= \frac{6.1}{6.1 + 0.8} = 0.884$$

8.
$$z_{vg} = \frac{z_g}{\cos^3\gamma} = \frac{40}{\cos^3 20°} = 48.2$$

9.Lewis form factor for gear, $y_g = 0.154 - \dfrac{0.912}{48.2} = 0.135$

10.Tangential load, $F_t = \dfrac{10^3 P}{v_g} = \dfrac{10^3 \times 2}{0.8} = 2500$ N

11.Effective load, $F_{eff} = \dfrac{C_s F_t}{C_v} = \dfrac{1.5 \times 2500}{0.884} = 4242$ N

12.Beam strength of gear tooth, $F_b = \sigma_{dg}\ b\pi y_g\ m_x$

For phosphor bronze gear, $\sigma_{dg} = \dfrac{245}{3} = 81.67$ MPa

$$F_b = 81.67 \times b \times \pi \times 0.135 \times 5.32 = 184.27\ b$$

13.Equate $F_b = F_{eff}$ to get the value of face width.

$$b = 4242/184.27 = 23 \text{ mm}$$

$$q = d_w / m_x = 29.23/5.32 = 5.5$$

$$b = 2m_x (q + 1)^{1/2} = 2 \times 5.32\ (5.5 + 1)^{½} = 27 \text{ mm}$$

or $= 0.73\ d_w = 0.73 \times 29.23 = 21.34$ mm

We adopt **b = 25 mm and m$_n$ = 5 mm.**

14. Wear load, $F_w = d_g\ bK$

Load stress factor, K = 0.685 for hardened steel on phosphor bronze and lead angle between 10° – 25°.

$$F_w = 212.8 \times 25 \times 0.685 = 3644.2 \text{ N} < F_{eff}$$

Therefore, hardness of the gear has to be increased.

15.Rubbing velocity, $v_r = \dfrac{v_w}{\cos\gamma} = \dfrac{2.2}{\cos 20°} = 2.34$ m/s

Coefficient of friction, $\mu = \dfrac{0.0422}{(v_r)^{0.28}}$

$$= \frac{0.0422}{(2.34)^{0.28}} = 0.033$$

$$\tan\phi_n = \tan\phi_t \cos\gamma = \tan 20° \cos 20° = 0.342$$

$$\phi_n = 18.9°$$

16. Efficiency, $\eta = \dfrac{\cos\phi_n - \mu\tan\gamma}{\cos\phi_n + \mu\cot\gamma}$

$$= \frac{\cos 18.9° - 0.033 \tan 20°}{\cos 18.9° + 0.033 \cot 20°}$$

$$= 0.900 \text{ or } 90\%$$

17. Heat generated, $H_g = (1 - \eta)\ P = (1 - 0.9) \times 2 = 0.2$ kW or 200 Watts

Heat dissipated, $H_d = h_{cr}\ A_c\ .\ \Delta t$

Taking $h_{cr} = 10$ W.m^2 . °C and $\Delta t = 50$°C

Area required, $A_c = \dfrac{200}{10 \times 50} = 0.40 \text{ m}^2$

Example 20.2 A worm shaft is supplied with 2 kW at 740 rpm. The number of starts for threads on the worm are 4 with a 60 mm pitch circle diameter. The worm wheel has 30 teeth with a 5 mm axial module. The normal pressure angle is 20°. Calculate the efficiency of the worm gear set and power lost in friction.

Solution.

Given : P = 2 *k*W, n_w = 740 *rpm*, z_w = 4, d_w = 60 mm, z_g = 30, m_x = 5 mm, ϕ_n = 20°

1.Lead, $p_z = \pi m_x z_w = \pi \times 5 \times 4 = 20\pi\ mm$

2.Lead angle, $\gamma = \tan^{-1}\left[\frac{p_z}{\pi d_w}\right] = \tan^{-1}\left[\frac{20\pi}{60\pi}\right]\tan^{-1}\left(\frac{1}{3}\right) = 18.435°$

3. Rubbing velocity, $v_r = \dfrac{\pi\, d_w n_w}{60 \times 10^3 \times \cos\gamma}$

$$= \frac{\pi \times 60 \times 740}{60 \times 10^3 \times \cos 18.435°} = 2.45 \text{ m/s}$$

4. Coefficient of friction, $\mu = \dfrac{0.0422}{(v_r)^{0.28}} = \dfrac{0.0422}{(2.45)^{0.28}} = 0.0328$

5. Efficiency, $\eta = \dfrac{\cos\phi_n - \mu\tan\gamma}{\cos\phi_n + \mu\cot\gamma}$

$$= \frac{\cos 20° - 0.0328 \tan 18.435°}{\cos 20° + 0.0328 \cot 18.435°}$$

$= 0.89468$ or 89.47%

6.Power lost in friction = $(1 - \eta)$P = (1 – 0.89468) × 2 = 0.21064 kW or 210.64 Watts.

Example 20.3 A worm gear box with an effective surface area of 2 m² is operating in still air with an overall heat transfer coefficient of 15 W/m²C. The temperature rise of lubricating oil is limited to 45°C. The worm gears are designated as 1/30/10/8. The worm shaft is rotating at 1440 rpm and normal pressure angle is 20°. Calculate the power transmitting capacity based on thermal considerations.

Solution.

Given : $A_c = 2m^2$, h_{cr} = 15 W/m²°C, Δt = 45°C, z_w = 1, z_g = 30, q = 10, m_x = 8 mm, n_w = 1440 *rpm*, ϕ_n = 20°

1.Lead angle, $\gamma = \tan^{-1}\left(\frac{z_w}{q}\right) = \tan^{-1}\left(\frac{1}{10}\right) = \tan^{-} 0.1 = 5.71°$

2. $d_w = m_x\, q = 8 \times 10 = 80$ mm

3. $v_r = \dfrac{\pi\, d_w n_w}{60 \times 10^3 \times \cos\gamma}$

$= \pi \times 80 \times 1440 / (60 \times 10^3 \times \cos 5.71°) = 6.06$ m/s

4. $$\mu = 0.025 + 3.281\frac{v_r}{10^3} = 0.025 + \frac{3.281 \times 6.06}{10^3} = 0.045$$

5. Efficiency, $$\eta = \frac{\cos\phi_n - \mu\tan\gamma}{\cos\phi_n + \mu\cot\gamma}$$

$$= \frac{\cos 20° - 0.045\tan 5.71°}{\cos 20° + 0.045\cot 5.71°}$$

$$= 0.6734 \text{ or } 67.34\%$$

6. Power transmitting capacity,

$$P = \frac{h_{cr}A_c \cdot \Delta t}{10^3 \times (1-\eta)}$$

$$= \frac{15 \times 2 \times 45}{10^3 \times (1-0.6734)} = 20.1335 \text{ kW}$$

Example 20.4 A worm – down type 2/60/10/6 worm gear pair transmits 10 kW worm shaft rotating at 1440 rpm to the worm gear shaft. The worm is left hand type and rotates in an anticlockwise direction as viewed from the right side. The coefficient of friction between the worm and worn gear teeth is 0.05 and the normal pressure angle is 20°. Determine the following:

(a) Components of tooth forces acting on the worm and worm gear.

(b) Efficiency of worm gear pair.

(c) Power lost in friction

Solution.

Given : $z_w = 2$, $z_g = 60$, $q = 10$, $m_x = 6$ mm, P = 10 kW, $n_w = 1440$ *rpm*, $\mu = 0.05$, $\phi_n = 20°$

Diameter of worm, $$d_w = m_x \cdot q = 6 \times 10 = 60 \text{ mm}$$

Lead angle, $$\gamma = \tan^{-1}\left(\frac{z_w}{q}\right) = \tan^{-1}\left(\frac{2}{10}\right) = 11.31°$$

Pitch line velocity of worm, $$v_w = \frac{\pi d_w n_w}{60 \times 10^3} = \frac{\pi \times 60 \times 1440}{60 \times 10^3} = 4.524 \text{ m/s}$$

(a) Tangential force on worm, $$F_{tw} = \frac{P}{v_w} = \frac{10{,}000}{4.524} = 2210.433 \text{ N}$$

Axial force on worm, $$F_{aw} = F_{tw}\left[\frac{\cos\phi_n\cos\gamma - \mu\sin\gamma}{\cos\phi_n\sin\gamma + \mu\cos\gamma}\right]$$

$$= 2210.433\left[\frac{\cos 20°\cos 11.31° - 0.05\sin 11.31°}{\cos 20°\sin 11.31° + 0.05\cos 11.31°}\right]$$

$$= 8636.74 \text{ N}$$

$$\text{Radial force on worm, } F_{rw} = F_{tw}\left[\frac{\sin\phi_n}{\cos\phi_n \sin\gamma + \mu\cos\gamma}\right]$$

$$= 2210.433\left[\frac{\sin 20^\circ}{\cos 20^\circ \sin 11.31^\circ + 0.05\cos 11.31^\circ}\right]$$

$$= 3240.25 \text{ N}$$

$$F_{tg} = F_{aw} = 8636.74 \text{ N}$$

$$F_{ag} = F_{tw} = 2210.433 \text{ N}$$

$$F_{rg} = F_{rw} = 3240.25 \text{ N}$$

(*b*) $$\text{Efficiency, } \eta = \frac{\cos\phi_n - \mu\tan\gamma}{\cos\phi_n + \mu\cot\gamma}$$

$$= \frac{\cos 20^\circ - 0.05\tan 11.31^\circ}{\cos 20^\circ + 0.05\cot 11.31^\circ}$$

$$= 0.78146 \text{ or } 78.146\%$$

(*c*) Power lost in friction, $P_f = (1-\eta)P = (1 - 0.78146) \times 10 = 2.1854$ kW.

Example 20.5 A double start worm made of a case hardened steel (Sut = 700 MPa) is meshing with worm gear made of phosphor bronze (Sut = 240 MPa). The gear pair is required to transmit 7.5 kW from an electric motor running at 1500 rpm to a machine running at 60 rpm. The service factor is 1.5 and the factor of safety required is 2.0. The face width of the worm is 0.73 time the pitch circle diameter of worm. The worm gear wear factor is 0.685 MPa and the diametral quotient is 10. The normal pressure angle is 20o. Design the gear pair.

Solutions

Given : $z_w = 2$, $S_{up} = 700$ MPa, $S_{ug} = 240$ MPa, P = 7.5 kW, $n_w = 1500$ *rpm*, $n_g = 60$ *rpm*, $C_s = 1.5$, FOS = 2.0, $b = 0.73d_w$, K = 0.685 MPa, $q = 10$, $\phi_n = 20^\circ$, $\mu = 0.04$

Beam strength of worm gear teeth :

$$\sigma_{dg} = \frac{240}{3} = 80 \text{ MPa}$$

$$d_w = m_x \cdot q = 10m_x \text{ mm}$$

$$b = 0.73\, d_w = 7.3\, m_x \text{ mm}$$

$$i = \frac{n_w}{n_g} = \frac{1500}{60} = 25$$

$$\frac{z_g}{z_w} = 25,\ z_g = 25 \times 2 = 50$$

For $$\phi_n = 20^\circ$$

Lewis form factor, $$y_g = 0.154 - \frac{0.912}{z_g} = 0.154 - \frac{0.912}{50} = 0.1358$$

$$\tan\gamma = \frac{z_w}{q} = \frac{2}{10} = 0.2$$

$$\gamma = 11.31°$$

$$m_n = m_x \cos\gamma = m_x \cos 11.31° = 0.9806\, m_x \text{ mm}$$

$$F_b = \sigma_{ag} b m_n \pi y_g$$

$$= 80 \times 7.3\, m_x \times 0.9806\, m_x \times \pi \times 0.1358$$

$$= 244.32\, m_x^2 \times N$$

Wear strength of worm gear :

$$F_w = d_g\, bK$$

$$d_g = m_x \cdot z_g = m_x \cdot 50 = 50 m_x \text{ mm}$$

$$F_w = 50 m_x \times 7.3 m_x + 0.685 = 250\, m_x^2 \text{ N}$$

We find that $F_b < F_w$. Therefore, the worm gear shall be designd against bending failure.

Effective load on worm gear :

Efficiency, $$\eta = \frac{\cos\phi_n - \mu\tan\gamma}{\cos\phi_n + \mu\cot\gamma}$$

$$= \frac{\cos 20° - 0.04 \tan 11.31°}{\cos 20° + 0.04 \cot 11.31°}$$

$$= 0.9518\%$$

Output power, $$P_o = P_i \times \eta = 7.5 \times 0.9518 = 7.1382\ kW$$

$$v_g = \frac{\pi d_g n_g}{60 \times 10}$$

$$= \frac{\pi \times 50 m_x \times 60}{60 \times 10^3} = 0.157\, m_x \text{ m/s}$$

$$F_{tg} = \frac{P_o}{v_g} = \frac{7.1382 \times 10^3}{0.157\, m_x} = \frac{45466}{m_x} \text{ N}$$

$$C_v = \frac{6.1}{6.1 + v_g}$$

$$C_v = \frac{6.1}{6.1 + v_g} \text{ for generated teeth by hobbing}$$

$$F_{eff} = \frac{C_s F_{tg}}{C_v}$$

$$= 1.5 \times \left(\frac{45466}{m_x}\right) \times \frac{(6.1 + 0.157\, m_x)}{6.1}$$

$$= \frac{11180\,(6.1 + 0.157/m_x)}{m_x}$$

Calculate axial module :

$$F_b = FOS \times F_{eff}$$

$$244.32\ m_x^2 = \frac{2.0 \times 11180\,(6.1 + 0.157\,m_x)}{m_x}$$

$$244.32\ m_x^3 = 22360\ (6.1 + 0.157\ m_x)$$

$$= 136396 + 3510\ m_x$$

$$f(m_x) = m_x^3 - 14.37\ m_x - 558.3 = 0$$

m_x, mm	5	6	8	9	10
$f(m_x)$	– 505.15	– 428.52	– 161.26	41.37	298

We adopt $m_x = 9$ mm

Designation of worm gear pair is : 2/50/10/9

Dimensions of worm and worm gear :

$$m_x = 9\text{ mm},\ z_w = 2,\ z_g = 50,\ d_w = 90\text{ mm},\ d_g = 450\text{ mm}$$

$$p_x = \pi m_x = 28.27\text{ mm}$$

$$p_z = p_x \cdot z_w = 56.54\ mm$$

$$\gamma = 11.31°$$

$$b = 7.3\ m_x = 65.7\text{ mm}$$

Length of worm, L_w $= \pi m_x \left(4.5 + \frac{z_g}{50}\right) = \pi \times 9 \left(4.5 + \frac{450}{50}\right) = 121.5$ mm

$$h_a = 1\ m_x = 9\text{ mm}$$

$$h_f = 1.2\ m_x = 10.8\text{ mm}$$

$$a = \frac{d_w + d_g}{2} = \frac{90 + 450}{2} = 270\text{ mm}$$

EXERCISES

20.1 A worm gear pair 2/40/10/8 consists of worm gear made of phosphor bronze (S_{ut} = 245 MPa) and worm made of case hardened alloy steel (S_{ut} = 700 MPa). The coefficient of friction between worm and worm gear teeth is 0.03 and the normal pressure angle is 20°. The worm gear wear factor is 0.825 MPa. The overall heat transfer coefficient for the gear pair is 20 W/m^2 C°. The permissible temperature rise for the lubricating oil is above the atmospheric temperature is 50°C. The worm rotates at 750 rpm. The service factor is 1.5. Determine the input power rating based upon (*a*) beam strength (*b*) wear strength, and (*c*) thermal considerations.

The external surface area of housing is given by :

$$A_c = 1.14 \times 10^{-4} \times a^{1.7} \text{ m}^2$$

where a = centre distance, mm

20.2 A worm gear pair is used to transmit 5 kW from an electric motor running at 900 rpm to a machine. The worm is quadruple threaded and the worm gear has 48 teeth. The axial module is 8 mm and the diametral quotient is of the worm is 10. The face width of worm gear is 60 mm. The worm is made of case hardened steel 10C4 and the worm gear is centrifugally cast of phosphor bronze (S_{ut} = 240 MPa). The wear load factor for worm gear is 0.6 MPa. The transverse pressure angle is 21.4°. The coefficient of friction between worm and worm gear is teeth 0.04. The service factor is 1.25. The external surface area of the housing is 1.4 m^2. The overall heat transfer coefficient is 18 W/m^2 . C°. Determine : (*a*) factor of safety against bending failure (*b*) factor of safety againt wear failure, and (*c*) temperature rise of the lubricating oil above atmospheric temperature.

20.3 A pair of worm and worm gear is designated as : 3/60/10/6. The worm is transmitting 7.5 kW at 1440 rpm to the worm gear. The coefficient of friction is 0.15 and the normal pressure angle is 20°. Determine the components of gear tooth forces acting on the worm and worm gear.

20.4 The worm shaft of a worm gear pair is supplied 2.5 kW at 750 *rpm*. The number of starts of threads on worm are 4 with a 60 mm pitch cirle diameter. The worm gear has 36 threads with a 6 mm module. The normal pressure angle is 20°. Calculate the efficiency of the worm gear pair and the power lost in friction.

20.5 A worm gear housing with an effective surface area of 1.5 m^2 is operating in still air with overall heat transfer coefficient of 15 W/m^2. C°. The worm gears are designated as : 1/30/10/8. The 4 worm shaft is rotating at 1440 rpm and the normal pressure angle is 20°. Calculate the power transmitting capacity based on thermal considerations.

SLIDING BEARINGS

Chapter 21

21.1 INTRODUCTION

A bearing may be defined as a machine member whose function is to support and retain another moving member. The part of the shaft inside the bearing is called the journal. Bearings may be grouped into two distinct classes: sliding bearings and rolling or anti — friction bearings. In sliding bearings, a lubricant film is introduced between the relatively moving surfaces. In rolling bearings, rolling components are introduced between the surfaces. A journal bearing is a sliding contact bearing which gives lateral support to the rotating shaft. A thrust bearing carries the axial load of a rotating shaft.

Anti — friction or rolling bearings have either point or line contact. A rolling contact bearing consists of four parts: inner and outer races, a rolling element like a ball, roller or needle and a cage which holds the rolling elements together and spaces them evenly around the periphery.

21.2 SLIDING BEARINGS

21.2.1 Types of sliding bearings

The sliding bearings may be classified into two groups.

1. Hydrodynamic bearings, and
2. Hydrostatic bearings.

In the hydrodynamic bearings, the load supporting fluid film is created by the shape and relative motion of the sliding surfaces. In the case of hydrostatic bearings, the load supporting fluid film, separating the two surfaces, is created by an external source, like a pump.

21.2.2 Working principles of bearings

(a) Hydrodynamic Bearing

The principle of working of hydrodynamic bearing is illustrated in Fig. 21.1. The shaft is at rest initially and the contact between the journal and bearing is at the bottom most point, as shown

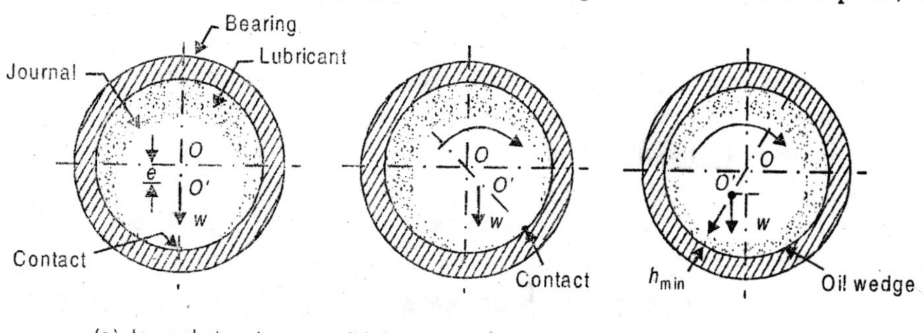

Fig. 21.1 Hydrodynamic Lubrication

in Fig. 21.1 (*a*). The surface of the journal and bearing contact during at rest. As the journal starts rotating the contact point between journal and bearing will shift opposite to the direction of rotation of the journal due to frictional forces, as shown in Fig. 21.1 (*b*). As the speed of journal is further increased, it will force the lubricant into the wedge—shaped region between the journal and the bearing and the journal will be lifted up. The minimum clearance between journal and bearing will shift in the direction of rotation of the journal, as shown in Fig. 21.1 (*c*). As more and more lubricating fluid is forced into the wedge — shaped clearance space, pressure is generated to full value to generate the hydrodynamic action. The pressure distribution around the periphery of the journal is shown in Fig. 21.2.

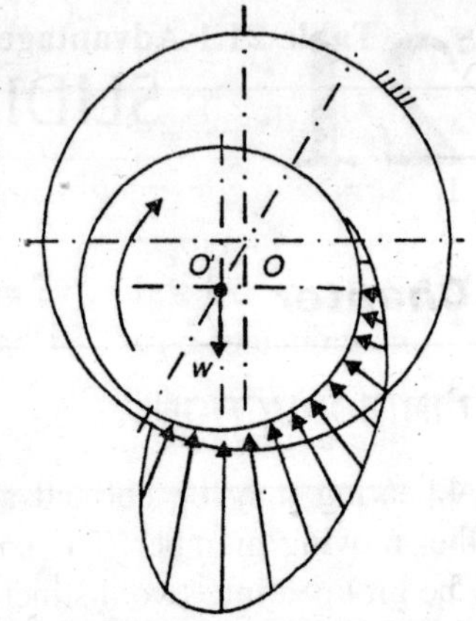

Fig. 21.3 Pressure distribution in hydrodynamic bearing

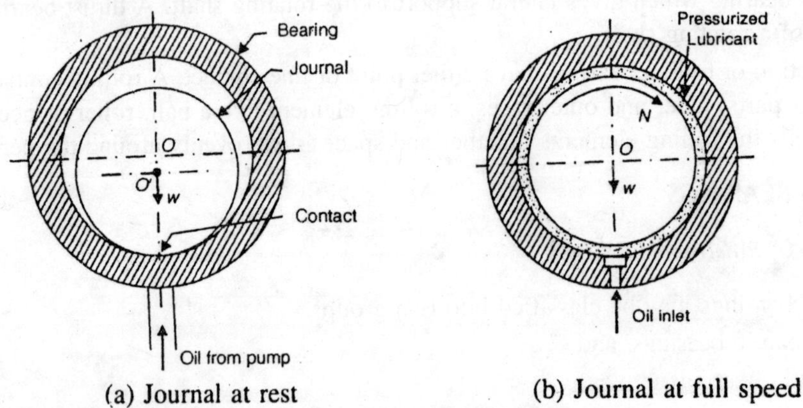

(a) Journal at rest (b) Journal at full speed

Fig. 2.3 Hydrostaticbearing

(*b*) Hydrostatic Bearing

The principle of working of a hydrostatic bearing is shown in Fig. 21.3. Initially, the journal rests on the bearing surface, as shown in Fig. 21.3 (*a*). As the high pressure fluid is admitted into the clearance space, it forces the journal and bearing surfaces to separate out, as shown in Fig. 21.3 (*b*). The hydrostatic bearings are used where pressures are very high like vertical turbo-generators, centrifuges and ball mills, etc.

21.3 CONDITIONS FOR HYDRODYNAMIC LUBRICATION

Conditions for hydrodynamic lubrication to be possible are :

1. The lubricant is incompressible and obeys Newton's law of viscosity.
2. The viscosity of lubricant is constant.
3. The inertia forces in the oil film are negligible.
4. The effect of curvature of film with respect to film thickness is neglected.
5. The film is so thin that the pressure is constant across the film thickness.
6. There is wedge — shaped clearance space available for the lubricant.

21.4 ADVANTAGES, LIMITATIONS AND APPLICATIONS OF HYDRODYNAMIC BEARINGS

The advantages, limitations and applications of hydrodynamic bearings are given in Table 21.1.

Table 21.1 Advantages, limitations and applications of hydrodynamic bearings.

Advantages	Limitations	Applications
1. Simple in construction and easy to maintian.	1. Unsiotable for low and extremely high speeds.	1. I.C. engines.
2. Low initial and maintenance cost.	2. High frictional loss at low speeds.	2. Large centrifugal pumps.
3. No auxiliary equipment like pump is required.	3. Poor positional accuracy.	3. Hydraulic turbines.
4. Reliable during operation.	4. High starting friction.	4. Guide ways of machine tools.
5. Low power loss.	5. Unsuitable for frequent starts and stops due to breakage of fluid film.	

21.5 DYNAMIC VISCOSITY

Dynamic viscosity of lubricant,

$$\mu = \left(0.22\,\text{SUS} - \frac{180}{\text{SUS}}\right)\rho \times 10^{-6}\ \text{N.s/m}^2 \qquad ...(21.1)$$

Where SUS = viscosity in Saybolt's Universal Seconds

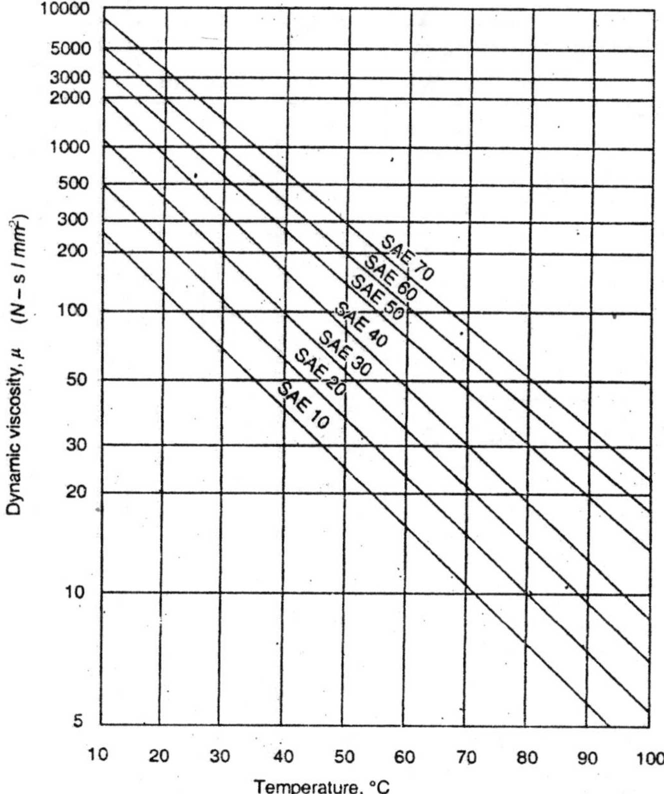

Fig. 21.4 Viscosity-temperature relationship

ρ = density, kg/m^3

$= \rho_{15.6} - 0.63\ (t_f - 15.6)$ kg/m^3 ...(21.2)

t_f = oil film temperature, °C

1cP = 10^{-2}P

1cP = 10^{-2}Pa.s or mPa.s

1 kg/m.s = 1 Pa.s

The effect of temperature on dynamic viscosity is shown in Fig. 21.4 for a range of SAE oils. The specific gravity of oils at 15.6°C is given in the Table 21.2.

Table 21.2 Specific gravity of oils at 15.6°C

Oil	Type of oil	Specific grvity at 15.6°C
SAE 10	Turbine oil (oil ring) light service and high speed	0.889
SAE 20	Light transmission for gears, automobile oil	0.925
SAE 30	Automobile oil	0.926
SAE 40	Medium transmission oil for large generators, motors, steam turbines, heavy speed gears, heavy motor oil, automobile oil	0.927
SAE 50	Light cylinder oil	0.890
SAE 60	Airplane oil	0.893
SAE 70	Gear oil	0.915

The properties of lubricants are given in Table 21.3.

Table 21.3 Properties of Lubricants

Puprpose	*SAE Number*	*Density kg/m³ at 15.5°C*	*Pour point °C*	*Flash point °C*	*Viscosity, index*	*Viscosity, Saybolt seconds at*			*Uses*
						38°C	*54°C*	*99°C*	
Transmis-sion gear oil	75	900	–23	193	121	220	–	50	Cobination pinion, reduction gear units, enclosed reduction gear sets.
	80	934	–32	185	78	320	–	52	
	90	930	–23	232	91	1300	–	100	
	140	937	–18	260	82	3350	–	160	
	250	–	–15	254.5	83	5660	–	220	
Automo-tive oil	10	870	–36	210	102	190	105	46	Automotive, truck and marine reciprocating engines, very heavcy duty oil used in Diesel engines.
	20	885	–23	227	96	330	153	54	
	30	891	–20	238	92	530	220	64	
	40	890	–18	240.5	90	800	255	77	
	50	902	–12	254.5	90	1250	–	97	
	60	–	–	–	80	–	–	115	
	70	–	–	260	80	–	–	137	
Aircraft engine oil		858	–65	111	87	43		33	Turbojet engines. Various reciprocating aircraft engines.
		864	–62	146	79	58		35	
		876	–18	215.5	106	350		57	
		884	–18	224	96	514		64	
		887	–18	232	95	829		80	
		892	–18	249	95	1240		99	
		892	–7	318	96	1711		120	

Puprpose	*SAE Number*	*Density kg/m³ at 15.5°C*	*Pour point °C*	*Flash point °C*	*Viscosity index*	*Viscosity, Saybolt seconds at*			*Uses*
						38°C	*54°C*	*99°C*	
Turbine grade oil								44	Direct connected turbines, electric motors. Land geared turbines, electric motors. Marine propulsion geared turbines.
Light		872	–18	210	109	150		53	
Medium		877	–12	235	105	300		62	
Heavy		885	–12	243	10	460			
Steam cylinder oil		895	1.4	210	101	1800		130	Rail road and stationary steam engine cylinders, enclosed gears.
		910	1.5	211	107	3750		210	
		940	15.5	343	103	6470		300	
Hydraulic oils									Hydraulic fluids for most indoor industrial hydraulic equipments. Heavier loads, higher temperature. Aircraft hydraulic system.
Light		887	–42	181	64	150		42	
Medium		805	–26	207	66	310		50	
Heavy		901	–12	267	70	910		74	
Extra-low temperature		844	–28	110	226	74		43	
Refrigerating machine oil		895	–45.5	146	53	72	3	6	Ammonia compressor
		898	–37	165.5	22	195		43	
		909	–29	182	34	235		45	
		902	–23	190.5	35	335		49	
Machine tools and general purpose oil		881	–4	177	80	105		39	All general prupose lubrication, machine tools.
		898	–4	199	80	205		46	
		915	–12	185	83	305		49	
		915	–15	199	25	520		59	
		890	–9	235	80	930		80	

21.6 HYDRODYNAMIC THEORY OF LUBRICATION

The theory of hydrodynamic lubrication was presented by Osborne Reynold. Reynold's equation is based on the following assumptions :

1. The lubricant obeys Newton's law of viscosity.
2. The lubricant is incompressible.
3. The viscosity of lubricant is constant.
4. The inertia forces in the oil film are negligible.
5. The effect of curvature of film with respect to film thickness is neglected.
6. The film is so thin that the pressure is constant across the film thickness.
7. The journal and bearing are rigid.
8. The lubricant is supplied continuously.

21.6.1 Reynold's Equation

Nomenclature

r_1 = bearing radius, mm

r = journal radius, mm

ε = eccentricity

c = $r_1 - r$, radial clearance

ε = $\dfrac{e}{c}$, eccentricity ratio

μ = dynamic viscosity of lubricant, NS /mm^2

h = film thickness

N = journal speed in rpm

n = journal speed, rps

W = radial load on the journal, N

l = length of bearing, mm

d = journal diameter, mm

p = $\dfrac{W}{LD}$, unit bearing pressure, N /mm^2

ϕ = attitude angle, deg

H_g = heat generated

H_d = heat dissipated

c_p = specific heat of lubricant, J/kg°C

m = mass of lubricating oil passing through the bearing, kg/s

ρ = density of oil, kg/m^3

Δt = temperature rise of lubricant, °C

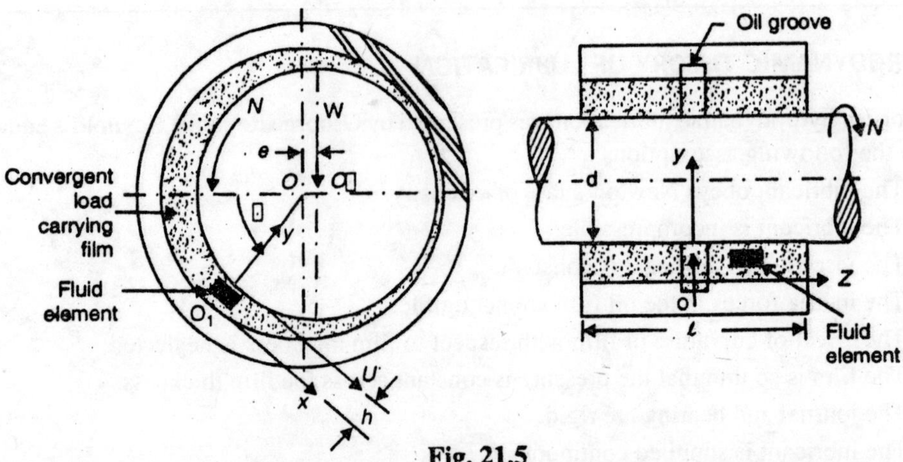

Fig. 21.5

Consider an elementary fluid element of dimensions dx, dy and dz in the xyz space as shown in Fig. 21.5 (*a*) and (*b*). Let

x, y, z = axes in the direction of motion, radial direction, and parallel to journal axis respectively.

u, v, w = velocities in x, y and z direction respectively.

τ_x, τ_z = shear stresses along x- and z– directions respectively.

The forces acting on the element in the x–direction are shown in Fig. 21.6. considering equilibrium of forces, we get

$$\left(p + \frac{\partial p}{\partial x}.dx\right).dydz + \tau_x dxdz = p \times dydz + \left(\tau_x + \frac{\partial \tau_x}{\partial y}.dy\right).dxdz$$

Simplifying, we get

$$\frac{\partial p}{\partial x}.dxdydz = \frac{\partial \tau_x}{\partial y}.dydxdz$$

Since $dxdydz \neq 0$, therefore

$$\frac{\partial \tau_x}{\partial y} = \frac{\partial p}{\partial x} \qquad ...(1)$$

According to Newton's law of viscosity,

$$\tau_x = \mu\left(\frac{\partial u}{\partial y}\right) \qquad ...(2)$$

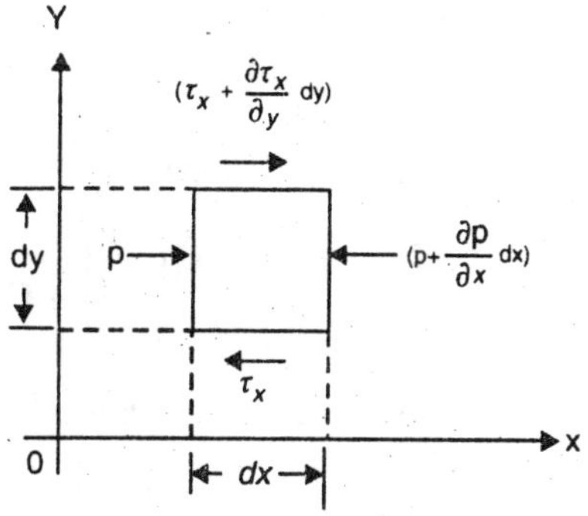

Fig. 21.6 Equilibrium of forces in x-direction

Substituting in Eq. (1), we have

$$\frac{\partial^2 u}{\partial y^2} = \left(\frac{1}{\mu}\right)\left(\frac{\partial p}{\partial x}\right) \qquad ...(3)$$

Integrating twice, we get

$$u = \left(\frac{1}{\mu}\right)\left(\frac{\partial p}{\partial x}\right)\left(\frac{y^2}{2}\right) + C_1 y + C_2 \qquad ....(4)$$

where C_1 and C_2 are constants of integration to be evaluated from the boundary conditions,

$$u = 0 \text{ for } y = 0$$

$$u = U \text{ for } y = h$$

Substituting boundary conditions in Eq. (4) we get

$$C_2 = 0$$

$$C_1 = \frac{U}{h} - \left(\frac{1}{\mu}\right)\left(\frac{\partial p}{\partial x}\right)\left(\frac{h}{2}\right)$$

Eq. (4) becomes,

$$u = \frac{Uy}{h} + \left[\frac{1}{2\mu}\right]\left(\frac{\partial p}{\partial x}\right)(y^2 - hy) \qquad ...(5)$$

The forces acting in the z-direction are shown in Fig. 21.7. Considering the equilibrium of forces, we have

$$\frac{\partial \tau_z}{\partial y} = \frac{\partial p}{\partial z} \quad ...(6)$$

According to Newton's law of viscosity,

$$\tau_z = \mu\left(\frac{\partial w}{\partial y}\right) \quad ...(7)$$

Substituting in Eq. (6), we have

$$\therefore \quad \frac{\partial^2 w}{\partial y^2} = \left(\frac{1}{\mu}\right)\left(\frac{\partial p}{\partial z}\right) \quad ...(8)$$

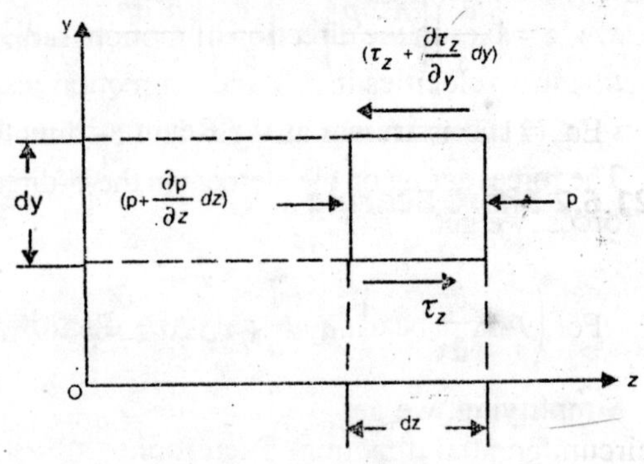

Fig. 21.7 Equilibrium of forces in z-direction

Integrating twice, we get

$$w = \left(\frac{1}{\mu}\right)\left(\frac{\partial p}{\partial z}\right)\left(\frac{y^2}{2}\right) + C_3 y + C_4 \quad ...(9)$$

where C_3 and C_4 are constants of integration to be evaluated from the boundary conditions,

$$w = 0 \text{ for } y = 0 \text{ and } y = h$$

Substituting boundary conditions in Eq. (9), we get

$$C_4 = 0$$

$$C_3 = -\left(\frac{1}{\mu}\right)\left(\frac{\partial p}{\partial z}\right)\left(\frac{h}{2}\right)$$

Eq. (9) becomes,

$$w = \left[\frac{1}{2\mu}\right]\left(\frac{\partial p}{\partial z}\right)(y^2 - hy) \quad ...(10)$$

The continuity equation for incompressible fluid flow is given by,

$$\frac{\partial u}{\partial x} + \frac{\partial v}{\partial y} + \frac{\partial w}{\partial z} = 0$$

Since there is no flow in the y–direction, we get

$$\frac{\partial u}{\partial x} + \frac{\partial w}{\partial z} = 0 \quad ...(11)$$

Integrating Eq. (11) w.r.t y from 0 to h, we get

$$\int_0^h \frac{\partial u}{\partial x} . dy + \int_0^h \frac{\partial w}{\partial z} dy = 0$$

Or

$$\left(\frac{\partial}{\partial x}\right)\int_0^h u \, dy + \left(\frac{\partial}{\partial z}\right)\int_0^h w \, dy = 0$$

Solving, we get

$$\left(\frac{\partial}{\partial x}\right)\left(\frac{h^3 \partial p}{\partial x}\right) + \left(\frac{\partial}{\partial z}\right)\left(\frac{h^3 \partial p}{\partial z}\right) = 6\mu U \left(\frac{\partial h}{\partial x}\right) \quad ...(21.3)$$

Eq. (21.3) is known as the Reynold's equation.

21.6.2 Short Bearing

For a short bearing $\left(\frac{l}{d} < 1\right)$ the oil flow in the axial direction is much greater than in the circumferential direction. Therefore. $\frac{\partial p}{\partial x} << \frac{\partial p}{\partial z}$ and first term in Eq. (21.3) on the LHS can be neglected as compared to the second term. Thus

$$\frac{\partial^2 p}{\partial z^2} = \left(\frac{6\mu U}{h^3}\right)\frac{\partial h}{\partial x} \quad ...(12)$$

Where l = length of the bearing

The solution of Eq. (12) may be obtained as :

$$p = \left[\frac{3\mu U}{h^3}\right]\left(\frac{\partial h}{\partial x}\right)\left(z^2 - \frac{l^2}{4}\right)$$

Substituting $dx = rd\theta$, we get

$$p = \left[\frac{3\mu U}{h^2 r}\right]\left(\frac{dh}{d\theta}\right)\left(z^2 - \frac{l^2}{4}\right) \quad ...(13)$$

The geometry of the bearing is shown in Fig. 21.8

O = bearing centre

O′ = journal centre

OO′ = e = eccentricity

OA = r_1 = bearing radius

O′F = r = journal radius

O′A = $r + h$

AF = h = oil film thickness

∠BO′A = θ

Fig. 21.8 Geometry of bearing

Draw OE ⊥ O′A.

Now $O'A = r + h = O'E + EA \approx e\cos\theta + r_1$

$$h = e\cos\theta + (r_1 - r)$$
$$= e\cos\theta + c$$

Then $h = c(1 + \varepsilon\cos\theta)$ ...(14)

Now h is minimum at $\theta = \pi$.

$$h_{min} = c(1 - \varepsilon) \quad ...(21.4)$$

Differentiating Eq. (14) w.r.t. θ, we get

$$\frac{dh}{d\theta} = -\varepsilon\, c \sin\theta$$

Eq. (13) becomes,

$$p = \left[\frac{\mu U}{rc^2}\right]\left[\frac{3\varepsilon \sin\theta}{(1+\varepsilon\cos\theta)^3}\right]\left[\frac{l^2}{4} - z^2\right] \quad ...(21.5)$$

For moderately to heavily loaded bearings, it can be assumed that the load is carried by a film extending from $\theta = 0$ to π. The forces in the x – and y – directions can be obtained as,

$$W_x = \left(\frac{\mu U l^3}{c^2}\right)\left[\frac{\varepsilon^2}{(1-\varepsilon^2)^{3/2}}\right] \quad ...(21.6)$$

$$W_y = \left(\frac{\mu U l^3}{c^2}\right)\left[\frac{\pi\varepsilon^2}{4\,(1-\varepsilon^2)^{3/2}}\right] \quad ...(21.7)$$

$$W = \left[W_x^2 + W_y^2\right]^{1/2}$$

$$= \left[\frac{4\mu U l^3 \varepsilon}{c^2\,(1-\varepsilon^2)^2}\right][16\,\varepsilon^2 + \pi^2\,(1-\varepsilon^2]^{1/2} \quad ...(21.8)$$

Or

$$\left[\frac{W}{l\mu U}\right]\left(\frac{c}{r}\right)^2\left(\frac{c}{l}\right)^2 = \varepsilon\,\frac{[16\,\varepsilon^2 + \pi^2\,(1-\varepsilon^2)]^{1/2}}{(1-\varepsilon^2)^2} \quad ...(21.9)$$

Sommerfeld Number,

$$S = \left[\frac{l\mu U}{\pi W}\right]\left(\frac{r}{c}\right)^2$$

$$= \left[\frac{\mu N}{60\,p}\right]\left(\frac{r}{c}\right)^2 \quad ..(21.10)$$

Substituting Eq. (21.10) in (21.9), we get

$$S\left(\frac{l}{d}\right)^2 = \frac{(1-\varepsilon^2)^2}{\pi\varepsilon\,\{16\,\varepsilon^2 + \pi^2\,(1-\varepsilon^2)\}^{1/2}} \quad ...(21.11)$$

The distribution of load on the bearing surface is shown in Fig. 21.2.

Attitude angle. It is the angle which the line of centres makes with the direction of the load. It is denoted by ϕ.

$$\tan\phi = \frac{W_y}{W_x} = \frac{-\pi\,(1-\varepsilon^2)^{1/2}}{4\,\varepsilon} \quad ..(21.12)$$

Frictional force : $$F = \left(\frac{\mu U l d}{c}\right)\left[\frac{2\pi}{(1-\varepsilon^2)^{3/2}}\right] \quad ...(21.13)$$

21.6.3 Long Bearing

In case of a long bearing $\left(\frac{l}{d} > 1\right)$ the change in pressure in the axial direction is small as compared to the change in pressure in the circumferential direction. Thus the Reynold's equation reduces to,

$$\frac{\partial^2 p}{\partial x^2} = \left(\frac{6\mu U}{h^3}\right)\frac{\partial h}{\partial x} \quad ...(21.14)$$

Solving Eq. (21.14), it can be shown that the film thickness h_1 where the pressure is maximum is given by,

$$h_1 = c\left[\frac{1-\varepsilon^2}{2+\varepsilon^2}\right] \quad ...(21.15)$$

The resultant force is given by,

$$\left[\frac{W}{\mu U l}\right]\left(\frac{c}{r}\right)^2 = \frac{12\,\pi\varepsilon}{(2+\varepsilon^2)(1-\varepsilon^2)^{1/2}} \quad ...(21.16)$$

Sommerfeld Number, $$S = \frac{(2+\varepsilon^2)(1-\varepsilon^2)^{1/2}}{12\,\pi^2\,\varepsilon} \quad ...(21.17)$$

Tangential frictional force,

$$F = \left(\frac{\mu U l d}{c}\right)\left[\frac{4\pi\,(1+2\varepsilon^2)}{(2+\varepsilon^2)(1-\varepsilon^2)^{1/2}}\right] \quad ...(21.18)$$

Coefficient of friction, f $$= \frac{F}{W} = \left[\frac{1+2\varepsilon^2}{3\,\varepsilon}\right]\left(\frac{c}{r}\right) \quad ...(21.19)$$

Angle θ for maximum pressure is given by, $\cos\theta = \dfrac{-3\,\varepsilon}{(2+\varepsilon)^2}$ ...(21.20)

Film pressure at any angle θ is,

$$p_\theta = \left(\frac{3\mu U d}{c^2}\right)\left[\frac{\varepsilon\sin\theta\,(2+\varepsilon\cos\theta)}{(2+\varepsilon^2)(1+\varepsilon\cos\theta)^2}\right] \quad ...(21.21)$$

21.6.4 Bearing Pressure

$$p = \frac{W}{ld} \text{ N/mm}^2 \quad ...(21.22)$$

where l = length of journal, mm

d = journal diameter, mm

21.6.5 Rubbing Velocity

$$\text{Rubbing velocity, } U = \frac{\pi d N}{60 \times 10^3} \text{ m/s} \quad ...(21.23)$$

21.6.6 Bearing Modulus

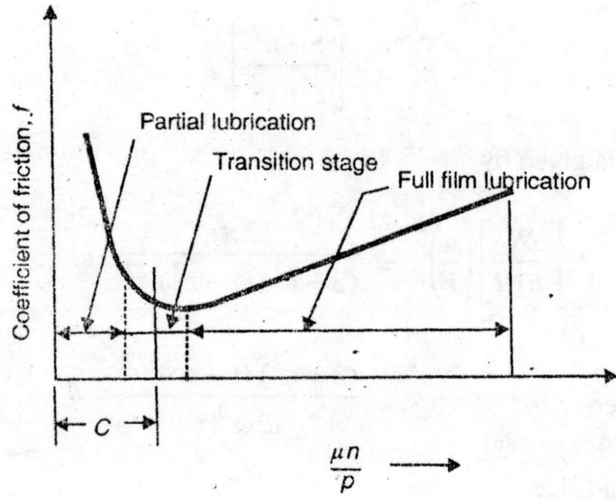

Fig. 21.9 Coefficient of friction v's $\frac{\mu n}{p}$

When the coefficient of friction 'f' for any bearing having film lubrication is plotted against $\left(\frac{\mu n}{p}\right)$, a curve similar to that shown in Fig. 21.9 is obtained. For any bearing there is a combination of μ, n and p that results in minimum coefficient of friction, indicated by C in the figure. Values of $\left(\frac{\mu n}{p}\right)$ greater than C indicate that the bearing may operate with complete film lubrication. At values less than C the oil film shall break down resulting into metal – to – metal contact and consequent higher friction and wear. This value of $\left(\frac{\mu n}{p}\right)$ at the breaking point is termed the bearing modulus.

When the bearing is operated at or near this value, slight decrease in speed or increase in pressure may be accompanied by large increase in friction, heating and wear. To prevent this, the bearing should operate at values of $\left(\frac{\mu n}{p}\right)$ at least three times the minimum value of C specified for journal and bearing materials.

Bearing modulus, $$C = \left[\frac{\mu n}{p}\right] \times 10^{-6} \quad ...(21.24)$$

Where n = journal speed, rps

p = bearing pressure, Pa

μ = dynamic viscosity of oil, Pa.s

$\left(\frac{\mu N}{p}\right) \geq 3C$ for minimum coefficient of friction

$\geq 15C$ for severe loading

The values of bearing modulus for various journal–bearing combinations are given in Table 21.4.

Table 21.4 Values of bearing modulus for various journal–bearing combinations

Journal material	Bearing material	$C = \frac{\mu N}{p^*}$
Hardened and ground steel	Babbit	2.85
Machined soft steel	Babbit	3.60
Hardened and ground steel	Plastic Bronze	4.30
Machined soft steel	Plastic Bronze	5.00
Hardened and ground steel	Rigid Bronze	5.70
Machined soft steel	Rigid Bronze	7.00

μ = Pa.s or N.s/m^2, N = rpm, p = N/mm^2

21.6.7 Clearance Ratio

It is the ratio of radial clearance between journal and bearing and radius of journal.

$$\frac{c}{r} = 0.001 \text{ for plastic bearing} \quad ...(21.25)$$

$$= 0.0015 \text{ for hard bearing}$$

Design data for journal bearings is given in Table 21.5.

Table 21.5 Journal bearing design practice

Machinery	Bearing	Maximum pressure p N/mm^2	Diametral clearance ratio $\frac{c}{d}$	Ratio $\frac{l}{d}$	Dynamic viscosity μ N.s/m^2	Bearing modulus (minimum) $C = \frac{\mu N}{p}$
Automobile and aircraft engines	Main	5.5 – 12	-	0.8–1.8	0.007 – 0.008	2.21
	Crank pin	10 – 24	-	0.7 – 1.4		1.45
	Wrist pin	16 – 34	-	1.5–2.2		1.16
Gas and oil engines (four stroke)	Main	4.8 – 8.2	0.001	0.6–2.0	0.020–0.065	2.90
	Crank pin	9.6 – 12.4	< 0.001	0.5–1.5		1.45
	Wrist Pin	12.4 – 15.2	< 0.001	1.5–2.0		0.72
Gas and oil engines (two stroke)	Main	3.4 – 5.5	0.001	0.6–2.0	0.020–0.060	3.62
	Crank pin	6.9 – 10	< 0.001	0.6 – 1.5		1.74
	Wrist pin	8.2 – 12.4	< 0.001	1.5–2.0		1.45
Marine steam engines	Main	3.4	< 0.001	0.7 – 1.5	0.030	2.90
	Crank pin	4.1	< 0.001	0.7–1.2	0.040	2.17
	Wrist pin	10.3	< 0.001	1.2 – 1.7	0.030	1.45
Stationary slow speed steam engines	Main	2.8	< 0.001	1.0 – 2.0	0.060	2.90
	Crank pin	10.3	< 0.001	0.9 – 1.3	0.080	0.87
	Wrist pin	12.4	< 0.001	1.2 – 1.5	0.060	0.72
Stationary high speed steam engines	Main	1.8	< 0.001	1.5–3.0	0.015	3.50
	Crank pin	4.1	< 0.001	0.9–1.5	0.030	0.84
	Wrist pin	12.4	< 0.001	1.3–1.7	0.025	0.70
Steam locomotives	Driving axle	3.8	0.001	1.6–1.8	0.100	4.35
	Crank pin	13.8	< 0.001	0.7–1.1	0.040	0.72
	Wrist pin	27.6	< 0.001	0.8–1.3	0.030	0.72
Reciprocating pumps and compressors	Main	1.8	< 0.001	1.0–2.2	0.030–0.080	4.35
	Crank pin	4.1	< 0.001	0.9–1.7		2.90
	Wrist pin	6.9	< 0.001	1.5–2.0		1.45
Railway cars	Axle	3.0	0.001	1.8–2.0	0.100	7.25
Steam turbines	Main	0.7–1.9	0.001	1.0–2.0	0.002–0.016	14.50
Generators, motors, centrifugal pumps	Rotor	0.7–1.4	0.0013	1.0–2.0	0.025	29.00

Machinery	Bearing	Maximum pressure p N/mm²	Diametral clearance ratio $\frac{c}{d}$	Ratio $\frac{l}{d}$	Dynamic viscosity μ N.s/m²	Bearing modulus (minimum) $C = \frac{\mu N}{p}$
Gyroscope	Rotor	5.9	0.0013	-	0.030	7.97
Transmission	Light, fixed	0.2	0.001	2.0–3.0		14.50
		1.0	0.001	2.0–3.0		4.35
shafting	Self-aligning	1.0	0.001	2.5–4.0	0.025–0.060	4.35
Cotton mill	Spindle	0.007	0.005	-	0.002	14.50
Machine tools	Main	2.1	0.001	1.0–1.4	0.040	5.80
Punching and	Main	27.5	0.001	1.0–2.0	0.100	-
shearing machines	Crank pin	55.0	0.001	1.0–2.0	0.100	-
Rolling mills	Main	20.6	0.0015	1.1 – 1.5	0.050	1.45

*μ in Pa.s, N in rpm and p in N/mm²

21.6.8 Sommerfeld Number

$$S = \left[\frac{\mu N}{60p}\right]\left(\frac{r}{c}\right)^2 \times 10^{-6} \text{ without side leakage} \quad ...(21.26)$$

Where p = pressure, MPa, N = rpm, μ = Pa.s, c = clearance mm, r = journal radius, mm

21.6.9 Coefficient of Friction

Petroff's equation for S > 150 and lightly loaded bearings,

$$f = \left(\frac{2\pi^2}{10^6}\right)\left[\frac{\mu N}{60\,p}\right]\left(\frac{r}{c}\right) \quad ...(21.28)$$

Where
p = pressure, Pa
r = journal radius, mm
c = radial clearance, mm
N = rpm
μ = dynamic viscosity of lubricating oil, Pa.s

McKee's equation for well lubricated and lightly loaded bearings, S > 150,

$$f = \left(\frac{0.326}{10^6}\right)\left[\frac{\mu N}{60p}\right]\left(\frac{r}{c}\right) + \Delta f \qquad ...(21.29)$$

Where Δf = a correction factor for end leakage

$$= 0.002 \text{ for } 0.75 < \frac{l}{d} < 2.8$$

The values for correction factor for end leakage are given in Table 21.6.

Table 21.6 Correction factor for end leakage, Δf

$\frac{l}{d}$	Δf	$\frac{l}{d}$	Δf
0.20	0.150	0.80	0.0018
0.30	0.012	1.00	0.0020
0.40	0.010	1.25	0.0022
0.50	0.005	1.50	0.0025
0.60	0.004	2.00	0.0027
0.70	0.003	2.50	0.0030

For medium to heavily loaded bearings.

$$f = \left[\frac{1+2\varepsilon^2}{3\varepsilon}\right]\left(\frac{c}{r}\right) \qquad ...(21.30)$$

For partially lubricated bearings having incomplete fluid film, Louis illaner equation for coefficient of friction,

$$f = \left(\frac{C_1 C_2}{8533}\right)\left(\frac{p_a}{U}\right)^{1/4} \qquad ...(21.31)$$

Where C_1 and C_2=constants (Table 21.7 and 21.8)

U = rubbing velocity, m/s

p_a = average pressure, Pa

$$= \frac{p_{max}}{2}$$

With a certain class of machinery the product of unit pressure and the rubbing velocity is a constant. Table 21.9 gives permissible values for various types of journals.

Table 21.7 Values of circumstance constant, C_1

Lubrication	Workmanship	Attendance	Operating condition	Value of C_1
Oil bath or flooded	High grade	First class	Clean and protected	1
Oil, free drop (constant feed)	Good	Fairly good	Favourable (ordinary condition)	2
Oil cup or grease (intermediate feed)	fair	Poor	Exposed to dirt, grit or other unfavourable conditions	4

Table 21.8 Values of type constant, C_2

Type of bearing	Value of C_2
Rotating journals, such as rigid bearings and crank pins.	1
Oscillating journals, such as rigid wrist pins and pintle blocks.	1
Rotating bearings lacking ample rigidity, such as eccentrics and the like	2
Rotating flat surfaces lubricated from the centre to the circumference, such as annular step or pivot bearings.	2
Sliding flat surfaces wiping over the guide ends, such as reciprocating crosshead shoes.	
Long guides	2
Short guides	3
Sliding or wiping surfaces lubricated from the periphery or outer wiping edge, such as marine thrust bearings and worm gears	3–4
Long power screw nuts and similar wiping parts over which it is difficult to effect a uniform distribution of lubricant or load	4–6

Table 21.9 Permissible *pv* values.

Class of bearing or journal	*pv* value	Class of bearing or journal	*pv* value
Axles, locomotives	4.20	Generators, motors	1.75
Railway cars	4.20	Line shaft	0.88
Aircraft crank pins	87.60	Reducing gears	3.50
Gas engine crank pins	14.00	Machine tools	0.35
High speed steam engine crank pins	14.00	Steam turbines	35.00
Air craft crank shaft and main bearings	70.10	Mill shafting	0.43
Diesel engine crank shaft and main bearing	35.00		

* p in N.mm^2 and v in m/s

21.6.10 Operating Pressure

Victor Tatarinoff equation for permissible pressure for film lubrication,

$$p = \left(\frac{1.33\mu n}{10^8}\right)\left(\frac{d}{c}\right)^2\left[\frac{l}{l+d}\right] \qquad ...(21.32)$$

Where p = pressure, MPa
μ = dynamic viscosity, mPas
n = speed of journal, rps

General Electric Company's formula (for motor and generator bearings),

$$p = 0.622\ U^{1/3}\ MPa \qquad ....(21.33)$$

where U = journal speed, m/s

Moore's equation,

$$p = 0.726\ U^{1/2}\ MPa \qquad ....(21.34)$$

Louis Illmer equation,

$$p = 4.65\left[\frac{U}{t-121.5}\right]^{1/3} \ MPa \qquad ....(21.35)$$

where U = velocity, m/s
t = temperature, °C

21.6.11 Minimum Oil Film Thickness

h_{min} = 0.01875 mm for ordinary babbitted bearings
= 0.0025 mm for fine-finished bronze bearings
= 0.075 to 0.125 mm for large steel shafts of fans, turbo-generators, etc.
= 0.00025D, D in mm ...(21.36a)

$$h_{min} = c\,(1-\varepsilon) \qquad ...(21.36b)$$

$$h_{min} \geq C_p\,(R_{ab} + R_{aj}) \qquad ....(21.37)$$

where R_{ab}, R_{aj} = roughness of bearing and journal respectively (Table 21.10)
C_p = 4.5 for ground finish
= 7.0 for super finish
= 10.0 for loose abrasive lapped

Table 21.10 Bearing and journal surface roughness, μm

Types of surface	Surface roughness
High precision journal	0.125 – 0.2
High precision bearing	0.125 – 0.5
Medium precision journal	0.4
Medium precision bearing	0.8
Commercial grinding	0.4 – 1.6
Broaching, reaming, boring, finished turning	0.8 – 1.6

Table 21.11 gives values of Sommerfeld number for h_{min}/c and l/d ratio

Table 21.11 Sommerfeld number for h_{min}/c and l/d ratio

l/d	h_{min}/c								
	1.0	0.2	0.1	0.067	0.05	0.04	0.033	0.0286	0.025
0.1	0	43.5	8.76	4.32	2.50	1.79	1.37	1.10	0.962
0.2	0	18.9	4.39	2.20	1.32	0.970	0.775	0.649	0.581
0.3	0	11.6	2.96	1.52	0.952	0.714	0.585	0.503	0.452
0.4	0	8.27	2.22	1.18	0.770	0.585	0.495	0.438	0.396
0.5	0	6.21	1.81	1.00	0.680	0.537	0.452	0.406	0.369
0.6	0	5.12	1.60	0.909	0.640	0.510	0.436	0.390	0.357
0.8	0	3.98	1.40	0.833	0.599	0.483	0.416	0.377	0.346
1.0	0	3.41	1.29	0.787	0.568	0.467	0.403	0.366	0.338
1.2	0	3.12	1.21	0.752	0.540	0.450	0.394	0.357	0.330
1.5	0	2.86	1.12	0.715	0.526	0.435	0.380	0.347	0.321
2.0	0	2.53	1.06	0.681	0.505	0.420	0.370	0.339	0.314
∞	0	2.07	0.922	0.611	0.411	0.394	0.352	0.322	0.301

21.6.12 Heat Generated and Dissipated

(*a*) Heat generated in bearing,

$$H_g = f p A U \text{ Watts} \quad ...(21.38)$$

(*b*) Heat dissipated through bearing.

(*i*) Surface exposed in terms of projected area of journal

$$H_d = k_b A (t_b - t_a) \times 10^{-6} \text{ Watts} \quad ...(21.39)$$

Where t_b = bearing temperature, °C

$= t_a + 0.5 (t_f - t_a)$

t_f = oil film temperature

t_a = surrounding air temperature, °C

k_b = energy dissipation coefficient, W/m^2 °C (Table 21.12)

A = ld = projected area of journal, mm^2

Table 21.12 Energy dissipation coefficient, k_b, W/m^2 °C

Bearing condition	k_b
1. Simple pillow block bearing	
(*a*) Quiet air	145 – 245
(*b*) air at 2.5 m/s	490–770
2. Heavy duty self–aligning ring oil lubricated	
(*a*) Quiet air	290 – 390
(*b*) Air at 2.5 m/s	880–1230

(*ii*) Based on area of housing.

$$H_d = K_1 A (t_b - t_a) \text{ J/s} \quad \text{...(21.40)}$$

where $K_1 (t_b - t_a)$ = heat radiating capacity of bearing, W/mm^2 (Table 21.13)

Table 21.13. Heat radiating capacity of bearing, $K_1 (t_b - t_a)$, W/mm^2

	Δt, °C					
	10	15	20	25	30	35
$K_1 (t_b - t_a)$ Ventilated	142×10^{-6}	214×10^{-6}	286×10^{-6}	361×10^{-6}	443×10^{-6}	531×10^{-6}
$K_1 (t_b - t_a)$ Unventilated	69×10^{-6}	116×10^{-6}	163×10^{-6}	204×10^{-6}	252×10^{-6}	301×10^{-6}

(*iii*) Based on projected area of bearing

$$\dot{H}_d = K_2 ld (t_b - t_a) \text{ J/s} \quad \text{...(21.41)}$$

where $K_2 (t_b - t_a)$ = heat radiating capacity of bearing, W/mm^2 (Table 21.14

Table 21.14 Heat radiating capacity of bearing, $K_2 (t_b - t_a)$, W/mm^2

	Δt, °C					
	10	15	20	25	30	35
$K_2 (t_b - t_a)$ Ventilated	284×10^{-5}	428×10^{-5}	567×10^{-5}	672×10^{-5}	886×10^{-5}	1060×10^{-5}
$K_2 (t_b - t_a)$ Unventilated	137×10^{-5}	232×10^{-5}	325×10^{-5}	417×10^{-5}	502×10^{-5}	600×10^{-5}

Heat dissipated is also given by,

$$H_d = ld\left[(t_b - t_a) + 18\right]^2 / K_3 \text{ J/s} \quad \text{...(21.42)}$$

K_3 = 2.6223×10^6 for well ventilated bearings

= 4.6513×10^6 for unventilated bearings

21.6.13 Volume Flow Rate of Coolant

Heat carried away by lubricating oil,

$$H_c = \rho Q C_p (t_0 - t_i)$$

Volume flow required, ...(21.43)

$$Q = \frac{(H_g - H_d) \times 10^3}{\rho C_p (t_0 - t_i)} m^3/s$$

ρ = density of coolant, kg/m^3 ...(21.44)

= 860 kg/m^3 for most of the oils

C_p = thermal capacity of coolant, J/kg °C

= 1.86 J/kg°C for most of the oils

t_i, t_0 = inlet and outlet temperature of coolant.

21.7 Design Considerations

The variables used in the design of hydrodynamic journal bearing are :

1. Design variables—These are those variables whose values are either given or are under the control of the designer. The design variables are :

(*i*) Viscosity, μ

(*ii*) Unit bearing pressure, p

(*iii*) Journal speed, n

(*iv*) Bearing dimensions, l, d, c.

2. Performance variables—These are those variables which control the performance of the bearing. The performance variables are :

(*i*) Coefficient of friction, f

(*ii*) Temperature rise, ΔT.

(*iii*) Lubricant flow rate, Q

(*iv*) Minimum oil film thickness, h_{min}

21.7.1 Selection of Design Variables

The design variables may be selected as described below :

1. Length of diameter ratio, l/d

Long bearings ($l/d > 1$) reduce the side leakage and require less oil flow rate. They are used where boundary lubrication is present. They have high load carrying capacity and low coefficient of friction. However, as the oil flow rate is less in such bearings, the temperature rise is more. They are more susceptible to metal – to – metal contact at the edges when the shaft is deflected under the load.

The short bearings ($l/d < 1$) result in greater flow of oil out of the ends, thus keeping the bearing cool. However, they have high coefficient of friction and low load carrying capacity. In practice, l/d ratio varies from 0.5 – 2.0. It is advisable to keep $l/d \approx 1.0$.

2. Unit Bearing Pressure, p

The permissible value of unit bearing pressure depends upon the bearing meterial, the nature of load, and the operating temperature. The permissible values of unit pressure for different applications are given in Table 21.15 and for different bearing meterials in Table 21.16.

Table 21.15 Permissible unit bearing pressure for different applications.

Application	Unit bearing pressure, *p*, MP*a*
(*a*) Uniform loads :	
1. Electric motors	0.8 – 1.5
2. Steam turbines	0.8 – 1.5
3. Gear reducers	0.8 – 1.5
4. Centrifugal pumps	0.6 – 1.2
(*b*) Moderate shock loads :	
1. Air compressors	
• Main bearings	1 – 2
• Connecting rod bearings	2 – 4
(*c*) Heavy shock loads :	
1. Diesel engines	
• Main bearings	6 – 12
• Connecting rod bearings	8 – 15
2. Automotive gasoline engines	
• Main bearings	4 – 5
• Connecting rod bearigs	10 – 15

Table 21.16 Permissible unit bearing pressure for different bearing materials.

Bearing materials	Unit bearing pressure, p, MPa
1. Lead base Babbit	4.2 – 5.6
2. Tin base Babbit	5.6 – 7.0
3. Copper – Lead (55 : 45)	14.0 – 21.0
4. Copper – Lead – Tin (72:25:3)	21.0 – 28.0
5. Silver	35.0
6. Bronze	70.0

3. Radial clearance ratio, *c*/*r*

The radial clearance should be small to provide the necessary velocity gradient. However, this requires fine surface finish, clean lubricating oil and rigid bearing mounting. This increases the initial and maintenance cost of the bearing. Increase in radial clearance increases the oil flow rate and decreases the load carrying capacity. The permissible values for radial clearance ratio are given in Table 21.17.

Table 21.17 Radial clearance for bearings.

Bearings	Material	Radial clearance ratio, c/r
1. Precision bearings	Babits	0.0010 – 0.0015
2. General machinery bearings	Copper–Lead	0.0015 – 0.0025
3. Rough service bearings	Aluminium alloy	0.003 – 0.004

4. Minimum oil film thickness, h_{min}

The minimum oil film thickness depends upon surface finish, viscosity of lubricating oil, and nature of load. The lower limit on the minimum oil film thickness to avoid metal–to–metal contact is given by,

$$h_{min} = (0,0002 - 0.0003)\ r \qquad ....(21.45)$$

The optimum values of h_{min}/c are given in Table 21.18.

Table 21.18 Optimum values of h_{min}/c for full journal bearing

$\frac{l}{d}$	h_{min}/c for maximum load	h_{min}/c for minimum friction or minimum power loss
∞	0.66	0.6
1	0.53	0.3
0.5	0.43	0.12
0.25	0.27	0.03

21.8 LUBRICATION AND LUBRICANTS

21.8.1 Lubrication

Lubrication is the science of reducing friction by the application of a lubricant between two rubbing surfaces having relative motion. There are two types of lubrication systems used for bearings.

1. Thin film lubrication—Thin film lubrication is defined as a condition of lubrication where the lubricant film is relatively thin and there is partial metal–to–metal contact. This mode of lubrication is seen in door hinges and machine tool slides. It is also called boundary lubrication. The conditions resulting in thin film lubrication are excessive load, insufficient surface area or oil supply, low speed and misalignment.

2. Thick film lubrication. When the bearing and the journal is completely separated from each other by the lubricant film, then it is called thick film lubrication. Since there is no contact between the surfaces, the properties of surface, like surface finish, have little or no influence on the performance of the bearing. The resistance to relative motion between the journal and the bearing arises from the viscous resistance of the lubricant. Therefore, the performance of the bearing is affected by the viscosity of the lubricant. Thick film lubrication could be hydrodynamic or hydrostatic. Thick film lubrication is provided in steam turbines, centrifugal compressors, crank shaft bearings, etc.

21.8.2 Functions of a Lubricant

The main functions of a lubricant are :

1. To reduce power loss and wear between rubbing surfaces.
2. To carry away the heat generated due to frictional work converted.
3. To provide protection against corrosion.

1.8.3 Properties of a Lubricant

The desirable properties of a lubricant are :

1. It should be available in a wide range of viscosities.
2. There should be little change in viscosity of the oil with change in temperature.
3. The oil should be chemically stable with bearing material and atmosphere at all temperatures encountered in its application.
4. The oil should have sufficient specific heat to carry away frictional heat without abnormal rise in temperature.
5. It should be commercially available at a reasonable cost.

21.8.4 Lubricants

Both mineral and synthesis oils are used as a bearing lubricant. Mineral oils are basically hydrocarbons, which may be classified as follows :

Paraffins—They contains significant amounts of waxy hydrocarbons but little or no asphaltic matter.

Naphatheline—It contains asphaltic matter in least volatile fractions, but little or no wax.

Mixed base—It contains both waxy and asphaltic materials.

Plain mineral oils are used in many units and systems for the lubrication of bearings where their oxidation ability, operating temperature range, ability to prevent wear, etc. are adequate. Special chemicals or additives are added to many oils to improve their properties.

Synthesis oils, most commonly used are : di-ester, silicones, polyglycol, chlorinated diphenyl, polyphenyl ether, fluorocarbon and silicate ester, etc.

A grease may be defined as a solid to semi-fluid lubricant consisting of a dispersion of a thickening agent in a lubricating fluid. The thickening agent may consist of a soap, a clay or a dyestuff. The lubricating fluid is usually a mineral oil, a diester or a silicone. They should be used for operating speeds below 1 to 2 m/s.

The solid fluids are used when fluids are undesirable and ineffective. The solid lubricants may be classified as follows :

Lamellar solids : MoS_2, WS_2, Graphite, TaS_2 and CaF_2.

Polymers : FEP, PTFCE,PTFE, Nylons, Acetals and Polyurethanes

Soft metals : Pb, Au, Ag, Sn, In

Oxides : MoO_3, PbO/SiO_2, B_2O_3/PbS

Salts : Na_2WO_4, $PbMoO_4$

21.8.5 Lubrication Methods

The various methods of lubrication methods are given in Table 21.19

Table 21.19 Methods of liquid lubricant supply for plain bearings.

Method of supply	Main characteristics	Applications
1. Hand oiling	Non-automatic, irregular, low initial cost, high maintenance cost.	Low speed, cheap journal bearings.
2. Drip and wick feed	Non-automatic, adjustable, moderately efficient, cheap	Journals in some machine tools, axles
3. Ring and collar feed	Automatic, reliable, efficient, fairly cheap, mainly horizontal bearings.	Journals in pumps, blowers, large electric motors
4. Bath and splash Lubrication	Automatic, reliable efficient, oil–tight housing required, high initial cost.	Thrust bearings, bath only, engines process machinery, general appliances.
5. Pressure fed	Automatic, positive and adjustable, reliable and efficient, high initial cost.	High speed and heavily loaded journal and thrust bearings in machine tools, engines and compressors.

21.9 BEARING MATERIALS

21.9.1 Characteristics of a good Journal Bearing Material

The desirable properties of a good journal bearing material are :

1. Compressive strength-The bearing material should have high compressive strength to withstand the maximum pressure in order to prevent undue deformation of the bearing.
2. Fatigue strength-It should have sufficient fatigue strength to withstand repeated loads without developing surface cracks.
3. Comformability–It should accommodate shaft deflections and bearing inaccuracies without excessive wear and heating.

4. Structure-It should have duplex structure with hard particles supported in a soft matrix.
5. Embeddability–It should accommodate small particles of dust, grit, etc. without scoring the journal.
6. Bondability–It should easily make bonds with other materials.
7. Corrosion resistance–It should not corrode under the action of the lubricating oil.
8. Thermal conductivity–It should have high thermal conductivity to dissipate generated heat.
9. Thermal expansion–It should have low coefficient of thermal expansion so that there is no undue change in the clearance during operation under heavy loads.
10. Cost and availabiliy–It should be easily available at a reasonable cost.

21.9.2 Bearing Materials

The commonly used materials for sliding contact bearings are ·

1. Babbits (white metals)
   (i) Lead–Base Babbits
   (ii) Tin–Base Babbits
2. Bronze
   (i) Gun Metal
   (ii) Phosphor Bronze
3. Copper–Lead alloy
4. Aluminium alloy
5. Silver
6. Cast iron
7. Sintered metal (porous metals)
8. Non-metallic materials.
   (i) Rubber
   (ii) Plastics
   (iii) Carbon–graphite
   (iv) Teflon
   (v) Ceramics and cermets
   (vi) Glass
   (vii) Synthetic sapphires and jewels.

The journal bearing materials, their characteristics, limitations and applications are given in Table 21.20.

Table 21.20 Journal bearing materials.

Material	Characteristics	Limitations	Applications
1. Wood (maple lignum, vital)	Self-lubricating, low cost, long life	Lighter loads at high speeds, under 65°C	Conveyors
2. Cast iron	Low friction, low cost	Not over 3.4 MPa and 40 m/min speed	Refrigerator, compressors, cam-shafts, light transmission
3. Steel	Low cost	Light loads, upto 45 m/min speeds	Guides
4. Bronze bushing	Low cost, simple construction	Loads upto 20 Mpa. Speeds upto 270 m/min	All equipment
5. Heavy babbit liner on steel or cast iron	Long life, low friction, good lubrication	Steady loads under 6.7 MPa	Motor, turbine shafting
6. Light liner on steel or bronze backing	Heavy duty, general purpose, good for dynamic loads	About 3.4 MPa loads and speeds upto 1200 m / min	Gas and diesel engines, compressors
7. Rubber	Low friction, resists abrasion, shock absorbent, long life	About 0.5 MPa, needs water lubrication	Marine propellers, pumps and turbines
8. Carbon graphite needed, light	No lubrication needed, light duty applications	Under 455° C, 3.2 MPa at low speeds	Electric motors, meters, conveyors
9. Moulded plastic	Low friction, clean	Low loads at high speeds machinery	Dairy, textile and food
10. Sintered bearing, steel bronze	Self lubricating	Low loads, high speeds	All inaccessible places, typewrite electric fans, instruments, piston rings.

21.10 Bearing Housing

The bearing is supported in a housing for its proper support and mounting. The different types of journal bearings used are :

1. Solid journal bearing
2. Bushed bearing
3. Split bearing or Plummer block.

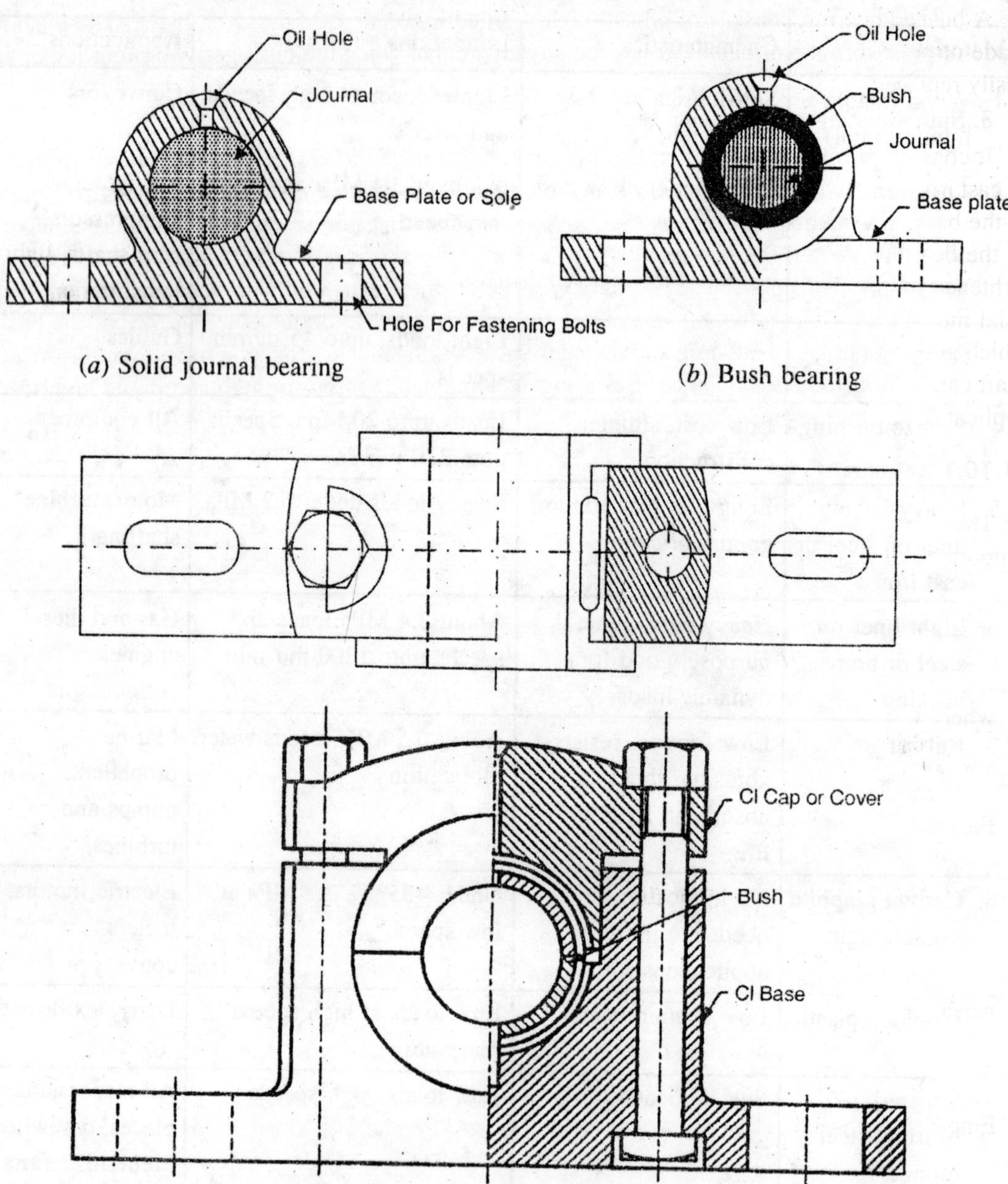

(*a*) Solid journal bearing

(*b*) Bush bearing

(*c*) Split bearng or Plummer block

Fig. 21.10 Types of journal bearings

The various types of bearings are shown in Fig. 21.10

1. Solid Journal Bearing

It consists of a cast iron block with hole for the journal. The lower portion of the block is extended to form a base plate which is fixed to the frame or foundation with the help of bolts. An oil hole is provided at the top for lubrication purpose. This type of bearing is used for low speed and light load applications. It has simple construction and is of low cost. No provision is made for adjustment of the bearing in case of wear. The shaft has to be inserted axially into the bearing.

2. Bushed Bearing

A bushed bearing consists of a bush press - fitted in a hole of a cast iron block. The bush is made of brass, bronze or babbit. It has the advantage that when the bush gets worn out it can be easily replaced.

3. Split Bearing or Plummer Block

It consists of a base made of cast iron, a split bush made of bearing material and a cap made of cast iron. The two halves of the bush are held together by tightening a cap with the help of bolts to the base. The adjustment for wear is provided by inserting thin shims between the two halves of the bearings. When the bottom wears out, one of the shims is removed and again the cap is tightened by means of bolts. The two halves of the bush are provided with collar to prevent their axial movement. Their rotation is prevented by providing snugs on the outer surface of the bush which goes into holes in the base and the cap. Such bearings have provision for wear adjustment, shaft can be inserted perpendicular to bearing axis and can be used for high speed and high load applications. However their cost is high.

21.10.1 *Bearing Cap*

The cap is designed by assuming it to be a simply supported beam loaded at the centre and supported at the bolt centres. The thickness of the bearing cap, shown in Fig. 21.11, is given by,

$$t = \left[\frac{3WL}{2\sigma l}\right]^{\frac{1}{2}} \qquad (21.46)$$

where L = distance between bolt centres

l = bearing length – oil hole diameter

σ = permissible bending stress for the cap material

Based upon deflection, the cap thickness is given by,

$$t = \left[\frac{WL^3}{4EI\delta}\right]^{\frac{1}{3}} \qquad ...(21.46)$$

Fig. 21.11 Bearing cap.

where I = moment of inertia of cap cross – section

δ = deflection of cap ≈ 0.0254 mm for medium size cap

= 0.0127 mm for small size cap

Empirically, $t = 0.08d + 2.5$ mm for small size brass and bronze bearing caps ... (21.48a)

$= 0.2d$ mm for medium size C.I. bearing caps ...(21.48b)

$= 0.202d + 3.2$ mm for babbit bearing cap ...(21.48c)

E = Young's modulus of elasticity of cap material.

21.10.2 *Oil Grooves and Rings*

The oil grooves are provided in the bearing to ensure supply of lubricating oil in the bearing. Three types of oil grooves provided in the bearing are shown in Fig. 21.12.

1. Circumferential oil groove. This divides the bearing into two half-length bearings. It reduces the load carrying capacity of the bearing. Such bearings are used for connecting rod ends and crankshaft of automobile engines.

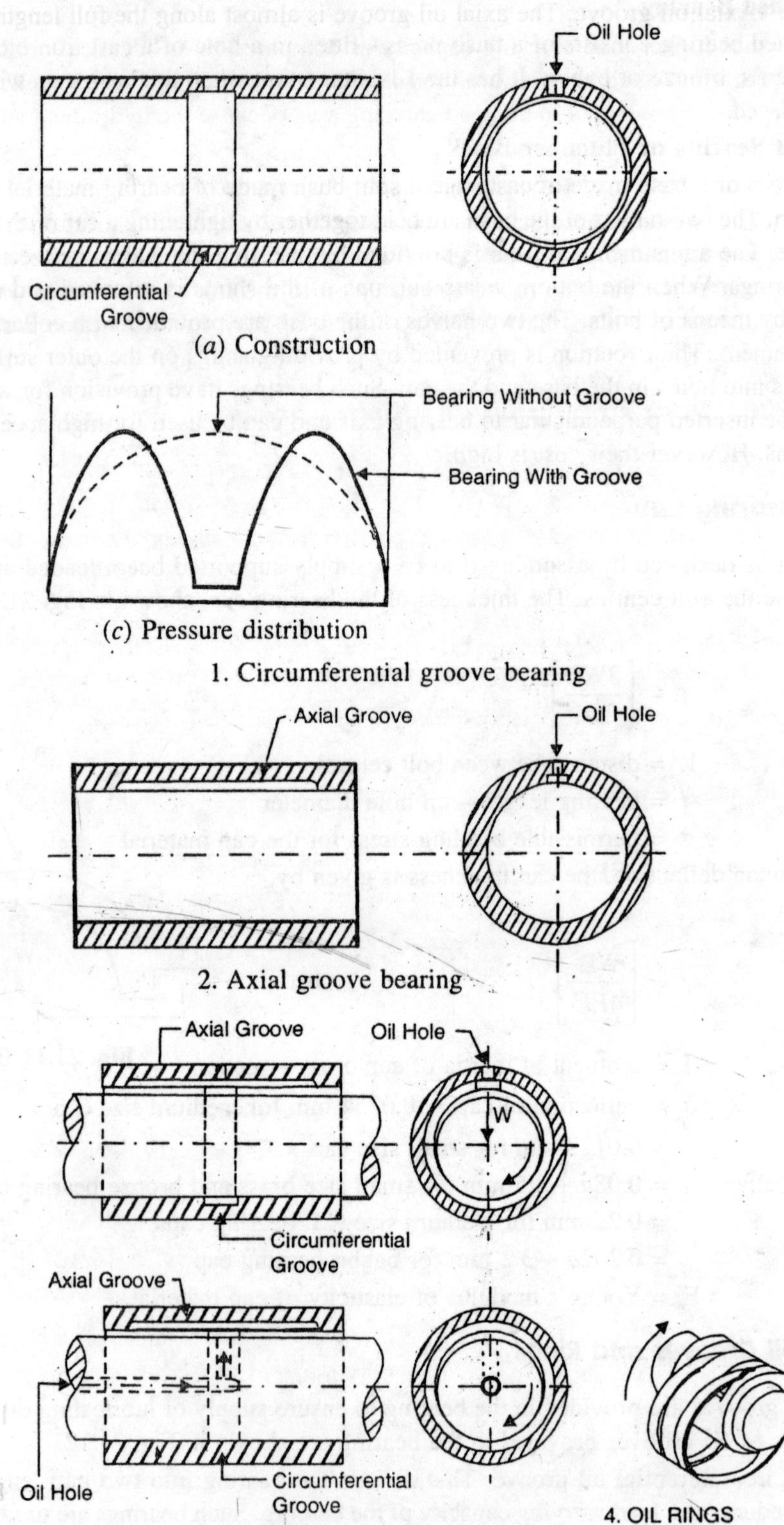

(*a*) Construction

(*c*) Pressure distribution

1. Circumferential groove bearing

2. Axial groove bearing

3. Combination of grooves.

4. OIL RINGS

2. Axial oil groove. The axial oil groove is almost along the full length of the bearing. It has higher load carrying capacity as compared to an identical circumferential groove bearing. Such grooves are used in gear boxes and high speed applications. Oil groove should not be cut in the load carrying area because the hydrodynamic pressure drops to nearly zero at the groove.
3. Combination of circumferential and axial oil grooves.
4. Oil rings are approximately twice the shaft diameter which rest on the upper side of the journal and dipping into an oil reservoir below the bearing. Rotation of the journal causes the rings to rotate and carry oil to the top of the journal where axial grooves distribute it. They have the disadvantage as they do not function at slow and very high speeds.

21.11 PROCEDURE FOR DESIGN OF JOURNAL BEARING

1. For the given machinery, select l/d ratio for journal, unit pressure p, radial clearance ratio c / r, coefficient of friction, dynamic viscosity and bearing modulus.
2. For the load to be carried by the bearing, find length and diameter of journal for the unit pressure selected above.
3. Calculate the sommerfeld number.
4. Check the viscosity and unit pressure.
5. Calculate the minimum oil film thickness.
6. Calculate the coefficient of friction.
7. Calculate heat generated.
8. Calculate heat dissipated by assuming reasonable lubricant film temperature.
9. Calculate the coolant rate for a feasible temperature rise.

Example 21.1 A journal bearing has a diameter and length of 80 mm and 120 mm respectively to support a radial load of 15 kN. The shaft revolves at 900 rpm. The bearing surface temperature is not to exceed 60° C where the room temperature is 30°C. The radial clearance ratio is 0.001. The viscosity of lubricating oil is 10 Pa.s at the operating temperature. Calculate the amount of artificial cooling required.

Solution :

Given : W =15 kN, l = 120 mm, d = 80 mm, N = 900 rpm, t_b = 60°C, t_a = 30°C,

$$\mu = 10 \text{ Pa.s}, \quad \frac{c}{r} = 0.001$$

Unit pressure, $$p = \frac{W}{ld} = \frac{15 \times 10^3}{120 \times 80} = 1.5625 \text{ N/mm}^2$$

Coefficient of friction, $$f = \left(\frac{0.326}{10^6}\right)\left[\frac{\mu N}{60p}\right]\left(\frac{r}{c}\right) + \Delta f$$

For $l/d = \frac{120}{80} = 1.5$ $\Delta f = 0.0025$ (Table 21.4)

$$f = \left(\frac{0.326}{10^6}\right)\left[\frac{10 \times 900}{60 \times 1.5625}\right]\left(\frac{1}{0.001}\right) + 0.0025$$

$$= 0.00563$$

Rubbing speed, $U = \dfrac{\pi d N}{60 \times 10^3} = \dfrac{\pi \times 80 \times 900}{60 \times 10^3} = 3.77 \text{ m / s}$

Heat generated, $H_g = f\,WU = 0.00563 \times 15 \times 10^3 \times 3.77 = 318.37$ Watts

Heat dissipated, $H_d = K_1 \Delta t \text{ (ld)}$

Now $\Delta t = 60 - 30 = 30°C$

For $\Delta t = 30°C$, $K_1 \Delta t = 0.000443 \text{ W / mm}^2$ for ventilated bearing (Table 21.7)

$H_d = 0.000443 \times 120 \times 80 = 80.05$ Watts

Artificial cooling required, $H_c = H_g - H_d = 318.37 - 80.05 = 238.32$ Watts

Example 21.2 The following data is given for a full journal bearing :

Radial load = 35 kN, diameter of journal = length of journal = 80 mm, speed of journal = 3000 rpm, radial clearance = 0.15 mm, inlet temperature = 40° C, specific heat of lubricant = 1760 J / kg.°C, oil used SAE – 30 of specific gravity at 15°C = 0.926.

Calculate the power lost in friction and requirement of oil flow.

Solution.

Given : W = 35 kN, $l = d = 80$ mm, N = 3000 rpm, c = 0.15 mm, $t_i = 40°C$

$cC_p = 1760$ J / kg. °C, specific gravity of oil at 15°C = 0.926

Let film temperature be, $t_f = 60°C$

Density at 15°C = 0.926 × 1000 = 926 kg / m³

Density of oil at 60°C, $\rho = \rho_{15} - 0.63\,(t_f - 15)$

$= 926 - 0.63\,(60 - 15) = 897.65 \text{ kg / m}^3$

From Fig. 21.4, dynamic viscosity, $\mu = 29$ mPa.s

Unit bearing pressure, $p = \dfrac{W}{ld} = \dfrac{35 \times 10^3}{80 \times 80} = 21.47 \text{ N / mm}^2$

Sommerfeld number, $S = \left[\dfrac{\mu N}{60 p}\right]\left(\dfrac{r}{c}\right) \times 10^{-6}$

$$= \left[\frac{29 \times 10^{-3} \times 3000}{60 \times 21.47}\right]\left(\frac{40}{0.15}\right)^2 \times 10^{-6}$$

$$= 0.019$$

Coefficient of friction, $f = \left(\frac{0.326}{10^6}\right)\left[\frac{\mu N}{60p}\right]\left(\frac{r}{c}\right) + \Delta f$

For $l/d = \frac{80}{80} = 1.5,\ \Delta f = 0.0020$ (Table 21.5)

$$f = \left(\frac{0.326}{10^6}\right)\left[\frac{29 \times 10^{-3} \times 3000}{60 \times 21.47}\right]\left(\frac{40}{0.15}\right) + 0.0020$$

$$= 0.002024$$

Rubbing speed, $U = \frac{\pi d N}{60 \times 10^3} = \frac{\pi \times 80 \times 3000}{60 \times 10^3} = 12.566$ m/s

Heat generated, $H_g = f W U = 0.002024 \times 35 \times 10^3 \times 12.566 = 890.25$ Watts

Heat dissipated, $H_d = \rho Q C_p . \Delta t$

$$Q = \frac{890.25 \times 60}{897.65 \times 1760 \times (60 - 40)}$$

$$= 1.69 \times 10^{-3} \text{ m}^3/\text{min}$$

Example 21.3 Design a bearing and journal for the main bearing of a four – stroke gas engine to support a radial load of 4.5 kN when the shaft is revolving at 600 rpm. Assume suitable data.

Solution.

Given : W = 4.5 kN, N = 600 rpm

Let the oil film temperature; $t_f = 70°C$

For the main bearing of a four – stroke gas engine, the range of dynamic viscosity,

$$\mu = 0.020 \text{ to } 0.065 \text{ Pa.s}$$

Let us assume, $\mu = 0.03$ Pa.s.

From Fig. 21.4, we find that SAE 40 at 70° C gives,

$$\mu \approx 0.03 \text{ Pa.s}$$

Specific gravity of SAE 40 oil at 15.6°C = 0.927 (Table 21.2)

Density, $\rho = 0.927 \times 1000 = 927$ kg / m^3

Now $\rho_{70} = \rho_{15.6} - 0.63\,(t_f - 15.6)$

$= 927 - 0.63\,(70 - 15.6) = 892.73$ kg / m^3

Bearing modulus for the main bearing of gas engine is,

$$C = \left(\frac{\mu N}{p}\right)_{min} = 2.90.$$

Clearance ratio, $\frac{c}{r}$ = 0.001

$$\frac{l}{d} \text{ ratio} = 0.6 - 2.0$$

Design bearing modulus = 3C = 3 × 2.90 = 8.70

$$\rho = \frac{\mu N}{C} = \frac{0.03 \times 600}{8.7} = 2.07 \text{ N / mm}^2$$

Projected area of journal, $A = \frac{W}{p} = \frac{4.5 \times 10^3}{2.07} = 2174 \text{ mm}^2$

Let l / d = 1.25

$$1.25\, d^2 = 2174$$
$$d = 41.7 \approx 42 \text{ mm}$$
$$l = 52.5 \text{ mm}$$

Rubbing speed, $U = \frac{\pi d N}{60 \times 10^3} = \frac{\pi \times 42}{60 \times 10^3} = 1.32 \text{ m / s}$

Operating pressure, $p = 0.726\, U^{1/2}$

$$= 0.726 \times (1.32)^{1/2} = 0.834 \text{ N / mm}^2.$$

The actual pressure is more than the operating pressure. Hence, the bearing shall function satisfactorily.

21.12 RAIMONDI AND BOYD METHOD

There is no exact solution to Reynold's equation for a journal bearing of finite length. In the Raimondi and Boyd method, the performance of the journal bearing is expressed in terms of dimensionless parameters. The geometry of the bearing is shown in Fig. 21.13.

h_o = minimum film thickness, mm

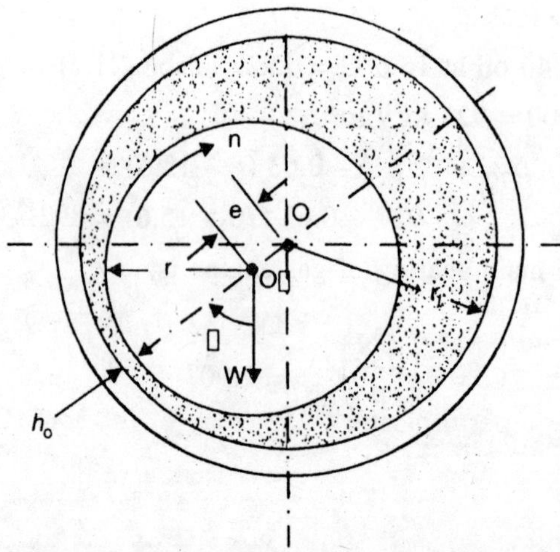

Fig. 21.13 Bearing geometry for Raimondi and Boyd method

$$\varepsilon = \frac{e}{c} = 1 - \frac{h_0}{c}, \text{ eccentricity ratio}$$

$$\bar{h} = \frac{h_0}{c} = \text{minimum film thickness variable}$$

$$\frac{l}{d} = \text{length to diameter ratio}$$

$$S = \left(\frac{\mu n}{p}\right)\left(\frac{r}{c}\right)^2, \text{ Sommerfeld number}$$

$$\bar{f} = \left(\frac{r}{c}\right) f = \text{coefficient of friction variable (CFV)}$$

$$P_f = \frac{fWU}{10^3}, \text{ frictional power, kW}$$

$$U = \frac{\pi dN}{60 \times 10^3}, \text{ journall tangential speed, } m/s$$

$$\frac{p}{p_{max}} = \text{maximum film pressure ratio}$$

$$\bar{Q} = \frac{Q}{rcnl}, \text{ flow variable (FV)}$$

$$Q = \text{flow of lubricant } mm^3/s$$

$$Q_s = \text{side leakage of lubricant, } mm^3/s$$

$$\frac{Q_s}{Q} = \text{flow ratio}$$

$$\frac{p}{p_{max}} = \text{maximum film pressure ratio}$$

$$\frac{\rho C_p \Delta t}{p} = \text{temperature rise variable}$$

The dimensionless static performance parameters for full journal bearing with side flow are given in Table 21.21.

Table 21.21 Dimensionless static performance parameter for full journal bearing with side flow

$\frac{l}{d}$	ε	$\bar{h}=\frac{h_0}{c}$	s	ϕ	$\bar{f}=f\left(\frac{r}{c}\right)$	$\bar{Q}=\frac{Q}{rcnl}$	$\frac{Q_s}{Q}$	$\frac{\rho c_p \Delta t}{H_g}$	$\frac{p}{p_{max}}$
∞	0	1.0	∞	70.92	∞	π	0	∞	-
	0.1	0.9	0.240	69.10	4.80	3.03	0	19.9	0.826
	0.2	0.8	0.123	67.26	2.57	2.83	0	11.4	0.814
	0.4	0.6	0.0626	61.94	1.52	2.26	0	8.47	0.764
	0.6	0.4	0.0389	54.31	1.20	1.56	0	9.73	0.667
	0.8	0.2	0.021	42.22	0.961	0.760	0	15.9	0.495
	0.9	0.1	0.115	31.63	0.756	0.411	0	23.1	0.358
	0.97	0.03			-	-	-	0	- -
	1.0	0	0	0	0	0	0	0	
1	0	1.0	∞	85	∞	π	0	∞	-
	0.1	0.9	1.33	79.5	26.4	3.37	0.150	106	0.540
	0.2	0.8	0.631	74.02	12.8	3.59	0.280	52.1	0.529
	0.4	0.6	0.264	63.10	5.79	3.99	0.497	24.3	0.484
	0.6	0.4	0.121	50.58	3.22	4.33	0.680	14.2	0.415
	0.8	0.2	0.0446	63.24	1.70	4.62	0.842	8.00	0.313
	0.9	0.1	0.0188	26.45	1.05	4.74	0.919	5.16	0.247
	0.97	0.03	0.00474	15.47	0.514	4.82	0.973	2.61	0.152
	1.0	0	0	0	0	–	1.0	0	0
0.5	0	1.0	∞	88.5	∞	π	0	∞	-
	.01	0.9	4.31	81.62	85.6	3.43	0.173	343.0	0.523
	0.2	0.8	2.03	74.94	40.9	3.72	0.318	164.0	0.506
	0.4	0.6	0.779	61.45	17.0	4.29	0.552	68.6	0.441
	0.6	0.4	0.319	48.14	8.10	4.85	0.730	33.0	0.365
	0.8	0.2	0.0923	33.31	3.26	5.41	0.874	13.4	0.267
	0.9	0.1	0.0313	23.66	1.60	5.69	0.939	6.66	0.206
	0.97	0.03	0.00609	13,75	0.610	5.88	0.980	2.56	0.126
	1.0	0	0	0	0	–	-	1.0	0
0.25	0	1.0	∞	89.5	∞	π	0	∞	-
	0.1	0.9	16.2	82.31	322	3.45	0.180	1287.0	0.515
	0.2	0.8	7.57	75.18	153.0	3.76	0.330	611.0	0.489
	0.4	0.6	2.83	60.86	61.1	4.37	0.567	245.0	0.415
	0.6	0.4	1.07	46.72	26.7	4.99	0.746	107.0	0.334
	0.8	0.2	0.261	31.04	8.80	5.60	0.884	34.4	0.240
	0.9	0.1	0.0736	21.85	3.50	5.91	0.945	14.1	0.180
	0.97	0.03	0.0101	12.22	0.922	6.12	0.984	3.73	0.108
	1.0	0	0	0	0	-	1.0	0	0

Example 21.4 A full journal bearing operating under a steady load has the following specifications :

Journal diameter = 60 mm, bearing length = 60 mm, radial load = 2.8 kN, jounral speed = 1020 rpm, radial clearance = 0.05 mm, oil viscosity = 80×10^{-9} N.s / mm^2, oil density = 860 kg / m^3, specific heat of oil = 1760 J / kg.°C.

Using Raimondi and Boyd method, determine (*a*) Sommerfeld number, (*b*) power loss in friction, (*c*) temperature rise if heat generated is entirely carried by oil, and (*d*) minimum film thickness, and its location.

Solution.

Given : $W = 2.8$ kN, $l = d = 60$ mm, $n = 1020$ rpm, $c = 0.05$ mm, $\mu = 80 \times 10^{-9}$ N.s/mm^2, $\rho = 860$ kg / m^3, $C_p = 1760$ J / kg. °C.

Pressure, $$p = \frac{W}{ld} = \frac{2800}{60 \times 60} = 0.778 \text{ N/mm}$$

Sommerfeld number, $$S = \left(\frac{\mu n}{p}\right)\left(\frac{r}{c}\right)^2$$

$$= \frac{80 \times 10^{-9} \times 1020}{[60 \times 0.778]} = 0,63$$

Now $$\frac{l}{d} = \frac{60}{60} = 1 \text{ and } S = 0.63$$

Coefficient of friction variable, $$\bar{f} = f\left(\frac{r}{c}\right) = 12.8$$

Coefficient of friction, $$f = \frac{12.8 \times 0.05}{30} = 0.0213$$

Minimum film thickness variable, $$\bar{h} = \frac{h_0}{c} = 0.8$$

Minimum film thickness, $$h_0 = 0.05 \times 0.8 = 0.04 \text{ mm}$$

Flow variable, $$Q = \frac{Q}{rcnl} = 3.59$$

$$Q = \frac{3.59 \times 30 \times 0.05 \times 1020 \times 60}{60} = 5492.7 \text{ mm}^3/\text{s}$$

Frictional power, $$P_f = (2\pi n) . \frac{(fWr)}{10^6}$$

$$= \frac{2\pi \times 1020 \times 0.0213 \times 2800 \times 30 \times 10^{-3}}{60 \times 10^3} = 0.1911 \text{ kW}$$

Heat generated, $H_g = 0.1911$ kJ / s

Heat carried away by oil, $H_c = \rho Q C_p . \Delta t$.

$= 860 \times 5492.7 \times 10^{-9} \times 1760 \times \Delta t = 8.3137 \times \Delta t$

For heat balance, $H_c = H_g$

$8.3137 \times \Delta t = 191.1$

$\Delta t = 23°C$

Example 21.5 The following data refer to a full hydrodynamic journal bearing :

Radial load = 4.5 kN

Journal diameter = 60 mm

Bearing length = 60 mm

Journal speed = 1440 rpm

Radial clearance = 60 microns

Lubricant viscosity = 25 mPa.s

Lubricant density = 860 kg/m^3

Lubricant specific heat = 1.76 kJ / kg. °C

Assuming that total heat generaxed in the bearing is carried by the total oil flow in the bearing, calculate (*a*) minimum oil film thickness (*b*) coefficient of friction (*c*) power lost in friction (*d*) total lubricant flow rate (*e*) side leakage, and (*f*) temperature rise.

Solution.

Given : W = 4.5 kN, d = 60 mm, l = 60 mm, N = 1440 rpm or n = 24 rps, c = 0.06 mm, μ = 25 mPa.s ρ = 860 kg / m^3, c_p = 1.76 kJ / kg. °C

Bearing pressure, $$p = \frac{W}{ld} = \frac{4500}{60 \times 60} = 1.25 \text{ N / mm}^2$$

Clearance ratio, $$\frac{c}{r} = \frac{0.06}{30} = 0.002$$

Sommerfeld number, $$S = \left(\frac{\mu n}{p}\right)\left(\frac{r}{c}\right)^2$$

$$= \left(25 \times 10^{-9} \times \frac{24}{1.25}\right)\left(\frac{1}{0.002}\right)^2 = 0.12$$

For $\frac{l}{d} = \frac{60}{60} = 1.0$ and S = 0.12, from Table 21.21, we have

l/d	S	h_{min}/c	(r/c)f	Q/(rcnl)	Q_s/Q
1.0	0.0446	0.2	1.70	4.62	0.842
0.121	0.4	3.22	4.33	0.680	

By linear interpolation, we have

$$\frac{h_{min}}{c} = 0.3974, \left(\frac{r}{c}\right) f = 3.20, \frac{Q}{rcnl} = 4.334, \frac{Q_s}{Q} = 0.682$$

(*a*) Minimum oil film thickness,

$$h_{min} = 0.06 \times 0.3974 = 0.0238 \text{ mm}$$

(*b*) Coefficient of friction, $$f = 3.20 \times \frac{0.06}{30} = 0.0064$$

(*c*) Power lost in friction, $$P_f = \frac{fWU}{10^3} kW$$

$$= 0.0064 \times 4500 \times \pi \times 60 \times \frac{24}{10^6} = 0.13 \text{ kW}$$

(*d*) Total lubricant flow rate, $Q = 4.334 \times 30 \times 0.06 \times 24 \times 60$

$= 11233.73 \text{ mm}^3/\text{ s or } 6.74 \times 10^{-4} \text{ m}^3 / \text{min or } 0.674\ lpm$

(*e*) Side leakage, $Q_s = 0.682 \times 0.674 = 0.4597\ l\text{ pm}$

(*f*)Temperature rise considering side leakage,

$$P_f = \rho Q C_p \Delta t\ [1 - 0.5\ Q_s / Q] / 10^9$$

$$0.13 = 860 \times 11233.73 \times 1.76 \times \Delta t \times [1 - 0.5 \times 0.682] / 10^9$$

$$\Delta t = 11.6°\text{C},$$

EXERCISES

21.1 Design a bearing and journal to support a load of 5 kN at 600 rpm using a hardened steel journal and a bronze – backed babbit bearing. An abudnance of oil is supplied by means of oiling rings. The oil viscosity is 250 SUS at 37.8° and the specific gravity is 0.90 at 15.6°C.

21.2 A pressure fed bearing with an annular central groove has a lengh of 50 mm and diameter of 50 mm. The bearing supports a load of 5 kN at a journal speed of 1200 rpm. The radial clearnace is 0.025 mm and the oil is SAE 30. Determine (*a*) the minimum oil film thickness, and (*b*) the equilibrium temperature.

21.3 A journal bearing for a steam turbine is required to carry 5 kN of load. The diameter of journal = 60 mm, length of journal = 75 mm, speed = 1500 rpm, radial clearance ratio = 0.001, ambient temperature = 15.5°C, lubricating oil = SAE 10, oil film temperature = 60°C. Determine the heat generated and heat dissipated.

21.4 A 10 kW steam turbine running at a speed of 1800 rpm has two bearings. The temperature of oil is 70°C when SAE 40 oil is used. Assuming suitable data, determine the dimensions of journal, bearing and cap.

21.5 The following data refer to a full hyrodynamic journal bearing :
Radial load = 10 kN, journal speed = 1500 rpm, bearing length = 60 mm, 1/d ratio = 1.0, radial clearance = 20 microns, eccentricity = 15 microns, specific gravity of lubricant = 0.86, specific heat of lubricant = 2.0 kJ / Kg.°C. Determine (*a*) minimum oil film thickness (*b*) coefficient of friction (*c*) power lost in friction (*d*) viscosity of lubricant in mPa.s (*e*) total flow rate of lubricant (*f*) side leakage, and (*g*) average temperature, if make up oil is suppllied at 25°C.

ROLLING CONTACT BEARINGS

Chapter

22. 1 INTRODUCTION

The bearings in which the contacting surfaces have rolling contact, are called rolling contact bearings. Due to low friction offered by rolling contact bearings they are known as anti — friction bearings. The coefficient of friction in rolling cantact bearing is about 0.001.

22.2 TYPES OF ROLLING CONTACT BEARINGS

The rolling contact bearings may be classified as follows:

1. Ball bearings
   - (*i*) Radial ball bearings
   - (*ii*) Angular contact ball bearings
   - (*iii*) Thrust ball bearings

   Various types of ball bearings are shown in Fig. 22.1
2. Roller bearings
   - (*i*) Cylindrical roller bearings
   - (*ii*) Needle roller bearings
   - (*iii*) Typer roller bearings
   - (*iv*) Spherical roller bearings

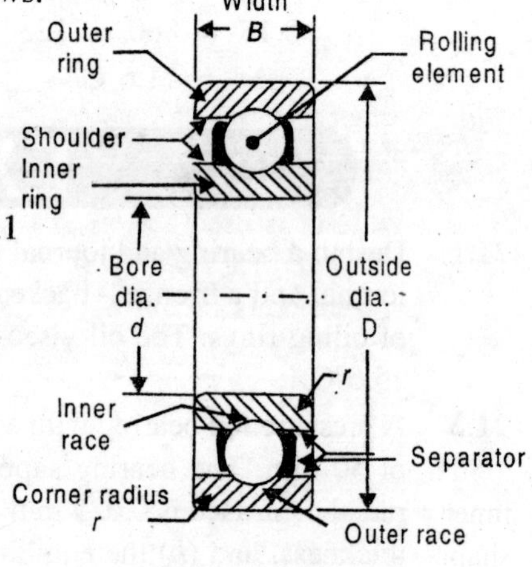

(a) Deep groove ball bearing

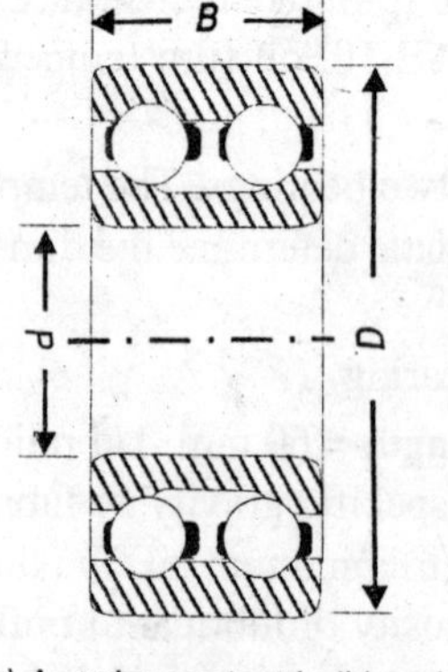

(c) Angular contact ball bearing (Double row)

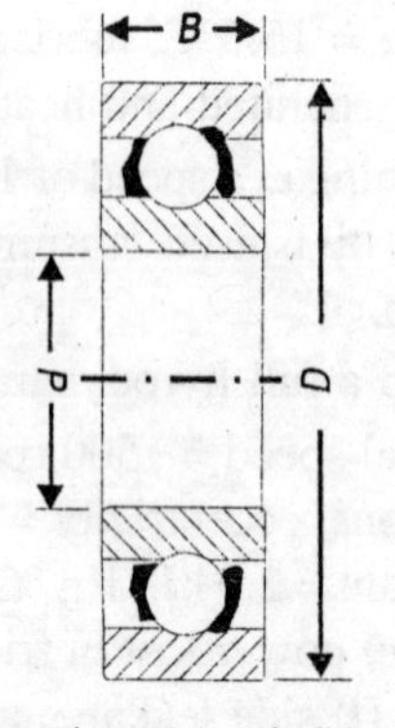

(b) Angular contact ball bearing (Single row)

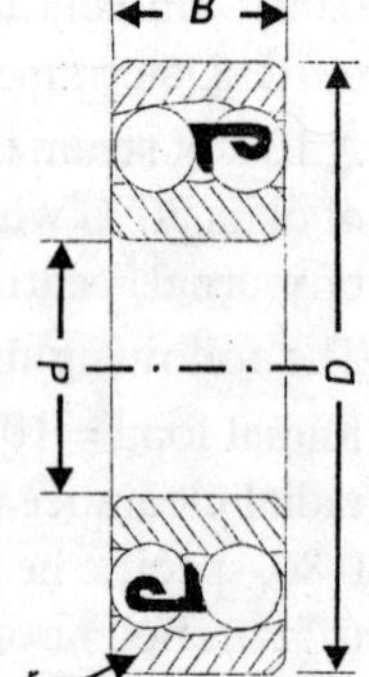

(d) Self aligning ball bearing

Fig. 22.1 Ball bearings.

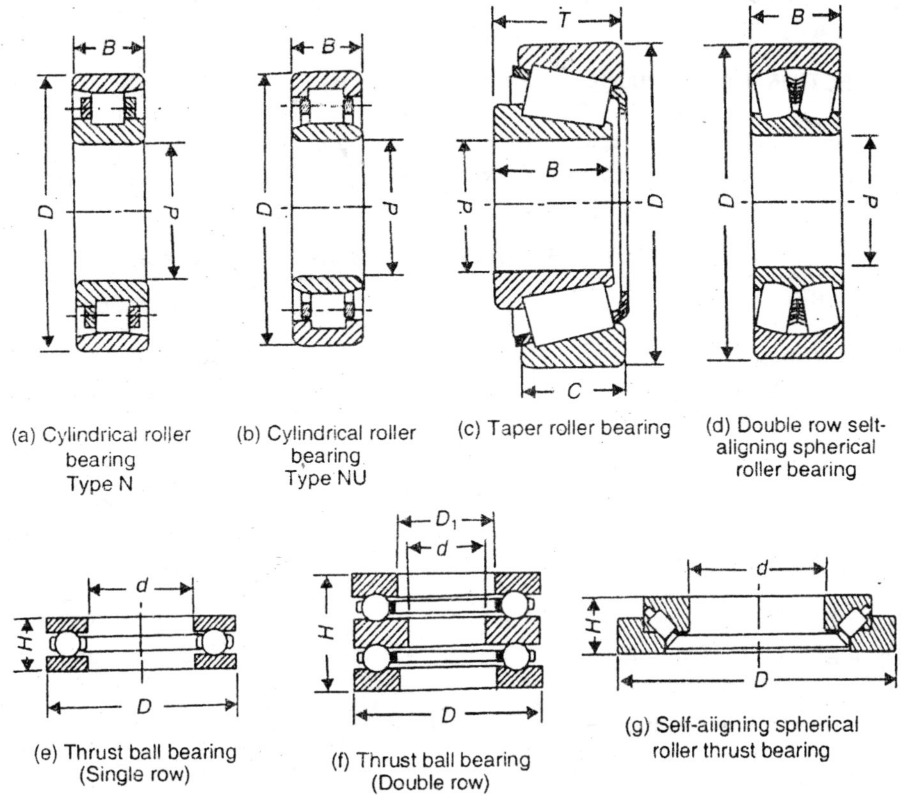

(a) Cylindrical roller bearing Type N

(b) Cylindrical roller bearing Type NU

(c) Taper roller bearing

(d) Double row selt-aligning spherical roller bearing

(e) Thrust ball bearing (Single row)

(f) Thrust ball bearing (Double row)

(g) Self-aiigning spherical roller thrust bearing

Fig. 22.2 Roller bearings.

Various types of roller bearings are shown in Fig. 22.2.

The most common type of ball bearing is the deep groove ball bearing. In this bearing, the inner race is placed eccentric with respect to the outer race and the balls are inserted in the crescent shaped space. There is a point contact between the balls and the races. They have a high radial load carrying capacity and moderate thrust load capacity.

Cylindrical roller bearing consists of short rollers, which are positioned and guided by a cage. The line contact between rollers and races increases the rigidity and load carrying capacity. They have a low coefficient of friction, high load carrying capacity and high rigidity.

In angular contact bearings, one side of the outer race is cut to insert the balls. This permits the bearing to take the thrust load in only one direction. They are mounted in pairs to take the thrust load in both directions.

The self-aligning ball bearings consists of two rows of balls, which roll on a common spherical surface in the outer race. In this case, the inner race can freely adjust itself to the angular misalignment of the shaft. They can compensate angular misaligment between the axes of shaft and housing.

Spherical roller bearing consists of two rows of spherical rollers, which run on a common spherical surface in the outer race. They can carry relatively high radial and thrust loads.

The taper roller bearing consists of rolling elements that are frustums of a conc. They are arranged in such a way that all their elements and and axes intersect at a common apex point on the axis of the bearing. They have high radial and thrust load carrying capacity. They are mainly used to take thrust loads.

Needle roller bearings are useful where radial space is limited. They have high radial load carrying capacity.

22.3 SELECTION OF BEARINGS

The selection of the type of bearing for a particular application depends upon the requirement of the situation and the characteristics of the bearing. The guide lines for selecting a proper type of bearing are given in Table 22.1.

Table 22.1 Selection guide for anti-friction bearings.

Bearing type	Applications
1. Ball bearings	Low and medium radial loads
2. Roller bearings	Heavy loads and larges shaft diameters
3. Self-aligning ball bearings and Spherical roller bearings	Where misalignement between the the axes of the shaft and housing is likely to exist.
4. Thrust ball bearings	Medium thrust loads
5. Cylindrical roller thrust bearings	Heavy thrust loads
6. Deep groove ball bearings, angular contact bearings, and spherical roller bearings	Combind radial and thrust loads
7. Deep groove ball bearings, angular contact bearings, andcylindrical roller bearings	High speed applications
8. Double row cylindrical roller bearings, taper roller bearings	Rigidity of machine tool spindles
9. Deep groove ball bearings	Household application and less noise

22.4 ADVANTAGES AND DISADVANTAGES OF ROLLING CONTACT BEARINGS

The advantages and disadvantages of rolling bearings as compared to journal bearings are given in Table 22.2

Table 22.2 Advantages and disadvantages of rolling as compared to journal bearings.

Advantages	Disadvantages
1. Low starting frictional torque.	Requires greater diametral space.
2. Ease of lubricatio..	Higher initial cost.
3. Occupies lesser space	Noisier in normal operation.
4. Can support radial and thrust loads.	Finite life due to eventual failure by fatigue
5. Gives early warning of impending failure.	Lesser capacity to withstand shocks.
6. Readily replaceable as stock items.	Higher coefficient of friction at higher speeds.
7. Close tolerances permit their use in precision machinery.	Components are not normally strictly changeable.
8. Wide versatility with respect to mounting.	

22.5 TYPES OF FAILURES IN ROLLING CONTACT BEARINGS

The prominent failure of bearings is by fatigue. The failure of bearings appears as spalling or flaking of race - ways or rolling elements due to contact stresses. Due to the repeated contact the bearing will have a limited life. As pure rolling does not occur in the bearing and is always accompanied by sliding, small particles of metal are removed, which is called spalling failure. Failure of the bearing may take place due to the following reasons:

1. Faulty mounting with high preload.
2. Indentation of races and rolling elements.
3. Chatter marks due to vibration.
4. Entry of foreign particles.
5. Faulty lubrication resulting into overheating.
6. Corrosion.
7. Presence of electic voltage.

22.6 COMPARISON BETWEEN BALL AND ROLLER BEARINGS

The comparison between ball and roller bearings is given in Table 22.3

The 22.3 Comparison between ball and roller bearings

SI. No	Parameter	Ball bearing	Roller bearing
1.	Rolling element	Spherical balls	Cylindrical, taper or spherical rollers
2.	Nature of contact	Point contact	Line contact
3.	Load carrying capacity	Low	High
4.	Redial dimensions	More	Less
5.	Axial dimensions	Less	More

22.7 APPLICATIONS OF ROLLING CONTACT BEARINGS

1. Industrial and automotive gear boxes.
2. Electric motors.
3. Machine tool spindles.
4. Small size centrifugal pumps.
5. Automobile front and rear axles.

22.8 DESIGNATION OF ROLLING CONTACT BEARINGS

The basic plan for designation of rolling contact bearings is given in Table 22.4

Table 22.4 Basic plan for designation of rolling contact bearings

Bearing Type		Bearing Series		Bore	
First Number	Type of Bearing	Second Number	Bearing series	Third Number	Bore 'd' mm
6	Deep - groove ball bearings	0	Extra light	00	10

Bearing Type		Bearing Series		Bore	
First Number	Type of Bearing	Second Number	Bearing series	Third Number	Bore 'd' mm
2	Self - aligning bearings	2	Light	01	12
3	Double - row angular contact ball bearings	3	Medium	02	15
30 31 32	Taper roller bearings	4	Heavy	03	17
NU 2	Cylindrical roller bearings			04 and onwards	5× number

22.9 LOAD AND LIFE RATINGS

Basic Static Load Rating. The basic static load rating of radial ball bearings is defined as that radial load which corresponds to a total permanent deformaton of rolling element and race at the most heavily stressed contact of 0.0001 of the ball or roller diameter.

In single — row angular - contact ball bearings, the basic static load rating relates to the radial component of that load which causes a purely radial displacement of the bearings in relation to each other.

Basic Dynamic Load Rating. The basic dynamic load rating of radial ball bearings is that constant stationary radial load which a group of apparently identical ball bearings with sationary outer rings can endure for a rating life of one million revolutions of the inner rings.

In single — row angular - contact bearings, the dynamic load rating relates to the radial component of the load which results in a purely radial displacement of the bearing rings in relation to each other.

Load ratings, if given for specific speeds, are to be based on a rating life of 500 hours. This is equivalent to running the bearing for 500 hours at

$33\frac{1}{3}$ rpm or a total of 10^6 revolutions.

Life of Ball Bearings. The life of radial ball bearings is the number of revolutions(or hours, at some given constant speed) that the bearing runs before the first evidence of fatigue develops in the material of either ring or in a ball.

The rating life of a group of apparently identical radial ball bearings, thrust ball bearings, radial roller bearings or thrust roller bearings is defined as the number of revolutions(or hours at some given constant speed) that 90% of a group of apparently identical bearings will complete or exceed before the first evidence of fatigue develops. Rating life is commonly referred to as L_{10} life.

Equivalent Load. It is that constant stationary radial load which, if applied to a bearing with rotating inner and stationary outer ring, would give the same life as that which the bearing would attain under the actual conditions of load and rotation.

22.10 BASIC STATIC LOAD RATINGS

1. Radial Ball Bearings

The static equivalent load P_0 in N for radial ball bearings under combined action of radial and thrust loads shall be the greatest magnitude of those obtained from the two following equations :

$$P_o = X_o F_r + Y_o F_a \qquad ....(21.1a)$$

And

$$P_o = F_r \qquad ... (22.1b)$$

where X_o = a radial factor

Y_o = a thrust factor

F_r = radial load, N

F_a = axial load, N

For values of X_o and Y_o, refer to Table 22.5.

2. Radial Roller Bearings

The static equivalent load P_o in N for radial roller bearings under combined action of radial and thrust loads shall be the greatest magnitude of those obtained from the two following equations :

$$P_o = X_o F_r + Y_o F_a \qquad ....(22.2a)$$

and

$$P_o = F_r \qquad ...(22.2b)$$

X_o = a radial factor

Y_o = a thrust factor

F_r = radial load, N

F_a = axial load, N

For values of X_o and Y_o, refer to Table 22.5.

3. Thrust Ball Bearings

The static equivalent load P_{oa} in N for thrust ball barings with contact angle ≠ 90° under combined action of radial an thrust loads shall be given by the following formula :

$$P_{oa} = F_a + 2.3\, F_r \tan \alpha \qquad ...(22.3)$$

4. Thrust Roller Bearings

The static equivalent load P_{oa} in N for thrust roller bearings with contact angle ≠ 90° under combined action of radial and thrust loads shall be given by the following formula :

$$P_{oa} = F_a + 2.3\, F_r \tan \alpha \qquad ...(22.4)$$

22.11 BASIC DYNAMIC LOAD AND LIFE RATINGS

1. Radial Ball Bearings

The equivalent dynamic load P in N for radial and angular contact ball bearings of conventional type, except filling slot type bearing, under combined action of constant radial and constant thrust loads shall be given by the following formula :

$$P = XVF_r + YF_a \qquad \text{....(22.5)}$$

Where X = radial factor

V = rotation factor

Y = thrust factor

For values of X, V and Y, refer to Table 22.6.

Table 22.5 Values for Factors X_o and Y_o for Radial Ball and Roller Bearings.

Type of Bearing	*Single Row Bearings*[1]		*Double Row Bearings*[2]	
	X_o	Y_o	X_o	Y_o
Radial Ball Bearings :				
Radial contact groove ball bearings[3]	0.6	0.5	0.6	0.5
Angular contact groove ball bearing[4]				
$\alpha = 20°$	0.5	0.42	1	0.84
$\alpha = 25°$	0.5	0.38	1	0.76
$\alpha = 30°$	0.5	0.38	1	0.66
$\alpha = 35°$	0.5	0.29	1	0.58
$\alpha = 40°$	0.5	0.26	1	0.52
Self-aligning ball bearings	0.5	$0.22 \cot \alpha$	1	$0.44 \cot \alpha$
Radial roller bearings : Self-aligning and tapered roller bearings, $\alpha \neq 0°$	0.5	$0.22 \cot \alpha$	1	$0.44 \cot \alpha$

IS : 3823 (Part-I, II)–1966

Note :

1. P_o is always $\geq F_r$
2. Double row bearings are presumed to be symmetrical.
3. Permissible maximum value of $\frac{F_a}{C_o}$ depends on the bearing design (groove depth and internal clearance).
4. For two similar single row-angular contact ball bearings, mounted 'face-to-face' or 'back-to-back' use the value of X_o and Y_o which apply to a double-row angular contact ball bearings. For two or more similar single row angular contact bearings mounted 'in tandem', use the values of X_o and Y_o which apply to single row angular contact ball bearing.

Table 22.6 Value of Factors V., X and Y

Type of Bearing	$\frac{iF_a}{C_o}$	$\frac{F_d}{Zd_{br}^2}$	$\frac{F_a}{C_o}$	$\frac{F_a}{izd_{br}^2}$	e^1	*V* — *In relation to the load, the inner ring is*		X^1 — *Single Row*[2]	*Double Row*[3]		*Single Row*[3]	*Double Row*[3]	
						Rotati	*Station*	$\frac{F_a}{VF_r} > e$	$\frac{F_a}{VF_r} \le e$	$\frac{F_a}{VF_r} > e$	$\frac{F_a}{VF_r} > e$	$\frac{F_a}{VF_r} > e$	$\frac{F_a}{VF_r} > e$
(1)	(2)	(3)	(4)	(5)	(6)	(7)	(8)	(9)	(10)	(11)	(12)	(13)	(14)
Ball Bearings :			0.014	0.018	0.19						2.30		2.30
			0.028	0.035	0.22						1.99		1.99
Non			0.056	0.070	0.26						1.71		1.71
Filling	—	—	0.084	0.110	0.28		+	0.56	1	0.56	1.55	0	1.55
slot			0.110	0.140	0.30	1.0	1.2				1.45		1.45
assembly,			0.170	0.210	0.34						1.31		1.31
Radial			0.280	0.350	0.38						1.15		1.15
contact			0.420	0.530	0.42						1.04		1.04
groove ball bearing			0.560	0.700	0.44						1.00		1.00
Ball Bearing* $\alpha = 5°$	0.014	0.018	—	—	0.23							2.78	3.74
	0.028	0.036			0.26			For this type, use X, Y and *e* values applicable to single row, non-filling slot assembly radial contact, groove ball bearing		For this type, use X, Y and *e* values applicable single row, non-filling slot assembly radial contact, groove ball bearing		2.40	3.23
	0.056	0.070			0.30							2.07	2.78
	0.085	0.110			0.34		+					1.87	2.52
	0.110	0.140	—	—	0.36	1.0	1.2		1		0.78	1.75	2.36
	0.170	0.210			0.40							1.58	2.13
	0.280	0.350			0.45							1.39	1.87
	0.420	0.530			0.50							1.26	1.69
	0.560	0.700			0.52							1.21	1.63
Angular contact groove $\alpha = 15°$ $\alpha = 10°$	0.014	0.018			0.29						1.88	2.18	3.06
	0.029	0.035			0.32						1.71	1.98	2.78
	0.057	0.070			0.36						1.52	1.76	2.47
	0.086	0.110			0.38		+				1.41	1.63	2.29
	0.110	0.140	—	—	0.40	1.0	1.2	0.46	1	0.75	1.34	1.55	2.18
	0.170	0.210			0.44						1.23	1.42	2.00
	0.290	0.350			0.49						1.10	1.27	1.79
	0.430	0.530			0.54						1.01	1.17	1.64
	0.570	0.700			0.54						1.00	1.16	1.63

Type of Bearing	$\frac{iF_a}{C_o}$	$\frac{F_d}{Zd_{br}^2}$	$\frac{F_a}{C_o}$	$\frac{F_a}{izd_{br}^2}$	e[1]	V: In relation to the load, the inner ring is		X[1]: Single Row[2]	X[1]: Double Row[3]		Y[1]: Single Row[3]	Y[1]: Double Row[3]	
						Rotating	Stationart	$\frac{F_a}{VF_r} > e$	$\frac{F_a}{VF_r} \le e$	$\frac{F_a}{VF_r} > e$	$\frac{F_a}{VF_r} > e$	$\frac{F_a}{VF_r} > e$	$\frac{F_a}{VF_r} > e$
(1)	(2)	(3)	(4)	(5)	(6)	(7)	(8)	(9)	(10)	(11)	(12)	(13)	(14)
	0.015	0.018			0.38						1.47	1.65	2.39
	0.029	0.035			0.40						1.40	1.57	2.28
	0.058	0.070			0.43	+					1.30	1.46	2.11
	0.087	0.110	—	—	0.46	1.0	1.2	0.44*	1	0.72	1.23	1.38	2.00
	0.120	0.140			0.47						1.19	1.34	1.93
	0.170	0.210			0.50						1.12	1.26	1.82
	0.290	0.350			0.55						1.02	1.14	1.66
	0.440	0.530			0.56						1.00	1.12	1.63
	0.580	0.700			0.56						1.00	1.12	1.63
α = 20°					0.57	1	1.2	0.43	1	0.70	1.00	1.09	1.63
α = 25°					0.68	1	1.2	0.41	1	0.67	0.87	0.92	1.41
α = 30°	—	—	—	—	0.80	1	1.2	0.39	1	0.63	0.76	0.78	1.24
α = 35°					0.95	1	1.2	0.37	1	0.60	0.66	0.66	1.07
α = 40°					1.14	1	1.2	0.35	1	0.57	0.57	0.55	0.93
Self-alig-ning ball bear-ngs	—	—	—	—	1.5 tan α	1	1	0.40	1	0.65	0.4 cot α	0.42 cot α	0.65 cot α
Single rowradial con-tact separable ballbear-ings (magneto bearigns)	—	—	—	—	0.2	1	1	0.5	—	—	2.5	—	—

IS : 3824 (Part-1)—1966

$d_{br} = d_b$ = dia of ball, mm; $d_{br} = d_r$ = dia of roller, mm; C_o = static basic load rating.

1. Value of X, Y and e for a load or a contact angle other than those shown in the table, are obtained by inner interpolation.
2. For a single row bearings, when $\frac{F_a}{VF_r} \le e$, use X = 1 and Y = 0.
3. Double row bearings are presumed to be symmetrical.

(*a*) When calculating the equivalent load for a unit consisting of two similar single row angular contact ball bearings in duplex mounting, 'face-to-face' or 'back-to-back', the units shall be considered as one of double row angular contact ball bearings.

(*b*) When calculating the equivalent load for a unit consisting of two or more single row radial or angular contact ball bearings 'in tandem', X, Y values for single row ball bearings shall be used.

\+ Because experimental data are incomplete, no correct value of factor V for radial and angular contact groove ball bearings with innering stationary in relation to the load can be stated. The values shown in the able are, however well on the safe side.

* Permissible maximum value of $\frac{F_o}{C_o}$ - depends upon the bearing design.

Table 22.7 Values of X and Y for Thrust Ball and Roller Bearings.

Types of bearing	X			Y			e
	Single direction bearing*	Double direction direction		Single Direction bearing	Double direction bearing		
	$\frac{F_a}{F_r} < e$	$\frac{F_a}{F_r} \leq e$	$\frac{F_a}{F_r} > e$	$\frac{F_a}{F_r} > e$	$\frac{F_a}{F_r} \leq e$	$\frac{F_a}{F_r} > e$	
Thrust ball $\alpha = 45°$	0.66	1.18	0.66	1.0	0.59	1.0	1.125
Thrust ball $\alpha = 60°$	0.92	1.90	0.92	1.0	0.54	1.0	2.17
Thrust ball $\alpha = 75°$	1.66	3.89	1.66	1.0	0.52	1.0	4.67
Thrust ball $\alpha = 90°$**	—	—	—	—	—	—	—
Self-aligning and tapered thrust roller bearing $\alpha \neq 90°$**	tan α	1.5 tan α	tan α	1.0	0.67	1.0	1.5 tan α

* Double direction bearings are presumed to be symmetrical.

** For $\alpha = 90°$, X = 0, Y = 1.

IS : 3824 (Part III, IV)—1966

Table 22.8 Deep Groove Ball Bearings Series 02 (Fig. 22.1 a)

Bearing No.		*Boundary Dimension mm*				*Basic Load Rating kN*		*Limiting speed rev./min*
BIS	*SKF*	*Bore d*	*Outside dia D*	*Width B*	*Corner radius*	*Static C_0*	*Dynamic C*	$\times 10^3$
10BC02	6200	10	30	9		2..24	3.9	24
12BC02	6201	12	23	10		3.1	5.3	22
15BC02	6202	15	35	11	1	3.55	6	19
17BC02	6203	17	40	12		45	7.35	17
20BC02	6204	20	47	14		6.2	9.8	15
25BC02	6205	25	52	15	1.5	6.95	10.8	12
30BC02	6206	30	62	16		10	15	10

Bearing No.		*Boundary Dimension mm*				*Basic Load Rating kN*		*Limiting speed rev./min × 10^3*
BIS	*SKF*	*Bore d*	*Outside dia D*	*Width B*	*Corner radius*	*Static C_0*	*Dynamic C*	
35BC02	6007	35	72	17		13.7	19.6	9
40BC02	6208	40	80	18		16.6	23.6	8.5
45BC02	6209	45	85	19	2.0	18.6	25.5	7.5
50BC02	6210	50	90	20		19.6	27	7
55BC02	6211	55	100	21		25	33.5	6.3
60BC02	6212	60	110	22		28	36.5	6
65BC02	6213	65	120	23	2.5	34	43	5.5
70BC02	6214	70	125	24		37.5	47.5	5
75BC02	6215	75	130	25		40.5	51	4.8
80BC02	6216	80	140	26		45	54	4.5
85BC02	6217	85	150	28	3.0	53	64	4.3
90BC02	6218	90	160	30		62	73.5	3.8
95BC02	6219	95	170	32		69.5	83	3.6
100BC02	6220	100	180	34		78	95	3.4
105BC02	6221	105	190	36	3.5	90	102	3.2
110BC02	6222	110	200	38		100	112	3
120BC02	6224	120	215	40		100	112	2.8

Table 22.9 Deep Groove Ball Bearing Series 03

Bearing No.		*Boundary Dimension mm*				*Basic Load Rating kN*		*Limiting speed rev./min × 10^3*
BIS	*SKF*	*d*	*D*	*B*	*r*	*Static C_0*	*Dynamic C*	
10BC03	6300	10	35	11	1	3.75	6.2	20
12BC03	6301	12	37	12		4.65	7.5	19
15BC03	6302	15	42	13	1.5	5.4	8.8	17
17BC03	6303	17	47	14		6.55	10.4	16
20BC03	6304	20	52	15		7.8	12.2	13
25BC03	6305	25	62	17	2	11.4	17.3	11
30BC03	6306	30	72	19		14.6	21.6	9

Bearing No.		*Boundary Dimension mm*				*Basic Load Rating kN*		*Limiting speed rev./min × 10^3*
BIS	*SKF*	*d*	*D*	*B*	*r*	*Static C_0*	*Dynamic C*	
35BC03	6307	35	80	21		18	25.5	8.5
40BC03	6308	40	90	23	2.5	22.4	31.5	7.5
45BC03	6309	45	100	25		30	40.5	6.7
50BC03	6310	50	110	27		36	47.5	6.3
55BC03	6211	55	120	29	3	41.5	55	5.6
60BC03	6312	60	130	31		48	63	5
65BC03	6313	65	140	33		56	71	4.8
70BC03	6314	70	150	35	3.5	63	80	4.5
75BC03	6315	75	160	37		72	86.5	4.3
80BC03	6316	80	170	39		80	95	3.8
85BC03	6317	85	180	41		90	102	3.6
90BC03	6318	90	190	43		98	110	3.4
95BC03	6319	95	200	45	4	110	118	3.2
100BC03	6320	100	215	47		132	134	3.6
105BC03	6321	105	225	49		143	140	2.8
110BC03	6322	110	240	50		166	156	2.6
120BC03	6324	120	260	55		166	160	2.4

Table 22.10 Deep Groove Ball Bearings Series 04

Bearing No.		*Boundary Dimensions mm*				*Basic Load Rating kN*		*Limiting speed rev./min × 10^3*
BIS	*SKF*	*d*	*D*	*B*	*r*	*Static C_0*	*Dynamic C*	
17BC04	6403	17	62	17	2	11.8	17.6	12
20BC04	6404	20	72	19		16.6	23.6	10
25BC04	6405	25	80	21		19.6	27.5	9
30BC04	6406	30	90	23	2.5	24	33.5	8.5
35BC04	6407	35	100	25		31	42.5	7
40BC04	6408	40	110	27	3	36.5	49	6.7
45BC04	6409	45	120	29		45.5	58.5	6
50BC04	6410	50	130	31		52	67	5.3

Bearing No.		*Boundary Dimensions mm*				*Basic Load Rating kN*		*Limiting speed rev./min*
BIS	*SKF*	*d*	*D*	*B*	*r*	*Static* C_0	*Dynamic C*	$\times 10^3$
55BC04	6411	55	140	33		63	76.5	5
60BC04	6412	60	150	35	3.5	69.5	83	4.8
65BC04	6413	65	160	37		78	91.5	4.5
70BC04	6414	70	180	42		104	110	3.8
75BC04	6415	75	190	45	4	114	118	3.6
80BC04	6416	80	200	48		125	125	3.4
85BC04	6417	85	210	52	5	124	134	3.2
90BC04	6418	90	225	54		146	140	3

Table 22.11 Cylindrical Roller Bearings Series 02. (Fig. 22.2 a)

Bearing No.		*Boundary Dimension mm*				*Basic Load Rating kN*		*Limiteing speed rev./min*
BIS	*SKF*	*d*	*D*	*B*	*r*	*Static* C_0	*Dynamic C*	$\times 10^3$
20RN02	N204	20	47	14		7.35	13.4	15
25RN02	N205	25	52	15	1.5	8.8	15.8	12
30RN02	N206	30	92	16		12	20.4	10
35RN02	N207	35	72	17		17.6	29	9
40RN02	N208	40	80	18	2	24	38	8.5
45RN02	N209	45	85	19		25.5	40	7.5
50RN02	N210	50	90	20		27.5	42.5	7
55RN02	N211	55	100	21		34	51	6.3
60RN02	N212	60	110	22	2.5	43	62	5.6
65RN02	N213	65	120	23		51	72	5.3
70RN02	N214	70	125	24		51	72	5
75RN02	N215	75	130	25		63	88	4.8
80RN02	N216	80	140	26	3	68	96.5	4.5
85RN02	N217	85	150	28		78	110	4.3
90RN02	N218	90	160	30		100	134	3.8
95RN02	N219	95	170	32		112	1500	03.6
100RN02	N220	100	180	34	3.5	125	166	3.4
110RN02	N222	110	200	38		166	200	3
120RN02	N224	120	215	40		183	236	2.8

Table 22.12 Cylindrical Roller Bearings Series 03

Bearing No.		*Boundary Dimensions mm*			*Basic Load Rating kN*		*Limiting speed rev./min* $\times 10^3$
BIS	*SKF*	*d*	*D*	*B*	*Static* C_0	*Dynamic* *C*	
17RN03	N303	17	47	14	8.65	15.6	14
25RN03	N305	25	62	17	15	26	9.5
30RN03	N306	30	72	19	20	34	8.5
35RN03	N307	35	80	21	27	43	8
40RN03	N308	40	90	23	32.5	51	6.7
45RN03	N309	45	100	25	41.5	65	6.3
50RN03	N310	50	110	27	52	80	5.6
55RN03	N311	55	120	29	67	100	5
60RN03	N312	60	130	31	76.5	112	4.8
65RN03	N313	65	140	33	85	125	4.5
70RN02	N314	70	150	35	102	146	4
75RN03	N315	75	160	37	125	176	3.8
80RN03	N316	80	170	39	125	176	3.6
85RN03	N317	85	180	41	146	204	3.4
90RN03	N318	90	190	43	160	220	3.2
95RN03	N318	95	200	45	190	220	3.6
100RN03	N320	100	215	47	220	250	2.8
110RN03	N322	110	240	50	290	375	2.4
120RN03	N324	120	260	55	340	440	2.2

Table 22.13 Single Direction Thrust Ball Bearings Series 11 (Fig. 22.2 e)

Bearing No.		*Boundary Dimension mm*			*Basic Load Rating kN*		*Limiting speed rev./min* $\times 10^3$
BIS	*SKF*	*d*	*D*	*H*	*Static* C_0	*Dynamic* *C*	
10TA11	51100	10	24	9	8.8	6.7	7
12TA11	51101	12	26	9	19	6.95	7
15TA11	51102	15	28	9	11.2	7.2	6.3
17TA11	51103	17	30	9	12.2	7.5	6.3
20TA11	51104	20	35	10	16.6	9.8	5.6
25TA11	51105	25	42	11	22.8	12.2	4.8
30TA11	51106	30	47	11	26.5	12.9	4.5
35TA11	51107	35	52	12	30	13.4	4.3
40TA11	51108	40	60	13	40	18	3.8

Bearing No.		Boundary Dimensions mm			Basic Load Rating kN		Limiting speed rev./min $\times 10^3$
BIS	SKF	d	D	H	Static C_0	Dynamic C	
45TA11	51109	45	65	14	45	18.6	3.4
50TA11	51110	50	70	14	50	19.6	3.4
55TA11	51111	55	78	16	62	23.6	3
60TA11	51112	60	85	17	71	27.5	2.6
65TA11	51113	65	90	18	78	28.5	2.4
70TA11	51114	70	95	18	88	32.5	2.4
75TA11	51115	75	100	19	98	34	2.2
80TA11	51116	80	105	19	102	34.5	2
85TA11	51117	85	110	19	106	34.5	2
90TA11	51118	90	120	22	120	39	1.8
100TA11	51120	100	135	25	173	57	1.7
110TA11	51122	110	145	25	190	58.5	1.6
120TA11	51124	120	155	25	250	68	1.6
130TA11	51126	130	170	30	285	80	1.4
140TA11	51128	140	180	31	305	81.5	1.3
150TA11	51130	150	190	31	325	83	1.2
160TA11	51132	160	200	31	345	86.5	1.2
170TA11	51134	170	215	34	415	104	1.1
180TA11	51136	180	225	34	430	104	1
190TA11	51138	190	240	37	530	132	0.95
200TA11	51140	200	250	37	550	132	0.95

H = width

Table 22.14 Single Direction Thrust Ball Bearing Series 14

Bearing No.		Boundary Dimensions mm			Basic Load Rating kN		Limiting speed rev./min $\times 10^3$
BIS	SKF	d	D	H	Static C_0	Dynamic C	
24TA14	51405	25	60	24	63	40	2600
30TA14	51406	30	70	28	98	52	2200
35TA14	51407	35	80	32	112	63	1800
40TA14	51408	40	90	36	146	80	1700
45TA14	51409	45	100	39	173	93	1600
50TA14	51410	50	110	43	204	106	1500
55TA14	51411	55	120	48	265	137	1300

Bearing No.		Boundary Dimensions mm			Basic Load Rating kN		Limiting speed rev./min × 10^3
BIS	SKF	d	D	H	Static C_0	Dynamic C	
60TA14	51412	60	130	51	285	143	1200
65TA14	51413	65	140	56	360	166	1000
70TA14	51414	70	150	60	400	180	950
75TA14	51415	75	160	65	500	208	900
80TA14	51416	80	170	68	550	224	850
85TA14	51417	85	180	72	610	236	850
90TA14	51418	90	190	77	670	250	800
100TA14	51420	100	210	85	865	305	700

Table 22.15 Taper Roller Bearings Series 322 (Fig.22.2c)

Bearing No.		Boundary Dimensions mm				Basic Load Rating kN			Limiting speed rev./min × 10^3
BIS	SKF	d	D	B	C	Static C_0	Dynamic C		
30RK322	32206	30	62	20	17	21.25	34	43	6.3
35RK322	32207	35	72	23	19	24.25	45	56	5.3
40RK322	32208	40	80	23	19	24.75	50	64	4.8
45RK322	32209	45	85	23	19	24.75	56	66	4.5
50RK322	32210	50	90	23	19	24.75	57	69.5	4.3
55RK322	32211	55	100	25	21	26.75	75	90	3.8
60RK322	32212	60	110	28	24	39.75	91.5	108	3.4
65RK322	32213	65	120	31	27	32.75	112	129	3
70RK322	32214	70	125	31	27	33.25	118	134	2.8
75RK322	32215	75	130	31	27	33.25	120	137	2.6
80RK322	32216	80	140	33	28	35.25	137	160	2.4
85RK322	32217	85	150	36	30	38.5	163	183	2.2
90RK322	32218	90	160	40	34	42.5	193	216	2
95RK322	32219	95	170	43	37	45.5	220	240	1.9
100RK322	32220	100	180	46	39	49	250	270	1.8
105RK322	32221	105	190	50	43	53	290	310	1.8
110RK322	32222	110	200	53	46	56	325	340	1.7
120RK322	32224	120	215	58	50	61.5	340	355	1.6
130RK322	32226	130	230	64	54	67.75	405	425	1.5
140RK322	32228	140	250	68	58	71.75	480	490	1.4
150RK322	32230	150	270	73	60	77	640	630	1.2
160RK322	32232	160	290	80	67	84	780	750	1.1

C = cup length, T = cone length

The approximate value of rating life L_{10} for ball bearings, except filling slot bearings, is given by

$$L_{10} = \left(\frac{C}{P}\right)^3 \text{ millions of revolutions} \qquad \text{...(22.6)}$$

For static and dynamic basic load rating C_0 and C respectively of deep groove ball bearings, refer to Table 22.8 to 22.10.

2. Radial Roller Bearings

The equivalent dynamic load P in N for self-aligning and tapered roller bearings of conventional type under combined action of constant radial and constant thrust loads shall be given by the following formula :

$$P = XVF_r + YF_a \qquad \text{...(22.7)}$$

where X = radial factor

V = rotation factor

Y = thrust factor

For values of X, V and Y, refer to Table 22.7

The approximate value of rating life L_{10} for radial roller bearings is given by,

$$L_{10} = \left(\frac{C}{P}\right)^{10/3} \text{ millions of revolutions} \qquad \text{...(22.8)}$$

For static and dynamic basic load rating C_0 and C respectively of cylindrical roller bearings, refer to Table 22.11 and 22.12.

3. Thrust Ball Bearings

The equivalent dynamic thrust load P_a in N for thrust ball bearings with $\alpha \neq 0°$ under combined action of constant radial and constant thrust loads shall be given by the following formula :

$$P_a = XF_r + YF_a \qquad \text{...(22.9)}$$

Where X = radial factor

Y = thrust factor

For values of X, Y refer to Table 22.7

The rating life L_{10} of thrust ball bearings is given by,

$$L_{10} = \left(\frac{C_a}{P_a}\right)^{10/3} \text{ millions of revolutions} \qquad \text{...(22.10)}$$

For static and dynamic basic load rating C_0 and C respectively of single direction thrust ball bearings, refer to Table 22.13 and 22.14

4. Thrust Roller Bearings

The equivalent dynamic thrust load P_a in N for thrust roller bearings with $\alpha \neq 0°$ under combined action of constant radial and constant thrust loads shall be given by the following formula :

$$P_a = XF_r + YF_a \qquad \text{...(22.11)}$$

where X = radial factor

Y = thrust factor

For values of X, Y refer to Table 22.7

The rating life L_{10} of thrust ball bearings is given by,

$$L_{10} = \left(\frac{C_a}{P}\right)^{10/3} \text{ millions of revolutions} \qquad ...(22.10)$$

For static and dynamic basic load rating C_0 and C respectively of thrust roller bearings, refer to Table 22.15

22.12 SERVICE FACTOR

The forces acting on the bearing are calculated by considering the equilibrium of horizontal and vertical forces. These equations do not take into account the effect of service conditions prevailing in the bearing. To account for this, service factor C_a is used to multiply the theoretical load. The values of C_s are given in Table 22.16.

Equivalent load, $$P_e = C_s (XVF_r + YF_a) \qquad ..(22.13)$$

Table 22.16 Service factor for rolling cantact bearing, C_s

Type of load	C_s
Constant or steady	1.0
Light shocks	1.5
Moderate shocks	2.0
Heavy shocks	3.0

22.13 VARIABLE LOADS

Rearranging Eq. (22.6) for ball bearings, we have

$$L_{10} = \left(\frac{C}{P}\right)^3 \left[\frac{10^6}{60 \times n}\right]$$

Where n = rpm

$$C^3 = \frac{60nL_{10}P^3}{10^6} \qquad ...(22.14)$$

If during a fraction of cycle r_i the speed n_i and equivalent load P_i are constant, then

$$C^3 = \left(\frac{60L_{10}}{10^6}\right) \sum_{i=1}^{m} r_i n_i \, (C_s P)_i^3 \qquad ...(22.15)$$

where C_s = bearing service factor during fraction of the cycle

$$\sum_{i=1}^{m} r_i = 1$$

Equivalent steady load under fluctuating load,

$$P = \left(\frac{2P_{max} + P_{min}}{3}\right) \quad ...(22.16)$$

For roller barings, we have

$$C^{10/3} = \left(\frac{60L_{10}}{10^6}\right) \sum_{i=1}^{m} r_i n_i \, (C_s P)_i^{10/3} \quad ...(22.17)$$

22.14 BEARING LIFE

While selecting the proper size of a bearing, it is necessary to specify the expected life of the bearing for the given application. The information regarding the life of the bearing is statistical in nature and is generally vague. The life of the bearing in hours for various machines is given in

Bearing life in hours, $$L_h = \frac{10^6 L}{60n} \quad ...(22.18)$$

Where n = rpm and L = life in millions of revolutions.

The ebaring life for vehicle applications is given in Table 22.17.

Table 22.17 Bearing life for vehicle applications

Type of vehicle	Life in million revolutions
Automobile cars	50
Trucks	100
Rail road cars	1000

Baering life for industrial applications given in Table 22.18. Loading ratio $\frac{C}{P}$ for different lives in millions of revolutions for ball and roller bearings is given in Table 22.19 and 22.20 respectively.

22.15 SELECTING BEARING FROM MANUFACTURER'S CATALOGUE

The following procedure may be adopted to select the bearing from manufacturer's catalogue :

1. Calculate the radial and thrust loads acting on the bearing and determine the diameter of the journal.
2. Select the type of bearing for the given application.
3. Determine the static and dynamic load capacity C_0 and C of the bearing respectively from the manufacturer's catalogue.
4. Calculate $\frac{F_a}{F_r}$ and $\frac{F_a}{C_0}$ ratios.
5. Determine the value of 'e'. Then determine the values of X and Y.
6. Calculate the equivalent dynamic load from Eq. (22.5), (22.7), (22.9) or (22.11)
7. Calculate the life of the bearing from Eq. (22.10), (22.12),(22.15) or (22.17)

Table 22.18 Guide to Values of life Requirements for Different Classes of Machines

Type of Machine	Life in Hours of Operation
Instruments and apparatus that are only infrequently used **Ex. :** Demonstration apparatus, devices for operation of sliding doors	500
Aircrafts engines	500—2,000
Machines for service of short duration or intermittent operation, where service interruptions are of minor importance. **Ex. :** Hand tools, lifting tackle in machinery shops, hand driven machines in general, farm machinery, assemblycranes charging machines, foundry cranes, household machine Machines for intermittlent service where dependable operation is of great importance	4,000—8,000 8,000—12,000
Ex. : Auxiliary machines in power stations, conveying equipment in production lines, elevators, general cargo cranes, machine tools less frequently used	
Machines for 8-h service that are not always fully utilized **Ex. :** Stationary electric motor, gear drives for general purposes Machines for 8-h service that are fully utilized **Ex. :** Machines in general in the mechanical industries, cranes for continuous services, blowers, jack shafts	12,000—20,000 20,000—30,000
Machines for continuous operation (24-h services) **Ex. :** Separators, compressors, pumps, main lines shafting, rollerbeds and conveyor rollers, mine hists, stationary electric motors.	40,000—60,000
Machines for 24-h service where dependability is of great importance **Ex. :** Pulp and paper machines, public power stations, mine pumps,public pumping stations machiens for continuous service aboard ship,	1000,000—2000.000

Table 22.19 Loading ratio C/P for different expressed in millions of revolutions : ball bearings

Life in millions of revolutions L_n	$\frac{C}{P}$	*Life in million of revolutions* L_n	$\frac{C}{P}$	*Life in millions of revolutions* L_n	$\frac{C}{P}$	*Life in millions of revolutions* L_n	$\frac{C}{P}$
0.5	0.793	80	4.31	600	8.43	3200	14.7
0.75	0.909	90	4.48	650	8.66	3400	15.0
1.0	1	100	4.64	700	8.88	3600	15.3
1.5	1.14	120	4.93	750	9.09	3800	15.6
2	1.26	140	5.19	800	9.28	4000	15.9
3	1.44	160	5.43	850	9.47	4500	16.5
4	1.59	180	5.65	900	9.65	5000	17.1
5	1.71	200	5.85	950	9.83	5500	17.7
6	1.82	220	5.04	1000	10	6000	18.2
8	2	240	6.21	1100	10.3	6500	18.7
10	2.15	260	6.38	1200	10.6	7000	19.1
12	2.29	280	6.54	1300	10.9	7500	19.6
14	2.41	300	6.69	1400	11.2	8000	20.0
16	2.52	320	6.84	1500	11.4	8500	20.4
18	2.62	340	6.98	1600	11.7	9000	20.8
20	2.71	360	7.11	1700	11.9	9500	21.2
25	2.92	380	7.24	1800	12.2	10000	21.5
30	3.11	400	7.37	1900	12.4	12000	22.9
35	3.27	420	7.49	2000	12.6	14000	24.1
40	3.42	440	7.61	2200	13.6	12000	25.1
45	3.56	460	7.72	2400	13.4	18000	26.2
50	3.68	480	7.83	2600	13.8	20000	27.1
60	3.91	500	7.94	2800	14.1	25000	29.2
70	4.12	550	8.19	3000	14.4	30000	31.1

Table 22.20 Loading ratio C/P for different lives expressed in millions of revolutions : Roller bearings

Life in millions of revolutions L_n	$\frac{C}{P}$	*Life in millions of revolutions* L_n	$\frac{C}{P}$	*Life in millions of revolutions* L_n	$\frac{C}{P}$	*Life in millions of revolutions* L_n	$\frac{C}{P}$
0.5	0.812	80	3.72	600	6.81	3200	11.3
0.75	0.917	90	3.85	650	6.98	3400	11.5

Life in millions of revolutions L_n	$\frac{C}{P}$	*Life in millions of revolutions* L_n	$\frac{C}{P}$	*Life in millions of revolutions* L_n	$\frac{C}{P}$	*Life in millions of revolutions* L_n	$\frac{C}{P}$
1.0	1	100	3.98	700	7.14	3600	11.7
1.5	1.13	120	4.20	750	7.29	3800	11.9
2	1.24	140	4.40	800	7.43	4000	12.0
3	1.39	160	4.58	850	7.56	4500	12.5
4	1.52	180	4.75	900	7.70	5000	12.9
5	1.62	200	4.90	950	7.82	5500	13.2
6	1.71	220	5.04	1000	7.94	6000	13.6
8	1.87	240	5.18	1100	8.17	6500	13.9
10	2.00	260	5.30	1200	8.39	7000	14.2
12	2.11	280	5.42	1300	8.59	7500	14.5
14	2.21	300	5.54	1400	8.79	8000	14.8
16	2.30	320	5.64	1500	8.97	8500	15.1
18	2.38	340	5.75	1600	9.15	9000	15.4
20	2.46	360	5.85	1700	9.31	9500	15.6
25	2.63	380	5.94	1800	9.48	10000	15.8
30	2.77	400	6.03	1900	9.63	12000	16.7
35	2.91	420	6.12	2000	9.78	14000	17.5
40	3.02	440	6.21	2200	10.1	16000	18.2
45	3.13	460	6.29	2400	10.3	18000	18.9
50	3.23	480	6.37	2600	10.6	20000	19.5
60	3.42	500	6.45	2800	10.8	25000	20.9
70	3.58	550	6.64	3000	11.0	30000	22.0

22.16 BEARING LIFE FOR HIGHER RELIABILITY

The definition of bearing life is based upon 10% failure rate or 90% survival rate in many applications, failure rates other than 10% are required. To determine the rating life at other failure rates than 10%, the rating life L_{10} may be multiplied by the life adjustment reliability factor, K_L, which can be obtained from Fig. 22.3.The median life corresponding to 5×10^6 revolutions is given by

$$\text{Median life} = 5\ L_{10} \qquad \text{... (22.19)}$$

The realibility R is defined as,

$$R = \frac{\text{Number of bearings which have successfully completed L million revolutions}}{\text{Total number of bearings under test}}$$

The relationship between bearing life and reliability is given by a statistical curve, known as Weibull distribution. For Weibull distribution.

$$R = \exp.\left[-\left(\frac{L}{a}\right)^b\right] \quad \text{... (22.20)}$$

where R = reliability in fraction

L = corresponding life

a, b = constants

Rearranging the above equation, we have

$$\left(\frac{1}{R}\right) = \exp.\left[\left(\frac{L}{a}\right)^b\right]$$

Or
$$ln\left(\frac{1}{R}\right) = \left(\frac{L}{a}\right)^b \quad \text{... (1)}$$

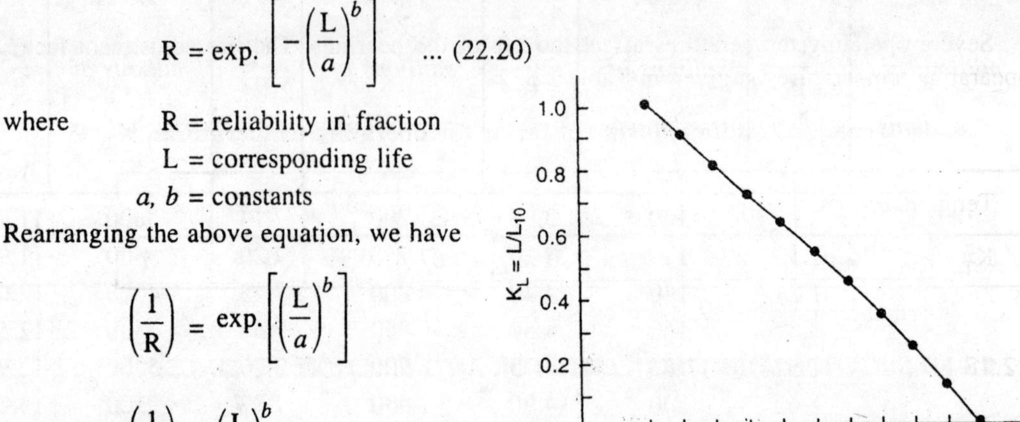

Fig. 22.3 Life adjustment reliability factor

For R = 90% corresponding to L_{90}, we have

$$Ln\left(\frac{1}{R_{90}}\right) = (L_{90})^b \quad \text{... (2)}$$

Dividing Eq. (1) by Eq. (2), we have

$$\left(\frac{L}{L_{10}}\right) = \left[\frac{\ln\left(\frac{1}{R}\right)}{\ln\left(\frac{1}{R_{90}}\right)}\right]^{1/b} \quad \text{... (22.21)}$$

$$= K_L$$

The values of 'a' and 'b' are : $a = 6.84$, $b = 1.17$ for,

$$L_{50} = 5L_{90} \quad \text{... (22.22)}$$

Where L_{50} = median life or life which 50% of the bearings will complete or exceed before fatigue failure. Eq. (22.21) is used for selecting the bearing life when the reliability is other than 90%.

K_L = life adjustment factor for reliability (see Table 22.21)

Table 22.21 Life adjustment factor for reliability, K_L

Reliability	0.90	0.91	0.92	0.93	0.94	0.95	0.96	0.97	0.98	0.99	0.995	0.999
K_L	1.0	0.928	0.855	0.779	0.701	0.618	0.531	0.437	0.332	0.208	0.131	0.0448

In a system, if there are number of bearings, the individual reliability of each bearing should be fairly high. If there are n bearings in the sytem, each having the same reliability R, then the reliability of the complete system is given by,

$$R_s = R^n \quad \text{... (22.23)}$$

Where R_s = probability of one out of n bearings failing during its life time.

22.17 LIFE ADJUSTMENT FACTOR FOR OPEATING TEMPERATUARE

Severe operating temperatures affect the life of the bearings. The life adjustment factor for opearating temperatures is given in Table 22.22.

Table 22.22 Life adjustment factor for operating temperatures, K_T

Temperature, °C	100	125	150	175	200	225	250
K_T	0.98	0.95	0.91	0.87	0.80	0.74	0.71

Actual bearing life, $L_a = K_L K_T L_{10}$ ... (22.24)

22.18 LUBRICATION AND LUBRICANTS FOR ANTI-FRICTION BEARINGS

The purpose of lubrication of anti—friction bearings are :

1. Reduce the friction between balls and races.
2. Dissipation of frictional heat.
3. Prevention of corrosion and protection of bearing from dirt and other foreign particles.

There are two types of lubricants used for anti—friction bearings : oil and grease. Compared with grease, oils offer the following advantages :

1. It is more effective in carrying frictional heat.
2. Easy to feed into contact area of the bearing under load.
3. More effective in flushing out dirt, corrosion and foreign particles from the bearing.

The advantages offered by grease lubricants are :

1. Simple housing design.
2. Less maintenance cost.
3. Better sealing against rust.
4. Less possibility of leakage.

The guidelines for selecting the lubricant are :

1. When the temperature is less than 100°C, grease is suitable, while lubricating oils are preferred where the temperature exceeds 100°C.
2. When the product of bore (in mm) × speed (in rpm) is below 200,000, grease is suitable.
3. Grease is suitable for low and moderate loads and oils for heavy duty applications.
4. Oils are used for central lubricating system.

The following methods are commonly used for oil lubrication :

(*i*) Jet oil lubrication.

(*ii*) Circulating oil lubrication.

(*iii*) Oil splash lubrication

(*iv*) Oil bath lubrication

(*v*) Drop feed lubrication

(*vi*) Air—oil mist lubricate

22.19 MOUNTING OF BEARINGS

Mounting of bearings is the installation of bearing on the shaft and in the housing. The basic principles in bearing mountings are :

1. The displacement of inner and outer race in an axial direction should be restricted.
2. The thrust load must be transmitted by the bearing to the housing.
3. The assembly and disassembly should be easy and fool proof.

The following methods are used for mounting of bearings :

1. Fixed–fixed bearing mounting arrangement :

Such an arrangment is shown in Fig. 22. The inner races of the bearings are backed up against the shaft shoulders, spacers of hub of the mechanical element like gear or pulley. The outer races are held in positon by the cover plates, which are bolted to the housing. The adjustable spacers are used to permit the axial free play of the shaft to be adjusted to the desired amount. Each bearing takes the axial thrust directed towards it. This method cannot be used when the distance between the bearings is large and temperature of surroundings is high.

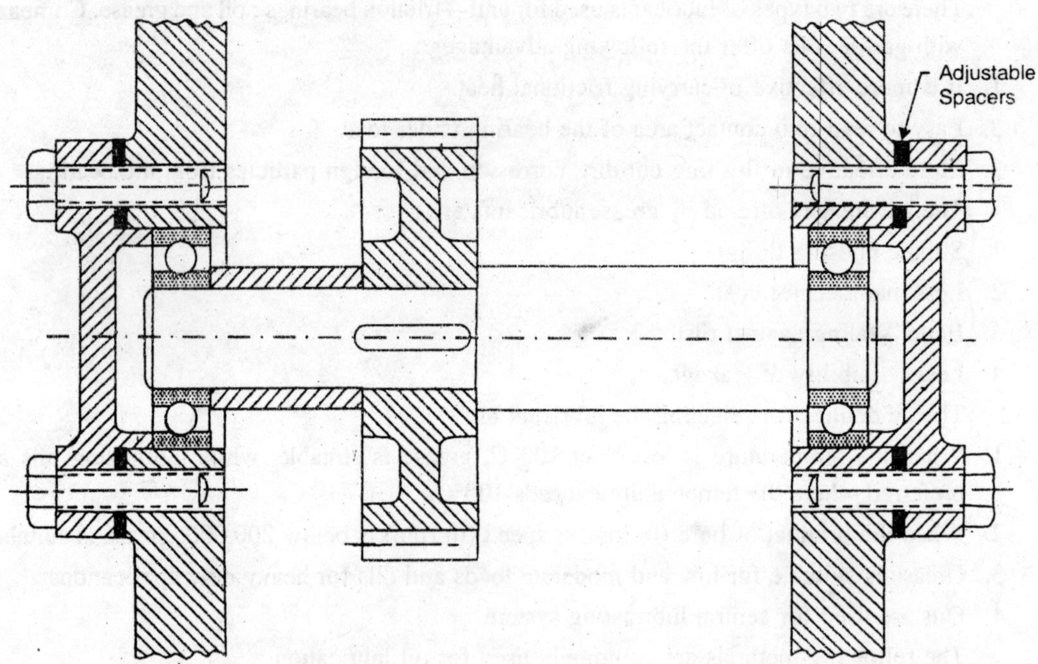

Fig. 22.4 Fixed — Fixed Bearing Mounting

2. Fixed–float bearing mounting arrangement :

This type of bearing mounting arrangment is shown in Fig. 22.5 The inner races are backed up against the shaft shoulders and are held in position by lock – nuts threaded on the to the shaft. The outer race of the left side bearing is held in position by the cover plate and the outer race of the right side bearing floats in the housing. The left side bearing takes thrust in both directions.

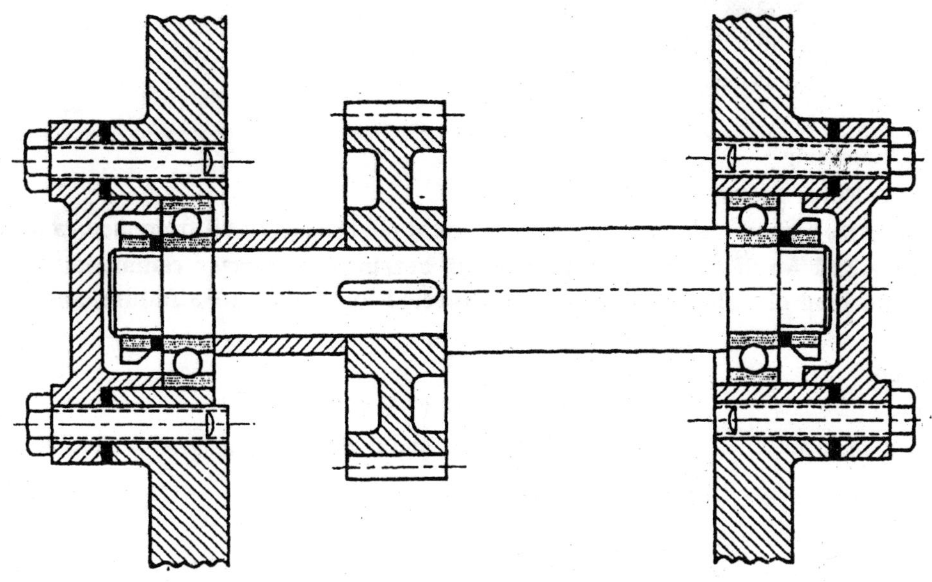

Fig. 22.5 Fixed-Float Bearing Mounting

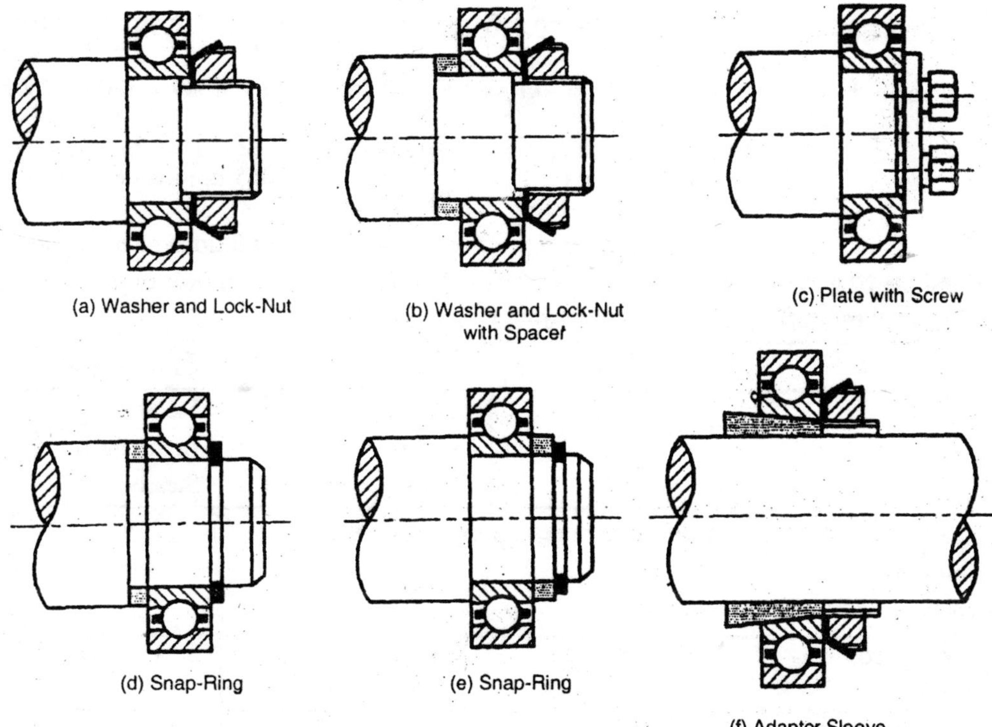

Fig. 22.6 Mounting of Bearing on Shaft

3. Alternate methods of bearing mounting :

The alternate methods of bearing mounting are shown in Fig. 22.6

(*i*) Washer and lock–nut

(*ii*) Washer and lock–nut with spacer

(*iii*) Plate fastened with screws

(*iv*) Snap ring

(*v*) Adaptor sleeve.

Gaskets and oil seals may be used to prevent the leakage of oil from the bearings.

Example 22.1 An SKF 6201 deep groove ball bearing is to carry a radial load of 2400 N and a thrust load of 1200 N. Calculate the expected life for (*a*) 10% failure, and (*b*) 59% survival.

Solution.

Given : $F_r = 2400$ N, $F_a = 1200$ N

The static capacity, $C_o = 2940$ N, and dynamic capacity, C = 5250 N

$$\frac{F_a}{F_r} = \frac{1200}{2400} = 0.5$$

$$\frac{F_a}{C_o} = \frac{1200}{2940} = 0.408$$

By interpolation, the value of $e = 0.416$, and $\frac{F_a}{F_r} > e$, so that X = 0.56, Y = 1.049

Equivalent load, $P = XF_r + YF_a = 0.56 \times 2400 + 1.049 \times 1200 = 2602.8$ N

(*a*) $L_{10} = \left(\frac{C}{P}\right)^3 = \left(\frac{5250}{2602,8}\right)^3 = 8.2065$ millions of revolutions

(*b*) $L_{50} = 5L_{10} = 5 \times 8.2065 = 41.03$ millions of revolutions.

Example 22.2 A single – row deep – groove ball bearing has a dynamic capacity of 42500 N and operates on the following work cycle :

Radial load, N	Duration	Speed, rpm	Type of load
5,000	0.25	500	Steady
10,000	0.50	700	Light shock
7,000	0.25	400	Steady

Calculate the life of the bearing in hours for 10% failures.

Solution.

The service factors are taken from

Radial load, $P = F_r$, N	Fraction of cycle, *r*	Speed, *n* *rpm*	Service factor, C_S
5,000	0.25	500	1.0
10,000	0.50	700	1.5
7,000	0.25	400	1.0

$$C^3 = \left(\frac{60L_{10}}{10^6}\right)\sum_{i=1}^{m} r_i n_i \,(C_s P)_i^3$$

$$= \left(\frac{60\,L_{10}}{10^6}\right)\ [(0.25 \times 500\,(1.0 \times 5000)^3 + 0.5 \times 700\,(1.5 \times 10{,}000)^3 + 0.25 \times$$

$$400\,(1.0 \times 7{,}000)^3]$$

$$(42500)^3 = 7.38705 \times 10^{10} \times L_{10}$$

$$L_{10} = 1039.2 \text{ hours}$$

Example 22.3 A ball bearing subjected to a radial load of 10 kN is expected to have a life of 10,000 hours at 1500 rpm with a reliability of 95%. Calculate the dynamic load capacity of the bearing for a reliability of 90%.

Solution.

$$L_{95} = \frac{60nL_h}{10^6} = \frac{60 \times 1500 \times 10{,}000}{10^6} = 900 \text{ millions of revolution}$$

$$\frac{L_{95}}{L_{90}} = \left[\frac{\ln\left(\frac{1}{0.95}\right)}{\ln\left(\frac{1}{0.9}\right)}\right]^{1/1.17}$$

$$= 0.5405$$

$$L_{90} = \frac{900}{0.5405} = 16622.086 \text{ millions of revolutions}$$

$$C = P(L_{90})^{1/3} = 10{,}000\,(16622.086)^{1/3} = 118526 \text{ N.}$$

Example 22.4 A double direction 60° taper roller bearing has a static capacity of 57 kN and dynamic capacity of 69.5 kN. It is required to carry a thrust load of 10 kN and radial load of 5 kN. Calculate the life of the bearing for 90% survival.

Solution.

For double direction 60° taper roller breaing, $e = 2.17$

$$\frac{F_a}{F_r} = \frac{10}{5} = 2 < e$$

$$X = 1.90,\ Y = 0.54$$

Equivalent thrust load,

$$P_a = XF_r + YF_a$$

$$= 1.90 \times 5 + 0.54 \times 10 = 14.9 \text{ kN}$$

$$L_{10} = \left(\frac{C}{P_a}\right)^{10/3} = \left(\frac{69.5}{14.9}\right)^{10/3} = 169.562 \text{ millions of revolution.}$$

Example 22.5 A deep groove ball bearing having 60 mm bore and rotating at 1440 rpm is subjected to a radial load of 3 kN an axial load of 2 kN. The radial and thrust factors are 0.56 and 2.0 respectively. The service factor is 1.2. If the expected rating life is 20,000 hours, calculate the required basic dynamic capacity of the bearing and select the bearing from manufacturer's catalogue.

Solution.

Given : d = 60 mm, n = 1440 *rpm*, F_r = 3 *k*N, F_a = 2*k*N, X = 0.56, Y = 2.0, C_s = 1.2,

L_h = 20,000 hours, V = 1.0 for inner race rotating

Equivalent dynamic load,

$$P = (XVF_r + YF_a)\,C_s = (0.56 \times 1 \times 3 + 2 \times 2)\,1.2 = 6.816\ kN$$

$$L_{10} = \frac{L_h \times 60 \times n}{10^6} = \frac{20{,}000 \times 60 \times 1440}{10^6} = 1728 \text{ millions of revolutions}$$

$$L_{10} = \left(\frac{C}{P}\right)^{\frac{1}{3}}$$

$$1728 = \left(\frac{C}{6816}\right)^3$$

$$C = 81792\ N$$

From Table 22. C = 83 *k*N for 60BC04 (SKF 6412) deep groove ball bearing.

We select this bearing.

Example 22.6 A single row deep – groove ball bearing is subjected to the following work cycle :

Sr. No.	Fraction of cycle, r	Radial load F_r N	Trust load F_a N	Speed n *rpm*	Service factor C_s
1.	$\frac{1}{10}$	2500	1800	450	2.5
2.	$\frac{1}{10}$	2000	1500	500	1.5
3.	$\frac{1}{5}$	1200	1600	600	2.0
4.	$\frac{3}{5}$	1500	2000	750	1.5

If the desired rating life is 15,000 hours, select a suitable bearing. Inner race rotates and radial and axial load factors are 1.0 and 1.5 respectivey.

Solution.

Given : L_h =15,000, hours, V = 1.0, X = 1.0, Y = 1.5

$$P = (XVF_r + YF_a)\,C_s$$

$$C^3 = \left(\frac{60L_h}{10^6}\right) \sum_{i=1}^{m} r_i n_i\,(C_s P)_i^3$$

$$C^3 = \left(\frac{60 \times 15000}{10^6}\right)\left[\left(\frac{1}{10}\right) \times 450 \times \{2.5\,(1.0 \times 1.0 \times 25{,}000 + 1.5 \times 1800)\}^3\right.$$

$$+ \left(\frac{1}{10}\right) \times 500 \times \{1.5\,(1.0 \times 1.0 \times 2{,}000 + 1.5 \times 1{,}500)\}^3$$

$$+ \left(\frac{1}{5}\right) \times 600 \times \{2.0\,(1.0 \times 1.0 \times 1200 + 1.5 \times 1600)\}^3$$

$$\left. + \left(\frac{1}{5}\right) \times 750 \times \{1.5\,(1.0 \times 1.0 \times 1500 + 1.5 \times 2000)\}^3\right]$$

$$= 1.8247 \times 10^{14}$$

$$C = 56718.93 \text{ N or } 56.72\ k\text{N}$$

C = 64 kN of 85BC02 (or SKF 6217) single row deep groove ball bearing

EXERCISES

22.1 Select a suitable bearing for the following operating conditions :
Speed = 1440 *rpm*, radial load = 7.5 *k*N, thrust load = 6 *k*N, bore = 45 mm, life = 500 hours.
Load is steady and inner race rotates.

22.2 Select a single – row deep groove ball bearing for a radial load of 5 *k*N and thrust load of 6 *k*N, operating at a speed of 1500 *rpm,* for an average life of 6 years (300 working days per year) running for 12 hours per day. Assume light shocks.

22.3 Select a suitable rolling contact bearing for the following operating conditions :
Bore = 50 mm, speed = 600 rpm, radial load = 3 *k*N, thrust load = 1 *k*N, bearing life for 95% reliablity = 15000 hours, operating temperature = 50°C, inner race rotates.

22.4 A ball bearing operates on a work cycle as given below :

Sr. No.	% of cycle	Radial load, F_r *k*N	Thrust load F_a *k*N	Speed n *rpm*	Service factor C_s
1.	30	3	2	750	1.25
2.	40	8	5	1500	1.5
3.	30	5	3	900	1.5

The basic dynamic capacity of the bearing is 36.5 kN. Determine :

(*a*) Rating life of bearing in hours

(*b*) Average speed of rotation, and

(*c*) Life of earing with 95% reliability

22.5 A single – row deep – groove ball bearing operates the following work cycle :

Sr. No.	% of cycle	Radial load $F_n kN$	Thrust load F_a kN	Radial factor X	Thrust factor Y	Race rotating	Speed n rpm	Serice factor C_s
1.	50	3.0	1.5	0.56	1.4	Inner	750	1.5
2.	20	2.5	1.0	0.56	1.6	Outer	1500	2.0
3.	30	2.0	0.5	0.56	1.4	outer	750	1.25

If the expected life of the bearing is 20,000 hours with a reliability of 95%, determine the basic dynamic load rating of the bearing.

If there are 6 such bearings in a system, what is the probability that all bearings will survive for 20,000 hours ?

23 FLYWHEEL

Chapter

23.1 INTRODUCTION

A flywheel is a heavy body rotating about its axis. It acts as a reservior of energy which is stored in the form of kinetic energy. The extra energy is stored during the idle stroke of the driven machinery and released during the working stroke. Thus the flywheel controls the fluctuation of speed during each cycle of the driven machinery. In the case of a prime mover, the flywheel absorbs the excess energy during the expansion stroke and release during the suction, compression and exhaust strokes.

23.2 COEFFICIENT OF FLUCTUATION OF SPEED

The difference between the maximum and minimum speeds during a cycle is called the maximum fluctuation of speed. The ratio of maximum fluctuation of speed to the mean speed is called coefficient of fluctuation of speed.

Let ω_{max} = maximum angular speed

ω_{min} = minimum angular speed

$$\omega_{mean} = \left(\frac{1}{2}\right)(\omega_{max} + \omega_{min}) = \text{mean angular speed}$$

Coefficient of fluctuation of speed,

$$C_s = \frac{\omega_{max} - \omega_{min}}{\omega_{mean}} \quad \text{... (23.1}a\text{)}$$

$$= \frac{N_{max} - N_{min}}{N_{mean}} \quad \text{... (23.1}b\text{)}$$

where N = speed in rpm

Refer to Table 23.1 for values of C_s for various machinery.

Table 23.1 Coefficient of fluctuation of speed, C_s

Machine type	Coefficient of fluctuation of speed C_s
Punching machine, Pressing machine	0.2 – 0.07
Pumps	0.05 – 0.03
Machine tools	0.03 – 0.025
Textile and paper machinery	0.025

Machine type	Coefficient of fluctuation of speed C_s
Spinning machinery	0.02
DC generator (lighting)	0.007
AC generator	0.003
DC generator	0.01
Belt driven DC generator	0.014
AC generator direct drive	0.006
Belt driven AC generator	0.008
Automobile (idling)	0.2
Automobile (normal speed)	0.1
Weaving machine	0.025
Flour mill machinery	0.02
Reciprocating pumps and compressors	0.04

23.3 COEFFICIENT OF FLUCTUATION OF ENERGY

It is defined as the ratio of maximum fluctuation of energy to the work done per cycle.

Maximum fluctuation of energy,

$$\Delta E = \text{Maximum energy} - \text{Minimum energy}$$

Coefficient of fluctuation of energy,

$$C_e = \frac{\Delta E}{\text{Work done per cycle}} \quad \text{... (23.2)}$$

$$\text{Work done per cycle} = T_{mean} \times \theta \quad \text{... (23.3}a\text{)}$$

Where T_{mean} = mean torque

$$= \frac{\text{Power transmitted in watts}}{\text{angular speed}}$$

$$= \frac{P}{\omega}$$

$$\omega = \frac{2\pi N}{60} \text{ rad/s}$$

$$\text{Work done per cycle is also} = \frac{60P}{n} \quad \text{... (23}b\text{)}$$

where n = number of working strokes per minute

= N, in case of steam engines and two stroke internal combustion engines

$= \frac{n}{2}$, in case of four stroke internal combustion engines

Fluctuation of kinetic energy,

$$\Delta E = C_e E. \quad \text{... (23.4)}$$

Where C_e = coefficient of fluctuation of energy (see Table 23.2 and 23.3)

Let I = moment of inertia of flywheel

Change of kinetic energy,

$$\Delta E = \frac{1}{2}\left(\omega_{max}^2 - \omega_{min}^2\right)$$

$$= \frac{1}{2}\left(\omega_{max} + \omega_{min}\right)\left(\omega_{max} - \omega_{min}\right) \quad \text{... (23.5)}$$

Table 23.2 Coefficient of fluctuation of energy for steam engines, C_e

Cut off percentage	Coefficient of fluctuation of energy, C_e		
	Single cylinder	Two cylinder	Three cylinder
10	0.35	0.088	0.040
20	0.33	0.082	0.037
40	0.31	0.078	0.034
60	0.29	0.072	0.032
80	0.28	0.070	0.031
100	0.27	0.068	0.030

Table 23.3 Coefficient of fluctuation of energy for I.C. engines, C_e

Type of engine	Number of cylinders	Angle between crank (*s*) Deg	Coefficient of fluctuation of energy C_e	
			Four stroke	Two Stroke
Single acting	1	360	2.35 – 2.40	0.95 – 1.00
Single acting	2	180	1.50 – 1.60	0.20 – 0.25
Vertical	3	120	0.60 – 0.75	0.15 – 0.18
Vertical	4	180 and 90	0.15 – 0.20	0.075 – 0.10
Vertical	6	180 and 60	0.10 – 0.12	0.016 – 0.02

23.4 ENERGY STORED IN A FLYWHEEL

Consider a flywheel as shown in Fig. 23.1

Let m = mass of flywheel, kg

K = radius of gyration of flywheel about its axis of rotation.

I = mass moment of inertia of the flywheel about its axis of rotation, kg.m^2

= mk^2

N_{max}, N_{min} = maximum and minimum speeds during a cycle, rpm

$\omega_{max}, \omega_{min}$ = maximum and minimum angular speeds during a cycle, rad/s

ω_{mean} = mean angular speed during a cycle

Mean kinetic energy of flywheel,

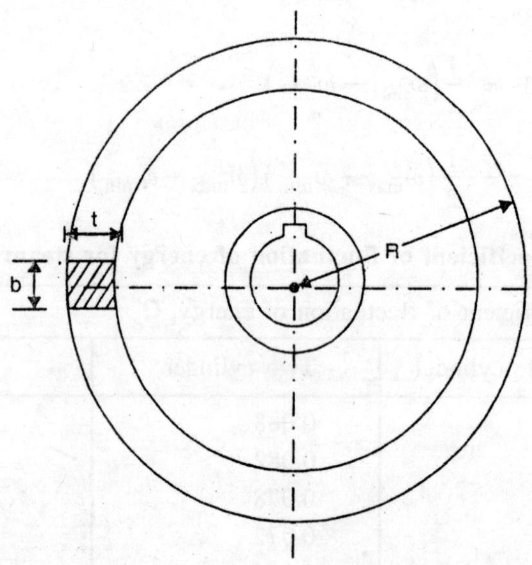

Fig. 23.1 Flywheel

$$E = \left(\frac{1}{2}\right) I\, \omega_{mean}^2$$

$$= \left(\frac{1}{2}\right) \times m K^2 \times \omega_{mean}^2$$

Maximum fluctuation of energy as the speed changes from ω_{max} to ω_{min},

$$\Delta E = \text{Maximum K.E.} - \text{Minimum K.E.}$$

$$= \left(\frac{1}{2}\right) \times I \times \left[\omega_{max}^2 - \omega_{min}^2\right]$$

$$= \left(\frac{1}{2}\right) \times I \times (\omega_{max} + \omega_{min})(\omega_{max} - \omega_{min})$$

$$= I \times \omega_{mean} (\omega_{max} - \omega_{min})$$

$$= \frac{I \times \omega^2{}_{mean} (\omega_{max} - \omega_{min})}{\omega_{mean}}$$

$$= 2EC_s \quad \text{... (23.6)}$$

For a flywheel having very small thickness of rim as compared to the rim diameter, $K \approx R$.

Therefore, $I = mR^2$ ... (23.7)

Mass of flywheel, $m = 2\pi R \times bt \times \rho$ ... (23.8)

where b = width of rim

t = thickness of rim

ρ = density of rim material

= 7260 kg/m^3 for cast iron

= 7800 kg/m^3 for steel

Generally $= \frac{b}{t} = 2$

23.5 TYPES OF FLYWHEELS

The flywheels can be of the following types :

1. Solid disc type
2. Rimmed type with either arms or solid web.

Solid disc type flywheels are rarely used because they have less capacity of storing energy. Rimmed types of flywheels with arms are preferred because they can store more energy. Small rimmed type flywheels are manufactured with solid web or holes drilled in the web.

1. Solid disc type flywheel

Moment of inertia of a solid disc of radius R and mass *m*,

$$I = \frac{mR^2}{2} \quad \text{... (23.9)}$$

Mass of the disc, $m = \pi R^2 t\rho$ ... (23.10)

where t = thickness of disc

ρ = density of disc material

2. Rimmed flywheel

The arms, hub and the shaft contribute only 10% of the total energy stored by a rimmed flywheel. The rim contributes approximately 90% of the required moment of inertia.

Moment of inertia of the rim,

$$I_r = KI$$

where k = a factor = 0.9

I = moment of inertia of the entire flywheel

$$I_r = m_r R^2 \quad \text{... (23.11)}$$

where m_r = mass of rim

R = mean radius of rim

The limiting mean rim velocity is kept below 30 m/s for flywheels made of grey cast iron. When the velocity exceeds this limit, there is a possibility of bursting due to centrifugal force.

$$R < 30/\omega_{mean} \quad \text{... (23.12)}$$

For limiting rim velocity, refer to Table 23.4.

Table 23.4 Limiting rim velocity of flywheel

Machine type	Rim velocity, m/min
Cast iron wheels under 100 kg	1500
Large machinery	2000
Flywheel without blow holes	3500
Cast steel wheel	3500
Large flywheel for steel mills	4500
Cast iron flywheel for automobiles	3100
Cast steel flywheel for automobiles	6200

23.6 Design of Belt – Pulley Flywheel

A rimmed type flywheel is shown in Fig. 23.2

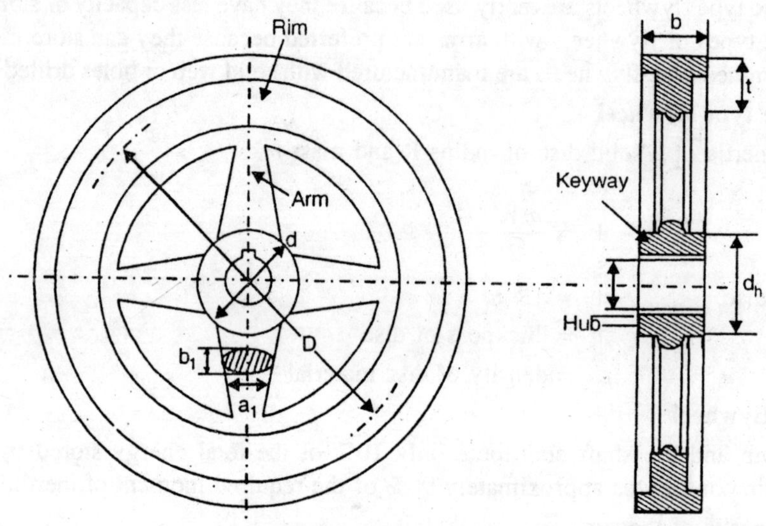

Fig. 23.2 Rimmed type flywheel

1. Rim size

Let b = width of flywheel rim

= belt width + 40 mm

t = thickness of rim ... (23.13)

$$\frac{b}{t} = 0.5 - 1.5 \text{ for belt – pulley flywheel} \quad \text{... (23.14}a\text{)}$$

$$= 2 - 4 \text{ for a rimmed flywheel} \quad ..(23.14b)$$

2. Shaft and hub size

$$P = \text{power transmitted, kW}$$

Torque on flywheel shaft, $T = \dfrac{60 \times 10^3 \times P}{2\pi N}$ N.m ... (23.15)

where N = rpm of flywheel

M = bending moment on the flywheel shaft

= weight of flywheel × overhang ... (23.16)

Diameter of flywheel shaft,

$$d = \left[\left\{\frac{16}{\pi \tau_d}\right\}\left\{(K_m M)^2 + (K_t T)\right\}^{\frac{1}{2}}\right]^{\frac{1}{3}} \quad ... (23.17)$$

where τ_d = design shear stress for shaft

K_m, K_t = shock and fatigue factors for bending and torsion respectively.

Diameter of hub, $d_h = 2d$ ... (23.18a)

Length of hub, $l_h = (2 \text{ to } 2.5)d$ ... (23.18b)

23.7 STRESSES IN FLYWHEEL RIM

The following stresses are induced in the rim of a flywheel :

(*i*) Tensile stress due to centrifugal force.

(*ii*) Tensile bending stress caused by the restraint of the arms.

(*iii*) Shrinkage stresses due to unequal rate of cooling of casting.

It is difficult to calculate these stresses.

(*i*) Tensile stress in flywheel rim due to centrifugal force

$$\sigma_t = \rho v^2 \quad ... (23.19)$$

where v = linear velocity of rim

$= R\omega$

(*ii*) Tensile bending stress due to restraint of the arms

It is assumed that each portion of the rim between a pair of arms behaves like a beam fixed at both ends and uniformly loaded, as shown in Fig. 23.3.

Length between fixed ends, $l = \dfrac{2\pi R}{n}$

Where n = number of arms

The uniformly distributed load 'w' per metre length will be equal to the centrifugal force between a pair of arms.

$$w = bt\rho\omega^2 R \text{ N/m}$$

Maximum bending moment,

$$M = \frac{wl^2}{12} = \left(\frac{bt\rho\omega^2 R}{12}\right)\left(\frac{2\pi R}{n}\right)^2$$

$$= \frac{bt\omega^2 R^3 \pi^2}{3n^2}$$

$$= \frac{\pi^2 bt\rho v^2 R}{3n^2}$$

Section modulus,

$$z = \frac{bt^2}{6}$$

Fig. 23.3 Bending stress due to restraint of arms

Bending stress in flywheel rim due to restraint of arms,

$$\sigma_b = \frac{M}{z} = \frac{2\pi^2 \rho v^2 R}{n^2 t} \quad \text{... (23.20)}$$

Where R = mean rim radius

It has been observed that the arms of a flywheel stretch about $\frac{3}{4}$ th of the amount necessary for free expansion. Therefore, the total stress in the rim,

$$\sigma_{total} = \left(\frac{3}{4}\right)\sigma_t + \frac{\sigma_b}{4} \quad \text{... (23.21a)}$$

$$= \left(\frac{3}{4}\right)\rho v^2 + \rho v^2 \left\{\frac{\pi^2 R}{2n^2 t}\right\}$$

$$= \rho v^2 \left[0.75 + \frac{4.935 R}{n^2 t}\right] \quad \text{... (23.21b)}$$

For a safe design, $\sigma_{total} < 38$ MPa for C.I.

23.8 STRESSES IN FLYWHEEL ARMS

The following stresses are induced in the flywheel arms :

(*i*) Tensile stress due to centrifugal force acting on the rim.

(*ii*) Bending stress due to torque transmitted from the rim to the shaft or vice – versa.

(*iii*) Shrinkage stresses due to unequal rate of cooling of casting.

It is difficult to calculate these stresses.

(*i*) Tensile stress due to centrifugal force

The tensile stress due to centrifugal force acting on the rim, assuming 75% effect, is given by,

$$\sigma_{t1} = \left(\frac{3}{4}\right)\sigma_t = \left(\frac{3}{4}\right)\rho v^2 \quad \text{... (23.22)}$$

(ii) Bending stress due to torque transmitted

Due to the torque transmitted from the rim to the shaft or vice – versa, the arms will be subjected to bending, because they are required to carry the full torque load.

Consider the arm as shown in Fig. 23.4. The arm may be assumed as a cantilever beam fixed at the hub and carrying a concentrated load at the free end at the rim.

Let R = mean radius of the rim

r = radius of hub / $\frac{d_h}{2}$

T = maximum torque transmitted by the shaft

Load at the mean radius of rim,

$$F = \frac{T}{R}$$

$$\text{Load in each arm} = \frac{T}{(nR)}$$

Maximum bending moment on the arm near the hub,

$$M = \frac{T(R-r)}{nR}$$

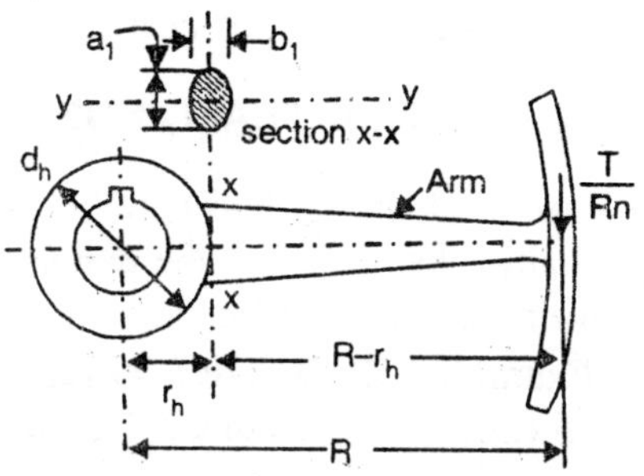

Fig. 23.4 Bending stress in arm

Bending stress in arm, $$\sigma_{b1} = \frac{T(R-r)}{nRz_y} \quad \text{... (23.23)}$$

Where z_y = section modulus of arms about y – axis

Total tensile stress in the arms near the hub,

$$\sigma = \sigma_{t1} + \sigma_{b1} \quad \text{... (23.24)}$$

For belt – pulley flywheel, bending stress at the hub due to belt tensions, considering that only half the number of arms are effective in transmitting belt tension,

$$\sigma_{b2} = \frac{(F_1 - F_2)(R-r)}{\left(\frac{nz_y}{2}\right)} \quad \text{... (23.25)}$$

Where F_1, F_2 = belt tensions on tight and slack sides respectively

Total stress, $$\sigma_{total} = \sigma_{b1} + \sigma_{b2} + \sigma_{t1} \quad \text{... (23.26)}$$

For a safe design, σ_{total} < 20 MPa for C.I. ... (23.27a)

< 60 MPa for steel ... (23.27b)

23.9 DESIGN OF FLYWHEEL ARMS

The cross – section of the arms is generally elliptical with major axis as twice the minor axis.

Let a_1 = major axis of arms near the hub

b_1 = minor axis of arms near the hub

Section modulus of arms about the y – axis, $z_y = \dfrac{\pi b_1 a_1^2}{32}$

For $b_1 = \dfrac{a_1}{2}$,

major axis of elliptical arm section

$$a_1 = \left[\frac{64 z_y}{\pi}\right]^{\frac{1}{3}} \quad \text{... (23.28)}$$

Cross section of arms near the rim = $\dfrac{2}{3}$ rd of that near the hub ... (23.29)

23.10 DESIGN OF SHAFT, HUB AND KEY

Size of shaft and hub may be determined as explained in art. 23.6. The diameter of hub is generally taken as twice the diameter of shaft. The length of hub is taken as 2 – 2.5 times the diameter of shaft or twice the rim width. A standard sunk key is used for the shaft and hub.

Example 23.1 Design a suitable flywheel for a 12 BHP, 2 – cylinder vertical 4 – stroke diesel engine coupled to a 7.5 kW A.C. generator. The normal speed of the engine is 900 rpm.

Solution.

Given : P = 7.5 kW, BHP = 12, N = 900 rpm

1.For a 2 – cylinder, 4 – stroke, I.C. engine, C_e = 1.5 – 1.6 (Table 23.2)

We take, C_e = 1.5

2.For A.C. generator directly coupled, C_s = 0.006 (Table 23.3)

3.Fluctuation of energy, $\Delta E = \dfrac{C_e \times 10^3 \times P \times 60}{N}$

$$= \frac{1.5 \times 10^3 \times 7.5 \times 60}{900} = 750 \text{ N.m}$$

4. We take v =25 m/s for cast iron flywheel. (v < 50 m/s for I.C. Engine flywheel) Table (23.4)

5. Mass of flywheel, $m = \dfrac{\Delta E}{C_s v^2} = \dfrac{750}{0.006 \times 25^2} = 200$ kg

6. Mass of rim, m_r = 0.9 m = 180 kg

7. Mean radius of flywheel,

$$R = \frac{60v}{2\pi N}$$

$$= \frac{60 \times 25}{\pi \times 900} = 0.265 \text{ m}$$

8. Size of rim

$$m_r = 2\pi R b t \rho$$

$$\frac{b}{t} = 0.5 - 1.5 \text{ for pulley type flywheel}$$

$$= 1.25$$

For C.I., $\rho = 7100 \text{ kg/m}^3$

$$180 = 2\pi \times 0.265 \times 1.25t^2 \times 7100$$

$$t = 110.4 \text{ mm}, b = 137.95 \text{ mm}$$

We adopt **b = 140 mm and t = 112 mm, R = 265 mm**

9. Flywheel shaft.

Let overhang of flywheel = 150 mm

Bending moment, M = weight of flywheel × overhang

$$= 200 \times 9.81 \times 0.15 = 294.3 \text{ N.m}$$

$$\text{Torque, T} = \frac{0.746 \times \text{BHP} \times 10^3 \times 60}{2\pi \text{N}}$$

$$= \frac{0.746 \times 12 \times 10^3 \times 60}{2\pi \times 900} = 94.9837 \text{ N.m}$$

Let $K_m = 1.5$ and $K_t = 1.5$, $\tau_d = 40$ MPa

$$\text{Diameter of shaft, } d = \left[\left\{\frac{16}{\pi\tau_d}\right\}\left\{(K_m M)^2 + (K_t T)^2\right\}^{\frac{1}{2}}\right]^{\frac{1}{3}}$$

$$= \left[\left\{\frac{16}{\pi \times 40}\right\}\left\{(1.5 \times 294.3)^2 + (1.5 \times 94.9837)^2\right\}^{\frac{1}{2}}\right]^{\frac{1}{3}} \times 10$$

$$= 38.94 \approx 40 \text{ mm}$$

10. Hub diameter, d_h = $2d$ = 80 mm

11. Hub length, l_h = $2d$ = 80 mm

12. The size of the flywheel is small. We shall use web for the flywheel with thickness,

$$t_w = \frac{b}{20} = \frac{140}{20} = 7 \text{ mm}$$

Example 23.2 A punching machine is required to produce 30 holes per minute of 20 mm diameter in a steel plate of 15 mm thickness having ultimate shearing strength of 340 MPa. The actual punching operation last for a period of 36° rotation of punching machine crankshaft. Mechanical efficiency of punching machine is 80%. It rotates at 240 rpm. Design a suitable flywheel for the punching machine.

Solution.

Given : d =20 mm, t = 15 mm, τ_u = 340 MPa, θ = 36°, η_{mech} = 0.8, N = 240 rpm

1. Punching force, F $= \pi d t \tau_u = \pi \times 20 \times 15 \times 340 = 320443$ N

2. Work done in punching one hole,

$$E_1 = 0.5F \times \text{distance travelled by punch}$$

$$= 0.5 \times 320443 \times 15 \times 10^{-3} = 2403.3 \text{ N.m}$$

3. Energy required per hole at flywheel shaft,

$$E = \frac{E_1}{\eta_{mech}} = \frac{2403.3}{0.8} = 3004.15 \text{ N.m}$$

4. Actual punching operation period

$$= \frac{\theta}{360} = \frac{36}{360} = 0.1 \text{ part of one revolution of crank}$$

5. Energy stored by the flywheel in the remaining period,

$$\Delta E = (1 - 0.1)\, E = 0.9 \times 3004.15 = 2703.73 \text{ N.m}$$

6. Coefficient of fluctuation of speed,

$$C_s = 0.07 - 0.2$$

We take, $C_s = 0.1$

7. Let v = 10 m/s

Rim diameter, $D = \frac{60v}{\pi N} = \frac{60 \times 10}{\pi \times 240}$

$$= 0.796 \approx 0.8 \text{ m or } 800 \text{ mm}$$

8. Mass of flywheel, $m = \frac{\Delta E}{C_s v^2} = \frac{2703.73}{0.1 \times 10^2} = 270.4$ kg

9. Mass of rim,

$$m_r = 0.9m = 0.9 \times 270.4 = 243.4 \text{ kg}$$

10. Rim size

Let $\frac{b}{t} = 1.25$

$$1.25t^2 = \frac{243.4}{\pi \times 0.8 \times 7100}$$

$$t = 104.5 \approx 105 \text{ mm}$$

$$b = 131.25 \approx 132 \text{ mm}$$

We adopt b = 132 mm, t = 105 mm, D = 800 mm

11. Flywheel shaft.

Energy required per minute at flywheel shaft = E × number of holes punched per minute

$$= 3004.15 \times 30 = 90124.5 \text{ N.m/min}$$

Torque on the flywheel shaft,

$$T = \frac{90124.5}{2\pi \times 240} = 59.766 \text{ N.m}$$

$$\text{Let overhang} = 150 \text{ mm}$$

$$\text{Bending moment, M} = 270.4 \times 9.81 \times 0.150 = 397.9 \text{ N.m}$$

$$\text{Let } K_m = 1.5,\ K_t = 2.0,\ \tau_d = 40 \text{ MPa}$$

$$\text{Shaft diameter, } d = \left[\left\{\frac{16}{\pi\tau_d}\right\}\left\{(K_m M)^2 + (K_t T)^2\right\}^{\frac{1}{2}}\right]^{\frac{1}{3}}$$

$$= \left[\left\{\frac{16}{\pi \times 40}\right\}\left\{(1.5 \times 397.9)^2 + (2.0 \times 59.766)^2\right\}^{\frac{1}{2}}\right]^{\frac{1}{3}} \times 10$$

$$= 42.6 \approx 45 \text{ mm}$$

12. Hub diameter, $d_h = 2d = 90$ mm

13. Hub length, $l_h = 2.25d \approx 100$ mm

14.The size of the flywheel is small. We shall use web for the flywheel with thickness,

$$t_w = \frac{b}{20} = \frac{132}{20} = 6.6 \approx 7 \text{ mm}$$

Example 23.3 Design a flywheel for a 4 – stroke, 4 – cylinder, 130 kW engine running at 375 rpm. The engine is directly coupled to an A.C. generator. Due to space restrictions the flywheel diameter should not be more than 1.25 m.

Solution.

Given : P =130 kW, N = 375 rpm. $D_{max} = 1.25$ m

1.Coefficient of fluctuation of energy,

$$C_e = 0.15 - 0.20$$

$$\text{We take, } C_e = 0.16$$

2. Fluctuation of energy

$$\Delta E = \frac{60 C_s P}{N} = \frac{60 \times 0.16 \times 130 \times 10^3}{375} = 3328 \text{ N.m}$$

3.Let D = 1 m, giving $v = \dfrac{\pi \times 1 \times 375}{60} = 19.63$ m/s < 30 m/s.

Hence safe for C.I. flywheel.

4.For directly coupled A.C. generator,

$$C_e = 0.006$$

5. Mass of flywheel, $m = \dfrac{\Delta E}{C_e v^2} = \dfrac{3328}{0.006 \times (19.63)^2} = 1439.43$ kg

6. Mass of rim, $m_r = 0.9$ m = 1295.5 kg

7. Rim size.

$$\text{Let } b = 2t, \text{ then } 2t^2 = \frac{1295.5}{\pi \times 1 \times 7100}$$

$$t = 170 \text{ mm}, b = 340 \text{ mm}$$

Outer diameter of flywheel,

$$D_o = D + t = 1.17 \text{ m} < 1.25 \text{ m, the space available.}$$

8. Shaft design.

$$\text{Torque, T} = \frac{10^3 \text{ P} \times 60}{2\pi \text{N}} = \frac{10^3 \times 130 \times 60}{2\pi \times 375} = 3310.4 \text{ N.m}$$

$$\text{Let overhang} = 300 \text{ mm}$$

$$\text{Bending moment, M} = 1439.43 \times 9.81 \times 0.3 = 42323.24 \text{ N.m}$$

$$\text{Let } K_m = K_t = 1.5, \tau_d = 40 \text{ MPa}$$

$$\text{Shaft diameter, } d = \left[\left\{\frac{16}{\pi \tau_d}\right\}\left\{(K_m M)^2 + (K_t T)^2\right\}^{\frac{1}{2}}\right]^{\frac{1}{3}}$$

$$\left[\left\{\frac{16}{\pi \times 40}\right\}\left\{(1.5 \times 42323.24)^2 + (1.5 \times 3310.4)^2\right\}^{\frac{1}{2}}\right]^{\frac{1}{3}} \times 10$$

$$= 101 \approx 110 \text{ mm}$$

9. Hub diameter, $d_h = 2d = 220$ mm

10. Hub length, $l_h = 2.25d = 247.5 \approx 250$ mm

11. Number of arms, $n = 4$ for D < 450 mm

12. Check for rim stresses.

$$\text{Total stress in rim} = 0.75\sigma_t + 0.25\sigma_b$$

$$= \rho v^2 \left[0.75 + \frac{0.25\pi^2 D}{n^2 t}\right]$$

$$= 7100 \times (19.63)^2 \left[0.75 + \frac{0.25 \times \pi^2 \times 1}{16 \times 0.170}\right] \times 10^{-6}$$

$$= 4.53 \text{ MPa} < 38 \text{ MPa}$$

Hence safe.

13. Arms cross – section.

Required section modulus of arms,

$$z_y = \frac{T K_t (D - d)}{n D_{\sigma_{b1}}}$$

$$= \frac{3310.4 \times 10^3 \times 1.5(1000-110)}{4 \times 1000 \times 14}$$

$$= 78918 \text{ mm}^3$$

Major axis of arms cross - section,

$$a_1 = \left[\frac{64 z_y}{\pi}\right]^{\frac{1}{3}}$$

$$= \left[\frac{64 \times 78918}{\pi}\right]^{\frac{1}{3}} = 117 \approx 120 \text{ mm}$$

$$\text{Minor axis, } b_1 = \frac{a_1}{2} = 60 \text{ mm}$$

14. Key size.

Rectangular key : width, $w = \frac{d}{4} = \frac{110}{4} \approx 30$ mm, thickness $= \frac{d}{6} \approx 20$ mm

Example 23.4 An Otto cycle engine develops 50 kW at 140 rpm with 70 explosions per minute. The fluctuation of speed should not exceed 0.5 % of the mean speed on either side. Design a suitable cast iron rim section having width four times the depth so that the hoop stress does not exceed 4.5 MPa. Assume that the flywheel stores $\frac{16}{15}$ times the energy stored by the rim and that the work done during power stroke is 1.5 times the work done during the cycle. Take density of rim material as 7200 kg/m^3.

Solution.

Given : P = 50 kW, N = 140 rpm, n = 70, σ_t = 4.5 MPa, ρ = 7200 kg/m^3

1. **Fluctuation of energy**

$$\text{Power transmitted, } P = \frac{2\pi N T_{mean}}{60 \times 10^3}$$

$$T_{mean} = \frac{50 \times 60 \times 10^3}{2\pi \times 140} = 3410.5 \text{ N.m}$$

$$\text{Number of explosions, } n = 70 = \frac{N}{2}$$

Therefore, it is a four stroke engine. The T – θ diagram is shown in Fig. 23.5

Work done per cycle $= T_{mean} \times \theta = 3410.5 \times 4\pi = 42857.6$ N.m

Work done during power stroke = 1.5 × 42857.6 = 64286.4 N.m

The work done during power stroke is shown by the triangle ABC in which base AC = π radians and height FB = T_{max}

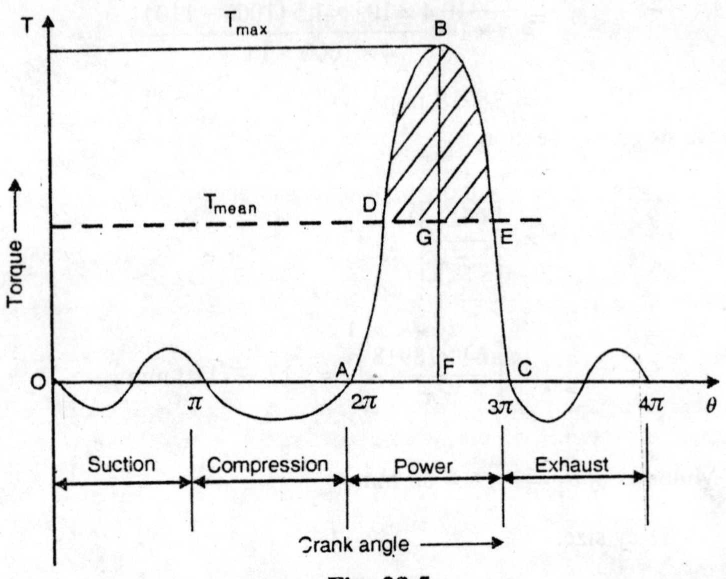

Fig. 23.5

$$\text{Work done during power stroke} = \left(\frac{1}{2}\right) T_{max} \times \pi$$

$$= 64286.4$$

$$T_{max} = 40926 \text{ N.m}$$

Height above the mean torque line,

$$GB = FB - FG$$

$$= T_{max} - T_{mean}$$

$$= 40926 - 3410.5 = 37515.5 \text{ N.m}$$

Maximum fluctuation of energy, ΔE, is represented by the area DBED. It being a parabola, its area is proportional to the square of the height of parabola. Therefore,

$$\frac{\text{Area DBED}}{\text{area ABCA}} = \left(\frac{GB}{FB}\right)^2$$

$$\Delta E = 64286.4 \times \left(\frac{37515.5}{40926}\right)^2 = 54018.4 \text{ N.m}$$

2. Mean diameter of flywheel

Hoop stress in flywheel rim,

$$\sigma_\theta = \rho v^2$$

$$v = \left(\frac{4.5 \times 10^6}{7200}\right)^{\frac{1}{2}} = 25 \text{ m/s}$$

$$\frac{\pi DN}{60} = v$$

$$D = \frac{25 \times 60}{\pi \times 140} = 3.41 \text{ m}$$

3. Size of rim

Width of rim, $b = 4t$

Area of cross section, $A = bt = 4t^2$

$N_{max} - N_{min} = 2 \times 0.05 = 1\%$ of mean speed N_{mean}

Coefficient of fluctuationof speed,

$$C_s = \frac{N_{max} - N_{min}}{N_{mean}}$$

$$= 0.01$$

Maximum fluctuation of energy,

$$\Delta E = 2\, C_s E$$

$$= 2 \times 0.01 \times E = 0.02\, E$$

$$= 54018.4$$

$$E = 2700920 \text{ N.m}$$

$$\text{Energy of rim, } E_r = \left(\frac{15}{16}\right) E \; \frac{15 \times 2700920}{16} = 2532112.5 \text{ N.m}$$

Now

$$E_r = \left(\frac{1}{2}\right) m_r v^2$$

$$\text{Mass of rim, } m_r = \frac{2 \times 2532112.5}{(25)^2} = 8102.76 \text{ kg}$$

$$m_r = A \times \pi D \times \rho$$

$$= 4t^2 \times \pi \times 3.41 \times 7200$$

$$= 8102.76$$

$$t = 0.162 \text{ m or } 162 \text{ mm}$$

$$b = 4t = 648 \text{ mm}$$

Example 23.5 Design a cast iron flywheel for a four – stroke I.C. Engine developing 200 kW at 240 rpm. The hoop stress developed in the rim is 5 MPa. The total fluctuation of speed is to be limited to 2.5% of mean speed. The work done during the power stroke is 30% more than the average work done during the whole cycle. The maximum torque on the shaft is twice the mean torque. The density of cast iron is 7200 kg/m^3.

Solution.

Given : P =200 kW, N = 240 rpm, $\sigma_\theta = 5$ MPa, $N_{max} - N_{min} = 0.025\, N_{mean}$, $T_{max} = 2T_{mean}$,

$\rho = 7200$ kg/m^3

1. **Fluctuation of energy**

$$T_{mean} = \frac{P \times 10^3 \times 60}{2\pi \times N} = \frac{200 \times 10^3 \times 60}{2\pi \times 240} = 7957.75 \text{ N.m}$$

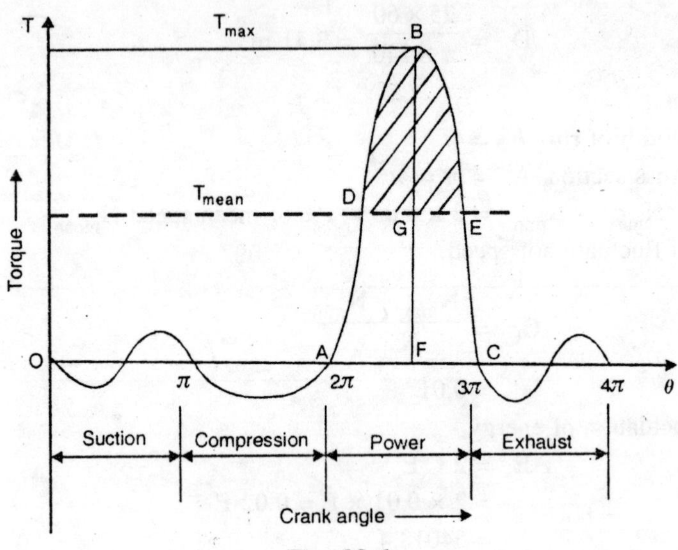

Fig. 23.6

Work done per cycle $= T_{mean} \times \theta = 5957.75 \times 4\pi = 100{,}000$ N.m

Work done during power stroke $= 1.3 \times 100{,}000 = 130{,}000$ N.m

The $T - \theta$ diagram is shown in Fig. 23.6

The work done during the power stroke is represented by the triangle ABC in which base

$$AC = \pi \text{ radians, and height } FB = T_{max}$$

$$\text{Work done during power stroke} = \left(\frac{1}{2}\right) \times \pi \times T_{max} = 130{,}000$$

$$T_{max} = 82760.6 \text{ N.m}$$

Height above the mean torque line,

$$GB = FB - FG$$
$$= T_{max} - T_{mean}$$
$$= 82760.6 - 7957.75 = 74802.82 \text{ N.m}$$

Maximum fluctuation of energy, ΔE, is represented by the area DBED. It being a parabola, its area is proportional to the square of the height of parabola. Therefore,

$$\frac{\text{Area DBED}}{\text{area ABCA}} = \left(\frac{GB}{FB}\right)^2$$

$$\Delta E = 130{,}000 \times \left(\frac{74802.82}{82760.6}\right)^2 = 106201.8 \text{ N.m}$$

2.Mean diameter of flywheel

Hoop stress in flywheel rim,

$$\sigma_\theta = \rho v^2$$

$$v = \left(\frac{5 \times 10^6}{7200}\right)^{\frac{1}{2}} = 26.35 \text{ m/s}$$

$$\frac{\pi DN}{60} = v$$

$$D = \frac{26.35 \times 60}{\pi \times 240} = 2.097 \text{ m}$$

3. Mass of rim

Coefficient of fluctuation of speed,

$$C_s = \frac{N_{max} - N_{min}}{N_{mean}}$$
$$= 0.025$$

Maximum fluctuation of energy,

$$\Delta E = 2\, C_s E$$
$$= 2 \times 0.025 \times E = 0.05\ E$$
$$= 106201.8$$
$$E = 2124036 \text{ N.m}$$

Now $$E = \left(\frac{1}{2}\right) m_r v^2$$

$$\text{Mass of rim, } m_r = \frac{2 \times 2124036}{(26.35)^2} = 6118.3 \text{ kg}$$

4. Rim size

Taking $b = 2t$

Area of rim, $a_r = 2t^2$

$m_r = A_r \times \pi D \times \rho = 2t^2 \times \pi \times 2.097 \times 7200$

$= 6118.3$

Thickness of rim, $t = 0.254$ m or 254 mm

Width of rim, $b = 2t = 508$ mm

5. Shaft and hub size

$$T_{max} = 2T_{mean}$$
$$= 2 \times 7957.75 = 15915.5 \text{ N.m}$$

$$\left(\frac{\pi}{16}\right) d^3 \tau = T_{max}$$

Taking $\tau = 42$ MPa

$$d = \left[\frac{16 \times 15915.5 \times 10^3}{\pi \times 42}\right]^{\frac{1}{3}}$$

$$= 124.5 \approx 125 \text{ mm}$$

Diameter of hub, $d_h = 2d = 250$ mm

Length of hub, $l_h = b = 508$ mm

6. Arms size

Number of arms for D = 2.097 m are 6

Bending moment on the arms,

$$M = \frac{T_{max}(D - d_h)}{Dn}$$

$$= \frac{15915.5(2.097 - 0.25)}{2.097 \times 6}$$

$$= 2336.38 \text{ N.m}$$

Section modulus, $z_y = \left(\frac{\pi}{32}\right) b_1 a_1^2 = \left(\frac{\pi}{32}\right) \times 0.5a_1 \times a_1^2 = 0.049\, a_1^3$

Taking permissible bending stress in arms = 15 MPa

$$15 = \frac{2336.38 \times 10^3}{0.049 a_1^3}$$

Major axis of arms, $a_1 = 147 \approx 150$ mm

Minor axis of arms, $b_1 = 75$ mm

7. Key size

For a shaft of diameter 125 mm, the size of rectangular sunk key are :

Width, $w = 36$ mm

Thickness, $h = 20$ mm

Considering shearing failure of the key,

$$\frac{L \times w \times \tau \times d}{2} = T_{max}$$

$$L = \frac{15915.5 \times 10^3}{36 \times 42 \times \frac{125}{2}}$$

$$= 168.4 \approx 170 \text{ mm}$$

8. Check for maximum stress in rim

Total stress in rim, $\sigma = \rho v^2 \left[0.75 + \frac{4.935 R}{n^2 t}\right]$

$$= 7200 \times (26.35)^2 \left[0.75 + \frac{4.935 \times \left(\frac{2.097}{2}\right)}{6^2 \times 254}\right]$$

$$= 6.58 \text{ MPa} < 15 \text{ MPa}$$

Hence safe.

Example 23.6 A punching press is used to punch 30 holes per minute in a plate using 10 kN.m of energy per hole during each revolution. Each piercing takes 42% of the time needed to make one revolution. The punch receives power through a gear reduction unit which in turn is fed by an electric motor driven belt pulley 900 mm diameter and running at 240 rpm. Find the power of the electric motor if overall efficiency of the transmission unit is 85%. Design a cast iron flywheel to be sued with the punching machine for a coefficient of steadiness of 5. Space limitations restrict the diameter to 1.2 m.

Permissible shear stress in shaft material = 45 MPa

Permissible tensile strees for cast iron = 4 MPa

Density of cast iron = 7200 kg/m^3

Solution.

Given : Number of holes =30 per minute, energy consumed per hole = 10 kN.m, d_p = 900 mm,

N_p = 240 rpm, η = 85%, $C_s = \frac{1}{5} = 0.2$, D_{max} = 1.2 m, τ = 45 MPa, σ_t = 4MPa, ρ = 7200 kg/m^3

1. Power of electric motor

Energy used for piercing holes per minute = Number of holes pierced × energy used per hole

= 30 × 10,000 = 300,000 N.m/min

$$\text{Power required for the electric motor, P} = \frac{\text{Energy used per minute}}{60 \times \eta}$$

$$= \frac{300{,}000}{60 \times 0.85} = 5882 \text{ W} = 5.882 \text{ kW}$$

2. Fluctuation of energy

Total energy to be supplied during each revolution,

$$E_t = \frac{10{,}000}{0.85} = 11764.7 \text{ N.m}$$

$$\text{Velocity of belt, } v = \frac{\pi d_p N_p}{60 \times 10^3}$$

$$= \frac{\pi \times 900 \times 240}{60 \times 10^3} = 11.3 \text{ m/s}$$

$$\text{Net pull acting on the belt. } F = \frac{P}{v} = \frac{5882}{11.3} = 520.8 \text{ N}$$

$$\text{Time required to punch a hole} = \frac{0.42}{30} = 0.014 \text{ min}$$

Distance moved by belt during punching a hole =velocity of belt × time required to punch a hole

= 11.3 × 60 × 0.014 = 9.492 m

Energy supplied by belt during punching a hole,

$$E_b = \text{net pull in belt} \times \text{distance travelled by belt} = 520.8 \times 9.492 = 4943.4 \text{ N.m}$$

Energy to be supplied by the flywheel,

$$\Delta E = E_t - E_b = 11764.7 - 4943.4 = 6821.3 \text{ N.m}$$

3. Size of flywheel

$$\text{Angular velocity, } \omega_{mean} = \frac{2\pi N}{60} = \frac{2\pi \times 240}{60} = 25.13 \text{ rad/s}$$

Diameter of flywheel rim, D = 1.2 m

R = 0.6 m

$$\Delta E = m R^2 \omega_{mean}^2 C_s$$

$$6821.3 = m \times (0.6)^2 \times (25.13)^2 \times 0.2$$

$$m = 150 \text{ kg}$$

4. Cross – section of flywheel rim

Let thickness of rim = t and width of rim, $b = 2t$

Area of cross – section of rim,

$$A = 2t^2$$

$$\text{Mass of rim, } m = A \times \pi D \times \rho$$

$$150 = 2t^2 \times \pi \times 1.2 \times 7200$$

$$t = 0.052 \text{ m or } 52 \text{ mm}$$

$$b = 104 \text{ mm}$$

5. Shaft and hub size

$$\text{Mean torque, } T_{mean} = \frac{P \times 60}{2\pi N}$$

$$= \frac{5882 \times 60}{2\pi \times 240} = 234 \text{ N.m}$$

$$\text{Let } T_{max} = 2T_{mean} = 2 \times 234 = 468 \text{ N.m}$$

$$\left(\frac{\pi}{16}\right) d^3 \tau = T_{max}$$

$$d = \left[\frac{16 \times 468 \times 10^3}{\pi \times 45}\right]^{\frac{1}{3}} = 37.56 \approx 40 \text{ mm}$$

Diameter of hub, $d_h = 2d = 80$ mm

Length of hub, $l_h = b = 104$ mm

6. Arms size

For D = 1.2 m, $n = 6$

Bending moment on the arms,

$$M = \frac{T_{max}\left(D - d_h\right)}{Dn}$$

$$= \frac{468\,(1.2 - 0.08)}{1.2 \times 6} = 72.8 \text{ N.m}$$

$$\text{Section modulus, } z_y = \left(\frac{\pi}{32}\right) b_1 a_1^2 = \left(\frac{\pi}{32}\right) \times 0.5 a_1 \times a_1^2 = 0.049\, a_1^3$$

Permissible bending stress in arms = 4 MPa

$$4 = \frac{72.8 \times 10^3}{0.049 a_1^3}$$

Major axis of arms, a_1 = 71.88 ≈ 72 mm

Minor axis of arms, b_1 = 36 mm

7. Key size

For a shaft of diameter 40 mm, the size of rectangular sunk key are :

Width, w = 16 mm

Thickness, h = 10 mm

Considering shearing failure of the key,

$$\frac{L \times w \times \tau \times d}{2} = T_{max}$$

$$L = \frac{468 \times 10^3}{16 \times 45 \times \frac{40}{2}}$$

$$= 32.5 \approx 35 \text{ mm}$$

8. Check for maximum stress in rim

$$\text{Velocity of rim, } v = \frac{\pi DN}{60} = \frac{\pi \times 1.2 \times 240}{60} = 15 \text{ m/s}$$

$$\text{Total stress in rim, } \sigma = \rho v^2 \left[0.75 + \frac{4.935 R}{n^2 t}\right]$$

$$= 7200 \times (15)^2 \left[0.75 + \frac{4.935 \times \left(\frac{1.2}{2}\right)}{6^2 \times 0.052}\right]$$

$$= 3.78 \text{ MPa} < 4 \text{ MPa}$$

Hence safe.

EXERCISES

23.1 Design a cast iron flywheel for a four – stroke engine to develop 100 kW at 160 rpm. The work done in the power stroke is 1.25 times the average work done during the whole cycle. The mean diameter of the flywheel is 2.5 m and the fluctuation of speed is limited to 5% of the mean speed. The material density is 7200 kg/m^3. The permissible shear stress in the shaft material is 42 MPa and the bending stress in the arms of the flywheel is 20 MPa.

23.2 A punch press is fitted with a flywheel which can furnish 2500 N.m of energy during quarter of a revolution near the bottom dead centre while blanking a hole in sheet metal. The maximum speed of the flywheel during the operation is180 rpm and the speed decreases by 10% during the blanking stroke. The mean diameter of the rim is 2 m. Calculate the mass of the flywheel rim assuming that it contributes 90% of the energy requirements.

23.3 Design a flywheel for a punching machine having a capacity of punching 30 holes per minute of 25 mm diameter in a steel plate 20 mm thick. The ultimate shear strength for the material of the plate is 360 MPa. The actual punching operation takes place during 30° rotation of the punching machine crankshaft. This crankshaft is powered by a flywheel shaft through reduction gearing having a ratio of 1 : 8. Assume that the mechanical efficiency of the punching machine is 80% and during the actual punching operation the flywheel speed is reduced by 10%. The diameter of the flywheel is restricted to 0.8 m

23.4 A cast iron flywheel is to be designed for a single cylinder double acting steam engine which delivers 140 kW at 100 rpm. The maximum fluctuation of energy per revolution is 10% and total fluctuation of speed is 5% of the mean speed. The mean diameter of the flywheel rim is 2.5m Determine :

(*a*) Cross – section of rim, assuming that the hub and arms provide 5% of the rotational inertia of the wheel. The density of cast iron is 7200 kg/m^3 and permissible tensile stress is 15 MPa. Width is twice the thickness.

(*b*) Diameter of shaft, hub and key size. the permissible shear stress in shaft and key is 42 MPa.

(*c*) Cross – section of ellipitcal arms assuming major axis to be double the minor axis. The number of arms are 6.

Chapter 24 CLUTCHES AND BRAKES

24.1 CLUTCHES

A clutch is a device which is used to connect a driving shaft to a driven shaft when desired without stopping the driving shaft.

24.1.1 *Types of Clutches*

There are two types of clutches :

1. Positive action clutches. These are jaw type of clutches in which power is transmitted as long as the jaws are in contact. The part of the jaw on the driving shaft is rigidly fixed to the shaft whereas the other parts of the clutch may be made to slide on the driven shaft having splines.

2. Friction clutches. These clutches work on the basis of frictional forces developed between the surfaces of contact. Friction clutches may be of the axial engagement type or radial type. In the axial type of clutches, the engagement or disengagement of the clutch is achieved by an axial movement of the clutch parts attached to the driven shaft. Whereas in the radial type of clutches the contact pressure is applied in a direction perpendicular to the shaft axis. Plate or disc and cone type of clutches are of the axial type and tyre–pneumatic, centrifugal, over-running, block, expending ring and band clutches are of the radial type. The plate type of clutches may be either of the single plate or multi-plate friction clutches. Multi-plate friction clutches are used in small vehicles two and three wheelers whereas single plate clutches are used in heavy vehicles like trucks, buses, etc. Multi-plate clutches are wet clutches and operate in oil which acts as cushion and takes away the heat energy released due to rubbing action of friction plates.

24.1.2 *Design Considerations*

A clutch of good design should have the following qualities :

1. Ability to withstand and dissipate heat
2. Adequate torque capacity.
3. Long life.
4. Positive release.
5. Smooth engagement.
6. Low operating force.
7. Ease of repair.

Clutches should be designed for overload capacities of 75-100%. If the load is variable or subjected to shock, additional service factors should be used, as suggested in Table 24.1.

Table 24.1 Service factors for friction clutches

Type of service	Service factor excluding starting factor
Driving machine :	
Electric motor, steady load	1.0
Fluctuating load	1.5
Gas engine :	
Single cylinder	1.5
Multiple cylinder	1.0
Diesel engine :	
High speed	1.5
Large, slow speed	2.0
Driven machine :	
Generator, steady load	1.0
Fluctuating load	1.5
Blower	10
Compressor, depending on the number of cylinders	2.0 – 2.5
Pumps, centrifugal :	
Single acting	2.0
Double acting	1.5
Line shaft	1.5
Wood working machinery	1.75
Hoist, elevators, cranes, Shovels	2.0
Hammer mills, ball mills, crushers	2.0
Brick machinery	3.0
Rock crushers	3.0

24.1.3 Properties of Material for the lining of a Clutch

The lining material for clutches should have the following properties :

1. High coefficient of friction.
2. Unaffected by moisture and oil.
3. Ability to withstand high temperature caused by slippage.
4. High heat conductivity.
5. High resistance to wear and scoring.

Coefficient of friction and unit pressure for clutch facings are given in Table 24.2.

Table 24.2 Coefficient of friction and unit pressure for clutch facings

Facing Material	Coefficient of friction, μ	Unit pressure, MPa
Cast iron on cast iron :		
Dry	0.20	0.28–0.42
Oily	0.07	

Facing Material	Coefficient of friction, μ	Unit pressure, MPa
Cast iron on steel :		
Dry	0.30	0.28–0.42
Oily	0.10	
Leather on cast iron :		
Dry	0.50	0.07–0.084
Oily	0.15	
Cork on cast iron :		
Dry	0.35	0.00875–0.014
Oily	0.30	
Asbestos fabric :		
Dry	0.35–0.45	0.21–0.42
Oily	0.25	
Moulded asbestos	0.25–0.35	0.21–0.42
Wood on cast iron	0.30	0.172–0.35

24.1.4 Disc or Plate Clutch

Consider two friction surfaces maintained in contact by an axial force F_a, as shown in Fig 24.1.

Let T = torque transmitted by the clutch

p = intensity of axial pressure

r_1, r_2 = external and internal radii of friction surfaces respectively

r_m = mean radius of friction surface

μ = coefficient of friction

The two conditions prevailing on the clutch plate may be,

1. Uniform pressure, p = const.
2. Uniform wear, pr = const.

The first condition generally prevails in a new clutch and the second condition in a worm out clutch.

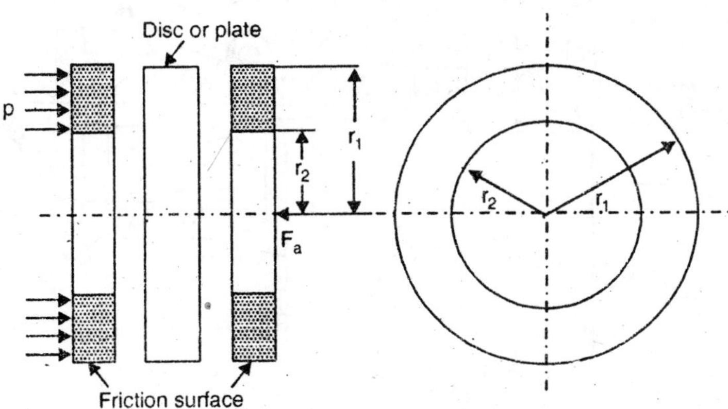

Fig. 24.1 Single Plate friction clutch

(*a*) For uniform pressure (p = const.), we have

Axial force, $$F_a = \pi\left(r_1^2 - r_2^2\right)p \qquad ...(24.1)$$

Frictional torque, $$T_f = \mu F_a r_m \qquad ...(24.2)$$

Where $$r_m = \left(\frac{2}{3}\right)\left[\frac{r_1^3 - r_2^3}{r_1^2 - r_2^2}\right]$$

(*b*) For uniform wear (pr = const.), we have

Axial force, $$F_a = 2\pi\, p_a r_2 (r_2 - r_1) \qquad ...(24.3)$$

$$T_f = \mu F_a r_m \qquad ....(24.4)$$

Where $$r_m = \frac{r_1 + r_2}{2} \qquad ...(24.5)$$

$$\frac{r_2}{r_1} = 0.6 - 0.7 \text{ for general machinery} \qquad ...(24.5)$$

$$= 0.748 \text{ for tractors}$$

24.1.5 MULTIPLE DISC CLUTCH

A multiple disc clutch is shown in Fig. 24.2

Let n_1 = number of discs on the driven shaft

$n = n_1 + n_2 - 1$ = number of pairs of contact surfaces ...(24.6)

Total friction torque acting on the clutch,

$$T_f = n\mu F_a r_m \qquad ....(24.7)$$

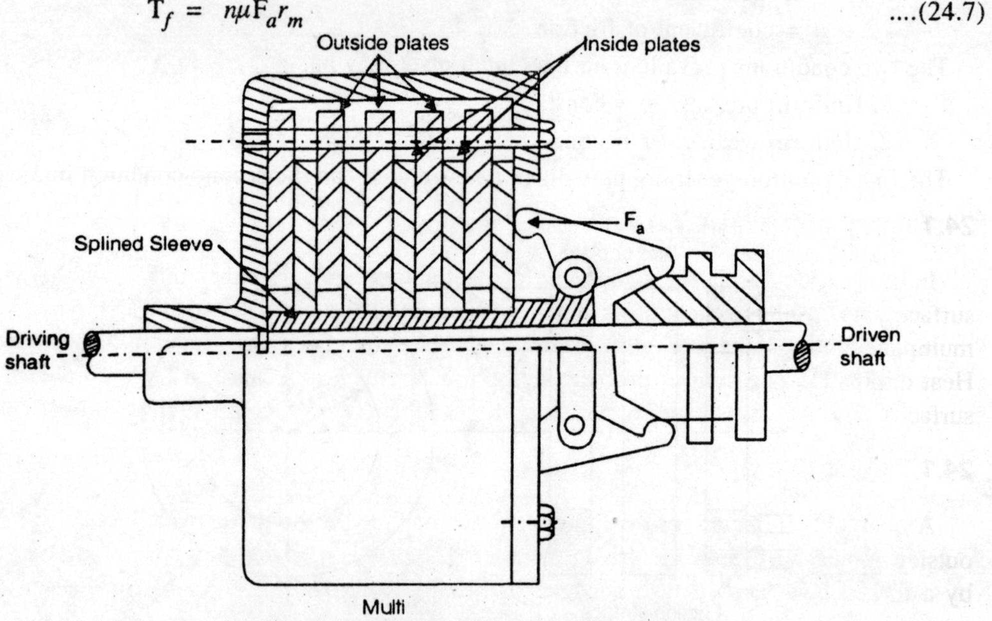

Fig. 24.2 Multi-Plate Friction clutch

Number of plates = 10 – 20 ...(24.8)

Thickness of core plate = 0.75 – 3.0 mm (See Table 24.3) ...(24.9)

Table 24.3 Thickness of core plates

Clutch diameter, mm	Upto 150 mm	150–300	300–450	450–460
Core plate thickness, mm	0.75	1.5	2.25	3.0

The multiplate clutches work wet in oily medium. The face of the wet clutch plate may be provided with radial slots, radial grooves or spiral grooves for proper circulation of oil through the clutch plate. The driving disc and clutch disc are shown in Fig. 24.3. The grooving practice for wet clutch plates are given in Table 24.4.

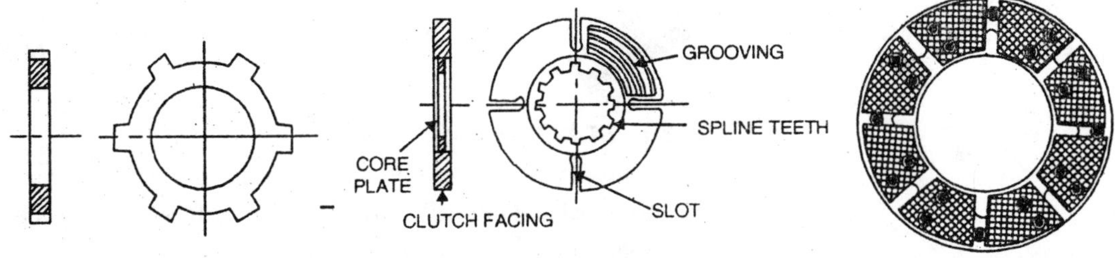

(*a*) Driving disc (*b*) Clutch disc (*c*) Slots on clutch facing

Fig. 24.3.

Table 24.4 Grooving practice for wet friction clutch plate

Grooves or Slots	Disc diameter, mm
No grooves	0–230
Spiral grooves	125–500
Radial slots	200–450
Radial grooves	230–450
Alternating radial grooves and slots	350–500

24.1.6 Dry and Wet Clutches

In case of single plate clutches, the contacting surfaces are either one or two. Due to large surface area available, heating is not a problem, and the clutch is of the dry type. In case of multiplate clutches, the work done during engagement and disengagement is converted into heat. Heat dissipation is a serious problem in multiplate clutches because of large number of friction surfaces. Cooling oil is used to dissipate this heat. Therefore, these clutches are wet clutches.

24.1.7 Cone Clutch

A cone clutch has an inside conical surface or face on the driver which exactly fits into the outside conical surface of the driven member. The driven member may be shifted along the shaft by a forked lever provided to engaged or disengage by shifting the two conical surfaces. The torque is transmitted from one shaft to the other due to frictional resistance. A cone clutch is shown in fig. 24.4.

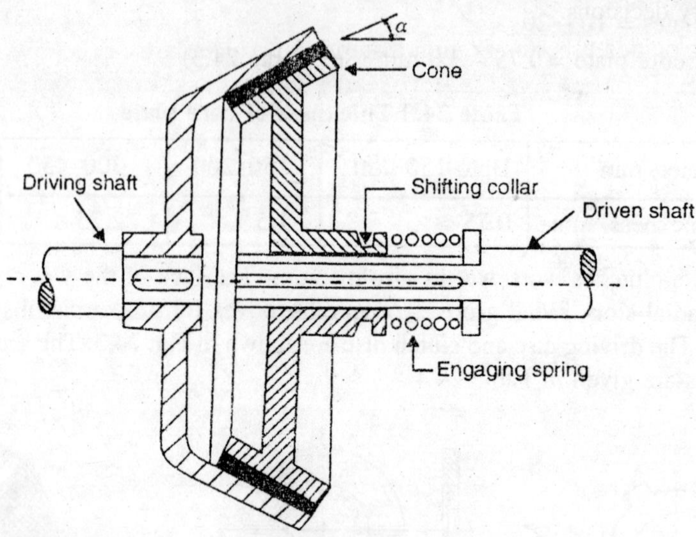

Fig. 24.4 Cone clutch

Let

p_n = intensity of normal pressure between contacting surfaces

r_1, r_2 = external and internal radii of friction surfaces respectively.

r_m = mean radius of friction surface.

μ = coefficient of friction

α = semi-cone angle of cone

b = width of the friction surface

(*a*) For uniform pressure, we have

$$F_a = \pi\left(r_1^2 - r_2^2\right) p_n \quad ...(24.10)$$

Frictional torque,

$$T_f = \mu F_a r_m \operatorname{cosec} \alpha \quad ...(24.11)$$

where

$$r_m = \left(\frac{2}{3}\right)\left[\frac{r_1^3 - r_2^3}{r_1^2 - r_2^2}\right] \quad ...(24.11)$$

(*b*) For uniform wear, we have

Total frictional torque,

$$T_f = \mu F_a r_m \operatorname{cosec} \alpha \quad ...(24.12)$$

Where

$$r_m = \frac{r_1 + r_2}{2}$$

Axial force required for engaging the clutch,

$$F_e = F_n (\sin \alpha + 0.25 \mu \cos \alpha) \quad ...(24.13)$$

Example 24.1 Design a single plate clutch of a vehicle to transmit 20 kW at a speed of 3000 rpm, allowing for 25% overload. The axial thrust is to be provided by 6 springs of about 20 mm mean coil diameter. The clutch may be made with 6 splines.

Solution.

Given : P = 20 kW, N = 3000 rpm, Overload = 25%, Number of springs = 6, D_m = 20 mm, number of splines = 6, n = 2

1. Preliminary decisions.

We assume moulded asbestos friction material for which in dry condition,

$$\mu = 0.2 - 0.5 \text{ and } p = 0.34 - 0.98 \text{ MPa (Table 24.2)}$$

We take $\mu = 0.3$, $p = 0.4$ MPa and $\frac{r_2}{r_1} = 0.7$

Shear stress in spring wire = 400 MPa

2. Torque transmitted,

$$T = \frac{1.25 \times 10^3 \times P \times 60}{2\pi N}$$

$$= \frac{1.25 \times 10^3 \times 20 \times 60}{2 \times \pi \times 3000} = 79.58 \text{ N.m}$$

3.Axial force, $F_a = 2\pi\, p_a\, r_2 (r_1 - r_2)$ for uniform wear

4. Frictional torque, $T_f = n\mu F_a r_m$

Where $r_m = \frac{r_1 + r_2}{2}$, for uniform wear

5. For $T = T_f$, we have

$$2 \times 0.3 \times 2\pi \times 0.4 \times r_2 \times \frac{\left(r_1^2 - r_2^2\right)}{2} = 79580$$

$$0.754 \times 0.7 r_1 \times r_1^2 (1 - 0.49) = 79580$$

$$r_1 = 66.6 \approx 70 \text{ mm}$$

$$r_2 = 49 \approx 50 \text{ mm}$$

$$r_m = \frac{70 + 50}{2} = 60 \text{ mm}$$

6. Mean peripheral speed,

$$v_m = \frac{2\pi r_m N}{60} = \frac{2\pi \times 60 \times 3000}{60 \times 10^3} = 18.85 \text{ m/s}$$

7. Axial force, $F_a = 2\pi\, p_a r_2 (r_1 - r_2)$

$$= 2\pi \times 0.40 \times 50 \times (70 - 50) = 2513 \text{ N}$$

8. Spring design,

Thrust to be exerted by the spring,

$$F = \frac{F_a}{\text{number of springs}}$$

$$= \frac{2513}{6} = 418.88 \approx 420 \text{ N}$$

Let spring index, $C = 5$

Wahl's factor, $$K_w = \frac{4C-1}{4C-4} + \frac{0.615}{C}$$

$$= \frac{4 \times 5 - 1}{4 \times 5 - 4} + \frac{0.615}{5} = 1.31$$

Wire diameter, $$d = \left[\frac{8FCK_w}{\pi \tau_d}\right]^{\frac{1}{2}}$$

$$= \left[\frac{8 \times 420 \times 5 \times 1.31}{\pi \times 400}\right]^{\frac{1}{2}} = 4.18 \approx 4.25 \text{ mm}$$

Mean coil diameter,

$$D_m = 5 \times 4.25 = 21.25 \text{ mm} \approx 20 \text{ mm}$$

Outside spring diameter $= D + d = 25.5$ mm

Number of active coils $= 6$ (say

Total coils, $i_t = i + 2 = 8$ for squared and ground ends

Maximum deflection, $$\delta_{max} = \frac{8FC^3 i}{Gd}$$

$$= \frac{8 \times 420 \times 5^3 \times 6}{78.45 \times 10^3 \times 4.25} = 7.56 \text{ mm}$$

Free length of spring,

$$l_f = i_t d + 1.15\,\delta_{max} = 8 \times 4.25 + 1.15 \times 7.56 = 42.7 \text{ mm}$$

9. Splines of clutch shaft.

Torque transmission capacity of a splined shaft,

$$T = 0.25 \text{ phli } (d_1 + d) = 0.5\text{phli } (d + h)$$

$$b = 0.250d,\ h = 0.1d \text{ when loaded}$$

where d = minor diameter of splined shaft

d_1 = major diameter of splined shaft

h = thickness of spline

i = number of splines

l = length of spline

p = allowable bearing pressure for splines

= 7 MPa, for splines to slide when loaded

$$T_{shaft} = \left(\frac{\pi}{16}\right) d^3 \tau$$

$$d = \left[\frac{16 \times 79580}{\pi \times 25}\right]^{\frac{1}{3}}$$

$$= 25.3 \approx 26 \text{ mm}$$

$$d_1 = d + 2h = 26 + 2 \times 0.1 \times 26 = 31.2 \approx 32 \text{ mm}$$

$$b = 0.250 \times 26 = 6.5 \text{ mm}$$

$$h = 0.1 \times 26 = 2.6 \approx 3 \text{ mm}$$

$$T = 0.25 \times 7 \times 3 \times 1 \times 6 \times (32 + 26) = 79580$$

$$l = 43.6 \approx 50 \text{ mm}$$

Example 24.2 Design a suitable clutch for a 150 cc scooter to develop 5.8 HP at 5500 rpm. At maximum torque it develops 4 HP at 3200 rpm. Space restricts the outer diameter of the clutch to 150 mm.

Solution.

Given : P_{max} = 5.8 HP, N = 5500 rpm, P = 4 HP at N = 3200 rpm, r_1 = 75 mm

1. Maximum torque, $$T_{max} = \frac{4 \times 0.746 \times 1000 \times 60}{2\pi \times 3200} = 24.905 \text{ N.m}$$

2. Design torque, $$T_d = 1.5\, T_{max} = 13.357 \text{ N.m}$$

3.Torque cpacity of clutch assuming uniform wear,

$$T_f = n\mu F_a r_m$$

Assume $$p = 0.08 \text{ MPa}, \mu = 0.1 \text{ for wet clutch,}$$

$$\frac{r_2}{r_1} = 0.6, r_1 = 75 \text{ mm}, r_2 = 45 \text{ mm}$$

$$r_m = \frac{75 + 45}{2} = 60 \text{ mm}$$

$$F_a = 2\pi p_a r_2 (r_1 - r_2)$$

$$= 2\pi \times 0.08 \times 45\ (75 - 45)$$

$$= 678.6 \text{ N}$$

4. Number of plates,

$$n \times 0.1 \times 678.6 \times 60 = 13357$$

$$n = 3.28 \approx 4$$

$$= n_1 + n_2 - 1$$

$$n_1 + n_2 = 5$$

$$n_1 = 3, n_2 = 2$$

5. Actual axial force,

$$F_a = \frac{T_d}{n\mu r_m} = \frac{13.357 \times 10^3}{4 \times 0.1 \times 60} = 556.54 \text{ N}$$

Actual pressure,

$$p = \frac{F_a}{2\pi r_2 (r_1 - r_2)}$$

$$= \frac{556.54}{2\pi \times (45 \times 30)} = 0.0656 \text{ MPa}$$

6. Shaft diameter,

$$d_s = \left[\frac{16T}{\pi\tau}\right]^{\frac{1}{3}}$$

$$= \left[\frac{16 \times 13357}{\pi \times 25}\right]^{1/3} = 13.96 \approx 14 \text{ mm}$$

7. Splines.

$b = 0.250d$, $h = 0.1\ d$ when loaded

where $d =$ d_s minor diametr of splined shaft

$d_1 =$ $d + 2h$, major diameter of splined shaft

$h =$ thickness of spline

$i =$ number of splines

$l =$ length of spline

$p =$ allowable bearing pressure for splines

$=$ 7 MPa, for spines to slide when loaded

$h = 0.1 \times 14 = 1.4 \approx 2$ mm

$d_1 = d + 2h = 14 + 2 \times 2 = 18$ mm

$b = 0.250 \times 14 = 3.5$ mm

$T = 0.25\ phli\ (d_1 + d) = 0.5\ phli\ (d + h)$

$= 0.25 \times 7 \times 2 \times 1 \times 6 \times (18 + 14) = 13357$

$l = 19.87 \approx 40$ mm allowing for movement of hub.

8. Design of spring

Thrust to be exerted by the spring, $F = \frac{F_a}{\text{number of springs}}$

$$= \frac{556.54}{3} = 185.5 \text{ N}$$

Let spring index, $C = 6$

Wahl's factor, $$K_w = \frac{4C - 1}{4C - 4} + \frac{0.615}{C}$$

$$= \frac{4 \times 6 - 1}{4 \times 6 - 4} + \frac{0.615}{6} = 1.2525$$

Wire diameter, $d = \left[\frac{8FCK_w}{\pi\tau_d}\right]^{\frac{1}{2}}$

$$= \left[\frac{8 \times 185.5 \times 6 \times 1.2525}{\pi \times 400}\right]^{\frac{1}{2}} = 2.98 \approx 3 \text{ mm}$$

Mean coil diamter, D = 6 × 3 = 18 mm

Outside spring diamter = D + d = 21 mm

Number of active coils = 5 (say)

Total coils, $i_t = i + 2 = 7$ for squared and ground ends

Maximum deflection, $\delta_{max} = \frac{8FC^3 i}{Gd} = \frac{8 \times 185.5 \times 6^3 \times 5}{724.45 \times 10^3 \times 3} = 6.81$ mm

Free length of spring, l_f $= i_t d + 1.15\delta_{max} = 7 \times 3 = 1.15 \times 6.81 = 224.8$ mm

Example 24.3 Desing a cone clutch with leather facing and cast iron mating surface t. transmit 15 kW at 750 rpm. The mean diameter is limited to 250 mm and the slope of the face is to be 15°. The clutch is to work under oily conditions.

Solution.

Given : P = 15 kW, N = 750 rpm, D_m = 250 mm, α = 15°

1. Preliminary decisions.

For leather on cast iron in oily conditions, coefficient of friction,

μ = 0.15, p = (0.07 – 0.084) MPa (Table 24.2)

We take, p = 0.08 MPa

2. Torque transmitted.

$$T_f = \frac{10^3 P \times 60}{2\pi N} = \frac{10^3 \times 15 \times 10^3 \times 60}{(2\pi \times 750)} = 191 \text{ N.m}$$

3. Shaft diameter

Taking permissible shear stress in shaft material, τ = 40 mPa

Shaft diameter, $d_s = \left[\frac{16 T_f}{\pi\tau}\right]^{1/3}$

$$= \left[\frac{16 \times 191 \times 10^3}{\pi \times 40}\right]^{1/3} = 34.9 \approx 40 \text{ mm (adopt)}$$

4. Torque transmitted, $T_f = \frac{\mu F_n D_m}{2}$

Normal force, $F_n = \frac{191 \times 10^3}{0.15 \times 125} = 10{,}187$ N

5. Axial force, $F_a = F_n (\sin \alpha + 0.25 \mu \cos \alpha)$

$= 10,187 (\sin 15° + 0.25 \times 0.15 \times \cos 15°)$

$= 3005.6$ N

6. Width of cone face, $$b = \frac{F_n}{\pi D_m p}$$

$$= \frac{10,187}{\pi \times 250 \times 0.08} = 162 \text{ mm}$$

7. Inside cone diamter, $D_i = D_m - b \sin \alpha$

$= 250 - 162 \sin 15° = 208$ mm

8. Outside cone diamter, $D_o = D_m + b \sin \alpha$

$= 250 + 162 \sin 15° = 292$ mm

9. Helical spring design.

Let spring index, C = 6 and allowable shear stress in spring wire, $\tau_d = 500$ MPa

Wahl's correction factor, $$K_w = \frac{4C-1}{4C-4} + \frac{0.615}{C}$$

$$= \frac{4 \times 6 - 1}{4 \times 6 - 4} + \frac{0.615}{6}$$

$= 1.2525$

Wire diameter, $$d = \left[\frac{8 F_a C K_w}{\pi \tau_d}\right]^{\frac{1}{2}}$$

$$= \left[\frac{8 \times 3005.6 \times 6 \times 1.2525}{\pi \times 500}\right]^{\frac{1}{2}}$$

$= 10.72 \approx 11$ mm

Mean coil diamter, $D = 6 \times 11 = 66$ mm

Let number of active turns, $i = 6$

Maximum deflection, $$\delta_{max} = \frac{8 F_a i D^3}{G d^4}$$

$$= \frac{8 \times 3005.6 \times 6 \times 66^3}{78.45 \times 10^3 \times 11^4}$$

$= 36$ mm

Free length of squared and ground ends,

$l_f = (i + 2) d + 1.15\delta_{max}$

$= (6 + 2) \times 11 + 1.15 \times 36 = 129.4$ mm

Example 24.4 A single plate (one side) clutch can transmit 718 N.m torque. Calculate the inner diamter of friction lining using uniform pressure theory using following parameters :

D_o = 300 mm, μ = 0.25, and p_{max} = 835 kPa. Check the capacity of clutch using uniform wear assumption.

Solution.

Given : T_f = 718 N.m, D_o = 300 mm, μ = 0.25, p_{max} = 825 kPa or 0.825 MPa, n = 1

(*i*) Uniform pressure.

$$\text{Axial force, } F_a = \left(\frac{\pi}{4}\right)\left(D_o^2 - D_i^2\right)p$$

$$\text{Torque transmitted, } M_t = \frac{n\mu F_a D_m}{2}$$

$$\text{Mean diamter, } D_m = \left(\frac{2}{3}\right)\left[\frac{D_o^3 - D_i^3}{D_o^2 - D_i^2}\right]$$

$$M_t = 1 \times \mu \times \left(\frac{\pi}{4}\right)\left(D_o^2 - D_i^2\right)p \times \left(\frac{1}{3}\right)\left[\frac{D_o^3 - D_i^3}{D_o^2 - D_i^2}\right]$$

$$= \mu \times \left(\frac{\pi}{12}\right)p \times \left(D_o^3 - D_i^3\right)$$

$$718 \times 10^3 = \left(\frac{0.25 \times \pi}{12}\right) \times 0.825 \times \left(300^3 - D_i^3\right)$$

$$D_i = 239.3 \approx 240 \text{ mm}$$

(*ii*) Uniform wear.

$$\text{Axial force, } F_a = \left(\frac{\pi}{2}\right)D_i\left(D_o - D_i\right)p$$

$$\text{Torque transmitted, } M_t = \frac{n\mu F_a D_m}{2}$$

$$\text{Mean diamter, } D_m = \left(\frac{1}{2}\right)\left[D_o + D_i\right]$$

$$M_t = 1 \times \mu \times \left(\frac{\pi}{2}\right)D_i\left(D_o - D_i\right)p \times \left(\frac{1}{4}\right)\left(D_o + D_i\right)$$

$$= \mu \times \left(\frac{\pi}{8}\right)p \times D_i\left(D_o^2 - D_i^2\right)$$

$$718 \times 10^3 = \left(\frac{0.25 \times \pi}{8}\right) \times 0.825 \times D_i \left(300^2 - D_i^2\right)$$

$$D_i \left(300^2 - D_i^2\right) = 8864834$$

We find the value of D_i by hit and trial method.

D_i, mm	240	235	230	225
$D_i \left(300^2 - D_i^2\right)$	1088834	692709	331834	5451

We take $D_i = 225$ mm

24.2 BRAKES

A brake is a device to retard or stop the motion of a machine. It absorbs either the kinetic or potential energy of a moving object. The absorbed energy is converted into heat energy, which has to be dissipated.

24.2.1 Types of Mechanical Brakes.

The various types of mechnical brakes are :

1. Single block or shoe brake
2. Pivoted block or shoe brake
3. Double block or shoe brake
4. Simple band brake
5. Differential band brake
6. Band and block brake
7. Internal expanding brake

We shall study single block brake, simple band brake and differential band brake.

24.2.2 Factors on which Capacity of a Brake Depends

Capacity of a brake depends upon the following factors :

1. Unit pressure between braking surfaces.
2. Coefficient of friction between contacting surfaces.
3. Velocity of drum.
4. Projected area of friction surfaces.
5. Ability of brake to dissipate heat generated.

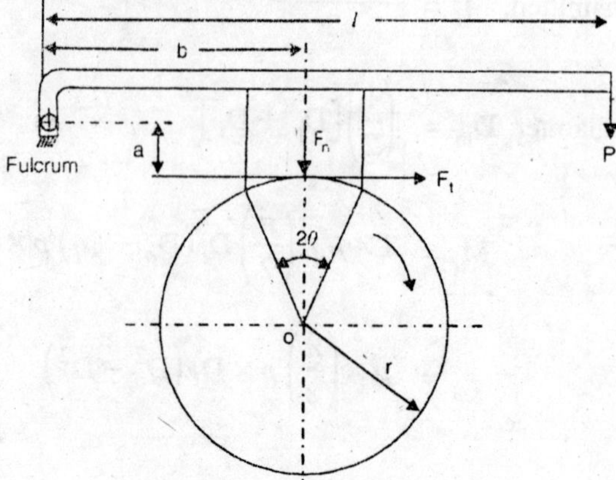

Fig. 24.5 Single block brake

24.2.3 Single Block Brake

A single block brake is shown in Fig. 24.5.

Let P = force applied at the end of the lever

F_n = normal reaction between wheel and block

r = radius of wheel

2θ = angle of contact between block and wheel

μ = coefficient of friction

F_t = tangential braking force

$F_t = \mu F_n$

Braking torque, $T_{br} = F_t \cdot r = \mu F_n \cdot r$

Let the line of action of F_t be at a distance 'a' below the fulcrum, and the wheel rotates clockwise, as shown in Fig. 24.5. The forces acting on the lever are shown in its free body diagram. Taking moments about the fulcrum, we have,

$$F_n \times b + F_t \times a = P \times l$$

$$F_n \times b + \mu \times F_n \times a = Pl$$

$$F_n = Pl\,(b + a\mu) \qquad \text{... (24.14)}$$

$$\text{Braking torque, } T_{br} = \frac{\mu P l r}{b + a\mu} \qquad \text{... (24.15)}$$

Similar results can be obtained when F_t is above the fulcrum and wheel rotates clockwise or anti-clockwise.

Self–energizing brake. When the moment of frictional force adds to the moment of the applied force, then frictional force helps to apply the brake. Such a brake is called self – energizing brake.

Self–locking brake. When the frictional force is great enough to apply the brake without the help of external force, then the brake is said to be of the self-locking type.

The coefficient of friction and allowable pressure for brakes is given in Table 24.5.

Table 24.5 Coefficient of friction and allowable bearing pressure for brakes

Materials of contact	Coefficient of friction			Allowable bearing pressure, MPa
	Dry	Greasy	Lubricated	
Cast iron on cast iron	0.20 – 0.15	0.10 – 0.06	0.10 – 0.05	1.03 – 1.72
Bronze on cast iron	—	0.10 – 0.06	0.17 – 0.05	0.55 – 0.83
Steel on cast iron	0.35 – 0.25	0.12 – 0.07	0.10 – 0.06	0.83 – 1.38
Wood on cast iron	0.35 – 0.10	0.12 – 0.08	—	0.41 – 0.62
Fibre on metal	—	0.20 – 0.10	—	0.07 – 0.08
Cork on metal	0.35	0.30 – 0.25	0.25 – 0.22	0.06 – 0.10
Leather on metal	0.5 – 0.3	0.20 – 0.15	0.15 – 0.12	0.07 – 0.28
Wired asbestos on metal	0.50–035	0.030 – 0.25	0.25 – 0.20	0.21 – 0.55
			0.25 – 0.20	0.38 – 2.07
short action				
Steel on cast iron, or cast iron on cast iron, short action	—	—	0.25 – 0.25	1.38 – 2.08

The working pressure for shoe material against cast iron is given in table 24.6.

Table 24.6 Working pressure for shoe material against Cast Iron

Rubbing velocity m/s	Pressure, MPa		
	Wood block	Asbestos fabric	Asbestos blocks
1	0.549	0.687	1.103
2	0.451	0.549	1.035
3	0.343	0.412	0.896
4	0.245	0.275	0.687
5	0.173	0.206	0.483
6	0.173	0.206	0.275

24.2.4 Simple Band Brake

A simple band brake is shown in Fig. 24.6.

Let F_1, F_2 = tension in the band on the tight and slack side respectively

θ = angle or wrap

μ = coefficient of friction between band and drum

r = radius of drum

Now $\frac{F_1}{F_2} = \exp(\mu\theta)$

Braking torque,

$$T_{br} = (F_1 - F_2)\, r \quad ...\ (24.16)$$

Taking moments about the fulcrum, we have

$Pl = F_1.b$ for cw rotation of drum

$= F_2.b$ for ccw rotation of drum

Let w = width of band

t = thickness of band

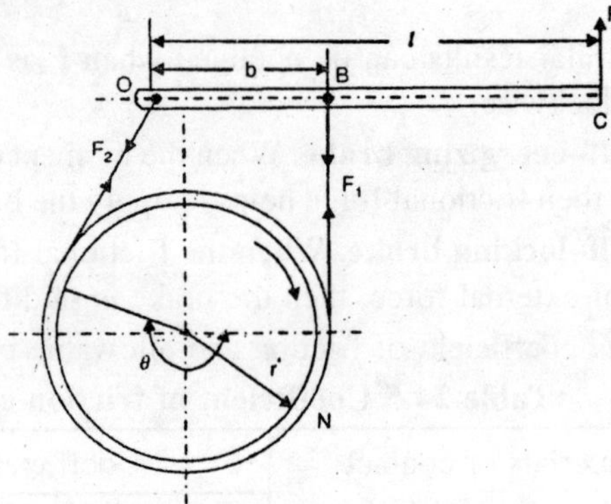

Fig. 24.6 Simple band brake

σ_t = permissible tensile strength of band material

$$F_1 = \sigma_t wt \quad ...\ (24.17)$$

24.2.5 Differential Band Brake

Consider the differential band brake shown in Fig. 24.7.

Considering the forces acting on the lever, we have

$P.l + F_1.b = F_2.a$, for cw rotation of drum

$Pl = F_2a - F_1b \ ...\ (24.18)$

And $Pl + F_2b = F_1a$, for ccw rotation of drum

$Pl = F_1a - F_2b \ ...\ (24.19)$

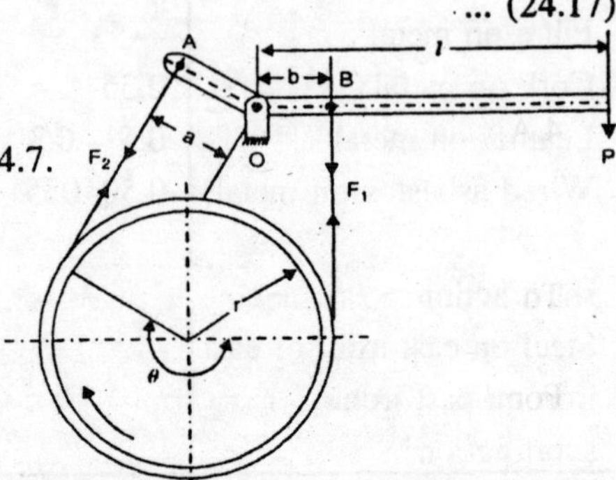

Fig. 24.7 Differential band brake

We find from Eq. (24.13) that P will become zero or negative, *i.e.* the brake will become self – locking, when, $F_2 a - F_1 b \leq 0$

Or $$\frac{F_2}{F_1} \leq \frac{b}{a} \text{ for cw rotation of drum}$$

And $$\frac{F_1}{F_2} \leq \frac{b}{a} \text{ a for ccw rotation of drum}$$

Example 24.5 Design a differential band brake for a winch lifting a load of 50 kN by means of a rope wound round a barrel of 450 mm diamter.

Solution.

Given: W =50 kN, D_b = 450 mm

1. Preliminary decisions

Let angle of wrap, θ = 240°, coefficient of friction, μ = 0.3

Diamter of rope drum D > D_b = 600 mm (say)

The brake is shown in Fig. 24.8.

2. Ratio of tensions,

$$\frac{F_1}{F_2} = \exp(\mu\theta)$$

$$= \exp\left(\frac{0.3 \times \pi \times 240}{180}\right)$$

$$= 3.51$$

3. Braking torque,

$$T_{br} = (F_1 - F_2)\, D/2$$

$$= \frac{WD_b}{2}$$

$$F_1 - F_2 = \frac{50 \times 10^3 \times 450}{600} = 37{,}500 \text{ N}$$

$$F_2 = \frac{37{,}500}{2.51} = 14940 \text{ N}$$

$$F_1 = 52439 \text{ N}$$

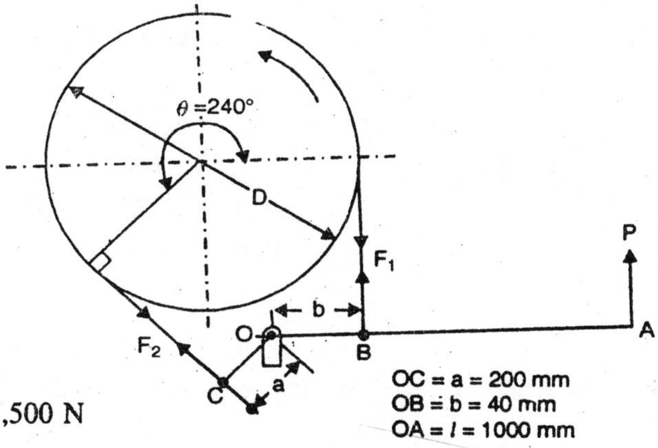

Fig. 24.8

4. Applied effort at the end of lever.

$$P = F_2 \frac{[a - b\exp(\mu\theta)]}{l}$$

To avoid self – locking, let $a = 1.5 \times b \exp(\mu\theta) \approx 5b$

For a = 200 mm, $b = \frac{200}{5}$ = 40 mm

Let $l = 1000$ mm

$$P = \frac{14940\,(200 - 40 \times 3.51)}{1000} \approx 600 \text{ N}$$

5. Band width and thickness.

$$\text{Band width, } w = \frac{F_1}{pr} = \frac{52439}{0.7 \times 300} = 249 \approx 250 \text{ mm}$$

$$\text{Band thickness, } t = \frac{F_1}{w\,\sigma_t} = \frac{52439}{250 \times 50} = 4.2 \approx 4.5 \text{ mm}$$

6. Lever.

Maximum bending moment in the lever occurs near the fulcrum,

$$M = F_2 \,.\, a = \frac{14940 \times 200}{1000} = 2988 \text{ N.m}$$

Taking $\sigma_t = 1000$ MPa for steel lever, and depth, $d = 3$ times thickness, h

$$\text{Section modulus of lever, } z = \frac{h \times (3h)^2}{6} = \frac{3h^3}{2}$$

$$z = \frac{M}{\sigma_t}$$

$$\frac{3h^3}{2} = \frac{2988 \times 10^3}{100}$$

$$h = 27.1 \approx 28 \text{ mm}$$

$$d = 84 \text{ mm}$$

7. Pins

$$\text{Fulcrum reaction, } R = \left[(52439)^2 + (14940 \cos 60°)^2 + (14940 \sin 60°)^2\right]^{\frac{1}{2}} = 61290 \text{ N}$$

Pin at O and A

$$\frac{l_p}{d_p} = 2 \text{ and } p_b = 10 \text{ MPa}$$

$$2d_p \times d_p \times 10 = 61290$$

$$d_p = 55 \text{ mm and } l_p = 110 \text{ mm}$$

Boss diamter, $d_p = 1.5 d_p \approx 85$ mm

Pin at B

$$2d_p \times d_p \times 10 = 14940$$

$$d_p = 28 \text{ mm and } l_p = 56 \text{ mm}$$

Boss diamter, $d_p = 45$ mm

8.Check of the lever at the fulcrum.

$$\text{Section modulus, } z = \frac{\left[\dfrac{100\left(85^3 - 60^3\right)}{12}\right]}{\left(\dfrac{85}{2}\right)}$$

$$= 85870 \text{ mm}^3 > 29880 \text{ mm}^3$$

Hence safe.

Example 24.6 A band brake acts on the $\frac{3}{4}$th of circumference of a drum of 450 mm diameter which is keyed to the shaft. The band brake provides a braking torque of 225 N.m. One end of the band is attached to a fulcrum pin of the lever and the other end to a pin 100 mm from the fulcrum. If the operationg force is applied at 500 mm from the fulcrum and the coefficient of friction is 0.25, find the operating force when the drum rotates in anti – clockwise direction.

If the brake lever and the pins are to be male of mild steel having permissible stress in tension and crushing as 70 MPa and for shear 56 MPa, design the shaft, key, lever and pins. The bearing pressure between pin and lever arm may be taken as 8 MPa.

Solution.

Given : D = 450 mm, T_{br} = 225 N.m, OB = 100 mm, L = 500 mm, μ = 0.25, τ = 56 MPa, $\sigma_t = \sigma_c = 70$ MPa

(*a*)The band brake is shown in Fig. 24.9. The operating force P will act upwards to apply the brake when the drum rotates ccw.

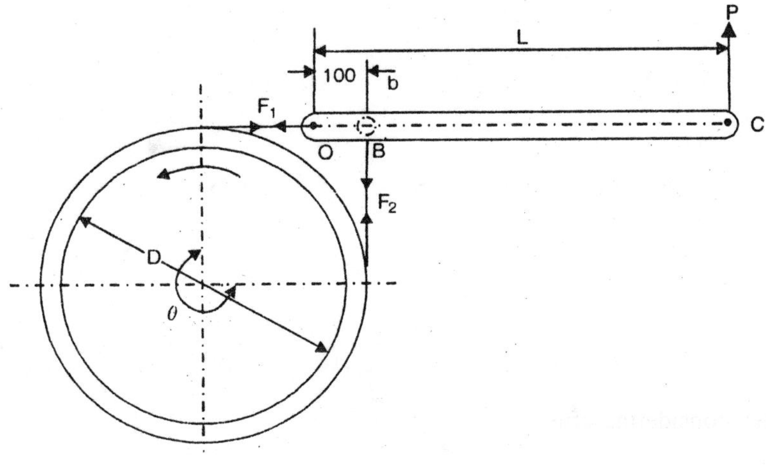

Fig. 24.9

1.Angle of wrap, $\theta = \frac{3 \times 360}{4} = 270°$ or $\frac{3\pi}{2}$ *rad*

2. Ratio of tensions, $\frac{F_1}{F_2} = \exp(\mu\theta) = \exp\left(\frac{0.25 \times 3\pi}{2}\right) = 3.25$

3. Braking torque, $T_{br} = (F_1 - F_2)\, D/2$

$$F_1 - F_2 = \frac{2 \times 225 \times 10^3}{450} = 1000 \text{ N}$$

$$F_2 = \frac{1000}{2.25} = 444.45 \text{ N}$$

$$F_1 = 1444.45 \text{ N}$$

4.Taking moments about fulcrum, we get

$$P = \frac{F_2 \times OB}{500} = \frac{444.45 \times 100}{500} = 88.9 \text{ N}$$

5. Shaft design.

$$\text{Braking torque, } T_{br} = \left(\frac{\pi}{16}\right) d_s^3\, \tau$$

$$\text{Shaft diameter, } d_s = \left[\frac{225 \times 10^3 \times 16}{\pi \times 56}\right]^{\frac{1}{3}} = 27.35 \approx 30 \text{ mm}$$

6. Key size.

$$\text{Width of key, } w = \frac{d_s}{3} = \frac{30}{3} = 10 \text{ mm}$$

$$\text{Thickness of key, } t = \frac{d_s}{4} = \frac{30}{4} = 7.5 \approx 8 \text{ mm}$$

Length of key, considering shearing of key,

$$l = \frac{2\,T_{br}}{w\,\tau\,d_s} = \frac{2 \times 225 \times 10^3}{(10 \times 56 \times 30)} = 27 \text{ mm}$$

Length of key considering crushing of key,

$$l = \frac{4\,T_{br}}{t d_s \sigma_c} = \frac{4 \times 225 \times 10^3}{80 \times 70 \times 30} = 54 \text{ mm}$$

We adopt, $l = 54$ mm

7. Lever.

$$\text{Let } t_1 = \text{thickness of lever}$$

$$b = \text{width of lever}$$

Bending moment on the lever, M = PL = 88.9 × 500 = 44450 N.mm

$$\text{Section modulus of lever, } z = \frac{t_1 b^2}{6} = \frac{b^3}{12} \text{ for } b = 2t_1$$

$$z = \frac{M}{\sigma_t}$$

$$\frac{b^3}{12} = \frac{44450}{70}$$

$$b = 20 \text{ mm}$$

$$t_1 = 10 \text{ mm}$$

8. Pins.

$$\text{Let } d_p = \text{diameter of pins}$$

$$l_p = 1.25 d_p = \text{length of pins}$$

$$p_b = 8 \text{ MPa}$$

Considering bearing of pins, we have

$$F_1 = d_p l_p p_b$$

$$1444.45 = 1.25\, d_p^2 \times 8$$

$$d_p = 12 \text{ mm}$$

$$l_p = 15 \text{ mm}$$

Check of pins in double shear,

$$1444.45 = 2 \times \frac{\pi}{4} \times 12^2 \times \tau_p$$

$$\tau_p = 6.4 \text{ MPa, which is quite safe.}$$

Check of pins in bending,

$$\text{Maximum bending moment, } M_{max} = \frac{5 F_1 l_p}{24} = \frac{5 \times 144.45 \times 15}{24} = 4515 \text{ N.mm}$$

$$\text{Section modulus, } \quad z_p = \frac{\pi d_p^3}{32} = \frac{\pi \times 12^3}{32} = 170 \text{ mm}^3$$

$$\text{Bending stress} = \frac{M_{max}}{z_p} = \frac{4514}{170} = 26.5 \text{ MPa} < 70 \text{ MPa.}$$

9. Eye hole in lever.

$$\text{Thickness of each eye, } t_e = \frac{l_p}{2} = \frac{15}{2} = 7.5 \text{ mm}$$

$$\text{Outer diameter of eye, } d_e = 2d_p = 2 \times 12 = 24 \text{ mm}$$

10. Bush size in the eye.

$$\text{Thickness of bush} = 3 \text{ mm}$$

$$\text{Diameter of hole in the lever} = d_p + 2 \times 3 = 12 + 6 = 18 \text{ mm}$$

$$\text{Outer diameter of boss} = 2d_p = 24 \text{ mm}$$

$$\text{Length of boss} = l_p = 15 \text{ mm}$$

11. Check of lever at the fulcrum in bending.

$$M = PL = 88.9 \times 500 = 44450 \text{ N.mm}$$

$$z = \frac{\left[\dfrac{15\left(24^3 - 18^3\right)}{12}\right]}{12} = 833 \text{ mm}^3$$

$$\sigma_b = \frac{44450}{833} = 53.3 < 70 \text{ MPa. Hence safe.}$$

EXERCISES

24.1 A clutch is to be designed for a lathe. Caclulate the number of plates and the operating force required for the clutch to transmit a torque of 50 N.m under normal operating conditions. The clutch should slip under 200% of rated torque. The space limitations restrict the diameter of the friction surfaces to 120 mm and 80 mm. The clutch is to operate in an oily atmosphere.

24.2 Design a single plate dry friction clutch for the following data :

Power transmitted = 30 kW

Speed = 1500 rpm

Ratio of mean radius to radial width of friction faces = 5

Assume any other missing data suitably.

24.3 Design a cone clutch with leather facing and cast iron mating surface to transmit 10 kW at 750 rpm. The mean diameter is limited to 300 mm and the slope of the face is to be 15°. The clutch is to work under oily conditions.

24.4 In a band brake, the drum diameter is 900 mm and the band thickness is 6 mm. The brake facing has a coefficient of friction of 0.25. The arc of contact is 240°. The brake drum is attached to a hoisting drum that sustains a rope load of 10 kN. The operating force has a moment arm of 1.5 m and the band is attached 150 mm from the fulcrum point. Determine (*a*) the force required to just support the load, (*b*) the required force when the direction of rotation is reversed, and (*c*) the width of steel band, limiting its tensile strength to 45 MPa.

24.5 In a differential band brake the angle of contact is 250°. The brake is to sustain a torque of 500 N.m. The coefficient of friction for brake lining is 0.26. The drum diameter of the brake is 450 mm. The distances of tight side and slack side bands from the fulcrum are 30 mm and 120 mm respectively. The length of the lever is 500 mm. Determine the band size, diameter of fulcrum pin and cross – section of lever.

INTERNAL COMBUSTION ENGINE PARTS

25.1 INTRODUCTION

A line diagram of the principal parts of an internal combustion engine are shown in Fig. 25.1. The main parts to be designed are :

1. Cylinder along with liner, cylinder head and water jacket
2. Piston, piston rings and gudgeon pin.
3. Connecting rod.
4. Crank shaft and crank.

The cylinder retains the working fluid and guides the piston. In order to eliminate wear of the cylinder and for cheaper and easy replacement a liner is used. To provide cooling space for the cylinder a jacket is provided. A liner which does not experience and direct contact with the engine cooling medium is called a dry liner. A wet liner experiences direct contact with the engine cooling medium.

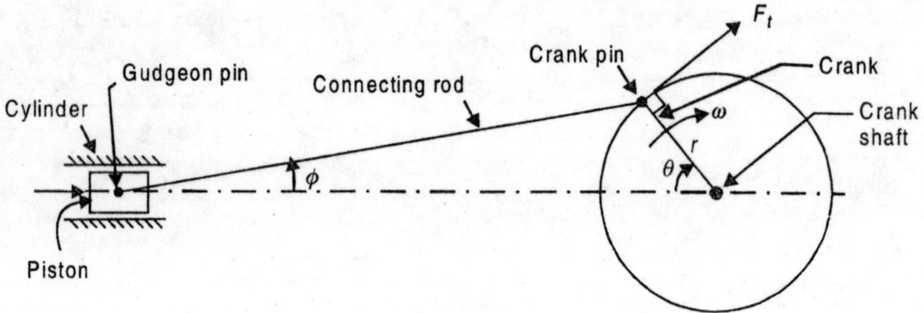

Fig. 25.1 Line diagram of an I.C. Engine

Piston for single - acting engine is of the trunk type and act as cross head to guide the small end of the connecting rod. The piston also transmits large amount of heat from the combustion chamber to the cylinder walls. The piston rings help in reducing friction force between piston and cylinder and also act as seal for the gases.

The main function of the connecting rod is to transmit the push and pull from the piston pin to the crank pin. The connecting rod transmits the gas and inertia forces to the crankshaft.

The crankshaft converts the reciprocating motion of the piston into rotary motion or vice-versa. The crankshaft may be either of the side type or centre type.

25.2 DESIGN OF CYLINDER

The cylinder may be designed by treating it as a thin or a thick cylinder depending upon the ratio of wall thickness to the mean radius. The wall thickness based upon the thin cylinder theory is given by,

$$t = \frac{pd}{2\sigma_\theta} + c \qquad \text{...(25.1)}$$

Where d = cylinder bore, p = maximum internal pressure (Table 25.1)

σ_θ = permissible hoop stress for the cylinder material

= 35 – 70 MPa for cast iron

= 70 – 100 MPa for alloy steel

c = allowance for reboring (Table 25.2)

Table 25.1 Cylinder bore diameters for internal combustion engines

Preferred bore diameter (mm). 30, 32, 34, 36, 38, 40, 44, 46, 48, 50, 52, 54, 56, 58, 60, 65, 70, 75, 80, 85, 90, 95, 100, 105, 110, 115, 120, 125, 130, 135, 140, 145, 150, 155, 160, 165, 170, 175, 180, 185, 190, 195, 200, 210, 220, 230, 240, 250, 260, 270, 280, 290, 300, 310, 320, 330, 340, 360.

Table 25.2 Reboring allowance for I.C. engine cylinders

Cylinder dia, mm	75	100	150	200	250	300	350	400	450	500
c, mm	1.5	2.4	4.0	6.3	8.0	9.5	11.0	12.5	12.5	12.5

The thickness of the cylinder wall based on the thick cylinder theory may be determined as follows :

1. Maximum principal theory,

$$t = \left(\frac{d}{2}\right)\left[\left\{\frac{\sigma_\theta + p}{\sigma_\theta - p}\right\}^{1/2} - 1\right] \qquad \text{...(25.2)}$$

This theory is suitable for brittle materials, such as cast iron or cast steel.

2. Maximum strain theory,

(*a*) Clavarino's Equation

When the cylinder ends are closed.

$$t = \left(\frac{d}{2}\right)\left[\left\{\frac{\sigma_\theta + (1-2\nu)\,p}{\sigma_\theta - (1+\nu)\,p}\right\}^{\frac{1}{2}} - 1\right] \qquad \text{...(25.3)}$$

(*b*)Birnie's Equation

For a cylinder with open ends,

$$t = \left(\frac{d}{2}\right)\left[\left\{\frac{(\sigma_\theta + (1-\nu)\,p}{(\sigma_\theta - (1+\nu)\,p}\right\}^{\frac{1}{2}} - 1\right] \qquad \text{...(25.4)}$$

This theory is suitable for ductile materials, such as mild steel or alloy steel.

Empirically, the thickness of cylinder wall is given by,

$$t = 0.045\,d + 1.5 \text{ mm}$$

25.2.1 Design of Liner and Water Jacket Wall

Empirically, the thickness of dry liner is given by :

$$t_1 = (0.030 - 0.035)\, d \qquad \text{...(25.6)}$$

Thickness of water jacket wall, $t_2 = 0.032\, d + 1.5$ mm ...(25.7)

Water space between the outer cylinder wall and inner jacket wall,

$$t_3 = 0.08\, d + 6.5 \text{ mm}$$

The hardness of cylinder liners should be as given in Table 25.3.

Table 25.3 Hardness of cylinder liners for I.C. engines

Wall thickness of liner castings, mm	Brinell hardness, N / mm^2
From 8 upto 25	210 – 290
Above 25 upto 50	200 – 280
Above 50	190 – 270

25.2.2 Mean Effective Pressure in the Cylinder

Power developed, $P = p_m\, LAN / 10^3$ kW ...(25.9)

Where p_m = mean effective pressure, N / mm^2

L = length of stroke, m

$A = \left(\frac{\pi}{4}\right) d^2$ = area of piston, mm^2

N = number of strokes per second

L / d ratio is generally 1.5 to 2.

25.2.3 Cylinder Flange and Head

The cylinder flange is integral with the cylinder. It should be made thicker than the cylinder wall with a generous fillet at the flange and the cylinder wall.

Thickness of cylinder flange, $t_f = (1.2 - 1.4)\, t$ ...(25.10)

The cylinder head may be approximated as a flat circular plate loaded uniformly. Its thickness,

$$t_c = C_1\, d \left(\frac{p}{\sigma_t}\right)^{\frac{1}{2}} \qquad \text{...(25.11)}$$

Where C_1 = a constant = 0.3 – 0.4

σ_t = allowable tensile stress in plate material

= 34.42 MPa for C.I.

= 55 MPa for nickel C.I.

= 42 MPa for Al alloys

= 82.4 MPa for low carbon cast steel

25.2.4 Diameter of Studs or Bolts

The cylinder head is fitted on the cylinder with the help of number of bolts or studs. The material used for the studs is nickel steel or nickel - chromium steel.

$$\text{Diameter of studs, } d_s = d\left(\frac{p}{n\sigma_t}\right)^{\frac{1}{2}} \qquad ...(25.12)$$

Where d_s = diameter of studs at the root of thread, mm (> 16 mm)

p = combustion pressure, N / mm^2

n = number of studs

$$= \left(\frac{d}{100}+4\right) \text{ to } \left(\frac{d}{50}+4\right) \qquad ...(25.13)$$

d = cylinder bore, mm

σ_t = allowable tensile stress for stud material, MPa

Pitch circle diameter of studs or bolts, $d_p = d + 3d_s$ ...(25.14)

Outside diameter of flange or cylinder head, $d_0 = d_p + 2d_s + 12$ mm ...(25.15)

Pitch of studs, $p_s = (19 - 28)\sqrt{d_s}$

Example 25.1 A four - stroke diesel engine is developing 6 kW brake power at 1000 rpm. The indicated mean effective pressure is 0.4 MPa and mechanical efficiency is 80%. Determine (*a*) cylinder bore and length, (*b*) thickness of cylinder wall and cylinder head, (*c*) thickness of liner, and (*d*) size of studs for the cylinder head.

Solution.

Given : BP = 6 kW, N = 1000 rpm or n = 500 strokes / min, $p_{im} = 0.4$ MPa, $\eta_{mech} = 0.8$

(*a*)Bore and length of cylinder :

Let L / d = 1. 5

$$\text{Indicated power, IP} = \frac{BP}{\eta_{mech}} = \frac{6}{0.8} = 25.5 \text{ kW}$$

$$IP = \frac{P_{im}\,LAn}{60}$$

$$25.5 \times 10^3 = 0.4 \times 1.5d \times 10^{-3} \times \frac{\pi}{4} \times d^2 \times \frac{500}{60}$$

$$d^3 = 1909.86 \times 10^3$$

Bore, $d = 124 \approx 125$ mm

Stroke, L = 187.5 mm

Taking clearance on both sides of cylinder to be 15% of stroke, length of cylinder = 215 mm

(*b*) Thickness of cylinder

We assume alloy steel for the cylinder material having allowable tensile strength, $\sigma_t = 80$ MPa

Assuming the cylinder to be a thin pressure, vessel, thickness of cylinder wall,

$$t = \frac{p_{max}\,d}{2\sigma_t} + c$$

$$p_{max} = (9 \text{ to } 10)\ p_{mean} = 9 \times 0.4 = 3.6 \text{ MPa}$$

Reboring allowance, c = 3.2 mm (see Table 25.2)

$$\sigma_t = 50 \text{ MPa for C.I.}$$

$$t = \frac{3.6 \times 125}{2 \times 50} + 3.2 = 7.7 \text{ mm}$$

$$\frac{d}{t} = \frac{125}{6} \approx 21, \text{ hence, the cylinder can be taken as thin.}$$

Empirically, $t = 0.045\ d + 1.5 = 0.045 \times 125 + 1.5 = 25.125$ mm

We adopt, $t = 25$ mm

Thickness of cylinder flange, $t_f = 1.3t = 1.3 \times 25 = 32.5$ mm

Thickness of water jacket, $t_j = 0.032\ d + 1.5 = 0.032 \times 125 + 1.5 = 5.5 \approx 6$ mm

Thickness of water space, $t_w = 10$ mm

Thickness of cylinder head :

Assuming the cylinder head to be made of nickel C.I. having $\sigma_t = 55$ MPa,

Thickness of cylinder head,

$$t_h = 0.35\ d \left(\frac{p_{max}}{\sigma_t}\right)^{\frac{1}{2}}$$

$$= 0.35 \times 125 \left(\frac{3.6}{55}\right)^{\frac{1}{2}} = 11.2 \approx 12 \text{ mm}$$

(c)Thickness of liner, $t_1 = 0.032\ d = 0.032 \times 125 = 4$ mm

(*d*)Number of studs, $n = \frac{d}{50} + 4 = \frac{125}{50} + 4 \approx 6$

Material selected for studs is alloy steel 40 Ni3 having $\sigma_t = 70$ MPa

Core diameter of studs,

$$d_c = d\left[\frac{p_{max}}{n\sigma_t}\right]^{\frac{1}{2}}$$

$$= 125\left[\frac{3.6}{6 \times 70}\right]^{\frac{1}{2}} = 11.6 \text{ mm}$$

Nominal diameter of studs, $d_s = \frac{d_c}{0.84} = 13.8$ mm

We adopt, $d_s = 16$ mm (min)

25.3 PISTON

Pistons for single - acting engines are of the trunk type. They consist of a head to carry the cylinder pressure, a skirt to act as a bearing for the connecting rod side thrust, a piston pin to connect it to the connecting rod, and piston rings to seal the cylinder. They are open at one end. Fig. 25.2 shows the parts of a trunk piston. Pistons are made of cast iron and cast steel for large stationary engines and cast iron and aluminium alloys (duralium) for small high speed engines.

(*a*)Strength Considerations

The head may be treated as a flat plate with a uniform load and rigidly supported at the outer edge. The head thickness is given by,

$$t_n = 0.433d \left[\frac{p}{\sigma_t}\right]^{\frac{1}{2}} \quad ...(25.17)$$

where σ_t = permissible stress in tension for the piston material (Table 25.4)

Table 25.4 Allowable tensile stress for piston materials

Material	Allowable tensile stress, MPa
Cast iron	35–40
Alloy cast iron	50–70
Aluminium alloy	50–70
Steel	80–90

Empirically, $t_h = 0.032\,d + 1.5$ mm ...(25.18)

Thickness of wall under the rings, $t_r = t_b$ ...(25.19)

Length of piston, $l = (1 - 1.5)\,d$ ...(25.20)

Distance between centre of piston pin and centre of skirt = $(0.02 - 0.04)\,d$ ...(25.21)

(*b*)Heat Transfer Considerations

Head thickness for heat flow,

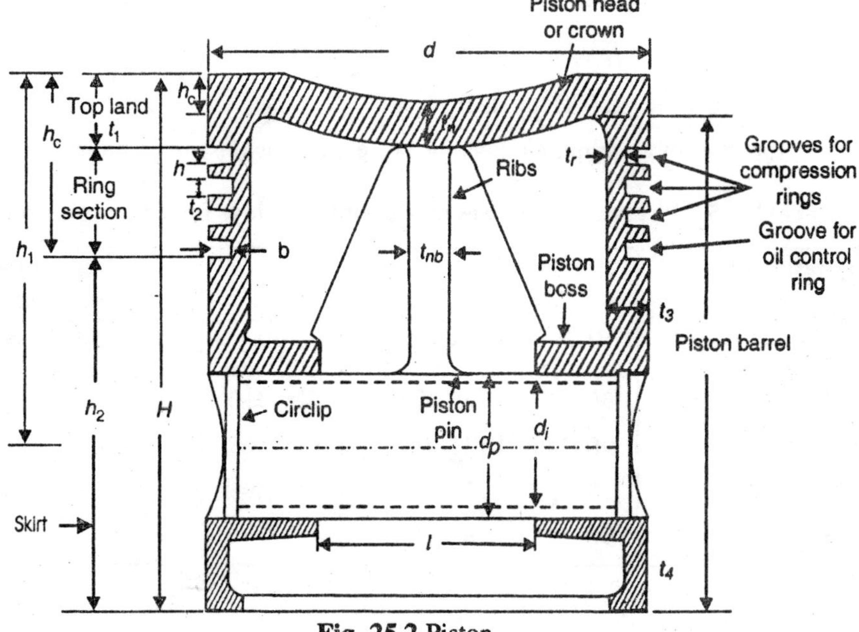

Fig. 25.2 Piston

$$t_h = \frac{Hd^2}{16h_c(T_c - T_e)\,A}$$

$$= \frac{H}{12.56h_c(T_c - T_e)} \quad ...(25.22)$$

Where H = heat flow through the head, Watts

= KCW × Brake power per cylinder ...(25.23)

K = fraction of heat absorbed by the piston ≈ 0.05

C = higher heat value of fuel used in kJ/kg = 41840 kJ/kg

W = weight of fuel used in kg/brake power in kW per hour

d = cylinder diameter, m

A = piston head area, m²

h_c = heat conduction factor, W/m^2.°c per meter length

= 46.5 W/m^2. °C per meter length for grey cast iron

= 162.73 W/m^2 °C per meter length for aluminium.

= 51 W/m^2. °C per meter length for steel

T_c = temperature at centre of head, °C

T_e = temperature at edge of head, °C

$T_c - T_e$ = 220°C for grey cast iron

= 75°C for aluminium alloys

= 250°C for steel

Physical properties of aluminium alloys for I.C. engine pistons are given in Table 25.5.

Table 25.5 Physical properties of aluminium alloys for I.C. engine pistons

Alloy designation		Hardness	Tensile strength, MPa		Coefficient of thermal expansion
Casting	Forging	BHN	Chill casting	Forging	(20° – 200°C) per °C × 10^{-6}
2285	24850	90-130	225-275	345–410	23–24
4658	59582	90–140	125–140	195–365	20.5–21.5
4928–A	49285	90–125	175–215	225–295	18.5–19.5
4928-B	–	90–125	165–205	–	17–18

25.3.1 Piston Rings

Radial thickness of piston ring

$$t_{pr} = d\left(\frac{3p_w}{\sigma_{br}}\right)^{1/2} \qquad ...(25.24)$$

where t_{pr} = radial thickness of piston rings, mm

p_w = radial wall pressure between the piston ring and cylinder wall, MPa

= (0.03 – 0.06) MPa

σ_{br} = allowable bending stress for piston ring

= (90 – 120) MPa for cast iron

Typical materials for piston rings are given in Table 25.6

Width of rings, $t_2 = \frac{d}{20}$ for concentric rings ...(25.25*a*)

$= \frac{d}{225.5}$ opposite to the joint of eccentric rings ...(25.25*b*)

$= \frac{d}{55}$ at the joint of eccentric rings ...(25.25*c*)

Width of lands on piston between the piston rings, $h = (0.75 - 1.0)\, t_2$ ...(25.26)

Width of top land, $h_0 = (1 - 1.25)\, t_h$ ...(25.27)

Table 25.6 Typical materials for piston rings

Material symbol	Material type	Modulus of elasticity, GPa	Minimum ring tensile strength, MPa	Minimum bending strength, MPa	Minimum hardness
A_1	Grey cast iron, as cast	100	250	350	200 BHN
A_2	Spun grey cast iron, as cast	115	250	350	230 BHN
A_3	Alloyed cast iron, heat treated	120	360	500	HV5 – 400
A_4	Alloyed cast iron, heat treated	115	310	1100	260 BHN
A_5	Alloyed cast iron, heat treated	145	400	910	270 BHN
A6	S.G. Iron, heat treated	160	600	1300	270 BHN
S_1	Steel	200	–	–	38 – 44 HRC

The thickness of the ring section must be modified to maintain equivalent area across the ring grooves. The area under the grooves should equal the edge area of the head since the same amount of heat is flowing. For that reason,

$$\pi d t_h = \pi (d_r - t_r)\, t_r$$

Or

$$t_r = 0.5\left[d_r \pm (d_r^2 - 4dt_h)^{1/2}\right] \qquad ...(25.28)$$

Where t_r = thickness under the ring grooves, mm

d_r = diameter at the bottom of the ring grooves, mm

= $d - (2b + 0.006\,d + 0.5$ mm) at the compression rings

= $d - (2b + 0.006\,d + 1.5$ mm) at the oil grooves

b = radial depth of the ring section, mm

Gap between free ring ends, $g_r = (3 - 4)\,t_{pr}$ ...(25.29)

Small engines usually have three compression rings and one oil ring. Side thrust from the connecting rod is transmitted through the piston bosses to the skirt, and a stiffening rib should be provided at the centre line of the boss and should extend around the skirt to distribute this load and prevent distortion of the skirt. There should be gap or clearance between the ends of the rings to account for expansion of the rings to avoid distortion.

25.3.2 Piston Skirt

The portion of the piston barrel below the rings section up to the open end is called piston skirt. It acts as a bearing surface for connecting rod side thrust. Its length should be such that the bearing pressure between the piston skirt and cylinder wall, due to side thrust, should not exceed 0.3 MPa for low speed engines and 0.5 MPa for high speed engines.

Maximum side thrust,

$$F_s = (0.05 - 0.10) \times \text{maximum gas force}$$

$$= (0.05 - 0.10) \times \left(\frac{\pi}{4}\right) d^2\, p_{max} \quad ...(25.30)$$

Now $$F_s = h_2\, d\, p_{bc} \quad ...(25.31)$$

Where h_2 = length of piston skirt

p_{bc} = permissible bearing pressure between piston skirt and cylinder wall

25.3.3 Length of Piston

Length of piston = width of top land + total width of piston rings + total width of lands + length of piston skirt

$$= t_1 + zt_2 + (z - 1)\,h + h_2 \quad ...(25.32)$$

The empirical proportions of piston are given in Table 25.7.

Table 25.7 Empirical proportions of I.C. engine piston

Dimension	Petrol engine	Diesel engine
Piston head thickness, t_h	$(0.05 - 0.10)\,d$	$(0.01 - 0.2)\,d$
Piston length, H	$(0.8 - 1.3)\,d$	$(0.10 - 1.7)\,d$
Height of top part of piston, h_1	$(0.45 - 0.75)\,d$	$(0.6 - 1.0)\,d$
Skirt length, h_2	$(0.6 - 0.8)\,d$	$(0.7 - 1.1)\,d$
Top land, t_1	$(0.06 - 0.12)\,d$	$(0.11 - 0.2)\,d$
Radial clearance of ring, Δt	0.7 – 0.95 mm	0.7 – 0.95 mm
Length of pin, l_p	$(0.78 - 0.93)\,d$	$(0.8 - 0.93)\,d$
Reinforcing ribs	$\left(\frac{1}{3} - \frac{1}{2}\right) t_h$	$\left(\frac{1}{3} - \frac{1}{2}\right) t_h$

25.3.4 Piston Clearances

Adequate clearances should be provided between the piston and the liner due to the following reasons :

1. Thermal expansion of piston.
2. Distortion of piston under load.

To prevent piston seizure during operation, the diameters of piston crown and skirt are provided clearances between the liner walls and the piston in the cold state. The clearances are provided as given below :

Clearance on piston crown,

$$\Delta c = (0.006 \text{ to } 0.008)\, d, \text{ for aluminium piston} \quad ...(25.33\ a)$$

$$= (0.004 \text{ to } 0.006)\, d, \text{ for cast iron piston} \quad ...(25.33\ b)$$

Clearance on skirt,

$$\Delta s = (0.001 \text{ to } 0.002)\, d, \text{ for both aluminium and cast iron pistons} \quad ...(25.34)$$

Where d = piston diameter

The clearances in the hot state can be determined as follows :

$$\Delta c' = d\,[1 + \alpha_{cyl}\,(T_{cyl} - T_0)] - d_c\,[1 + \alpha_p\,(T_c - T_0)] > (0.002 \text{ to } 0.0025)\, d \quad ...(25.35)$$

$$\Delta s' = d\,[1 + \alpha_{cyl}\,(T_{cyl} - T_0)] - d_s\,[1 + \alpha_p\,(T_c - T_0)] > (0.0005 \text{ to } 0.0015)\, d \quad ...(25.36)$$

Where α_{cyl} = coefficient of expansion for material of cylinder liner

α_p = coefficient of expansion for material of piston

$$= 22 \times 10^{-6} \text{ for aluminium, } \frac{1}{K}$$

$$= 11 \times 10^{-6} \text{ for cast iron, } \frac{1}{K}$$

For water cooled engines,

$$T_{cyl} = 385 \text{ K}, T_c = 473 \text{ to } 723 \text{ K}, T_s = 403 \text{ to } 473 \text{ K}, T_0 = 293 \text{ K}$$

For air cooled engines,

$$T_{cyl} = 450 \text{ K}, T_c = 573 \text{ to } 873 \text{ K}, T_s = 483 \text{ to } 613 \text{ K}$$

25.3.5 Piston (or Gudgeon) Pin

The piston pin should be designed for the maximum combustion pressure or inertia force of the piston, whichever is larger. The piston pin is usually hollow to reduce the weight and is often tapered on the inside, the smallest diameter being at the centre of the pin. The piston pin bearing area should be about equally divided between bearing in the connecting rod and in the piston. Thus, the length of the pin in the connecting rod bearing will be about 0.45 times the piston diameter, allowing for end clearance for the pin, etc.

As shown in Fig. 25.3, the piston is assumed to be uniformly loaded for the length l, with supports at the centres of the bosses at both ends, making l', the length between the supports.

$$l' = l + 0.5\,(d - l) = \frac{(l + d)}{2}$$

Maximum bending moment,

$$M_{max} = \left(\frac{F_g}{2}\right)\left(\frac{l'}{2} - \frac{l}{4}\right) = \frac{F_g d}{8} \quad ...(25.33)$$

Where $F_g = p_{max} \times$ area of cylinder

Resisting moment,

$$M_r = \left(\frac{\pi}{32}\right)\left[\frac{d_0^4 - d_i^4}{d_0}\right]\sigma \quad ...(25.34)$$

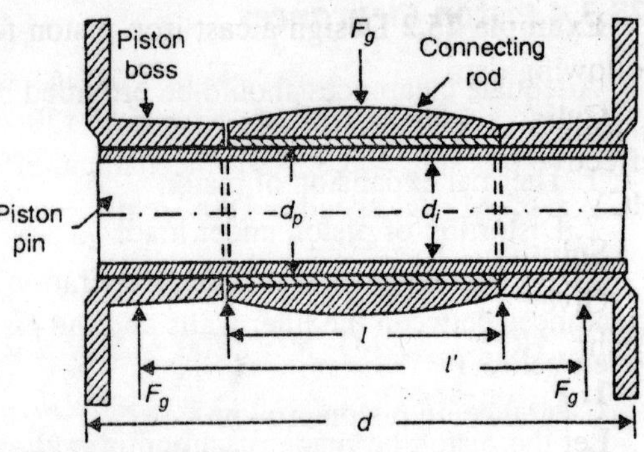

Fig. 25.3 Piston)or gudgeon) pin

Where d_i, d_0 = inside and outside diameters of the pin

σ = permissible bending stress for the pin material.

Eqs. (25.33) and (25.34) must be equated to obtain the values of d_i and d_0.

The bearing pressure between the piston pin and the connecting rod bearing is given by,

$$F_g = l d_p p_b \quad ...(25.35)$$

Where L = length of the piston pin in connecting rod bearing

d_p = diameter of piston pin

p_b = permissible bearing pressure between the piston pin and connecting rod bearing (Table 25.8)

Table 25.8 Permissible bearing pressure between piston pin and connecting rod bearing

Bearing material	p_b, MPa
Aluminium alloys	5-10
Cast iron	10-15
Babbit	15-20
Bronze	20-25
Needle roller bearing	40-60

Shear stress in piston pin, $$\tau = \frac{2F_g}{\pi\left(d_p^2 - d_i^2\right)} \quad ...(25.36)$$

The piston pin is made of plain carbon steels containing nickel, chromium and molybdenum. The pin is hardened and ground. The bush is made of bronze.

The specifications for piston pins are given in Table 25.9

Table 25.9 Specifications for Piston Pins for I.C. Engines

Steel type	Material	Symbol	Tensile strength, Mpa	Core hardness, HRC
Case hardening	15Cr3	A	325	15–35
	17Mn5Cr4	B	432	15–35
	15C8	C	233	15-35
	15Ni5Cr4Mo1	D	540	25–45
	13Ni13Cr3	E	460	25–45
Nitriding	40Cr7Al10Mo2	F	325	25–45

Example 25.2 Design a cast iron piston for a single–acting four–stroke diesel engine for the following data :

Cylinder bore = 100 *mm*, stroke = 130 *mm*, maximum pressure = 6 MP*a*, indicated mean effective pressure = 0.8 MP*a*, mechanical efficiency = 80%, fuel consumption = 0.2 kg/ BP.hr., HCV = 42 MJ/kg, speed = 1800 *rpm*

Solution.

Given : d = 100 *mm*, l = 130 *mm*, p_{max} = 6 MP*a*, p_{im} = 0.8 MP*a*, η_{mech} = 0.8, m_f = 0.2 kg/ BP.hr., HCV = 42 MJ, N = 1800 *rpm* or n = 900 strokes/min.

1. Piston head :

Let the piston be made of aluminium alloy having allowable tensile stress, (Table 25.4)

$$\sigma_t = 60 \text{ MPa}$$

Thickness of piston head based on strength considerations,

$$t_h = 0.433d\left[\frac{p_{max}}{\sigma_t}\right]^{1/2}$$

$$= 0.433 \times 100\left[\frac{6}{60}\right]^{1/2} = 13.7 \approx 14 \text{ mm}$$

$$\text{Indicated power, IP} = \frac{p_{im}\,\text{LAn}}{60} = 0.8 \times 130 \times 10^{-3} \times \frac{\pi}{4} \times 100^2 \times \frac{900}{60} = 12.252 \text{ kW}$$

Brake power, BP = 12.252 × 0.8 = 9.8 kW

Heat flow through the head,

$$\text{H} = \text{C} \times \text{HCV} \times m_f \times \text{BP}$$

$$= 0.05 \times 42 \times 10^6 \times \left(\frac{0.2}{3600}\right) \times 9.8 = 1143 \text{ W}$$

Thickness of cylinder head based upon heat flow considerations,

$$t_h = \frac{\text{H}}{12.56\, h_c\left(\text{T}_c - \text{T}_e\right)}$$

$$= \frac{1143}{12.56 \times 46.4 \times 220} = 8.9 \approx 9 \text{ mm}$$

We adopt, t_h = 14 mm

2. Ribs

Number of ribs = 4

$$\text{Thickness of ribs, } t_{rib} = \frac{t_h}{3} \text{ to } \frac{t_h}{2} = 4.67 \text{ to } 7 \text{ mm}$$

We adopt, t_{rib} = 6 mm

3. Rings

There shall be three compression rings and one oil ring.

Thickness of rings, $t_{pr} = d\left[\frac{3pr}{\sigma_{br}}\right]^{1/2}$

$$= 100\left[\frac{3\times 0.035}{60}\right]^{1/2} = 4.2 \approx 4 \text{ mm}$$

Depth of piston rings $= (0.7 \text{ to } 1.0)\ t_{pr} = 3$ mm

4.Proportions of Piston

Piston crown thickness,

t_h = 14 mm (already calculated)

Piston height, H = $1.0\,d$ = 100 mm

Height of top part of piston,

$h_1 = 0.6d = 60$ mm

Skirt length, $h_2 = 0.8d = 800$ mm

Top land, $t_1 = 0.15\text{d} = 15$ mm

Radial clearance of ring,

$\Delta t = 0.8$ mm

Length of pin, $l_p = 0.85d = 85$ mm

5. Piston pin

Maximum gas force on the piston,

$$F_g = \left(\frac{\pi}{4}\right)d^2\ p_{\max} = \left(\frac{\pi}{4}\right)\times 100^2 \times 6 = 47124 \text{ N}$$

Let d_p = outside diameter of pin, mm

l = length of pin in the bush of the small end of the connecting rod, mm

= $0.45d$ = 45 mm

p_b = bearing pressure at the small end of connecting rod

$\approx$ 25 MPa

Load on the pin during bearing pressure, $F_b = p_b d_p\,(l_p - l)$

Now $F_b = F_g$

$25 \times d_p \times (85 - 45) = 47123$

$d_p = 48 \approx 50$ mm

Inside diameter, $d_i = 0.6d_p = 30$ mm

Maximum bending moment,

$$M_{\max} = \frac{F_g d}{8}$$

$$= \frac{47124\times 100}{8}$$

$$= 589050 \text{ N.mm}$$

$$M_{max} = \left(\frac{\pi}{32}\right)\left[\frac{d_p^4 - d_i^4}{d_0}\right]\sigma_b$$

$$589050 = \left(\frac{\pi}{32}\right)\left[\frac{50^4 - 30^4}{50}\right]\sigma_b$$

$$\sigma_b = 55.15 \text{ MPa} < 140 \text{ MPa}$$

Maximum shear stress in pin due to double shear,

$$\tau_{max} = \frac{2F_g}{\pi\,(d_p^2 - d_i^2)}$$

$$= \frac{2 \times 47124}{\pi\,(50^2 - 30^2)}$$

$$= 18.75 \text{ MPa} < (60 - 70) \text{ MPa}$$

Hence safe.

25.4 CONNECTING ROD

The forces acting on the connecting rod are :

1.Force due to gas pressure and the piston inertia.

2.Force due to friction of piston rings and piston.

3.Inertia of the connecting rod.

4.Frictional force in the big and small end bearings.

Maximum force due to gas pressure,

$$F_g = \left(\frac{\pi}{4}\right)d^2 p_e \quad \text{...(25.37)}$$

Where p_e = explosion pressure

Force due to piston inertia, $F_i = M\omega^2 r\left[\cos\theta + \left(\frac{1}{n}\right)\cos 2\theta\right]$ ...(25.38)

Where M = mass of (piston and piston rings + piston pin + $\frac{1}{3}^{rd}$ of connecting rod)

$\approx 4.15 \times 10^{-6}\, d^3$ kg for cast iron pistons

$\approx 1.78 \times 10^{-6}\, d^3$ for aluminium pistons

d = bore diameter, mm

ω = angular speed, rad./s

r = crank radius, mm

l = length of connecting rod, mm

θ = crank angle

$$n = \frac{l}{r}$$

The force in the connecting rod is maximum when the crank angle is 90°. At that instant, the force in the connecting rod,

$$F_c = \frac{nF_a}{(n^2 - \sin^2\theta)^{1/2}} \qquad \text{...(25.39)}$$

Where $F_g = \left(\frac{\pi}{4}\right) d_p^2 \, p_a$

p_a = net pressure along the cylinder

Force of friction due to piston rings and piston,

$F_f = \pi h d i p_r \mu$

Where h = axial width of rings

i = number of rings

p_r = pressure of rings

μ = coefficient of friction ≈ 0.05 – 0.10

The inertia force of the connecting rod has two components: longitudinal and transverse. The longitudinal component is accounted for by taking 1/3rd of its mass at the piston pin and 2/3rd at the crank pin. The transverse components acts on every part of the rod. The rod may be assumed of uniform cross-section without any appreciable error, in which case the resultant inertia force due to the acceleration of the oscillating rod will act at a ditance $\left(\frac{2}{3}\right) l$ from the piston pin. At the crank pin the force per unit length will have a magnitude given by [see Fig. 25.4 (*a*)],

$$F = \rho A \omega^2 r \sin(\theta + \phi)$$

Where ρ = density of connecting rod material

A = cross – sectional area of connecting rod

F will be maximum when the crank and connecting rod are at right angles, *i.e.* $\theta + \phi = 90°$.

$$F_{max} = \rho A \omega^2 r$$

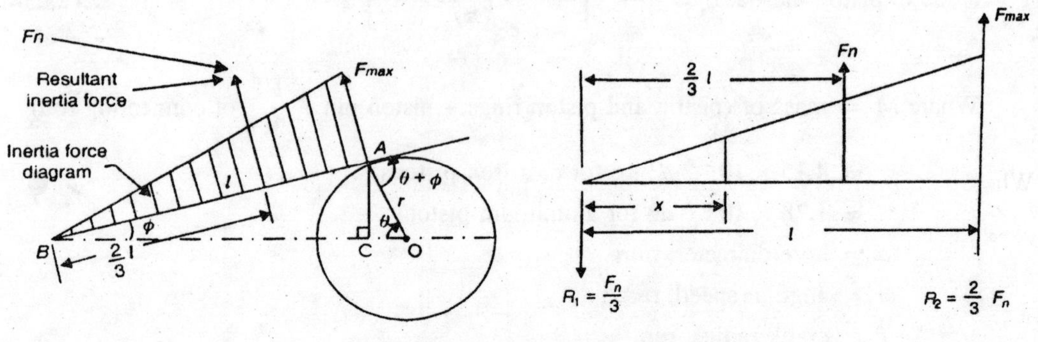

Fig. 25.4 Forces on connecting rod

The resultant normal force acting on the conecting rod then becomes,

$$F_n = \left(\frac{1}{2}\right) F_{max}\, l \qquad \text{...(25.41)}$$

The reactions at the small and big end are [see Fig. 25.4 (*b*)],

$$R_1 = \frac{F_n}{3} \text{ and } R_2 = \frac{2F_n}{3}$$

The bending moment at a distance from x from the small end is,

$$M_x = \frac{F_n \times x}{3} - \left(\frac{1}{2}\right) F_{max} \times \frac{x}{l} \times x \times \frac{x}{3}$$

$$= \left[\frac{F_n x}{3l^2}\right](l^2 - x^2)$$

For bending moment to be maximum $\frac{dM_x}{dx} = 0$

$$\left[\frac{F_n}{3l^2}\right](l^2 - 3x^2) = 0$$

$$x = \frac{l}{\sqrt{3}}$$

$$M_{max} = \frac{2F_n l}{9\sqrt{3}} = 0.1283\, F_n l \qquad \text{...(25.42)}$$

Maximum bending stress,

$$\sigma_{max} = \frac{M_{max}}{z} \qquad \text{...(25.43)}$$

Where z = section modulus of the rod.

Due to the gas force, the rod is subjected to buckling. The buckling load is given by the Rankine-Gordon's formula,

$$P_e = \frac{\sigma_c A}{\left[1 + a\left(\frac{l_e}{K}\right)^2\right]} \qquad \text{...(25.44)}$$

Where σ_c = ultimate compressive strength of rod material

l_e = equivalent length of the rod

a = constant = $\frac{1}{7500}$

K = radius of gyration of the rod about the buckling axis.

In the plane of motion of the rod, the ends are direction–free at the crank and gudgeon pins and the rod is, therefore, freely hinged for buckling about the neutral axis $x - x$, as shown in Fig. 25.5. In the plane perpendicular to the plane of motion, *i.e.* $y - y$ the rod has fixed ends due to constraining effects of the bearings at the crank and gudgeon pins.For buckling about $y - y$, the rod is therefore four times as strong as for buckling about $x - x$. To achieve this condition for a rectangular rod [Fig. 25.5 (*a*)] the ratio of width of depth should be approximately 0.25. For an I-section rod the necessary condition for the rod to be equally resistant to buckling in either plane is,

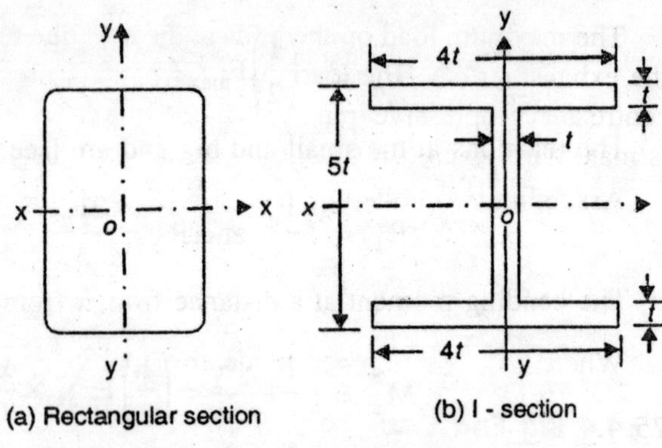

Fig. 25.5 Proportions of connecting rod

$$4I_{yy} = I_{xx}$$

Or $$4K^2_{yy} = K^2_{xx} \qquad \text{....(25.45)}$$

The proportions of the rod as shown in Fig. 25.5 (*b*) satisfy the condition given by Eq. (25.45) approximately. These dimensions may be taken at the middle of the rod. At the piston end the depth may be taken as 0.75 to 0.9 times and at the crank pin end 1.1 to 1.25 times.

25.4.1 Dimensions of Small End

Let l_p = length of piston pin

d_p = diameter of piston pin

p_b = permissible bearing pressure ≈ 10 – 14 N/mm^2

$$\frac{l_p}{d_p} = 1.5 \text{ to } 2$$

The force on the small end, $F_a = l_p d_p p_b$ ....(25.46)

25.4.2 Dimensions of Big End

Let l_c = length of crank pin

d_c = diameter of crank pin

p_b = permissible bearing pressure = 5 – 10 N/mm^2

$$\frac{L_c}{d_c} = 1.25 \text{ to } 1.5$$

The force on the small end,

$$F_a = l_c d_c p_b \qquad \text{....(25.47)}$$

25.4.3 Bolts for Big End

The maxmum load on the bolts and cap is due to the inertia force at the top dead centre during the exhaust stroke. The load due to explosion pressure and also the inertia load at bottom dead centre are compressive in nature and do not affect the bolts. The inertia force given by Eq. (25.38) is maximum when $\theta = 0°$. The bolts are under repeated but not alternating stresses.

Assuming two bolts, we have

$$2 \times \left(\frac{\pi}{4}\right) d_b^2 \sigma_t = F_i \quad ...(25.48)$$

Where d_b = diameter of bolts

25.4.4 Big End Cap

The cap for the big end is designed as a beam freely supported at the bolt centres and loaded in a manner intermediate between uniformly distributed and centrally concentrated loading. The maximum bending moment is given by,

$$M_{max} = \frac{F_{ip} l_0}{6} \quad ....(25.49)$$

Where l_0 = distance between bolt centres

$= d_c + 2 \times$ thickness of linear ($\approx 0.05d + 1$ mm) $+ d_b +$ clearance (≈ 1.5 mm)

If h = cap thickness, and l_c = width of cap = length of bearing, then

$$M_{max} = \frac{\sigma_b \times l_c h^2}{6} \quad ...(25.50)$$

The main parts of a connecting rod are shown in Fig. 25.6

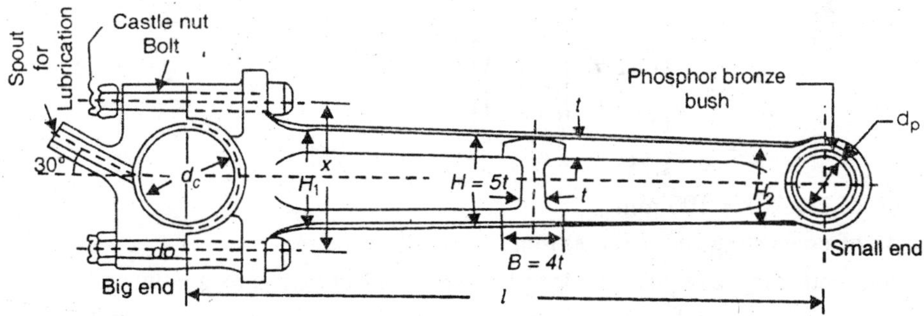

Fig. 25.6 Connecting rod

Example 25.3 Design a connecting rod for a four-stroke diesel engine for the following data :

Speed of engine = 1800 rpm, Maximum pressure = 3.5 MPa

Piston diameter = 100 mm, Mass of reciprocating parts per cylinder = 2 kg

Length of connecting rod = 400 mm, Stroke = 200 mm, Compression ratio = 6 : 1

Factor of safety = 6

Length of diameter ratio of big end = 1.25, Length to diameter ratio for small end = 2

Bearing pressure for big end = 10 MPa, Bearing pressure for small end = 15 MPa
Density or rod material = 7800 kg/m^3, Allowable stress in the bolts = 60 MPa
Allowable stress in the cap = 80 MPa, Cross – section of rod = 1 – section

Rankine constants = – 320 MPa and $\frac{1}{7500}$

Solution.

Given : N = 1800 rpm, n = 900 strokes/min, p_{max} = 3.5 MPa, d = 100 mm, R = 2 kg, l = 400 mm, L = 200 mm, *i.e.* r = 100 mm, r_c = 6, FOS = 6, $\frac{l_c}{d_c}$ = 1.25, $\frac{l_s}{d_s}$ = 2, p_{bc} = 10 MPa, p_{bs} = 15 MPa, ρ = 7800 kg/m^3, σ_b = 60 MPa, σ_{cap} = 80 MPa, σ_c = 320 MPa, $a = \frac{1}{7500}$

1. Gas force, $F_g = \left(\frac{\pi}{4}\right) d^2 p_{max}$

$$= \left(\frac{\pi}{4}\right) \times 100^2 \times 3.5 = 27489 \text{ N}$$

2. Cross-section of Rod

Buckling load, $P_e = F_g \times \text{FOS} = 27489 \times 6 = 164934$ N

The rod is of symmetrical I – section with width, B = 4t and depth, H = 5t, where t = thickness of flanges and web, This gives area of cross – section equal to 11t^2, $I_{xx} = 3.2\, I_{yy}$.

Radius of gyration about x-axis, $K_{xx} = \left(\frac{I_{xx}}{A}\right)^{1/2}$

$$= \left[\left(\frac{419t^4}{12}\right)\left(\frac{1}{11t^2}\right)\right]^{1/2}$$

$$= 1.78t$$

Length of crank, r = 100 mm

Length of connecting rod, l = 400 mm

For both ends hinged, equivalent length of the connecting rod as a column, $l_e = l$ = 400 mm

Rankine–Gordon formula for buckling of rod gives,

$$P_e = \frac{\sigma_c A}{\left[1 + a\left(\frac{l_e}{K_{xx}}\right)^2\right]}$$

$$164934 = \frac{320 \times 11t^2}{\left[1 + \left(\frac{1}{7500}\right)\left(\frac{400}{1.78t}\right)^2\right]}$$

$$46.85625 = \frac{t^4}{t^2 + 6.733}$$

$$t^4 - 46.85625t^2 - 315.5 = 0$$

$$t^2 = \frac{46.85625 \pm (2195.5 + 1262)^{1/2}}{2} = 52.8284$$

$$t = 25.268 \text{ mm}$$

$$\text{Adopt } t = 25.5 \text{ mm}$$

Thickness of flange and web of *l*-section,

$$t = 25.5 \text{ mm}$$

Width of section, $B = 4t = 30$ mm

Depth of section, $H = 5t = 325.5$ mm

These are the dimensions at the middle of the connecting rod. The width B is kept constant throughout the length of the rod, but the depth H is varied.

Depth near the big end, $H_1 = 1.2\text{ H} = 45$ mm

Depth near the small end, $h_2 = 0.85\text{ H} = 31.875 \approx 32$ mm

3. Design of Big End

The big end is designed for the maximum gas force.

$$\text{Load on the crankpin, } p_{bc}d_c l_c = 10 \times 1.25 \times d_c^2 = 12.5d_c^2$$

$$12.5d_c^2 = 27489$$

$$d_c = 46.895 \approx 48 \text{ mm}$$

$$l_c = 58.62 \approx 60 \text{ mm}$$

4. Design of Small End

The small end is designed for the maximum gas force.

$$\text{Load on the piston pin, } p_{bs}d_s l_s = 15 \times 2 \times d_s^2 = 30d_s^2$$

$$30d_b^2 = 27489$$

$$d_s = 30.27 \approx 31 \text{ mm}$$

$$l_s = 62 \text{ mm}$$

5. Bolts for Big End Cap

The bolts for the big end are designed for the inertia forces of the reciprocating parts at the top dead centre on the exhaust stroke.

$$F_i = R\omega^2 r\left(1 + \frac{r}{l}\right)$$

$$= 2\left(2\pi \times \frac{1800}{60}\right)^2 \times 0.1 \times \left(1 + \frac{100}{400}\right)$$

$$= 8882.6 \text{ N}$$

Force on the bolts $= 2\left(\frac{\pi}{4}\right) d_b^2 \sigma_b$

$$8882.6 = 2\left(\frac{\pi}{4}\right) d_b^2 \times 60$$

Nominal diameter of bolts, $d_b = 9.7 \approx 10$ mm

6. Big End Cap

Let t_c = thickness of cap
b_c = width of cap
= length of crankpin
= $l_c = 60$ mm

The cap is designed as a beam freely supported at the cap bolt centres and loaded by inertia force at the top dead centre of the exhaust stroke. The load is assumed to be in between uniformly distributed and centrally concentrated load. Therefore,

$$M_c = \frac{F_i x}{6}$$

Where x = distance between the bolt centres
= diameter of crankpin + 2 × thickness of bearing liner + nominal diameter of bolt + clearance
$= d_c + 2 \times 3 + d_b + 3$
$= 48 + 6 + 10 + 3 = 67$ mm

$$M_c = 8882.6 \times \frac{67}{6} = 99189 \text{ N.mm}$$

Section modulus for cap, $z_c = \frac{b_c t_c^2}{6} = \frac{60 \times t_c^2}{6} = 10t_c^2$

$$M_c = \sigma_{cap} \times z_c$$

$$99189 = 80 \times 10t_c^2$$

$$t_c = 11.1 \approx 11.5 \text{ mm}$$

7. Check for the rod against bending stresses due to inertia force.

Mass of connecting rod per metre length,

$$m_1 = Al\rho = 11t^2 \times l \times \rho$$

$$= 11 \times (25.5)^2 \times 10^{-6} \times 0.4 \times 7800 = 1.93 \text{ kg/m}$$

Maximum bending moment, $M_{max} = \frac{m\,\omega^2 r\,l}{9\sqrt{3}}$

$$= \frac{m_1\,\omega^2 r\,l^2}{9\sqrt{3}}$$

$$= \frac{1.93\,(2\pi \times 1800/60)^2\,(0.1)\,(0.4)^2}{9\sqrt{3}}$$

$$= 70.384 \text{ N.m}$$

$$= 70384 \text{ N.mm}$$

Section modulus, z_{xx} $= \dfrac{I_{xx}}{2.5\,t}$

$$= \frac{419\,t^4}{12}\left[\frac{1}{2.5\,t}\right]$$

$$= 13.97\,t^3$$

$$= 13.97\,(25.5)^3 = 5893.6 \text{ mm}^3$$

Bending stress, $(\sigma_b)_{max} = \dfrac{M_{max}}{z_{xx}} = \dfrac{70384}{5893.6} = 11.94$ MPa < 80 MPa. Hence safe.

25.5 CRANKSHAFT

A crankshaft is used to convert reciprocating motion of the piston into rotary motion or vice versa. The crankshaft consist of the following parts :

1. Shaft parts which revolve in the bearings.
2. The crankpins to which big ends of the connecting rod are connected.
3. The crank arms or webs (called cheeks) which connect the crankpins and the shafts.

25.5.1 Types of Crankshafts

Depending upon the position of the crank,, the crankshafts are of two types, as shown in Fig. 25.7.

1. Side (or overhung) crankshaft, and
2. Centre crankshaft.

A crankshaft with onlyone side crank or centre crank is called a single throw crankshaft. The crankshaft with two side cranks or two or more centre cranks is called a multi-throw crankshaft.

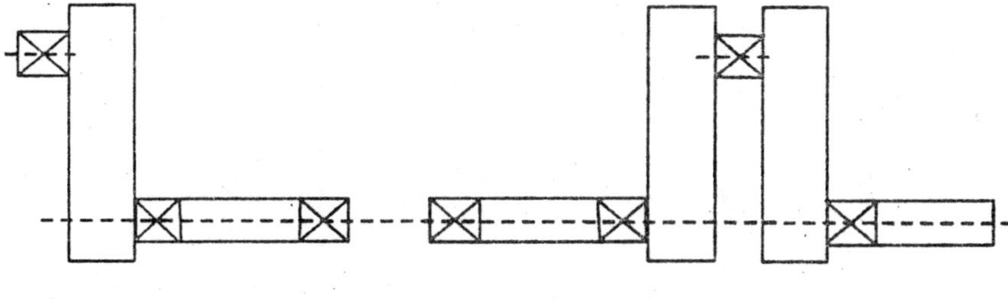

(a) Side Crankshaft (b) Centre Crankshaft

Fig. 25.7 Types of crankashofts

25.5.2 Materials and Manufacturing Methods

The materials for crankshaft are given in Table 25.10.

Table 25.10 Materials for crankshaft

Materials	Applications
1. 40C8, 55C8, 60C4	Industrial engines
2. 20Mn2, 27Mn2, 37Mn2	Transport engines
3. 35Ni1Cr60, 40Ni2Cr1Mo28	Aero engines.

The crankshafts are manufactured by drop forging or casting process. The surface of the crankpin is hardened by case carburizing, nitriding or induction hardening.

25.5.3 Stresses in Crankshaft

The stresses induced in a crankshaft are :

1.Bending stresses due to gas forces, and

2.Shear stresses due to torsional moment on the shaft.

The crankshaft is subjected to shock and fatigue loads. The failure of a crankshaft is caused by a progressive fracture due to repeated bending or reversed torsional stresses.

25.5.4 Design of Centre Crankshaft

The crankshaft is designed for two crank positions : (1) When the crank is at dead centre, it is subjected to maximum bending moment,, and (2) when the crank is at an angle, it is subjected to maximum twisting moment.

1. Crank at maximum bending moment position :

At this position the bending moment on the shaft is maximum and the twisting moment is zero. A single throw centre crankshaft with three bearings is shown in Fig. 25.8. Let

d = cylinder bore or piston diameter, mm

p_{max} = maximum pressure on the piston, MPa

W = weight of the flywheel acting downwards, N

$T_1 + T_2$ = resultant belt tension acting horizontally, N

Thrust in the connecting rod = Gas force on the piston acting horizontally

$$F_g = \left(\frac{\pi}{4}\right) d^2\ p_{max}$$

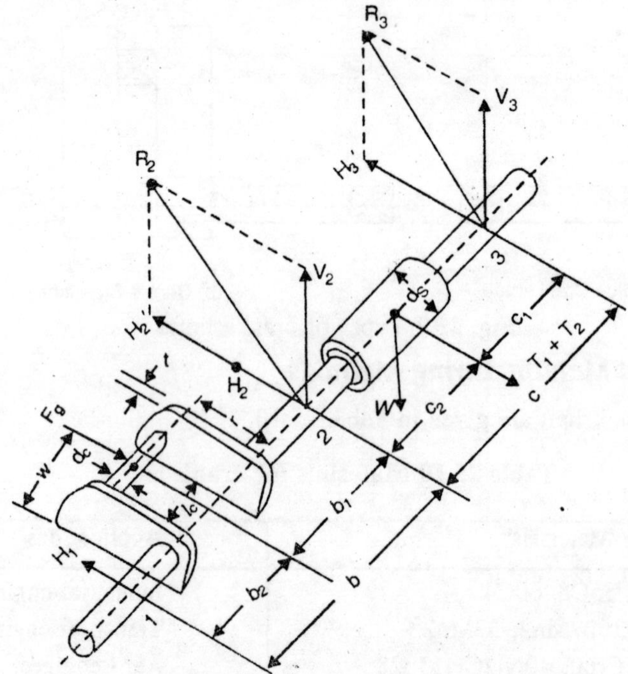

Fig. 25.8 Centre crankshaft at dead centre

Horizontal reactions on bearings 1 and 2 are,

$$H_1 = \frac{F_g\, b_1}{b} \text{ and } H_2 = \frac{F_g\, b_2}{b}$$

Vertical reactions due to weight W of the flywheel on the bearings 2 and 3 are,

$$V_2 = \frac{W.c_1}{c}, \text{ and } V_3 = \frac{W.c_2}{c}$$

Horizontal reactions due to belt tensions at bearings 2 and 3 are,

$$H'_2 = \frac{(T_1 + T_2)\, c_1}{c}, \; H'_3 = \frac{(T_1 + T_2)\, c_2}{c}$$

Resultant forces at bearings 2 and 3 become,

$$R_2 = [(H_2 + H'_2)^2 + V^2_2]^{1/2}$$
$$R_3 = [H^2_3 + V^2_3]^{1/2}$$

(i) Design of crankpin

Let d_c = diameter of crankpin, mm

l_c = length of crankpin, mm

σ_b = allowable bending stress for the crankpin, MPa

Bending moment at the centre of crankpin, $M_c = H_1 b_2$

Also
$$M_c = \left(\frac{\pi}{32}\right) d^3{}_c \sigma_b = H_1\, b_2 \quad ...(1)$$

We can determine d_c from Eq. (1)

Considering bearning of the crankpin, the length of crankpin is,

$$l_c = \frac{F_g}{d_c\, p_b} \quad ...(2)$$

where p_b = permissible bearing pressure.

(ii) Design of left hand crank web

The crank web is subjected to eccentric loading. The crank web is subjected to direct compressive stress and bending stress due to gas force.

Empirically, the thickness of the crankweb is,

$$t = (0.4 - 0.6)\, d_s$$
$$= (0.22 - 0.32)\, d$$
$$= 0.65\, d_c + 6.35 \text{ mm}$$

where d_s = diameter of shaft, mm

d = bore diameter, mm

d_c = crankpin diameter, mm

width of the crank web, $w = 1.125\, d_c + 12.7$ mm

Maximum bending moment on the crankweb, M

$$M = H_1 \left(b_2 - \frac{l_2}{2} - \frac{t}{2}\right)$$

Section modulus of web, $z_w = \frac{wt^2}{6}$

Bending stress in web, $\sigma_{wb} = \frac{M}{z_w} = \frac{6H_1 (b_2 - l_c/2 - t/2)}{wt^2}$

Direct compressive stress in web, $\sigma_{wd} = \frac{H_1}{(wt)}$

Total stress in the web, $\sigma_w = \sigma_{wb} + \sigma_{wd}$

For a safe design, $\sigma_w < \sigma_{allow}$

(iii) Design of right hand crank web

The dimensions of the right hand crank web are made the same as that of the left hand crank web from balancing point of view.

(iv) Design of shaft under the flywheel

Let d_s = diameter of shaft, mm

Bending moment due to the weight of the flywheel,

$$M_f = V_3 . c_1$$

Bending moment due to belt tensions,

$$M_b = H'_3 . c_1$$

Resultant bending moment, $M_s = [M^2_f + M^2_b]^{1/2}$

$$= \left(\frac{\pi}{32}\right) d_s^3 \sigma_b \quad \text{....(3)}$$

Where σ_b = allowable bending stress for the shaft, MPa

We can determine the value of d_s from Eq. (3).

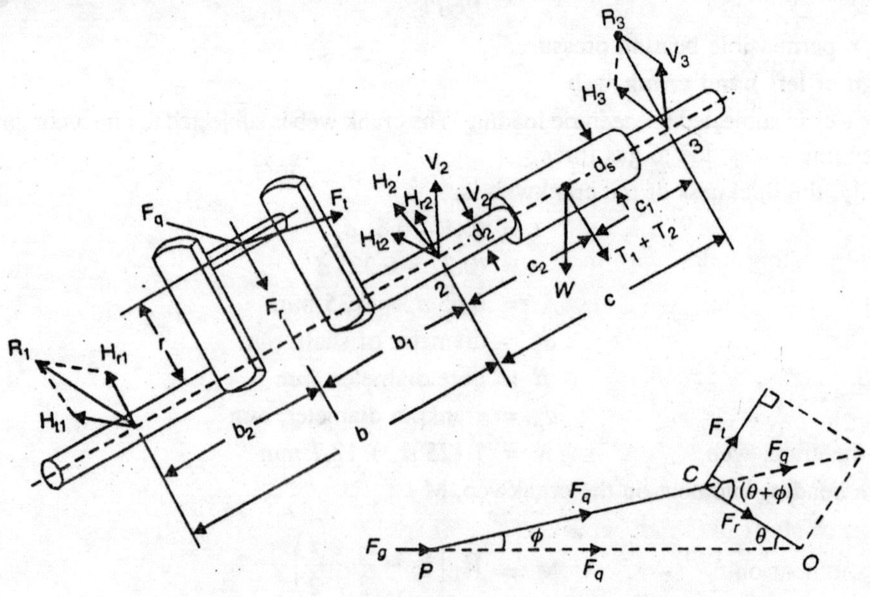

Fig. 25.9 *(a)* Crank at an angle of maximum twisting moment

2. Crank at maximum twisting moment position.

The twisting moment on the crank shaft is maximum when the tangential force on the crank is maximum. It occurs at an angle of 25°–30° the TDC for a petrol engine and 30°–40° for a diesel engine. The crankshaft m its maximum twisting moment position is shown in Fig. 25.9. Let p_0 be the pressure on the piston at this position, then the gas force on the piston is,

$$F_g = \left(\frac{\pi}{4}\right) d^2 \, p_0$$

Thrust on the connecting rod,

$$F_q = \frac{F_g}{\cos \phi}$$

Where ϕ = angle of inclination of the connecting rod with the line of stroke.

The tangential force F_t is the component of F_q perpendicular to the crank. The other component is the radial force F_r along the crank.

Tangential force, $F_t = F_q \sin (\theta + \phi)$

Radial force, $F_r = F_q \cos (\theta + \phi)$

The tangential force causes twisting whereas radial force causes bending of the shaft. Reactions at bearings 1 and 2 due to F_t are,

$$H_{t1} = \frac{F_t . b_1}{b}, \text{ and } H_{t2} = \frac{F_t . b_2}{b}$$

Reactions due to radial force at bearings 1 and 2 are,

$$H_{r1} = \frac{F_r . b_1}{b}, \text{ and } H_{r2} = \frac{F_r . b_2}{b}$$

The reactions at the bearings 2 and 3 due to flywheel and belt tensions will be the same as in case 1.

(i) Design of crankpin

Bending moment at the centre of crankpin,

$$M_c = H_{r1} . b_2$$

Twisting moment on the crankpin,

$$T_c = H_{t1} . r$$

Equivalent twisting moment on the crankpin,

$$T_e = [M^2_c + T^2_c]^{1/2}$$

$$= \left(\frac{\pi}{16}\right) d_c^3 \tau \qquad ...(4)$$

Where τ = allowable shear stress in the crankpin

From Eq. (4), the diameter of the crankpin can be determined.

(ii) Design of shaft under the flywheel

Bending moment on the shaft, $M_s = R_3 . c_1$

Twisting moment on the shaft, $T_s = F_t . r$

Equivalent twisting moment on the shaft,

$$T_e = \left[M_s^2 + T_s^2\right]^{1/2}$$

$$= \left(\frac{\pi}{16}\right) d_s^3 \tau \qquad \text{...(5)}$$

where τ = allowable shear in the shaft

From Eq. (5), the diameter of the shaft can be determined.

(iii) Design of shaft at the junction of right hand crank arm

Let d_{s1} = diameter of the shaft at the junction of right hand crank arm.

Bending moment at the junction of right hand crank arm,

$$M_{s1} = R_1 (b_2 + l_c / 2 + t/2) - F_q (l_c/2 + t/2)$$

Twisting moment at the junction of the right hand crank arm,

$$T_{s1} = F_r . r$$

Equivalent twisting moment at the junction of right hand crank arm,

$$T_e' = \left[M_{s_1}^2 + T_{s_1}^2\right]^{1/2}$$

$$= \left(\frac{\pi}{16}\right) d_{s1}^3 \tau \qquad \text{....(6)}$$

Where τ = allowable shear stress in the shaft.

From Eq. (6), d_{s1} can be determined.

(iv) Design of right hand crank web

The right hand crank web is subjected to (1) bending stresses due to the tangential and radial components of F_q, (2) direct compressive stress due to F_r, and (3) torsional stress.

Bending moment due to radial component of F_q is,

$$M_r = H_{r2} (b_1 - l_c/2 - t/2) \qquad \text{....(7)}$$

$$= \left(\frac{wt^2}{6}\right) . \sigma_{br} \qquad \text{....(7)}$$

where σ_{br} = bending stress in the radial direction.

Bending moment due to tangential component of F_q is,

$$M_t = F_t \left(r - \frac{ds_1}{2}\right)$$

$$= \left(\frac{tw^2}{6}\right) . \sigma_{bt} \qquad \text{....(8)}$$

Where σ_{bt} = bending stress in the tangential direction

Direct compressive stress, $$\sigma_d = \frac{F_r}{2wt} \qquad \text{...(9)}$$

Maximum compressive stress, $\sigma_c = \sigma_{br} + \sigma_{bt} + \sigma_d$ ...(10)

Twisting moment on the arm, $$T = H_{t1}\left(b_2 + \frac{l_c}{2}\right) - \frac{F_t \times l_c}{2}$$

$$= H_{t2}\left(b_1 - \frac{l_c}{2}\right)$$

Shear stress in the arm, $$\tau = \frac{T}{z_p} = \frac{4.5\,T}{wt^2} \quad ...(11)$$

Maximum combined stress, $$(\sigma_c)_{max} = \frac{\sigma_c}{2} + \left(\frac{1}{2}\right)[\sigma_c^2 + 4\tau^2]^{1/2} \quad ...(12)$$

It should be less than the permissible compressive stress.

(v) Design of left hand crank web.

The dimensions for the left hand crank web may be kept the same as the right hand crank web from balancing point of view.

(vi) Design of crankshaft bearings

The bearing 2 is the most heavily loaded and should be checked for bearing pressure.

$$R_2 = \frac{F_g}{2} + \frac{W}{2} + \frac{(T_1 + T_2)}{2}$$

Bearing pressure, $$p_{b2} = \frac{R_2}{l_2 . d_{s1}}$$

where l_2 = length of bearing 2.

25.5.5 Side Crankshaft

1. Crank at dead centre position.

At this position the bending moment on the shaft is maximum and the twisting moment is zero. A single throw centre crankshaft with three bearings is shown in Fig. 25.10. Let

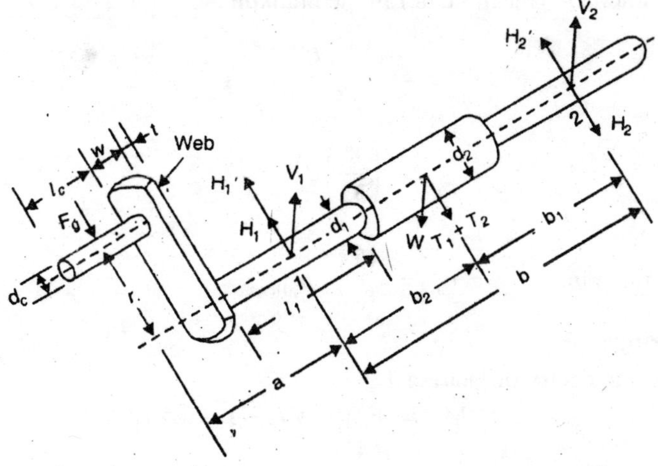

Fig. 25.10 Side crankshaft at dead centre

$$d = \text{cylinder bore or piston diameter, mm}$$
$$p_{max} = \text{maximum pressure on the piston, MPa}$$
$$W = \text{weight of the flywheel acting downwards, N}$$
$$T_1 + T_2 = \text{resultant belt tension acting horizontally, N}$$

Thrust in the connecting rod = Gas force on the piston acting horizontally

$$F_g = \left(\frac{\pi}{4}\right) d^2\ p_{max}$$

Horizontal reactions on bearings 1 and 2 are,

$$H_1 = \frac{F_g\ (a+b)}{b} \text{ and } H_2 = F_g \frac{a}{b}$$

Vertical reactions due to the weight W of the flywheel on the bearings 1 and 2 are,

$$V_1 = \frac{W.b_1}{b} \text{ and } V_2 = \frac{W.b_2}{b}$$

Horizontal reactions due to belt tensions at bearings 1 and 2 are,

$$H'_1 = \frac{(T_1 + T_2)\ b_1}{b}, \text{ and } H'_2 = \frac{(T_1 + T_2)\ b_2}{b}$$

(i) Design of crankpin

Let d_c = diameter of crankpin, mm

l_c = length of crankpin, mm

p_b = safe bearing pressure for the crankpin, MPa

Considering bearing of the crankpin, the length of the crankpin is,

$$F_g = l_c\ d_c\ p_b \qquad \text{...(1)}$$

$$\frac{l_c}{d_c} = 0.6 \text{ to } 1.5$$

The crankpin is checked under bending. The crankpin acts as antilever and the load is assumed to be uniformly distributed.

Maximum bending moment (mean value) in the crankpin,

$$M = \frac{3\ F_g\ l_c}{4}$$

Section modulus, $z_c = \left(\frac{\pi}{32}\right) d_c^3$

Bending stress in crankpin, $\sigma_b = \frac{M}{z_c} < \sigma_{allow}$ ....(2)

(ii) Design of bearings

Bending moment at the centre of bearing 1,

$$M_1 = F_g\ (0.75\ l_c + t + 0.5\ l_1)$$

Where t = thickness of crank web

$$= (0.45 - 0.75)\, d_c$$

$$l_1 = \text{length of bearing 1}$$

$$= (1.5 - 2)\, d_c$$

Now $$M_1 = \left(\frac{\pi}{32}\right) d^3{}_1\, \sigma_b \qquad ...(3)$$

Thus diameter of bearing 1 can be determined.

Bearing 2 may be made of the same size as bearing 1.

(iii) Design of crank web

The crank web is subjected to a bending moment and a direct compressive stress.

Bending moment on crank web, $M = F_g\,(0.75\, l_c + 0.5\, t)$

Section modulus, $$z_w = \frac{wt^2}{6}$$

Bending stress, $$\sigma_b = \frac{M}{z_w}$$

Direct compressive stress, $$\sigma_d = \frac{F_g}{w.t}$$

Total stress, $$\sigma_t = \sigma_b + \sigma_d < \sigma_{allow} \qquad ... (4)$$

***(iv)* Design of shaft under the flywheel**

Let d_s = diameter of shaft, mm

Horizontal bending moment at the flywheel location due to piston gas force,

$$M_1 = F_g\,(a + b_2) - H_1 \,.\, b_2 = H_2 \,.\, b_1$$

Horizontal bending moment,
at the flywheel location due to belt tensions.

$$M_2 = H_1' \,.\, b_2 = H_2' \,.\, b_1 = (T_1 + T_2)\frac{b_1 b_2}{b}$$

Total horizontal bending moment, $M_h = M_1 + M_2$

Vertical bending moment due to flywheel weight,

$$M_v = V_1 \,.\, b_2 = V_2 \,.\, b_1 = \frac{W b_1 b_2}{b}$$

Resultant bending moment. $$M_R = \left[M_h^2 + M_v^2\right]^{1/2}$$

$$= \left(\frac{\pi}{32}\right) d_s^3\, \sigma_b \qquad ... (5)$$

Where σ_b = allowable bending stress for that shaft, MPa

We can determine the value of d_s from Eq. (5).

2. Crank at maximum twisting moment position.

The twisting moment on the crank shaft is maximum when the tangential force on the crank is maximum. It occurs at an angle of 25° – 30° the TDC for a petrol engine and 30° – 40° for diesel engine. The crankshaft in its maximum twisting moment position is shown in Fig. 25.11. Let p_0 be the pressure on the piston at this position, then the gas force on the piston is,

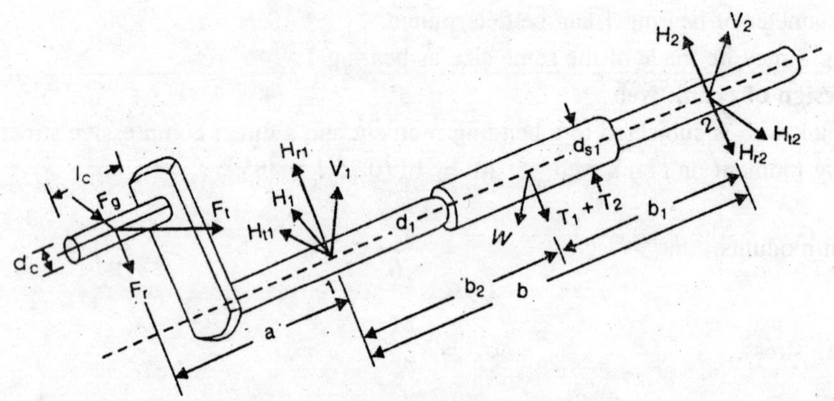

Fig. 25.11 Crank at an angle of maximum twisting moment

$$F_g = \left(\frac{\pi}{4}\right) d^2 p_0$$

Thrust on the connecting rod,

$$F_q = \frac{F_g}{\cos\phi}$$

Where ϕ = angle of inclination of the connecting rod with the line of stroke.

The tangential force F_t is the component of F_q perpendicular to the crank. The other component is the radial force F_r along the crank.

$$\text{Tangential force, } F_t = F_q \sin(\theta + \phi)$$

$$\text{Radial force, } F_r = F_q \cos(\theta + \phi)$$

The tangential force causes twisting whereas radial force causes bending of the shaft. Reactions at bearings 1 and 2 due to F_t are,

$$H_{t1} = F_t \cdot \frac{(a+b)}{b}, \text{ and } H_{t2} = F_t \cdot \frac{a}{b}$$

Reactions due to radial force at bearings 1 and 2 are,

$$H_{r1} = F_r \cdot \frac{(a+b)}{b} \text{ and } H_{r_2} = F_r \cdot \frac{a}{b}$$

The reactions at the bearings 1 and 2 due to flywheel and belt tensions will be the same as in case 1.

(i) Design of crank web

The most critical section of the crank web is where it joins the shaft. It is subjected to (1) bending stresses due to the tangential force F_t and radial force F_r, (2) direct compressive stress due to F_r, and (3) torsional shear stress due to twisting moment due to F_t.

Bending moment due to tangential component F_t is,

$$M_t = F_t\left(r - \frac{d_1}{2}\right)$$

$$= \left(\frac{tw^2}{6}\right).\sigma_{bt} \qquad \text{... (6)}$$

Where σ_{bt} = bending stress in the tangential direction.

d_1 = diameter of bearing 1

Bending moment due to radial component F_r is,

$$M_r = F_r\,(0.75\,l_c + 0.5t)$$

$$= \left(\frac{wt^2}{6}\right).\sigma_{br} \qquad \text{... (7)}$$

Where σ_{br} = bending stress in the radial direciton.

Direct compressive stress, $\sigma_d = \frac{F_r}{wt}$ ... (8)

Maximum compressive stress, $\sigma_c = \sigma_{br} + \sigma_{bt} + \sigma_d$ ... (9)

Twisting moment on the arm due to F_t,

$$T = F_t\,(0.75\,l_c + 0.5t)$$

Shear stress in the arm, $\tau = \frac{T}{z_p} = \frac{4.5\,T}{wt^2}$ ... (10)

Maximum combined stress, $(\sigma_c)_{max} = \frac{\sigma_c}{2} + \left(\frac{1}{2}\right)[\sigma_c^2 + 4\tau^2]^{\frac{1}{2}}$ ... (11)

It should be less than the permissible compressive stress.

(ii) Design of shaft at the junction of crank

Let d_{s_1} = diameter of the shaft at the junction of right hand crank arm.

Bending moment at the junction of crank.

$$M = F_g\,(0.75\,l_c + t)$$

Twistiny moment on the shaft,

$$T = F_t\,.\,r$$

Equivalent twisting moment at the junction of right hand crank arm,

$$T_e = [M^2 + T^2]^{1/2} \qquad \text{... (12)}$$

$$= \left(\frac{\pi}{16}\right) d_{s_1}^3\,\tau$$

Where τ = allowable shear stress in the shaft.

From Eq. (12), d_{s_1} can be determined.

(iii) Design of shaft under the flywheel

Horizontal bending moment on the shaft due to gas force,

$$M_1 = F_g (a + b_2) - [H_{v_1}^2 + H_{ti}^2]^{\frac{1}{2}} b_2$$

Horizontal bending moment at the flywheel location due to belt tensions,

$$M_2 = H_1' . b_2 = H_2' . b_1 + (T_1 + T_2)\frac{b_1 b_2}{b}$$

Total horizontal bending moment, $M_h = M_1 + M_2$

Vertical bending moment, due to flywheel weight,

$$M_v = V_1 \cdot b_2 = V_2 b_1 = b_1 b_2 / b$$

Resultant bending moment, $$M_R = \left[M_h^2 + M_v^2\right]^{\frac{1}{2}}$$

Twisting moment, $$T = F_t . r$$

Equivalent twisting moment on th shaft,

$$T_e = \left[M_R^2 + T^2\right]^{\frac{1}{2}}$$

$$= \left(\frac{\pi}{16}\right) d_s^3 \tau \qquad \text{... (13)}$$

Where τ = allowable shear stress in the shaft

From Eq. (13) the diameter of the shaft can be determined.

Example 25.4 Design a crankshaft for a single – acting four – stroke single cylinder diesel engine for the following data :

Bore = 450 mm, stroke = 600 mm, engine speed = 600 rpm, mean effective pressure = 0.5 MPa, maximum combustion pressure = 3 MPa, weight of flywheel used as a pulley = 50 kN, total belt pull = 6 kN.

When the crank has turned through 40° from the TDC, the pressure on the piston is 1 MPa and the torque on the crank is maximum. The ratio of length of connecting rod to crank radius is 5.

Solution.

Given : d = 450 mm, L = 600 mm or r = 300 mm, p_m = 0.5 MPa, p_{max} = 3 MPa, W = 50 kN,

$T_1 + T_2 = 6\ kN$, $\theta = 40°$, $p_0 = 1$ MPa, $\frac{l}{r} = 5$

Preliminary decisions : The crankshaft shall be made of plain carbon steel having allowable stresses in bending = 56 – 75 MPa, and in shear from 31 – 42 MPa.

Thrust in the connecting rod = Gas force on the piston acting horizontally

$$F_g = \left(\frac{\pi}{4}\right) d^2 p_{max}$$

$$= \left(\frac{\pi}{4}\right) \times 450^2 \times 3 = 477129.4 \text{ N}$$

Let $b = 2d = 900$ mm

$$b_1 = b_2 = \frac{b}{2} = 450 \text{ mm}$$

Horizontal reactions on bearings 1 and 2 are,

$$H_1 = H_2 = \frac{F_g b_1}{b} = \frac{477129.4 \times 450}{900} = 238564.7 \text{ N}$$

Let $c_1 = c_2 = \frac{c}{2}$

Vertical reactions due to the weight W of the flywheel on the bearings 2 and 3 are,

$$V_2 = V_3 = W \cdot \frac{c_1}{c} = \frac{50}{2} = 25 \text{ } kN$$

Horizontal reactions due to belt tensions at bearings 2 and 3 are,

$$H_2' = H_3' = (T_1 + T_2)\frac{c_1}{c} = \frac{6}{2} = 3 \text{ } kN$$

(*i*) Design of crankpin

Let d_c = diameter of crankpin, mm

l_c = length of crankpin, mm

σ_b = allowable bending stress for the crankpin, MPa

Bending moment at the centre of crankpin,

$$M_c = H_1 b_2 = 238564.7 \times 450 = 107354 \text{ kN.mm}$$

Also $$M_c = \left(\frac{\pi}{32}\right) d_c^3 \sigma_b$$

Let $\sigma_b = 75$ MPa, then

$$d_c = 244.3 \approx 245 \text{ mm}$$

Considering bearing of the crankpin, the length of the crankpin is,

$$l_c = \frac{F_g}{d_c p_b}$$

where p_b = permissible bearing pressure = 10 MPa

$$l_c = \frac{477129.4}{245 \times 10} = 194.74 \approx 195 \text{ mm}$$

(ii) Design of left hand crank web

The crank web is subjected to eccentric loading. The crank web is subjected to direct compressive stress and bending stress due to gas force.

Empirically, the thickness of the crankweb is,

$$t = 0.65\, d_c + 6.35 \text{ mm}$$
$$= 0.65 \times 245 + 6.35 = 165.6 \approx 166 \text{ mm}$$

Width of the crank web, $w = 1.125\, d_c + 12.7$ mm

$$= 1.25 \times 245 + 12.7 = 318.95 \approx 320 \text{ mm}$$

Maximum bending moment on the crankweb, M

$$M = H_1\left(b_2 - \frac{l_c}{2} - \frac{t}{2}\right)$$

$$= 238564.7\left(450 - \frac{195}{2} - \frac{166}{2}\right) = 64293.2 \text{ kN.mm}$$

Section modulus of web, $z_w = \dfrac{wt^2}{6} = \dfrac{320 \times 166^3}{6} = 1469653 \text{ mm}^3$

Bending stress in web, $\sigma_{wb} = \dfrac{M}{z_w} = \dfrac{64293.2 \times 10^3}{1469653} = 43.75$ MPa

Direct compressive stress in web,

$$\sigma_{wd} = \frac{H_1}{wt} = \frac{238564.7}{(320 \times 166)} = 4.49 \text{ MPa}$$

Total stress in the web, $\sigma_w = \sigma_{wb} + \sigma_{wd} = 43.75 + 4.49 = 48.24 \text{ MPa} < 75$ MPa

For a safe design, $\sigma_w < \sigma_{allow}$

(iii) Design of right hand crank web

The dimensions of the right hand crank web are made the same as that of the left hand crank web from balancing point of view.

(iv) Design of shaft under the flywheel

Let d_s = diameter of shaft, mm

$$l_1 = l_2 = l_3 = 2\left(\frac{b}{2} - \frac{l_c}{2} - t\right)$$

$$= 2\left(450 - \frac{195}{2} - 166\right) = 373 \text{ mm}$$

Let width of flywheel = 300 mm

$$c = 373 + 300 = 673 \text{ mm}$$

Allowing for clearance etc., let c = 800 mm

$$c_1 = c_2 = \frac{c}{2} = 400 \text{ mm}$$

Bending moment due to the weight of the flywheel,

$$M_f = V_3 \cdot c_1 = 25 \times 400 = 10{,}000 \text{ kN.mm}$$

Bending moment due to belt tensions,

$$M_b = H_3' \cdot c_1 = 3 \times 400 = 1200 \text{ kN . mm}$$

$$\text{Resultant bending moment, } M_s = \left[M_f^2 + M_b^2\right]^{\frac{1}{2}}$$

$$= [10{,}000^2 + 1200^2]^{1/2} = 10071.74 \text{ kN.mm}$$

$$= \left(\frac{\pi}{32}\right) d_s^3 \, \sigma_b$$

Where σ_b = allowable bending stress for the shaft, MPa = 42 MPa

$$d_s = 134.67 \approx 135 \text{ mm}$$

2. Crank at maximum twisting moment position.

The twisting moment on the crank shaft is maximum when the tangential force on the crank is maximum. It occurs at an angle at 25° – 30° the TDC for a petrol engine and 30° – 40° for a diesel engine. The crankshaft in its maximum twisting moment position is shown in Fig. 25.10. Let p_0 be the pressure on the piston at this position, then the gas force on the piston is,

$$F_g = \left(\frac{\pi}{4}\right) d^2_{p_0}$$

$$= \left(\frac{\pi}{4}\right) \times 450^2 \times 1 = 159043 \text{ N}$$

$$\sin\phi = \frac{\sin\theta}{\left(\frac{l}{r}\right)} = \frac{\sin 40°}{5} = 0.12856$$

$$\phi = 25.386°$$

Where ϕ = angle of inclination of the conneting rod with the line of stroke

Thrust on the connecting rod,

$$F_q = \frac{F_g}{\cos\phi} = \frac{159043}{\cos 25.386°} = 160374 \text{ N}$$

The tangential force F_t is the component of F_q perpendicular to the crank. The other component is the radial force F_r along the crank.

Tangential force, $F_t = F_q \sin(\theta + \phi) = 159043 \sin(40 + 25.386)° = 117045$ N

Radal forc, $F_r = F_q \cos(\theta + \phi) = 159043 \cos 425.386° = 107681$ N

The tangential force causes twisting whereas radial force causes bending of the shaft. Reactions at bearings 1 and 2 due to F_t are,

$$H_{t1} = H_{t2} = F_t \cdot \frac{b_1}{b} = \frac{117045 \times 450}{900} = 58522.5 \text{ N}$$

Reactions due to radial force at bearings 1 and 2 are

$$H_{t1} = H_{t2} = F_r \cdot \frac{b_1}{b} = \frac{107681}{2} = 53840.5 \text{ N}$$

The reactions at the bearings 2 and 3 due to flywheel and belt tensions will be the same as in case 1.

***(i)* Design of crankpin**

Bending moment at the centre of crankpin,

$$M_c = H_{t_1} \cdot b_2 = 53840.5 \times 450 = 24228.225 \ k\text{N.mm}$$

Twisting moment on the crankpin,

$$T_c = H_{t_1} \cdot r = 58522.5 \times 300 = 17556.75 \ k\text{N.mm}$$

Equivalent twisting moment on the crankpin,

$$T_e = [M_c^2 + T_c^2]^{\frac{1}{2}}$$

$$= [(24228.225)^2 + (17556.75)^2]^{\frac{1}{2}} = 29920668 \text{ N.mm}$$

$$= \left(\frac{\pi}{16}\right) d_c^3 \, \tau$$

Where τ = allowable shear stress in the crankpin = 35 MPa

$$d_c = 163.3 \approx 164 \text{ mm}$$

***(ii)* Design of shaft under the flywheel**

Bending moment on the shaft, $M_s = R_3 \cdot c_1 = 10071740 \ k\text{N.mm}$

Twisting moment on the shaft, $T_s = F_t \cdot r = 117045 \times 300 = 35113500$ N.mm

Equivalent twisting moment on the shaft,

$$T_e = [M_s^2 + T_s^2]^{\frac{1}{2}}$$

$$= 36529410 \text{ N.mm}$$

$$= \left(\frac{\pi}{16}\right) d_s^3 \, \tau$$

Where τ = allowable shear stress in the shaft = 35 MPa

$$= \left(\frac{\pi}{16}\right) (d_s)^3 \times 35$$

$$d_s = 174 \text{ mm} > 135 \text{ mm calculated earlier.}$$

We adopt d_s = 174 mm

***(iii)* Design of shaft at the junction of right hand crank arm**

Let d_{s1} = diameter of the shaft at the junction of right hand crank arm.

Resultant force at bearing 1,

$$R_1 = \left[H_{t1}^2 + H_{t1}^2\right]^{1/2} = [(58522.5)^2 + 53840.5)^2]^{1/2} = 79521.6 \text{ N}$$

Bending moment at the junction of right hand crank arm,

$$M_{s1} = R_1\left(b_2 + \frac{l_c}{2} + \frac{t}{2}\right) - F_q\left(\frac{l_c}{2} + \frac{t}{2}\right)$$

$$= 79521.6\left(450 + \frac{195}{2} + \frac{166}{2}\right) - 159043\left(\frac{195}{2} + \frac{166}{2}\right)$$

$$= 21431098 \text{ N.mm}$$

Twisting moment at the junction of the right hand crank arm,

$$T_{s1} = F_t \cdot r = 117045 \times 300 = 35113500 \text{ N.mm}$$

Equivalent twisting moment at the junction of right hand crank arm,

$$T_e = \left[M_{s1}^2 = T_{s1}^2\right]^{½} = [(21431098)^2 + (35113500)^2]^{½} = 41136964 \text{ N.mm}$$

$$= \left(\frac{\pi}{16}\right) d_{s^1}^3\ \tau$$

Where τ = allowable shear stress in the shaft = 42 MPa

$$d_{s1} = 170.86 \approx 171 \text{ mm.}$$

From Eq. (6), d_{s1} can be determined.

***(iv)* Design of right hand crank web**

The right hand crank web is subjected to (1) bending stresses due to the tangential and radial components of F_q, (2) direct compressive stress due to F_r, and (3) torsional stress.

Bending moment due to radial component of F_q is,

$$M_r = H_{r2}\left(b_1 - \frac{l_2}{2} - \frac{t}{2}\right)$$

$$= 53840.5\left(450 - \frac{195}{2} - \frac{166}{2}\right) = 14510015 \text{ N.mm}$$

$$= \left(\frac{wt^2}{6}\right) \cdot \sigma_{br}$$

$$= \left(\frac{320 \times 166^2}{6}\right)\sigma_{br} = 1469653\ \sigma_{br}$$

$$\sigma_{br} = 9.87 \text{ MPa}$$

Where σ_{br} = bending stress in the radial direction.

Bending moment due to tangential component of F_q is,

$$M_t = F_t\left(r - \frac{d_{s1}}{2}\right)$$

$$= 117045\left(300 - \frac{171}{2}\right) = 25106152 \text{ N.mm}$$

$$= \left(\frac{tw^2}{6}\right) . \sigma_{bt}$$

$$= \left(\frac{166 \times 320^2}{6}\right)\sigma_{bt} = 2833067\ \sigma_{bt}$$

$$\sigma_{bt} = 8.86 \text{ MPa}$$

Where σ_{bt} = bending stress in the tangential direction.

Direct compressive stress, $\sigma_d = \dfrac{F_r}{2wt} = \dfrac{107681}{2 \times 320 \times 166} = 1.014$ MPa

Maximum compressive stress,

$$\sigma_c = \sigma_{br} + \sigma_{bt} + \sigma_d$$

$$= 9.87 + 8.86 + 1.014 = 18.73 \text{ MPa}$$

Twisting moment on the arm,

$$T = H_{t1}\left(b_2 + \frac{l_c}{2}\right) - F_t \times \frac{l_c}{2}$$

$$= H_{t2}\left(b_1 - \frac{l_c}{2}\right)$$

$$= 58522.5\left(450 - \frac{195}{2}\right) = 20629181 \text{ N.mm}$$

Shear stress in the arm,

$$\tau = \frac{T}{z_p} = \frac{4.5T}{wt^2}$$

$$= \frac{20629181 \times 4.5}{320 \times 166^2} = 10.53 \text{ MPa}$$

Maximum combined stress,

$$(\sigma_c)_{max} = \frac{\sigma_c}{2} + \left(\frac{1}{2}\right)\left[\sigma_c^2 + 4\tau^2\right]^{\frac{1}{2}}$$

$$= \frac{18.73}{2} + \left(\frac{1}{2}\right)\left[(18.73)^2 + 4 \times (10.53)^2\right]^{\frac{1}{2}}$$

$$= 25.46 \text{ MPa} < 60 \text{ MPa}$$

Hence safe.

(*v*) Design of left hand crank web

The dimensions for the left hand crank web may be kept the same as for the right hand crank web from balancing point of view.

(*vi*) Design of crankshaft bearings

The bearing 2 is the most heavily loaded and should be checked for bearing pressure.

$$R_2 = \frac{F_g}{2} + \frac{W}{2} + \frac{T_1 + T_2}{2}$$

$$= \frac{159043}{2} + \frac{50,000}{2} + \frac{6,000}{2}$$

$$= 107521.5 \text{ N}$$

Bearing pressure,

$$p_{b2} = \frac{R_2}{l_2 . d_{s1}}$$

$$= \frac{107521.5}{(373 \times 171)}$$

$$= 1.68 \text{ MPa} < 6 \text{ MPa}$$

Example 25.5 Design a side crankshaft for a diesel engine of 300 mm piston diameter and 350 mm stroke. The weight of the flywheel is 36 kN and the explosion pressure is 2.5 MPa. The gas pressure at the maximum torque is 1 MPa, when the crank angle is 40° from IDC. The connecting rod is 4.5 times the crank radius.

Solution.

Given : $d = 300$ mm, L = 350 mm or $r = 175$ mm, W = 36 kN, $p_{max} = 2.5$ MPa, $p_o = 1$ MPa,

$\theta = 40°$, $\frac{l}{r} = 4.5$

1. Crank at dead centre position.

At this position the bending moment on the shaft is maximum and the twisting moment is zero. Thrust in the connecting rod = Gas force on the piston acting horizontally

$$F_g = \left(\frac{\pi}{4}\right) d^2 p_{max}$$

$$= \left(\frac{\pi}{4}\right) \times 300^2 \times 2.5 = 176715 \text{ N}$$

(*i*) Design of crankpin

Let d_c = diameter of crankpin, mm

$$l_c = \text{length of crankpin, mm}$$

$$p_b = \text{safe bearing pressure for the crankpin MPa} = 10 \text{ MPa}$$

Considering bearing of the crankpin, the length of the crankpin is

$$F_g = l_c d_c p_d$$

$$\frac{l_c}{d_c} = 0.6 \text{ to } 1.5 = 0.8$$

$$0.8 d_c^2 \times 10 = 176715$$

$$d_c = 148.6 \approx 150 \text{ mm}$$

$$l_c = 120 \text{ mm}$$

The crankpin is checked under bending. The crankpin acts as antilever and the load is assumed to be uniformly distributed.

Maximum bending moment (mean value) in the crankpin,

$$M = \frac{2F_g l_c}{4} = 0.75 \times 176715 \times 120 = 15904350 \text{ N.mm}$$

Section modulus,

$$z_c = \left(\frac{\pi}{32}\right) d_c^3 = \left(\frac{\pi}{32}\right) \times 150^3 = 331340 \text{ mm}^3$$

Bending stress is crankpin,

$$\sigma_b = \frac{M}{z_c}$$

$$= \frac{15904350}{331340} = 48 \text{ MPa} < \sigma_{allow} \text{ of } 60 \text{ MPa}$$

***(ii)* Design of bearings**

Bending moment at the centre of bearing 1,

$$M_1 = F_g (0.75 l_c + t + 0.5 l_1)$$

Where t = thickness of crank web

$$= (0.45 - 0.75)\, d_c = 0.6\, d_c = 0.6 \times 150 = 90 \text{ mm}$$

$$l_1 = \text{length of bearing 1}$$

$$= (1.5 - 2)\, d_c = 1.8\, d_c = 1.8 \times 150 = 270 \text{ mm}$$

$$M_1 = 176715 (0.75 \times 120 + 90 + 0.5 \times 270) = 55665225$$

N.mm

Now

$$M_1 = \left(\frac{\pi}{32}\right) d_1^3 \sigma_b, \sigma_b = 60 \text{ MPa}$$

$$d_1 = 211.4 \approx 212 \text{ mm}$$

Bearing 2 may be made of the same size as bearing 1.

***(iii)* Design of crank web**

The crank web is subjected to a bending moment and a direct compressive stress.

Bending moment on crank web, $M = F_g (0.75\, l_c + 0.5\, t)$

$= 176715\,(0.75 \times 120 + 0.5 \times 90)$

$= 23856525$ N.mm

Section modulus, $$z_w = \frac{wt^2}{6} = \frac{w \times 90^2}{6} = 1350\, w$$

Bending stress, $$\sigma_b = \frac{M}{z_w} = \frac{23856525}{(1350\, w)} = \frac{17671.5}{w} \text{ MPa}$$

Direct compressive stress, $$\sigma_d = \frac{F_g}{w.t.} = \frac{176715}{90\, w} = \frac{1963.5}{w} \text{ MPa}$$

Total stress, $$\sigma_t = \sigma_b + \sigma_d$$

$$= \frac{(17671.5 + 1963.5}{w} = 60$$

$$w = 3225.25 \approx 328 \text{ mm}$$

***(iv)* Design of shaft under the flywheel**

Let d_s = diameter of shaft, mm

Horizontal bending moment at the flywheel location due to piston gas force,

$$M_1 = F_g\,(a + b_2) - H_1.b_2 = H_2.b_1$$

Let us assume that width of flywheel = 200 mm and $l_1 = l_2 = 200$ mm.

Allowing some clearance, $$b = 200 + \frac{(l_1 + l_2)}{2} + 20$$

$$= 200 + 200 + 20 = 420 \text{ mm}$$

$$a = 0.75\, l_c + t + 0.5 l_1 = 0.75 \times 120 + 90 + 0.5 \times 200 = 280 \text{ mm}$$

$$b_1 = \frac{b}{2} = 210 \text{ mm}$$

Horizontal reactions on bearings 1 and 2 are,

$$H_1 = \frac{F_g\,(a + b)}{b} = \frac{176715\,(280 + 420)}{420} = 294525 \text{ N}$$

$$H_2 = \frac{F_g a}{b} = \frac{176715 \times 280}{420} = 117810 \text{ N}$$

Vertical reactions due to the weight W of the flywheel on the bearings 1 and 2 are,

$$V_1 = V_2 = W.\frac{b_1}{b_2} = \frac{36}{2} = 18 \text{ kN}$$

$$M_1 = 176715\,(280 + 210) - 294525 \times 210 = 24740100 \text{ N.mm}$$

Horizontal bending moment at the flywheel location due to belt tensions,

$$M_2 = H_1'.b_2 = H_2'.b_1 = \frac{(T_1 + T_2)\, b_1\, b_2}{b} = 0 \text{ as there is no belt pull}$$

Total horizontal bending moment,

$$M_h = M_1 + M_2 = 24740100 \text{ N.mm}$$

Vertical bending moment due to flywheel weight,

$$M_v = V_1.b_2 = V_2.b_1 = \frac{Wb_1b_2}{b} = \frac{36 \times 210 \times 210}{420} = 3780 \text{ kN.mm}$$

Resultant bending moment,

$$M_R = \left[M_h^2 + M_v^2\right]^{1/2}$$

$$= \left[(24740.1)^2 + 3780)^2\right]^{1/2} = 250225.2 \text{ kN.mm}$$

$$= \left(\frac{\pi}{32}\right) d_s^3 \sigma_b$$

Where σ_b = allowable bending stress for the shaft, MPa = 60 MPa

$$d_s = 161.9 \approx 162 \text{ mm}$$

But d_s should be more than d_1 = 212 mm. Let us take d_s = 230 mm

2. Crank at maximum twisting moment position

The twisting momenton on the crank shaft is maximum when the tangential force on the crank is maximum. It occurs at an angle of 25° – 30° the TDC for a petrol engine and 30° – 40° for a diesel engine. Let p_o be the pressure on the piston at this position, then the gas force on the piston is;

$$F_g = \left(\frac{\pi}{4}\right) d^2 p_o = \left(\frac{\pi}{4}\right) \times 300^2 \times 1 = 70686 \text{ N}$$

$$\sin\phi = \frac{\sin\theta}{\left(\frac{l}{r}\right)} = \frac{\sin 40^\circ}{4.5} = 0.14284$$

$$\phi = 8.21^\circ$$

where ϕ = angle of inclination of the connectng rod with the line of stroke.

Thrust on the connecting rod,

$$F_q = \frac{F_g}{\cos\phi} = \frac{70686}{\cos 8.23} = 71418 \text{ N}$$

The tangential force F_t is the component of F_q perpendicular to the crank. The other component is the radial force F_r along the crank.

Tangential force, $F_t = F_q \sin(\theta + \phi) = 70686 \sin(40 + 8.23)^\circ = 52719$ N

Radial force, $F_r = F_q \cos(\theta + \phi) = 70686 \cos 48.23^\circ = 47087$ N

The tangential force causes twisting whereas radial force causes bending of the shaft. Reactions at bearings 1 and 2 due to F_t are

$$H_{t1} = \frac{F_t.(a+b)}{b} = \frac{52719\,(280 + 420)}{420} = 87865 \text{N}$$

$$H_{t2} = \frac{F_t.a}{b} = \frac{52719 \times 280}{420} = 35146 \text{ N}$$

Reactions due to radial force at bearings 1 and 2 are,

$$H_{r1} = \frac{F_r.(a+b)}{b} = \frac{47087\,(280+420)}{420} = 78478\text{N}$$

$$H_{r2} = \frac{F_r.a}{b} = \frac{47087 \times 280}{420} = 31391 \text{ N}$$

The reactions at the bearing 1 and 2 due to flywheel will be the same as in case 1.

(*i*) Design of crank web

The most critical section of the crank web is where it joins the shaft. It is subjected to (1) bending stresses due to the tangential force F_t and radial force F_r, (2) direct compressive stress due to F_r, and (3) torsional shear stress due to twisting moment due to F_t.

Bending moment due to tangential component F_t is,

$$M_t = F_t\left(r - \frac{d_1}{2}\right) = 52719\left(175 - \frac{212}{2}\right) = 3637611\text{N.mm}$$

$$z = \left(\frac{tw^2}{6}\right) = \frac{90 \times 328^2}{6} = 1613760 \text{ mm}^3$$

σ_{bt} = bending stress in the tangential direction.

$$= \frac{M_t}{z} = \frac{3637611}{1613760} = 2.254 \text{ MPa}$$

d_1 = diameter of bearing 1

Bending moment due to radial component F_r is,

$$M_r = F_r\,(0.75\, l_c + 0.5\, t) = 47087\,(0.75 \times 120 + 0.5 \times 90)$$

$$= 6356745 \text{ N.mm}$$

$$z = \left(\frac{wt^2}{6}\right) = \frac{328 \times 90^2}{6} = 442800 \text{ mm}^3$$

$$\sigma_{br} = \frac{M_r}{z} = \frac{6356745}{442800} = 14.36 \text{ MPa}$$

Where σ_{br} = bending stress in the radial direction.

Direct compressive stress,

$$\sigma_d = \frac{F_r}{wt} = \frac{47087}{328 \times 90} = 1.595 \text{ MPa}$$

Maximum compressive stress,

$$\sigma_c = \sigma_{br} + \sigma_{bt} + \sigma_d$$

$$= 14.36 + 2.254 + 1.595 = 18.21 \text{ MPa}$$

Twisting moment on the arm due to F_t,

$$T = F_t (0.75\, l_c + 0.5t)$$
$$= 52719\,(0.75 \times 120 + 0.5 \times 90) = 7117065 \text{ N.mm}$$

Shear stress in the arm,

$$\tau = \frac{T}{z_p} = \frac{4.5T}{wt^2}$$

$$= \frac{4.5 \times 7117065}{328 \times 90^2} = 12.05 \text{ MPa}$$

$$\text{Maximum combined stress, } (\sigma_c)_{max} = \frac{\sigma_c}{2} + \left(\frac{1}{2}\right)\left[\sigma_c^2 + 4\tau^2\right]^{\frac{1}{2}}$$

$$= \frac{18.21}{2} + \left(\frac{1}{2}\right)\left[(18.21)^2 + 4(12.05)^2\right]^{\frac{1}{2}}$$

$$= 24.021 \text{ MPa} < 60 \text{ MPa}$$

***(ii)* Design of shaft at the junction of crank**

Let d_{s1}= diameter of the shaft at the junction of right hand crank arm.

Bending moment at the junction of crank,

$$M = F_g\,(0.75\, l_c + t) = 70686\,(0.75 \times 120 + 90) = 12723480 \text{ N.mm}$$

Twisting moment on the shaft,

$$T = F_t.r = 52719 \times 175 = 9225825 \text{ N.mm}$$

Equivalent twisting moment at the junction of right hand crank arm,

$$T_e = \left[M^2 + T^2\right]^{\frac{1}{2}} = \left[(12723480)^2 + (9225825)^2\right]^{\frac{1}{2}} = 15716322 \text{ N.mm}$$

$$= \left(\frac{\pi}{16}\right) d_{s1}^3\, \tau$$

Where τ = alowable shear stress in the shaft = 30 MPa

$$d_{s1} = 230 \text{ mm}$$
$$\tau = 6.57 \text{ MPa} < 30 \text{ MPa}$$

Hence safe.

***(iii)* Design of shaft under the flywheel**

Horizontal bending moment on the shaft due to gas force,

$$M_h = F_g\,(a + b_2) - \left[H_{r1}^2 + H_{ti}^2\right]^{\frac{1}{2}} b_2$$
$$= 70686\,(280 + 210) - [(78478)^2 + (87865)^2]^{1/2} \times 210$$
$$= 9896166 \text{ N.mm}$$

Vertical bending moment due to flywheel weight,

$$M_v = V_1.b_2 = V_2.b_1 = \frac{Wb_1\, b_2}{b} = \frac{36 \times 210 \times 210}{420} = 3780 \text{ kN.mm}$$

Resultant bending moment,

$$M_R = \left[M_h^2 + M_v^2\right]^{\frac{1}{2}}$$

$$= \left[(9896166)^2 + (37800000^2\right]^{\frac{1}{2}} = 10593512 \text{ N.mm}$$

Twisting moment, $T = F_t r = 52719 \times 175 = 9225825$ N.mm

Equivalent twisting moment on the shaft,

$$T_e = \left[M_R^2 + T^2\right]^{\frac{1}{2}}$$

$$= \left[(10593512)^2 + (9225825)^2\right]^{\frac{1}{2}}$$

$$= 14047717 \text{ N.mm}$$

$$= \left(\frac{\pi}{16}\right) d_s^3 \tau$$

where τ = allowable shear stress in the shaft = 30 MPa

$d_s = 133.6 \approx 134$ mm

We adopt $d_s = 230$ mm

EXERCISES

25.1 A four – stroke diesel engine has the following specifications :

Brake power	=	6 kW
Speed	=	1000 rpm
Indicated mean effective pressure	=	0.30 MPa
Mechanical efficiency	=	8%

Determine : (*a*) Bore and length of cylinder

(*b*) Thickness of cylinder head

(*c*) size of studs for the cylinder head.

25.2 Design a cast iron piston for a single acting 4 – stroke diesel engine for the following data :

Cylinder bore	=	100 mm
Stroke	=	120 mm
Maximum gas pressure	=	4.5 MPa
Indicated mean effective pressure	=	0.8 MPa
Mechanical efficiency	=	80%
Fuel consumption	=	0.16 kg/brake power per hour
Higher calorific value of fuel	=	42 MJ/kg
Speed	=	1800 rpm

25.3 Design a connecting rod for an I.C. engine running at 1800 rpm and developing maximum pressure of 3 MPa for the following data :

Piston diameter	=	120 mm
Mass of reciprocating parts per cylinder	=	2.5 kg
Length of connecting rod	=	400 mm
Stroke	=	200 mm
Compression ratio	=	6
Desired factor of safety	=	6
Big end bearing :		
Length of diameter ratio	=	1.4
Permissible bearing pressure	=	10 MPa
Small end bearing :		
Length to diameter ratio	=	2
Permissible bearing pressure	=	15 MPa
Density of connecting rod material	=	7800 kg/m^3
Allowable stress in bolts	=	65 MPa
Allowable stress in cap	=	85 MPa
Rankin – Gordon formula constants :		

$$\sigma_c = 320 \text{ MPa}, a = \frac{1}{7500}$$

Assume missing data suitably.

25.4 Design a plan carbon steel centre crankshaft for a single cylinder 4– stroke single cylinder I.C. engine for the following data :

Bore	=	400 mm
Stroke	=	600 mm
Speed	=	240 rpm
Mean effective pressure	=	0.45 MPa
Maximum combustion pressure	=	2 MPa
Weight of flywheel pulley	=	40 kN
Total belt pull	=	6 kN

When the crank has turned through 30° from TDC, the pressure on the piston is 0.9 MPa and the torque on the crank is maximum. The ratio of connecting rod length to crank radius is 4.5.

Assume missing data suitably.

26 LUBRICATION IN TRANSMISSION SYSTEM

Chapter

26.1 FUNCTIONS OF LUBRICATING SYSTEM

The important functions of a lubricating system are the following :

1. Lubrication. The main function of the lubricating system is to keep the moving parts sliding freely past each other and, thus, reduce the engine friction and wear.

2. Cooling. To keep the surfaces cool by taking away a part of their heat through the oil passing over them. This requires oil to possess good oxidation stability.

3. Cleaning. To keep the bearings and piston rings clean of the products of wear and combustion by washing them away.

4. Sealing. The lubricating oil must form a good seal between piston rings and cylinder walls. Therefore, it must possess adequate viscosity stability.

5. Reduction of noise. Lubrication reduces the noise of the engine and other sliding parts.

26.2 DESIRABLE PROPERTIES OF A LUBRICATING OIL

The desirable properties of a lubricant are :

1. It should be available in a wide range of viscosities.
2. There should be little change in viscosity of the oil with change in temperature.
3. The oil should be chemically stable at all temperatures encountered in its application.
4. The oil should have sufficient specific heat to carry away frictional heat without abnormal rise in temperature.
5. It should be commercially available at a reasonable cost.

26.3 LUBRICATING SYSTEMS

There are two types of lubricating systems :

1. Splash lubricating system, and
2. Pressure feed lubricating system.

1. Splash Lubricating System

The arrangement of splash lubricating system is shown in Fig. 26.1. This method is generally used for a vertical engine with a closed crankcase. The sump is located at the bottom of the crankcase. When the engine crankshaft rotates, the big end of the connecting rod splashes oil by centrifugal action. The connecting rod big end has a hollow pipe called a scoop which is fitted to the bearing cap and pointed towards the direction of rotation of the crankshaft. The lubricating oil passing through the scoop lubricate the big end bearing and gudgeon pin bearing. All other parts are lubricated by the splash. Excess oil is collected in the troughs and are provided with overflows and collected in the main sump. The level of the oil in the trough is maintained constant. The dripping from the cylinders is also collected in the sump. The oil from the sump is recirculated with the help of a pump.

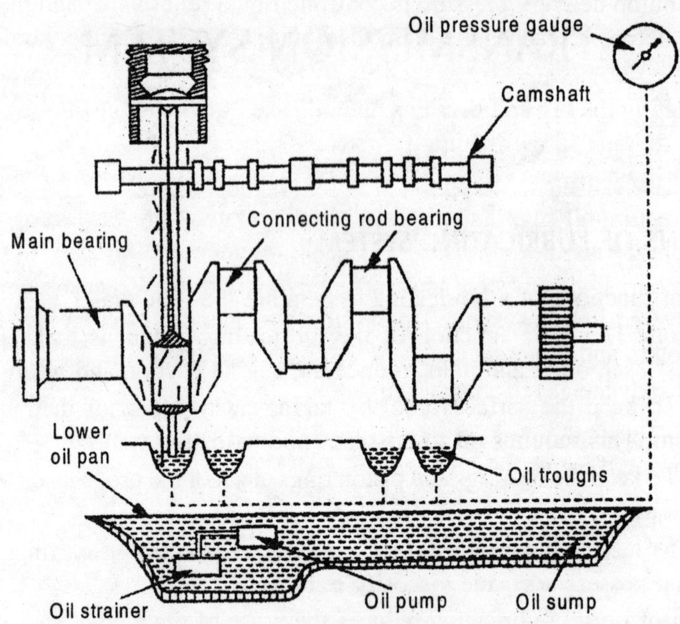

Fig. 26.1 Splash lubrication system.

The limitations of this system are : (*i*) Inability to regulate the quantity of oil splashed against the cylinder wall. (*ii*) Inability to keep the oil from getting past the piston head into the combustion chamber, burning with the fuel and passing out with exhaust gases.

2. Pressure Feed Lubricating System

This system is shown in Fig. 26.2. Such a system supplies oil under pressure directly to the connecting rod bearings, crankshaft bearings, valve gear and to the camshaft drive. Indirect supplies reach the cylinder walls, gudgeon pin, the distributor and pump drives.

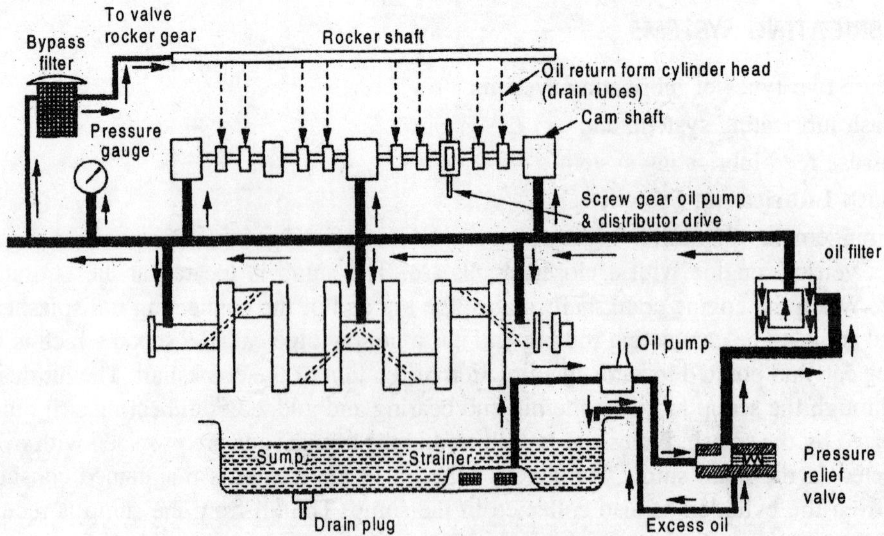

Fig. 26.2 Wet Sump high pressure Lubrication System.

Oil is carried in the sump and circulated by the gear pump which sucks from the sump through a strainer. The pump delivery pressure is controlled by a relief valve and the oil passes through a very fine filter before it reaches the main distributor gallery from the various bearings, surfaces and gears.

After lubricating the big end bearings, the oil is fed to the gudgeon pins through the oilway in the connecting rod and further squirted into the cylinder wall.

3. Charge Lubricating System

It does not require oil-filter and oil sump. In this system, the lubricating oil is pre-mixed with the petrol which carries the lubricating oil in the cylinder to lubricate the piston and cylinder. Most of the oil burns with the fuel and carried away with the exhaust gases. The lubricating oil cannot be recovered in this system. This system is generally used for two stroke spark ignition engines of scooters and motorcycles.

The main disadvantages of this system are :

(*i*) Carbon deposits due to burning of oil on the spark plug,

and (*ii*) Non-recovery of the oil used.

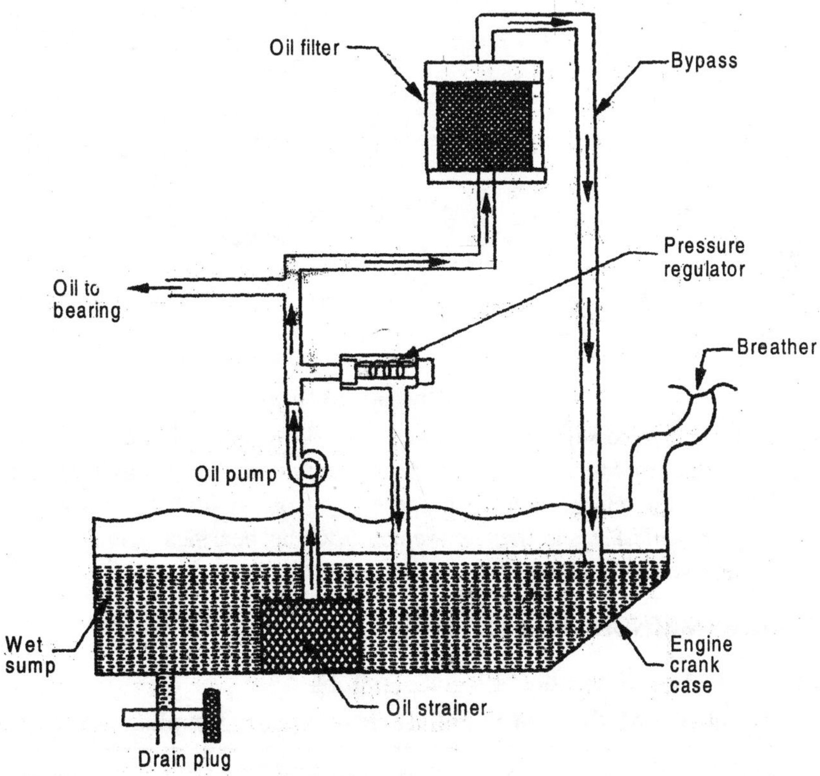

Fig. 26.3 Wet Sump System.

4. Wet Sump Lubricating System

The wet sump lubricating system is shown in Fig. 26.3. In this system, the sump is always full of oil. Oil is drawn from the sump by an oil pump through an oil strainer. A pressure relief valve is provided which automatically maintains the delivery pressure constant. If the pressure exceeds than the predetermined pressure, the valve opens and allows some of the oil to return to the sump and relieves the oil pressure in the system. The oil from the pump goes to the bearings and part of it passes through the filter which removes solid particles from the oil. As all the oil is not passed through the filter, the system is known as by-pass filtering system. The advantage of this system is that a clogged filter will not restrict the flow of oil to the engine.

5. Dry Sump Lubricating System.

In the dry pump lubricating system shown in Fig. 26.4, the sump does not contain oil and all the oil required for lubrication remains in circulation. High speed racing cars and military jeeps use this type of system.

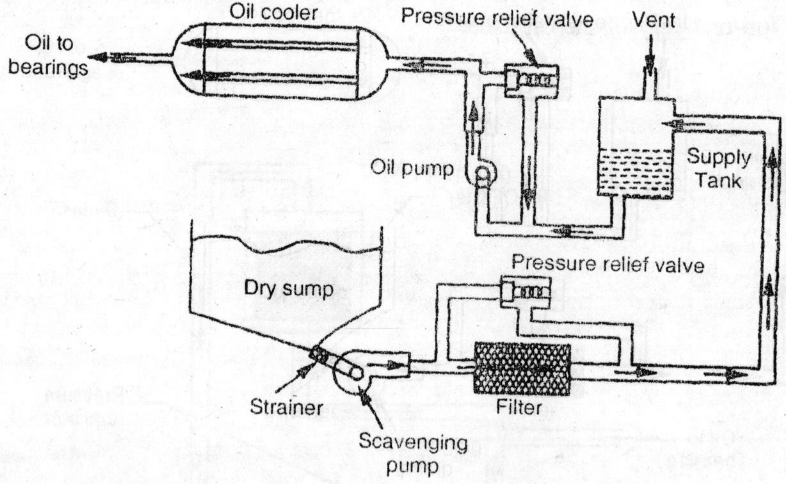

Fig. 26.4 Dry Sump System.

An auxiliary tank is used to supply the oil to the main bearings with the help of the pump. The oil returns back to the tray and then returned to the auxiliary tank by scavenging pump. If the filter gets clogged, the pressure relief valve opens permitting oil to flow by passing the filter and reaches the supply tank. The oil is then recirculated to the bearings from the supply tank. A separate oil cooler is used to cool the oil.

26.4 LUBRICATING METHODS FOR BEARINGS

The various methods of lubrication are given in Table 26.1.

Table 26.1 Methods of liquid lubricant supply for plain bearings.

Method of supply	Main characteristics	Applications
1. Hand oiling	Non-automatic, irregular, low initial cost, high maintenance cost.	Low speed, cheap general bearings.

Method of supply	Main characteristics	Applications
2. Drip and wick feed	Non-automatic, adjustable, moderately efficient, cheap	General in some machine tools, axles
3. Ring and collar feed	Automatic, reliable, efficient, fairly cheap, mainly horizontal bearings.	Journals in pumps, blowers, large electric motors.
4. Bath and splash Lubrication	Automatic, reliable, efficient, oil-tight housing required, high initial cost.	Thrust bearings, bath only, engines process machinery, general appliances.
5. Pressure feed	Automatic, positive and adjustable, reliable and efficient, high initial cost.	High speed and heavily loaded general and thrust bearings in machine tools, engines and compressors.

26.5 LUBRICATION IN GEAR TRANSMISSION SYSTEM

For the satisfactory performance and durability of the gears, proper lubrications in the gear teeth is essential. The main function of the lubricant is to separate the liquid surfaces and remove heat from the gears. The relative motion between two teeth in ??? is a combination of rolling and sliding motion. This combination of motion, increases the lubrication problems. Splash and pumped lubrication are the two methods by which lubricant is usually applied to a gear system. In splash lubrication, the gears are enclosed in a box and dipped in a bath of lubricating oil. In the pumped lubrication system, a jet of oil is directed towards the meshing teeth. For low velocities, under 15 m/s, the splash system is used, whereas for velocities above 25 m/s, pump feed lubrication is preferred.

Gears are lubricated by grease, straight mineral oils or extreme pressure lubricants. Grease is used as a lubricant for hand-operated gear mechanisms or drives working in low speeds and low loads working intermittently. For enclosed gear drives operating, under high loads and normal speeds, straight mineral oils provide satisfactory lubrication. Small gears operating at low speeds (below 1200 rpm) and lightly loaded should be lubricated by SAE 80 grade mineral oils. Gears operating at low speeds and heavily loaded should be lubricated with SAE 90 grade mineral oil at normal ambient temperatures and SAE 140 grade mineral oil at high temperatures. For heavy duty applications, EP lubricants are used. EP lubricants are mineral oils with additives. They are used in the automobile gear box and heavy industrial gear boxes. Synthetic oils should be used for aircraft turbine engines.

26.6 LUBRICANTS

Both mineral and synthesis oils are used as lubricants. Mineral oils are basically hydrocarbons, which may be classified as follows :

Paraffins. They contains significant amounts of waxy hydrocarbons but little or no asphaltic matter.

NaphAthalene. It contains asphaltic matter in least volatile fractions, but little or no wax.

Mixed base. It contains both waxy and asphaltic materials.

Plain mineral oils are used in many units and systems for the lubrication of bearings where their oxidation ability, operating temperature range, ability to prevent wear, etc. are adequate. Special chemicals or additives are added to many oils to improve their properties.

Synthesis oils, most commonly used are : di-ester, silicones, polyglycol, chlorinated diphenyl, polyphenyl, fluorocarbon and silicate ester, etc.

A grease may be defined as a solid to semi-fluid lubricant consisting of a dispersion of a thickening agent in a lubricating fluid. The thickening agent may consist of a soap, a clay or a dyestuff. The lubricating fluid is usually a mineral oil, a di-ester or a silicone. They should be used for operating speeds below 1 to 2 m / s.

The solid fluids are used when fluids are undesirable and ineffective. The solid lubricants may be classified as follows :

Lamellar solids	:	MoS_2, WS_2, Graphite, TaS_2 and CaF_2.
Polymers	:	FEP, PTFCE, PTFE, Nylons, Acetals and Polyurethanes
Soft metals	:	Pb, Au, Ag, Sn, In
Oxides	:	MoO_3, PbO/SiO_2, B_2O_3/PbS
Salts	:	Na_2WO_4, $PbMoO_4$.

EXERCISES

26.1 What are the functions of lubricating system ?

26.2 What are the desirable properties of a lubricating oil ?

26.3 What are the lubricating systems ? Explain one of the methods.

26.4 Write short notes on the following :

(*a*) Splash lubricating system.

(*b*) Pressure feed lubricating system.

26.5 Differentiate between wet sump and dry sump lubricating system.

26.6 Describe briefly the lubricating methods for :

(*a*) Bearings, and

(*b*) Gear transmission system.

MULTI-CHOICE QUESTIONS

2 SELECTION OF ENGINEERING MATERIALS

Chapter

1. Small percentage of boron is added to steel to :
(*a*) increase hardenability
(*b*) reduce machinability
(*c*) increase wear resistance
(*d*) increase endurance strength.

2. Which of the following display properties similar to that of steel ?
1. Black-heart cast iron
2. White-heart cast iron
3. Gray cast iron
4. Pig iron

(*a*) 1 and 2 only (*b*) 3 and 4 only
(*c*) 2 and 4 only (*d*) 1 and 3 only.

3. Match List I with List II and select the correct answer using the codes given below the lists :

List I (Steel type)	List II (Product)
A. Mild steel	1. Screw driver
B. Tool steel	2. Commercial beams
C. Medium carbon steel	3. Crane hooks
D. High carbon steel	4. Blanking dies

Codes :	A	B	C	D
(*a*)	1	4	3	2
(*b*)	2	4	1	3
(*c*)	1	3	4	2
(*d*)	2	4	3	1

4. Consider the following statements :
1. grinding
2. coating
3. hammer peening

Of the above statements :
(*a*) 1 and 2 are correct
(*b*) 2 and 3 are correct
(*c*) 1 and 3 are correct
(*d*) 1, 2 and 3 are correct.

5. The main alloying elements in high speed Steel in order of increasing proportion are :
(*a*) Vanadium, chromium, tungsten
(*b*) Tungsten, titanium, vanadium
(*c*) Chromium, titanium, vanadium
(*d*) Tungsten, chromium, titanium.

6. Duralumin Alloy contains aluminium and copper in the ratio of :

	%Al	%Cu
(*a*)	94	4
(*b*)	90	8
(*c*)	88	10
(*d*)	86	12.

7. Consider the following statements :
1. From design considerations, it is always advantageous to place cast iron ribs on the tension side rather than on the compression side.
2. Cast iron is an excellent choice for machine tool guides and frames
3. Cast iron parts have a low notch sensitivity.

Which of these statements are correct ?
(*a*) 1, 2 and 3 (*b*) 2 and 3
(*c*) 1 and 3 (*d*) 1 and 2.

8. Consider the following fits :
1. I.C. engine cylinder and piston
2. Ball bearing outer race and housing
3. Ball bearing inner race and shaft

Which of the above fits are based on the shaft basis system ?
(*a*) 1 and 2 (*b*) 2 and 3
(*c*) 1 and 3 (*d*) 1, 2 and 3.

9. The damping capacity of materials used for machine members should be :
(*a*) negligible (*b*) low
(*c*) high
(*d*) any one of the above.

10. A material capable of absorbing large amount of strain energy is called :
(*a*) ductile (*b*) shock proof
(*c*) resilient (*d*) tough.

11. Which process will increase the fatigue strength of parts ?
(*a*) polishing (*b*) shot-peening
(*c*) decarburisation (*d*) electroplating.

12. The endurance limit of a material is :
(*a*) greatly decreased by poor surface finish
(*b*) improved due to rough surface
(*c*) independent of surface conditions
(*d*) decreased due to surface defects.

3 DESIGN FOR STATIC LOAD

Chapter

13. A transmission shaft subjected to bending loads must be designed on the basis of :
(*a*) maximum normal stress theory
(*b*) maximum shear stress theory
(*c*) maximum normal stress and maximum shear stress theories
(*d*) fatigue strength.

14. Design of shafts made of brittle materials is based on :
(*a*) Guest's theory
(*b*) Rankine's theory
(*c*) St. Venant's theory
(*d*) Von Mises theory.

15. A solid shaft of diameter 'D' caries a twisting moment that develops maximum shear stress τ. If the shaft is replaced by a hollow one of outside diameter 'D' and inside diameter D/2, then the maximum shear stress will be :
(*a*) 1.067 τ (*b*) 1.143 τ
(*c*) 1.333 τ (*d*) 2 τ.

16. A circular shaft can transmit a torque of 5 kNm. If the torque is reduced to 4 kNm then the maximum value of bending moment that can be applied to the shaft is :
(*a*) 1 kNm (*b*) 2 kNm
(*c*) 3 kNm (*d*) 4 kNm.

17. If σ_y is the yield strength of a particular material, then the distortion energy theory is expressed as :
(*a*) $(\sigma_1 - \sigma_2)^2 + (\sigma_2 - \sigma_3)^2 + (\sigma_3 - \sigma_1)^2 = 2\,\sigma_y^{\,2}$
(*b*) $(\sigma_{12} + \sigma_{22} + \sigma_{32}) - 2v\,(\sigma_1\,\sigma_2 + \sigma_2\,\sigma_3 + \sigma_3\,\sigma_1) = \sigma\,y^2$
(*c*) $(\sigma_1 - \sigma_2)^2 + (\sigma_2 - \sigma_3)^2 + (\sigma_3 - \sigma_1)^2 = 3\,\sigma_y^{\,2}$
(*d*) $(1 - 2v)\,(\sigma_1 + \sigma_2 + \sigma_3)^2 = 2\,(1 + v)\,\sigma_y^{\,2}$.

18. For a two-dimensional state $(\sigma_1\,\sigma_2)$, $\sigma_1 > 0$, $\sigma_2 < 0$ the designed values are most conservative if which one of the following failure theories were used ?
(*a*) Maximum principal strain theory
(*b*) Maximum distortion energy theory
(*c*) Maximum shear stress theory
(*d*) Maximum principal stress theory.

19. 30C8 steel has its yield strength of 400 N/mm^2 and modulus of elasticity of 2×10^5 MPa. Assuming the material to obey Hooke's law up to yielding, what is its proof resilience ?
(*a*) 0.8 N/mm^2 (*b*) 0.4 N/mm^2
(*c*) 0.6 N/mm^2 (*d*) 0.7 N/mm^2.

20. Autofrettage is a method of :
(*a*) improving the transmission of power from engine to the axle in automobiles
(*b*) improving the strength of welded joints

(*c*) pre-stressing the cylinders before putting them in service
(*d*) improving the strength of riveted joints.

21. A body is subjected to a pure tensile stress of 100 units. What is the maximum shear produced in the body at some oblique plane due to the above ?
(*a*) 100 units (*b*) 75 units
(*c*) 50 units (*d*) 0 unit.

22. Autofrettage is a method of :
(*a*) Joining thick cylinders
(*b*) Relieving stresses from thick cylinders
(*c*) Pre-stressing thick cylinders
(*d*) Increasing the life of thick cylinders.

23. A hollow shaft of the same cross-sectional area and material as that of a solid shaft, transmits :
(*a*) Same torque
(*b*) Lesser torque
(*c*) More torque
(*d*) Cannot be predicted without more data.

24. Angle of twist of a shaft of diameter '*d*' is inversely proportional to :
(*a*) d (*b*) d^2
(*c*) d^3 (*d*) d^4.

25. A loaded semi-infinite flat plate is having an elliptical hole $\left(\frac{A}{B}=2\right)$ in the middle as shown in the figure.

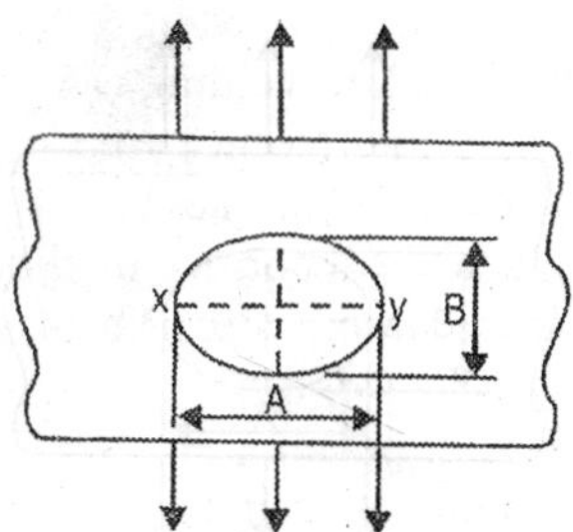

The stress concentration factor at points either X or Y is :
(*a*) 1 (*b*) 3
(*c*) 5 (*d*) 7.

26. In a semi-infinite flat plate shown in the figure, the theoretical stress concentration factor K_t for an elliptical hole of major axis $2a$ and minor axis $2b$ is given by :

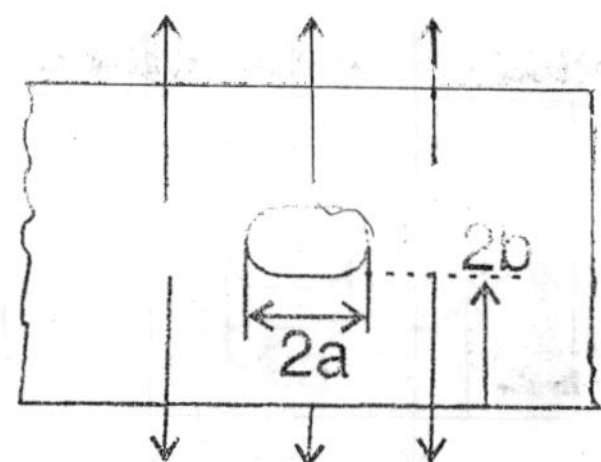

(*a*) $K_t = \frac{a}{b}$ (*b*) $K_t = 1 + \frac{a}{b}$
(*c*) $K_t = 1 + \frac{2b}{a}$ (*d*) $K_t = 1 + \frac{2a}{b}$.

27. Match List-I (Loaded cylindrical bar) with List-II (Type of fracture) and select the correct answer using the codes given below the lists :

List-I **Loaded cylindrical bar**	**List-II** **Type of fracture**
A. CI bar subjected to tension	1. Cup and Cone
B. CI bar under torsion	2. Granular helicoidal
C. MS bar under torsion perpendicular to axis	3. Granular plain perpendicular to axis
D. MS bar under tension	4. Smooth plain perpendicular to axis
	5. Granular plain at 45° to axis

Codes :

	A	B	C	D
(*a*)	3	2	4	1
(*b*)	5	1	3	2
(*c*)	3	1	4	2
(*d*)	5	2	3	1.

28. Stress-concentration is caused due to :
(*a*) material discontinuities
(*b*) load discontinuities
(*c*) geometric discontinuities
(*d*) all of the above.

29. The maximum stress-concentration factor in a wide plate with a circular hole under uniaxial tension is :
(*a*) 2 (*b*) 3
(*c*) 4 (*d*) 5.

30. Stress-concentration under static loading is to be accounted for in design for :
(a) ductile materials
(b) brittle materials
(c) both ductile and brittle materials
(d) neither ductile nor brittle materials.

4 DESIGN FOR VARIABLE LOAD

Chapter

31. The design calculations for members subject to fluctuating loads with the same factor of safety yield the most conservative estimates when using :
(a) Gerber relation
(b) Soderberg relation
(c) Goodman relation
(d) None of the above.

32. A shaft is subjected to a maximum bending stress of 80 N/mm^2 and maximum shearing stress equal to 30 N/mm^2 at a particular section. If the yield point in tension of the material is 280 N/mm^2, and maximum shear stress theory of failure is used, then the factor of safety obtained will be :
(a) 2.5 (b) 2.8
(c) 3.0 (d) 3.5.

33. A shaft subjected to fluctuating loads for which the normal torque (T) and bending moment (M) are 1000 N-m add 500 N-m respectively. If the combined shock and fatigue factor for bending is 1.5 and combined shock and fatigue factor for torsion is 2, then the equivalent twisting moment for the shaft is :
(a) 2000 N-m (b) 2050 N-m
(c) 2100 N-m (d) 2136 N-m.

34. In the figure shown, if the line AB represents Goodmon criterion of failure, then Soderberg criterion could be represented by line :

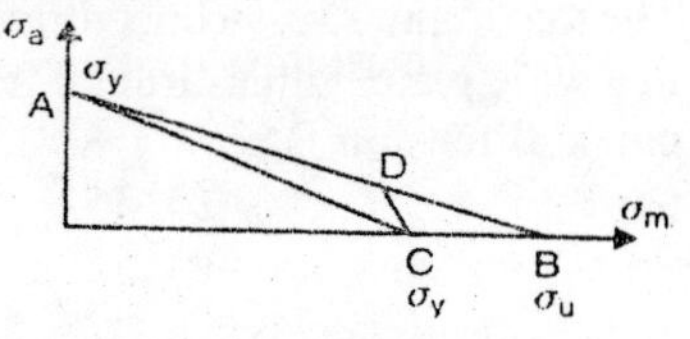

(a) AD (b) DB
(c) DC (d) AC.

35. In designing a shaft for variable loads, the S.N. diagram can be drawn by :
(a) Joining the S_{ut} at 0 cycles and S_e at 10^6 cycles by a straight line on an S.N. graph
(b) Joining the 0.9 S_{ut} at 1000 cycles and S_e at 10^6 cycles by a straight line on a log S-log N graph
(c) Joining the 0.9 S_{ut} at 1000 cycles and S_e at 10^6 cycles by a straight line on an S-N graph
(d) Joining the S_{ut} at 1000 cycles and 0.9 S_e at 10^6 cycles by a straight line on a log S-log N graph.
(S_{ut} stands for ultimate tensile strength and S_e for the endurance limit).

36. How is the tyre of a railway wheel fitted ?
(a) By seam welding
(b) By fillet welding
(c) By interference fit
(d) None of the above.

37. Select the incorrect method to increase the endurance strength of parts :
(a) cold working
(b) shot peening
(c) understressing
(d) surface decarburisation.

38. Coaxing is a procedure to increase the fatigue strength of parts by :
(a) overstressing
(b) understressing
(c) understressing and then increasing the stress in small increments
(d) surface treatment.

39. Stress-concentration under dynamic loading is to be accounted for in design for :

(*a*) ductile materials
(*b*) brittle materials
(*c*) both ductile and brittle materials
(*d*) neither ductile nor brittle materials.

40. In terms of theoretical stress-concentration factor (K_t) and fatigue stress concentration factor (K_f), the notch sensitivity '*q*' is expressed as :

(*a*) $\frac{K_{f-1}}{K_{t-1}}$ (*b*) $\frac{K_{f-1}}{K_{t+1}}$

(*c*) $\frac{K_{t-1}}{K_{f-1}}$ (*d*) $\frac{K_{f+1}}{K_{t+1}}$.

41. The S-N curve for steel becomes asymptotic nearly at :

(*a*) 10^3 cycles (*b*) 10^4 cycles

(*c*) 10^6 cycles (*d*) 10^9 cycles.

5 RIVETED JOINTS

Chapter

42. The most efficient riveted joint possible is one which would be as strong in tension, shear an bearing as the original plates to be joined. But this can never be achieved because :

(*a*) rivets cannot be made with the same material
(*b*) rivets are weak in compression
(*c*) there should be at least one hole in the plate reducing its strength
(*d*) clearance is present between the plate and the rivet.

43. Which one of the following structural joints with 10 rivets and same size of plate and material will be the most efficient ?

(*a*)

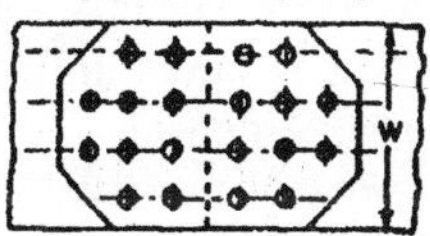

(*b*)

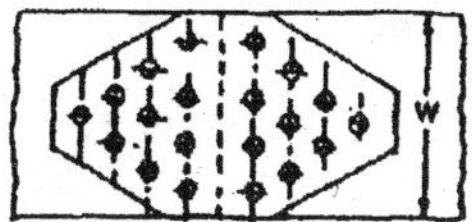

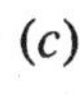

(*c*)

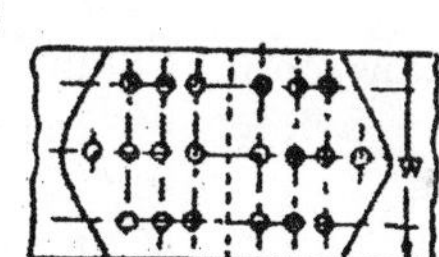

(*d*)

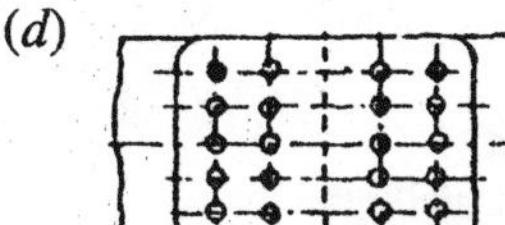

44. Match List I with List II and select answer using the codes given below the Lists :

List I (Types of joints)	List II (An element of the joint)
A. Riveted joint	1. Pin
B. Welded joint	2. Strap
C. Bolted joint	3. Lock washer
D. Knuckle joint	4. Fillet

Codes	A	B	C	D
(*a*)	4	3	2	1
(*b*)	2	3	4	1
(*c*)	2	4	3	1
(*d*)	2	4	1	3.

45. Match List I with List II and select the correct answer using the code give below the lists :

List I (Application)	List II (Joint)
A. Boiler shell	1. Cotter joint
B. Marine shaft coupling	2. Knuckle joint
C. Crosshead and Piston rod	3. Riveted joint
D. Automobile gear box (gears to shaft)	4. Bolted joint

Code :

	A	B	C	D
(*a*)	1	4	2	5
(*b*)	3	5	1	4
(*c*)	1	5	2	4
(*d*)	3	4	1	5.

46. The margin in riveted joints is kept as :
(*a*) equal to rivet diameter
(*b*) 1.25 *d*
(*c*) 1.5 *d* (*d*) 2 *d*.

47. If tearing efficiency of a riveted joint is 80%, then the ratio of diameter of rivet to pitch is :
(*a*) 0.2 (*b*) 0.25
(*c*) 0.4 (*d*) 0.5.

48. Rivet holes in plates should be made by :
(*a*) punching and reaming
(*b*) drilling
(*c*) punching
(*d*) torch cutting.

49. Unwin's formula for rivet diameter *d* in terms of plate thickness '*t*' is :
(*a*) $d = 1.2\sqrt{t}$ (*b*) $d = 6\sqrt{t}$
(*c*) $d = 2\sqrt{t}$ (*d*) $d = 5\sqrt{t}$.

50. In an eccentric riveted joint, that rivet shall experience the maximum stress that is :
(*a*) nearer to the load line
(*b*) farthest away from the load line
(*c*) at the C.G. of the rivet group
(*d*) anywhere.

51. For a boiler, the longitudinal riveted joint should be a :
(*a*) lap joint (*b*) butt joint
(*c*) lozenge joint (*d*) anyone.

52. For a boiler, the circumferential riveted joint should be a :
(*a*) lap joint (*b*) butt joint
(*c*) diamond joint (*d*) anyone.

53. Design of riveted joints for material storage tanks should be based on :
(*a*) strength and rigidity
(*b*) leakage
(*c*) joint efficiency
(*d*) environmental conditions.

54. Match List-I (Different systems) with List-II (Associated terminology) and select the correct answer using the codes given below the lists :

List-I (Different systems)	List-II (Associated terminology)
A. Riveted joints	1. Nipping
B. Welded joints	2. Angular movement
C. Leaf springs	3. Fullering
D. Knuckle joints	4. Fusion

Codes :

	A	B	C	D
(*a*)	3	2	1	4
(*b*)	1	2	3	4
(*c*)	1	4	3	2
(*d*)	3	4	1	2

55.

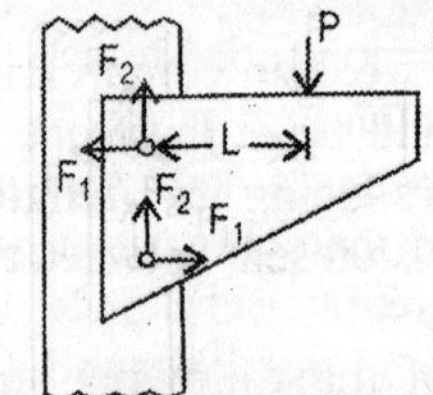

A riveted joint has been designed to support an eccentric load P. The load generates value of F_1 equal to 4 kN and F_2 equal to 3 kN. The cross-sectional area of each rivet is 500 mm^2. Consider the following statements :
1. The stress in the rivet is 10 N/mm^2
2. The value of eccentricity L is 100 mm
3. The value of load P is 6 kN
4. The resultant force in each rivet is 6 kN.

Which of these statements are correct ?
(*a*) 1 and 2 (*b*) 2 and 3
(*c*) 3 and 4 (*d*) 1 and 3.

6 WELDED JOINTS

Chapter

56. The permissible stress in a fillet weld is 100 N/mm^2. The fillet weld has equal leg lengths of 15mm each. The allowable shearing load on weldment per cm length of the weld is :

(*a*) 22.5 kN (*b*) 15.0 kN
(*c*) 10.6 kN (*d*) 7.5 kN.

57. The following two figures show welded joints (*xxx* indicated welds, for the same load and same dimensions of plate and weld.)

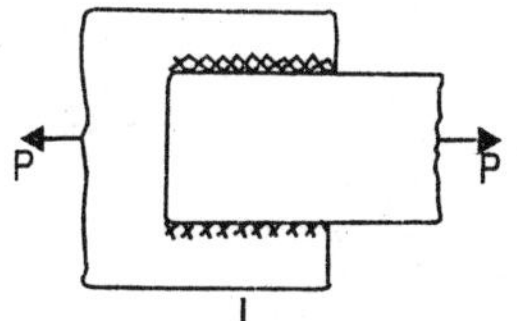

I

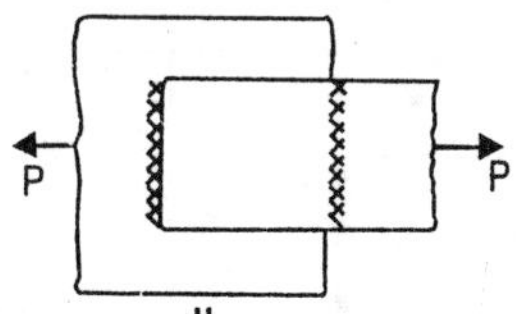

II

The joint shown in :

(*a*) Fig. I is better because the weld is in shear and the principal stress in the weld is not in line with P.
(*b*) Fig. I is better because the load transfer from the tie bar to the plate is not direct.
(*c*) Fig. II is better because the weld is in tension and safe stress of weld in tension is greater than that in shear.
(*d*) Fig. II is better because it has less stress concentration.

58. A double fillet welded joint with parallel fillet weld of length L and leg B is subjected to a tensile force P. Assuming uniform stress distribution, the shear stress in the weld is given by :

(*a*) $\frac{\sqrt{2}.P}{B.L}$ (*b*) $\frac{P}{2.B.L}$

(*c*) $\frac{P}{\sqrt{2}B.L.}$ (*d*) $\frac{2P}{B.L}$.

59. In the welded joint shown in the given figure, if the weld at B has thicker fillets than that at A, then the load carrying capacity P, of the joint will :

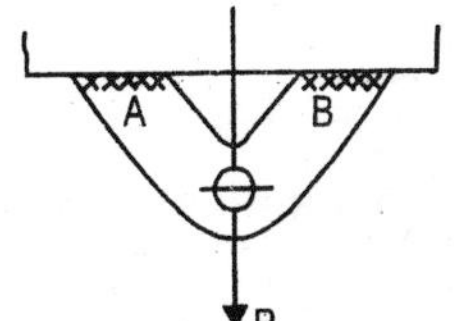

(*a*) increase
(*b*) decrease
(*c*) remain unaffected
(*d*) exactly get doubled.

60. A butt welded joint, subjected to tensile force P is shown in the given figure, l = length of the weld (in mm) h = throat of the butt weld (in mm) and H is the total height of weld including reinforcement. The average tensile stress σ_t in the weld is given by :

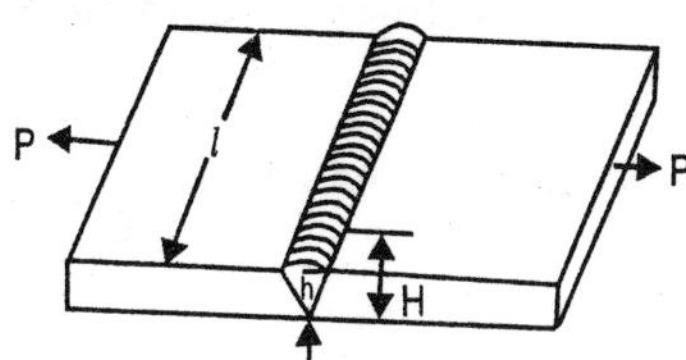

(*a*) $\sigma_t = \frac{P}{Hl}$ (*b*) $\sigma_t = \frac{P}{hl}$

(*c*) $\sigma_t = \frac{P}{2hl}$ (*d*) $\sigma_t = \frac{2P}{hl}$.

61. Two metal plates of thickness '*t*' and width '*w*' are joined by a fillet weld of 45° as shown in given figure. When subjected to a pulling force 'F', the stress induced in the weld will be :

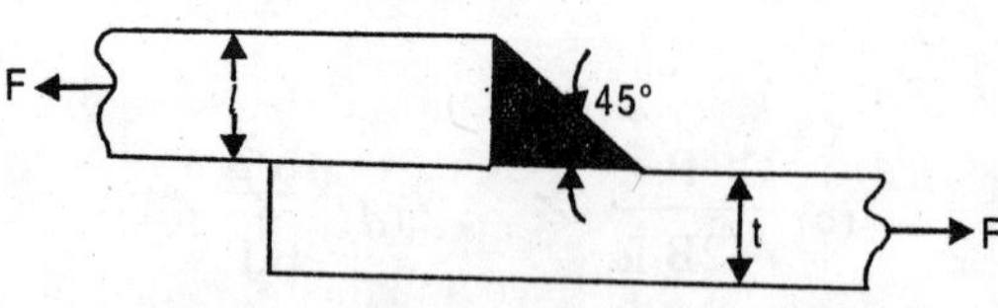

(*a*) $\frac{F}{wt \sin 45°}$ (*b*) $\frac{F}{wt}$

(*c*) $\frac{F \sin 45°}{wt}$ (*d*) $\frac{2F}{wt}$.

62. Consider the following statements.

A splined shaft is used for

1. Transmitting power
2. Holding a flywheel rigidly in position
3. Moving axially the gear wheels mounted on it
4. Mounting V-belt pulleys on it.

Of these statements

(*a*) 2 and 3 are correct
(*b*) 1 and 4 are correct
(*c*) 2 and 4 are correct
(*d*) 1 and 3 are correct.

63. A single parallel fillet weld of total length L and weld size *h* subjected to a tensile load P, will have what design stress ?

(*a*) Tensile and equal to $\frac{P}{0.707\, Lh}$

(*b*) Tensile and equal to $\frac{P}{Lh}$

(*c*) Shear and equal to $\frac{P}{0.707\, Lh}$

(*d*) Shear and equal to $\frac{P}{Lh}$.

64. A circular rod of diameter *d* is welded to a flat plate along its circumference by fillet weld of thickness *t*. Assuming τ_w as the allowable shear stress for the weld material, what is the value of the safe torque that can be transmitted ?

(*a*) $\pi d^2 . t.\tau_w$ (*b*) $\frac{\pi d^2}{2} t\tau_w$

(*c*) $\frac{\pi d^2}{2\sqrt{2}} t\tau_w$ (*d*) $\frac{\pi d^2}{\sqrt{2}} t\tau_w$.

65. A circular solid rod of diameter *d* welded to a rigid flat plate by a circular fillet weld of throat thickness *t* is subjected to a twisting moment T. The maximum shear stress induced in the weld is :

(*a*) $\frac{T}{\pi t d^2}$ (*b*) $\frac{2T}{\pi t d^2}$

(*c*) $\frac{4T}{\pi t d^2}$ (*d*) $\frac{2T}{\pi t d^3}$.

66.

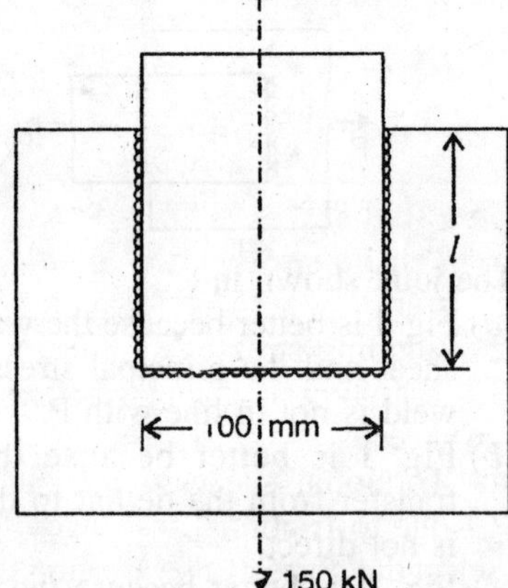

Two plates are joined together by means of single transverse and double parallel fillet welds as shown in figure given above. If the size of fillet is 5 mm and allowable shear load per mm is 300 N, what is the approximate length of each parallel fillet ?

(*a*) 150 mm (*b*) 200 mm
(*c*) 250 mm (*d*) 300 mm.

7 SCREWED JOINTS

Chapter

67. The number of slots in a 25 mm castle nut is :
(a) 2 (b) 4
(c) 6 (d) 8.

68. Which of the following is true about following specifications Hex Bolt M 16 NL – IS : 1363 :
(a) 16 eye bolts used on electric motors
(b) a stress relieved bolt for connecting rod
(c) a bolt with a nut
(d) a bolt with a nut and a lock nut.

69. Two rigid plates are clamped by means of bolt and nut with an initial force N. After tightening, a separating force P (P < N) is applied to the lower plate, which in turn acts on nut. The tension in the bolt after this is :
(a) (N + P) (b) (N – P)
(c) P (d) N.

70. Which of the following stresses are associated with the tightening of a nut on a stud ?
1. Tensile stresses due to stretching of stud.
2. Bending stresses of stud.
3. Transverse shear stresses across threads.
4. Torsional shear stresses in threads due to frictional resistance.

Select the correct answer using the codes given below :

Codes :
(a) 1, 2 and 3 (b) 1, 2 and 4
(c) 2, 3 and 4 (d) 1, 3 and 4.

71. A bolt of uniform strength can be developed by :
(a) keeping the core diameter of threads equal to the diameter of unthreaded portion of the bolt
(b) keeping the core diameter smaller than the diameter of the unthreaded portion
(c) keeping the nominal diameter of threads equal the diameter of unthreaded portion of the bolt
(d) one end fixed and the other end free.

72. Which of the following stresses are associated with the tightening of nut on a bolt ?
(a) Tensile stress due to the stretching of bolt
(b) Bending stress due to the bending of bolt
(c) Crushing and shear stresses in threads
(d) Torsional shear stress due to frictional resistance between the nut and the bolt.

Select the correct answer using the codes given below :
(a) 1, 2 and 4 (b) 1, 2 and 3
(c) 2, 3 and 4 (d) 1, 3 and 4.

73. A screw thread specified by M 20 × 2.5 C as per BIS thread system means :
(a) Metric thread of 20 mm nominal diameter and 2.5 mm pitch having coarse tolerance
(b) Metric thread of 20 mm root diameter and 2.5 mm pitch having coarse tolerance
(c) Metric thread of fine class having 20 mm root diameter and 2.5 mm pitch
(d) Metric thread of 20 mm shank diameter and 2.5 mm thread depth with coarse tolerance.

74. For the bracket bolted as shown in the figure, the bolts will develop :

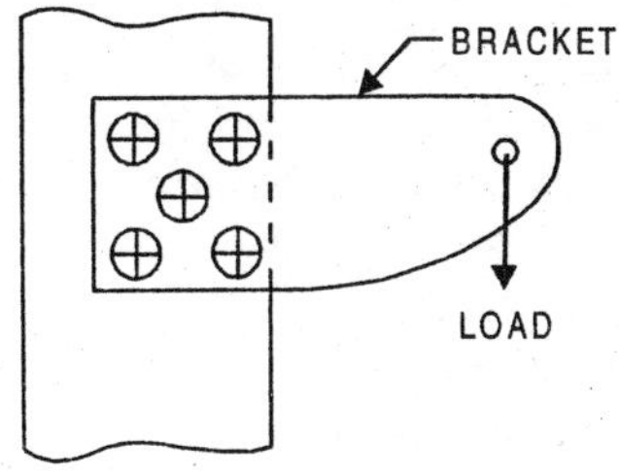

(*a*) Primary tensile stresses and secondary shear stresses
(*b*) Primary shear stresses and secondary shear stresses
(*c*) Primary shear stresses and secondary tensile stresses
(*d*) Primary tensile setresses and secondary compressive stresses.

75. Stresses in a screw, thread are estimated by considering the thread to be :
(*a*) long cantilever beam projecting from the pitch cylinder
(*b*) long cantilever beam projecting from the root cylinder
(*c*) short cantilever beam projecting from the root cylinder
(*d*) short cantilever beam projecting from the pitch cylinder.

76. How can shock absorbing capacity of a bolt be increased ?
(*a*) By tightening it properly
(*b*) By increasing the shank diameter
(*c*) By grinding the shank
(*d*) By making the shank diameter equal to the core diameter of thread.

77. Stresses in a screw thread are estimated by considering the thread to be :
(*a*) long cantilever beam projecting from the pitch cylinder
(*b*) long cantilever beam projecting from the root cylinder
(*c*) short cantilever beam projecting from the root cylinder
(*d*) short cantilever beam projecting from the pitch cylinder.

78. What is the safe static tensile load for a M 36 × 4C bolt of mild steel having yield stress of 280 MPa and a factor of safety 1.5 ?
(*a*) 285 kN (*b*) 190 kN
(*c*) 142.5 kN (*d*) 95 kN.

79. The hemispherical end of a pressure vessel is fastened to the cylindrical portion of the pressure vessel with the help of gasket, bolts and lock nuts. The bolts are subjected to :
(*a*) Tensile stress
(*b*) Compressive stress
(*c*) Shear stress
(*d*) Bearing stress.

80. For bolts of uniform strength, the shank diameter is made equal to :
(*a*) Major diameter of threads
(*b*) Pitch diameter of threads
(*c*) Minor diameter of threads
(*d*) Nominal diameter of threads.

81. Match List-I (Items in joints) with List-II (Type of failure) and select the correct answer using the codes given below the lists :

List-I (Items in joints)	**List-II (Type of failure)**
A. Bolts in bolted joints of engine cylinder cover plate	1. Double transverse shear
B. Cotters in cotter joint	2. Torsional shear
C. Rivets in lap joints	3. Single transverse shear
D. Bolts holding two flanges in a flange coupling	4. Tension

Codes :

	A	B	C	D
(*a*)	4	1	3	2
(*b*)	4	2	3	1
(*c*)	3	1	4	2
(*d*)	3	2	4	1

82. In a bolted joint two members are connected with an axial tightening force of 2200 N. If the bolt used has metric threads of 4 mm pitch, the torque required for achieving the tightening force is :
(*a*) 0.7 N.m (*b*) 1.0 N.m
(*c*) 1.4 N.m (*d*) 2.8 N.m.

83. The function of a washer is to :
(*a*) provide cushioning effect
(*b*) provide bearing area
(*c*) absorb shocks and vibrations
(*d*) provide smooth surface.

84. When a nut is tightened on a bolt by placing a washer below it, then the stress in the bolt is :
(*a*) compression (*b*) tension
(*c*) shear (*d*) complex.

85. The following type of threads are most suitable for applications involving high stresses in one direction only :
(*a*) Iso metric thread (*b*) Acme thread
(*c*) Square thread (*d*) Buttress thread.

86. Which of the following is a permanent fastening ?
(*a*) bolts (*b*) keys
(*c*) cotters (*d*) rivets.

87. The included angle of acme thread is :
(*a*) 60° (*b*) 55°
(*c*) 47.5° (*d*) 29°.

88. The fatigue strength of a bolt can be increased by :
(*a*) increasing shank diameter
(*b*) increasing threaded diameter
(*c*) decreasing shank diameter
(*d*) making threaded and shank diameter same.

89. Resilience of a bolt can be increased by increasing its :
(*a*) length
(*b*) shank diameter
(*c*) threaded diameter
(*d*) head size.

90. Metal-to-metal joint is used for applications involving :
(*a*) high temperature
(*b*) high pressure
(*c*) severe vibrations
(*d*) stress raisers.

91. Eight bolts are to be selected for fixing the cover plate of a cylinder subjected to a maximum load of 980.175 kN. If the design stress for the bolt material is 315 N/mm^2, what is the diameter of each bolt ?
(*a*) 10 mm (*b*) 22 mm
(*c*) 30 mm (*d*) 36 mm.

8 POWER SCREWS

Chapter

92. Stress in a screw thread are estimated by considering the thread to be :
(*a*) long cantilever beam projecting from the pitch cylinder
(*b*) long cantilever beam projecting from the root cylinder
(*c*) short cantilever beam projecting from the root cylinder
(*d*) short cantilever beam projecting from the pitch cylinder.

93. To ensure self-locking in a screw jack it is essential that helix angle is :
(*a*) larger than friction angle
(*b*) smaller than friction angle
(*c*) equal to friction angle
(*d*) such as to give maximum efficiency in lifting.

94. The frictional torque for square thread at mean radius while raising load is given by :
(W = load, R = Mean Radius, ϕ = Angle of friction, α = Helix angle)
(*a*) $WR_0 \tan(\phi - \alpha)$
(*b*) $WR_0 \tan(\phi + \alpha)$
(*c*) $WR_0 \tan\alpha$
(*d*) $WR_0 \tan\phi$.

95. Consider the following statements regarding power screws :
1. The efficiency of a self-locking screw cannot be more than 50%
2. If the friction angle is less than the helix angle of the screw, then the efficiency will be more than 50%
3. The efficiency of ACME (trapezoidal) thread is less than that of square thread.

Of these statements
(*a*) 1, 2 and 3 are correct
(*b*) 2 and 3 are correct
(*c*) 1 and 3 are correct
(*d*) 1 and 2 are correct.

96. The load cup of a screw jack is made separate from the head of the spindle to :
(*a*) enhance the load carrying capacity of the jack
(*b*) reduce the effort needed for lifting the working load
(*c*) reduce the value of frictional torque to be countered for lifting the load
(*d*) prevent the rotation of load being lifted.

97. While designing a screw in a screw jack against buckling failure, the end conditions for the screw are taken as :
(*a*) both the ends fixed
(*b*) both the ends hinged
(*c*) one end fixed and other end hinged
(*d*) one end fixed and the other end free.

98. The maximum efficiency of a self-locking screw is :
(*a*) 50% (*b*) 70%
(*c*) 75% (*d*) 80%.

99. What is the efficiency of a self-locking power screw ?
(*a*) 70% (*b*) 60%
(*c*) 55% (*d*) < 50%.

100. In case of power screws, what is the combination of materials used for the screw and the nut ?
(*a*) Cast iron screw and mild steel nut
(*b*) Carbon steel screw and phosphor bronze nut
(*c*) Cast iron screw and cast iron nut
(*d*) Aluminium screw and alloy steel nut.

101. Consider the case of a square- threaded screw loaded by a nut as shown in the given figure. The value of the average shearing stress of the screw is given by (symbols have the usual meaning)

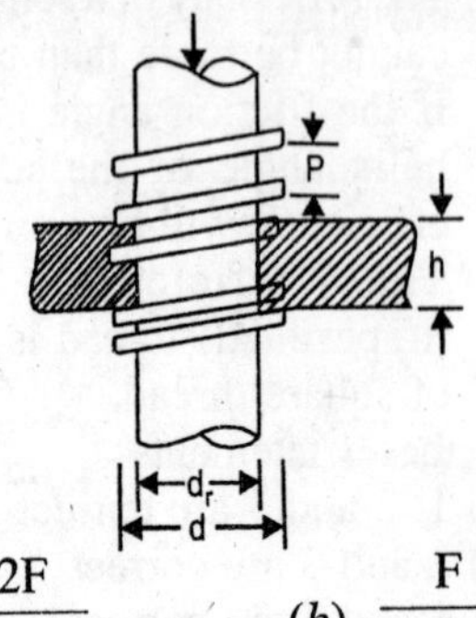

(*a*) $\frac{2F}{\pi d_r h}$ (*b*) $\frac{F}{\pi d_r h}$
(*c*) $\frac{2F}{\pi dh}$ (*d*) $\frac{F}{\pi dh}$.

102. The diameter of tommy bar for a screw jack is designed for :
(*a*) Bending moment due to effort applied
(*b*) Torque on the tommy bar due to effort applied
(*c*) A percentage of axial loads
(*d*) Some axial loads coupled with transverse loads.

103. The following parameters are to be calculated while designing a screw jack :
1. Core diameter of screw.
2. Torque required to rotate the screw.
3. Principal stresses.
4. Height of the nut.
The correct sequence of the calculation of these parameters is :
(*a*) 1, 2, 4, 3 (*b*) 1, 2, 3, 4
(*c*) 2, 1, 3, 4 (*d*) 2, 1, 4, 3.

104. Power screws are used to produce uniform, slow and powerful motion such as required in presses, jacks and other machinery. 'V' threads are usually *not* used for this application due to low efficiency.
This is because :
(*a*) Profile angle is zero
(*b*) Profile angle is moderate
(*c*) Profile angle is large
(*d*) There is difficulty in manufacturing the profile.

105. Which one of the following is the value of helix angle for maximum efficiency of a square threaded screw ? ($\phi = \tan^{-1} \mu$)
(*a*) $45° + \phi$ (*b*) $45° - \phi$
(*c*) $45° - \phi/2$ (*d*) $45° + \phi/2$.

106. The maximum efficiency of a screw jack having square threads with friction angle ϕ is :
(*a*) $\frac{1-\tan(\phi/2)}{1+\tan(\phi/2)}$ (*b*) $\frac{1-\tan\phi}{1+\tan\phi}$
(*c*) $\frac{1-\sin\phi}{1+\sin\phi}$ (*d*) $\frac{1-\sin(\phi/2)}{1+\sin(\phi/2)}$.

107. Ball bearing type of screws are found in following applications :
(*a*) screw jack
(*b*) machine tools
(*c*) crane
(*d*) steering mechanism.

108. In order to obtain a bolt of uniform strength :
(*a*) increase shank diameter
(*b*) increase the diameter of threaded portion
(*c*) drill an axial hole through head upto threaded portion so as to make shank area equal to the root area of the threads
(*d*) provide threads on the whole bolt length.

109. Turnbuckle has :
(*a*) right hand threads on both ends
(*b*) left hand threads on both ends
(*c*) left hand threads on one end and right hand threads on the other
(*d*) threads in the middle only.

110. Eye bolts are used for :
(*a*) foundations
(*b*) locking devices
(*c*) power transmission
(*d*) lifting purposes.

111. The threads on the nut and bolt are of the :
(*a*) same hand
(*b*) different hand
(*c*) different pitch
(*d*) same or different hand.

112. Threads used on a bench vice are :
(*a*) metric (*b*) square
(*c*) acme (*d*) trapezoidal.

113. If shear stress in the nut is half of the tensile stress in a bolt, then nut length would be equal to :
(*a*) diameter of bolt
(*b*) 0.5 × diameter of bolt
(*c*) 1.5 × diameter of bolt
(*d*) 2 × diameter of bolt.

114. The maximum efficiency of a screw jack having square threads and friction angle of 30° will be :
(*a*) 11% (*b*) 22%
(*c*) 33% (*d*) 44%.

115. For a lifting screw to be of the overhauling type :
(*a*) Helix angle (α) = friction angle (ϕ)
(*b*) $\alpha < \phi$
(*c*) $\alpha > \phi$
(*d*) $\alpha = 2\phi$.

9 TRANSMISSION SHAFTS

Chapter

116. Which one of the following loadings is considered for design of axles ?
(*a*) Bending moment only
(*b*) Twisting moment only
(*c*) Combined bending moment and torsion
(*d*) Combined action of bending moment; twisting moment and axial thrust.

117. Match List I with List II and select the correct answer using the codes given below the lists :

List I	List II
A. Crank shaft	1. Supports the revolving parts and transmits torque.
B. Wire shaft	2. Transmits motion between shafts where it is not possible to effect a rigid coupling between them.
C. Axle	3. Converts linear motion into rotary motion.
D. Plain shaft	4. Supports only the revolving parts.

Codes :	A	B	C	D
(*a*)	3	2	1	4
(*b*)	4	2	3	1
(*c*)	3	2	4	1
(*d*)	1	4	2	3.

118. In the assembly design of shaft, pulley and key, the weakest member is :
(*a*) Pulley (*b*) Key
(*c*) Shaft (*d*) None.

119. When a shaft transmits power through gears, the shaft experiences :
(*a*) torsional stresses alone
(*b*) bending stresses alone
(*c*) constant bending and varying torsional stresses
(*d*) varying bending and constant torsional stresses.

120. In power transmission shafts, if the polar moment of inertia of a shaft is doubled, then what is the torque required to produce the same angle of twist ?
(*a*) $\frac{1}{4}$ of the original value
(*b*) $\frac{1}{2}$ of the original value
(*c*) Same as the original value
(*d*) Double the original value.

121. While transmitting the same power by a shaft, if its speed is doubled, what should be its new diameter if the maximum shear stress induced in the shaft remains same ?
(*a*) $\frac{1}{2}$ of the original diameter
(*b*) $\frac{1}{\sqrt{2}}$ of the original diameter
(*c*) $\sqrt{2}$ of the original diameter
(*d*) $\frac{1}{(2)^{\frac{1}{3}}}$ of the original diameter.

122. The diameter of a shaft is increased from 30 mm to 60 mm, all other conditions remaining unchanged. How many times is its torque carrying capacity increased ?
(*a*) 2 times (*b*) 4 times
(*c*) 8 times (*d*) 16 times.

123. For obtaining the maximum shear stress induced in the shaft shown in the given figure, the torque should be equal to :

(*a*) T
(*b*) $Wl + T$
(*c*) $\left[(W\,l)^2 + \left(\frac{wL}{2}\right)^2\right]^{\frac{1}{2}}$
(*d*) $\left[\left\{W\,l + \frac{wL^2}{2}\right\}^2 + T^2\right]^{\frac{1}{2}}$.

124. Which one of the following statement is correct ? Shafts used in heavy duty speed reducers are generally subjected to :
(*a*) bending stress only
(*b*) shearing stress only
(*c*) combined bending and shearing stresses
(*d*) bending, shearing and axial thrust simultaneously.

125. Choose the correct statement :
(*a*) cold rolling produces a stronger shaft than hot rolling
(*b*) hot rolling produces a stronger shaft than cold rolling
(*c*) both cold and hot rolling produce equally strong shaft
(*d*) shafts are not made by rolling process.

126. A hollow shaft of outer dia 40 mm and inner dia of 20 mm is to be replaced by a solid shaft to transmit the same torque at the same maximum stress. What should be the diameter of the solid shaft ?
(*a*) 30 mm
(*b*) 35 mm
(*c*) $10 \times (60)^{1/3}$ mm
(*d*) $10 \times (20)^{1/3}$ mm.

Chapter 10 KEYS AND COUPLINGS

127. Consider the following operations :
1. Cutting key ways on shafts
2. Cutting external screw threads.
3. Cutting teeth of spur gears.
4. Cutting external splines.

Those which can be performed with milling cutters would include

(*a*) 1 and 2 (*b*) 2, 3 and 4
(*c*) 1 and 3 (*d*) 1, 2, 3 and 4.

128. Match List I with List II and select the correct answer using the codes given below the lists :

List I (Name)	List II (Type)
A. Oldham coupling	1. Joins collinear shafts and is rigid type.
B. Flange coupling	2. Joins non-collinear shafts and is adjustable.
C. Universal coupling	3. Joins collinear shafts and engages and disengages them during motion.
D. Friction coupling	4. Compensates peripheral shafts, longitudinal and angular shifts of shafts.

Codes :

	A	B	C	D
(*a*)	2	1	4	3
(*b*)	3	2	1	4
(*c*)	1	4	2	3
(*d*)	2	3	1	4

129. Match List-I (Type of keys) with List-II (Characteristics) and select the correct answer using the codes given below the Lists :

List-I	List-II
A. Woodruff key	1. Loese fitting, light duty
B. Kennedy key	2. Heavy duty
C. Feather Key	3. Self-aligning
D. Flat key	4. Normal industrial use.

Codes :

	A	B	C	D
(*a*)	2	3	1	4
(*b*)	3	2	1	4
(*c*)	2	3	4	1
(*d*)	3	2	4	1

130. Which of the following stresses are associated with the design of pins in bushed pin-type, flexible coupling ?
1. Bearing stress
2. Bending stress
3. Axial tensile stress
4. Transverse shear stress

Select the correct answer using the codes given below :

(*a*) 1, 3 and 4 (*b*) 2, 3 and 4
(*c*) 1, 2 and 3 (*d*) 1, 2 and 4.

131. The shearing area of a key of length 'L' breadth 'b' depth 'h' is equal to

(*a*) $b \times h$ (*b*) $L \times h$
(*c*) $L \times b$ (*d*) $L \times (h/2)$

132. Match List-1 (connecting shaft) with List-II (Couplings) and select the correct answer using the codes given below the lists :

List-I Connecting shaft	List-II Couplings
A. In perfect alignment	1. Oldham coupling
B. With angular misalignment of 10°	2. Rigid coupling
C. Shafts with parallel misalignment	3. Universal joint
D. Where one of the shafts may undergo more deflection with respect to the other	4. Pin type flexible coupling.

Codes :

	A	B	C	D
(a)	2	1	3	4
(b)	4	3	1	2
(c)	2	3	1	4
(d)	4	1	3	2.

133.

The key shown in the above figure is a
(*a*) Barth key (*b*) Kennedy key
(*c*) Lewis key (*d*) Woodruff key.

134. Match List-I (keys) with List-II (Characteristics) and select the correct answer using the codes given below the lists :

List-I	List-II
A. Saddle key	1. Strong in shear and crushing
B. Woodruff key	2. Withstands tension in one direction
C. Trangent key	3. Transmission of power through frictional resistance
D. Kennedy key	4. Semi-circular in shape.

Codes :

	A	B	C	D
(a)	3	4	1	2
(b)	4	3	2	1
(c)	4	3	1	2
(d)	3	4	2	1

135. A key connecting a flange coupling to a shaft is likely to fail in
(*a*) Shear (*b*) tension
(*c*) torsion (*d*) bending.

136. Which key is preferred for the condition where a large amount of impact torque is to be transmitted in both direction of rotation.
(*a*) Woodruff key (*b*) Feather key
(*c*) Gib-head key (*d*) Tangent key.

137. The bolts in a rigid flanged coupling connecting two shafts transmitting power are subjected to
(*a*) shear force and bending moment
(*b*) axial force
(*c*) torsion
(*d*) torsion and bending.

138. The coupling used to connect two shafts with large angular misalignment is
(*a*) a flange coupling
(*b*) an Oldham's coupling
(*c*) a flexible bush coupling
(*d*) a hook's joint.

139. What is sunk key in the form of a segment of a circular disc of uniform thickness, known as ?
(*a*) Feather key (*b*) Kennedy key
(*c*) Woodruff key (*d*) Saddle key.

140. It two parallel shafts are to be connected and the distance between the axes of shafts is small and variable, then one would need to use.
(*a*) a clutch
(*b*) a universal joint
(*c*) an Oldham's coupling
(*d*) a knuckle joint.

141. Which one of the following statements is not correct ?
(*a*) Hooke's joint is used to connect two rotating co-planar, non-intersecting shafts.
(*b*) Hooke's joint is used to connect two rotating co-planar, intersecting shafts.
(*c*) Oldham's coupling is used to connect two parallel rotating shafts.
(*d*) Hooke's joint is used in the steering mechanism for automobiles.

142. Which one of the following statements is correct ? While designing a parallel sunk key it is assumed that the distribution of force along the length of the key.
(*a*) varies linearly
(*b*) is uniform throughout
(*c*) varies exponentially, being more at the torque input end
(*d*) varies exponentially, being less at torque output end.

143. A pulley is connected to a power transmission shaft of diameter *d* by means of rectangular sunk key of width *w* and length *l*. The width of the key is taken as ¼. For full power transmission, the shearing strength of the key is equal to the torsion shearing strength of the shaft. The ratio of the length of the key to the diameter of the shaft (l/d) is

(*a*) $\frac{\pi}{4}$ (*b*) $\frac{\pi}{\sqrt{2}}$

(*c*) $\frac{\pi}{2}$ (*d*) π.

144. A square key of side $d/4$ is to be fitted on a shaft of diameter d and in the hub of a pulley. If the material of the key and shaft is same and the two are to be equally strong in shear, what is the length of the key ?

(*a*) $\frac{\pi d}{2}$ (*b*) $\frac{2\pi d}{3}$

(*c*) $\frac{3\pi d}{4}$ (*d*) $\frac{4\pi d}{5}$.

145. Consider the following statements in respect of flexible couplings :

1. The flanges of flexible coupling are usually made of grey cast iron FG200.
2. In the analysis of flexible coupling, it is assumed that the power is transmitted by the shear resistance of the pins.
3. Rubber bushes with brass lining are provided to absorb misalignment between the two shafts.

Which of the statements given above are correct ?

(*a*) 1, 2 and 3 (*b*) Only 1 and 2
(*c*) Only 2 and 3 (*d*) Only 1 and 3.

146. Match List-I (Coupling) with List-II (Purpose) and select the correct answer using the codes given below the lists :

List-I (Coupling)	**List-II (Purpose)**
A. Muff coupling	1. To transmit power between two parallel shafts.
B. Flange coupling	2. To transmit power between two intersecting shafts with flexibility.
C. Oldham's coupling	3. For rigid connection between two aligned shafts for power transmission.
D. Hooke's joint.	4. For flexible connection between two shafts with some misalignment for transmitting power.

Codes :

	A	B	C	D
(*a*)	1	4	3	2
(*b*)	3	4	2	1
(*c*)	3	2	1	4
(*d*)	1	2	3	4

147. Square key of side "$\frac{d}{4}$" each and length l is to transmit torque "T" from the shaft of diameter "d" to the hub of a pulley. Assuming the length of the key to be equal to the thickness of the pulley, the average shear stress developed in the key is given by

(*a*) $\frac{4T}{ld}$ (*b*) $\frac{16T}{ld^2}$

(*c*) $\frac{8T}{ld^2}$ (*d*) $\frac{16T}{\pi d^3}$.

148. Consider the following statements in respect of flexible couplings :

1. The flanges of flexible coupling are usually made of grey cast iron FG200.
2. In the analysis of flexible coupling, it is assumed that the power is transmitted by the shear resistance of the pins.
3. Rubber bushes with brass lining are provided to absorb misalignment between the two shafts.

Which of the statements given above are correct ?

(*a*) 1, 2 and 3 (*b*) Only 1 and 2
(*c*) Only 2 and 3 (*d*) Only 1 and 3.

149. Which one of the following statements is not correct ?

(*a*) Hooke's joint is used to connect two rotating co-planar, non-intersecting shafts.
(*b*) Hooke's joint is used to connect two rotating co-planar, intersecting shafts.
(*c*) Oldham's coupling is used to connect two parallel rotating shafts.
(*d*) Hooke's joint is used in the steering mechanism for automobiles.

150. Oldham's coupling is used to connect two shafts which

(*a*) have lateral misalignment
(*b*) have two axes intersecting
(*c*) have lateral movement during power transmission
(*d*) change axes during power transmission.

151. Match List I with List II and select the correct answer using the code given below the lists :

List-I Key/spline	List-II Application
A. Gib head key	1. Self aligning
B. Woodruff key	2. Facilitates removal
C. Parallel key	3. Mostly used
D. Spline	4. Axial movement possible

Code :

	A	B	C	D
(*a*)	1	2	3	4
(*b*)	1	2	4	3
(*c*)	2	1	3	4
(*d*)	2	1	4	3.

152. Which of the following keys is usually strong in failure under shear and crushing ?

(*a*) rectangular (*b*) flat
(*c*) square (*d*) circular.

153. Splined shafts are commonly used in

(*a*) machine tools (*b*) bicycles
(*c*) aircraft (*d*) automobiles.

154. Which of the following keys transmits power through frictional resistance only ?

(*a*) woodruff (*b*) Kennedy
(*c*) sunk (*d*) saddle.

155. Splined shafts are used in applications requiring

(*a*) less stress-concentration due to deep keyways
(*b*) high torque transmission
(*c*) axial relative movement between shaft and hub
(*d*) quick engagement and disengagement.

156. For two parallel shafts, the distance between whose axes is small and variable, which coupling will be used ?

(*a*) hydraulic coupling
(*b*) universal joint
(*c*) flange coupling
(*d*) Oldham's coupling.

157. In flange coupling, the two flanges are coupled together by means of bolts fitted in holes which are

(*a*) reamed (*b*) machined
(*c*) threaded (*d*) gasketed.

158. Muff coupling is used to join two shafts which have

(*a*) lateral misalignment
(*b*) intersecting axes at small angle
(*c*) aligned axes
(*d*) to absorb shocks.

159. Universal coupling is used to join two shafts which have

(*a*) lateral misalignment
(*b*) intersecting axes at small angle
(*c*) aligned axes
(*d*) to absorb shocks.

Chapter 12 COTTER AND KNUCKLE JOINTS

160. Match List-I with List-II and select the correct answer using the code given below the Lists :

List-I (Parts to be Joined)	List-II (Types of Joint)
A. Two rods having relative axial motion	1. Pin Joint
B. Strap end of the connecting rod	2. Knuckle Joint
C. Piston rod and cross head	3. Gib and Cotter Joint
D. Links of four-bar chain	4. Cotter Joint

Codes :

	A	B	C	D
(*a*)	1	3	4	2
(*b*)	2	4	3	1
(*c*)	1	4	3	2
(*d*)	2	3	4	1

161. In a gib and cotter joint, the gib and cotter are subjected to
(*a*) single shear only
(*b*) double shear only
(*c*) single shear and crushing
(*d*) double shear and crushing.

162. Match List-I with List-II and select the correct answer using the codes given below the lists :

List-I (Type of Joint)	List-II (Mode of joining members)
A. Cotter joint	1. Connects two rods or bars permitting small amount of flexibility.
B. Knuckle joint	2. Rigidly connects two members
C. Turn buckle	3. Connects two rods having thread ends
D. Riveted joint	4. Permanent fluid-tight joint between two flat pieces
	5. Connects two shafts and transmits torque

Codes :

	A	B	C	D
(*a*)	5	1	3	2
(*b*)	2	1	3	4
(*c*)	5	3	2	4
(*d*)	2	3	1	4

163. Match List-I with List-II and select the correct answer using the codes given below the lists :

List-I (Different types of detachable joints)	List-II (Specific use of these detachable joints)
A. Cotter joint	1. Tie rod of a wall crane
B. Knuckle joint	2. Suspension bridges
C. Suspension link joint	3. Diagonal stays in boiler
D. Turn buckle (adjustable joint).	4. Cross-head of a steam engine.

Codes :

	A	B	C	D
(*a*)	4	2	3	1
(*b*)	4	3	2	1
(*c*)	3	2	1	4
(*d*)	2	1	4	3.

164. The piston rod and the crosshead in a steam engine are usually connected by means of :
(*a*) Cotter joint (*b*) Knuckle joint
(*c*) Ball joint (*d*) Universal joint.

165. Match List-I (Machine element) with List-II (Cause of failure) and select the correct answer using the codes given below the lists :

List-I	List-II
A. Axle	1. Shear stress
B. Cotter	2. Tensile/ compressive stress
C. Connecting rod	3. Wear
D. Journal bearing	4. Bending stress

Codes :

	A	B	C	D
(*a*)	1	4	2	3
(*b*)	4	1	2	3
(*c*)	4	1	3	2
(*d*)	1	4	3	2.

166. In a gib and cotter joint, the gib and cotter are subjected to :
(*a*) single shear only
(*b*) double shear only
(*c*) single shear and crushing
(*d*) double shear and crushing.

167. The spigot of a cotter joint has a diameter of D and carries a slot for cotter. The permissible crushing stress is x times the permissible tensile stress for the material of spigot where $x > 1$. The joint carries an axial load P. Which one of the following equations will give the diameter of the spigot ?

(*a*) $D = 2\sqrt{\frac{P}{\pi\sigma_t}\frac{x-1}{x}}$

(*b*) $D = 2\sqrt{\frac{P}{\pi\sigma_t}\frac{x+1}{x}}$

(*c*) $D = \frac{2}{\pi}\sqrt{\frac{P}{\sigma_t}\frac{x+1}{x}}$

(*d*) $D = \frac{2P}{\pi\sigma_t}\sqrt{(x+1)}$.

168. In a cotter joint, the width of the cotter at the centre is 50 mm and its thickness is 12 mm. The load acting on the cotter is 60 kN. What is the shearing stress developed in the cotter ?
(*a*) 120 N/mm^2 (*b*) 100 N/mm^2
(*c*) 75 N/mm^2 (*d*) 50 N/mm^2.

169. Match List-I with List-II and select the correct answer using the code given below the Lists :

List-I (Parts to be Joined)	List-II (Type of Joint)
A. Two rods having relative axial motion	1. Pin Joint
B. Strap end of the connecting rod	2. Knuckle Joint
C. Piston rod and cross head	3. Gib and Cotter Joint
D. Links of four-bar chain.	4. Cotter Joint.

Code :

	A	B	C	D
(*a*)	1	3	4	2
(*b*)	2	4	3	1
(*c*)	1	4	3	2
(*d*)	2	3	4	1.

170. Match List-I with List-II and select the correct answer using the code given below the Lists :

List-I (Parts to be Joined)	List-II (Types of Joint)
A. Two rods having relative axial motion	1. Pin Joint
B. Strap end of the connecting rod	2. Knuckle Joint
C. Piston rod and cross head	3. Gib and Cotter Joint
D. Links of four-bar chain.	4. Cotter Joint.

Codes :

	A	B	C	D
(*a*)	1	3	4	2
(*b*)	2	4	3	1
(*c*)	1	4	3	2
(*d*)	2	3	4	1.

171. In a gib and cotter joint, the gib and cotter are subjected to :
(*a*) single shear only
(*b*) double shear only
(*c*) single shear and crushing
(*d*) double shear and crushing.

13 PIPE JOINTS

Chapter

172. Following type of pipe joint is mostly used for pipes carrying water at low pressure :
(*a*) socket (*b*) nipple
(*c*) union (*d*) spigot and socket.

173. Which of the following type of pipe joints would be suitable for pipes carrying steam ?
(*a*) flanged (*b*) threaded
(*c*) expansion (*d*) bell and spigot.

14 MECHANICAL SPRINGS

Chapter

174. The ends of the leaves of semi-elliptical leaf spring are made triangular in plan in order to :
(*a*) Obtain variable I in each leaf
(*b*) permit each leaf to act as a overhanging beam
(*c*) have variable bending moment in each leaf
(*d*) Make $\frac{M}{1}$ constant throughout the length of the leaf.

175. Cracks in helical springs used in Railway carriages usually start on the inner side of the coil because of the fact that :
(*a*) it is subjected to a higher stress than the touter side
(*b*) it is subjected to a higher cyclic loading than the outer side
(*c*) it is more stretched than the outer side during the manufacturing process
(*d*) it has a lower curvature than the outer side.

176. In the calculation of induced shear stress in helical springs, the Wahl's correction factor used to take care of :
(*a*) combined effect of transverse shear stress and bending stress in the wire.
(*b*) combined effect of bending stress and curvature of the wire
(*c*) combined effect of transverse shear stress and curvature of the wire.
(*d*) combined effect of torsional shear stress and transverse shear stress in the wire.

177. Wahl stress factor takes into account :
(*a*) direct shear stress
(*b*) torsional shear stress
(*c*) wire curvature effect
(*d*) direct shear and wire curvature effect.

178. An open coiled helical spring is subjected to an axial force, the wire of the spring is subjected to :
(*a*) direct shear only
(*b*) combined shear and bending only
(*c*) combined shear, bending and twisting
(*d*) combined shear and twisting only.

179. Spring index may be defined as :
(*a*) ratio of coil diameter to wire diameter
(*b*) load required to produce unit deflection
(*c*) as the capability of the spring to store energy
(*d*) ratio of wire diameter to coil diameter.

180. The deflection of a spring designed for uniform strength and of uniform breadth b and thickness t under load W is :
(a) $\pi Wl^3/2Ebt^3$ (b) $Wl^3/2Ebt^3$
(c) $Wl^3/4Ebt^3$ (d) $Wl^3/8Ebt^3$.

181. The shearing stress on the inner side as compared to the outer side of a helical spring is :
(a) more
(b) less
(c) equal
(d) depends upon coil diameter.

182. Wahl factor accounts for :
(a) direct shear
(b) curvature of coils
(c) direct shear and curvature of coils
(d) helix angle.

183. If a spring is cut into two springs, the stiffness of cut spring will be :
(a) same (b) half
(c) double (d) four times.

184. Concentric helical springs should be wound :
(a) in the same direction
(b) with opposite hand helices
(c) in any direction
(d) depending on the load.

185. Allowable shear stress for helical spring wire with increase in diameter of wire :
(a) decreases
(b) increases
(c) remains same
(d) unpredictable.

15 FLAT AND V-BELTS

Chapter

186. In a belt drive, if the pulley diameter is doubled keeping the tension and belt width constant, then it will be necessary to :
(a) increase the key length
(b) increase the key depth
(c) increase the key width
(d) decrease the key length.

187. The following data refers to an open belt drive :

	Pulley A	Pulley B
Purpose	Driving	Driven
Diameter	450 mm	750 mm
Angle of contact	$\theta_A = 150$	$\theta_B = 210°$
Coefficient of friction between belt and pulley	$f_A = 0.36$	$f_B = 0.22$

The ratio of tensions may be calculated using the relation $(T_1/T_2) = \exp(z)$ where z is :
(a) $f_A\theta_A$
(b) $f_B\theta_B$
(c) $(f_A + f_B)(\theta_B + \theta_B)/4$
(d) $(f_A\theta_A + f_B\theta_B)/2$.

188. In a multiple V belt drive, when a single belt is damaged, it is preferable to change the complete set to :
(a) reduce vibration
(b) reduce slip
(c) ensure uniform loading
(d) ensure proper alignment.

189. In flat belt drive, if the slip between the driver and the belt is 1%, that between belt and follower is 3% and driver and follower pulley diameters are equal, then the velocity ratio of the drive will be :
(a) 0.99 (b) 0.98
(c) 0.97 (d) 0.96.

190. When belt drive is transmitting maximum power :
(a) effective tension is equal to centrifugal tension.
(b) effective tension is half of centrifugal tension.
(c) driving tension on slack side is equal to the centrifugal tension.
(d) driving tension on tight side is twice the centrifugal tension.

191. In the case of a vertical belt pulley drive with T_c as centrifugal tension and T_o as the initial tension, the belt would tend to hang clear of the power pulley when :

(*a*) $T_c < T_o$ (*b*) $T_c = T_o/3$
(*c*) $T_c > T_o$ (*d*) $T_c = T_o/2$.

192. Given that T_1 and T_2 are the tension on the tight and slack of the belt respectively, the initial tension of the belt taking centrifugal tension T_c is equal to :

(*a*) $\frac{T_1 + T_2 + T_c}{3}$

(*b*) $\frac{T_1 + T_2 + 2T_c}{2}$

(*c*) $\frac{T_1 + T_2 + 3T_c}{2}$

(*d*) $\frac{T_1 + T_2 + 3T_c}{2}$

193. Which one of the following belts should not be used above 40° C ?

(*a*) Balata belt (*b*) Rubber belt
(*c*) Fabric belt (*d*) Synthetic belt.

194. Centrifugal tension in belts is :

(*a*) Useful because it maintains some tension even when no power is transmitted
(*b*) Not harmful because it does not take part in power transmitted
(*c*) Harmful because it increases belt tension and reduces the power transmitted
(*d*) A hypothetical phenomenon and does not actually exist in belts.

195. Which one of the following statements with regard to belt drives is NOT correct ?

(*a*) Increase in the angle of wrap of the belt enables more power transmission
(*b*) Maximum power is transmitted when the centrifugal tension is three times the tight side tension
(*c*) Wide and thin belt is preferable for better life than a thick and narrow one
(*d*) Crown is provided on the pulley to make the belt run centrally on the pulley.

196. In case of belt drives, the effect of the centrifugal tension is to :

(*a*) cause the belt to leave the pulley and increase the power to be transmitted
(*b*) cause the belt to stay on the pulley and increase the power to be transmitted
(*c*) reduce the driving power of the belt
(*d*) stretch the belt in longitudinal direction.

197. Consider the following statements in case of belt drives :

1. Centrifugal tension in the belt increases the transmitted power.
2. Centrifugal tension does not affect the driving tension.
3. Maximum tension in the belt is always three times the centrifugal tension.

Which of the statements given above is/are correct ?

(*a*) 1, 2 and 3 (*b*) 2 and 3 only
(*c*) 1 and 3 only (*d*) 1 only.

198. In case of belt drives, the effect of the centrifugal tension is to :

(*a*) cause the belt to leave the pulley and increase the power to be transmitted
(*b*) cause the belt to stay on the pulley and increase the power to be transmitted
(*c*) reduce the driving power of the belt
(*d*) stretch the belt in longitudinal direction.

199. Belt slip may occur due to :

(*a*) heavy load
(*b*) loose belt
(*c*) driving pulley too small
(*d*) all of the above.

200. Creep in belt drive is due to :

(*a*) material of the pulley
(*b*) material of the belt
(*c*) size of the driven pulley
(*d*) unequal extensions and contractions due to varying tension.

201. For maximum power to be transmitted by a belt, the velocity of belt should be such that the ratio of tight side tension to centrifugal tension should be :

(*a*) 1 (*b*) 2
(*c*) 3 (*d*) 4.

202. The phenomenon of elongation and contraction of the belt is known as :
(*a*) slip (*b*) creep
(*c*) fatigue (*d*) elasticity.

203. In multi-V belt transmission, if one of the belts is broken, we have to change the :
(*a*) broken belt
(*b*) broken belt and its adjacent belts
(*c*) all the belts
(*d*) there is no need of changing the broken belt.

204. In a V - belt drive, the belt touches :
(*a*) at the bottom
(*b*) at the sides only
(*c*) both at the bottom and the sides
(*d*) could touch anywhere.

205. The minimum angle of contact for a V - belt is :
(*a*) 110° (*b*) 120°
(*c*) 140° (*d*) 160°.

206. The maximum angle of inclination of a flat belt drive is :
(*a*) 50° (*b*) 60°
(*c*) 70° (*d*) 80°.

207. The length of a V - belt is designated as :
(*a*) inside length
(*b*) outside length
(*c*) pitch length
(*d*) length at neutral axis.

Chapter

16 CHAINS AND ROPES

208. Sources of power loss in a chain drive are given below :
1. Friction between chain and sprocket teeth.
2. Overcoming the chain stiffness.
3. Overcoming the friction in shaft bearing.
4. Frictional resistance to the motion of the chain in air or lubricant.

The correct sequence of descending order of power loss due to these sources is :
(*a*) 1, 2, 3, 4 (*b*) 1, 2, 4, 3
(*c*) 2, 1, 3, 4 (*d*) 2, 1, 4, 3.

209. Give that p = chain pitch, c = centre distance
N, n = number of teeth on large and small sprocket respectively,
the length of chain in terms of pitches can be approximated by.

(*a*) $\frac{2c}{p}$

(*b*) $\frac{2c}{p}+(\text{N}+n)/2$

(*c*) $\frac{2c}{p}+[(\text{N}+n)/2\pi]^2\,\frac{p}{c}$

(*d*) $\frac{2c}{p}+(\text{N}+n)/2p+\left[\frac{\text{N}-n}{2p}\right]^2\frac{p}{c}$.

210. When compared to a rod of the same diameter and material, a wirerope :
(*a*) is less flexible
(*b*) has a much smaller load carrying capacity
(*c*) does not provide much warning before failure
(*d*) provides much greater time for remedial action before failure.

211. Effective stress in wire ropes during normal working is equal to the stress due to :
(*a*) axial load plus stress due to bending.
(*b*) acceleration / retardation of masses plus stress due to bending.
(*c*) axial load plus stress due to acceleration / retardation.
(*d*) bending plus stress due to acceleration / retardation.

212. Match List-I (Application) with List-II (Drive element) and select the correct answer using the codes given below the lists :

List-I	List-II
A. Automobile differential	1. Flat belt
B. Bicycle	2. V-belt
C. Planning machine	3. Chain drive
D. Radiator fan of automobile	4. Gear drive.

Codes :

	A	B	C	D
(*a*)	4	3	1	2
(*b*)	1	3	4	2
(*c*)	4	2	1	3
(*d*)	1	2	4	3.

213. Consider the following types of stresses in respect of a hoisting rope during acceleration of load :

1. Direct stress due to weight hoisted and weight of the rope.
2. Bending stress due to bending of rope over the sheave.
3. Stresses due to initial tightening.
4. Acceleration stresses.

Which of these are the correct types of stresses induced in a hoisting rope during acceleration of load ?

(*a*) 1, 3 and 4 (*b*) 2, 3 and 4
(*c*) 1, 2 and 4 (*d*) 1, 2 and 4.

214. Given that W = weight of load handled, W_r = weight of rope and f = acceleration, the additional load in ropes of a hoist during starting is given by :

(*a*) $F_a = \left(\frac{W - W_r}{g}\right) f$

(*b*) $F_a = \left(\frac{W + W_r}{g}\right) f$

(*c*) $F_a = \frac{W}{g} f$

(*d*) $F_a = \frac{W_r}{g} f$.

215. A 6 × 19 rope implies that there are :

(*a*) 6 wires in each strand and 19 strands in the rope
(*b*) 6 strands and 19 wires in each strand
(*c*) 6 large diameter wires and 19 small diameter wires
(*d*) 19 large diameter wires and 6 small diameter wires.

216. Which one of the following drives is used for a constant velocity ratio, positive drive with large centre distance between the driver and driven shafts ?

(*a*) Gear drive (*b*) Flat belt drive
(*c*) Chain drive (*d*) V-belt drive.

217. A chain drive is used for :

(*a*) short distances
(*b*) longer distances
(*c*) medium distances
(*d*) distance is not a criteria for chain drive.

218. For roller chain drive with sprocket having 10 teeth, the velocity of the driven shaft with respect to that of drive will be approximately :

(*a*) same
(*b*) 5% above
(*c*) 5% below
(*d*) 5% above to 5% below.

219. Match List-I (Device) with List-II (Component/Accessory) and select the correct answer using the codes given below the Lists :

List-I (Device)	List-II (Component/ Accessory)
A. Lifting machine	1. Idler of jockey pulley
B. Fibre rope drive	2. Sun wheel
C. Differential gear	3. Sheave
D. Belt drive	4. Power screw.

Codes :

	A	B	C	D
(*a*)	4	3	1	2
(*b*)	3	4	1	2
(*c*)	4	3	2	1
(*d*)	3	4	2	1

220. A wire rope is designated as 6 × 19 standard hoisting. The number 6 × 19 represent :

(*a*) diameter in millimeter × length in metre
(*b*) diameter in centimeter × length in metre

(*c*) number of strands × number of wires in each strand
(*d*) number of wires in each strand × number of strands.

221. Wire ropes are used for applications experiencing :
(*a*) low speed and low tension
(*b*) low speed and high tension
(*c*) high speed and low tension
(*d*) high speed and high tension.

222. In cross or regular lay ropes :
(*a*) direction of twist of wires in strands is opposite to the direction of twist of the strands
(*b*) direction of twist of strands and wires is the same
(*c*) wires in two adjacent strands are twisted in opposite direction
(*d*) wires are not twisted.

223. In parallel lay ropes :
(*a*) direction of twist of wires in strands is opposite to direction of twist of strands
(*b*) direction of twist of wires and strands is same
(*c*) wires in two adjacent strands are twisted in opposite directions
(*d*) wires are not twisted.

224. In reverse laid ropes :
(*a*) direction of twist of wires in strands is opposite to direction of twist of strands
(*b*) direction of twist of wires and strands is same
(*c*) wires in two adjacent strands are twisted in opposite directions
(*d*) wires are not twisted.

225. 6 × 19 ropes means a wire rope containing :
(*a*) 6 strands and 19 wires in each strand
(*b*) 19 strands and 6 wires in each strand
(*c*) 6 strands of 19 m length
(*d*) 19 strands of 6 m length.

226. Silent chain is made of
(*a*) links and blocks
(*b*) links, pins, bushings and rollers
(*c*) three or more roller chains
(*d*) inverted tooth over - lapping links.

227. The number of teeth in a chain sprocket should be :
(*a*) odd
(*b*) even
(*c*) either even or odd
(*d*) maximum possible.

228. The following type of chain is used in motor cycles :
(*a*) steel roller (*b*) bush roller
(*c*) precision bush (*d*) silent.

SPUR, HELICAL AND BEVEL GEARS

229. Which of the following is/are used for cutting internal gears ?
1. Gear hobber 2. Gear shaper
3. Rack cutter 4. Jig borer.
Select the correct answer using the codes given below :

Codes :
(*a*) Only 1 and 2 (*b*) Only 2 and 3
(*c*) Only 1 and 4 (*d*) Only 2.

230. Match List-I with List-II and select the correct answer using the codes given below the lists :

List-I (Standard tooth forms)	**List-II (Advantages or disadvantages)**
A. 20° and 50° system	1. Results in lower loads on bearing

List-I (Standard tooth forms)	List-II (Advantages or disadvantages)
B. $14\frac{1°}{2}$ stub-tooth system	2. Broadest at the base and strongest in bending
C. 25° Full depth system	3. Obsolete
D. 20° Full depth system	4. Standards for new applications.

Codes :

	A	B	C	D
(*a*)	4	3	2	1
(*b*)	3	1	2	4
(*c*)	3	2	1	4
(*d*)	4	2	3	1.

231. Which one of the following pairs is not correctly matched ?
(*a*) Positive drive .... Belt drive.
(*b*) High velocity ratio .... Worm gearing.
(*c*) To connect non-parallel and non-intersecting shafts .... Spiral gearing.
(*d*) Diminished noise and smooth operation .... Helical gears.

232. The gear train usually employed in clocks is a :
(*a*) reverted gear train
(*b*) simple gear train
(*c*) sun and planet gear
(*d*) differential gear.

233. A spur gear transmitting power is connected to the shaft with a key of rectangular section. The type(s) of stresses developed in the key is/are :
(*a*) shear stress alone.
(*b*) bearing stress alone.
(*c*) both shear and bearing stresses.
(*d*) shearing, bearing and bending stresses.

234. In the formulation of Lewis equation for toothed gearing, it is assumed that tangential tooth load F, acts on the :
(*a*) Pitch point
(*b*) Tip of the tooth
(*c*) Root of the tooth
(*d*) Whole face of the tooth.

235. Internal gears can be made by :
(*a*) hobbing
(*b*) shaping with pinion cutter
(*c*) shaping with rack cutter
(*d*) milling.

236. The curve traced by a point on the circumference of a circle which rolls along the inside of a fixed circle, is known as :
(*a*) epicycloid (*b*) hypocycloid
(*c*) cardiod (*d*) involute.

237. Match List-I with List-II and select the correct answer using the codes given below the lists :

List-I	List-II
A. Unwin's formula	1. Bearings
B. Wahl factor	2. Rivets
C. Reynold's equation	3. Gears
D. Lewis form factor	4. Springs.

Codes :

	A	B	C	D
(*a*)	1	4	2	3
(*b*)	2	4	1	4
(*c*)	1	3	2	4
(*d*)	2	4	1	3.

238. What is the value of pressure angle generally used for involute gears ?
(*a*) 35° (*b*) 30°
(*c*) 25° (*d*) 20°.

239. Match List-I (Properties) with List-II (Units) and select the correct answer using the codes given below the Lists :

List-I	List-II
A. Dynamic viscosity	1. Pa
B. Kinematic viscosity	2. m^2/s
C. Torsional stiffness	3. Ns/m^2
D. Modulus of rigidity	4. Nm
	5. N/m

Codes :

	A	B	C	D
(*a*)	3	2	4	1
(*b*)	5	2	4	3
(*c*)	3	4	2	3
(*d*)	5	4	2	1.

240. Which type of gears are used for shaft axes having an offset ?
(*a*) Mitre gears
(*b*) Spiral bevel gears
(*c*) Hypoid gears
(*d*) Zerol gears.

241. Which one of the following statements is correct ?
Certain minimum number of teeth on the involute pinion are necessary in order to :
(*a*) provide an economical design
(*b*) avoid interference
(*c*) reduce noise in operation
(*d*) overcome fatigue failure of the teeth.

242. Match List-I (Phenomenon) with List-II (Significant Parameters/Phenomenon) and select the correct answer using the codes given below the Lists :

List-I (Phenomenon)	List-II (Significant Parameters/ Phenomenon)
A. Interference fit	1. Viscosity index
B. Cyclic loading	2. Interference
C. Gear meshing	3. Notch sensitivity
D. Lubricating of bearings	4. Induced compressive stress.

Codes :

	A	B	C	D
(*a*)	3	4	1	2
(*b*)	4	3	2	1
(*c*)	3	4	2	1
(*d*)	4	3	1	2.

243. Consider the following statements pertaining to the basic Lewis equation for the strength design of spur gear teeth :
1. Single pair of teeth participates in power transmission at any instant.
2. The tooth is considered as a cantilever beam of uniform strength.
3. Loading on the teeth is static in nature.
4. Lewis equation takes into account the inaccuracies of the tooth profile.
5. Meshing teeth come in contact suddenly.

Which of the statements given above are correct ?
(*a*) 1, 3, 4 and 5 (*b*) 1, 2, 3, and 4
(*c*) 1, 2, and 3 (*d*) 2, 4 and 5.

244. Which type of gear will be used for non intersecting perpendicular shafts ?
(*a*) helical gears
(*b*) worm gears
(*c*) hypoid gears
(*d*) herringbone gears.

245. Best profile of gears to withstand resistance to wear is :
(*a*) 14.5° full depth involute tooth
(*b*) 20° involute stub tooth
(*c*) 20° full depth involute tooth
(*d*) 14.5° stub tooth.

246. Gears are best mass produced by :
(*a*) hobbing (*b*) milling
(*c*) shaping (*d*) casting.

247. The locus of a point on a thread unwound from a cylinder will be :
(*a*) a straight line (*b*) a circle
(*c*) an involute (*d*) a cycloid.

248. Spring driven watches and clocks utilize :
(*a*) involute gears (*b*) cycloid gears
(*c*) spur gears (*d*) epicyclic gears.

249. The velocity ratio between pinion and gear in a gear drive is 2.3, the module of teeth is 2.0 mm and sum of number of teeth on pinion and gear is 99. What is the centre distance between pinion and the gear ?
(*a*) 49.5 mm (*b*) 99 mm
(*c*) 148.5 mm (*d*) 198 mm.

250. Compared to gears with 20° pressure angle involute full depth teeth, those with 20° pressure angle and stub teeth have :
1. Smaller addendum
2. Smaller dedendum
3. Smaller tooth thickness
4. Greater bending strength.

Which of the statements given above are correct ?
(*a*) 1, 2 and 3 (*b*) 1, 2 and 4
(*c*) 1, 3 and 4 (*d*) 2, 3 and 4.

251. Common contact ratio of a pair of spur pinion and gear is :
(*a*) less than 1.0 (*b*) equal to 1
(*c*) between 2 and 3 (*d*) greater than 3.

252. The gears employed for connecting two non-intersecting and non-parallel, *i.e.* non-coplaner shafts are :
(*a*) Bevel gears (*b*) Spiral gears
(*c*) Helical gears (*d*) Mitre gears.

253. In heavy duty gears drives, proper heat treatment of gears is necessary in order to ;

(*a*) avoid interference
(*b*) prevent noisy operation
(*c*) minimize wear of gear teeth
(*d*) provide resistance against impact loading on gear teeth.

254. Consider the following statements :
1. A stub tooth has a working depth larger than that of a full-depth tooth.
2. The path of contact for involute gears is an arc of a circle.

Which of the statements given above is/ are correct ?
(*a*) Only 1 (*b*) Only 2
(*c*) Both 1 and 2 (*d*) Neither 1 nor 2.

255. Consider the following statements :
A 20° stub tooth system is generally in spur gears as it results in
1. Stronger teeth
2. Lesser number of teeth on the pinion.
3. Lesser chances of surface fatigue failure.
4. Reduction of interference.

Which of the above statements are correct ?
(*a*) 1, 2 and 4 (*b*) 3 and 4
(*c*) 1 and 3 (*b*) 1, 2, 3 and 4.

256. Consider the following specifications of gears, A, B, C and D :

Gears	A	B	C	D
Number of teeth	20	60	20	60
Pressure angle	14.5°	14.5°	20°	14.5°
Module	1	3	3	1
Material	Steel	Brass	Brass	Steel

Which of these gears form the pair of spur gears to achieve a gear ratio of 3 ?
(*a*) A and B (*b*) A and D
(*c*) B and C (*d*) C and D.

257. Minimum number of teeth for involute rack and pinion arrangement for pressure angle of 20° is :
(*a*) 18 (*b*) 20
(*c*) 30 (*d*) 34.

258. In gears, interference takes place when
(*a*) the tip of a tooth of mating gear digs into the portion between base and root circles
(*b*) gear do not move smoothly in the absence of lubrication.
(*c*) pitch of the gear is not same.
(*d*) gear teeth are undercut.

259. Match List–I with List-II and select the correct answer using the codes given below the Lists :

List–I (Types of Gears)	List–II (Characteristics)
A. Helical gearing	1. Zero axial thrust
B. Herringbone gearing	2. Non-interchangeable
C. Worm gearing	3. Skew shafts
D. Hypoid gearing	4. Parallel shafts.

Codes :	A	B	C	D
(*a*)	4	1	3	2
(*b*)	3	2	4	1
(*c*)	4	2	3	1
(*d*)	3	1	4	2

260. In involute gears, the pressure angle is :
(*a*) dependent on the size of the teeth
(*b*) dependent on the size of the gears
(*c*) always constant
(*d*) always variable.

261. Consider the following expressions for a spur gear.

1. Addendum $= \dfrac{1}{P}$

2. Clearance $= \dfrac{P}{20}$

3. Centre distance for internal gears $= \dfrac{z_1 + z_2}{2P}$

4. Dedendum $= \dfrac{1.157\, p}{\pi}$

Of these expressions
(*a*) 1, 2, 3 and 4 are correct
(*b*) 1 and 2 are correct
(*c*) 1, 2 and 3 are correct
(*d*) 1, 2 and 4 are correct.

262. Which one of the following is true for involute gears ?
(*a*) Interference is inherently absent
(*b*) Variation in centre distance of shafts increase radial force
(*c*) A convex flank is always in contact with concave flank
(*d*) Pressure angle is constant throughout the teeth engagement.

263. Consider the following statements :
The form factor of a spur gear teeth depends upon the
1. number of teeth
2. pressure angle
3. addendum modification
4. circular pitch.
Of these statements :
(*a*) 1 and 3 are correct
(*b*) 2 and 4 are correct
(*c*) 1, 2 and 3 are correct
(*d*) 1 and 4 are correct.

264. Consider the following statements :
1. Small interference
2. Strong tooth
3. Low production cost
4. Gear with small number of teeth.
Those characteristics which are applicable to stub 20° involute system would include
(*a*) 1 alone (*b*) 2, 3 and 4
(*c*) 1, 2 and 3 (*d*) 1, 2, 3 and 4.

265. When two shafts are neither parallel nor intersecting, power can be transmitted by using :
(*a*) a pair of spur gears
(*b*) a pair of helical gears
(*c*) an Oldham's coupling
(*d*) a pair of spiral gears.

266. In a single reduction, a large velocity ratio is required. The best transmission is :
(*a*) Spur gear drive
(*b*) Helical gear drive
(*c*) Bevel gear drive
(*d*) Worm gear drive.

267. A certain minimum number of teeth is to be kept for a gear wheel :
(*a*) so that the gear is of a good size
(*b*) for better durability
(*c*) to avoid interference and undercutting
(*d*) for better strength.

268. The limiting wear load of spur gear is proportional to (where E_p = Young's modulus of pinion material; E_g = Young's modulus of gear material)
(*a*) $(E_p + E_g)^{-1}$ (*b*) $\frac{E_p + E_g}{E_p E_g}$
(*c*) $1 + \frac{E_p}{E_g}$ (*d*) $1 + \frac{E_g}{E_p}$.

269. Cycloidal tooth profile of a cycloidal gear tooth is a combination of :
(*a*) hypocycloid and involute of a circle
(*b*) hypocycloid and epicycloids
(*c*) epicycloids and involute of a circle
(*d*) straight line and epicycloids.

270. Match List–I with List–II and select the correct answer using the codes given below the Lists :

List–I (Type of Gears)	List–II (Characteristics)
A. Helical gearing	1. Non-interchangeable
B. Herringbone gearing	2. Zero axial thrust
C. Worm gearing	3. Quiet motion
D. Hypoid gearing reduction	4. Extreme speed

Codes :	A	B	C	D
(*a*)	1	2	3	4
(*b*)	3	2	1	4
(*c*)	3	1	4	2
(*d*)	3	2	4	1

271. Match List–I with List–II and select the correct answer using the codes given below the Lists :

List–I (Gears)	List–II (Configurations)
A. Spur	1. Connecting two non-parallel or intersecting but coplanar shafts
B. Bevel	2. Connecting two parallel and coplanar shafts and teeth inclined to the axis of the gear wheel
C. Helical	3. Connecting two parallel and coplanar shafts using teeth inclined to the axis of the gear wheel

List–I (Gears)	List–II (Configurations)
D. Mitre	4. Connecting two shafts whose axes are mutually perpendicular to each other.

Codes :	A	B	C	D
(*a*)	2	4	3	1
(*b*)	3	1	2	4
(*c*)	2	1	3	4
(*d*)	3	4	2	1

272. Match List–I (Type of gear failure) with List–II (Reasons) and select the correct answer using the codes given below the Lists :

List–I (Type of gear failure)	List–II (Reasons)
A. Scoring	1. Oil film breakage
B. Pitting	2. Yielding of surface under heavy loads
C. Scuffing	3. Cyclic loads causing high surface stress
D. Plastic flow	4. Insufficient lubrication

Codes :	A	B	C	D
(*a*)	2	1	3	4
(*b*)	2	3	1	4
(*c*)	4	3	1	2
(*d*)	4	1	3	2

273. Match List–I with List–II and select the correct answer using the codes given below the Lists :

List–I (Type of gears)	List–II (Arrangement of shafts)
A. Bevel gears	1. Non-parallel off-set shafts
B. Worm gears	2. Non-parallel intersecting shafts
C. Herringbone shafts	3. Non-parallel and non-intersecting shafts
D. Hypoid gears	4. Parallel shafts

Codes :	A	B	C	D
(*a*).	4	2	1	3
(*b*)	2	3	4	1
(*c*)	3	2	1	4
(*d*)	1	3	4	2

274. The surface of the gear tooth below the pitch surface is called :

(*a*) addendum (*b*) dedendum
(*c*) flank (*d*) face.

275. The path of contact in involute gears is ;

(*a*) a straight line (*b*) parabola
(*c*) curved path (*d*) circle.

276. For two meshing gears, their

(*a*) number of teeth must be same
(*b*) addendum must be same
(*c*) dedendum must be same
(*d*) module must be same.

277. Match List I with List II and select the correct answer using the code given below the lists ;

List I	List II
A. Worm gear	1. Imposes no thrust load on the shaft
B. Spur gear	2. No transmit power between two non-intersecting shafts which are perpendicular to each other
C. Herringbone gear	3. To transmit power when the shafts are at right angles to one another.

Code :

	A	B	C	D
(*a*)	1	2	3	4
(*b*)	2	3	1	4
(*c*)	1	2	4	3
(*d*)	2	3	4	1.

20 WORM GEARS

Chapter

278. The lead angle of a worm is 32.5′. Its helix will be :
(*a*) 22.5° (*b*) 45°
(*c*) 67.5° (*d*) 90°.

279. Consider the following statements regarding improvement of efficiency of worm gear drive :
1. Efficiency can be improved by increasing the spiral angle of worm thread to 45° or more.
2. Efficiency can be improved by adopting proper lubrication.
3. Efficiency can be improved by adopting worm diameter as well as practicable to reduce sliding between worm-threads and wheel teeth.
4. Efficiency can be improved by adopting convex tooth profile both for worm and wheel.

Which of the statements given above are correct ?
(*a*) 1, 2 and 3 (*b*) 1, 2 and 4
(*c*) 2, 3 and 4 (*d*) 1, 3 and 4.

21 SLIDING BEARINGS

Chapter

280. In a flat collar pivot bearing, the moment due to friction is to (r_1 and r_2 are the outer and inner radii respectively)
(*a*) $\frac{r_1^2 - r_2^2}{r_1 - r_2}$ (*b*) $\frac{r_1^2 - r_2^2}{r_1 + r_2}$
(*c*) $\frac{r_1^3 - r_2^3}{r_1^2 - r_2^2}$ (*d*) $\frac{r_1^3 - r_2^3}{r_1 - r_2}$.

281. A friction circle is drawn when a journal rotates in bearing. Its radius depends on the coefficient of friction and the
(*a*) magnitudes of the forces on the journal
(*b*) angular velocity of the journal
(*c*) clearance between the journal and the bearing
(*d*) radius of the journal.

282. In a journal bearings, the radius of the friction circle increases with the increase in :
(*a*) load
(*b*) radius of the journal
(*c*) speed of the journal
(*d*) viscosity of the lubricant.

283. Match List–I and List–II and select the correct answer using the codes given below the Lists :

List–I (Bearing materials)	List–II (Properties)
A. Babbits	1. Porous
B. Bronze	2. Good Embeddability
C. C.I.	3. Suitable for high loads and low speeds
D. Sintered powdered metal	4. Runs well with C.I. journals

Codes :

	A	B	C	D
(*a*)	2	3	4	1
(*b*)	3	2	1	4
(*c*)	2	3	1	4
(*d*)	3	2	4	1

284. Which one of the following sets of parameters should be monitored for determining safe operation of journal bearing ?

(*a*) Oil pressure, bearing metal temperature and bearing vibration
(*b*) Bearing vibration, oil pressure and speed of shaft
(*c*) Bearing metal temperature and oil pressure
(*d*) Oil pressure and bearing vibration.

285. A full journal bearing having clearance to radius of ratio $\frac{1}{100}$, using a lubricant with $\mu = 28 \times 10^{-1}$ Pas support the shaft journal running at N = 2400 r.p.m. If bearing pressure is 1.4 MPa, the Sommerfeld number is :

(*a*) 8×10^{-3} (*b*) 8×10^{-5}
(*c*) 0.48 (*d*) 0.48×10^{-2}.

286. A sliding contact bearing is operating unstable condition. The pressure developed in film is *a* when the journal rotates at N r.p.m. The dynamic viscosity of lubricant is μ of effective coefficient of friction between bear and journal of diameter D is *f*. Which one the following statements is correct for bearing ?

(*a*) *f* is directly proportional to μ and *p*
(*b*) *f* is directly proportional to μ and N
(*c*) *f* is inversely proportional to p and D
(*d*) *f* is directly proportional to μ and inversely proportional to N.

287. In a collar thrust bearing, the number of collars have been doubled while maintaining co-efficient of friction and axial thrust same. It will result in

(*a*) Same friction torque and same bearing pressure
(*b*) Double friction torque and half bearing pressure
(*c*) Double friction torque and same bearing pressure
(*d*) Same friction torque and half bearing pressure.

288. The radius of the friction circle in a journal bearing is dependent on co-efficient of friction and the :

(*a*) Angular velocity of the journal
(*b*) Radius of the journal
(*c*) Magnitudes of the forces on the journal
(*d*) Journal and bearing clearance.

289. Consider the following statements :
Radius of friction circle for a journal bearing depends upon

1. coefficient of friction
2. radius of the journal
3. angular speed of rotation of the shaft

Which of the statements given above are correct ?

(*a*) 1, 2 and 3
(*b*) Only 1 and 2
(*c*) Only 2 and 3
(*d*) Only 1 and 3.

290. It is seen from the curve that there is a minimum value of the coefficient of friction (*m*) for a particular value of the Bearing Characteristic Number denoted by *a*. What is this value of the Bearing Characteristic Number called ?

(*a*) McKee Number
(*b*) Reynolds Number
(*c*) Bearing Modulus
(*d*) Sommerfeld Number.

291. In hydrostatic bearings :

(*a*) the oil film pressure is generated only by the rotation of the journal.
(*b*) the oil film is maintained by supplying oil under pressure
(*c*) external supply of lubricant is not required.
(*d*) grease is used for lubrication.

292. Antifriction bearings are :

(*a*) journal bearing
(*b*) sleeve bearing
(*c*) foot step bearing
(*d*) ball and roller bearing.

293. Bearing modulus is defined at the value of coefficient of friction to be :

(*a*) maximum (*b*) minimum
(*c*) zero (*d*) any.

294. Bearing modulus is defined by :

(a) $\frac{\mu n}{p}$ (b) $\frac{\mu}{n p}$

(c) $\frac{\mu p}{n}$ (d) $\frac{p}{\mu n}$.

where μ = dynamic viscosity, n = rev./s, p = pressure.

295. The clearance ratio for hydrodynamic bearing is :

(a) 0.1 (b) 0.01
(c) 0.001 (d) 0.0001.

296. As compared to spur gears, helical gears :

(a) run more smoothly
(b) run with more vibrations and noise
(c) run exactly alike
(d) consume more power.

297. Which type of gear will be used to give minimum axial thrust ?

(a) helical gears
(b) herringbone gears
(c) hypoid gears
(d) worm gears.

298. Which of the following journal bearing material is self lubricating ?

(a) cast iron (b) babbit
(c) graphite (d) sintered.

299. Consider the following statements :

For a journal rotating in a bearing under film lubrication conditions, the frictional resistance is :

1. Proportional to the area of contact.
2. Proportional to the viscosity of lubricant.
3. Proportional to the speed of rotation.
4. Independent of the pressure.

Which of the statements given above are correct ?

(a) 1, 2, 3 and 4 (b) 1 and 4 only
(c) 2, 3 and 4 only (d) 2 and 3 only.

300. Increase in values of which of the following results in an increase of the co-efficient of friction in a hydrodynamic bearing ?

(a) Viscosity of the oil
(b) Clearance between shaft and bearing
(c) Shaft speed.
(d) Shaft diameter.

301. Babbit lining is used on brass/bronze bearings to :

(a) Increase bearing resistance
(b) Increase compressive strength
(c) Provide anti-friction properties
(d) Increase wear resistance.

302. Satisfactory hydrodynamic film in a journal bearing is formed when :

(a) Journal speed is low, unit pressure on the bearing is high and viscosity of lubricant used is low.
(b) Journal speed is low, unit pressure on the bearing is low and viscosity of lubricant used is high.
(c) Journal speed is high, unit pressure on the bearing is high and viscosity of lubricant used is high.
(d) Appropriate combination of journal speed, unit pressure on bearing and lubricant viscosity exists resulting in low coefficient of friction.

303. In an oil-lubricated bearing, coefficient of friction between the journal and the bearing.

(a) Remains constant at all speeds
(b) Is minimum at zero speed and increases monotonically with increase in speed
(c) Is maximum at zero speed and decrease monotonically with increase in speed
(d) Becomes minimum at an optimum speed and then increases with further increase in speed.

304. What is the main advantage of hydrodynamic bearing over roller bearing ?

(a) Easy to assemble
(b) Relatively low price
(c) Superior load carrying capacity at higher speeds
(d) less frictional resistance.

305. In sliding contact bearings, a positive pressure can be built up and a load supported by a fluid only by the use of :

(a) Diverging film
(b) Converging-diverging film
(c) Converging film
(d) Flat film.

306. Which one of the curves shown below represents the characteristic of a hydrodynamically Lubricated journal bearing ?
(*a*) 1
(*b*) 2
(*c*) 3
(*d*) 4.

307. Consider the following statements regarding boundary lubrication conditions in hydrodynamic journal bearings ?
(*a*) It is an undesirable condition
(*b*) It always occurs during starting and stopping of the journal.
(*c*) It necessitates frequent repair of the journal.
(*d*) This condition can be completely eliminated by using high viscosity lubricant.

308. When the lubricating oil is forced between the journal and bearing by external pressure, the bearing is known as :
(*a*) Zero film bearing
(*b*) Boundary lubricated bearing
(*c*) Hydrostatic bearing
(*d*) Hydrodynamic bearing.

309. The performance of a hydrodynamic journal bearing is a function of :
(*a*) Reynold's number
(*b*) Sommerfeld number
(*c*) Speed
(*d*) Bearing modulus.

310. The most suitable bearing for carrying very heavy loads with slow speed is :
(*a*) Hydrodynamic bearing
(*b*) Ball bearing
(*c*) Roller bearing
(*d*) Hydrostatic bearing.

311. Starting friction is low in :
(*a*) Hydrostatic bearing
(*b*) Hydrodynamic bearing
(*c*) Mixed (or semi-fluid) lubrication
(*d*) Boundary lubrication.

312. In thick film hydrodynamic journal bearings, the coefficient of friction :
(*a*) Increases with increase of load
(*b*) Is independent of load
(*c*) Decreases with increase of load
(*d*) May increase or decrease with increase in load.

313. The SI unit of kinematic viscosity is :
(*a*) m^2/s
(*b*) kg (m/s)
(*c*) m/s^2
(*d*) m^3/s^2.

314. To restore stable operating condition in a hydrodynamic journal bearing, when it encounters higher magnitude loads,
(*a*) Oil viscosity is to be decreased
(*b*) Oil viscosity is to be increased
(*c*) Oil viscosity index is to be increased
(*d*) Oil viscosity index is to be decreased.

315. In a hydrodynamic journal bearing there is :
(*a*) A very thin film of lubricant between the journal and the bearing such that there is contact between the journal and the bearing.
(*b*) A thick film of lubricant between the journal and the bearing.
(*c*) No lubricant between the journal and the bearing.
(*d*) A forced lubricant between the journal and the bearing.

316. The bearing characteristic number in a hydrodynamic bearing does not depend on :
(*a*) Coefficient of friction
(*b*) Dynamic viscosity of lubricant
(*c*) Speed of journal
(*d*) Pressure of lubricant film.

317. The bearing characteristic number in a hydrodynamic bearing depends on :
(*a*) Length, width and load
(*b*) Length, width and speed
(*c*) Viscosity, speed and load
(*d*) Viscosity, speed and bearing pressure.

318. Match List–I with List–II and select the correct answer using the code given below the lists :

List–I (Requirements)	List–II (Types)
A. High temperature service	1. Teflon bearing
B. High load	2. Carbon bearing
C. No lubrication	3. Hydrodynamic bearing
D. Bushings	4. Sleeve bearing

Codes :

	A	B	C	D
(*a*)	1	2	3	4
(*b*)	4	1	2	3
(*c*)	2	1	3	4
(*d*)	2	3	1	4

319. Match List–I with List–II and select the correct answer using the code given below the Lists :

List–I (Terminology)	List–II (Relevant terms)
A. Interference	1. Arc of approach, arc of recess, circular pitch
B. Dynamic load on tooth	2. Lewis equation
C. Static load	3. Minimum number of teeth on pinion
D. Contact ratio	4. Inaccuracies in tooth profile

Codes :

	A	B	C	D
(*a*)	3	4	1	2
(*b*)	1	2	3	4
(*c*)	4	3	2	1
(*d*)	3	4	2	1

22 ROLLING CONTACT BEARINGS

Chapter

320. Removal of metal particles from the raceway of a rolling contact bearing is a kind of failure of bearing known as :
(*a*) pitting (*b*) wearing
(*c*) spalling (*d*) scuffing.

321. Which one of the following statements is NOT true of rolling contact bearing ?
(*a*) The bearing characteristic number is given by ZN/*p* where Z is the absolute viscosity of the lubricant, N is the shaft speed and *p* is the bearing pressure.
(*b*) Inner race of a radial ball bearing has an interference fit with the shaft and rotates along with it.
(*c*) Outer race of bearing has an interference fit with bearing housing and does not rotate
(*d*) In some cases, the inner race is stationary and outer race rotates.

322. In a single row deep groove ball-bearing, cages are needed to :
(*a*) Separate the two races
(*b*) Separate the balls from the inner race
(*c*) Separate the outer race from the balls.
(*d*) Ensure that the balls do not cluster at one point and maintain proper relative angular positions.

323. Jewel bearing are used in :
(*a*) contaminated atmosphere
(*b*) low-torque applications
(*c*) fully immersed in water condition
(*d*) high seed shafts.

324. A ball-bearing is characterized by basic static capacity = 11000 N and dynamic capacity = 18000 N. This bearing is subjected to equivalent static load = 5500 N. The bearing loading ratio and life in million revolutions respectively are :
(*a*) 3.27 and 52.0 (*b*) 3.27 and 35.0
(*c*) 2.00 and 10.1 (*d*) 1.60 and 4.1.

325. The rated life of a ball bearing varies inversely as which one of the following ?
(*a*) load (*b*) $(\text{load})^2$
(*c*) $(\text{load})^3$ (*d*) $(\text{load})^{3.33}$.

326. A suitable bearing for combined radial and axial loads is :
(*a*) single row angular contact ball bearing
(*b*) spherical roller bearing
(*c*) cylindrical roller bearing
(*d*) taper roller bearing.

327. Average values of effective co-efficients of friction for bearings are described below :
1. Spherical ball bearing–f_1,
2. Cylindrical roller bearing–f_2,
3. Taper roller bearing–f_3,
4. Stable (thick film) sliding contact bearing–f_4.

Which one of the following sequences is correct ?
(*a*) $f_1 < f_2 < f_3 < f_4$
(*b*) $f_1 < f_2 < f_4 < f_3$
(*c*) $f_2 < f_1 < f_3 < f_4$
(*d*) $f_2 < f_4 < f_2 < f_3$.

328. Match List–I (*Applications*) with List–II (*Choice of Bearings*) and select the correct answer using the codes given below the lists :

List–I (Applications)	List–II (Choice of Bearings)
A. Granite table of a coordinate measuring machine	1. Hydrodynamic bearing
B. Headstock spindle of a lathe	2. Deep groove ball bearing
C. Crank shaft of a diesel engine	3. Hydrostatic bearing
D. Armature of 0.5 kW induction motor	4. Taper roller bearing

Codes :

	A	B	C	D
(*a*)	1	4	3	2
(*b*)	3	2	1	4
(*c*)	1	2	3	4
(*d*)	3	4	1	2

329. Which one of the following statements is correct ?
Antifriction bearings are
(*a*) sleeve bearings
(*b*) gas lubricated bearings
(*c*) ball and roller bearings
(*d*) journal bearings.

330. To carry a large axial in a flat collar bearing, a number of collars is provided to :
(*a*) reduce frictional torque
(*b*) increase frictional torque
(*c*) decrease intensity of pressure
(*d*) increase intensity of pressure.

331. Deep Groove ball bearing are used for :
(*a*) Heavy thrust load only
(*b*) Small angular displacements of shafts
(*c*) Radial load at high speed
(*d*) Combined thrust and radial loads at high speed.

332. The life of a ball bearing at a load of 10 kN is 8000 hours. Its life in hours if the load is increased to 20 kN, keeping other conditions same, as
(*a*) 4000 (*b*) 2000
(*c*) 1000 (*d*) 500.

333. The basic load rating of a ball bearing is :
(*a*) The maximum static radial load that can be applied without causing any plastic deformation of bearing components.
(*b*) The radial load at which 90% of the group of apparently identical bearings run for one million revolutions before the first evidence of failure.
(*c*) The maximum radial load that can be applied during operation without any plastic deformation of bearing components.
(*d*) A combination of radial and axial loads that can be applied without any plastic deformation.

334. The life of a ball bearing is inversely proportional to :
(*a*) $(\text{Load})^{1/3}$ (*b*) $(\text{Load})^3$
(*c*) $(\text{Load})^{10/3}$ (*d*) $(\text{Load})^2$.

335. Spherical roller bearings are normally used :
(*a*) For increased radial load
(*b*) For increased thrust load
(*c*) When there is less radial space
(*d*) To compensate for angular misalignment.

336. Which one of the following correctly relates the life of a roller bearing in millions of revolutions (L), the load P and the basic dynamic load rating for one million revolutions (C) ?

(*a*) $L = \left(\frac{C}{P}\right)^3$ (*b*) $L = \left(\frac{C}{P}\right)^{10/3}$

(*c*) $L = \left(\frac{C}{P}\right)^{1/3}$ (*d*) $L = \left(\frac{C}{P}\right)^{1/3}$

337. If $k_1 = 3$ for ball bearings and $k_2 = \frac{10}{3}$ for roller bearings, which one of the following correctly states the load (P)–Life (L) relationship for roller contact bearings ?

(*a*) $\frac{L_1}{L_2} = \left(\frac{P_1}{P_2}\right)_1^k$

(*b*) $\frac{L_2}{L_1} = \left(\frac{P_1}{P_2}\right)_1^{1(k-1)}$

(*c*) $\frac{L_2}{L_1} = \left(\frac{P_1}{P_2}\right)_2^k$

(*d*) $\frac{L_2}{L_1} = \left(\frac{P_1}{P_2}\right)_2^{(k-1)}$

338. Which bearing is preferred for oscillating conditions ?
(*a*) Double row roller bearing
(*b*) Angular contact single row ball bearing
(*c*) Taper roller bearing
(*d*) Needle roller bearing.

339. If the load on a ball bearing is halved, its life :
(*a*) Remains unchanged
(*b*) Increases two times
(*c*) Increases four times
(*d*) Increases eight times.

340. The rating life of a group of apparently identical ball bearings is defined as the number of revolutions exceeded before the first evidence of fatigue crack by :
(*a*) 100% of the bearings of the group
(*b*) 95% of the bearings of the group
(*c*) 90% of the bearings of the group
(*d*) 66.66% of the bearings of the group.

341. Ball bearings are provided with a cage :
(*a*) To reduce friction
(*b*) To maintain the balls at a fixed distance apart
(*c*) To prevent the lubricant from flowing out
(*d*) To facilitate slipping of balls.

342. Taper roller bearings can take :
(*a*) Radial load only
(*b*) Axial load only
(*c*) Both radial and axial loads and the ratio of these being less than unity
(*d*) Both radial and axial loads and the ratio of these being greater than unity.

343. Which of the following types of bearings is employed in shafts of gearboxes of automobiles ?
(*a*) Hydrodynamic journal bearings
(*b*) Multi-lobed journal bearings
(*c*) Anti-friction bearings
(*d*) Hybrid journal bearings.

344. The rolling element bearings are :
(*a*) Hydrostatic bearings
(*b*) Squeeze film bearings
(*c*) Anti-friction bearings
(*d*) Grease lubrication bearings.

345. Consider the following statements about anti-friction bearings :
1. They have low starting and low running friction at moderate speeds
2. They have high resistance to shock loading.
3. They can carry both radial and thrust loads.
4. Their initial cost is high.
5. They can accommodate some amount of misalignment of a shaft.

Of these statements.

(*a*) 1, 2, 3 and 4 are correct
(*b*) 1, 3 and 4 are correct
(*c*) 1, 4 and 5 are correct
(*d*) 1, 2, 3 and 5 are correct.

346. Match List–I (Type of Bearings) with List–II (Type of Load) and select the correct answer using the code given below the Lists :

List–I (Type of Bearings)	List–II (Type of Load)
A. Deep Groove bearing	1. Radial load
B. Tapered roller bearing	2. Radial and axial load
C. Self alignment bearing	3. Mainly radial load with shaft misalignment
D. Thrust bearing	4. Mainly axial load

Code :

	A	B	C	D
(*a*)	1	2	3	4
(*b*)	3	4	1	2
(*c*)	1	4	3	2
(*d*)	3	2	1	4

347. Consider the following pairs of types of bearings and applications :

1. Partial journal bearing .... Rail wagon axles
2. Full journal bearing............ Diesel engine crankshaft
3. Radial bearing .... Combined radial and axial loads

Which of these pairs is / are correctly matched

(*a*) 1 alone (*b*) 1 and 2
(*c*) 2 and 3 (*d*) 1, 2 and 3

348. Match List-I (Bearings) with List-II (applications) and select the correct answer using the codes given below the Lists :

List-I	List-II
A. Journal bearing	1. Electric motors
B. Thrust bearing	2. Watches
C. Conical pivot bearing	3. Marine engines
D. Ball bearing	4. Swivelling chairs

Code :

	A	*B*	*C*	*D*
(*a*)	3	4	1	2
(*b*)	4	2	1	3
(*c*)	3	4	2	1
(*d*)	4	2	3	1

349. Match List-I with List-II and select the correct answer using the code given below the Lists :

List-I	List-II
A. Automobile wheel mounting on axle	1. Magneto bearing
B. High speed grinding wheel	2. Angular contact bearing
C. I.C. Engine connecting rod	3. Taper roller bearing
D. Leaf spring eye mounting	4. Hydrodynamic journal bearing
	5. Sintered metal bearing
	6. Teflon/Nylon bush

Codes :

	A	B	C	D
(*a*)	1	2	4	5
(*b*)	3	2	4	6
(*c*)	2	3	6	4
(*d*)	1	4	2	6

350. Match list-I (Bearings) with List-II (applications) and select the correct answer using the codes given below the Lists :

List-I	List-II
A. Cylindrical roller	1. Radial load
B. Ball bearing	2. Machine needs frequent dismantling and assembling
C. Taper roller bearing with lesser thrust	3. Radial loads
D. Angular contact ball bearing	4. Shock loads
	5. Axial expansion of shaft due to rise in temperature.

Codes :

	A	B	C	D
(a)	4	3	1	5
(b)	1	3	2	5
(c)	4	1	2	3
(d)	5	4	1	3.

351. Match List-I (Bearings) with List-II (Type of load) and select the correct answer using the codes given below the Lists :

List-I	List-II
A. Hydrodynamic journal bearing	1. High radial and thrust load combined
B. Rectangular hydro-static bearing	2. Radial load only
C. Taper roller bearing	3. Thrust load only
D. Angular contact ball bearing	4. Medium to low radial and thrust combined.

Codes :

	A	B	C	D
(a)	2	3	1	4
(b)	4	1	3	2
(c)	2	1	3	4
(d)	4	3	1	2.

352. Match List-I (Bearings) with List-II (Purpose) and select the correct answer using the codes given below the Lists :

List-I	List-II
A. Ball bearings	1. Heavy loads with oscillatory motion
B. Tapered roller bearings	2. Light loads
C. Spherical roller bearings	3. Carrying both radial and thrust loads
D. Needle roller bearings	4. Self-aligning property

Codes :

	A	B	C	D
(a)	4	1	3	2
(b)	2	1	4	3
(c)	2	3	1	4
(d)	2	3	4	1.

353. Match List-I with List-II and select the correct answer using the codes given below the lists :

List-I (Type of Anti-friction bearing)	List-II (Specific use)
A. Self-aligning ball bearing	1. For pure axial load
B. Taper roller bearing	2. For hinged condition
C. Deep groove ball bearing	3. For pure radial load
D. Thrust ball bearing	4. For axial and radial load.

Codes :

	A	B	C	D
(a)	2	1	3	4
(b)	3	4	2	1
(c)	2	4	3	1
(d)	3	1	2	4.

Chapter 23 FLYWHEEL

354. In which one of the following is a flywheel generally employed ?

(a) Lathe

(b) Electric motor

(c) Punching machine

(d) Gearbox.

355. The safe rim velocity of a flywheel is influenced by the :

(*a*) Centrifugal stresses
(*b*) Fluctuation of energy
(*c*) Fluctuation of speed
(*d*) Mass of the flywheel.

356. What is the value of the radius of gyration of disc type flywheel as compared to rim type flywheel for the same diameter ?
(*a*) $\sqrt{2}$ times (*b*) $1/\sqrt{2}$ times
(*c*) 2 times (*d*) $\frac{1}{2}$ times.

357. Which one of the following statements is correct ?
If the size of a flywheel in a punching machine is increased
(*a*) Then the fluctuation of speed and fluctuation of energy will both decrease.
(*b*) Then the fluctuation of speed will decrease and the fluctuation of energy will increase.
(*c*) Then the fluctuation of speed will increase and the fluctuation of energy will decrease.
(*d*) Then the fluctuation of speed and fluctuation of energy both will increase.

358. If speed of a flywheel is halved, then for the same performance, its weight has to be increased
(*a*) Twice (*b*) Four times
(*c*) Eight times (*d*) $\sqrt{2}$ times.

359. The weight of a flywheel is proportional to :
(*a*) speed (N) (*b*) N^2
(*c*) $N^{\frac{1}{2}}$ (*d*) $\frac{1}{N^2}$.

Chapter 24 CLUTCHES AND BRAKES

360. Match List-I with List-II and select the correct answer using the codes given below the lists :

List-I	List-II
A. Single-plate friction clutch	1. Scooters
B. Multi-plate friction clutch	2. Rolling mills
C. Centrifugal clutch	3. Trucks
D. Jaw clutch	4. Mopeds.

Codes :

	A	B	C	D
(*a*)	1	3	4	2
(*b*)	1	3	2	4
(*c*)	3	1	2	4
(*d*)	3	1	4	2.

361. In the multiple disc clutch, if there are 6 discs on the driving shaft and 5 discs on the driven shaft, then the number of pairs of contact surfaces will be equal to :
(*a*) 11 (*b*) 12
(*c*) 10 (*d*) 22.

362. Consider the following statements regarding a centrifugal clutch :
1. It need not be unloaded before engagement.
2. It enables the prime mover to start up under no-load conditions.
3. It picks up the load gradually with the increase in speed.
4. It will not slip to the point of destruction.
5. It is very useful when the power unit has a low starting torque.

Which of these are the advantages of a centrifugal clutch ?
(*a*) 1, 2 and 4 (*b*) 1, 3 and 5
(*c*) 2, 3 and 5 (*d*) 1, 3, 4 and 5.

363. In a multi-plate clutch with a number of outer discs and n_1 number of inner discs, the number of pairs of active surfaces is :
(*a*) $n_i + n_o$ (*b*) $n_i + n_o + 1$
(*c*) $n_i + n_o - 1$ (*d*) $n_i + n_o - 2$.

364. Which one of the following is not a friction clutch ?
(*a*) Disc or plate clutch
(*b*) Cone clutch
(*c*) Centrifugal clutch
(*d*) Jaw clutch.

365. On the motors with low starting torque, the type of the clutch to be used is :
(*a*) Multiple-plate clutch
(*b*) Cone clutch
(*c*) Centrifugal clutch
(*d*) Single-plate clutch with both sides effective.

366. In a multiple disc clutch if n_1 and n_2 are the number of discs on the driving and driven shafts, respectively, the number of pairs of contact surfaces will be :
(*a*) $n_1 + n_2$ (*b*) $n_1 + n_2 - 1$
(*c*) $n_1 + n_2 + 1$ (*d*) $\frac{n_1 + n_2}{2}$.

367. Which one of the following is the correct expression for the torque transmitted by a conical clutch of outer radius R, inner radius *r* and semi-cone angle α, assuming uniform pressure ? (Where W = total axial load and μ = coefficient of friction)
(*a*) $\frac{\mu W(R+r)}{2\sin\alpha}$
(*b*) $\frac{\mu W(R+r)}{3\sin\alpha}$
(*c*) $\frac{2\mu W(R^3 - r^3)}{3\sin\alpha(R^2 - r^2)}$
(*d*) $\frac{3\mu W(R^3 - r^3)}{4\sin\alpha(R^2 - r^2)}$.

368. In a band brake, the ratio of tight side band tension to the tension on the slack side is 3. If the angle of overlap of band on the drum is 180°, the coefficient of friction required between drum and the band is :
(*a*) 0.20 (*b*) 0.25
(*c*) 0.30 (*d*) 0.35.

369. The brake commonly used on train boggies is :
(*a*) Internal expanding
(*b*) Band brake
(*c*) Band and block brake
(*d*) Shoe brake.

370. Which of the following clutches is positive type ?
(*a*) Cone (*b*) Disc
(*c*) Jaw (*d*) Centrifugal.

371. The type of clutch used in mopeds is :
(*a*) Disc (*b*) Jaw
(*c*) Cone (*d*) Centrifugal.

Chapter 25

INTERNAL COMBUSTION ENGINE PARTS

372. Consider the following design considerations :
1. Tensile failure 2. Creep failure
3. Bearing failure 4. Shearing failure
5. Bending failure.
The design of the pin of a rocker arm of an I.C. engine is based on :
(*a*) 1, 2 and 4 (*b*) 1, 3 and 4
(*c*) 2, 3 and 5 (*d*) 3, 4 and 5.

373. Consider the following statements :
Cam followers are generally classified according to :
1. The nature of its motion
2. The nature of its surface in contact with the cam
3. The speed of the cam.
Which of the statements given above are correct ?
(*a*) 1, 2 and 3 (*b*) Only 1 and 2
(*c*) Only 2 and 3 (*d*) Only 1 and 3.

ANSWERS

1 (a) 2 (a) 3 (d) 4 (c) 5 (a) 6 (a) 7 (b) 8 (b) 9 (c) 10 (d)

11 (b) 12 (a) 13 (a) 14 (b) 15 (a) 16 (c) 17 (a) 18 (d) 19 (b) 20 (c)

21 (c) 22 (c) 23 (b) 24 (d) 25 (c) 26 (d) 27 (a) 28 (d) 29 (b) 30 (b)

31 (b) 32 (b) 33 (d) 34 (d) 35 (b) 36 (c) 37 (d) 38 (c) 39 (c) 40 (a)

41 (c) 42 (c) 43 (b) 44 (c) 45 (b) 46 (c) 47 (a) 48 (b) 49 (b) 50 (a)

51 (b) 52 (a) 53 (a) 54 (d) 55 (d) 56 (c) 57 (c) 58 (c) 59 (a) 60 (b)

61 (a) 62 (d) 63 (c) 64 (b) 65 (b) 66 (b) 67 (c) 68 (d) 69 (a) 70 (b)

71 (a) 72 (d) 73 (a) 74 (b) 75 (c) 76 (d) 77 (c) 78 (b) 79 (a) 80 (c)

81 (a) 82 (c) 83 (b) 84 (b) 85 (d) 86 (d) 87 (d) 88 (d) 89 (a) 90 (b)

91 (c) 92 (c) 93 (b) 94 (b) 95 (a) 96 (d) 97 (d) 98 (a) 99 (d) 100 (b)

101 (b) 102 (a) 103 (c) 104 (c) 105 (c) 106 (c) 107 (b) 108 (c) 109 (c) 110 (d)

111 (a) 112 (c) 113 (a) 114 (c) 115 (c) 116 (a) 117 (c) 118 (b) 119 (d) 120 (d)

121 (d) 122 (c) 123 (d) 124 (d) 125 (a) 126 (c) 127 (c) 128 (a) 129 (d) 130 (d)

131 (c) 132 (b) 133 (a) 134 (d) 135 (a) 136 (d) 137 (a) 138 (d) 139 (c) 140 (c)

141 (a) 142 (b) 143 (c) 144 (a) 145 (d) 146 (b) 147 (b) 148 (d) 149 (a) 150 (a)

151 (c) 152 (c) 153 (d) 154 (d) 155 (c) 156 (d) 157 (a) 158 (c) 159 (b) 160 (d)

161 (d) 162 (b) 163 (b) 164 (a) 165 (b) 166 (d) 167 (b) 168 (d) 169 (d) 170 (d)

171 (d) 172 (a) 173 (c) 174 (d) 175 (a) 176 (c) 177 (d) 178 (d) 179 (a) 180 (b)

181 (a) 182 (c) 183 (c) 184 (b) 185 (a) 186 (c) 187 (a) 188 (c) 189 (d) 190 (d)

191 (c) 192 (b) 193 (b) 194 (c) 195 (b) 196 (c) 197 (b) 198 (c) 199 (d) 200 (d)

201 (b) 202 (b) 203 (c) 204 (b) 205 (b) 206 (c) 207 (a) 208 (a) 209 (d) 210 (d)

211 (a) 212 (a) 213 (c) 214 (b) 215 (b) 216 (c) 217 (d) 218 (c) 219 (c) 220 (c)

221 (b) 222 (a) 223 (d) 224 (c) 225 (a) 226 (d) 227 (a) 228 (b) 229 (d) 230 (a)

231 (a) 232 (a) 233 (c) 234 (b) 235 (b) 236 (b) 237 (d) 238 (d) 239 (a) 240 (d)

241 (b) 242 (b) 243 (c) 244 (c) 245 (a) 246 (a) 247 (c) 248 (c) 249 (b) 250 (c)

251 (c) 252 (b) 253 (c) 254 (c) 255 (a) 256 (a) 257 (a) 258 (a) 259 (a) 260 (c)

261 (d) 262 (d) 263 (c) 264 (d) 265 (d) 266 (d) 267 (c) 268 (b) 269 (d) 270 (d)

271 (c) 272 (c) 273 (b) 274 (c) 275 (a) 276 (d) 277 (b) 278 (c) 279 (d) 280 (c)

281 (d) 282 (b) 283 (a) 284 (a) 285 (a) 286 (b) 287 (b) 288 (b) 289 (b) 290 (c)

291 (b) 292 (d) 293 (b) 294 (a) 295 (c) 296 (a) 297 (b) 298 (d) 299 (c) 300 (b)

301 (d) 302 (d) 303 (d) 304 (b) 305 (c) 306 (c) 307 (a) 308 (c) 309 (b) 310 (d)

311 (c) 312 (c) 313 (a) 314 (b) 315 (b) 316 (a) 317 (d) 318 (d) 319 (d) 320 (a)

321 (d) 322 (d) 323 (b) 324 (b) 325 (d) 326 (d) 327 (a) 328 (b) 329 (c) 330 (c)

331 (d) 332 (c) 333 (b) 334 (b) 335 (a) 336 (b) 337 (d) 338 (d) 339 (d) 340 (c)

341 (b) 342 (c) 343 (c) 344 (c) 345 (c) 346 (a) 347 (b) 348 (c) 349 (b) 350 (c)

351 (a) 352 (d) 353 (c) 354 (c) 355 (a) 356 (b) 357 (b) 358 (b) 359 (d) 360 (d)

361 (a) 362 (a) 363 (c) 364 (d) 365 (c) 366 (b) 367 (c) 368 (d) 369 (d) 370 (c)

371 (d) 372 (d) 373 (b)

SUBJECT INDEX

W